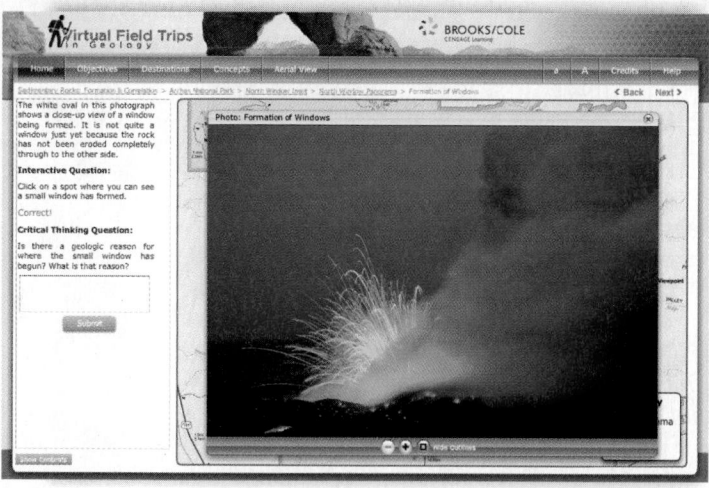

Bringing the field to you.

The Virtual Field Trips are concept-based modules that teach students geology by using famous locations throughout the United States. Sideling Hill Syncline, Grand Canyon, and Hawaii Volcanoes National Parks are included, as well as many others. Designed to be used as homework assignments or lab work, the modules use a rich array of multimedia to demonstrate concepts. High definition videos, images, animations, quizzes, and Google Earth layers work together in **Virtual Field Trips** to bring the concepts to life.

Packaged with EARTH

- Sedimentary Rocks
- Geologic Time
- Hydrothermal Activity
- Running Water
- Desert Environments

Additional Bundle Options*

- Metamorphism and Metamorphic Rocks
- Mass Wasting Processes
- Glaciers and Glaciation
- Igneous Rock Textures
- Earthquakes and Seismicity
- Volcano Types
- Plate Tectonics
- Mineral Resources
- Ground water
- Shorelines and Shoreline Processes

*For availability and pricing information, contact your local Brooks/Cole representative.

VIRTUAL FIELD TRIPS:
LET'S GO!

GOALS OF THE TRIPS

1. **Engage students with panoramas, videos, photographs, and Google Earth.**
2. **Help teach geologic concepts in a clear way.**
3. **Hold students accountable for their learning, with critical thinking questions throughout.**
4. **Help get students excited about the Earth and its processes.**
5. **Have fun!**

Integration of Field Trips with the EARTH text

Students are directed to further their knowledge of key concepts using the Web-based Virtual Field Trips. The Virtual Field Trips sections in the book help students focus on what they should pay particular attention to while completing their field trip. Used as a road map to the field trip, the text sections include top objectives, integrated points from the text and field trip, photos, and critical thinking questions. Here is a list of the modules that are included in the EARTH package:

Sedimentary Rocks: Formation and Correlation

Found in: Chapter 3, Rocks
Locations: Arches and Capitol Reef National Parks
Main concepts covered:
- Sandstones, shales, and volcanic ash in the field
- Lithology and weathering profiles
- Bedding planes, strata, and unconformities
- Formation of arches in sandstones
- Story of the Delicate Arch

Geologic Time

Found in: Chapter 4, Geologic Time: A Story in the Rocks
Locations: Grand Canyon and Capitol Reef National Parks
Main concepts covered:
- Stratigraphic section and correlation of units
- Laws of relative dating (superposition, original horizontality, crosscutting relationships, and inclusions)
- Unconformities and missing time
- Ancient sedimentary environments and primary and secondary structures preserved in rocks (cross-bedding, ripple marks, trace fossils)
- Index fossils and correlation

Hydrothermal Activity

Found in: Chapter 8, Volcanoes and Plutons
Location: Yellowstone National Park
Main concepts covered:
- Hot springs and geysers (Old Faithful)
- Fumaroles and boiling mud pots
- Hydrothermal mineral deposits (travertine)
- Microorganisms adapting to extreme environments

Running Water

Found in: Chapter 11, Freshwater: Streams,
Lakes, Ground Water, and Wetlands
Location: Zion National Park
Main concepts covered:
- River erosion, cut banks, and slot canyons
- River deposition and point bars
- River abrasion and rounded pebbles
- Incised meanders and downcutting

Desert Environments

Found in: Chapter 14, Deserts and Wind
Location: Death Valley National Park
Main concepts covered:
- Geological features of an arid landscape
- Chemical sedimentary deposits and evaporites
- Formation and migration of sand dunes
- Features of an alluvial fan
- Borax mining in Death Valley

BROOKS/COLE
CENGAGE Learning™

EARTH
Graham R. Thompson and Jonathan Turk

Publisher: Yolanda Cossio

Earth Science Editor: Laura Pople

Developmental Editor: Anna Lustig

Assistant Editor: Samantha Arvin

Editorial Assistant: Kristina Chiapella

Media Editor: Alexandria Brady

Marketing Manager: Nicole Mollica

Marketing Coordinator: Kevin Carroll

Marketing Communications Manager: Belinda Krohmer

Content Project Manager: Hal Humphrey

Art Director: John Walker

Print Buyer: Karen Hunt

Rights Acquisitions Account Manager, Text:
   Bob Kauser

Rights Acquisitions Account Manager, Image:
   Deanna Ettinger

Production Service: Patti Zeman, Hespenheide Design

Cover Design: Red Hanger Design

Cover Image: © Robert Landau/Corbis

Compositor: Hespenheide Design

For product information and technology assistance, contact us at
**Cengage Learning Customer & Sales Support,
1-800-354-9706.**

For permission to use material from this text or product, submit all requests online at **www.cengage.com/permissions**. Further permissions questions can be e-mailed to **permissionrequest@cengage.com**

Library of Congress Control Number: 2009942999

Student Edition:
ISBN-13: 978-0-538-74099-9
ISBN-10: 0-538-74099-X

Brooks/Cole
20 Davis Drive
Belmont, CA 94002-3098
USA

Cengage Learning is a leading provider of customized learning solutions with office locations around the globe, including Singapore, the United Kingdom, Australia, Mexico, Brazil, and Japan. Locate your local office at **www.cengage.com/global**.

Cengage Learning products are represented in Canada by Nelson Education, Ltd.

To learn more about Brooks/Cole visit **www.cengage.com /brookscole**.
Purchase any of our products at your local college store or at our preferred online store **www.CengageBrain.com**.

Printed in the United States of America
1 2 3 4 5 6 7 14 13 12 11 10

Contents

Contents

COURTESY OF GRAHAM R. THOMPSON/JONATHAN TURK

Contents

© ANTON FOLTIN/SHUTTERSTOCK

Contents

COPYRIGHT AND PHOTOGRAPH BY DR. PARVINDER S. SETHI

CHAPTER 20

Climate 386

CHAPTER 21

Climate Change 402

UNIT 6 (ONLINE ONLY)
ASTRONOMY

CHAPTER 22 (Online Only)

Motions in the Heavens 426

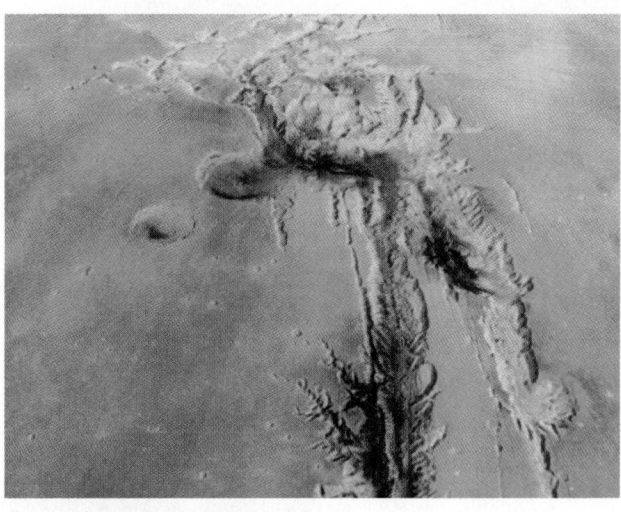

NASA/JPL

ANGLO-AUSTRALIAN TELESCOPE BOARD

1

EARTH SYSTEMS

Atmosphere

Hydrosphere

Geosphere

Biosphere

visit **4ltrpress.cengage.com**

> ### When one tugs at a single thing in nature, he finds it attached to the rest of the world.
>
> *John Muir*

Earth is sometimes called the water planet or the blue planet because azure seas cover more than two-thirds of its surface. Earth is the only planet or moon in the Solar System with rain falling from clouds and water that runs over the land to collect in extensive oceans. It is also the only body we know of that supports life.

1.1 The Earth's Four Spheres

Imagine walking along a rocky coast as a storm blows in from the sea. Wind whips the ocean into whitecaps, gulls wheel overhead, and waves crash onto shore. Before you have time to escape, blowing spray has soaked your clothes. A hard rain begins as you scramble over the rocks to your car. During this adventure, you have observed the four major spheres of Earth. The rocks and soil underfoot are the surface of the **geosphere**, or the solid Earth. The rain and sea are parts of the **hydrosphere**, the watery part of our planet. The gaseous layer above the Earth's surface is the **atmosphere**. Finally, you, the gulls, the beach grasses, and all other forms of life in the sea, on land, and in the air are parts of the **biosphere**, the realm of organisms.

You can readily observe that the atmosphere is in motion because clouds waft across the sky and wind blows against your face. In the biosphere, animals, and to a lesser extent plants, also move. Flowing streams, crashing waves, and falling rain are all familiar examples of motion in the hydrosphere. Although it is less apparent, the geosphere is also moving and dynamic. Vast masses of subterranean rock flow vertically and horizontally within the planet's interior, continents drift, and mountains rise and then erode into sediment. Throughout this book we will study many of these phenomena to learn which energy forces set matter in motion and how the motion affects the planet we live on. Figure 1.1 shows schematically all the possible interactions among the spheres.

Figure 1.2 shows that the geosphere is by far the largest of the four spheres. The Earth's radius is about 6,400 kilometers, more than 1.5 times the distance from New York to Los Angeles. Despite this great size, nearly all

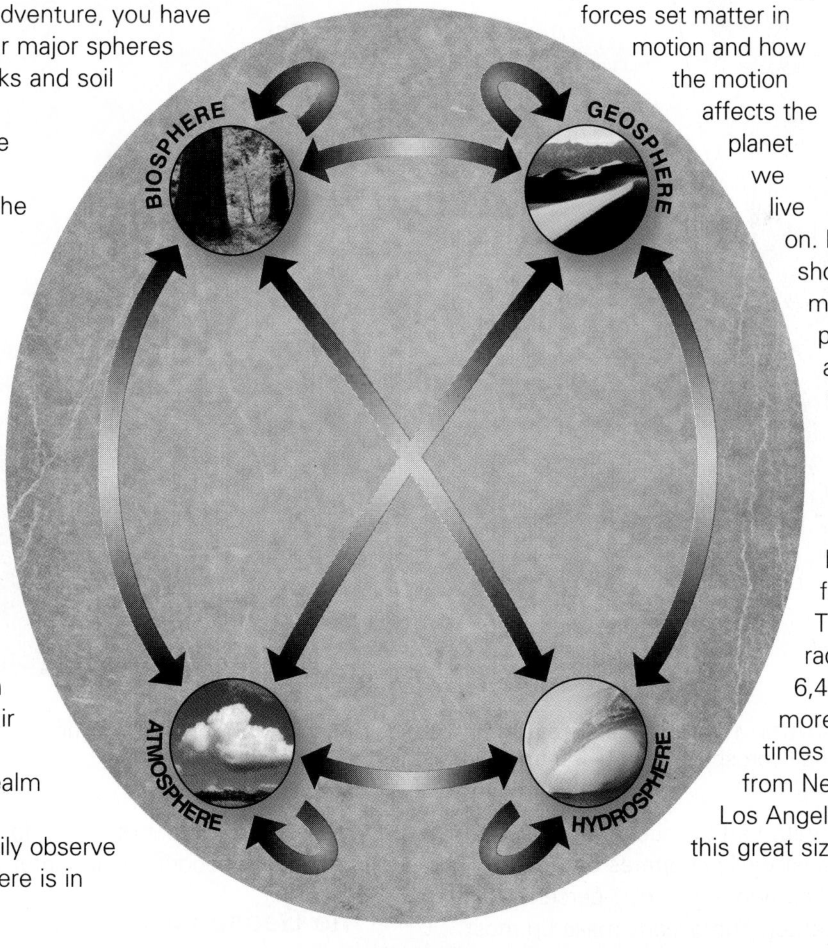

geosphere The solid Earth, consisting of the entire planet from the center of the core to the outer crust.

hydrosphere All of Earth's water, which circulates among oceans, continents, glaciers, and atmosphere.

atmosphere The gaseous layer above the Earth's surface, mostly nitrogen and oxygen, with smaller amounts of argon, carbon dioxide, and other gases. The atmosphere is held to Earth by gravity and thins rapidly with altitude.

biosphere The zone of Earth comprising all forms of life in the sea, on land, and in the air.

FIGURE 1.1 All of Earth's cycles and spheres are interconnected.

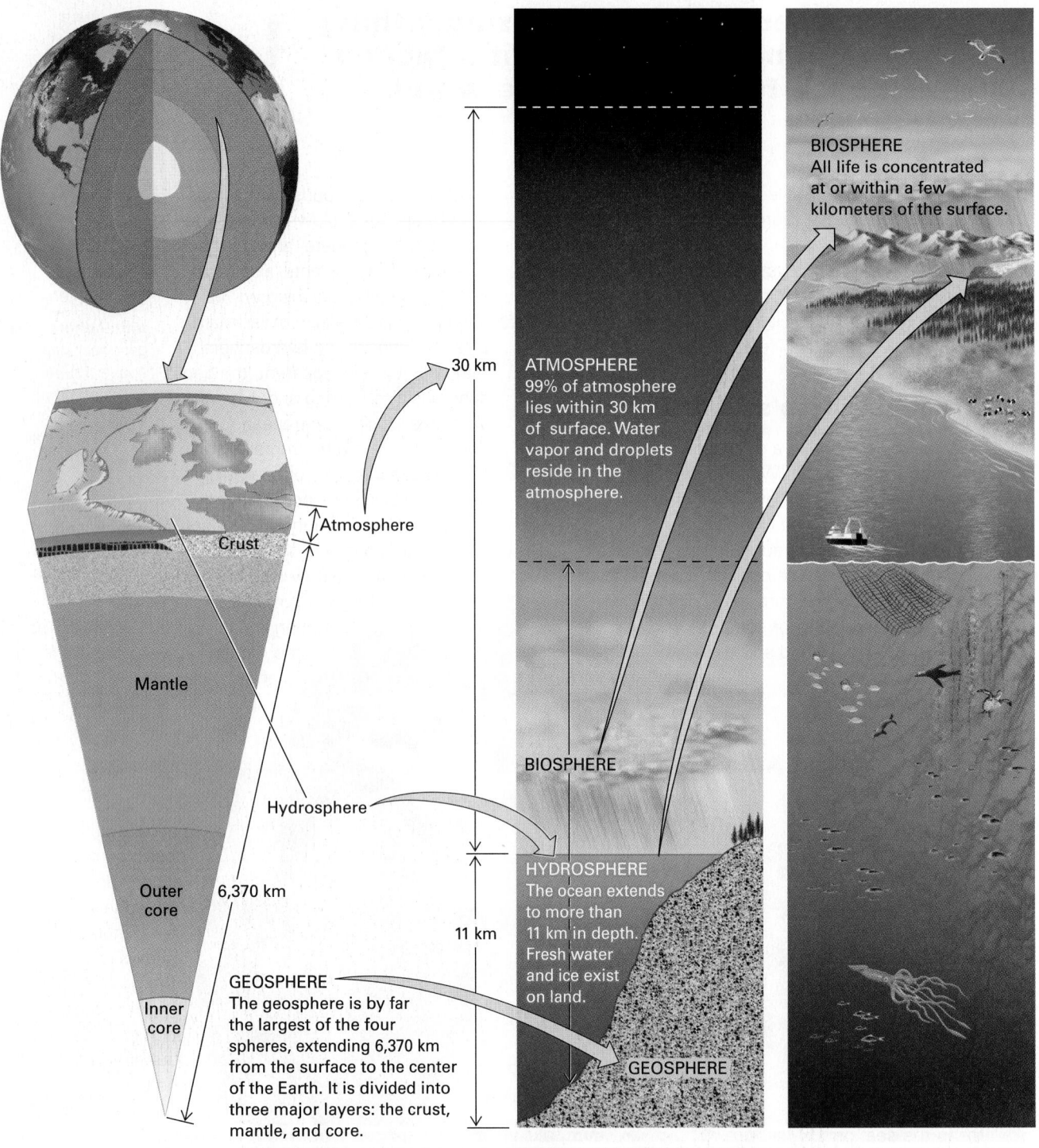

FIGURE 1.2 The geosphere is the largest component of Earth. It is surrounded by the hydrosphere, the biosphere, and the atmosphere.

of our direct contact with Earth occurs at or very near its surface. The deepest well penetrates only about 12 kilometers, less than two-tenths of 1 percent of the distance to Earth's center. The oceans make up most of the hydrosphere, and although it can extend as deep as 11 kilometers, the central ocean floor averages only 5 kilometers in depth. Most of Earth's atmosphere lies within 30 kilometers of the surface, and the biosphere is a thin shell about 15 kilometers thick.

The Geosphere

Our Solar System coalesced from a frigid cloud of dust and gas rotating slowly in space. The Sun formed as gravity pulled material toward the swirling center. At the

FIGURE 1.3 Earth's crust is made up of different kinds of rock. (A) A grand view of the sandstone, siltstones, limestones, and shales of the Grand Canyon, as seen from the South Rim. These sedimentary rocks are relatively soft, crumbly, and show a marked horizontal layering. Note people for scale. (B) The granite of Baffin Island in the Canadian Arctic is gray, hard, and strong.

core The dense, metallic, innermost region of Earth's geosphere, consisting mainly of iron and nickel. The outer core is molten, but the inner core is solid.

mantle The rocky, mostly solid layer of Earth's geosphere lying beneath the crust and above the core. The mantle extends from the base of the crust to a depth of about 2,900 kilometers.

crust The outermost layer of Earth's geosphere, about 7 to 70 kilometers thick and composed of relative low-density silicate rocks.

lithosphere The cool, rigid, outer part of Earth, about 100 kilometers thick, which includes the crust and the uppermost mantle.

same time, rotational forces spun material in the outer cloud into a thin disk. When the turbulence of the initial accretion subsided, small grains stuck together to form fist-sized masses. These planetary seeds then accreted to form rocky clumps, which grew to form larger bodies called *planetesimals*, 100 to 1,000 kilometers in diameter. Finally, the planetesimals collected to form the planets. This process was completed about 4.6 billion years ago.

As the Earth coalesced, the rocky chunks and planetesimals were accelerated by gravity so that they slammed together at high speeds. Particles heat up when they collide, so the early Earth warmed as it formed. Later, asteroids, comets, and more planetesimals crashed into the surface, generating additional heat. At the same time, radioactive decay heated the Earth's interior. As a result of these three processes, our planet became so hot that all or most of it melted soon after it formed.

Within the molten Earth, the denser materials gravitated toward the center, while the less dense materials floated toward the top, creating a layered structure. Today, the geosphere consists of three major layers: a dense metallic **core**, a less dense rocky **mantle**, and an even less dense surface **crust** (Figure 1.2).

The temperature of modern Earth grows hotter with depth. At its center, Earth is 6,000°C—as hot as the Sun's surface. The core is composed mainly of iron and nickel. The outer core is molten metal. However the inner core, although hotter yet, is solid because the great pressure compresses the metal to a solid state.

The mantle surrounds the core and lies beneath the crust. The physical characteristics of the mantle vary with depth. Near the surface, to a depth of about 100 kilometers, the outermost mantle is cool, strong, and hard. In contrast, the layer below 100 kilometers is so hot that the rock is weak, soft, *plastic* (a solid that will deform permanently), and flows slowly—like cold honey. Even deeper in Earth, pressure overwhelms temperature and the mantle rock becomes strong again.

The crust is the outermost layer, a thin veneer below a layer of soil and beneath the ocean water. It is composed almost entirely of solid rock. Even a casual observer sees that rocks of the crust are different from one another: some are soft, others hard, and they come in many colors, which you can see in Figure 1.3.

The uppermost mantle is relatively cool and its pressure is relatively low. Both factors combine to produce hard, strong rock similar to that of the crust. The outer part of Earth, including both the uppermost mantle and the crust, makes up the **lithosphere**. The lithosphere averages about 100 kilometers thick.

According to a theory developed in the 1960s, the lithosphere is broken into seven major segments called **tectonic plates**. These tectonic plates float on the weak, plastic, mantle rock beneath, and glide across Earth (Figure 1.4). For example, North America is currently drifting toward China about as fast as your fingernail grows. In a few hundred million years—almost incomprehensibly long on a human time scale but brief when compared with planetary history—Asia and North America may collide, crumpling the edges of the continents and building a giant mountain range. In later chapters we will learn how plate tectonic theory explains earthquakes, volcanic eruptions, the forma-tion of mountain ranges, and many other processes and events that have created our modern Earth and environment.

The Hydrosphere

The hydrosphere includes all of Earth's water, which circulates among oceans, continents, glaciers, and the atmosphere. Figure 1.5 shows the proportion of water in each of these areas. Oceans cover 71 percent of Earth and contain 97.5 percent of its water. Ocean currents transport heat across vast distances, altering global climate.

About 1.8 percent of Earth's water is frozen in glaciers. Although glaciers cover about 10 percent of Earth's land surface today, they covered much greater portions of the globe as recently as 18,000 years ago.

Only about 0.64 percent of Earth's total water exists on the continents as a liquid. Although this is a small proportion, freshwater is essential to life on Earth. Lakes, rivers, and clear, sparkling streams are the most visible reservoirs of continental water, but they constitute only 0.01 percent of Earth's water. In contrast, **ground water**, which saturates rock and soil of the upper few kilometers of the geosphere, is much more abundant and accounts for 0.63 percent of Earth's water. Only a minuscule amount, 0.001 percent, exists in the atmosphere, but because this water is so mobile, it profoundly affects both the weather and climate of our planet.

FIGURE 1.4 The lithosphere is composed of the crust and the uppermost mantle. It is a 100-kilometer-thick layer of strong rock that floats on the underlying plastic mantle. The lithosphere is broken into seven major segments, called tectonic plates, that glide horizontally over the plastic mantle at rates of a few centimeters per year. In the drawing, the thickness of the mantle and the lithosphere are exaggerated to show detail.

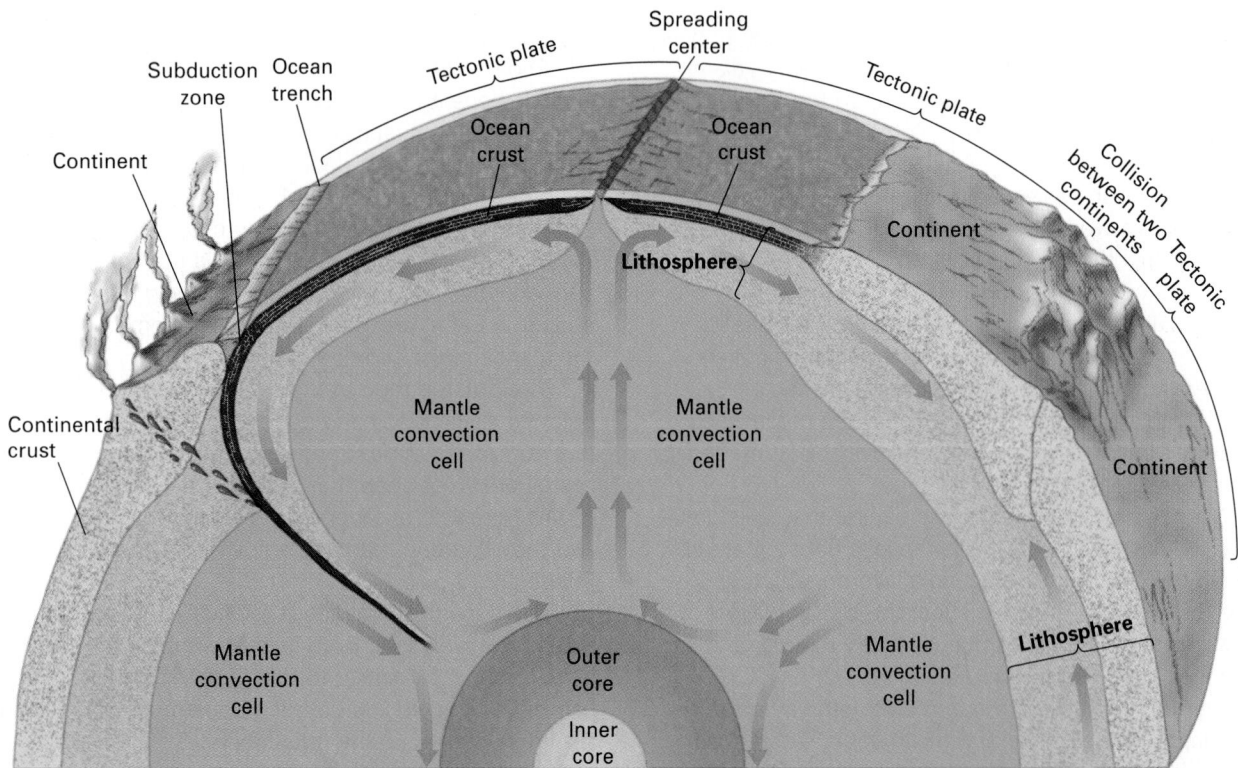

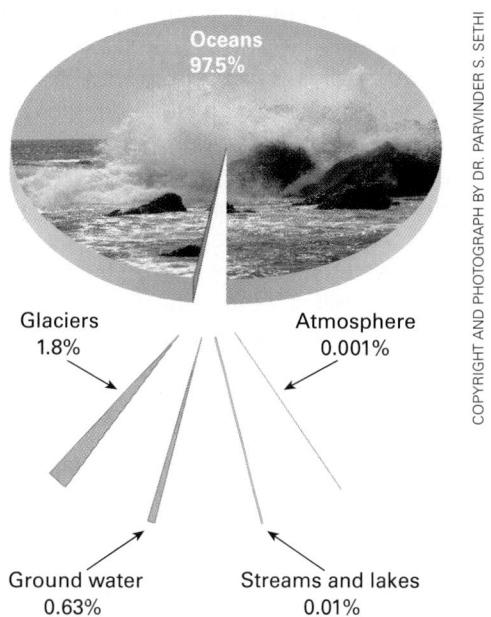

Oceans
97.5%

Glaciers
1.8%

Atmosphere
0.001%

Ground water
0.63%

Streams and lakes
0.01%

FIGURE 1.5 The oceans contain most of Earth's surface water. Most freshwater is frozen into glaciers. Most available freshwater is stored underground as ground water.

The Atmosphere

The atmosphere is a mixture of gases, mostly nitrogen and oxygen, with smaller amounts of argon, carbon dioxide, and other gases. It is held to Earth by gravity and thins rapidly with altitude. Ninety-nine percent is concentrated in the first 30 kilometers, but a few traces remain as far as 10,000 kilometers above Earth's surface.

The atmosphere supports life because animals need oxygen, and plants need both carbon dioxide and oxygen. In addition, the atmosphere supports life indirectly by regulating climate. Air acts as both a blanket and a filter, retaining heat at night and shielding us from direct solar radiation during the day. Wind transports heat from the equator toward the poles, cooling equatorial regions and warming temperate and polar zones.

The Biosphere

The biosphere is the zone that life inhabits. It includes the uppermost geosphere, the hydrosphere, and the lower parts of the atmosphere. Sea life concentrates near the surface, where sunlight is available. Plants also grow on Earth's surface, with roots penetrating a few meters into the soil. Animals live on the surface, fly a kilometer or two above it, or burrow a few meters underground. Large populations of bacteria live in rock to depths of as much as 4 kilometers, some organisms live on the ocean floor, and a few windblown microorganisms drift at heights of 10 kilometers or more. But

even at these extremes, the biosphere is a very thin layer at Earth's surface.

Plants and animals are clearly affected by Earth's environment. Organisms breathe air, require water, and thrive in a relatively narrow temperature range. Terrestrial organisms ultimately depend on soil, which is part of the geosphere. Less obviously, plants and animals also alter and form the environment they live in. For example, living organisms contributed to the evolution of the modern atmosphere.

1.2 Earth Systems

A **system** is any assemblage or combination of interacting components that form a complex whole. For example, the human body is a system composed of bones, nerves, muscles, stomach, and many other organs. Each organ is discrete, yet all the organs interact to produce a living human. Blood nurtures the stomach; the stomach helps provide energy to maintain the blood.

Systems are driven by the flow of matter and energy. Thus a person ingests food, which contains both matter and chemical energy, and inhales oxygen. Waste products are released through urine, feces, sweat, and exhaled breath. Some energy is used for respiration and motion, and the remainder is released as heat.

A single bacterium is a system, just as a person is. The human body contains hundreds of millions of bacteria that are essential to the functioning of metabolic processes such as digestion. Thus a larger system (the human body) is composed of many smaller ones, including bacteria.

In addition, humans interact with their environment. Imagine a Stone Age hunter-gatherer living in a small valley. All the plants and animals in this valley are components of an **ecosystem**, which is defined as a complex community of organisms and their environment, functioning as an ecological unit in nature. Therefore, to understand the human system, we must study smaller systems (e.g., bacteria) that exist within the body and also must appreciate how humans interact with larger ecosystems.

But we're not finished yet. Individual ecosystems interact with climate systems, ocean currents, and other Earth systems. Thus:

- The size of systems varies dramatically.
- Large systems contain numerous smaller systems.
- Systems interact with one another in complex ways.

system Any combination of interacting components that form a complex whole.

ecosystem A complex community of individual organisms, interacting with each other and with their physical environment, functioning as an ecological unit in nature.

As we have learned, Earth is composed of four major systems: geosphere, hydrosphere, atmosphere, and biosphere. Each of these large systems is further subdivided into a great many interacting smaller ones. For example, a single volcanic eruption is part of a system. Energy from deep within Earth melts rock, magma rises, and gases escape from deep Earth layers and react chemically with surface materials. But this volcanic eruption is driven by mantle convection and movements of tectonic plates that are components of Earth's interior, which is also a system. At the same time, ash spewed skyward from the volcanic eruption cools the Earth and becomes a component of the climate system. Looking at the eruption from yet another perspective, a volcanic mountain alters drainage patterns and erosion rates and is thus part of the hydrologic system. In this book, we will study systems of all sizes and illustrate many of the complex interactions among them.

Earth's surface systems—the atmosphere, hydrosphere, and biosphere—are ultimately powered by the Sun. Wind is powered by uneven solar heating of the atmosphere, ocean waves are driven by the wind, and ocean currents move in response to wind or temperature differences. Luckily for us, Earth receives a continuous influx of solar energy, and it will continue to receive this energy for another 5 billion years or so.

In contrast, Earth's interior is powered by the decay of radioactive elements and by residual heat from the primordial coalescence of the planet. We will discuss these sources in later chapters.

Several energy and material cycles are fundamental to our study of Earth systems. These include the rock cycle (Chapter 3), the hydrologic cycle or water cycle (Chapter 11), the carbon cycle (Chapter 17), and the nitrogen cycle (Chapter 17). During the course of these cycles, matter is always conserved—however, matter continuously and commonly changes form.

For example, ice melts to form water, and if more heat is added, water vaporizes to a gas. The process occurs in reverse when heat is removed. Thus, in the hydrologic cycle, water evaporates from the ocean into the atmosphere, falls to Earth as rain or snow, and eventually flows back to the oceans. Chemical reactions also change the physical properties of matter. Common table salt is a solid crystalline mineral. If you drop a crystal of salt in water, it dissolves, thus altering the properties of both the salt and the water.

Because matter exists in so many different chemical and physical forms, most materials occur in all four of Earth's major spheres: geosphere, hydrosphere, atmosphere, and biosphere. Water, for example, is chemically bound into clays and other minerals, as a component of the geosphere. It is the primary constituent of the hydrosphere. It exists in the atmosphere as vapor and clouds. And it is an essential part of all living organisms. As another example, thick layers of salt occur as chemical sedimentary rocks; large quantities of salt are dissolved in the oceans; salt aerosols are suspended in the atmosphere; and salt is an essential component of life. Thus, all the spheres continuously exchange matter and energy. In our study of Earth systems, we categorize the four separate spheres and numerous material cycles independently, but we also recognize that Earth materials and processes are all part of one integrated system (Figure 1.1).

© THE PRINT COLLECTOR / ALAMY

1.3 Time and Rates of Change in Earth Science

James Hutton was a gentleman farmer who lived in Scotland in the late 1700s. Although trained as a physician, he never practiced medicine and instead turned to geology. Hutton observed that a certain type of rock, called sandstone, is composed of sand grains cemented together. He also noted that rocks in the Scottish Highlands slowly decompose into sand and that streams carry sand into the lowlands. He inferred that sandstone is composed of sand grains that originated by the erosion of ancient cliffs and mountains.

Hutton tried to deduce how much time was required to form a thick bed of sandstone. He studied sand grains slowly breaking away from rock outcrops. He watched sand bouncing down streambeds. Finally he traveled to beaches and river deltas where sand was accumulating. By estimating the time needed for thick layers of sand to accumulate on beaches, Hutton concluded that sandstone must be much older than human history.

Hutton had no way to measure the magnitude of geologic time. However, modern geologists have learned that certain radioactive materials in rocks act as clocks to record the passage of time. Using these clocks and other clues embedded in Earth's crust, the moon, and in meteorites fallen from the Solar System, geologists estimate that Earth formed 4.6 billion years ago.

The primordial Earth was vastly different from our modern world. There was no crust as we know it today, no oceans, and the diffuse atmosphere was vastly different from the modern one. There were no living organisms.

No one knows exactly when or how the first living organisms evolved, but we know that life existed at least as early as 3.8 billion years ago, 800 million years after the planet formed. For the following 3.3 billion years, life evolved slowly, and although some multi-cellular organisms developed, most of the biosphere consisted of single-celled organisms. Organisms rapidly became more complex, abundant, and varied about 543 million years ago. The dinosaurs flourished between 225 million and 65 million years ago.

Homo sapiens and our direct ancestors have been on Earth for 5 to 7 million years, for only about one-tenth of 1 percent of the planet's history. In his book *Basin and Range*, John McPhee offers a metaphor for the magnitude of geologic time.[1] If the history of Earth were represented by the old English measure of a yard—the distance from the king's nose to the end of his outstretched hand—all of human history could be erased by a single stroke of a file on his middle fingernail. Figure 1.6 summarizes Earth history in graphical form. Geologists routinely talk about events that occurred millions or even billions of years ago. If you live on the east coast of North America, you may be familiar with the Appalachian Mountains, which started rising from a coastal plain nearly half a billion years ago. If you live in Seattle, the land you are standing on was an island in the Pacific between 80 and 100 million years ago, when dinosaurs walked Earth.

There are two significant consequences of the vast span of geologic time:

1. Events that occur slowly become significant. If a continent moves a few centimeters a year, the movement makes no noticeable alteration of Earth systems over decades or centuries. But over hundreds of millions of years, the effects are significant.
2. Improbable events occur regularly. The chances are great that a large meteorite won't crash into

1. John McPhee, *Basin and Range* (New York: Farrar, Straus & Giroux, 1982).

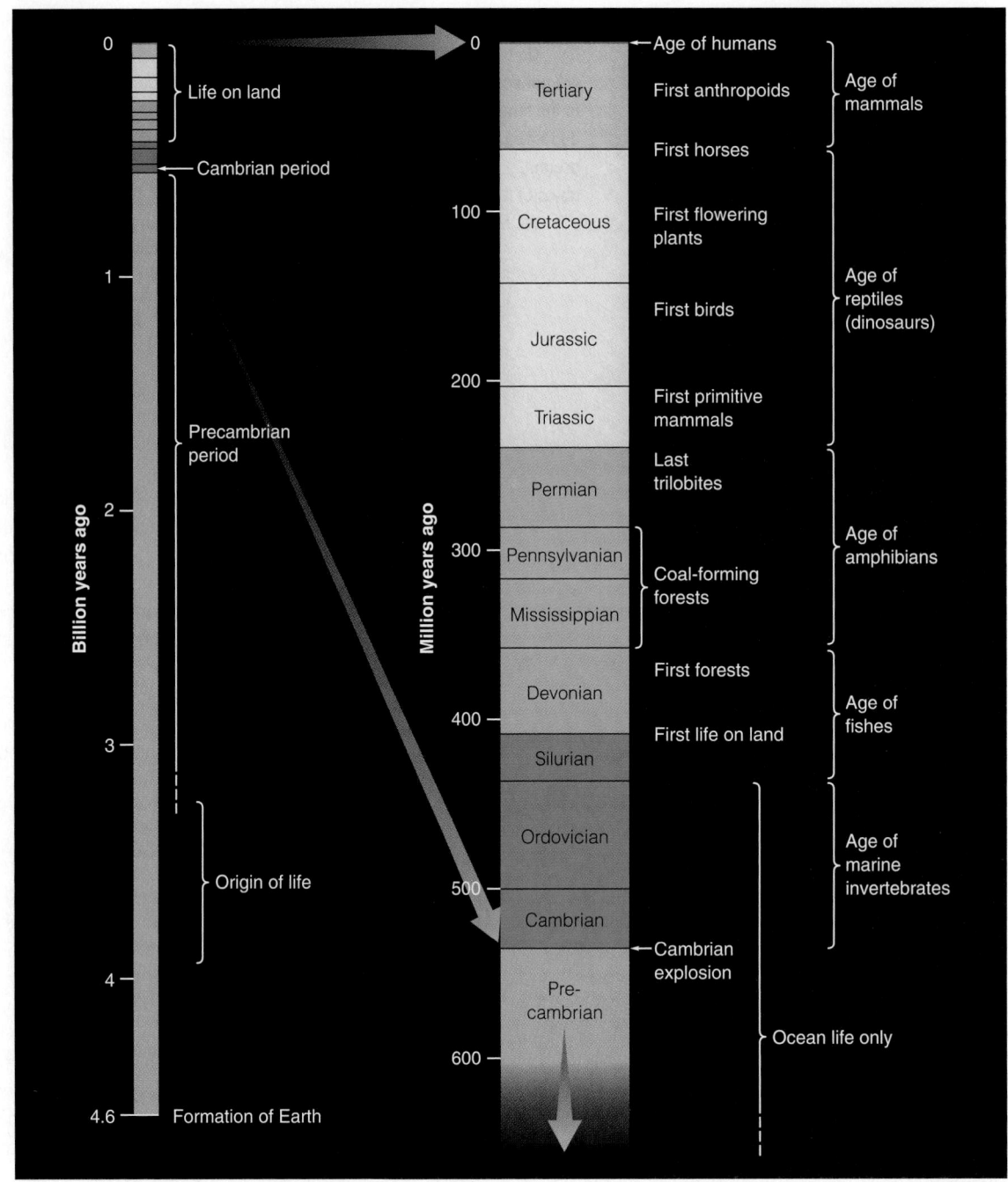

FIGURE 1.6 The geologic time scale. Note the great length of time before multicellular organisms became abundant, about 543 million years ago.

Earth tomorrow, or next year, or during the next century. But during the past 500 million years, several catastrophic impacts have occurred and they will probably occur sometime in the future.

James Hutton deduced that sandstone forms when rocks slowly decompose to sand, the sand is transported to lowland regions, and the grains cement together. This process occurs step by step—over

many years. Hutton's conclusions led him to formulate the principles now known as **gradualism** and **uniformitarianism**. The principle of gradualism states that geologic change occurs over long periods of time, by a sequence of almost imperceptible events. Uniformitarianism stipulates that geologic processes operating today also operated in the past. Thus scientists can explain events that occurred in the past by observing slow changes occurring today. Sometimes this idea is summarized in the statement: "The present is the key to the past."

However, not all geologic change is gradual. William Whewell, a geologist in the 19th century, argued that geologic change was sometimes rapid. He wrote that the geologic past may have "consisted of epochs of paroxysmal and catastrophic action, interposed between periods of comparative tranquility."[2] Earthquakes and volcanoes are examples of catastrophic events; Whewell argued for the principle of **catastrophism**—that occasionally huge catastrophes have altered the course of Earth history. He couldn't give an example of such an event, however, because they happen so infrequently that none had occurred within human history.

Today, geologists know that Hutton's gradualism and Whewell's catastrophism are both correct. Over the great expanses of geologic time, slow, uniform processes alter Earth. In addition, infrequent catastrophic events radically modify the path of slow change.

Gradual Change in Earth History

Tectonic plates creep slowly across Earth's surface. Since the first steam engine was built 200 years ago, North America has migrated 8 meters westward, the distance a sprinter can run in 1 second. Thus, the movement of tectonic plates is too slow to be observed without sensitive instruments. However, over geologic time, this movement alters the shapes of ocean basins, forms lofty mountains and plateaus, generates earthquakes and volcanic eruptions, and affects our planet in many other ways.

Catastrophic Change in Earth History

Chances are small that the river flowing through your city will flood this spring, but if you live to be 100 years old, you will probably see a catastrophic flood. In fact, destructive floods occur every year, somewhere on the planet. In recent years, three catastrophic geological events caused significant loss of human life: the tsunami in Southeast Asia, Hurricane Katrina, and the Pakistani/Kashmir earthquake. But even these catastrophes are minuscule compared with events that have occurred in the past.

When geologists study the 4.6 billion years of Earth history, they find abundant evidence of catastrophic events that are highly improbable in a human lifetime or even in human history. For example, clues preserved in rock and sediment indicate that giant meteorites have smashed into our planet, vaporizing portions of the crust, spreading dense dust clouds over the sky, and leaving large craters, such as the one in Figure 1.7.

Geologists have suggested that some meteorite impacts (much larger than that shown in Figure 1.7) have almost instantaneously driven millions of species into extinction. As another example, when the continental glaciers receded at the end of the last ice age, huge floods occurred that were massively larger than any floods in historic times. In addition, periodically throughout Earth history, catastrophic volcanic eruptions changed environmental conditions for life around the globe.

1.4 Threshold and Feedback Effects

Earth systems often change in ways that are difficult to predict. Two mechanisms that contribute to the challenges and complexities facing of earth science today are discussed next.

Threshold Effects

Imagine that some process gradually warmed the atmosphere. Imagine further that the summer temperature of coastal Greenland glaciers was −1.0°C. If gradual change warmed the atmosphere by 0.2°C, the glaciers would not be affected appreciably because ice melts at 0°C and the summer temperature would be −0.8°C, still below freezing. But now imagine that gradual warming brought summer temperature to 0.2°C. At this point, the system is at a threshold. Another small increment of warming would raise the temperature to the melting point of ice. As soon as ice starts to melt, large changes occur rapidly: glaciers recede and global sea level rises. A **threshold effect** occurs when the environment initially changes slowly (or not at all) in response to a small perturbation—but after the threshold is crossed, an additional small perturbation causes rapid and dramatic change.

If we don't understand a system thoroughly, it is easy to be deluded into a false sense of complacency. Imagine, in the case just described, that we didn't

gradualism A principle stating that geological change occurs as a consequence of slow or gradual accumulation of small events, such as the deposition of sand in a river delta to eventually form a large land mass. More recently, scientists studying biological evolution use the term to describe a theory of evolution that proposes that species change gradually in small increments.

uniformitarianism A principle stating that the geologic processes operating today also operated in the past.

catastrophism The principle stating that infrequent catastrophic geologic events alter the course of Earth history and modify the path of slow geologic time.

threshold effect A reaction whereby the environment initially changes slowly (or not at all) in response to a small perturbation, but after the threshold is crossed, an additional small perturbation causes rapid change.

---

2. Cited in Charles Lyell, *Principles of Geology*, vol. 2 (London: Murray, 1832), Quarterly Reviews 47, 103–123.

know the melting point of ice. If we observed that several years of incremental warming had no effect on the system, we might conclude that additional warming might also have no effect. But, in fact, the next small increment, which crosses the threshold, melts the ice.

Populations of plants, animals, and people also respond to thresholds. In 1912 scientists released 4 male and 21 female reindeer onto St. Paul Island in the Bering Sea. There were no predators on the island and hunting pressure was low, so the reindeer population increased dramatically. By 1938 approximately 2,000 reindeer grazed peacefully on the tundra vegetation of St. Paul Island. To an untrained eye, the introduction was successful because the reindeer were healthy and multiplying. But, the reindeer herd was overgrazing the available pasture—eating more food than was growing—heading toward a

threshold. Everything was fine as long as the reindeer could move across the island to find new sources of food, but when the entire island became overgrazed, suddenly there was nothing to eat—and the reindeer starved. Between 1938 and 1950 almost the entire herd starved to death, until only eight emaciated animals remained.

Feedback Mechanisms

A **feedback mechanism** occurs when a small perturbation in one component affects a different component of Earth systems, which amplifies the original effect and perturbs the system even more, leading to an even greater effect. The evolution of the Venusian climate is a perfect example of a feedback mechanism: Solar energy evaporated surface water. The water vapor trapped more heat in the atmosphere. The temperature rose and more water evaporated. When the water evaporated, dissolved carbon dioxide escaped into the atmosphere and this gaseous carbon dioxide trapped even more heat. The temperature rose further.

FIGURE 1.7 Meteor Crater, Arizona.

METEOR CRATER NATIONAL MONUMENT

1 mi

Chapter 1: Earth Systems

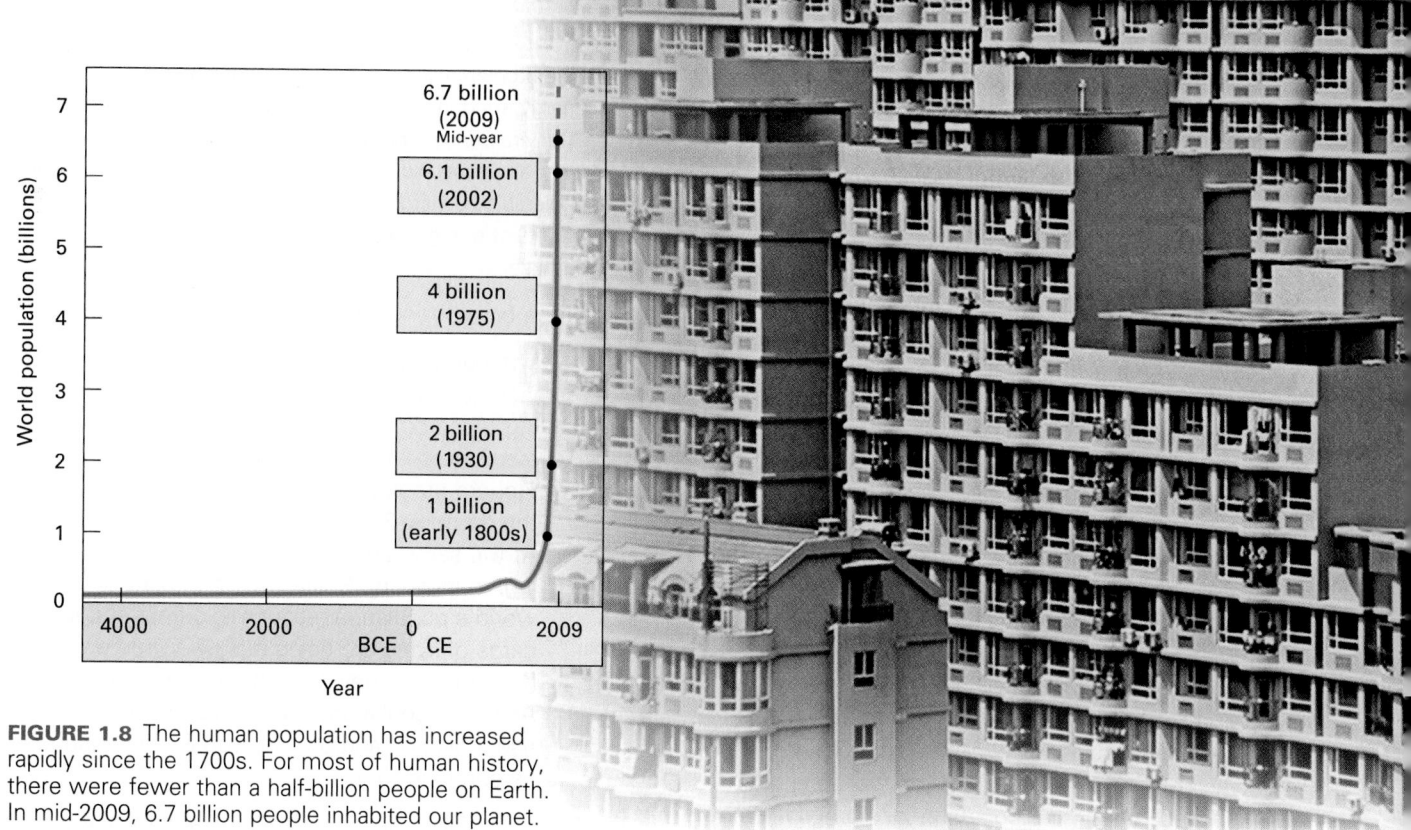

FIGURE 1.8 The human population has increased rapidly since the 1700s. For most of human history, there were fewer than a half-billion people on Earth. In mid-2009, 6.7 billion people inhabited our planet.

Scientists have observed many examples of feedback mechanisms in modern earth science. For example, in recent years, industrial emissions have increased the concentration of carbon dioxide in the atmosphere and the mean global atmospheric temperature has risen. Two separate studies published in September 2005 described specific effects: (1) Higher air temperatures in England caused soil bacteria to flourish. The bacteria consumed more food, and respiration increased, releasing even more carbon dioxide into the air. (2) Extremely hot weather in France during the summer of 2003 led to a reduction of plant photosynthesis, with a consequent reduction of the uptake of carbon dioxide by plants. Thus atmospheric carbon dioxide levels increased.

In both studies, emissions of industrial carbon dioxide led to warming, but warming caused natural systems to release even more carbon dioxide, which presumably leads to additional warming. No one fully understands the eventual ramifications of this feedback loop, but Chapter 21 explores what scientists do know.

1.5 Humans and Earth Systems

Ten thousand years ago, a mere eyeblink in Earth history, human population was low and technology was limited. Sparsely scattered bands hunted and gathered for their food, built simple dwellings, and warmed themselves with small fires. Human impact was not appreciably different from that of many other species on the planet. But with the invention of agriculture and

the growth of cities, problems arose in some cultures. In his recent book, *Collapse*, Jared Diamond documents several case histories where people settled in a region, built a society, and consumed resources faster than they were naturally replaced.[3] Then, like the reindeer on St. Lawrence Island, populations crashed. For example, at the height of the Mayan civilization between the 3rd and the 9th centuries, the central cities supported from 1,800 to 2,600 people per square mile. (In comparison, Los Angeles County averaged 2,345 people per square mile in 2000.) Then, due to a variety of human and environmental factors, the cities were abandoned, the population crashed, and an estimated 90 to 95 percent of the Mayans died. In a similar manner the Anasazi civilizations in the American Southwest disappeared in the 13th century, and Viking colonies starved to death in Greenland. All three of these human tragedies were exacerbated by climate change and human-to-human interactions, such as war. Nevertheless, Diamond points out that after thresholds were reached, the populations didn't just decline back to a sustainable level. Rather, they collapsed catastrophically.

In mid-2009 there were 6.7 billion people on Earth (Figure 1.8). Farms, pasture lands, and cities covered 40 percent of the land area of the planet. We diverted half

---

3. Jared Diamond, *Collapse: How Societies Choose to Fail or Succeed* (New York: Penguin Group, 2005).

of all the freshwater on the continents and controlled 50 percent of the total net terrestrial biological productivity. As the land and its waters have been domesticated and manipulated, other animal and plant species have perished. The species extinction rate today is 1,000 times more rapid than it was before the Industrial Revolution. In fact, species extinction hasn't been this rapid since a giant meteorite smashed into Earth's surface 65 million years ago, leading to the demise of the dinosaurs and many other species.

Emissions from our fires, factories, electricity-generating plants, and transportation machines have raised the atmospheric carbon dioxide concentration to its highest level in 420,000 years. Since the Industrial Revolution, Earth's mean annual temperature has increased by 0.8°C.

Roughly 20 percent of the human population lives in wealthy industrialized nations where food, shelter, clean drinking water, efficient sewage disposal, and advanced medical attention are all readily available. At the same time, 1.5 billion people in the impoverished regions of the world survive on less than $1 per day. One-eighth to one-half of the human population on this planet is malnourished, and 14 million infants or young children starve to death every year. If you were a member of a family in the Himalayan region of northern India—where you spent hours every day gathering scant firewood or cow dung for fuel, where meals consisted of an endless repetition of barley gruel and peas, where you brought your sheep and goats into the house during the winter so you could all huddle together for warmth—your experience with the Earth and its systems would be quite different from the way you probably experience things now.

Yet no matter where we live and regardless of whether we are rich or poor, our survival is interconnected with the flow of energy and materials among Earth systems. We affect Earth, and at the same time Earth affects us; it is the most intimate and permanent of marriages.

Predictions for the future are, by their very nature, uncertain. More than 200 years ago, in 1798, the Reverend Thomas Malthus argued that the human population was growing exponentially while food production was increasing much more slowly. As a result, humankind was facing famine, accompanied by "misery and vice."[4] But technological advancements have led to an exponential increase in food production, and today food is plentiful in many regions. At the same time, however, serious shortages exist, especially in parts of Africa.

Many scientists are deeply concerned about our future: John Holdren, professor at the John F. Kennedy School of Government, argues that:

[The problem is] not that we are running out of energy, food, or water but that we are running out of environment—that is, running out of the capacity of air, water, soil, and biota to absorb, without intolerable consequences for human well-being, the effects of energy extraction, transport transformation and use.[5]

Three broad trends are currently facing humanity:

1. The human population is large and continuing to increase—although the rate of increase is slowing. According to one prediction, population will stabilize at 7 billion by 2020. But other demographers predict that 14 billion people will walk the earth by 2100.
2. At the same time, extreme poverty is decreasing. In India and China, with one-third of the world's population, economic output and standards of living are rising rapidly. When people have money, they consume more resources.
3. As both population and per capita consumption rise, pollution and pressure on Earth's resources continue to rise.

Professor Diamond, quoted earlier, argues that we are in a "horse race" with the environment.[6] One horse, representing increased human consumption and the resulting environmental degradation and resource depletion, is galloping toward an unsustainable society with a possible catastrophic collapse. The other horse, powered by improved technology, increased awareness, and declining rates of population growth, leads us toward a happy, healthy, sustainable society. Which horse will win?

This book will not provide answers—because there are none—but it will provide a basis for understanding Earth processes so you can evaluate and appreciate issues concerning our stewardship of the planet. The accompanying table suggests an environmental "to-do" list.

4. Thomas Malthus, *An Essay on the Principle of Population*, 1st ed. (Penguin Classics, 1798).

5. John Holdren, "Energy: Asking the Wrong Question," *Scientific American* 286 (January 2002), 65–67.
6. Jared Diamond, *Collapse: How Societies Choose to Fail or Succeed* (New York: Penguin Group, 2005).

OUR ENVIRONMENTAL "TO-DO" LIST

1. Slow the increase in human population.

2. Reverse the buildup of greenhouse gases and the depletion of the ozone layer.

3. Slow the increase in human per capita resource consumption.

4. Conserve fossil fuels and firewood resources.

5. Conserve water and protect against causes of water shortages.

6. Prevent toxic chemicals from harming the environment.

7. Protect against soil erosion.

8. Halt the destruction of natural habitats such as forests, wetlands, prairies, and coral reefs.

9. Reduce our use of the world's photosynthetic capacity so that some is left over to support the growth of natural forests.

10. Protect against the loss of ocean fisheries.

11. Protect endangered species so as not to lose biodiversity.

12. Protect indigenous species, and keep invasive species from destroying ecosystems.

2

MINERALS

visit **4ltrpress.cengage.com**

Minerals are the building blocks of rock, and therefore the geosphere is composed of minerals. Minerals make up the rocks beneath your feet, the soil that supports plants, and the deep rock of Earth's mantle. Any thorough study of Earth must include an understanding of minerals. But it is not sufficient to study minerals isolated from the rest of the planet. Rather, we can learn more by observing the ways that minerals interact with other Earth systems.

Most minerals in their natural settings are harmless to humans and other species. In addition, many resources that are essential to modern life are produced from rocks and minerals: iron, aluminum, gold and all other metals, concrete, fertilizers, coal, and nuclear fuels. Both oil and natural gas are recovered from reservoirs in rock. The discoveries of methods to extract and use these Earth materials profoundly altered the course of human history. The Stone Age, the Bronze Age, and the Iron Age are historical periods named for the rock and mineral resources that dominated the development of civilization during those times. The Industrial Revolution occurred when humans discovered how to convert the energy stored in coal into useful work. The automobile

age occurred because humans found vast oil reservoirs in shallow rocks of Earth's crust and built engines to burn the refined fuels. The computer age relies on the movement of electrons through wafer-thin slices of silicon, germanium, and other materials that are also crystallized from minerals.

However, some minerals become harmful to humans and animals when crushed to dust during mining, quarrying, road building, and other activities. Some dissolve to release chemical elements such as lead, sulfur, and arsenic that are poisonous to plants and animals, including humans. In this chapter, we will consider the nature of minerals and then return to a discussion of minerals as natural resources and as environmental hazards.

2.1 What Is a Mineral?

Pick up any rock and look at it carefully. You will probably see small, differently colored specks, such as those you see in the close-up photo of granite in Figure 2.1. Each speck is a mineral. In the photo, the white grains are the mineral feldspar, the black ones are biotite, and the glassy-gray ones are quartz. A rock is a mixture of

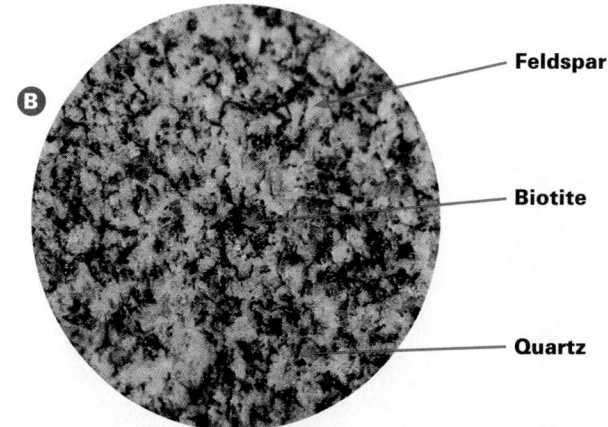

FIGURE 2.1 (A) The Bugaboo Mountains of British Columbia are made of granite. (B) Each of the differently colored grains in this close-up photo of granite is a different mineral. The white grains are feldspar, the black ones are biotite, and the glassy-gray ones are quartz. The granite sample shown here is about 4 centimeters across.

COPYRIGHT AND PHOTOGRAPH BY DR. PARVINDER S. SETHI

COURTESY OF GRAHAM R. THOMPSON/JONATHAN TURK

mineral A naturally occurring inorganic solid with a definite chemical composition and a crystalline structure.

element A substance that cannot be broken down into other substances by ordinary chemical means.

ion An atom with an electrical charge, either positive or negative.

cation A positively charged ion.

anion An ion that has a negative charge.

minerals. Most rocks contain two to five abundant minerals plus minor amounts of several others. Few rocks are made of only one mineral.

Although we can define minerals as the building blocks of rocks, such a definition does not tell us much about the nature of minerals. More precisely, a **mineral** is a naturally occurring inorganic solid with a definite chemical composition and a crystalline structure. *Chemical composition* and *crystalline structure* are the two most important properties of a mineral: They distinguish any mineral from all others. Before discussing these two properties, however, let's briefly consider the other properties of minerals that this definition describes.

Naturally Occurring

A synthetic diamond can be identical to a natural one, but it is not a true mineral, according to our definition, because a mineral must form by natural processes. Like diamonds, most other gems that occur naturally can be manufactured by industrial processes. Natural gems are more highly valued than manufactured ones. For this reason, jewelers should always tell their customers whether a gem is natural or artificial, by prefacing the name of a manufactured gem with the term *synthetic*.

TABLE 2.1 The Eight Most Abundant Chemical Elements in the Earth's Crust*

Element	Symbol	Weight Percent	Atom Percent	Volume Percent[†]
Oxygen	O	46.60	62.55	93.80
Silicon	Si	27.72	21.22	0.90
Aluminum	Al	8.13	6.47	0.50
Iron	Fe	5.00	1.92	0.40
Calcium	Ca	3.63	1.94	1.00
Sodium	Na	2.83	2.64	1.30
Potassium	K	2.59	1.42	1.80
Magnesium	Mg	2.09	1.84	0.30
	Totals	98.59	100.00	100.00

*Abundances are given in percentages by weight, by number of atoms, and by volume.
[†]These numbers will vary somewhat as a function of the ionic radii chosen for the calculations.
Source: From *Principles of Geochemistry* by Brian Mason and Carleton B. Moore. © 1982 by John Wiley & Sons, Inc.

Inorganic

Organic substances are composed mostly of carbon that is chemically bonded to hydrogen. *Inorganic* compounds do not contain carbon–hydrogen bonds. Although organic compounds can be produced in laboratories and by industrial processes, plants and animals create most of Earth's organic material. The tissue of most plants and animals is organic. When these organisms die, they decompose to form other organic substances. Both coal and oil form by the decay of plants and animals and are not minerals because of their organic properties. In addition, oil is not a mineral because it is not solid and has neither a crystalline structure nor a definite chemical composition.

Some material that organisms produce is not organic. For example, limestone, one of the most common sedimentary rocks, is usually composed of the shells of dead corals, clams, and similar marine organisms. Shells, in turn, are made of the mineral calcite. Although organisms produce calcite, the calcite is a mineral: an inorganic solid that formed naturally and has a definite chemical composition and crystalline structure.

Solid

All minerals are solids. Thus, ice is a mineral, but neither water nor water vapor is a mineral.

2.2 The Chemical Composition of Minerals

Minerals are the fundamental building blocks of rocks, but what are minerals made of? Minerals and all other Earth materials are composed of chemical elements. An **element** is a fundamental component of matter that cannot be broken into simpler particles by ordinary chemical processes. Most common minerals consist of a small number—usually two to five—of different chemical elements.

A total of 88 elements occur naturally in Earth's crust. However, eight elements—oxygen, silicon, aluminum, iron, calcium, sodium, potassium, and magnesium—make up more than 98 percent of the crust (Table 2.1). Each element is represented by a one- or two-letter symbol, such as O for oxygen and Si for silicon. In nature, most chemical elements have either a positive or negative electrical charge. For example, oxygen has a charge of negative 2 (–2) and silicon has a charge of plus 4 (+4). An atom with an electrical charge, whether it is positive or negative, is called an **ion**. A positively charged ion is a **cation**; an ion that has a negative charge is an **anion**. Ions with opposite charges are attracted to each other, like the positive end of a magnet attracts the negative end.

Recall that a mineral has a definite chemical composition. A substance with a definite chemical composition

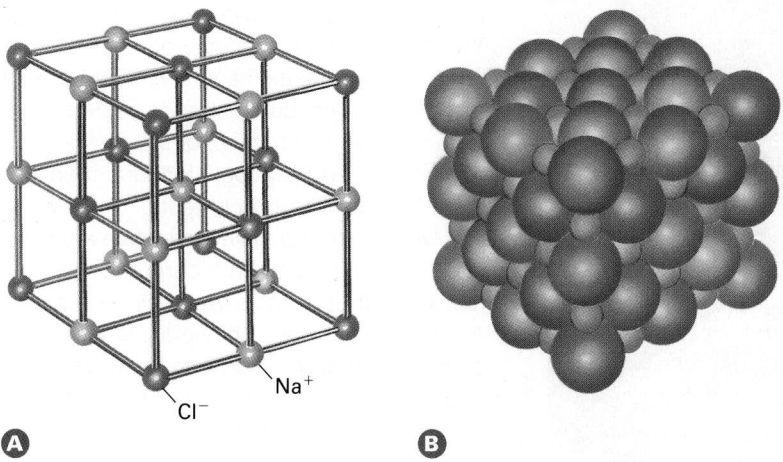

FIGURE 2.2 The orderly arrangement of sodium and chlorine ions in NaCl, or halite. The crystal model in (A) is exploded so that you can see into it; the ions are actually closely packed, as in (B).

is made up of chemical elements that are combined in definite proportions. Therefore, the composition can be expressed as a chemical formula, which is written by combining the symbols of the individual elements. A few minerals, such as gold and silver, consist of only a single element. Their chemical formulas, respectively, are Au (the symbol for gold) and Ag (the symbol for silver). Most minerals, however, are made up of two to five essential elements. For example, the formula of quartz is SiO_2: it consists of one atom of silicon (Si) for every two of oxygen (O). Quartz from anywhere in the Universe has that exact composition. If it had a different composition, it would be some other mineral. The compositions of some minerals, such as quartz, do not vary by even a fraction of a percent. The compositions of other minerals vary slightly, but the variations are limited.

The 88 elements that occur naturally in Earth's crust can combine in many ways to form many different minerals. In fact, more than 3,500 minerals are known. However, the eight abundant elements commonly combine in only a few ways. As a result, only nine **rock-forming minerals** (or mineral "groups") make up most rocks of Earth's crust. We will describe the rock-forming minerals in Section 2.5.

2.3 The Crystalline Nature of Minerals

Every mineral has a crystalline structure, and therefore every mineral is a crystal. A **crystal** is any solid element or compound whose atoms are arranged in a regular, periodically repeated pattern. The mineral halite (common table salt) has the composition NaCl: one sodium ion (Na$^+$) for every chlorine ion (Cl$^-$). Figure 2.2A is

an exploded view of the ions in halite. Figure 2.2B is more realistic, showing the ions in contact. In both sketches the sodium and chlorine ions alternate in orderly rows and columns intersecting at right angles. This orderly, repetitive arrangement of atoms is the **crystalline structure** of halite.

Most minerals initially form as tiny crystals that grow as layer after layer of atoms (or ions) are added to their surfaces. A halite crystal might grow, for example, as salty seawater evaporates from a tidal pool. At first, a tiny grain might form, similar to the sketch of halite in Figure 2.2. This model shows a halite crystal containing 125 atoms; it would be only about one-millionth of a millimeter long on each side, far too small to see with the naked eye. As evaporation continued, more and more sodium and chlorine ions would precipitate onto the faces of the growing crystal.

A **crystal face** is a flat surface that develops if a crystal grows freely in an uncrowded environment, such as halite growing in evaporating seawater. When a mineral grows freely like this, it commonly forms a symmetrical crystal with perfectly flat faces that reflect light like a mirror. In nature, however, mineral grains often impede the growth of adjacent crystals. For this reason, minerals rarely show perfect development of crystal faces. Figure 2.3 is a photomicrograph (a photo taken through a microscope) of a thin slice of granite in which the crystals fit like pieces of a jigsaw puzzle. This interlocking texture developed because some crystals grew around others as molten magma cooled and solidified.

FIGURE 2.3 Photomicrograph of a thin slice (0.03 mm) of a tourmaline-rich granite.

2.4 Physical Properties of Minerals

How does a geologist identify a mineral in the field? Chemical composition and crystal structure distinguish each mineral from all others. For example, halite always consists of sodium and chlorine in a one-to-one ratio, with the atoms arranged in a cubic fashion. But if you pick up a crystal of halite, you cannot see the atoms. You could identify a sample of halite by measuring its chemical composition and crystal structure by laboratory procedures, but such analyses are expensive and time-consuming. Instead, geologists commonly identify minerals by visual recognition, and then confirm the identification with simple tests.

Most minerals have distinctive appearances. Once you become familiar with common minerals, you will recognize them just as you recognize any familiar object. For example, an apple just looks like an apple, even though apples come in many colors and shapes. In the same way, quartz looks like quartz to a geologist. The color and shape of quartz may vary from sample to sample, but it still looks like quartz. Some minerals, however, look like others, so that their physical properties must be examined further to make a correct identification. For example, halite can look like calcite, quartz, and several other minerals, but halite tastes salty and scratches easily with a knife blade, two characteristics that distinguish it from the other minerals. Geologists commonly use properties such as crystal habit, cleavage, fracture, hardness, specific gravity, color, streak, and luster to identify minerals.

FIGURE 2.4 (A) Prismatic quartz grows as elongated crystals. (B) Massive rosy quartz shows no characteristic shape.

A

B

COPYRIGHT AND PHOTOGRAPH BY DR. PARVINDER S. SETHI

FIGURE 2.5 Mica has a single, perfect cleavage plane.

Ⓐ

Ⓑ

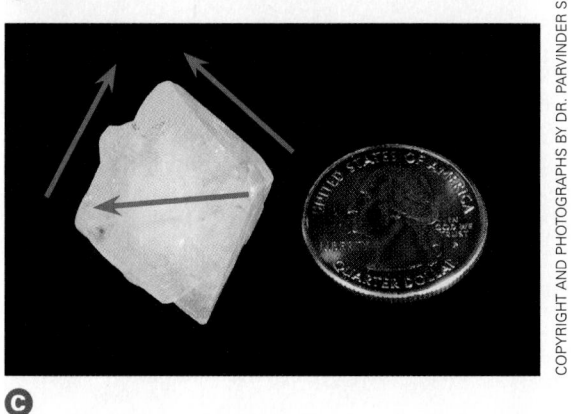

Ⓒ

COPYRIGHT AND PHOTOGRAPHS BY DR. PARVINDER S. SETHI

Crystal Habit

Crystal habit refers to the characteristic shape of an individual crystal and the manner in which aggregates of crystals grow. If a crystal grows freely, it develops a characteristic shape that the arrangement of its atoms controls. Some minerals occur in more than one habit. For example, Figure 2.4A shows quartz with a prismatic (pencil-shaped) habit, and Figure 2.4B shows massive quartz.

Cleavage

Cleavage is the tendency of some minerals to break along flat surfaces. The surfaces are planes of weak bonds in the crystal. Some minerals, such as mica, have one set of parallel cleavage planes (Figure 2.5). Others have two, three, or even four different sets, as shown by the three photos in Figure 2.6. Some minerals, like the micas, have excellent cleavage. You can peel sheet after sheet from a mica crystal as if you were peeling layers from an onion. Others have poor cleavage. Many minerals, such as quartz, have no cleavage at all because they have no planes of weak bonds. The number of cleavage planes, the quality of cleavage, and the angles between cleavage planes all help in mineral identification.

A flat surface created by cleavage and a flat crystal face can appear identical. However, a cleavage surface is duplicated when a crystal is broken, whereas a crystal face is not. So, if you are unsure of which type of flat surface you are looking at, break the sample with a hammer, unless, of course, you want to save it.

Fracture

Fracture is the manner in which a mineral breaks other than along planes of cleavage. Many minerals fracture into characteristic shapes. For instance, *conchoidal*

crystal habit The characteristic shape of an individual crystal, and the manner in which aggregates of crystals grow.

cleavage The tendency of some minerals to break along flat surfaces, which are planes of weak bonds in the crystal. When a mineral has excellent cleavage, sheet after sheet can be peeled from the crystal, like peeling layers from an onion.

fracture The manner in which minerals break, other than along planes of cleavage.

FIGURE 2.6 Some minerals have more than one cleavage plane. (A) Feldspar has two cleavages intersecting at right angles. (B) Calcite has three sets of cleavage planes. Because the three sets do not intersect at right angles, these cleaved crystals have the appearance of deformed boxes. (C) Fluorite has four cleavage planes. Each cleavage face has a parallel counterpart on the opposite side of the crystal, and thus, perfectly cleaved fluorite forms a double pyramid.

Chapter 2: Minerals   21

hardness The resistance of a mineral to scratching, controlled by the bond strength between its atoms.

Mohs hardness scale A scale, based on a series of 10 fairly common minerals and numbered 1 to 10 (from softest to hardest), used to measure and express the hardness of minerals.

specific gravity The weight of a substance relative to the weight of an equal volume of water.

fracture creates smooth, curved surfaces (Figure 2.7). It is characteristic of quartz and olivine. Some minerals break into splintery or fibrous fragments. Most fracture into irregular shapes.

Hardness

Hardness is the resistance of a mineral to scratching and is one of the most commonly used properties for identifying a mineral. It is easily measured and is a fundamental property of each mineral because it is controlled by the bond strength between the atoms in the mineral. Geologists commonly gauge hardness by attempting to scratch a mineral with a knife or other object of known hardness. If the blade scratches the mineral, the mineral is softer than the knife. If the knife cannot scratch the mineral, the mineral is harder.

To measure hardness more accurately, geologists use a scale based on 10 minerals, numbered 1 through 10 (Table 2.2). Each mineral is harder than those with lower numbers on the scale—so 10 (diamond) is the hardest and 1 (talc) is the softest. The scale is known as the **Mohs hardness scale**, after Friedrich Mohs, the Austrian mineralogist who developed it in the early 19th century.

As you can see from Table 2.2, the Mohs scale shows that a mineral scratched by quartz but not by orthoclase has a hardness between 6 and 7. Because the minerals of the Mohs scale are not always handy, it is useful to know the hardness values of common materials. A fingernail has a hardness of slightly more than 2, a pocketknife blade slightly more than 5, window glass about 5.5, and a steel file about 6.5. If you practice with a knife and the minerals of the Mohs scale, you can develop a "feel" for minerals with hardnesses of 5 and under by how easily the blade scratches them.

When testing hardness, it is important to determine whether the mineral has actually been scratched by the object or whether the object has simply left a trail of its own powder on the surface of the mineral. To check, rub away the powder trail and feel the surface of the mineral with your fingernail for the groove of the scratch. Fresh, unweathered mineral surfaces must be used in hardness measurements because weathering often produces a soft rind on minerals.

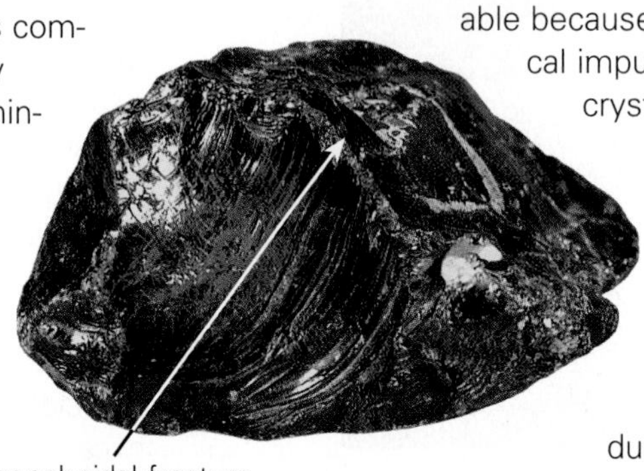

BRECK P. KENT

conchoidal fracture

FIGURE 2.7 Quartz typically fractures along smoothly curved surfaces, called conchoidal fractures. This sample is smoky quartz.

Specific Gravity

Specific gravity is the weight of a substance relative to that of an equal volume of water. If a mineral weighs 2.5 times as much as an equal volume of water, its specific gravity is 2.5. You can estimate a mineral's specific gravity simply by hefting a sample in your hand. If you practice with known minerals, you can develop a feel for specific gravity. Most common minerals have specific gravities of about 2.7. Metals have much greater specific gravities; for example, gold has the highest specific gravity of all minerals, 19. Lead is 11.3, silver is 10.5, and copper is 8.9.

Color

Color is the most obvious property of a mineral, and it is often used in identification. But color can be unreliable because small amounts of chemical impurities and imperfections in crystal structure can dramatically alter color. For example, corundum (Al_2O_3) is normally a cloudy, translucent, brown or blue mineral. Addition of a small amount of chromium can convert corundum to the beautiful, clear, red gem known as ruby. A small quantity of iron or titanium turns corundum into the striking blue gem called sapphire. Quartz occurs in many colors, including white, clear, black, purple, and red, as a result of tiny amounts of impurities and minor defects in the perfect ordering of atoms.

TABLE 2.2 Minerals of the Mohs Hardness Scale

Minerals of Mohs Scale	Common Objects with Similar Hardness
1. Talc	
2. Gypsum	Fingernail
3. Calcite	Copper penny
4. Fluorite	
5. Apatite	Knife blade; window glass
6. Orthoclase	Steel file
7. Quartz	
8. Topaz	
9. Corundum	
10. Diamond	

Streak

Streak refers to the color of the fine powder of a mineral. It is observed by rubbing the mineral across a piece of unglazed porcelain known as a "streak plate." Many minerals leave a streak of powder with a diagnostic color on the plate. Streak is commonly more reliable for identification than the color of the mineral itself. For example, the mineral hematite (iron oxide) can occur both as a dull red mineral or in a shiny black form that closely resembles black mica—but both types will leave the same red powder on a streak plate.

Luster

Luster is the manner in which a mineral reflects light. A mineral with a metallic look, irrespective of color, has a metallic luster. Pyrite is a yellowish mineral with a metallic luster (Figure 2.8). As a result, it looks like gold and is commonly called *fool's gold*. The luster of nonmetallic minerals is usually described by self-explanatory words such as *glassy*, *pearly*, *earthy*, and *resinous*.

Other Properties

Properties such as reaction to acid, magnetism, radioactivity, fluorescence, and phosphorescence can be characteristic of specific minerals. Calcite and some other carbonate minerals dissolve rapidly in acid, releasing visible bubbles of carbon dioxide gas. Minerals such as uranium, which contain radioactive elements, emit radioactivity that can be detected with a scintillometer. Fluorescent materials emit visible light when they are exposed to ultraviolet light. Phosphorescent minerals continue to emit light after the external stimulus ceases.

2.5 Mineral Classes and the Rock-Forming Minerals

Geologists classify minerals according to their chemical elements (Table 2.3). For example, the **silicates** all contain silicon and oxygen; the **carbonates** contain carbon and oxygen; and the **sulfides** contain sulfur. The **native elements** are a small class of minerals, including pure gold and silver, which consist of only a single element. Although more than 3,500 minerals are known in Earth's crust, only a small number—between 50 and 100— are common or valuable, and only nine rock-forming minerals make up most of the crust. These nine are important to geologists simply because they are the most common and abundant minerals. Seven of the rock-forming minerals are silicate minerals. The other two, calcite and dolomite, are carbonates.

Silicates

Silicates make up about 92 percent of Earth's crust. They are abundant for two reasons. First, silicon and oxygen are the two most plentiful elements in the crust. Second, silicon and oxygen combine readily.

To understand silicate minerals, remember four principles:

1. Every silicon atom in Earth's crust surrounds itself with four oxygen atoms. The bonds between silicon and its four oxygens are very strong.

FIGURE 2.8 Pyrite, or fool's gold, has a metallic luster.

streak The color of the fine powder of a mineral, usually obtained by rubbing the mineral on an unglazed, porcelain streak plate.

luster The quality and intensity of light reflected from the surface of a mineral.

silicates Minerals whose chemical elements include silicon and oxygen and whose crystal structures contain silicate tetrahedra. Also, all rocks composed principally of silicate minerals.

carbonates Minerals whose chemical elements include carbon and oxygen as a major part of their chemical composition; an example is calcite $(CaCO_3)$, containing CO_3^{-2} (the "carbonate atom").

sulfides Minerals whose chemical elements include sulfur (S) bonded to a metal ion; an example is pyrite (FeS_2).

native elements Minerals that consist of only one element and thus the element occurs in the native state (not chemically bonded to other elements). With the exception of atmospheric gases, only about 20 elements occur as native elements.

TABLE 2.3 Important Mineral Groups

Group	Member	Formula	Economic Use
Oxides	Hematite	Fe_2O_3	Ore of iron
	Magnetite	Fe_3O_4	Ore of iron
	Corundum	Al_2O_3	Gemstone; abrasive
	Ice	H_2O	Solid form of water
	Chromite	$FeCr_2O_4$	Ore of chromium
Sulfides	Galena	PbS	Ore of lead
	Sphalerite	ZnS	Ore of zinc
	Pyrite	FeS_2	Fool's gold
	Chalcopyrite	$CuFeS_2$	Ore of Copper
	Bornite	Cu_5FeS_4	Ore of copper
	Cinnabar	HgS	Ore of mercury
Sulfates	Gypsum	$CaSO_4 \cdot 2H_2O$	Plaster
	Anhydrite	$CaSO_4$	Plaster
	Barite	$BaSO_4$	Drilling mud
Native elements	Gold	Au	Electronics; jewelry
	Copper	Cu	Electronics
	Diamond	C	Gemstone; abrasive
	Sulfur	S	Sulfa drugs; chemicals
	Graphite	C	Pencil lead; dry lubricant
	Silver	Ag	Jewelry; photography
	Platinum	Pt	Catalyst
Halides	Halite	$NaCl$	Common salt
	Fluorite	CaF_2	Steel making
	Sylvite	KCl	Fertilizer
Carbonates	Calcite	$CaCO_3$	Portland cement
	Dolomite	$CaMg(CO_3)_2$	Portland cement
	Aragonite	$CaCO_3$	Portland cement
Hydroxides	Limonite	$FeO(OH) \cdot nH_2O$	Ore of iron; pigments
	Bauxite	$Al(OH)_3 \cdot nH_2O$	Ore of aluminum
Phosphates	Apatite	$Ca_5(F,Cl,OH)(PO_4)_3$	Fertilizer
	Turquoise	$CuAl_6(PO_4)_4(OH)_8 \cdot 4H_2O$	Gemstone
Silicates	Silicate minerals make up 92 percent of the Earth's crust. (Figure 2.10 summarizes the rock-forming minerals.)		

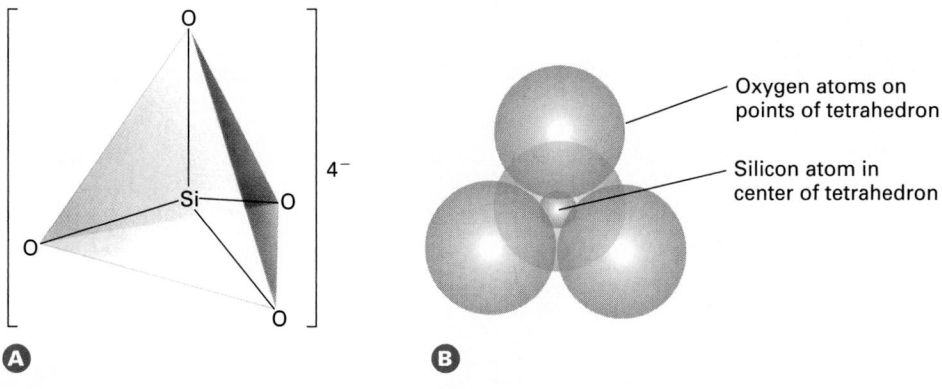

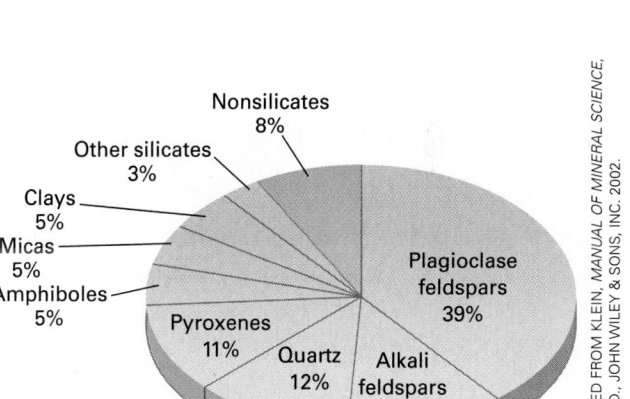

silicate tetrahedron The fundamental building block of all silicate minerals, a pyramid-shaped structure of a silicon ion bonded to four oxygen ions, $(SiO_4)^{4-}$.

Oxygen atoms on points of tetrahedron

Silicon atom in center of tetrahedron

FIGURE 2.9 The silicate tetrahedron consists of one silicon atom surrounded by four oxygen atoms. It is the fundamental building block of all silicate minerals. (A) Schematic representation. (B) Proportionally accurate model.

2. The silicon atom and its four oxygen atoms form a pyramid-shaped structure, called the **silicate tetrahedron**, with silicon in the center and oxygens at the four corners, as illustrated in the two panels of Figure 2.9. The silicate tetrahedron is the fundamental building block of all silicate minerals.

3. Most silicate tetrahedra combine with additional elements. Those elements are mostly other elements in Earth's crust: aluminum, iron, calcium, potassium, sodium, and magnesium. Quartz is the only common silicate that contains only silicon and oxygen.

4. Silicate tetrahedra commonly link together by sharing oxygen atoms to form chains, sheets, or three-dimensional networks as shown in Figure 2.10 (on page 26).

Rock-Forming Silicates

The silicate minerals fall into five groups, based on five ways in which tetrahedra link together. Each group contains at least one of the rock-forming minerals. Figure 2.10 on the next page displays these groups as well as the most common silicate minerals in each group.

Figure 2.11 shows that feldspar alone makes up about half of Earth's crust. Feldspars are divided into two main groups: plagioclase feldspar contains calcium, and alkali feldspar contains potassium or sodium (called *alkali elements*) instead of calcium. Orthoclase is a common alkali feldspar and is abundant in granite and other rocks of the continents. Rocks of the oceanic crust contain about 50 percent plagioclase. Plagioclase and orthoclase often look alike and can be difficult to tell apart.

Quartz comprises 12 percent of crustal rocks. It is widespread and abundant in most continental rocks but rare in oceanic crust and the mantle. Quartz is the only

FIGURE 2.11 The silicate minerals compose 92 percent of Earth's crust. Feldspar alone makes up about 50 percent of the crust, and pyroxene and quartz constitute another 23 percent.

MODIFIED FROM KLEIN, *MANUAL OF MINERAL SCIENCE*, 22ND ED., JOHN WILEY & SONS, INC. 2002.

common silicate mineral that contains no cations other than silicon; it is pure SiO_2. It has a ratio of one 4+ silicon for every two 2- oxygens, so the positive and negative charges neutralize each other perfectly without the addition of other cations.

Pyroxene makes up another 11 percent of the crust. The basalt rock of ocean crust is made up almost entirely of two minerals, plagioclase feldspar and pyroxene. Basalt and granite are described in Chapters 3 and 8.

Basalt commonly contains 1 or 2 percent olivine, but olivine is otherwise uncommon in Earth's crust. However, it makes up a large proportion of mantle rocks, and thus is an abundant and important mineral.

Amphibole, mica, and clay are the other rock-forming silicate minerals. Amphiboles and micas are common in granite and many other types of continental rocks but are rare in oceanic crust. Clay forms when

Class	Unit Composition	Arrangement of SiO$_4$ Tetrahedra	Mineral Examples
(A) Independent tetrahedra	$(SiO_4)^{4-}$		Olivine: 
(B) Single chains	$(SiO_3)^{2-}$	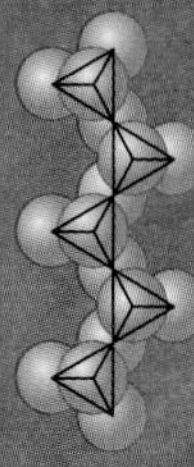	Pyroxene: 
(C) Double chains	$(Si_4O_{11})^{6-}$		Amphibole: 
(D) Sheet silicates	$(Si_2O_5)^{2-}$	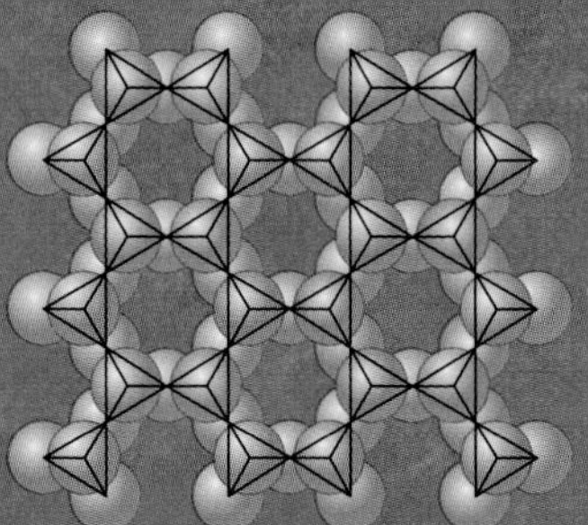	Mica: 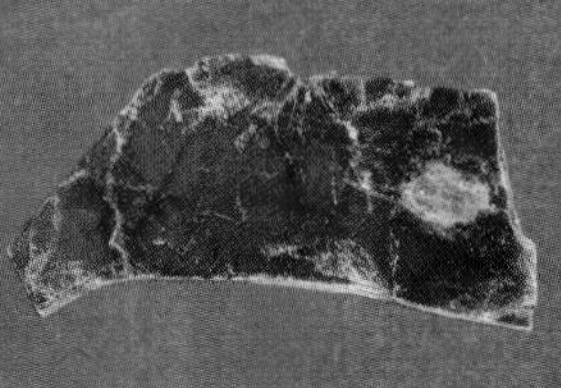
(E) Framework silicates	SiO_2	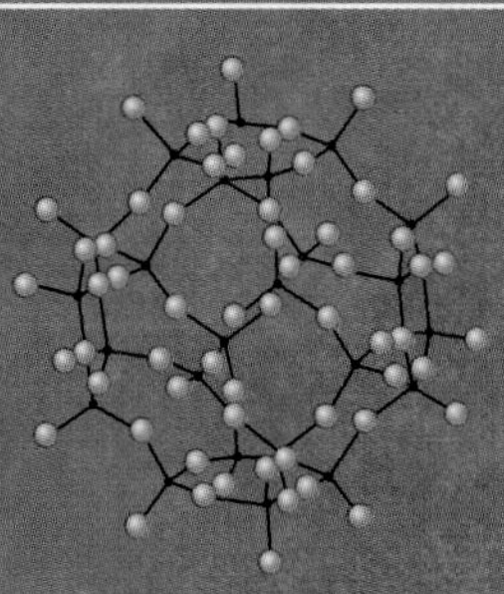	Quartz: 

OLIVINE, MICA, QUARTZ: COPYRIGHT AND PHOTOGRAPHS BY DR. PARVINDER S. SETHI; PYROXENE, AMPHIBOLE: JEFFREY SCOVILL; BLUE BACKGROUND: © NICHOLAS BELTON/ISTOCKPHOTO.COM

FIGURE 2.10 The five silicate structures are based on sharing of oxygen atoms among silicate tetrahedra. (A) Independent tetrahedra share no oxygen atoms. (B) In single chains, each tetrahedron shares two oxygens with adjacent tetrahedra, forming a chain. (C) A double chain is a pair of single chains that are cross-linked by additional oxygen sharing. (D) In the sheet silicates, each tetrahedron shares three oxygens with adjacent tetrahedra. (E) A three-dimensional silicate framework shares all four oxygens of each tetrahedron.

rain and atmospheric gases chemically decompose rocks of all types, and thus clay is common in soils. Clay minerals make up the rock called shale, which is the most abundant of all sedimentary rocks and is described in Chapter 3. Clay is the main raw material in the manufacture of ceramics such as plates and teacups.

Carbonates

Carbonate minerals are much less common than silicates in Earth's crust, but they are important rock-forming minerals because they form sedimentary rocks that cover large regions of every continent, as described in the following chapter. The shells and other hard parts of most marine organisms such as clams, oysters, and corals are made of carbonate minerals. These shells and skeletons accumulate to form the abundant sedimentary rock called limestone, described in Chapter 3. Calcite ($CaCO_3$) forms limestone, and dolomite ($CaMg(CO_3)_2$), makes up a similar rock also called dolomite, or dolostone. Limestone is mined as a raw ingredient of cement.

2.6 Commercially Important Minerals

Many minerals are commercially important even though they are not abundant. Our industrial society depends on metals such as iron, copper, lead, zinc, gold, and silver. **Ore minerals** are minerals from which metals or other elements can be profitably recovered. A few, such as native gold and native silver, are composed of

Dolomite

Calcite

ore minerals Minerals from which metals or other elements can be profitably recovered.

industrial minerals Rocks or minerals that have economic value exclusive of metal ores, fuels, and gems.

gem A mineral that is prized for its rarity and beauty rather than for industrial use.

a single element. However, most metals are chemically bonded to other elements. Iron is commonly bonded to oxygen. Copper, lead, and zinc are commonly bonded to sulfur to form sulfide ore minerals, such as the galena in Figure 2.12.

Industrial minerals are not metal ores, fuels, or gems but have economic value nonetheless. They are not considered "ore" because they are mined for purposes other than the extraction of metals. Halite is mined for table salt, and gypsum is mined for plaster and sheetrock. Apatite and other phosphorus minerals are sources of the phosphate fertilizers crucial to modern agriculture. Limestone is the raw material of cement. Native sulfur, used to manufacture sulfuric acid, insecticides, fertilizer, and rubber, is mined from the craters of dormant and active volcanoes, where it is deposited from gases emanating from the vents, like the one in Figure 2.13.

A **gem** is a mineral that is prized primarily for its rarity and/or beauty, although some gems, such as diamonds, are also used industrially. Depending on its value, a gem can be either *precious* or *semiprecious*. Precious gems include diamond, emerald, ruby, and

FIGURE 2.12 Galena (lead sulfide) is the most important ore of lead.

FIGURE 2.13 Yellow native sulfur is forming today in the vent of the Ollague volcano on the Chile–Bolivia border.

sapphire (Figure 2.14). Several varieties of quartz, including amethyst, agate, jasper, and tiger's eye, are semiprecious gems. Garnet, olivine, topaz, turquoise, and many other minerals can also occur as attractive semiprecious gems (Figure 2.15).

When you look at a lofty mountain or a steep cliff, you might not immediately think about the tiny mineral grains that form the rocks. Yet minerals are the building blocks of Earth. In addition, some minerals provide the basic resources for our industrial civilization.

As we will see throughout this book, Earth's surface and the human environment change as minerals react with the air and water, with other minerals, and with living organisms.

2.7 Harmful and Dangerous Rocks and Minerals

Most rocks and minerals in their natural states are harmless to humans and other organisms. That is not surprising because all of us evolved among the minerals and rocks that make up Earth, and species would not survive if they were poisoned by their surroundings. A few rocks and minerals, however, are harmful and dangerous. Asbestos is one such mineral; it is a

FIGURE 2.14 Sapphire is one of the most precious gems.

Unit 1: Earth Materials and Time

FIGURE 2.15 Topaz, a colorless crystal, is a popular semiprecious gem.

powerful *carcinogen*—a cancer-causing substance. Some rocks and minerals emit radon gas, a radioactive carcinogen. Other rocks and minerals contain toxic elements such as mercury, lead, arsenic, or sulfur.

In nature, most environmentally hazardous rocks and minerals are buried beneath the surface, where they are unavailable to plants and animals. Natural weathering and erosion expose them so slowly that they do little harm. Most minerals that contain toxic metals or other elements such as lead, mercury, and arsenic are also relatively insoluble. In their natural environments they weather so slowly that they release the toxic materials in low concentrations. However, if pollution controls are inadequate, mining, milling, and smelting can concentrate and release hazardous natural materials at greatly accelerated rates, poisoning humans and other organisms.

Silicosis and Black Lung

Most common minerals such as feldspar and quartz are harmless in their natural states in solid rock. However, if these minerals are ground to dust, they can enter the lungs and cause serious and even fatal inflammation and scarring of the lung tissue, called silicosis. In advanced cases, the lungs become inflamed and may fill with fluid, causing severe breathing difficulty and low blood oxygen levels. On-the-job exposure to silica dust can occur in mining, stonecutting, quarrying, building and road construction, working with abrasives, sandblasting, and other occupations and hobbies. Intense exposure to silica may produce silicosis in a year or less, but it usually takes at least 10 or 15 years of exposure before symptoms develop. Silicosis has become less common since the Occupational Safety and Health Administration (OSHA) instituted regulations requiring the use of protective equipment that limits the amount

FIGURE 2.16 Chrysotile asbestos is believed by some to be a less potent carcinogen, but is still being actively removed from public buildings.

of silica dust inhaled. Coal dust, inhaled by coal miners in large quantities before federal laws required dust suppression in coal mines, has similar effects and causes a disease called black lung.

Asbestos, Asbestosis, and Cancer

Asbestos is an industrial name for a group of minerals that crystallize as long, thin fibers. One type is a sheet silicate mineral called chrysotile, which has a crystal structure and composition similar to the micas and which forms tangled, curly fibers (Figure 2.16). The other type of asbestos includes four similar varieties of amphibole that crystallize as straight, sharply pointed needles. Chrysotile has been the more valuable and commonly used type of asbestos because the fibers are flexible, tough, and can be woven into fabric. It was considered to be a less harmful form because some believed its curly fibers would be more easily expelled from the lungs. However, recent studies show that all forms of asbestos are carcinogenic.[1]

Asbestos is commercially valuable because it is flameproof, chemically inert, and extremely strong. For example, chrysotile fibers are 8 times stronger than a steel wire of equivalent diameter. Asbestos has been

1. "Chrysotile Asbestos Fact Sheet," Environmental Information Association, 2009 (http://www.eia-usa.org/wp-content/post-files/ chrysotile-fact-sheet1.pdf); "Scientific and Medical Facts about Chrysotile Asbestos Released by the Environmental Information Association and the ADAO," *Medical News Today*, April 29, 2009 (http://www.medicalnewstoday.com/articles/147938.php); and "Chrysotile Asbestos Represents Little Health Hazard, Study Says," *Air Conditioning, Heating & Refrigeration News*, January 29, 1990 (http://www.accessmylibrary.com/article-1G1-8344361/chrysotile-asbestos-represents-little.html).

used to manufacture brake linings, fireproof clothing, insulation, shingles, tile, pipe, and gaskets but now is allowed only in brake pads, shingles, and pipe.

In the early 1900s, asbestos miners and others who worked with asbestos learned that prolonged exposure to the fibers caused an irreversible respiratory disease called asbestosis. Later, in the 1950s and 1960s, studies showed that asbestos also causes lung cancer and other forms of cancer. One reason that so much time passed before scientists recognized the cancer-causing properties of asbestos is that the disease commonly does not develop until decades after exposure to asbestos.

Although it is not clear how asbestos fibers cause cancer, it seems that the shapes of the crystals play an important role. Statistical studies also show that the sharp, pointed amphibole asbestos is a more potent carcinogen than the flexible chrysotile fibers, although both types cause cancer. In response to growing awareness of the health effects of asbestos, the Environmental Protection Agency (EPA) banned its use in building construction in 1978. However, the ban did not address the issue of what should be done with the asbestos already installed. In 1986, Congress passed a ruling called the Asbestos Hazard Emergency Response Act, requiring that all schools be inspected for asbestos. Public response has resulted in hasty programs to remove asbestos from schools and other buildings at a cost of billions of dollars. But what is the real level of hazard?

Most asbestos in buildings is the less potent chrysotile, woven into cloth or glued into a tight matrix, and often the surface has been further stabilized by painting. Therefore, the fibers are not free to blow around. The levels of airborne asbestos in most buildings are no higher than those in outdoor air. Some scientists argue that asbestos insulation, despite being carcinogenic, poses no health danger if left alone—but when the material is removed, it is disturbed and asbestos dust escapes. Not only are workers endangered, but airborne asbestos persists in the building for months after completion of the project.

Radon and Cancer

Radon is one of a series of elements formed by the radioactive decay of uranium. Uranium occurs naturally in small concentrations in several minerals and in all types of rock, but it concentrates in two abundant rocks—granite and shale—that are described in the next chapter. It is also found in soil that formed from granite and shale, and in construction materials made from those rocks, such as concrete and concrete blocks. Radon is itself radioactive, and it decays into other radioactive elements. Because of their radioactivity, radon and its decay products are carcinogenic, and because it is a gas, we breathe radon into our lungs.

Radon seeps from the ground into homes and other buildings, where it concentrates in indoor air and causes an estimated 5,000 to 20,000 cancer deaths per year among Americans. The risk of dying from radon-caused lung cancer in the United States is about 0.4 percent over a lifetime, much greater than the risk of dying from cancer caused by asbestos, pesticides, or other air pollutants and nearly as high as the risk of dying in an auto accident, from a fall, or in a fire at home.

Not all Americans are exposed to equal amounts of radon. Some homes contain very low concentrations of the gas; others have high concentrations. The variations in concentration are due to two factors: geology and home ventilation. Geology is important because some types of rocks, such as granite and shale, contain high concentrations of uranium and radon while others contain relatively little.

As radon forms by slow radioactive decay of uranium in bedrock, soil, or construction materials, it seeps into the basement of a home and circulates throughout the house. Radon concentrations are highest in poorly ventilated homes built on granite or shale, on soil derived from these rocks, and in homes constructed with concrete and concrete blocks containing these types of rocks. The highest home radon concentrations ever measured were found in houses built on the Reading Prong, a uranium-rich body of granite extending from Reading, Pennsylvania, through northern New Jersey and into New York. The air in one home in this area contained 700 times more radon than the EPA "action level"—the concentration at which the Environmental Protection Agency recommends that corrective measures be taken to reduce the amount of radon in indoor air.

People should ask two questions in regard to radon hazards: "What is the radon concentration in my home?" and "If it is high, what can be done about it?" Because radon is radioactive, it can be measured with a simple detector available at most hardware stores or from local government agencies for about $25. If the detector indicates excessive radon, three types of solutions can be implemented. The first is to extend a ventilation duct from the basement to the outside of the house. This solution prevents basement air from circulating through the house and also prevents radon from accumulating in the basement. The second solution is to ventilate the house so that indoor air is continually refreshed. However, this method allows hot air to escape and thus increases heating bills. A third solution is to pump outside air into the house to keep indoor air at a slightly higher pressure than the outside air. This positive pressure prevents gas from seeping from soil or bedrock into the basement.

It is impossible to avoid exposure to radon completely because it is everywhere—in outdoor air as well as in homes and other buildings. But it is relatively easy and inexpensive to minimize exposure and thus avoid a significant cause of cancer.

Acid Mine Drainage and Heavy Metals Contamination

Sulfide ore minerals are combinations of metals such as lead, zinc, copper, cadmium, mercury, and other heavy metals with sulfur. Some contain arsenic or other toxic elements in addition to the metals. These minerals are mined for their metals, which are essential to modern industrial societies.

However, mining and refining these minerals can create serious air and water pollution problems. When sulfide ore minerals are mined or refined without adequate pollution control, sulfur escapes into streams, ground water, and the atmosphere, where it forms hydrogen sulfide and sulfuric acid. Waterborne sulfides poison aquatic organisms, and atmospheric sulfur compounds contribute to acid precipitation.

Other toxic elements such as lead, mercury, cadmium, and arsenic can escape from mine wastes and smelters into the atmosphere and water. In addition, smelting and refining converts some poisonous elements, such as mercury, lead, and arsenic, from relatively insoluble to highly soluble forms, which are toxic. Most of the sulfur and toxic metals can be removed by pollution-control devices, which are required by law in the United States.

3

ROCKS

visit **4ltrpress.cengage.com**

> # It isn't the mountains ahead to climb that wear you out; it's the pebble in your shoe.
>
> *Muhammad Ali*

Earth is solid rock to a depth of 2,900 kilometers, where the mantle meets the liquid outer core. Even casual observation reveals that rocks are not all alike. The great peaks of the Sierra Nevada in California are hard, strong granite. The red cliffs of the Utah desert are soft sandstone. The top of Mount Everest is limestone containing clamshells and the remains of other small marine animals.

The marine fossils atop Mount Everest tell us that this limestone formed in the sea. What forces lifted the rock to the highest point of the Himalayas? Where did the vast amounts of sand in the Utah sandstone come from? How did the granite of the Sierra Nevada form?

All of these questions ask about the processes that formed the rocks and changed them throughout geologic history. In this chapter we will study rocks: how they form and what they are made of.

3.1 Rocks and the Rock Cycle

A rock is a solid aggregate of one or more minerals. Geologists group rocks into three categories on the basis of how the rocks form: igneous rocks, sedimentary rocks, and metamorphic rocks.

Earth's interior is hot and dynamic. The high temperature that exists within a few hundred kilometers of the surface can melt solid rock to form a molten liquid called **magma**, which then rises slowly toward Earth's surface. As it rises, the magma cools to become solid rock again. **Igneous rock** forms when magma solidifies.

All rocks may seem permanent and unchanging over a human lifetime, but this apparent permanence is an illusion created by our short observational time frame. Over geologic time, water and air attack rocks of all kinds at Earth's surface through the process called **weathering**, breaking them down into smaller particles. These particles—including gravel, sand, clay, and all other fragments weathered and eroded from rock—accumulate in loose, unconsolidated layers called **sediment**. Sand on a beach and mud on a lake bottom are examples of sediment. Weathering processes also form ions such as sodium and calcium dissolved in ground water and streams. Streams, wind, glaciers, and gravity then erode the sediment and dissolved ions and carry them downhill to deposit them at lower elevations. **Sedimentary rock** forms when sediment becomes cemented or compacted into solid rock. Marine organisms such as clams, oysters, and corals extract dissolved calcium from seawater. They combine the calcium with carbon dioxide that dissolves in seawater from Earth's

COPYRIGHT AND PHOTOGRAPHS BY DR. PARVINDER S. SETHI
PHOTO ALBUM/ISTOCKPHOTO.COM

igneous rock from magma, Hawaiian Volcanoes National Park

metamorphic rock from heat and pressure

magma Molten rock generated from below Earth's crust, from which igneous rock is formed.

igneous rock Rock that forms when magma rises to Earth's surface, cools, and solidifies.

weathering The decomposition and disintegration of rocks and minerals at Earth's surface by chemical and physical processes.

sediment Solid rock or mineral fragments that are transported and deposited by wind, water, gravity, or ice; that are weathered by natural forces, precipitated by chemical reactions, or secreted by organisms; and that accumulate in loose, unconsolidated layers.

sedimentary rock Rock formed when sediment becomes cemented or compacted through the process of lithification.

COPYRIGHT AND PHOTOGRAPH BY DR. PARVINDER S. SETHI

metamorphic rock A rock formed when igneous, sedimentary, or other metamorphic rocks recrystallize in response to elevated temperature, increased pressure, chemical change, and/or deformation.

rock cycle The sequence of events in which rocks are formed, destroyed, altered, and reformed by geological processes.

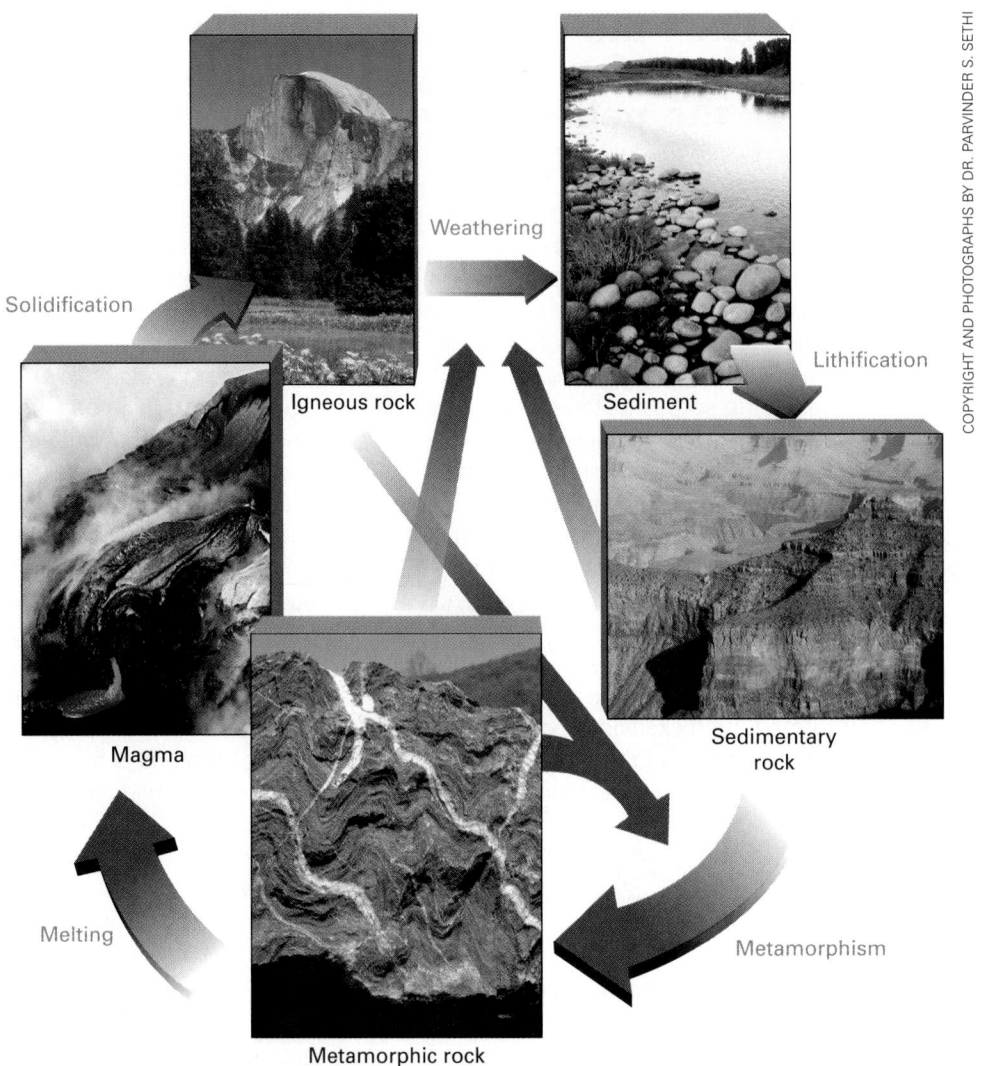

Solidification

Weathering

Lithification

Igneous rock

Sediment

Magma

Sedimentary rock

Melting

Metamorphism

Metamorphic rock

COPYRIGHT AND PHOTOGRAPHS BY DR. PARVINDER S. SETHI

FIGURE 3.1 The rock cycle shows that rocks change over geologic time. The arrows show paths that rocks can follow as they change.

atmosphere, and use it to form their shells and hard parts. After the organisms die, the remains of those shells and corals accumulate and become the sedimentary rock called limestone. Thus, limestone, one of the most common sedimentary rocks, forms by direct interactions among the geosphere, the hydrosphere, the biosphere, and the atmosphere.

A **metamorphic rock** forms when heat, pressure, or hot water alter any preexisting rock. For example, Earth's crust may slowly sink, forming a depression that may be hundreds of kilometers in diameter and thousands of meters deep. Sediment accumulates in the depression, becomes cemented, and forms sedimentary rock. As more sediment accumulates, it buries the deepest layers under a huge weight. The deep burial raises the temperature and pressure on those rocks and alters both the minerals and the texture to transform the sedimentary rock to metamorphic rock. This has been happening for the

past 150 million years where the Mississippi River has carried billions of tons of mud and sand to the northern part of the Gulf of Mexico, depositing a pile of sediment that now measures more than 12 kilometers in thickness in the deepest part of the basin.

No rock is permanent over geologic time; instead, all rocks undergo processes that change them from one of the three rock types to another. This continuous process is called the **rock cycle** (Figure 3.1). In the example of metamorphism just described, sediment has accumulated beneath the Mississippi Delta to form sedimentary rock. As the basin sank, new sediment buried the sedimentary rock to greater depths where rising temperature and pressure have converted it to metamorphic rock. If the temperature rises sufficiently at some future time, the metamorphic rock will melt to form magma. The magma will then rise and solidify to become igneous rock. Millions of years later, movement of Earth's crust might raise the igneous rock to the surface, where it will weather to form sediment. Rain and streams will then wash the sediment into a new basin, renewing the cycle.

The rock cycle does not follow a set order, and can take many different paths. The term "rock cycle" simply expresses the idea that rocks are not permanent but change over geologic time.

The Rock Cycle and Earth Systems Interactions

The rock cycle illustrates several types of Earth systems interactions—interactions among rocks, the atmosphere, the biosphere, and the hydrosphere. Rain and air, aided by acids and other chemicals secreted by plants, decompose solid rocks to form large amounts of clay and other sediment. During these processes, water and atmospheric gases react chemically to become incorporated into the clay. Thus, these processes transfer water and air from the atmosphere and hydrosphere to the solid minerals of the geosphere. More rain then washes the clay and other sediment into streams, which carry it to a sedimentary basin such as the Mississippi River delta on the edge of a continent. Recall from Chapter 1 that solar energy drives the hydrologic cycle. Thus, sunlight evaporates moisture to form rain, which in turn feeds flowing streams. But these same processes are also part of the rock cycle, illustrating once again that all Earth processes interact.

The rock cycle is also driven by Earth's internal heat. For example, when a sedimentary basin sinks under the weight of additional sediment, the deeper layers become heated and metamorphosed by Earth's heat. The same heat may melt the rocks to produce magma. But then the magma rises upward and perhaps even erupts onto Earth's surface from a volcano. In this way, heat is transferred from Earth's interior to the atmosphere.

Throughout this chapter, we will emphasize these and other interactions among the atmosphere, biosphere, hydrosphere, and rocks of the geosphere to illustrate the point that those systems continuously exchange both energy and material so that Earth functions as a single, integrated system.

3.2 Igneous Rocks

Magma: The Source of Igneous Rocks

If you drilled a well deep into the crust, you would find that Earth's temperature rises about 30°C for every kilometer of depth. Below the crust, temperature continues to rise, but not as rapidly. In the mantle between depths of 100 and 350 kilometers, the temperature is so high that rocks in some areas melt to form magma. The temperature of magma varies from about 600°C to 1,400°C, depending on its chemical composition and the depth at which it forms. As a comparison, an iron bar turns red-hot at about 600°C and melts at slightly over 1,500°C.

When rock melts, the resulting magma expands by about 10 percent, so it is less dense than the rock around it and therefore rises as it forms—much as a hot air balloon ascends in the atmosphere. When magma rises, it enters the cooler environment near Earth's surface, where it solidifies to form solid igneous rock.

Types of Igneous Rocks

Some igneous rock forms when magma solidifies within Earth's crust; other igneous rock is created when magma erupts onto the surface. When magma solidifies, it usually crystallizes to form minerals. The **texture** of a rock refers to the size, shape, and arrangement of its mineral grains, or crystals (Table 3.1). Some igneous rocks consist of mineral grains that are too small to be seen with the naked eye; others are made up of thumb-sized, or even larger, crystals.

Extrusive (Volcanic) Rocks

When magma rises all the way through the crust to erupt onto Earth's surface, it forms **extrusive igneous rock**, also called *volcanic rock*. **Lava**, a subcategory of extrusive igneous rock, is fluid magma that flows from a crack or a volcano onto Earth's surface. The term also refers to the rock that forms when lava cools and becomes solid. After lava erupts onto the relatively cool Earth surface, it solidifies rapidly—perhaps over a few days or years. Crystals form but do not have much time to grow. As a result, many volcanic rocks have fine-

texture The size, shape, and arrangement of mineral grains, or crystals, in a rock.

extrusive igneous rock Igneous rock formed from material that has erupted through the crust onto the surface of Earth; usually fine grained. Also called *volcanic rock.*

lava Fluid magma that flows onto Earth's surface from a volcano or fissure. Also, the rock formed by solidification of the same material.

TABLE 3.1 Igneous Rock Textures Based on Grain Size

Name of Texture	Grain Size
Glassy	No mineral grains (obsidian)
Very fine grained	Too fine to see with naked eye
Fine grained	Up to 1 millimeter
Medium grained	1 to 5 millimeters
Coarse grained	More than 5 millimeters
Porphyry	Relatively large grains in a finer-grained matrix

intrusive igneous rock A rock formed when magma solidifies within Earth's crust, without erupting to the surface; usually medium to coarse grained. Also called *plutonic rock*.

grained textures, with crystals too small to be seen with the naked eye. Basalt is a common very fine-grained volcanic rock.

In unusual circumstances, molten lava may solidify within a few hours of erupting. Because the magma hardens so quickly, the atoms have no time to align themselves to form crystals. As a result, the atoms are arranged in a random chaotic pattern, as happens in glass. Volcanic glass is called obsidian (Figure 3.2).

If magma rises slowly through the crust before erupting, some crystals may grow while most of the magma remains molten. If this mixture of magma and crystals then erupts onto the surface, the magma solidifies quickly, forming porphyry, a rock with the large crystals, called phenocrysts, embedded in a fine-grained matrix (Figure 3.3).

Intrusive (Plutonic) Rocks

When magma solidifies *within* the crust, without erupting to the surface, it is an **intrusive igneous rock**, also called *plutonic rock* (and the body of rock itself is called a *pluton*). The overlying rock insulates the magma like a thick blanket. The magma crystallizes slowly and the crystals grow over hundreds of thousands, or even millions, of years. As a result, most plutonic rocks are medium to coarse grained. Granite, the most abundant rock in continental crust, is a medium- or coarse-grained plutonic rock. The crystals in granite are clearly visible.

Many are a millimeter or so across, although some crystals may be much larger.

Naming and Identifying Igneous Rocks

Geologists use both minerals and texture to name igneous rocks. For example, any medium- or coarse-grained igneous rock consisting mostly of feldspar and quartz is called granite. Rhyolite also consists mostly of feldspar and quartz but is very fine grained. You can see the difference in the two panels of Figure 3.4. The same magma that solidifies slowly within the crust to form granite can also erupt onto Earth's surface to form rhyolite.

Like granite and rhyolite, most common igneous rocks are classified in pairs, each member of a pair containing the same minerals but having a different texture. The texture depends mainly on whether the rock is volcanic (fine grained) or plutonic (coarse grained).

Once you learn to identify the rock-forming minerals, it is easy to name a coarse-grained plutonic rock because the minerals are large enough to see. It is more difficult to name many volcanic rocks because the minerals are too small to identify. A field geologist often uses color to name a volcanic rock. Rhyolite is usually light in color: white, tan, red, and pink are common. Many andesite rocks are gray or green. Basalt is commonly black. The minerals in many volcanic rocks cannot be identified even with a microscope because of their tiny crystal sizes. In this case, definitive identification is based on chemical and X-ray diffraction analyses carried out in the laboratory.

FIGURE 3.2 Obsidian is natural volcanic glass. It contains no crystals. This sample is about 6 inches wide.

FIGURE 3.3 Porphyry is an igneous rock containing large crystals embedded in a fine-grained matrix. This is a rhyolite porphyry.

Common Igneous Rocks

Before proceeding with our discussion, you should become familiar with some terminology. Geologists commonly use the terms *basement rock*, *bedrock*, *parent rock*, and *country rock*. **Basement rock** is the igneous and metamorphic rock that lies beneath the thin layer of sediment and sedimentary rocks covering much of Earth's surface, thus forming the "basement" of the crust. **Bedrock** is the solid rock that lies beneath soil or unconsolidated sediments. It can be igneous, metamorphic, or sedimentary. **Parent rock** is any original rock before it is changed by weathering, metamorphism or other geological processes. The rock that is already in an area and is cut into by intrusive igneous rock or by a mineral deposit is called **country rock**.

granite

Granite and Rhyolite

Granite contains mostly feldspar and quartz. Small amounts of black biotite or hornblende often give it a speckled appearance. Granite (and metamorphosed granite) is the most common rock in continental crust. It is found nearly everywhere as basement rock, beneath the relatively thin veneer of sedimentary rocks and soil that covers most of the continents. Granite is hard and resistant to weathering; it forms steep, sheer cliffs in many of the world's great mountain ranges. Mountaineers prize granite cliffs for the steepness and strength of the rock.

As granitic magma (magma with the chemical composition of granite) rises through Earth's crust, some of it may erupt from a volcano to form rhyolite, while the remainder solidifies beneath the volcano, forming granite. Most obsidian forms from magma with a granitic (rhyolitic) composition.

Basalt and Gabbro

Basalt consists of approximately equal amounts of plagioclase feldspar and pyroxene. It makes up most of the oceanic crust, as well as huge basalt plateaus on continents. Gabbro is the plutonic counterpart of basalt; it is mineralogically identical but consists of larger crystals. Gabbro is uncommon at Earth's surface, although it is abundant in deeper parts of the oceanic crust, where basaltic magma crystallizes slowly.

Andesite and Diorite

Andesite is a volcanic rock intermediate in composition between basalt and granite. It is commonly gray or green

Grain size: it's a matter of time

rhyolite

FIGURE 3.4

and consists of plagioclase feldspar and dark minerals (usually biotite, amphibole, or pyroxene). It is named for the Andes Mountains, the volcanic chain on the western edge of South America, where it is abundant. Because it is volcanic, andesite is typically very fine grained. **Diorite** is the plutonic equivalent of andesite. It forms from the same magma as andesite and, consequently, often underlies andesitic mountain chains such as the Andes.

Peridotite and Komatiite

Peridotite is an *ultramafic* (rich in magnesium and iron) igneous rock that makes up most of the upper mantle but is rare in Earth's crust. It is coarse grained and composed of olivine and small amounts of pyroxene, amphibole, or mica, but no feldspar. Komatiite is the fine-grained, extrusive, equivalent of peridotite. Geologists think that it was the material of the earliest crust that formed as the molten planet cooled more than 4 billion years ago. Only a few traces of this primordial crust are known to exist today.

basement rock The older igneous and metamorphic rock that lies beneath the thin layer of sedimentary rocks and soil covering much of Earth's surface, forming the "basement" of the crust.

bedrock The solid rock that lies beneath soil or unconsolidated sediments; it can be igneous, metamorphic, or sedimentary.

parent rock Any original rock before it is changed by weathering, metamorphism, or other geological processes.

country rock The older rock already in an area, cut into by a younger igneous intrusion or mineral deposit.

lithification The process by which loose sediment is converted to solid rock.

precipitation A chemical reaction that produces a solid salt, called a *precipitate*, from a liquid solution.

3.3 Sedimentary Rocks

Over geologic time, the atmosphere, biosphere, and hydrosphere weather rock and convert it to clay, sand, gravel, and ions dissolved in water. Both chemical and physical processes weather rocks. Flowing water, wind, gravity, and glaciers erode the decomposed rock, transport it downslope, and deposit it on the seacoast or in lakes and river valleys. With time, the loose, unconsolidated sediment becomes compacted and cemented—a process called **lithification**—and forms sedimentary rock. Rivers carry dissolved ions into the sea, where they accumulate in the oceans. Marine organisms such as clams, oysters, and corals may then use these ions to form shells and other hard parts, which accumulate after the organisms die and then form limestone. Sedimentary rocks make up only about 5 percent of Earth's crust. However, because they form on Earth's surface, sedimentary rocks are widely spread in a thin veneer over underlying igneous and metamorphic rocks. As a result, they cover about 75 percent of continents. The formation and, in many cases, the lithification of sediment involve complex system interactions among Earth's spheres. As you read the descriptions in this section, pay special attention to the interplay among soil, air, water, and life that forms the rocks at Earth's surface.

To begin, sedimentary rocks are broadly divided into four categories:

1. Clastic sedimentary rock is composed of particles of weathered rocks, such as sand grains and pebbles, called *clasts*, which have been transported, deposited, and lithified. (The generic term *clastic* refers to any rocks that are composed of fragments of older rocks.) This category includes shale, siltstone, and sandstone. Clastic sedimentary rock makes up about 85 percent of all sedimentary rock (Figure 3.5).
2. Organic sedimentary rock consists of the lithified remains of plants or animals. Coal is an organic sedimentary rock made up of decomposed and compacted plant remains.
3. Chemical sedimentary rock forms by direct precipitation of minerals from solution. Rock salt, for example, forms when salt precipitates from evaporating seawater or saline lake water. (**Precipitation** is a chemical reaction that produces a solid salt, called a *precipitate*, from a liquid solution.)
4. Bioclastic sedimentary rock is composed of broken shell fragments and similar remains of

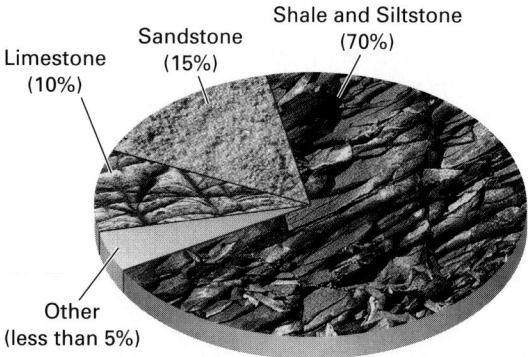

FIGURE 3.5 Earth's sedimentary rocks: Shale, siltstone, and sandstone are clastic rocks that make up about 85 percent of all sedimentary rocks. Limestone and some "other" sedimentary rocks make up less than 15 percent.

Limestone (10%)
Sandstone (15%)
Shale and Siltstone (70%)
Other (less than 5%)

living organisms. The fragments are clastic, but they have a biological origin. Most limestone formed from broken shells and is a bioclastic sedimentary rock.

Clastic Sedimentary Rocks

Clastic sediment is called gravel, sand, silt, or clay, in order of decreasing particle size. As clastic particles ranging in size from coarse silt to boulders tumble downstream, their sharp edges are worn off and they become rounded, like those in Figure 3.6. Finer silt and clay do not round effectively because they are so small and light that water and wind cushion them as they bounce along.

If you fill a measuring cup with sand (or some other clastic sediment), you can still add a substantial amount of water that occupies empty space—or

FIGURE 3.6 Boulders and rocks collide as they move downstream. The collisions wear away the sharp edges, producing rounded stones.

COPYRIGHT AND PHOTOGRAPH BY DR. PARVINDER S. SETHI

Unit 1: Earth Materials and Time

pore space—among the sand grains (Figure 3.7A). Commonly, sand and similar sediment have about 20 to 40 percent pore space.

As more sediment accumulates, the weight of overlying layers compresses the buried sediment. Some of the water is forced out, and the pore space shrinks (Figure 3.7B). This process is called **compaction**. If the grains have platy (flat and disk-like) shapes, as in clay and silt, compaction alone may lithify the sediment as the platy grains interlock like pieces of a puzzle.

As sediment is buried, water circulates through the pore space. This water commonly contains dissolved ions that precipitate in the pore spaces. When the ions precipitate, they form a *cement* that binds the clastic grains firmly together to form a hard rock (Figure 3.7C). Calcite, quartz, and iron oxides are the most common cements in sedimentary rocks.

The time required for lithification of sediment varies greatly. In some heavily irrigated areas of southern California, calcite has precipitated from irrigation water to cement soils within only a few decades. In the Rocky Mountains, calcite has cemented some glacial deposits that are less than 20,000 years old. In contrast, sand and gravel deposited in southwestern Montana about 30 million years ago can still be dug with a hand shovel.

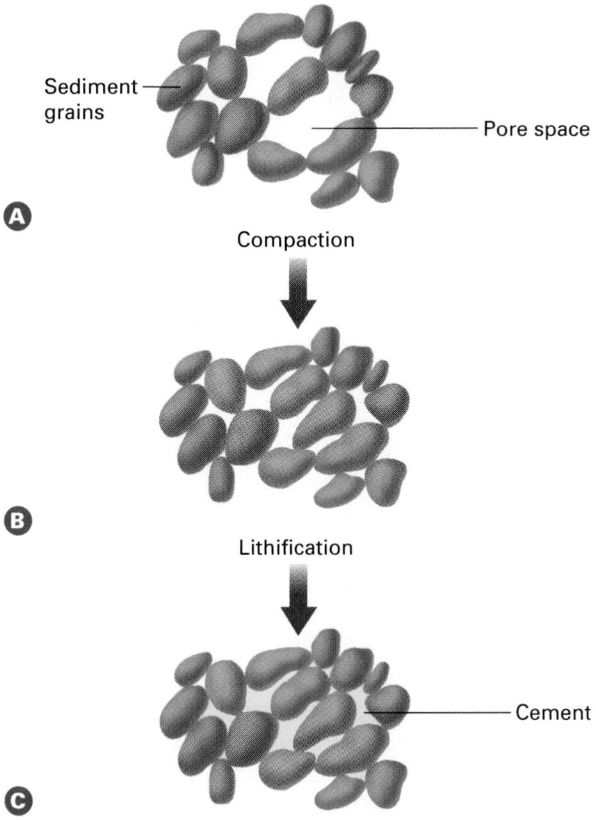

Conglomerate is lithified gravel. Each clast in a conglomerate is usually much larger than the individual mineral grains in the clast. In many conglomerates, the clasts may be fist sized or even larger. Therefore, the clasts retain most of the characteristics of the parent rock. If enough is known about the geology of an area where conglomerate is found, it may be possible to identify exactly where the clasts originated. For example, a granite clast probably came from nearby granite bedrock.

The next time you walk along a gravelly streambed, look carefully at the gravel. You will probably see sand or silt trapped among the larger clasts. In a similar way, most conglomerates contain fine sediment among the large clasts.

Sandstone consists of lithified sand grains (Figure 3.8). Granite, the most common rock in continents, is composed largely of quartz and feldspar. When granite and similar rocks weather, the feldspar reacts chemically with air and water to form tiny clay crystals. In contrast, quartz is resistant to chemical weathering and the unaltered quartz grains simply accumulate as sand in soil. Eventually rain erodes the sand, and streams carry it toward the seacoast. The sand grains collide as they bounce along on the bottom of a stream, wearing away the sharp edges and becoming rounded. Finally, the rounded sand grains accumulate on beaches. Over time, they become compacted and lithified to form sandstone. Most beach sands and most sandstones consist predominantly of rounded quartz grains.

Shale is a clastic sedimentary rock that consists mostly of tiny clay minerals and lesser amounts of quartz and organic particles. Shale typically splits easily along very fine layering. It is usually gray to black due to the presence

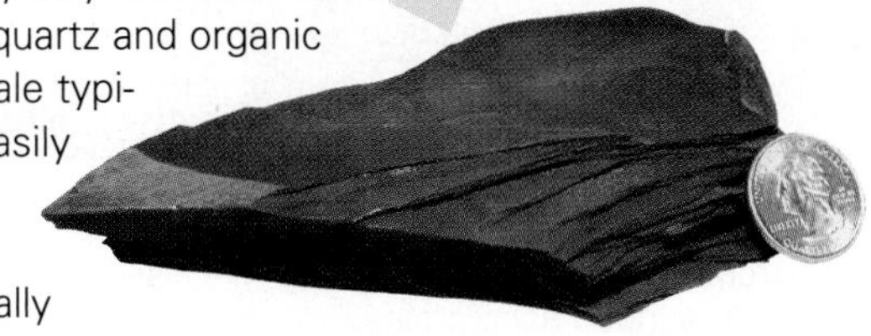

pore space The empty space between particles of rock, sediment, or soil.

compaction Tight packing of sedimentary grains, usually resulting from the weight of overlying sediment, causing weak lithification and a decrease in porosity.

conglomerate A clastic sedimentary rock that consists of lithified gravel.

shale A clastic sedimentary rock that consists of lithified tiny clay minerals and smaller amounts of quartz and organic particles. The organic material in shale is the source of most oil and natural gas.

Sediment grains

Pore space

A

Compaction

B

Lithification

Cement

C

FIGURE 3.7 (A) Pore space is the open space between sediment grains. (B) Compaction squashes the grains together, reducing the pore space and lithifying the sediment by interlocking the grains. (C) Cement fills the remaining pore space, lithifying the sediment by gluing the grains together.

COPYRIGHT AND PHOTOGRAPHS BY DR. PARVINDER S. SETHI

peat A loose, unconsolidated, brownish mass of partially decayed plant matter; a precursor to coal.

of partially decayed remains of plants and animals commonly deposited with clay-rich sediment. This organic material in shale is the source of most oil and natural gas. (The formation of oil and gas from this organic material is discussed in Chapter 5.)

Organic Sedimentary Rocks

Organic sedimentary rocks form by lithification of the remains of plants and animals. Chert is organic sedimentary rock composed of pure silica. There are two types of chert deposits: *bedded chert* occurs as sedimentary beds or layers, and *nodular chert* is found as irregularly shaped lumps called "nodules" in other sedimentary rocks (Figure 3.9). Microscopic examination of bedded chert often shows that it is made up of the remains of tiny marine organisms whose skeletons are composed of silica rather than calcium carbonate. Bedded chert forms as tiny marine organisms that float near the sea surface extract silica from seawater to form their skeletons. When they die, their remains sink to the sea floor to accumulate in layers, eventually forming chert. In contrast, some nodular chert appears to form by precipitation

FIGURE 3.9 Red nodules of chert in light-colored limestone. The nodules shown here are fist sized.

from silica-rich ground water, most often in limestone. Nodular chert, then, forms as a result of interactions between the hydrosphere and the geosphere. Chert was one of the earliest geologic resources. Flint, a dark gray to black variety, was frequently used for arrowheads, spear points, scrapers, and other tools chipped to hold a fine edge.

When plants die, their remains usually decompose by reaction with oxygen. However, in warm swamps and in other environments where plant growth is rapid, dead plants accumulate so rapidly that the oxygen is used up long before the decay process is complete. The partially decayed plant remains form **peat**. As peat is buried and compacted by overlying sediments, it is lithified and converts to coal, a hard, black, combustible rock. Coal formation is described in Chapter 5.

Chemical Sedimentary Rocks

Some common elements in rocks and minerals, such as calcium, sodium, potassium, and magnesium, dissolve during chemical weathering and are carried by ground water and streams to lakes or to the ocean. Streams that carry the salts to the ocean drain most lakes. However, some lakes, such as the Great Salt Lake in Utah, are landlocked. Streams flow into the lake, but no streams exit. As a result, water escapes only by evaporation. When the water evaporates, salts remain behind and the lake water steadily becomes saltier. Evaporites are rocks that form when evaporation concentrates the salts to the point at which they precipitate from solution (Figure 3.10). The same process can occur if ocean water is trapped in coastal or inland basins, where it can no longer mix with the open sea.

FIGURE 3.8 Sandstone is lithified sand. (A) Sandstone in Arches National Park, Utah. (B) A close-up photo shows the well-rounded sand grains.

carbonate rocks
Bioclastic sedimentary rocks composed of the carbonate minerals (minerals whose chemical elements include carbon and oxygen).

FIGURE 3.10 An evaporating lake precipitated thick salt deposits on the Salar de Uyuni, Bolivia. The salt polygons are about 2 meters across.

Bioclastic Sedimentary Rocks

Carbonate rocks are primarily made up of the carbonate minerals calcite and dolomite, described in Chapter 2. Calcite-rich rocks are called limestone, whereas rocks rich in dolomite are referred to as dolomite. Many geologists use the term *dolostone* for the rock name to distinguish it from the mineral dolomite.

Dissolved calcium ions are released into water during the chemical weathering of calcium-bearing rocks. The carbonate anion forms when atmospheric carbon dioxide gas dissolves in water. Seawater contains large quantities of both dissolved calcium and carbon dioxide. Clams, oysters, corals, some types of algae, and a variety of other marine organisms convert dissolved calcium and carbonate ions to shells and other hard body parts. When these organisms die, waves and ocean currents break the shells into small fragments. A rock formed by lithification of such sediment is called *bioclastic limestone*, indicating that it forms by both biological and clastic processes. Most limestones are bioclastic. The bits and pieces of shells appear as fossils in the rock, which you can see in detail in Figure 3.11.

Organisms that form limestone thrive and multiply in warm, shallow seas that are heated by the sun. Therefore, bioclastic limestone typically forms in shallow water along coastlines at low and middle latitudes. It also forms on continents where a rising sea level floods the land with shallow seas. Limestone makes up many of the world's great mountains; the summit of Mount Everest is made of limestone containing marine fossils. Leonardo da Vinci puzzled over the presence of fossils on mountaintops and was perhaps the first person to propose that Earth processes actively raise rocks from the sea floor to the tops of mountains.

Coquina is bioclastic limestone consisting wholly of coarse shell fragments cemented together. Chalk is a very-fine-grained, soft, white bioclastic limestone made of the shells and skeletons of microorganisms that float near the surface of the oceans. When they die, their remains sink to the bottom and accumulate to form chalk. The pale-yellow chalks of Kansas, the off-white chalks of Texas, and the gray chalks of Alabama remind us that all of these areas once lay beneath the sea.

Carbonate Rocks and Global Climate

Carbon dioxide is a greenhouse gas that traps heat in the atmosphere. Limestone is a hard, solid rock. Although it may seem counterintuitive, limestone rock is formed largely from carbon dioxide gas. Atmospheric

FIGURE 3.11 A close-up of shell fragments in limestone. Most limestone is composed of lithified shell fragments and other remains of marine organisms.

sedimentary structure Any feature of sedimentary rock formed during deposition or by later sedimentary processes—for example, layering, ripple marks, or fossils.

bedding Layering that develops as sediments are deposited; also called *stratification*.

cross-bedding A sedimentary structure in which wind or water deposits small beds at an angle to the main sedimentary layering.

carbon dioxide dissolves in seawater and then combines with dissolved calcium to form limestone rock, so formation of limestone removes carbon dioxide from both the seas and the air.

In Chapter 1 we mentioned briefly that Earth's outer shell is broken into several segments called *tectonic plates*. The plates float on the weak, plastic mantle below and glide across Earth, moving continents and oceans. You will learn in Chapter 6 that after a tectonic plate moves thousands of kilometers across Earth's surface, it eventually sinks deep into the mantle. In some instances, a sinking plate may carry limestone into the mantle, removing large amounts of carbon dioxide from Earth's surface and sequestering it in the deep mantle. In other cases, limestone on a sinking tectonic plate may become heated to the point that the carbon dioxide is released directly into the atmosphere in volcanic eruptions. In these ways the greenhouse gas is cycled between surface limestone, deep mantle rocks, and the atmosphere. Thus interactions between limestone and the carbon dioxide in the

atmosphere and that dissolved in the oceans are important determinants of global climate.

Sedimentary Structures

Nearly all sedimentary rocks contain **sedimentary structures**—features that developed during or shortly after deposition of the sediment. These structures help us understand how the sediment was transported and deposited. Because sedimentary rocks form at Earth's surface, sedimentary structures and other features of sedimentary rocks also contain clues about environmental conditions at Earth's surface when the rocks formed.

The most obvious and common sedimentary structure is **bedding**, or stratification—layering that develops as sediment is deposited (Figure 3.12). Bedding forms because sediment accumulates layer by layer. Nearly all sedimentary beds were originally horizontal because most sediment accumulates on nearly level surfaces.

Cross-bedding consists of small beds lying at an angle to the main sedimentary layers (Figure 3.13A). Cross-bedding forms in many environments where wind or water transports and deposits sediment. For example, wind heaps sand into parallel ridges called dunes and flowing water forms similar features called sand waves. Figure 3.13B shows that cross-beds are the layers formed by sand grains tumbling down the steep, downstream face of a dune or sand wave. Cross-bedding is common in sands deposited by wind, streams, and ocean currents and by waves on beaches.

FIGURE 3.12 Sedimentary beds exposed in Capitol Reef National Park, Utah.

COPYRIGHT AND PHOTOGRAPH BY DR. PARVINDER S. SETHI

Unit 1: Earth Materials and Time

A

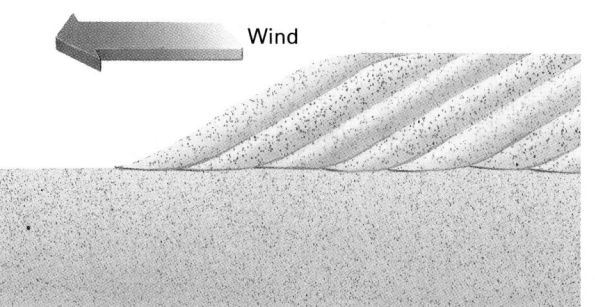

Wind

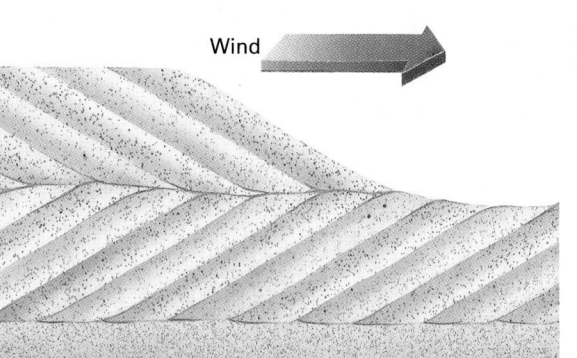

Wind

FIGURE 3.13 (A) Cross-bedding preserved in lithified ancient sand dunes in Zion National Park, Utah. (B) The development of cross-bedding as the prevailing wind direction changes.

ripple marks
Small, parallel ridges and troughs formed in sediment by wind, water currents, or waves, which are often preserved when the sediment is lithified.

mud cracks
Irregular polygonal fractures that develop when mud dries, forming patterns that may be preserved when the mud is lithified.

fossil The imprint, remains, or any other trace of a plant or animal preserved in rock.

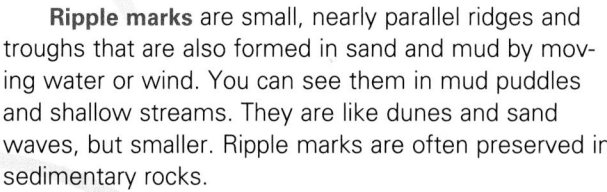

B

Ripple marks are small, nearly parallel ridges and troughs that are also formed in sand and mud by moving water or wind. You can see them in mud puddles and shallow streams. They are like dunes and sand waves, but smaller. Ripple marks are often preserved in sedimentary rocks.

Mud cracks are irregular polygonal cracks that form when mud shrinks as it dries. They indicate that the mud accumulated in shallow water that periodically dried up. For example, mud cracks are common on intertidal mud flats where sediment is flooded by water at high tide and exposed at low tide. The cracks often fill with sediment carried in by the next high tide and are commonly well preserved in rocks.

Occasionally, very delicate sedimentary structures are preserved in rocks. Geologists have found imprints of raindrops that fell on a muddy surface about a billion years ago and imprints of salt crystals that formed as a puddle of saltwater evaporated. Like mud cracks, raindrop and salt imprints show that the mud must have been deposited in shallow water that intermittently dried up.

Fossils are any remains or traces of a plant or animal preserved in rock—any evidence of past life. Fossils include remains of shells, bones, or teeth; whole bodies preserved in amber or ice; and a variety of tracks, burrows, and chemical remains. Fossils are discussed further in Chapter 4.

VIRTUAL FIELD TRIP

Sedimentary Rocks: Formation and Correlation

Ready to Go!

On this virtual field trip we will visit Arches and Capitol Reef national parks in Utah. The rocks that make up the brightly colored walls, arches, and towers are among the most common sedimentary rocks—sandstone, shale, and mudstone—described in Chapter 3, in Section 3.3 ("Sedimentary Rocks"). The scenic exposures allow us to see interesting and important features and principles common to all sedimentary rocks including sedimentary layering, the principle of superposition, and the principle of original horizontality. The origin of sedimentary layering is described under "Sedimentary Structures" in Chapter 3. The principles of superposition and original horizontality are described in Chapter 4, in Section 4.3 ("Relative Geologic Time"). Refer to these topics in Chapters 3 and 4 as you take this trip. We will also see examples of the processes that weather and erode all kinds of rocks, which are described in Section 10.1 ("Weathering and Erosion"), in Chapter 10 of this book.

GOALS OF THE TRIP

Some of the goals of this trip are to understand that:

1. Sedimentary rocks are layered, or "stratified," because sediments are deposited on the Earth's surface in successive layers.

2. Sedimentary rocks become progressively younger in the upward direction because each layer is deposited on top of older layers. This relationship is called the *principle of superposition*.

3. Sedimentary rocks form in horizontal layers. This relationship is called the *principle of original horizontality*. Tilting of the layering indicates that the rocks have been deformed by tectonic or other forces after the sediments were deposited.

4. The walls, arches, and towers of both parks are not permanent features; they are actively changing through both weathering and erosion.

5. Many of the sandstones of both parks are lithified windblown sand dunes. The processes that form the cross bedding in these sandstones is explained under "Sedimentary Structures" in Chapter 3.

FOLLOW-UP QUESTIONS

1. Explain how sedimentary rock layers can violate the principle of superposition (i.e., how can the older layers be stacked above younger layers.) Consult "Relative Geologic Time" in Chapter 4.

2. Describe two or three different geologic processes that can form sandstone. Consult "Sedimentary Structures" in Chapter 3.

3. What kinds of information would a geologist use to correlate sandstone layers found in Utah to layers found in Arizona? Consult Section 4.4 ("Unconformities and Correlation") in Chapter 4.

What to See When You Go

You can take this trip for its scenic beauty and to learn about the geology of these Utah parks before reading further in this book. However, your understanding of the geology will be greatly enhanced if you read the sections of Chapters 3, 4, and 10 noted here before taking the trip, or if you refer to these sections as you travel through Arches and Capitol Reef national parks on the field trip.

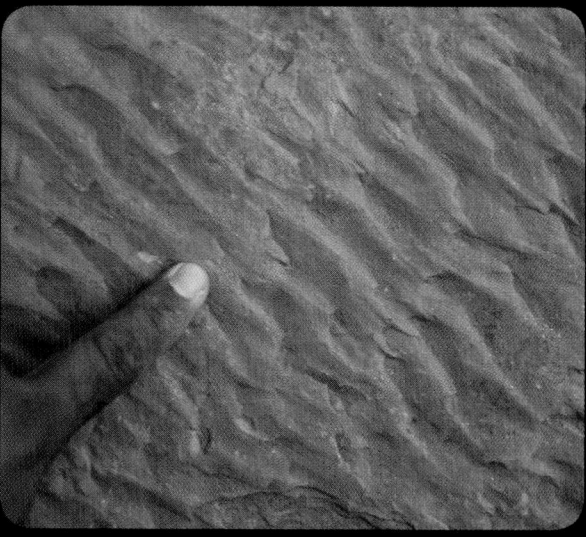

As you follow the field trip, notice that the sedimentary rocks show conspicuous layering, also called "bedding," and that there is clear evidence that the towers, cliffs, and arches are crumbling away by weathering and erosion.

As you learned under "Sedimentary Structures" in Chapter 3, sandstone consists of lithified sand grains. Here you can see how the sand grains formed asymmetric ripples as the sand was deposited by flowing water or by wind.

The "Sedimentary Structures" topic in Chapter 3 also explains that mudcracks in sand and shale formed when wet sediment deposited in water dried and shrank. They show that the water must have been shallow enough that the sediment was periodically exposed to air.

3.4 Metamorphic Rocks

A potter forms a delicate vase from moist clay. She places the soft piece in a kiln and slowly heats it to 1,000°C. As temperature rises, the clay minerals decompose. Atoms from the clay then recombine to form new minerals that make the vase strong and hard. The breakdown of the clay minerals, growth of new minerals, and hardening of the vase all occur without melting the solid materials.

Metamorphism (from the Greek words for "changing form") is the process by which rising temperature and pressure, or changing chemical conditions, transform rocks and minerals. Metamorphism occurs in solid rock, like the transformations in the vase as the potter fires it in her kiln. Small amounts of water and other fluids speed up the metamorphic mineral reactions, but the rock remains solid as it changes. Metamorphism can change any type of parent rock: sedimentary, igneous, or even another metamorphic rock.

A mineral that does not decompose or change in other ways, no matter how much time passes, is a "stable" mineral. Millions of years ago, weathering processes may have formed the clay minerals that today's potter used to create her vase. They were stable and had remained unchanged since they formed.

A stable mineral can become unstable when environmental conditions change. Three types of environmental change affect mineral stability and cause metamorphism: rising temperature, rising pressure, and changing chemical composition (usually caused by an influx of hot water). For example, when the potter put the clay in her kiln and raised the temperature, the clay minerals decomposed because they became unstable at the higher temperature. The atoms from the clay then recombined to form new minerals that were stable at the higher temperature. Like the clay, every mineral is stable only within a certain temperature range. In a similar manner, each mineral is stable only within a certain pressure range.

In addition, a mineral is stable only in a certain chemical environment. If hot water seeping through bedrock carries new chemicals to a rock, those chemicals may react with the original minerals to form different minerals that are stable in the new chemical environment. If hot water dissolves chemical components from a rock, new minerals may form for the same reason. Thus, water from the hydrosphere interacts with rock of the geosphere (at high temperatures) to create

new rocks, showing once again that Earth's spheres, while separate in one sense, interact continuously to shape our planet.

Metamorphism occurs because each mineral is stable only within a certain range of temperature, pressure, and chemical environment. If temperature or pressure rises above that range, or if chemicals are added or removed from the rock, the rock's original minerals may decompose and their components recombine to form new minerals that are stable under the new conditions.

Metamorphic Grade

The **metamorphic grade** of a rock is the intensity of metamorphism that formed it. Temperature is the most important factor in metamorphism, and therefore grade closely reflects the temperature of metamorphism. Because temperature increases with depth in Earth, a general relationship exists between depth and metamorphic grade (Figure 3.14). Low-grade metamorphism occurs at shallow depths, less than 10 kilometers beneath the surface, where temperature is no higher than 300 to 400°C. High-grade conditions are found deep within continental crust and in the upper mantle, 40 to 55 kilometers below Earth's surface. The temperature here is 600 to 800°C, close to the melting point

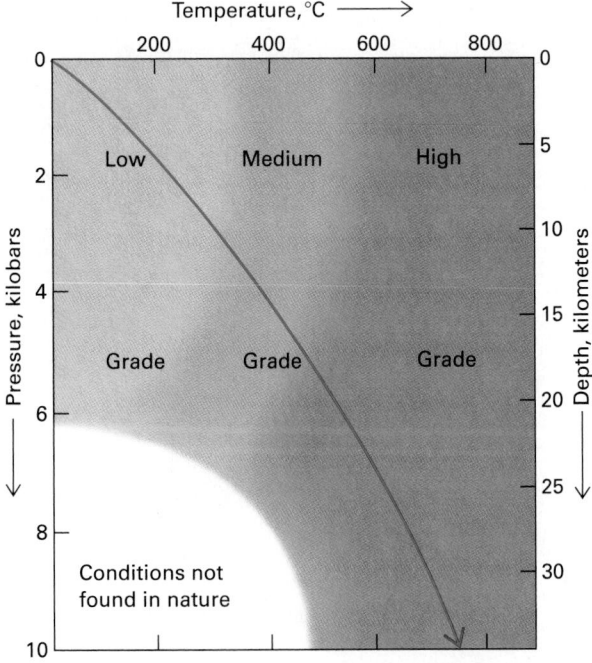

FIGURE 3.14 Metamorphic grade increases with depth because temperature and pressure rise with depth. The blue arrow traces the path of increasing temperature and pressure with depth in a normal part of the crust.

of rock. High-grade conditions can develop at shallower depths, however, in rocks adjacent to hot magma. For example, today metamorphic rocks are forming beneath Yellowstone Park, where hot magma lies close to Earth's surface.

Metamorphic Changes

Metamorphism commonly alters both the texture and mineral content of a rock.

Textural Changes

As a rock undergoes metamorphism, some mineral grains grow larger and others shrink. The shapes of the grains may also change. For example, fossils give the fossiliferous limestone shown in Figure 3.11 its texture. Both the fossils and the cement between them are made of small calcite crystals. If the limestone is heated, some of the calcite grains grow larger at the expense of others. In the process, the fossiliferous texture is destroyed. Metamorphism transforms limestone into a metamorphic rock called marble. Like the fossiliferous limestone, the marble is composed of calcite, but the texture is now one of large interlocking grains, and the fossils have vanished.

Micas are common metamorphic minerals that form when many different parent rocks undergo metamorphism. Micas are shaped like pie plates. When metamorphism occurs without deformation—without causing the rocks to change shape—the micas grow with random orientations, like pie plates flying through the air (Figure 3.15A). However, when tectonic force squeezes rocks as they are heated during metamorphism, the rock deforms into folds. When rocks are folded as mica crystals are growing, the micas grow with their flat surfaces perpendicular to the direction of squeezing. This parallel alignment of micas (and other minerals) produces the metamorphic layering called **foliation** (Figure 3.15B). Metamorphic rocks break easily along the foliation planes. This parallel fracture pattern is called **slaty cleavage**. In most cases, slaty cleavage cuts across original sedimentary bedding.

The foliation layers range from a fraction of a millimeter to a meter or more thick. Metamorphic foliation can resemble sedimentary bedding but is different in origin. Foliation results from alignment and segregation of metamorphic minerals during metamorphism; it forms at right angles to the forces acting on the rocks.

foliation The layering in micas and other minerals created by metamorphism.

slaty cleavage A metamorphic foliation producing a parallel fracture pattern that cuts across original sedimentary bedding in micas or other metamorphic rocks.

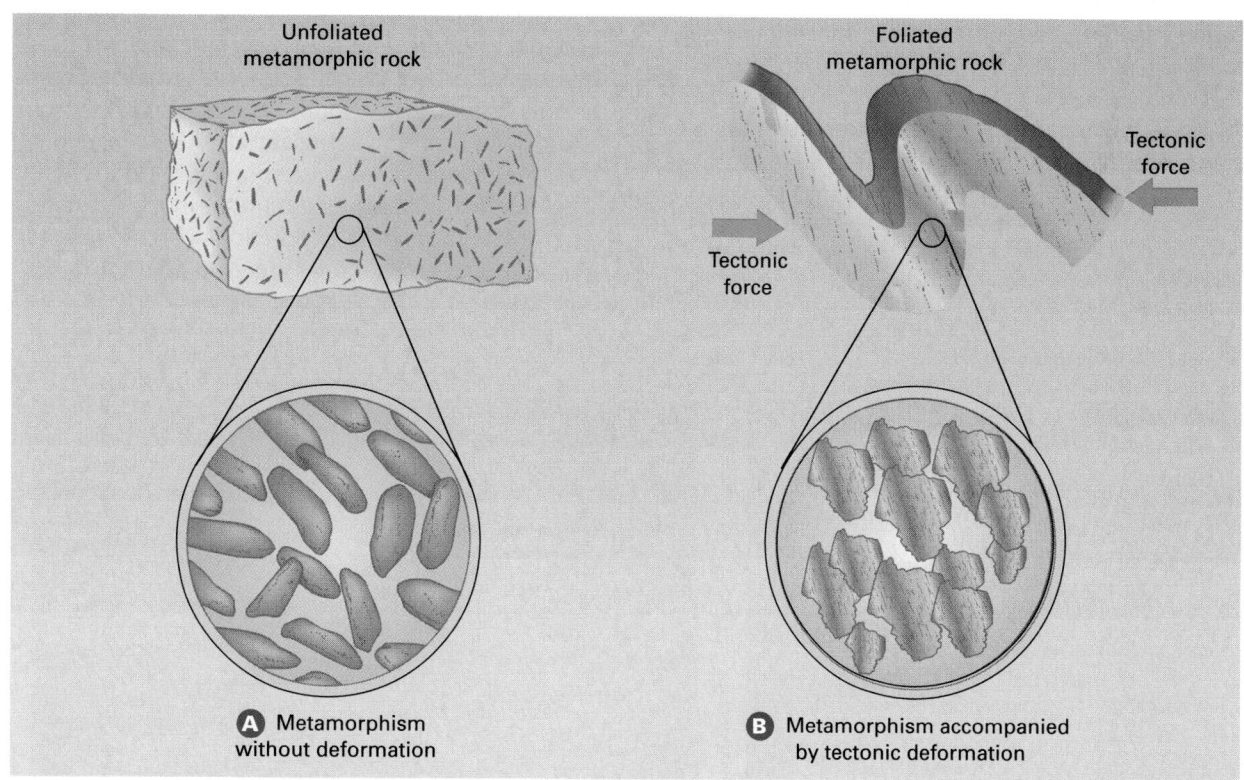

A Metamorphism without deformation

B Metamorphism accompanied by tectonic deformation

FIGURE 3.15 (A) When metamorphism occurs without deformation, platy micas grow with random orientations. (B) When deformation accompanies metamorphism, platy micas orient perpendicular to the force squeezing the rocks, forming foliated metamorphic rocks. The wavy lines represent original sedimentary layers that have been folded by the deformation.

Sedimentary bedding develops because sediments are deposited in a layer-by-layer process.

Mineralogical Changes

Sometimes, when a parent rock contains only one mineral, metamorphism transforms the rock into one composed of the same mineral, but with a coarser texture. Limestone converting to marble is one example of this generalization; both rocks consist of the mineral calcite, but their respective textures are very different. Another example is the metamorphism of quartz sandstone to quartzite, a rock composed of recrystallized quartz grains.

In contrast, metamorphism of a parent rock containing several minerals usually forms a rock with new and different minerals *and* a new texture. For example, a typical shale contains large amounts of clay, as well as quartz and feldspar. When heated, some of those minerals decompose, and their atoms recombine to form new minerals such as mica, garnet, and a different kind of feldspar. Figure 3.16 shows a rock called gneiss that formed when metamorphism altered both the texture and minerals of shale. If migrating fluids alter the chemical composition of a rock, new minerals invariably form.

Types of Metamorphism and Metamorphic Rocks

Recall that rising temperature, rising pressure, and changing chemical environment cause metamorphism. In addition, deformation resulting from movement of Earth's crust develops foliation and thus strongly affects the texture of a metamorphic rock. Four different geologic processes create these changes.

Contact Metamorphism

Contact metamorphism occurs where hot magma intrudes cooler rock of any type—sedimentary, metamorphic, or igneous. The highest grade metamorphic rocks form at the contact point, closest to the magma. Lower grade rocks develop farther out. A metamorphic halo around a pluton can range in width from less than a meter to hundreds of meters, depending on the size and temperature of the intrusion and the effects of water or other fluids (Figure 3.17). Contact metamorphism commonly occurs without deformation. As a result, the metamorphic minerals grow with random orientations—like the pie plates flying through the air— and the rocks develop no metamorphic layering.

Burial Metamorphism

Burial metamorphism results from burial of rocks in a sedimentary basin. A large river like the Mississippi carries massive amounts of sediment to the ocean every year, where it accumulates on a delta. Over tens or even hundreds of millions of years, the weight of the sediment becomes so great that the entire region sinks, just as a canoe settles when you climb into it. Younger sediment may bury the oldest layers to a depth of more than 10 kilometers in a large basin. Over time, temperature and pressure increase within the deeper layers until burial metamorphism begins.

Burial metamorphism is occurring today in the sediments underlying many large deltas, including the Mississippi River delta, the Amazon Basin on the east coast of South America, and the Niger River delta on the west coast of Africa. Like contact metamorphism, burial metamorphism occurs without deformation. Consequently, metamorphic minerals grow with random orientations and, like contact metamorphic rocks, burial metamorphic rocks are unfoliated.

FIGURE 3.16

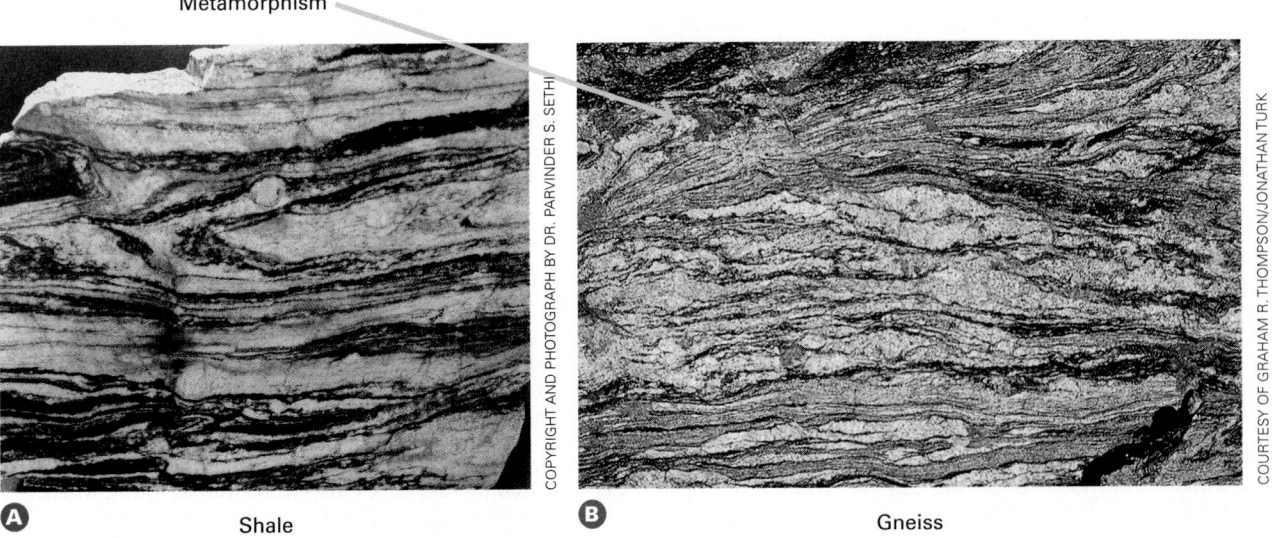

Metamorphism

(A) Shale

(B) Gneiss

COPYRIGHT AND PHOTOGRAPH BY DR. PARVINDER S. SETHI

COURTESY OF GRAHAM R. THOMPSON/JONATHAN TURK

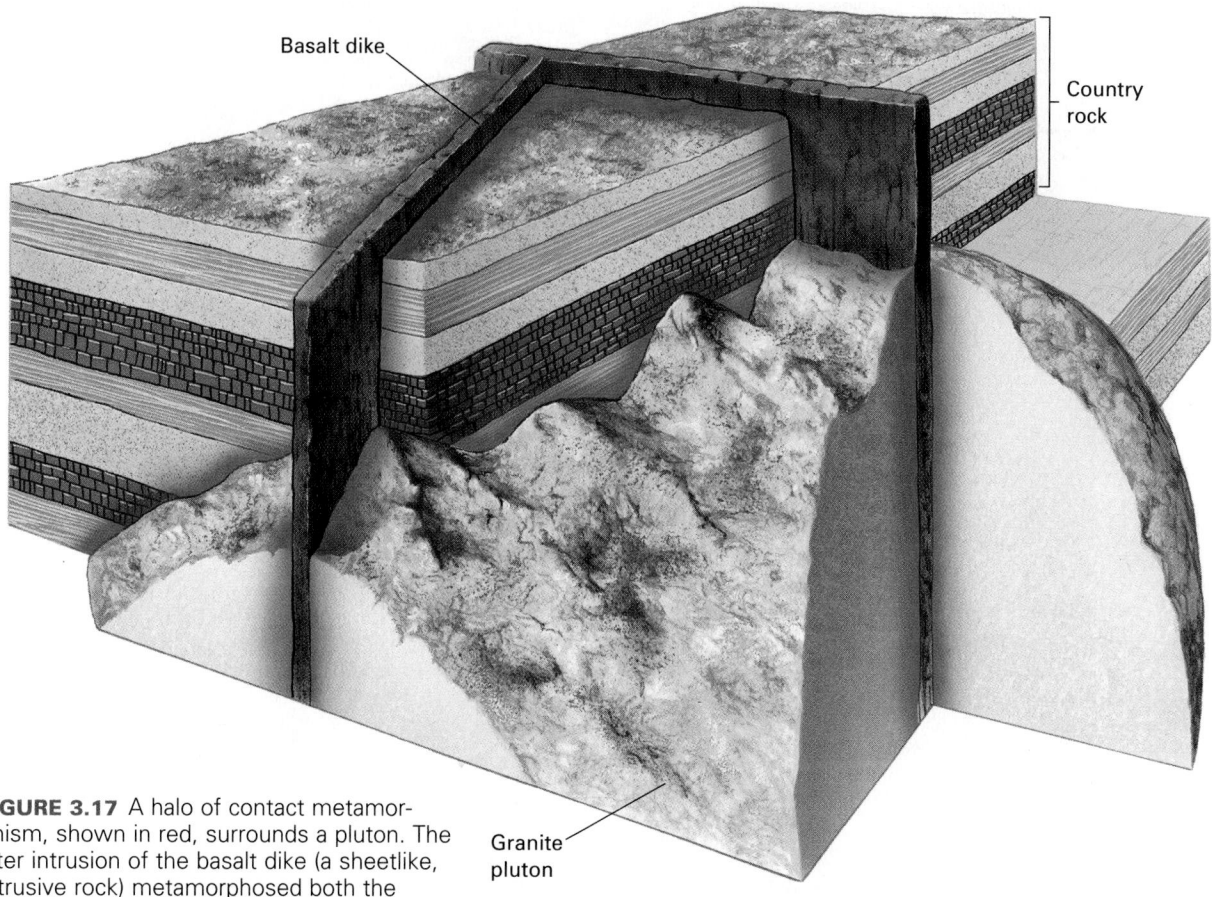

Basalt dike

Country rock

Granite pluton

FIGURE 3.17 A halo of contact metamorphism, shown in red, surrounds a pluton. The later intrusion of the basalt dike (a sheetlike, intrusive rock) metamorphosed both the pluton and the existing sedimentary rock, creating a second metamorphic halo.

Regional Dynamothermal Metamorphism

Regional dynamothermal metamorphism occurs where major crustal movements build mountains and deform rocks. The term *dynamothermal* simply indicates that the rocks are being deformed and heated at the same time. It is the most common and widespread type of metamorphism and affects broad regions of Earth's crust.

Magma rises and heats large portions of the crust in places where tectonic plates converge (see Chapter 6). The high temperatures cause new metamorphic minerals to form throughout the region. The rising magma also deforms the hot, plastic country rock as it forces its way upward. At the same time, the movements of the crust squeeze and deform rocks. As a result of all these processes acting together, regionally metamorphosed rocks are strongly foliated and are typically associated with mountains and igneous rocks. Regional dynamothermal metamorphism produces zones of foliated metamorphic rocks tens to hundreds of kilometers across.

Consider, for instance, the typical sequence of changes to shale as metamorphic grade increases. Shale is the most abundant type of sedimentary rock and consists of clay minerals, quartz, and feldspar. The mineral grains are too small to be seen with the naked eye and can barely be seen with a microscope. As regional metamorphism begins, clay minerals break down and mica and chlorite replace them. These new, platy minerals grow perpendicular to the direction of squeezing. As a result, the rock develops slaty cleavage and is called slate. With rising temperature and continued deformation, the micas and chlorite grow larger, and wavy or wrinkled surfaces replace the flat, slaty cleavage, creating phyllite, which has a silky or glossy appearance as opposed to the dullness of slate.

As temperature continues to rise, the mica and chlorite grow large enough to be seen by the naked eye, and foliation becomes very well developed. Rock of this type is called schist. Schist forms approximately at the transition from low to intermediate metamorphic grades.

At high metamorphic grades, light- and dark-colored minerals often separate into bands that are thicker than the layers of schist, to form a rock called gneiss (pronounced "nice"). At the highest metamorphic grade, the rock begins to melt, though only partially, forming small veins of granitic magma. When metamorphism wanes and the rock cools, the magma veins solidify to form migmatite, a mixture of igneous and metamorphic rock.

SHALE'S METAMORPHOSIS

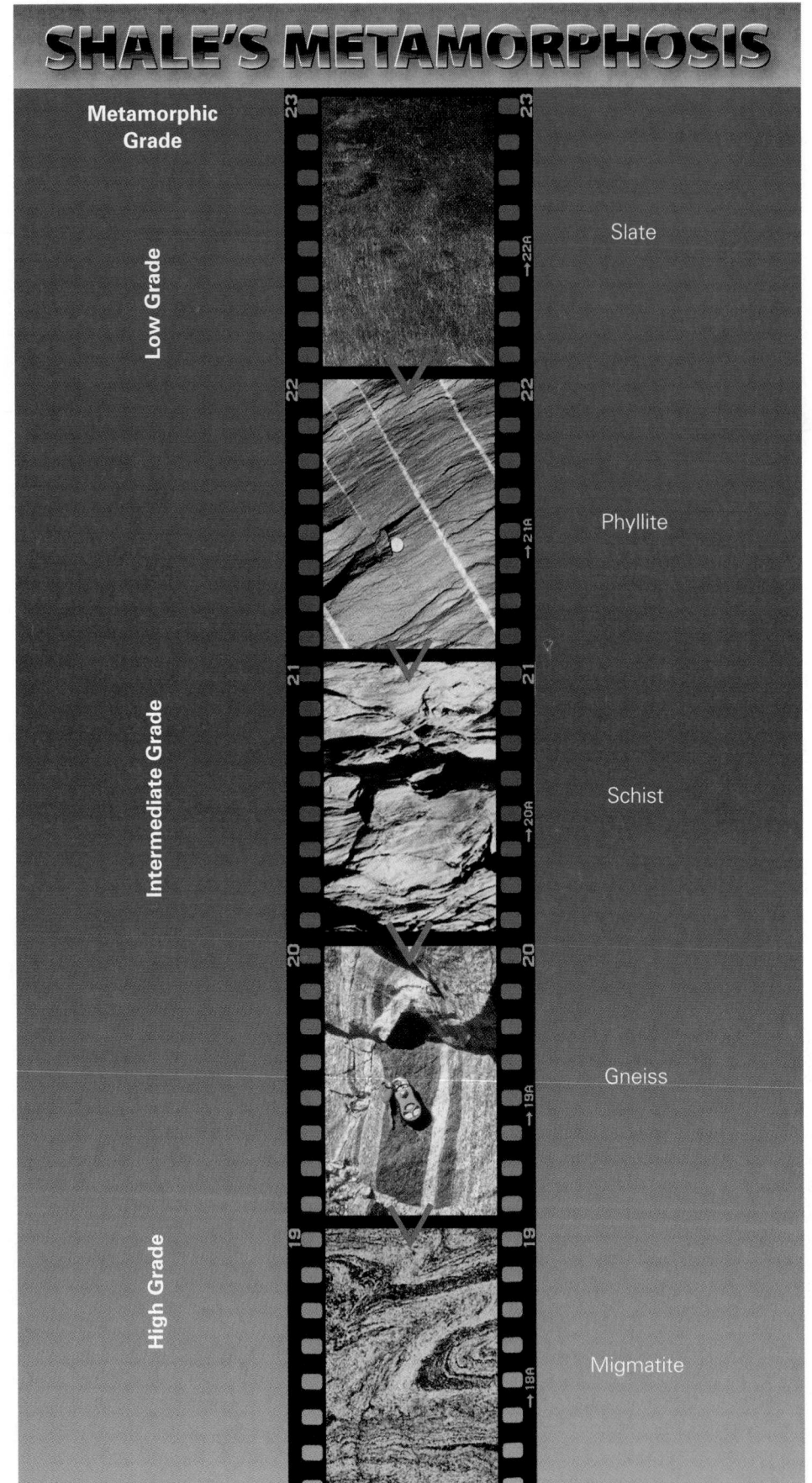

Metamorphic Grade

Low Grade

Intermediate Grade

High Grade

Slate

Phyllite

Schist

Gneiss

Migmatite

Unit 1: Earth Materials and Time

Shale is not the only example. For instance, quartz sandstone and limestone transform to foliated quartzite and foliated marble, respectively, during regional metamorphism.

Hydrothermal Metamorphism

Water is a chemically active fluid; it attacks and dissolves rocks and minerals. If the water is hot, it attacks minerals even more rapidly. Hydrothermal metamorphism (also called *hydrothermal alteration* or *metasomatism*) occurs when hot water and ions dissolved in the hot water react with a rock to change its chemical composition and minerals.

Most rocks and magma contain very low concentrations of metals such as gold, silver, copper, lead, and zinc. For example, gold makes up 0.0000002 percent of average crustal rock, while copper makes up 0.0058 percent and lead comprises 0.0001 percent. Although the metals are present in very low concentrations, hydrothermal solutions sweep slowly through vast volumes of country rock, dissolving and accumulating the metals as they go. The solutions then deposit the dissolved metals, where they encounter changes in temperature, pressure, or chemical environment (Figure 3.18). In this way, hydrothermal solutions scavenge and concentrate metals from average crustal rocks and then deposit them locally to form ore. Hydrothermal ore deposits are discussed further in Chapter 5.

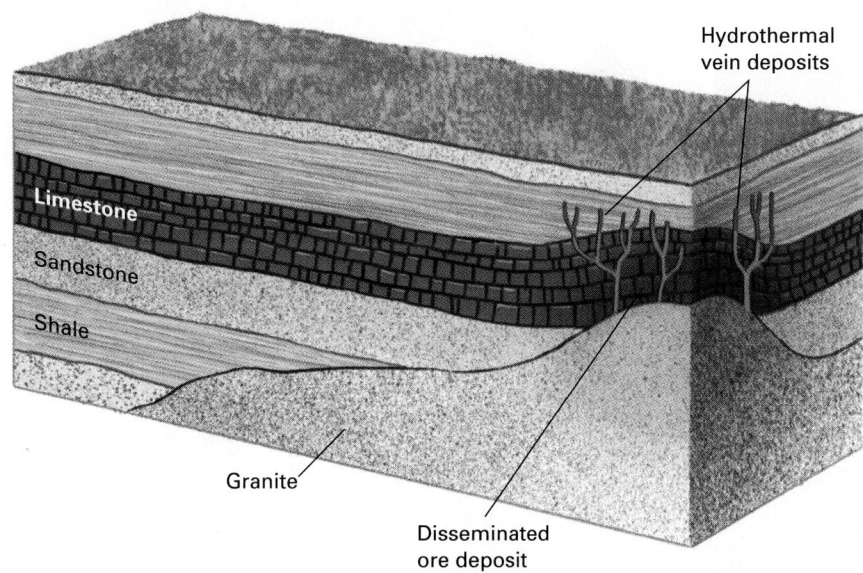

FIGURE 3.18 Hydrothermal ore deposits form when hot water dissolves metals from country rock and deposits them in fractures and surrounding country rock.

Common Igneous Rocks

Felsic	Intermediate	Mafic	Ultramafic
obsidian	andesite	basalt	peridotite
granite	diorite	gabbro	
rhyolite			

Common Sedimentary Rocks

Clastic	Bioclastic	Organic	Chemical
conglomerate	limestone	chert	evaporite
sandstone	dolomite (dolostone)	coal	
siltstone	coquina		
shale	chalk		

Common Metamorphic Rocks

Unfoliated	Foliated
marble	slate
quartzite	phyllite
	schist
	gneiss
	migmatite

4

GEOLOGIC TIME:
A STORY
IN THE ROCKS

visit **4ltrpress.cengage.com**

mass extinction A sudden, catastrophic event during which a significant part of all life-forms on Earth become extinct.

4.1 Earth Rocks, Earth History, and Mass Extinctions

In many locations, 19th century geologists found thick sequences of rock layers with abundant fossils that we now know represent millions, or hundreds of millions, of years of Earth history. But younger layers above the fossil-rich rocks contained few fossils or none at all. Looking even higher in the sequence of rock layers, they found abundant fossils again, but of organisms very different from those in the older rocks. Even more surprising, fossils of many of the most abundant animals and plants in lower layers were never seen again in younger rocks. Those organisms had simply disappeared from the face of Earth forever.

This fossil record suggests that sudden, catastrophic events had abruptly decimated life on Earth, and that new life-forms emerged following these **mass extinctions**. However, scientists did not immediately accept the idea that catastrophic events had caused sudden mass extinctions. Instead, they suggested that the rock record might not be complete. Maybe some rocks had been destroyed by erosion, or perhaps sedimentary rocks had not formed for a long period of time. According to this reasoning, the part of the record that was lost could have contained evidence for the gradual decline of the species that became extinct, and for the slow emergence of new species. Following conventional wisdom of the time, scientists concluded that the extinctions of old life-forms and emergence of new ones occurred slowly, as a result of gradually changing conditions.

But if pieces of the rock record were missing in some locations, they must exist in other places on Earth. Geologists searched for rocks to fill in the gaps and provide evidence for gradual extinctions—but they did not find them. Eventually, the evidence for near instantaneous extinctions became compelling. From studies of this type, scientists learned that at least six times in Earth history, catastrophic extinctions decimated life on Earth (Figure 4.1).

The most dramatic extinction occurred 248 million years ago, at the end of the Permian Period. At that time, 90 percent of all species in the oceans suddenly died out. On land, about two-thirds of reptile and amphibian species and 30 percent of insect species vanished.

One modern scientist exclaimed that this extinction event was "the closest life has come to complete extermination since its origin."[1]

The death of most life-forms at the end of Permian time left huge ecological voids in the biosphere. Ocean ecosystems changed as new organisms emerged in an environment free of predators and competition. On land, dinosaurs and many other terrestrial animals and plants emerged and proliferated.

About 180 million years later, at the end of Cretaceous times, another catastrophic extinction wiped out one-fourth of all species, including the dinosaurs. This mass extinction occurred 65 million years ago. Small mammals survived this disaster, facing a new world free of the efficient predators that had hunted them. The number of new mammal species increased rapidly after this extinction event, eventually leading to the evolution of humans.

What kinds of catastrophic events caused these sudden extinctions? Why would species that had survived for tens of millions of years instantaneously disappear? We consider three hypotheses for mass extinctions. Each of these hypotheses involves large-scale systems interactions among the biosphere, the geosphere, the atmosphere, and the hydrosphere. In the first hypothesis, described next, a fifth, extraterrestrial, system is seen as the trigger that initiated radical changes in all four Earth systems.

---

1. Douglas Erwin, *The Great Paleozoic Crisis* (New York: Columbia University Press, 1993), 187.

COPYRIGHT AND PHOTOGRAPH BY DR. PARVINDER S. SETHI

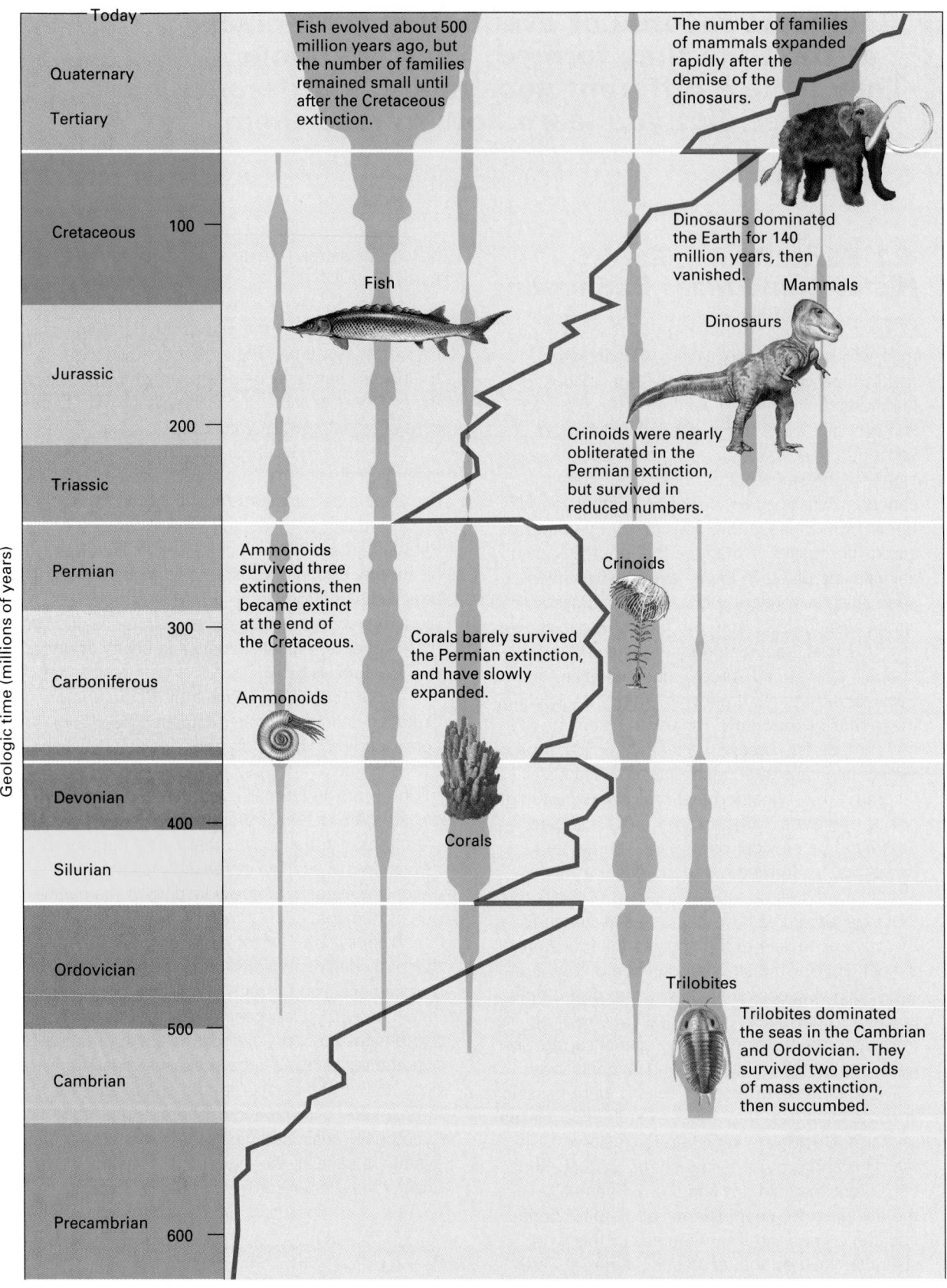

FIGURE 4.1 The red line shows that the number of families of organisms, indicated by the thickness of the blue-shaded areas, has varied throughout Earth history. Sudden leftward shifts of the red line indicate mass extinctions. The numbers in the left column indicate time in millions of years ago.

Text labels within the figure:

Geologic time (millions of years)

Today
Quaternary
Tertiary
Cretaceous — 100
Jurassic
— 200
Triassic
Permian
— 300
Carboniferous
Devonian
— 400
Silurian
Ordovician
— 500
Cambrian
Precambrian — 600

Fish evolved about 500 million years ago, but the number of families remained small until after the Cretaceous extinction.

Fish

The number of families of mammals expanded rapidly after the demise of the dinosaurs.

Dinosaurs dominated the Earth for 140 million years, then vanished.

Mammals

Dinosaurs

Crinoids were nearly obliterated in the Permian extinction, but survived in reduced numbers.

Ammonoids survived three extinctions, then became extinct at the end of the Cretaceous.

Crinoids

Corals barely survived the Permian extinction, and have slowly expanded.

Ammonoids

Corals

Trilobites

Trilobites dominated the seas in the Cambrian and Ordovician. They survived two periods of mass extinction, then succumbed.

Extraterrestrial Impacts

In 1977, the father-and-son team of Walter and Luis Alvarez were studying rocks that formed at the time the dinosaurs and so many other life-forms became extinct. In one layer, they found abundant dinosaur fossils. Just above it, they found very few fossils of any kind and no dinosaur fossils. Between these two rock layers, they found a thin, sooty, clay layer. They brought samples of the clay to the laboratory and found that it contained high concentrations of the element iridium. This discovery was surprising because iridium is rare in Earth rocks. Where did it come from?

Although rare in Earth's crust, iridium is more abundant in meteorites. Walter and Luis Alvarez suggested that 65 million years ago, a meteorite 10 kilometers in diameter hit the Earth with energy equal to 10,000 times that of today's entire global nuclear arsenal. The collision vaporized both the meteorite and Earth's crust at the point of impact, forming a plume of hot dust and gas that ignited fires around the planet. Soot from the global wildfires and iridium-rich meteorite dust rose into the upper atmosphere, circling the globe. The thick, dark cloud blocked out the Sun and halted photosynthesis for as much as a year. Surface waters froze, and many plants and animals died. This event, called the *terminal Cretaceous extinction*, killed off the dinosaurs and many other life-forms. Then the sooty, iridium-rich dust settled to Earth to form the distinctive clay layer.

Other scientists have found evidence of a huge meteorite crater that formed 65 million years ago in the Caribbean Sea north of the Yucatán Peninsula. Most scientists now agree with the meteorite impact hypothesis for the terminal Cretaceous extinction. Some scientists have suggested that an extraterrestrial impact also caused the Permian extinction, but the evidence is less compelling.[2]

Volcanic Eruptions

A volcanic eruption ejects gas and fine volcanic ash into the atmosphere. The ash acts as an umbrella that reflects sunlight back into space and thus causes climatic cooling. One volcanic gas, sulfur dioxide, forms small particles called *aerosols* that also reflect sunlight and cool Earth. Carbon dioxide is another gas emitted by volcanoes. It is a greenhouse gas that can cause warming of Earth's atmosphere. Thus, volcanoes erupt materials capable of causing both atmospheric cooling and warming. In many eruptions, the cooling overwhelms the warming effects. This cooling can occur rapidly—within a few weeks or months of the eruption.

Scientists have noted that some mass extinctions coincided with unusually high rates of volcanic activity. For example, massive flood basalts erupted onto Earth's surface in Siberia 248 million years ago—at the same time that the Permian extinction occurred. According to one hypothesis, explosive volcanic eruptions blasted massive amounts of volcanic ash and sulfur aerosols into the air. The ash and aerosols spread like a pall throughout the upper atmosphere, blocking out the Sun. Earth's atmospheric and oceanic temperatures plummeted, plants withered, and animals starved or froze to death. If this hypothesis is correct, the Permian extinction (and perhaps others) resulted from a classic Earth systems interaction, where a volcanic eruption occurring within the geosphere altered the composition of the atmosphere, with profound consequences for atmospheric and marine temperatures, and for life.

2. Douglas Erwin, *The Great Paleozoic Crisis* (New York: Columbia University Press, 1993).

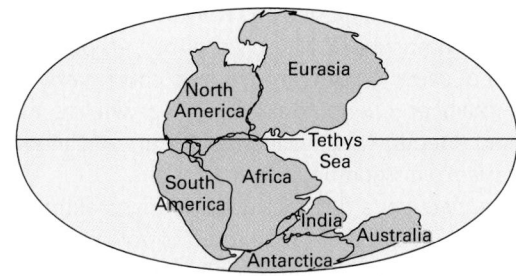

FIGURE 4.2 The assembly of all continents into the supercontinent Pangea during Permian time prevented vertical mixing of ocean water.

Supercontinents and Earth's Carbon Dioxide Budget

In the modern oceans, cold polar seawater sinks to the sea floor and flows as a deep ocean current toward the equator. This current transports oxygen to the deep ocean basins and forces the deep water to rise near the equator, where it is warmed. The current thus mixes both heat and gases throughout the oceans and the atmosphere.

Ocean currents are partially controlled by the positions of the continents. Recall from Chapter 1 that the continents slowly drift across the globe. Several times

in Earth history, all the continents have joined together to form one giant supercontinent. One such supercontinent assembled during Permian time (Figure 4.2). Computer calculations indicate that the assembly of all continents into a single land mass, and the consequent development of a single global ocean prevented mixing between the surface water and the ocean depths.[3]

Surface marine organisms absorb atmospheric carbon dioxide and use it to produce organic tissue. These organisms then die and settle to the sea floor. This process removes carbon dioxide from the atmosphere, converts the carbon to organic tissue, and transports it to the sea floor. Organisms living near the deep sea floor then consume the fallen litter and release carbon dioxide into the deep ocean water. Much of this gas dissolves in seawater and is held there by the pressure of overlying water. Modern ocean currents mix deep and shallow seawater, returning the carbon dioxide to the atmosphere.

However, because there was little vertical mixing in the Permian oceans, carbon accumulated in ever-increasing concentration in the deep oceans (Figure 4.3A). This process continued to remove carbon dioxide from the atmosphere and stored it in the deep oceans during late Permian time. But carbon dioxide is a greenhouse gas that absorbs heat and warms the atmosphere. As carbon dioxide was removed from the

FIGURE 4.3 (A) Global cooling in the late Permian Period. (B) The Permian extinction.

3. A. H. Knoll, R. K. Bambach, D. E. Canfield, and J. P. Grotzinger, "Comparative Earth History and Late Permian Mass Extinction," *Science* 273 (July 26, 1996), 452–457.

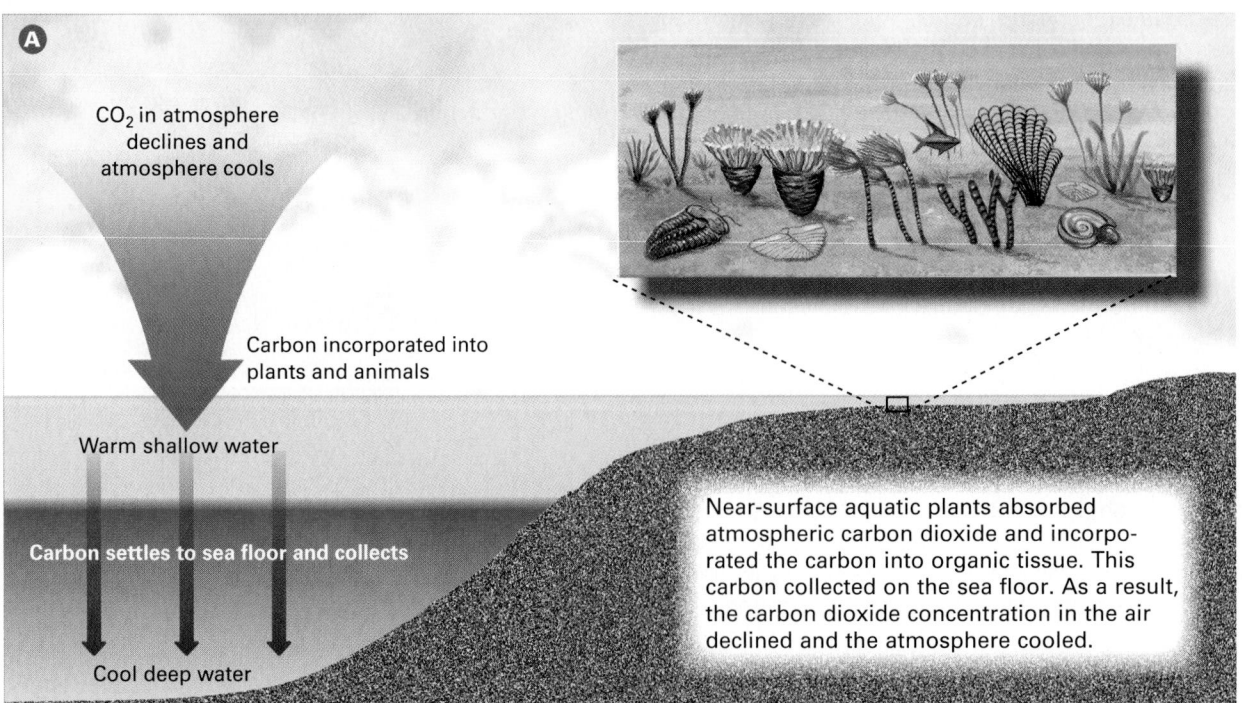

Near-surface aquatic plants absorbed atmospheric carbon dioxide and incorporated the carbon into organic tissue. This carbon collected on the sea floor. As a result, the carbon dioxide concentration in the air declined and the atmosphere cooled.

air and stored in deep ocean water, the atmosphere cooled. Some scientists have suggested that continental ice sheets formed as a result of the global cooling. Sea level fell as water evaporated from the oceans and accumulated on continental glaciers, exposing continental shelves and stressing populations of shallow-dwelling marine organisms.

The cool air cooled the surface of the sea. When water cools, it becomes denser. Eventually a threshold was reached at which polar surface water became denser than the deep water. At this point the cool, dense surface water sank. The sinking surface waters forced the carbon dioxide–rich deep water to rise to the sea surface, where it rapidly released massive amounts of carbon dioxide into the atmosphere (Figure 4.3B). According to this hypothesis, the carbon dioxide asphyxiated life both in the seas and on the continents, killing most life on Earth. Later, new organisms evolved when the carbon dioxide levels fell.

Each of these three hypotheses involves large-scale interactions among the atmosphere, the hydrosphere, the biosphere, and the geosphere, and the first also includes an extraterrestrial factor. The third hypothesis, if correct, is a classic example of both Earth systems feedback mechanisms and a threshold effect. The feedback process occurred as tectonic plate movement altered the shapes of the ocean basins. Ocean currents changed when the basins changed. The alteration in currents, in turn, changed atmospheric composition and unleashed a threshold event that rapidly killed most living organisms on Earth.

4.2 Geologic Time

While most of us think of time in terms of days or years, Earth scientists commonly refer to events that happened millions or billions of years ago. In Chapter 1 you learned that Earth is approximately 4.6 billion years old. Yet humans and our human-like ancestors have existed for only 5 to 7 million years, and recorded history is only a few thousand years old. How do scientists measure the ages of rocks and events that occurred millions or billions of years ago, long before recorded history or even human existence?

Scientists measure geologic time in two ways. **Relative age** measurement refers only to the order in which events occurred. Determination of relative age is based on a simple principle: In order for an event to affect a rock, the rock must exist first. Thus, the rock must be older than the event. This principle seems obvious, yet it is the basis of much geologic work. As you learned in Chapter 3, sediment normally accumulates in horizontal layers. However, the sedimentary rocks in Figure 4.4 are folded and tilted. We deduce that the folding and tilting occurred after the sediment accumulated. The order in which rocks and geologic features formed can usually be interpreted by such observation and logic.

Absolute age is age in years. Dinosaurs became extinct 65 million years ago. The terminal Permian extinc-

relative age An approach to measuring geologic time based on the order in which events occurred, but not measured in years.

absolute age Time measured in years.

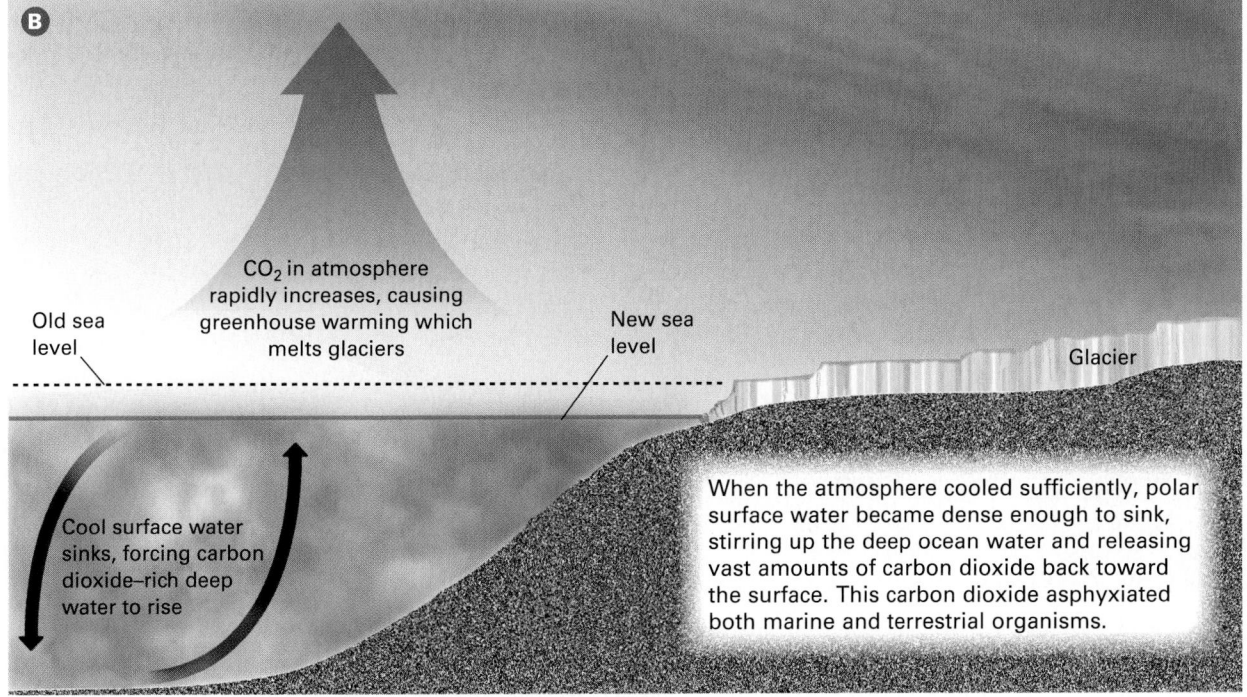

Old sea level

CO₂ in atmosphere rapidly increases, causing greenhouse warming which melts glaciers

New sea level

Glacier

Cool surface water sinks, forcing carbon dioxide–rich deep water to rise

When the atmosphere cooled sufficiently, polar surface water became dense enough to sink, stirring up the deep ocean water and releasing vast amounts of carbon dioxide back toward the surface. This carbon dioxide asphyxiated both marine and terrestrial organisms.

principle of original horizontality The principle that most sediment is deposited as nearly horizontal beds, and therefore most sedimentary rocks started out with nearly horizontal layering.

principle of superposition The principle that in any undisturbed layers of sediment or sedimentary rock, the age becomes progressively younger from bottom to top; younger layers always accumulate on top of older layers.

FIGURE 4.4 This limestone formed as horizontal layers 500 million years ago in a shallow sea. Then 400 million years later, tectonic forces uplifted and folded it in the Canadian Rockies, Alberta.

tion occurred 248 million years ago. The Teton Range in Wyoming began rising 6 million years ago. Absolute age tells us both the order in which events occurred and the amount of time that has passed since they occurred.

4.3 Relative Geologic Time

Absolute age measurements have become common only since the second half of the 20th century. Prior to that time, geologists used field observations to determine relative ages. Even today, with sophisticated laboratory processes available, most field geologists routinely use relative age measurement. Geologists use a combination of common sense and a few simple principles to determine the order in which rocks formed and changed over time.

The **principle of original horizontality** is based on the observation that sediment usually accumulates in horizontal layers (Figure 4.5A). If sedimentary rocks lie at an angle, as in Figure 4.5B, or if they are folded as you saw in Figure 4.4, we can infer that tectonic forces tilted or folded them after they formed.

The **principle of superposition** states that sedimentary rocks become younger from bottom to top (as long

FIGURE 4.5 (A) The principle of original horizontality tells us that most sedimentary rocks were oriinally deposited in horizontal layers, like these in San Juan River, Utah. (B) When we see tilted rocks such as these in Connecticut, we infer that they were tilted *after* the layers were deposited.

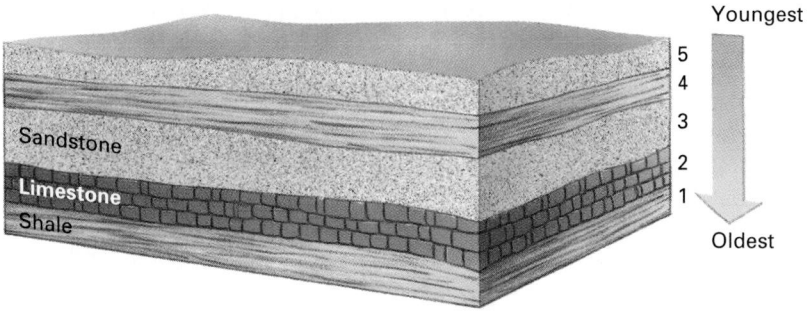

FIGURE 4.6 The principle of superposition. In a sequence of sedimentary beds, the oldest bed (labeled 1) is the lowest, and the youngest (labeled 5) is on top.

principle of crosscutting relationships The obvious principle that a rock or feature must first exist before anything can happen to it; thus, if an intrusion of rock cuts across an existing rock, the dike (intrusion) is younger.

evolution The biological theory that life-forms have changed in their physical and genetic characteristics over time.

as tectonic forces have not turned them upside down or thrust an older layer over a younger one). This is because younger layers of sediment always accumulate on top of older layers. In Figure 4.6, the sedimentary layers become progressively younger in the order 1, 2, 3, 4, and 5.

The **principle of crosscutting relationships** is based on the obvious fact that a rock must first exist before anything can happen to it. Figure 4.7 shows an igneous intrusion—called a *dike*—cutting across gneiss. Clearly, the dike must be younger than the gneiss. Figure 4.8 shows sedimentary rocks intruded by three granite dikes. Dike 2 cuts dike 1, and dike 3 cuts dike 2, so dike 2 is younger than 1, and dike 3 is the youngest. The sedimentary rocks must be older than all the dikes.

The theory of **evolution** states that life-forms have changed throughout geologic time. Species emerge, persist for awhile, and then become extinct. Recall from Chapter 3 that fossils are the remains and other traces of prehistoric life that are preserved in sedimentary rocks. They are useful in determining relative ages of rocks because different animals and plants lived at different times in Earth history. For example, trilobites (Figure 4.9) lived from 543 million to about 248 million years ago, and the first dinosaurs appeared about 248 million years ago.

The principle of superposition tells us that sedimentary rock layers become younger from bottom to top. If the rocks formed over a long time, different fossils appear and then vanish from bottom to top in the same order in which the organisms evolved and then became extinct. Rocks containing dinosaur bones must be younger than those containing trilobite remains.

FIGURE 4.9 Trilobites were marine creatures that lived in Earth's ancient seas before the time of the dinosaurs. That this trilobite fossil is found in slate tells us that the slate once lay in an aquatic environment.

igneous dike

FIGURE 4.7 An igneous dike cutting across gneiss. By the principle of crosscutting relationships, the dike must be younger than the gneiss.

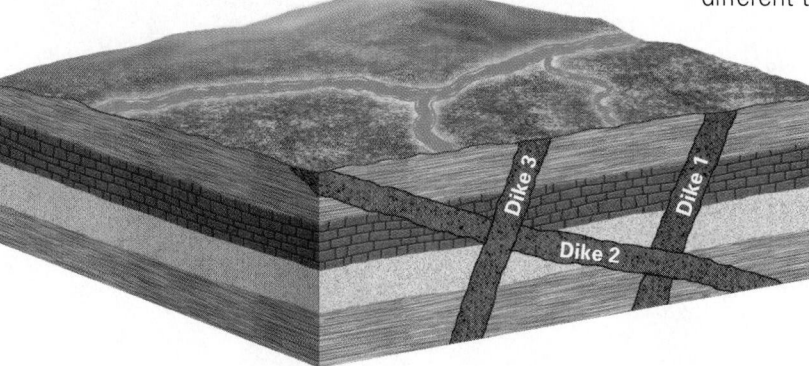

FIGURE 4.8 Three granite dikes cutting sedimentary rocks. The dikes become younger in the order 1, 2, 3. The sedimentary rocks must be older than all three dikes.

The **principle of faunal succession** states that species succeeded one another through time in a definite and recognizable order and that the relative ages of sedimentary rocks can therefore be recognized from their fossils.

Interpreting Geologic History from Fossils

Paleontologists study fossils to understand the history of life and evolution. The oldest known fossils are traces of bacteria-like organisms that lived about 3.5 billion years ago. A much younger fossil consists of the frozen and mummified remains of a Bronze Age man recently found in a glacier near the Italian–Austrian border.

Fossils also allow geologists to interpret ancient environments because most organisms thrive in specific environments and do not survive in others. For example, corals live in clear, warm, shallow seas. Yet today, fossil corals are abundant in rocks of many cold mountain environments such as the Canadian Rockies and the summit of Mount Everest. The fossil corals tell us that the rocks that lie exposed in these mountains were once submerged beneath clear, warm, shallow seas. Today, geologists understand that tectonic processes raised these marine rocks and their fossils to form some of the world's highest peaks.

FIGURE 4.10 The 2-kilometer-high walls of Grand Canyon are composed of sedimentary rocks lying on older igneous and metamorphic rocks. Their ages range from 200 million to nearly 2 billion years old. The 2-billion-year-old schist in the lower part of the photo lies beneath the horizontally bedded, younger sedimentary rocks. The red line indicates a nonconformity in the layers.

4.4 Unconformities and Correlation

The 2-kilometer-high walls of the Grand Canyon are composed of sedimentary rocks lying on older igneous and metamorphic rocks (Figure 4.10). Their ages range from about 200 million years to nearly 2 billion years. The principle of superposition tells us that the deepest sedimentary rocks are the oldest, and the rocks become younger as we climb up the canyon walls. However, no principle assures us that the rocks formed continuously from 2 billion to 200 million years ago. The rock record may be incomplete. Suppose that no sediment were deposited for a period of time, or that erosion removed some sedimentary layers before younger layers accumulated. In either case a gap would exist in the rock record. We know that any rock layer is younger than the layer below it, but without more information we do not know how much younger.

Unconformities

Layers of sedimentary rocks are **conformable** if they were deposited without detectable interruption. An **unconformity** represents an interruption in deposition, usually of long duration. During the interval when no sediment was deposited, some rock layers may have been eroded. Thus, an unconformity represents a long time interval for which no geologic record exists in that place. The lost record may involve hundreds of millions of years.

Several types of unconformities exist. In a **disconformity**, the sedimentary layers above and below the unconformity are parallel (Figures 4.11 and 4.12). A disconformity may be difficult to recognize unless

Nonconformity

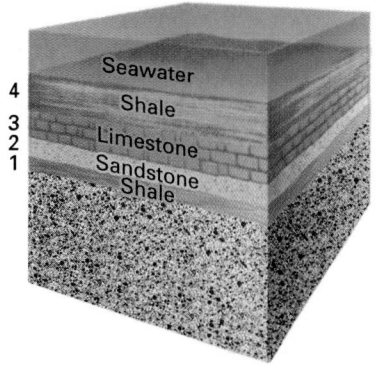

Sediment is deposited below
sea level.

A

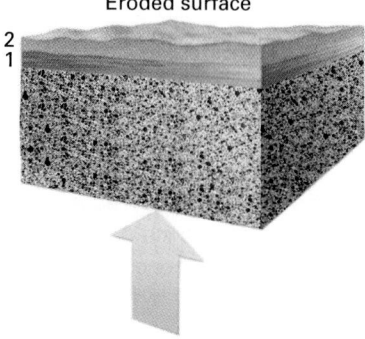

Rocks are exposed above sea level
and layers 4 and 3 are removed
by erosion.

B

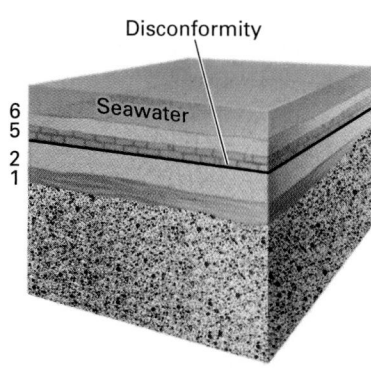

Rocks subside below sea
level and layers 5 and 6 are
deposited on the eroded surface.

C

FIGURE 4.11 A disconformity is
created by uplift and erosion, followed
by deposition of additional layers
of sediment. The disconformity
represents a gap in the rock record.

an obvious soil layer or erosional surface developed.
However, geologists can identify disconformities by
determining the ages of rocks, using methods based
on fossils and absolute dating, described later in this
chapter.

In an **angular unconformity**, tectonic activity tilted
older sedimentary rock layers and erosion planed them
off before the younger sediment accumulated (Figures
4.13 and 4.14). A **nonconformity** is an unconformity in
which sedimentary rocks lie on igneous or metamorphic
rocks. The nonconformity shown in Figure 4.10 represents a time gap of about 1 billion years.

Correlation

Ideally geologists would like to develop a continuous
history for each region of Earth by interpreting rocks
that formed in that place throughout geologic time.
However, there is no single place on Earth where rocks
formed and were preserved continuously. Erosion
removed some rock layers, and at times no rocks
formed. Consequently, the rock record in any one place
is full of unconformities, or historical gaps. To assemble
as complete and continuous a record as possible, geologists gather geological evidence from many localities
and match rocks of the same age from different localities in a process called **correlation** (Figure 4.15). But

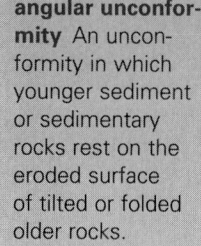

**angular unconfor-
mity** An uncon-
formity in which
younger sediment
or sedimentary
rocks rest on the
eroded surface
of tilted or folded
older rocks.

nonconformity A
type of unconformi-
ty in which layered
sedimentary rocks
lie on an erosion
surface cut into
igneous or meta-
morphic rocks.

correlation The
process of estab-
lishing the age
relationship of
rocks or geologic
features from dif-
ferent locations on
Earth; can be done
by comparing char-
acteristics of the
layers or the types
of fossils found in
those layers. There
are two types of
correlation: *time
correlation* (age
equivalence) and
*lithologic correla-
tion* (continuity of
the rock unit).

FIGURE 4.12 A dis-
conformity separates
parallel layers of
sandstone and over-
lying conglomerate
in Wyoming. Some
sandstone layers
were eroded away
before the conglom-
erate was deposited.

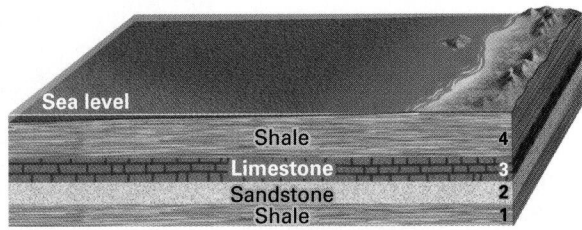

Sea level

Shale 4
Limestone 3
Sandstone 2
Shale 1

A Sediment is deposited below sea level.

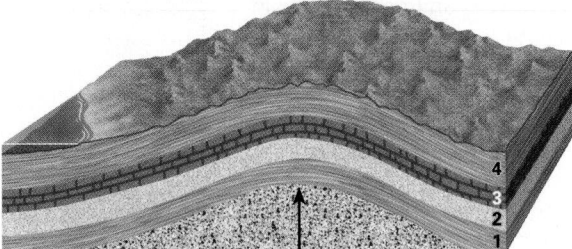

4
3
2
1

B Rocks are uplifted and tilted.

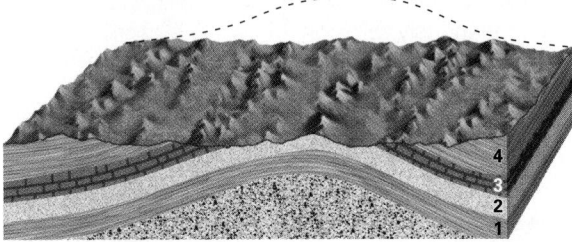

4
3
2
1

C Erosion exposes folded rocks.

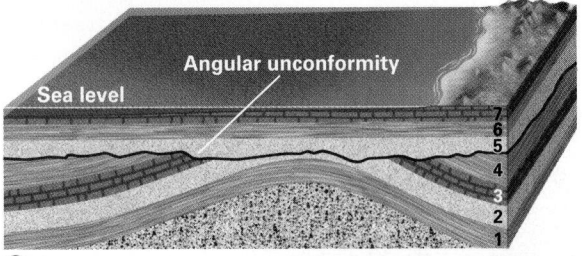

Angular unconformity

Sea level

7
6
5
4
3
2
1

D Rocks sink below sea level and new rocks are deposited.

FIGURE 4.13 An angular unconformity develops when older sedimentary rocks are folded and eroded before younger sediment accumulates.

how do we correlate rocks of the same age over great distances?

If you follow a single sedimentary bed from one place to another, then it is clearly the same layer in both places. But this approach is impractical over long distances or where rocks are not exposed. Another problem arises when time correlation is based on continuity of a sedimentary layer of a particular rock type, such as sandstone. When rocks are correlated for the purpose of building a geologic time scale, geologists want to show that certain rocks all formed at the same time. Suppose you are attempting to trace a beach sandstone that formed as sea level rose. The beach would have migrated inland over time and, as a result, the sandstone becomes younger in a landward direction (Figure 4.16).

Over a distance short enough to be covered in an hour or so of walking, the age difference may be unimportant. But if you traced a similar beach sand for hundreds of kilometers, its age may vary by millions of years. Thus, there are two kinds of correlation: *time correlation* refers to age equivalence, and *lithologic correlation* refers to continuity of a rock unit, such as the sandstone. The two are not always the same because some rock units, such as the sandstone, were deposited at different times in different places. To construct a record of Earth history and a geologic time scale, Earth scientists use several types of evidence to correlate rocks of the same age.

Angular unconformity

FIGURE 4.14 An angular unconformity near Capitol Reef National Park, Utah. The dark line shows the bottom of the angular unconformity.

Unit 1: Earth Materials and Time

One source is **index fossils**. Index fossils can accurately indicate the ages of sedimentary rocks because they were produced by an organism that (1) is abundantly preserved in rocks, (2) was geographically widespread, (3) existed as a species or genus for only a relatively short time, and (4) is easily identified in the field. Floating or swimming marine animals make the best index fossils because they spread rapidly and widely throughout the seas. The shorter the time span that a species existed, the more precisely the index fossil reflects the age of a rock.

In many cases, the presence of a single type of index fossil is sufficient to establish the age of a rock. More commonly, an assemblage of several fossils is used to date and correlate rocks. Figure 4.17 shows an example of how index fossils and fossil assemblages are used in correlation.

Another source used in correlation is the **key bed**—a thin, widespread, easily recognized sedimentary layer that was deposited rapidly and simultaneously over a wide area. Many volcanic eruptions eject great volumes of fine, glassy volcanic ash into the atmosphere. Wind carries the ash over great distances before it settles. Some historic ash clouds have encircled the globe. When the ash settles, the glass rapidly crystallizes to form a pale clay layer that is incorporated into sedimentary rocks. Such volcanic eruptions occur at a precise point in time, so the ash is the same age everywhere. The thin, sooty, iridium-rich clay layer deposited 65 million years ago from debris of a giant meteorite impact (described at the beginning of this chapter) is a classic example of a key bed.

index fossil A fossil that dates the layers where it is found because it came from an organism that is abundantly preserved in rocks, was widespread geographically, and existed as a species or genus for only a relatively short time.

key bed A thin, widespread, easily recognized sedimentary layer that can be used for correlation because it was deposited rapidly and simultaneously over a wide area.

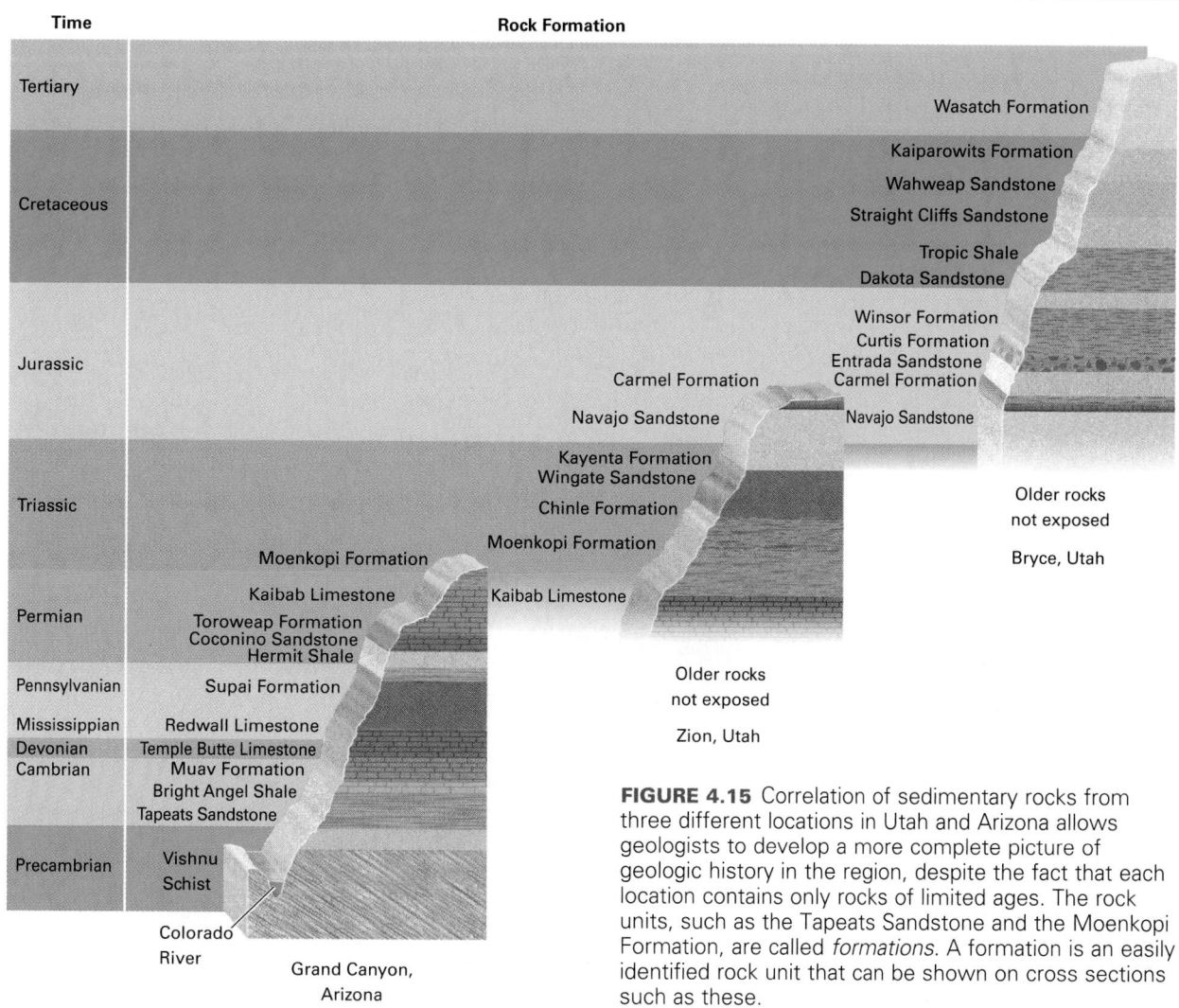

Time

Rock Formation

Tertiary

Wasatch Formation

Cretaceous

Kaiparowits Formation
Wahweap Sandstone
Straight Cliffs Sandstone
Tropic Shale
Dakota Sandstone

Winsor Formation
Curtis Formation
Entrada Sandstone
Carmel Formation

Jurassic

Carmel Formation

Navajo Sandstone

Navajo Sandstone

Triassic

Kayenta Formation
Wingate Sandstone

Chinle Formation

Moenkopi Formation

Older rocks
not exposed

Bryce, Utah

Moenkopi Formation

Kaibab Limestone

Kaibab Limestone

Permian

Toroweap Formation
Coconino Sandstone
Hermit Shale

Pennsylvanian

Supai Formation

Older rocks
not exposed

Zion, Utah

Mississippian

Redwall Limestone

Devonian
Cambrian

Temple Butte Limestone
Muav Formation
Bright Angel Shale
Tapeats Sandstone

Precambrian

Vishnu
Schist

Colorado
River

Grand Canyon,
Arizona

FIGURE 4.15 Correlation of sedimentary rocks from three different locations in Utah and Arizona allows geologists to develop a more complete picture of geologic history in the region, despite the fact that each location contains only rocks of limited ages. The rock units, such as the Tapeats Sandstone and the Moenkopi Formation, are called *formations*. A formation is an easily identified rock unit that can be shown on cross sections such as these.

isotopes Atoms of the same element that have the same number of protons but different numbers of neutrons.

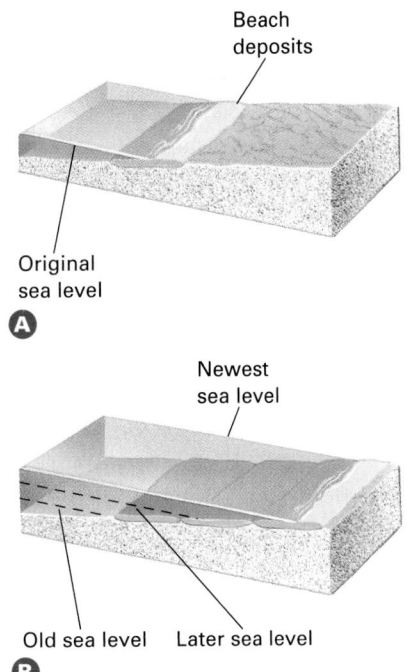

FIGURE 4.16 As sea level rises (A and B) or falls (C) slowly through time, a beach migrates laterally, forming a single sand layer that is of different ages in different localities.

Beach deposits

Original sea level

A

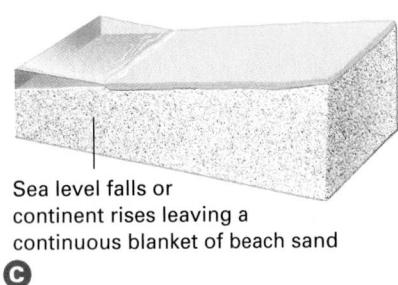

Newest sea level

Old sea level   Later sea level

B

Sea level falls or continent rises leaving a continuous blanket of beach sand

C

4.5 Absolute Geologic Time

Geologists have a challenging task—they are attempting to measure the absolute age of events that occurred before history was recorded. Measuring absolute age relies on (1) a process that occurs at a constant rate, such as Earth rotating once every 24 hours, and (2) some way to keep a cumulative record of that process, such as marking a calendar each time the Sun rises. Measurement of time with a calendar, a clock, an hourglass, or any other device depends on these two factors.

Geologists have found a natural process that occurs at a constant rate and accumulates its own record: it is the radioactive decay of elements that are present in many rocks. Thus, many rocks have built-in calendars.

To begin to understand radioactivity, you need a basic understanding of the atom. An atom consists of a small, dense nucleus surrounded by a cloud of electrons. A nucleus consists of positively charged *protons* and neutral particles called *neutrons*. All atoms of any given element have the same number of protons in the nucleus. However, the number of neutrons may vary. **Isotopes** are atoms of the same element with different numbers of neutrons. For example, all isotopes of potassium-40 contain 19 protons and 21 neutrons. Potassium-39 has 19 protons but only 20 neutrons.

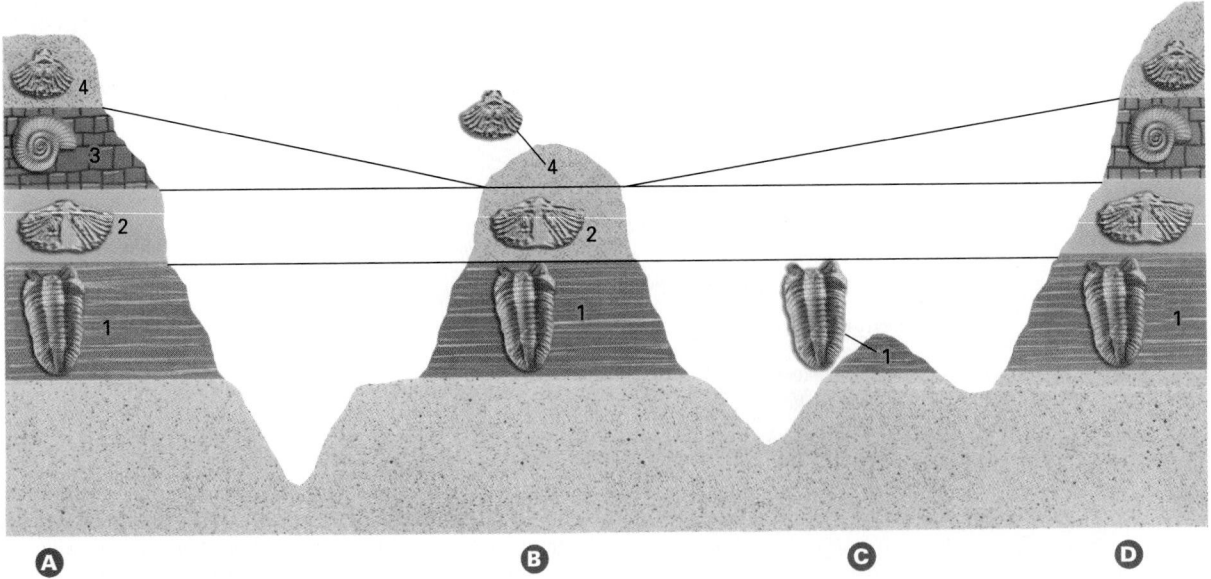

FIGURE 4.17 Index fossils demonstrate age equivalency of sedimentary rocks from widely separated localities. Sedimentary beds containing the same index fossils are interpreted to be of the same age. In this figure, the fossils show that at locality B, layer 3 is missing because layer 4 directly lies on top of layer 2. Either layer 3 was deposited and then eroded away before 4 was deposited at locality B, or layer 3 was never deposited here. At locality C, all layers above 1 are missing, either because of erosion or because they were never deposited.

Unit 1: Earth Materials and Time

Many isotopes are *stable*, meaning that they do not change with time. If you studied a sample of potassium-39 for 10 billion years, all the atoms would remain unchanged. Other isotopes are unstable or radioactive. Given time, their nuclei spontaneously decay.[4] Potassium-40 decays naturally to form two other isotopes: argon-40 and calcium-40 (Figure 4.18).

A radioactive isotope such as potassium-40 is known as a *parent isotope*. An isotope created by radioactivity, such as argon-40 or calcium-40, is called a *daughter isotope*.

Many common elements, such as potassium, consist of a mixture of radioactive and nonradioactive isotopes. With time, the radioactive isotopes decay but the nonradioactive ones do not. Some elements, such as uranium, consist only of radioactive isotopes. The amount of uranium on Earth slowly decreases as it decomposes to other elements, such as lead.

---

4. [4]Radioactive nuclei decay by three processes. In *electron capture*, the nucleus captures an electron. The electron combines with a proton to create a neutron. The decay of potassium-40 to argon-40 is an example of this process. In *beta emission*, a neutron ejects an electron from the nucleus, and the neutron becomes a proton. The ejected electron is called a beta particle. The decay of potassium-40 to calcium-40 is an example of this process. In a third type of decay, called *alpha emission*, a nucleus emits an alpha particle consisting of two protons and two neutrons. Uranium-238 emits eight alpha particles and six beta particles in a sequence of steps to become lead-206.

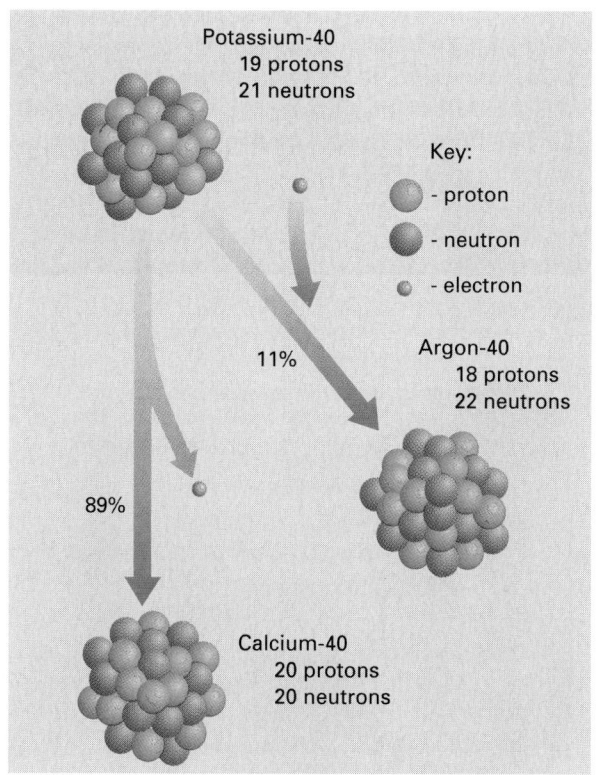

Radioactivity and Half-Life

If you watch a single atom of potassium-40, when will it decompose? This question cannot be answered, because any particular potassium-40 atom may or may not decompose at any time. But even a small sample of a material contains a huge number of atoms. Each atom has a certain probability of decaying at any time. Averaged over time, half of the atoms in any sample of potassium-40 will decompose in 1.3 billion years. The **half-life** is the time it takes for half of the atoms in a radioactive sample to decompose. The half-life of potassium-40 is 1.3 billion years. Therefore, if 1 gram of potassium-40 were placed in a container, 0.5 gram would remain after 1.3 billion years, 0.25 gram after 2.6 billion years, and so on. Each radioactive isotope has its own half-life; some half-lives are fractions of a second and others are measured in billions of years.

Radiometric Dating

Two properties of radioactivity are essential to the calendars in rocks. First, the half-life of a radioactive isotope is constant. It is easily measured in the laboratory and is unaffected by geologic or chemical processes, so radioactive decay occurs at a known, constant rate. Second, as a parent isotope decays, its daughter accumulates in the rock. The longer the rock exists, the more daughter isotope accumulates. The accumulation of a daughter isotope is analogous to marking off days on a calendar. Because radioactive isotopes are widely distributed throughout Earth, many rocks have built-in calendars that allow us to measure their ages. **Radiometric dating** is the process of determining the absolute ages of rocks, minerals, and fossils by measuring the relative amounts of parent and daughter isotopes.

Figure 4.19 shows the relationships between age and relative amounts of parent and daughter isotopes. At the end of one half-life, 50 percent of the parent

half-life The time it takes for half of the atoms of a radioactive isotope in a sample to decompose.

radiometric dating The process of measuring the absolute age of rocks, minerals, and fossils by measuring the concentrations of radioactive isotopes and their decay products.

FIGURE 4.18 Potassium-40 spontaneously decays either to argon-40 or to calcium-40. Eleven percent of the potassium-40 atoms decay to argon-40. In this case, the potassium-40 nucleus captures an electron. The electron combines with a proton to produce a neutron. This process creates an argon-40 nucleus with 18 protons and 22 neutrons. The other 89 percent of the potassium-40 nuclei decay to calcium-40 by expelling an electron from the nucleus. This process converts a neutron to a proton, producing a calcium-40 nucleus containing 20 protons and 20 neutrons.

FIGURE 4.19 As a radioactive parent isotope decays to a daughter, the proportion of parent decreases (blue line) and the amount of daughter increases (red line). The half-life is the amount of time required for half of the parent to decay to daughter. At time zero, when the radiometric calendar starts, a sample is 100 percent parent. At the end of one half-life, 50 percent of the parent has converted to daughter. At the end of two half-lives, 25 percent of the sample is parent and 75 percent is daughter. Thus, by measuring the proportions of parent and daughter in a rock, the rock's age in half-lives can be obtained. Because the half-lives of all radioactive isotopes are well known, it is simple to convert age in half-lives to age in years.

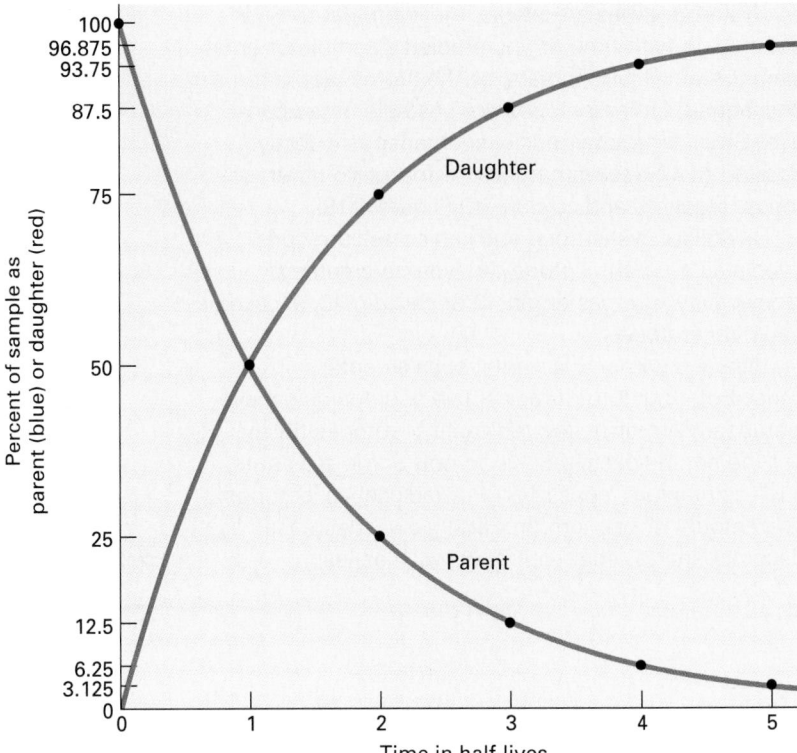

atoms have decayed to daughter. At the end of two half-lives, the mixture is 25 percent parent and 75 percent daughter. To determine the age of a rock, a geologist measures the proportions of parent and daughter isotopes in a sample and compares the ratio to a similar graph. Consider a hypothetical parent–daughter pair having a half-life of 1 million years. If we determine that a rock contains a mixture of 25 percent parent isotope and 75 percent daughter, Figure 4.19 shows that the age is two half-lives, or in this case 2 million years.

If the half-life of a radioactive isotope is short, that isotope gives accurate ages for young materials. For example, carbon-14 has a half-life of 5,730 years. Carbon-14 dating gives accurate ages for materials younger than 50,000 years. It is useless for older materials because by 50,000 years virtually all the

carbon-14 has decayed, and no additional change can be measured.

Isotopes with long half-lives give good ages for old rocks, but not enough daughter accumulates in young rocks to be measured. For example, rubidium-87 has a half-life of 47 billion years. In a geologically short period of time—10 million years or less—so little of its daughter has accumulated that it is impossible to measure accurately. Therefore, rubidium-87 is not useful for rocks younger than about 10 million years. The radioactive isotopes that are most commonly used for dating are summarized in Table 4.1.

TABLE 4.1 The Most Commonly Used Isotopes in Radiometric Age Dating

Isotope (Parent)	(Daughter)	Half-Life of Parent (Years)	Effective Dating Range (Years)	Minerals and Other Materials That Can Be Dated with This Isotope
Carbon-14	Nitrogen-14	5,730 ± 30	100–50,000	Anything that was once alive: wood, other plant matter, bone, flesh, or shells; also, carbon in carbon dioxide dissolved in ground water, deep layers of the ocean, or glacier ice
Potassium-40	Argon-40 Calcium-40	1.3 billion	50,000–4.6 billion	Muscovite Biotite Hornblende Whole volcanic rock
Uranium-238 Uranium-235 Thorium-232	Lead-206 Lead-207 Lead-208	4.5 billion 710 million 14 billion	10 million–4.6 billion	Zircon Uraninite and pitchblende
Rubidium-87	Strontium-87	47 billion	10 million–4.6 billion	Muscovite Biotite Potassium feldspar Whole metamorphic or igneous rock

4.6 The Geologic Column and Time Scale

As mentioned earlier, no single locality exists on Earth where a complete sequence of rocks formed continuously throughout geologic time. However, geologists have correlated rocks from many localities around the world to create the **geologic column** and **geologic time scale**, which together provide a nearly complete composite record of geologic time (Figure 4.20). The worldwide geologic column is frequently revised as geologic mapping continues.

Just as a year is subdivided into months, months into weeks, and weeks into days, geologic time is split into smaller intervals. The units are named, just as months and days are. The largest time units are **eons**, which are divided into **eras**. Eras are subdivided, in turn, into **periods**, which are further subdivided into **epochs**.

The geologic column and time scale were originally constructed on the basis of relative age determinations. When geologists developed radiometric dating, they added absolute ages to the column and time scale. Geologists commonly

geologic column A composite, columnar diagram that shows the sequence of rocks at a given place or region, arranged to show their position in the geologic time scale.

geologic time scale A chronological arrangement of geologic time subdivided into eons, eras, periods, and epochs.

eon The largest unit of geologic time. The most recent eon, the Phanerozoic Eon, is further subdivided into eras.

era A geologic time unit. Eons are divided into eras and, in turn, eras are subdivided into periods.

period A geologic time unit longer than an epoch and shorter than an era.

epoch The smallest unit of geologic time. Periods are divided into epochs.

FIGURE 4.20 The geologic column and geologic time scale is a record of geologic time divided into eons, eras, periods, and epochs. Time is given in millions of years (for example, 1,000 stands for 1,000 million, which is 1 billion). The column is not drawn to scale. We know relatively little about events that occurred during the early part of Earth's history. Therefore, the first 4 billion years are given relatively little space on this chart, while the more recent Phanerozoic Eon, which spans only 543 million years, receives proportionally more space.

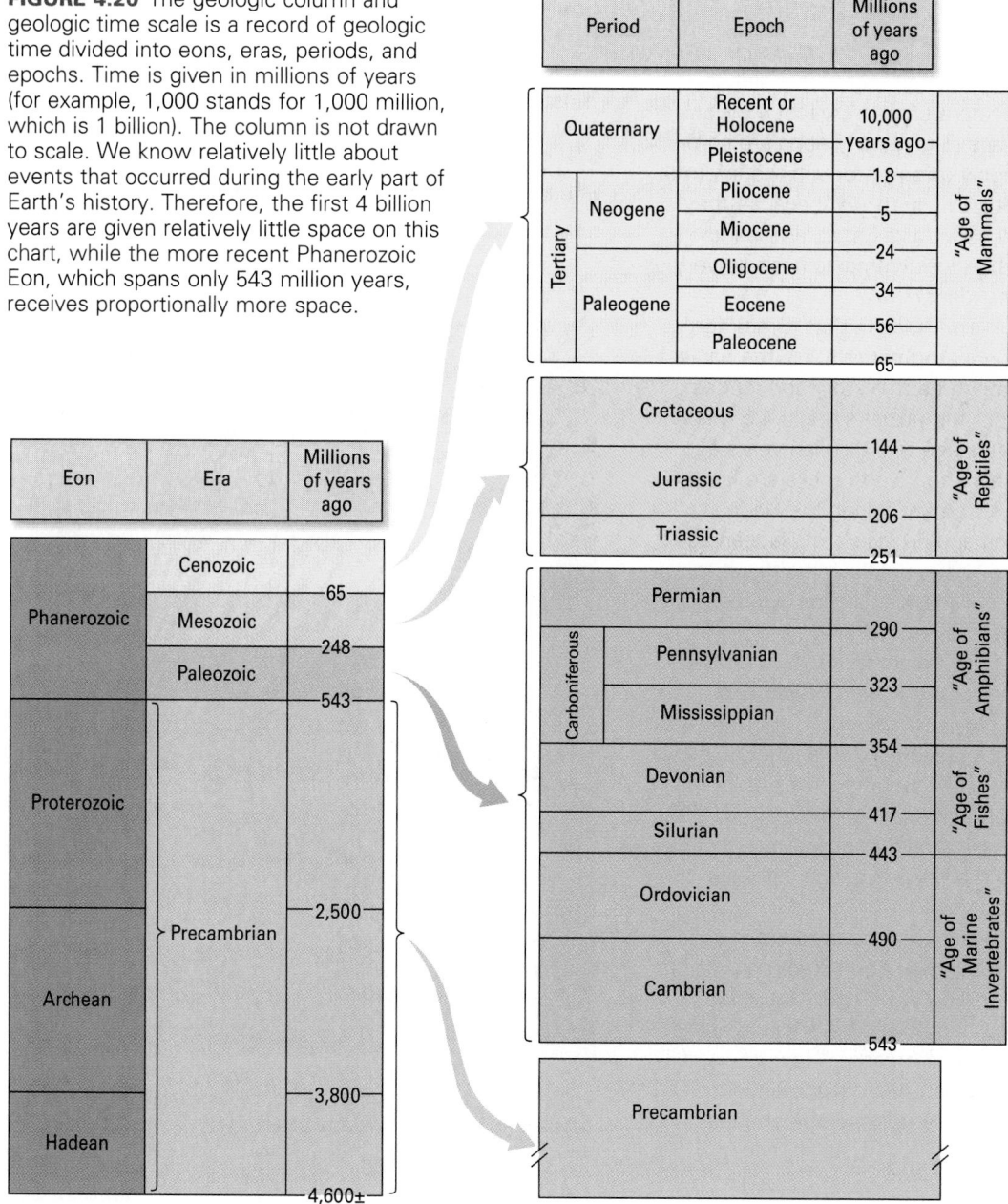

Precambrian A term referring to all of geologic time before the Paleozoic Era, encompassing approximately the first 4 billion years of Earth's history. Also refers to all rocks formed during that time.

Hadean Eon The earliest time in Earth's history, ranging from 4.6 billion years ago to 3.8 billion years ago.

Archean Eon A division of geologic time 3.8 to 2.5 billion years ago. The oldest known rocks formed at the beginning of, or just prior to, the start of the Archean Eon.

Proterozoic Eon The portion of geological time occurring during a period from 2.5 billion to 543 million years ago.

Phanerozoic Eon The most recent 543 million years of geologic time, including the present, represented by rocks that contain evident and abundant fossils.

use the time scale to date rocks in the field. Imagine that you are studying sedimentary rocks in the walls of Grand Canyon. If you find an index fossil or a key bed that has already been radiometrically dated by other scientists, you know the age of the rock you are studying, and you do not need to send the sample to a laboratory for radiometric dating.

The Earliest Eons of Geologic Time: Precambrian Time

The earliest eons—the Hadean, Archean, and Proterozoic Eons—often are not subdivided at all, even though together they constitute a time interval of 4 billion years. These early eons are commonly referred to with the informal term **Precambrian**, because they preceded the Cambrian Period, when fossil remains became very abundant.

The **Hadean Eon** (Greek for "beneath the Earth") is the earliest time in Earth history and ranges from the planet's origin 4.6 billion years ago to 3.8 billion years ago. Only a few Earth rocks are known that formed during the Hadean Eon and no fossils of Hadean age are known, making it difficult to subdivide the Hadean Eon based on fossils.

There are few fossils among the rocks of the **Archean Eon** (Greek for "ancient"), and they are not preserved well enough to allow for finely tuned subdivision of this eon that spanned from 3.8 to 2.5 billion years ago. The fossil record does indicate that life began on Earth 3.2 to 3.5 billion years ago, although the exact date is uncertain.

Diverse groups of fossils have been found in sedimentary rocks of the **Proterozoic Eon** (Greek for "earlier life"), 2.5 billion to 543 million years ago. The most complex are multicellular and have different kinds of cells arranged into tissues and organs. A few types of Proterozoic shell-bearing organisms have been identified, but shelled organisms did not become abundant until the Paleozoic Era.

The Phanerozoic Eon

Sedimentary rocks of the **Phanerozoic Eon**, which covers the most recent 543 million years of geologic time, contain abundant fossils (*phaneros* is Greek for "evident"). Four changes occurred at the beginning of Phanerozoic time that greatly improved the fossil record:

1. The number of species with shells and skeletons dramatically increased.
2. The total number of individual organisms preserved as fossils increased greatly.
3. The total number of species preserved as fossils increased greatly.
4. The average sizes of individual organisms increased.

Shells and skeletons are much more easily preserved than plant remains and soft body tissues. Thus, in rocks of earliest Phanerozoic time and younger, the most abundant fossils are the hard, tough shells and skeletons.

Subdivision of the Phanerozoic Eon into three eras is based on the most common types of life during each

WARD'S NATURAL SCIENCE ESTABLISHMENT

FIGURE 4.21 In Paleozoic time, the land was inhabited by reptiles and amphibians and covered by ferns, ginkgoes, and conifers.

era. Sedimentary rocks that formed during the **Paleozoic Era** (Greek for "old life") contain fossils of early life-forms, such as invertebrates, fishes, amphibians, reptiles, ferns, and cone-bearing trees (Figure 4.21). The Paleozoic Era ended abruptly about 248 million years ago when the terminal Permian mass extinction wiped out 90 percent of all marine species, two-thirds of reptile and amphibian species, and 30 percent of insect species.

The **Mesozoic Era** is most famous for the dinosaurs that roamed the land. Mammals and flowering plants also evolved during this era. The Mesozoic Era ended 65 million years ago with another mass extinction that wiped out the dinosaurs.

During the **Cenozoic Era** (Greek for "recent life"), mammals and grasses became abundant. Humans have evolved and lived wholly in the Cenozoic Era.

The eras of Phanerozoic time are subdivided into periods, the time unit most commonly used by geologists. Some of the periods are named after special characteristics of the rocks formed during that period. For example, the Cretaceous Period is named from the Latin word for chalk (*creta*), after chalk beds of this age in Africa, North America, and Europe. Other periods are named for the geographic localities where rocks of that age were first described. For example, the Jurassic Period is named for the Jura Mountains of France and Switzerland. The Cambrian Period is named for Cambria, the Roman name for Wales, where rocks of this age were first studied.

In addition to the abundance of fossils, another reason that details of Phanerozoic time are better known than those of Precambrian time is that many of the older rocks have been metamorphosed, deformed, and eroded. It is simple probability that the older a rock is, the greater the chance that metamorphism or erosion has destroyed the rock or the features that record Earth's history.

Paleozoic Era The part of geologic time occurring during a period from 543 to 248 million years ago. During this era invertebrates, fishes, amphibians, reptiles, ferns, and cone-bearing trees were dominant.

Mesozoic Era The part of geologic time roughly from 248 to 65 million years ago. Dinosaurs rose to prominence and became extinct during this era.

Cenozoic Era The latest of the four eras into which geologic time is subdivided, 65 million years ago to the present.

VIRTUAL FIELD TRIP

Geologic Time

GOALS OF THE TRIP

Some of the goals of this trip are to understand that:

1. Earth is about 4.6 billion years old.

2. Much of Earth history, including environments and geologic events, is recorded in rocks such as those in Grand Canyon and Capitol Reef national parks.

3. You can apply simple geologic tools and concepts such as the principles listed previously to understand how the rocks formed, the environmental conditions under which they formed, and the geologic events that affected the rocks after they formed.

4. You can also apply these principles to interpret the relative ages of rocks and geologic events.

Ready to Go!

On this virtual field trip we will visit both Grand Canyon National Park in Arizona, and Capitol Reef National Park in Utah. The scenery is spectacular and the rocks of both parks display evidence of the great length of geologic time. The rocks also illustrate the principles that geologists use to determine the relative ages of rocks and geologic events. Those principles include the "laws" of superposition, original horizontality, cross-cutting relationships, and faunal succession described in Section 4.3 ("Relative Geologic Time") in this chapter. Although early geologists dignified these concepts by calling them "laws" and "principles," they really are simple applications of common sense to an understanding of the ways in which rocks form. For example, a photo in "Relative Geologic Time" in this chapter shows an igneous dike cutting across gneiss. Common sense tells us that the gneiss must have been present before the dike intruded into it. Therefore the dike must be younger than the gneiss. This is an example of the principle of cross-cutting relationships. You may also wish to refer to Section 3.3 ("Sedimentary Rocks") in Chapter 3 for additional discussion of the processes that formed the rocks of these wonderful public parks.

FOLLOW-UP QUESTIONS

1. The principle of superposition tells us that the Bright Angle Shale in the Grand Canyon walls must be younger than the underlying Tapeats Sandstone. Refer to Section 4.5 ("Absolute Geologic Time") in Chapter 4 and consider how a geologist might measure the ages of the two formations in years rather than in relative time.

2. The narrative for this virtual field trip explains that the Tapeats Sandstone is a beach sand deposited as sea level rose during Cambrian time causing shallow seas to flood much of western North America. Refer to Section 4.4 ("Unconformities and Correlation") in this chapter and consider how, although it is laterally continuous, the Tapeats Sandstone can be older in one place and younger in another.

3. Refer to the sections titled "Unconformities and Correlation" and "Relative Geologic Time" in this chapter and consider how a geologist might use fossils to identify an unconformity.

What to See When You Go

The photos and videos in this field trip display the great beauty and awesome scale of the rocks and topographic features of Grand Canyon and Capitol Reef national parks. The narrative explains the geologic principles and concepts that are illustrated by the rocks of the parks and are explained in the section "Relative Geologic Time" in Chapter 4 and under "Sedimentary Structures" in Chapter 3. Notice especially the layered nature of sedimentary rocks, and ponder why the rock layers become progressively younger in the upward direction. Think also about why the layers are laterally continuous for many miles, and in some cases for hundreds of miles. What was the Earth's surface environment where and when these sediments were deposited? How did the sediment become hard rock?

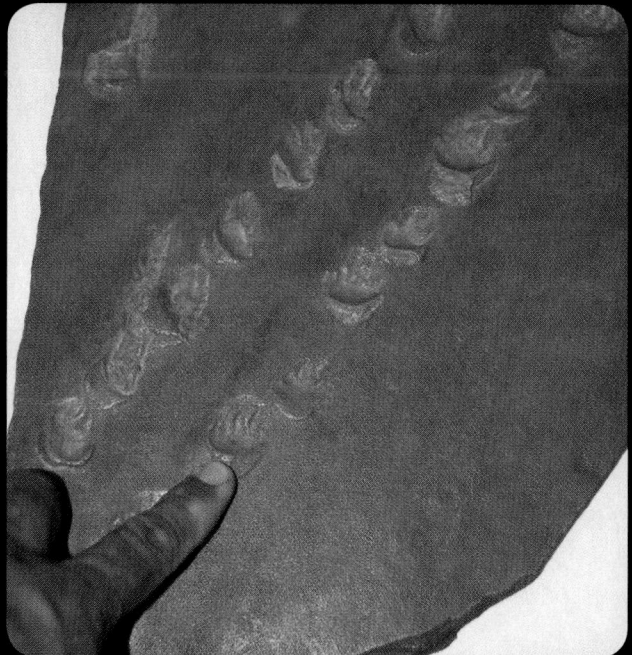

Consider the angular unconformity pictured below, which is in the walls of Grand Canyon. Review the section titled "Unconformities and Correlation" in this chapter and consider the sequence of events that formed the unconformity.

Fossils such as these (pictured below) are abundant in some of the sedimentary rocks of Grand Canyon and in many other sedimentary rocks around the world. Consider how a geologist might use fossils to correlate sedimentary layers over distances of hundreds of miles.

ALL PHOTOS FROM VIRTUAL FIELD TRIPS: COPYRIGHT DR. PARVINDER S. SETHI
MAP BACKGROUND: © ROBERT DANT/ISTOCKPHOTO.COM

5

GEOLOGIC RESOURCES

An offshore oil platform in the Gulf of Mexico.

{ **Fertile soil, level plains, easy passage across the mountains, coal, iron, and other metals imbedded in the rocks, and a stimulating climate, all shower their blessings upon man.** }

Ellsworth Huntington

mineral resources
Geologic resources including both metal ore and non-metallic minerals.

energy resources
Geologic resources, including petroleum, coal, natural gas, and nuclear fuels, used for heat, light, work, and communication.

nonmetallic mineral resources
Economically useful rocks or minerals that are not metals—such as salt, building stone, sand, and gravel.

Since human-like creatures emerged 5 to 7 million years ago, our use of geologic resources has become increasingly sophisticated. Early hominids used sticks and rocks as simple weapons and tools. Later prehistoric people used flint and obsidian to make more effective weapons and tools, and they used natural pigments to create elegant art on cave walls. About 8000 BCE, people learned to shape and fire clay to make pottery. Archaeologists have found copper ornaments in Turkey dating from 6500 BCE; 1,500 years later, Mesopotamian farmers used copper farm implements. Today, geologic resources provide the silicon chip that operates your computer, the titanium valves in a space probe, and the gasoline that powers your car.

We use two types of geologic resources: **mineral resources** and **energy resources**. Mineral resources include all useful rocks and minerals. As we will see in the sections that follow, many mineral resources are concentrated by processes that involve interactions among rock of the geosphere, atmospheric gases, and water from the hydrosphere. In turn, humans have used the mineral resources to create the industrial world that has altered our planet. The primary energy resources of the 21st century are coal, petroleum, and natural gas—all formed from the decayed remains of prehistoric plants and animals that have been altered by Earth systems processes.

5.1 Mineral Resources

Mineral resources include both metal ore and nonmetallic minerals. Recall from Chapter 2 that *ore* is rock sufficiently enriched in one or more minerals to be mined profitably. Geologists usually use the term to refer to metallic mineral deposits, and it is commonly accompanied by the name of the metal—for example, *iron ore* or *silver ore*.

Nonmetallic mineral resources refer to the useful rocks or minerals that are not metals—such as salt, building stone, sand, and gravel. When we think about "striking it rich" from mining, we usually think of gold. However, more money has been made mining sand and gravel than by mining gold. For example, in the United States in the year 2004, sand and gravel produced $6.2 billion in revenue, but gold produced only $3.2 billion. Sand and gravel are mined from stream and glacial deposits, sand dunes, and beaches. In turn, these nonmetallic resources are mixed with portland cement—a material produced by heating a mixture of crushed limestone and clay—to make concrete. Reinforced with steel, concrete is used to build roads, bridges, and buildings. Thus, reinforced concrete is one of the basic building materials of the modern world. In addition, many buildings are faced with stone—usually granite or limestone, although marble, slate, sandstone, and other rocks used for building are also mined from quarries cut into bedrock (Figure 5.1).

There are many important metals and other elements that are fundamental parts of our lives and the industries that produce many things in daily use. Some are familiar to us, such as iron, lead, copper, aluminum, silver, and gold. Others are less well known, such as molybdenum (rifle

magmatic processes Geologic processes that form ore deposits as liquid magma solidifies into igneous rock.

COURTESY OF GRAHAM R. THOMPSON/JONATHAN TURK

FIGURE 5.1 A quarryman in China splits a large granite block with a sledgehammer. After he splits the rock, the circular saws in the background will cut it into thin slabs for floors and walls.

barrels), tungsten (lightbulb filaments), and borax (soaps, antiseptics).

All mineral resources are nonrenewable: we use them up at a much faster rate than natural processes create them, although many can be recycled.

5.2 Ore and Ore Deposits

If you pick up any rock and send it to a laboratory for analysis, the report will probably show that the rock contains measurable amounts of iron, gold, silver, aluminum, and other valuable metals. However, the concentrations of these metals are so low in most rocks that the extraction cost would be much greater than the income gained by selling the metals. In certain locations, however, natural geologic processes have enriched metals many times above their normal concen-

trations. Table 5.1 shows that the concentration of a metal in ore may exceed its average abundance in ordinary rock by a factor—called the *enrichment factor*—of more than 100,000.

Successful exploration for new ore deposits requires an understanding of the processes that concentrate metals to form ore. For example, platinum concentrates in certain types of igneous rocks. Therefore, if you were exploring for platinum, you would focus on those rocks rather than on sandstone or limestone. The following topics describe some of the more common ore-forming processes.

With the exception of magmatic processes, which occur deep within the crust, the natural processes that concentrate ore minerals all involve interactions between rocks and minerals of the geosphere with water from the hydrosphere.

Magmatic Processes

Magmatic processes form mineral deposits as liquid magma solidifies to form an igneous rock. These processes create metal ores as well as some gems and nonmetallic mineral deposits including sulfur deposits and building stone.

Some large bodies of igneous rock, particularly those of *mafic* (basaltic; high in magnesium and iron) composition, solidify in layers. Each layer contains different minerals and is of a different chemical composition than is found in adjacent layers. Some of the layers may contain rich ore deposits. The layering can develop by at least two processes:

TABLE 5.1 Comparison of Concentrations of Specific Elements in Earth's Crust with Concentrations Needed to Operate a Commercial Mine

Element	Natural Concentration in Crust (% by Weight)	Concentration Required to Operate a Commercial Mine (% by Weight)	Enrichment Factor
Aluminum	8.0	24 to 32	3 to 4
Iron	5.8	40	6 to 7
Copper	0.0058	0.46 to 0.58	80 to 100
Nickel	0.0072	1.08	150
Zinc	0.0082	2.46	300
Uranium	0.00016	0.19	1,200
Lead	0.00010	0.20	2,000
Gold	0.0000002	0.0008	4,000
Mercury	0.000002	0.20	100,000

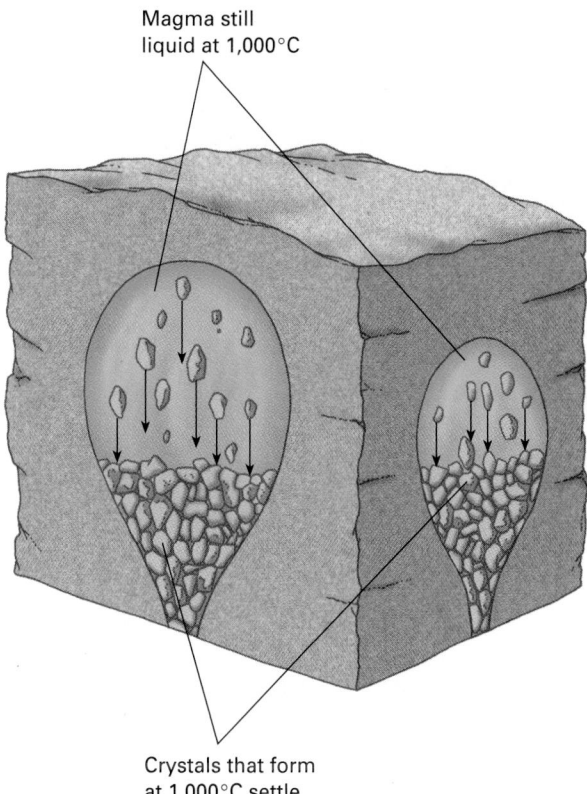

Magma still
liquid at 1,000°C

Crystals that form
at 1,000°C settle

FIGURE 5.2 The first crystals to form in cooling magma settle and concentrate near the bottom of the magma chamber. The crystals are magnified here for clarity. In actuality the crystals may be only a few millimeters in diameter, whereas the magma chamber may be many kilometers in width.

1. Cooling magma does not solidify all at once. Instead, higher-temperature minerals crystallize first, and lower-temperature minerals form later as the magma cools and temperature drops. Most minerals are denser than magma. Consequently, early-formed crystals may sink to the bottom of a magma chamber in a process called **crystal settling** (Figure 5.2). In some instances, ore minerals crystallize with other early-formed minerals and accumulate in layers near the bottom of a pluton.

2. Some large bodies of mafic magma crystallize from the bottom upward. Thus, early-formed ore minerals become concentrated near the base of the pluton.

The largest ore deposits found in mafic-layered plutons are the rich chromium and platinum reserves of South Africa's Bushveld intrusion. The pluton is about 375 by 300 kilometers in area—roughly the size of the state of Maine—and about 7 kilometers thick. The Bushveld deposits contain more than 20 billion tons of chromium and more than 10 billion grams of platinum, the greatest reserves in any known deposit on Earth.

Hydrothermal Processes

Hydrothermal processes—involving interactions between hot water or steam and rocks or minerals—are probably responsible for the formation of more ore deposits, and a larger total quantity of ore, than all other processes combined. To form a hydrothermal ore deposit, hot water (hence the roots *hydro* for "water" and *thermal* for "hot") dissolves metals from rock or magma. The metal-bearing solutions then seep through cracks or through permeable rock, where they precipitate to form an ore deposit.

Hydrothermal water comes from three sources—granitic magma, ground water, and the oceans:

1. Granitic magma contains more dissolved water than solid granite rock. Thus, the magma gives off hydrothermal water as it solidifies. For this reason, hydrothermal ore deposits are commonly associated with granite and similar igneous rocks.

2. Ground water can seep into Earth's crust, where it is heated and forms a hydrothermal solution. This is particularly true in volcanic areas where hot rock or magma heat ground water at shallow depths. For this reason, hydrothermal ore deposits are also common in volcanic regions.

3. In the oceans, hot, young basalt near the Mid-Oceanic Ridge heats seawater as it seeps into cracks in the seafloor.

Although water by itself is capable of dissolving minerals, most hydrothermal waters also contain dissolved salts, which greatly increase its ability to dissolve minerals. Therefore, hot, salty, hydrothermal water is a very powerful solvent, capable of dissolving and transporting metals.

Refer again to Table 5.1, which shows that tiny amounts of all metals are found in average rocks of the Earth's crust.

crystal settling A process in which the crystals that solidify first from a cooling magma settle to the bottom of the magma chamber because the minerals are more dense than magma; the ultimate result is a layered body of rock, each layer containing different minerals.

hydrothermal processes Geologic processes in which hot water or steam dissolves metals and minerals from rocks or magma; the solutions then seep through cracks before cooling, to create ore deposits.

© ROOCOUYT/SHUTTERSTOCK

hydrothermal vein deposit A rich, sheet-like mineral deposit that forms when dissolved minerals, precipitated from hot water solutions, fill a fault or other fracture.

disseminated ore deposit A large, low-grade hydrothermal deposit in which generally metal-bearing minerals are widely scattered throughout a rock body; not as concentrated as a hydrothermal vein.

submarine hydrothermal ore deposit An ore deposit that forms when hot seawater dissolves metals from sea-floor rocks and then, as it rises through the upper layers of oceanic crust, cools and precipitates the metals.

black smoker A jet of black water spouting from a fracture in the sea floor, commonly near the Mid-Oceanic Ridge. The black color is caused by precipitation of fine-grained metal sulfide minerals as the hydrothermal solutions cool on contact with seawater.

placer deposit A surface mineral deposit formed along streambeds, beneath waterfalls, or on beaches when water currents slow down and deposit high-density minerals.

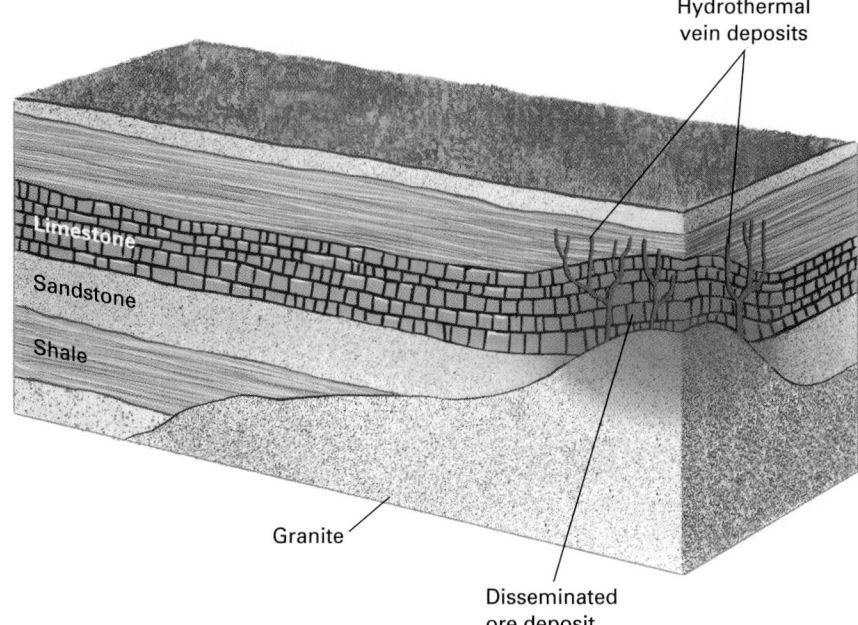

FIGURE 5.3 Hot water scavenges metals from country rock and deposits metallic minerals in ore-rich veins that fill fractures in bedrock. It also deposits low-grade disseminated metal ore in large volumes of rock surrounding the veins.

For example, gold makes up 0.0000002 percent of the crust, while copper makes up 0.0058 percent and lead 0.0001 percent. Although the metals are present in very low concentrations in country rock, hydrothermal solutions percolate through vast volumes of rock, dissolving and accumulating the metals. The solutions then deposit the metals, where they encounter changes in temperature, pressure, or chemical environment (Figure 5.3). In this way, hydrothermal solutions scavenge metals from large volumes of normal crustal rocks and then deposit them locally to form ore.

A **hydrothermal vein deposit** forms when dissolved metals precipitate in a fracture in rock. Ore veins range from less than a millimeter to several meters in width. A single vein can yield several million dollars worth of gold or silver. The same hydrothermal solutions may also soak into pores in country rock near the vein to create a large but much less concentrated **disseminated ore deposit**. Because they commonly form from the same solutions, rich ore veins and disseminated deposits are often found together. The history

of many mining districts is one in which early miners dug shafts and tunnels to follow the rich veins. After the veins were exhausted, later miners used huge power shovels to extract low-grade ore from disseminated deposits surrounding the veins.

In volcanically active regions of the seafloor, near the Mid-Oceanic Ridge and submarine volcanoes, seawater circulates through the hot, fractured oceanic crust. The hot seawater dissolves metals from the rocks and then, as it rises through the upper layers of oceanic crust, cools and precipitates the metals to form **submarine hydrothermal ore deposits**.

The metal-bearing solutions can be seen today as jets of black water, called **black smokers**, spouting from fractures in the Mid-Oceanic Ridge. The black color is caused by precipitation of fine-grained metal sulfide minerals as the solutions cool upon contact with seawater. The precipitating metals accumulate as chimneylike structures near the hot-water vent. Rich ore deposits form in such environments, but the cost to operate machinery beneath the sea is prohibitive.

Sedimentary Processes

Placer Deposits

Gold is denser than any other mineral. Therefore, if you swirl a mixture of water, gold dust, and sand in a gold pan, the gold falls to the bottom first. Differential settling also occurs in nature. Many streams carry silt,

Unit 1: Earth Materials and Time

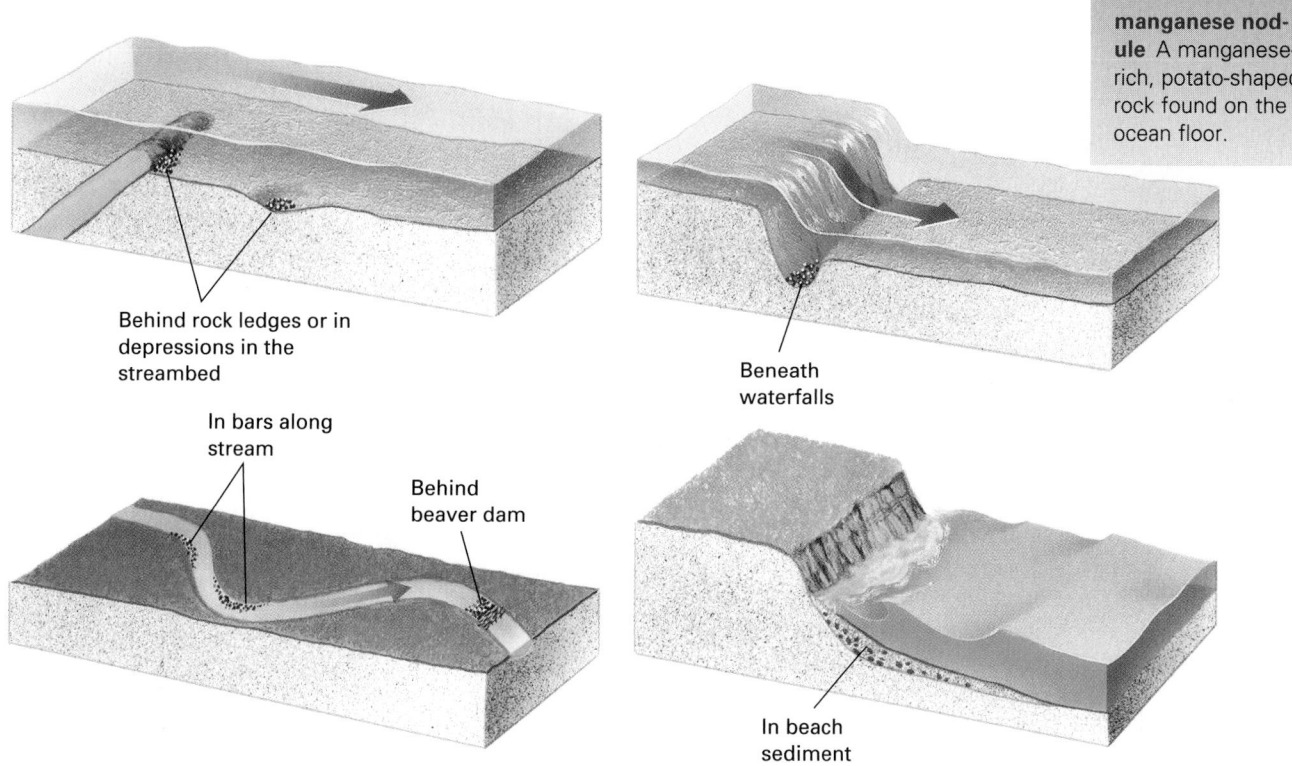

Behind rock ledges or in depressions in the streambed

In bars along stream

Behind beaver dam

Beneath waterfalls

In beach sediment

FIGURE 5.4 Placer deposits form where water currents slow down and deposit high-density minerals.

sand, and gravel with an occasional small grain of gold. The gold settles first when the current slows down. Over years, currents agitate the sediment and the dense gold works its way into cracks and crevices in the streambed. Thus grains of gold concentrate in gravel and bedrock cracks in the streambed, forming a **placer deposit** (Figure 5.4). Most of the prospectors who

rushed to California in the Gold Rush of 1849 searched for placer deposits.

Precipitates

Ground water dissolves minerals as it seeps through soil and bedrock. In most environments, ground water eventually flows into streams and then to the sea. Some of the dissolved ions, such as sodium and chloride, make seawater salty. In deserts, however, *playa lakes* develop with no outlet to the ocean. Water flows into the lakes but can escape only by evaporation. As

Manganese Nodules

About 25 to 50 percent of the Pacific Ocean floor is covered with golf ball to bowling ball sized **manganese nodules**. A typical nodule contains 20 to 30 percent manganese, 6 percent iron, about 1 percent each of copper and nickel, and lesser amounts of other metals. (Much of the remaining 60 to 70 percent consists of oxygen and other anions chemically bonded to the metals.) The metals are probably added to seawater by volcanic activity at the Mid-Oceanic Ridge, perhaps by the black smokers. Chemical reactions between seawater and sea floor sediment precipitate the dissolved metals to form the nodules.

A trillion or more tons of manganese nodules lie on the sea floor. They contain several valuable industrial metals that could be harvested without drilling or blasting. One can imagine undersea television cameras locating the nodules and giant vacuums scooping them up and lifting them to a ship. But, because the sea floor is a difficult environment in which to operate complex machinery, harvest of manganese nodules is not profitable at the present time.

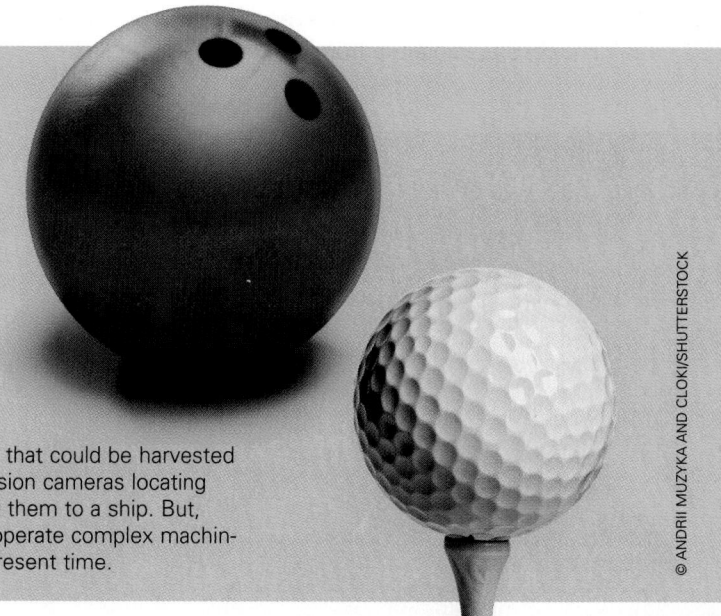

banded iron formation Iron-rich sedimentary rocks composed of alternating iron-rich and silica-rich layers; source of most of the world's supply of iron.

residual ore deposit A mineral deposit formed from relatively insoluble ions left in the soil near Earth's surface after most of the soluble ions were dissolved and removed by abundant water.

bauxite A gray, yellow, or reddish-brown rock, composed of a mixture of aluminum oxides and hydroxides, that formed as a residual deposit; the principle source of aluminum.

mineral reserves A term to describe the known supply of ore in the ground; can be used on a local, national, or global scale.

the water evaporates, the dissolved salts concentrate until they precipitate to form evaporite deposits (see Chapter 3). Evaporite deposits in desert lakes include table salt, borax, sodium sulfate, and sodium carbonate. These salts are used in the production of paper, soap, and medicines and for the tanning of leather.

Several times during the past 500 million years, shallow seas covered large regions of North America and all other continents. At times, those seas were so poorly connected to the open oceans that water did not circulate freely between them and the oceans. Consequently, evaporation concentrated the dissolved salts until they precipitated as marine evaporites. Periodically, storms flushed new seawater from the open ocean into the shallow seas, providing a new supply of salt. Thick marine evaporite beds, formed in this way, underlie nearly 30 percent of North America. Table salt, gypsum (used to manufacture plaster and sheetrock), and potassium salts (used in fertilizer) are mined extensively from these deposits.

Most of the world's supply of iron is mined from sedimentary rocks called **banded iron formations**, which are deposits composed of alternating iron-rich and silica-rich layers (Figure 5.5). These iron-rich rocks precipitated from the seas between 2.6 and 1.9 billion years ago, as a result of rising atmospheric oxygen concentrations.

Weathering Processes

In environments with high rainfall, the abundant water dissolves and removes most of the soluble ions from soil and rock near Earth's surface. This process leaves the relatively insoluble ions in the soil to form **residual ore deposits**. Both aluminum and iron have very low solubilities in water. **Bauxite**, the principal source of aluminum, forms as a residual deposit, and in some instances iron also concentrates enough to become ore. Most bauxite deposits form in warm, rainy, tropical, or subtropical environments where chemical weathering occurs rapidly. Thus bauxite ores are common in Jamaica, Cuba, Guinea, Australia, and parts of the southeastern United States (Figure 5.6).

USGA

FIGURE 5.6 This spheroidal texture is typical of bauxite, which is aluminum ore formed as a residual deposit by intense tropical weathering of aluminum-rich rocks.

5.3 Mineral Reserves

Mineral reserves are the known amount of ore in the ground. The term can refer to the amount of ore remaining in a particular mine, or it can be used on a global or national scale. Mining depletes mineral reserves by decreasing the amount of ore remaining in the ground—but reserves may also *increase* in two ways. First, geologists may discover new mineral deposits, thereby adding to the known amount of ore. Second, *subeconomic mineral deposits*—those in which the metal is not sufficiently concentrated to be mined at a profit—can become profitable if the price of that metal increases or if improvements in mining or refining technology reduce extraction costs.

In 1966 geologists estimated that global reserves of iron were about 5 billion tons.[1] At that time, world consumption of iron was about 280 million tons per year. Assuming that consumption continued at the 1966 rate, the global iron reserves identified in 1966 would have been exhausted in 18 years, and we would have run out of iron ore in 1984. But iron ore is still plentiful and cheap today because new and inexpensive methods of

COPYRIGHT AND PHOTOGRAPH BY DR. PARVINDER S. SETHI

FIGURE 5.5 Banded iron formations from Michigan.

---

1. Brian Mason, *Principles of Geochemistry*, 3rd ed. (New York: John Wiley & Sons, 1966), Appendix III.

FIGURE 5.7 Machinery extracts coal from an underground coal mine.

PEABODY COAL COMPANY

processing lower-grade iron ore were developed. Thus, large deposits that were subeconomic in 1966, and therefore not counted as reserves, are now ore.

The Geopolitics of Metal Resources

Earth's mineral resources are unevenly distributed, and no single nation is self-sufficient in all minerals. For example, almost two-thirds of the world's molybdenum reserves, and more than one-third of the lead reserves, are located in the United States. More than half of the aluminum reserves are found in Australia and

FIGURE 5.8 The Bingham Canyon, Utah, open-pit copper mine is the largest human-created excavation on Earth. It is 4 kilometers in diameter and 0.8 kilometer deep.

AGRICULTURAL STABILIZATION AND CONSERVATION SERVICE/USDA

Guinea. The United States uses about 30 percent of all aluminum produced in the world, yet it has no large bauxite deposits.

Five nations—the United States, Russia, South Africa, Canada, and Australia—supply most of the mineral resources used by modern societies. Many other nations have few mineral resources. For example, Japan has almost no metal or fuel reserves; despite its modern economy and high productivity, it relies entirely on imports for metals and fuel.

Developed nations consume most of Earth's mineral resources. While the United States, Japan, Germany, and Russia have traditionally been the largest consumers of these resources, China's industrial base is now growing so rapidly that it may outpace the developed world in consumption.

Currently, the United States depends on 25 other countries for more than half of its mineral resources. Some must be imported because we have no reserves of our own. We do have reserves of others, but we consume them more rapidly than we can mine them, or we can buy them more cheaply than we can mine them.

5.4 Mines and Mining

Miners extract both ore and coal (described in the following section) from underground mines and surface mines. A large **underground mine** may consist of tens of kilometers of interconnected passages that commonly follow ore veins or coal seams (Figure 5.7). The lowest levels may be several kilometers deep. In contrast, a **surface mine** is a hole excavated into Earth's surface. The largest human-created excavation on Earth is the open-pit copper mine at Bingham Canyon, Utah (Figure 5.8). It is 4 kilometers in diameter and 0.8 kilometer deep. The mine produced 263,700 tons of copper in 2004 and smaller amounts of gold, silver, and molybdenum. Most modern coal mining is done by large power shovels that extract coal from huge surface mines (Figure 5.9).

In the United States, the Surface Mining Control and Reclamation Act of 1977 requires mining companies to restore mined land, so it can be used for the same purposes for which it was used before mining began. In addition, a tax is levied to reclaim land that was mined before the law was enacted. Enforcement and compliance of environmental laws waxes and wanes with the political climate in Washington. Yet environmental awareness has increased dramatically over the past generation, and, overall, mining operations are less polluting today than they were when your parents were in their teens. One of the big challenges for

underground mine A mine consisting of subterranean passages that commonly follow ore veins or coal seams.

surface mine A hole excavated into Earth's surface for the purpose of recovering mineral or fuel resources.

FIGURE 5.9 A huge power shovel dwarfs a person standing inside the Navajo Strip Mine in New Mexico.

the future is to clean up old mines that were operated under lax or nonexistent environmental regulations of the past. In the United States, more than 6,000 unrestored coal and metal surface mines cover an area of about 90,000 square kilometers, almost as large as the state of Virginia. This figure does not include abandoned sand and gravel mines and rock quarries, which probably account for an even larger area.

Although underground mines do not directly disturb the land surface, some abandoned mines collapse, and occasionally buildings have fallen into the holes

FIGURE 5.10 This house tilted and broke in half as it sank into an abandoned underground coal mine.

(Figure 5.10). Over 800,000 hectares (2 million acres) of land in central Appalachia have settled into underground coal mine shafts.

Mining of both metal ores and coal also creates huge piles of *waste rock*—rock that must be removed to get at the ore or coal. If the waste piles are not treated properly, rain erodes the loose rock and leaches toxic elements such as arsenic, sulfur, and heavy metals from the piles, choking the streams with sediment and contaminating both stream water and ground water.

5.5 Energy Resources: Coal, Petroleum, and Natural Gas

Coal, petroleum, and natural gas are called **fossil fuels** because they formed from the remains of plants and animals. Fossil fuels are not only nonrenewable but also unrecyclable. When a lump of coal or a liter of oil (petroleum) is burned, the energy dissipates and is, for all practical purposes, lost. Thus our fossil fuel supply inexorably diminishes.

Coal

Coal (Figure 5.11) is a combustible rock composed mainly of carbon. Humans began using coal before they used petroleum and natural gas because coal is easily mined and can be burned without refining.

Coal-fired electric generating plants burn about 92 percent of the coal consumed in the United States. The remainder is used to make steel or to produce steam in factories. Although it is easily mined and abundant in many parts of the world, coal emits air pollutants that can be removed only with expensive control devices. It is still an abundant resource, with widespread availability projected until at least the year 2200.

FIGURE 5.11 Anthracite is a hard, compact variety of coal with the highest carbon count and lowest level of impurities. It is used today in hand-fired stoves and automatic stoker furnaces.

Large quantities of coal formed worldwide during the Carboniferous Period, between 360 and 286 million years ago, and later in Cretaceous and Paleocene times, when warm, humid swamps covered broad areas of low-lying land. When plants die in forests and grasslands, organisms consume some of the plant litter, and chemical reactions with oxygen and water decompose the remainder. As a result, little organic matter accumulates except in the topsoil. In some warm swamps, however, plants grow and die so rapidly that newly fallen vegetation quickly buries older plant remains. The

Unit 1: Earth Materials and Time

TABLE 5.2 Classification of Coal by Grade, Heat Value, and Carbon Content

Type	Color	Water (%)	Other Volatiles and Noncombustible Compounds (%)	Carbon (%)	Heat Value (BTU/lb)
Peat	Brown	75	10	15	3,000 to 5,000
Lignite	Dark brown	45	20	35	7,000
Bituminous (soft coal)	Black	5 to 15	20 to 30	55 to 75	12,000
Anthracite (hard coal)	Black	4	1	95	14,000

new layers prevent atmospheric oxygen from penetrating into the deeper layers, and decomposition stops before it is complete, leaving brown, partially decayed plant matter called *peat* (see Chapter 3). Commonly, peat is then buried by mud deposited in the swamp.

Plant matter is composed mainly of carbon, hydrogen, and oxygen and contains large amounts of water. During burial, rising pressure expels the water and chemical reactions release most of the hydrogen and oxygen. As a result, the proportion of carbon increases until coal forms (Figure 5.12). The grade of coal and the heat that can be recovered by burning coal can vary considerably depending on the carbon content (Table 5.2).

Petroleum

The word **petroleum** comes from the Latin for "rock oil" or "oil from the earth." The first commercial oil well was drilled in the United States in 1859, ushering in a new energy age. Crude oil, as it is pumped from the ground, is a gooey, viscous, dark liquid made up of thousands of chemical compounds. It is then refined to produce propane, gasoline, heating oil, and other fuels (Figure 5.13). Petroleum also is used to manufacture plastics, nylon, and other useful materials.

Formation of Petroleum

Streams carry organic matter from decaying land plants and animals to the sea and to some large lakes, and deposit it with mud in shallow coastal waters. Marine plants and animals die and settle to the seafloor, adding more organic matter to the mud. Over millions of years, younger sediment buries this organic-rich mud to depths of a few kilometers, where rising

fossil fuels Energy resources including petroleum, coal, and natural gas, which formed from the partially decayed remains of plants and animals; they are nonrenewable and unrecyclable.

petroleum A complex liquid mixture of hydrocarbons, formed from decayed plant and animal matter, that can be extracted from sedimentary strata and refined to produce propane, gasoline, and other fuels. Also called *crude oil* or simply *oil.*

Litter falls to floor of stagnant swamp.

Debris accumulates, barrier forms, decay is incomplete.

Sediment accumulates, organic matter is converted to peat.

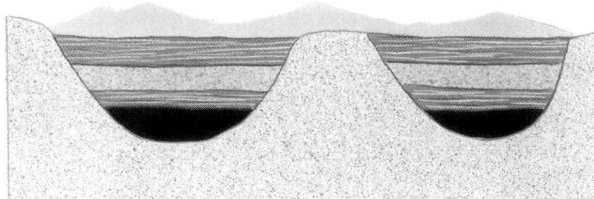
Peat is lithified to coal.

FIGURE 5.12 Peat and coal form as sediment buries organic litter in a swamp.

source rock The shale or other sedimentary rock in which oil or natural gas originates.

reservoir Oil-saturated permeable rock in which liquid petroleum or gas accumulates.

temperature and pressure convert the mud to shale. At the same time, the elevated temperature and pressure convert the organic matter to liquid petroleum that is dispersed throughout the rock (Figure 5.14). The activity of bacteria may enhance the process. Typically, petroleum forms in the temperature range of 50°C to 100°C.

The shale or other sedimentary rock in which oil originally forms is called the **source rock**. Oil dispersed in shale cannot be pumped from an oil well because shale is relatively *impermeable*; that is, liquids do not flow through it rapidly. But under favorable conditions, petroleum migrates slowly to a nearby layer of permeable rock—usually sandstone or limestone—where it can flow readily. Because petroleum is less dense than water or rock, it then rises through the permeable rock until it is trapped within the rock or escapes onto Earth's surface.

Many oil traps form where a layer of impermeable rock such as shale prevents the petroleum from rising further. The oil or gas then accumulates in a petroleum **reservoir**. An oil reservoir is not an underground pool or lake of oil. It consists of oil-saturated permeable rock that is like an oil-soaked sponge.

Petroleum Extraction, Transport, and Refining

To extract petroleum, an oil company drills a well into a reservoir. After the hole has been bored, the expen-

FIGURE 5.13 An oil refinery converts crude oil into useful products such as gasoline.

sive drill rig is removed and replaced by a pumper that slowly extracts the petroleum. Fifty years ago, many reservoirs lay near the surface and oil was easily pumped from shallow wells. But these reserves have been depleted, and modern oil wells are often a few kilometers or more deep.

Primary recovery techniques utilize the pressure in the reservoir that pushes oil into the well—but as oil is removed, the pressure decreases and the oil is said to be "inert." On average, more than half of the oil in a reservoir is too viscous to be pumped to the surface by conventional techniques and is left behind when the oil field has "gone dry." Recently, oil companies have developed methods of drilling horizontally through reservoirs, allowing access to vast amounts of oil left by

A

Organic-rich mud

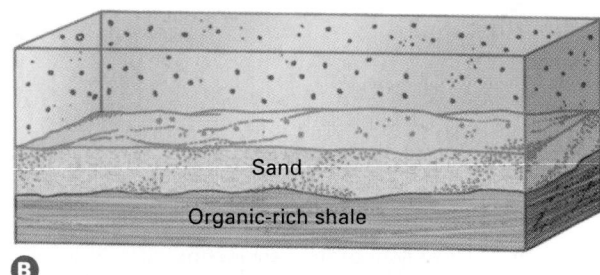

B

Sand

Organic-rich shale

FIGURE 5.14 (A) Organic matter from land and sea settles to the sea floor and mixes with mud. (B) Younger sediment buries this organic-rich mud. Rising temperature and pressure convert the mud to shale, and the organic matter to petroleum. (C) The petroleum is trapped in the reservoir by an impermeable cap rock.

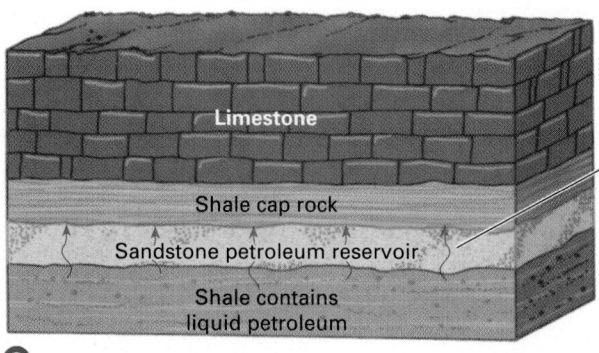

Limestone

Shale cap rock

Sandstone petroleum reservoir

Shale contains liquid petroleum

Petroleum rises into sandstone reservoir

C

FIGURE 5.15 An offshore, oil-drilling platform extracts oil from the continental shelf beneath a shallow sea.

SUN OIL

earlier wells. Additional oil can be extracted by **secondary and tertiary recovery techniques** that augment the energy in the reservoir by injecting water, detergent, pressurized gas, or some other fluid. Secondary methods are employed first, and when those are exhausted tertiary methods are used. In one simple secondary process, water is pumped into one well, called the "injection well." The pressurized water floods the reservoir, driving oil to nearby wells, where both the water and oil are extracted. At the surface, the water is separated from the oil and reused, while the oil is sent to the refinery. One tertiary process pumps detergent into the reservoir. The detergent dissolves the remaining oil and carries it to an adjacent well, where the petroleum is then recovered and the detergent recycled.

Because an oil well occupies only a few hundred square meters of land, most cause relatively little environmental damage. However, oil companies are now extracting petroleum from fragile environments such as the ocean floor and the Arctic tundra. To obtain oil from the seafloor, engineers build platforms on pilings driven into the ocean floor and mount drill rigs on these steel islands (Figure 5.15). Despite great care, accidents occur during the drilling and extraction of oil. When accidents occur at sea, millions of barrels of oil can spread throughout the waters, poisoning marine life and disrupting marine ecosystems. Significant oil spills have occurred in virtually all offshore drilling areas.

Natural Gas

Natural gas is an energy resource that forms in source rock or an oil reservoir when crude oil is heated above 100°C during burial. (Natural gas is mostly methane, CH_4, an organic molecule consisting of a single carbon atom bonded to four hydrogen atoms.) Consequently, many oil fields contain a mixture of oil, with natural gas floating above the heavier liquid petroleum. In other instances, the lighter, more mobile gas escaped into the atmosphere or was trapped in a separate underground reservoir.

Natural gas is used without refining for home heating, cooking, and to fuel large electrical generating plants. Because natural gas contains few impurities, it releases no sulfur or other pollutants when it burns, although, as with all fossil fuels, combustion of natural gas releases carbon dioxide, a greenhouse gas. This fuel has a higher net energy yield, produces fewer pollutants, and is less expensive to produce than petroleum. At current consumption rates, global natural gas supplies will last for 80 to 200 years.

Although most commercial natural gas is produced from petroleum fields, 7 percent of current U.S. gas production comes from coal seams, where both natural Earth heat and microbial activity slowly convert buried coal to **coal bed methane**, methane that is chemically bonded to coal. The coal bed methane reserves in the United States are estimated to be more than 700 trillion cubic feet (Tcf), although less than 100 Tcf may be economically recoverable.

Most coal beds have a high capacity to store water in small voids in the coal itself. As natural processes convert coal to methane, the gas dissolves in the coal bed ground water, where pressure of overlying rock keeps the methane in solution. Natural gas companies drill thousands of wells into the coal beds and pump the ground water to the surface, decreasing the pressure on the remaining water in the coal bed. The decreased pressure allows the methane to separate from the water. It is then piped to the surface, where it is compressed and sent to market.

Because they store so much water, coal beds are important ground water reservoirs for farmers and ranchers, especially in the arid and semiarid western United States, where extensive coal bed methane development is now occurring. Here, coal bed methane development has two serious impacts on regional agriculture and ecosystems. The extraction of so much water from the coal beds has depleted essential aquifers and lowered the water table over large areas. Secondly, coal bed water is commonly salty. After it is pumped to the surface, the saltwater can poison

secondary and tertiary recovery techniques Methods of extracting oil or natural gas by artificially augmenting the reservoir energy, as by injection of water, detergent, pressurized gas, or other fluid.

natural gas A mixture of naturally occurring light hydrocarbons composed mainly of methane, CH_4, that is used for home heating, cooking, and to fuel large electrical generating plants.

coal bed methane Methane that is chemically bonded to coal. The methane can be recovered by removing the ground water in a coal bed, which decreases the pressure and allows the methane to separate from the coal as a gas.

tar sands Sand deposits permeated with heavy oil and an oil-like substance called bitumen.

bitumen A thick, sticky, oil-like substance that permeates tar sands and can be converted to crude oil.

kerogen The waxy, solid organic material in oil shales that yields oil when the shales are heated and distilled; the precursor of liquid petroleum.

oil shale A kerogen-bearing sedimentary rock that yields liquid or gaseous hydrocarbons when heated.

streams and soils, rendering them useless for agriculture and wildlife. State and federal regulations on water extraction and disposal methods attempt to minimize these impacts.

5.6 Energy Resources: Tar Sands and Oil Shale

In 2008, nearly 85 percent of energy used in the United States came from petroleum, coal, and natural gas, which traditionally have been the cheapest fuels. However, the price of crude petroleum jumped from less than $40 to nearly $60 per barrel between 2004 and 2005. Prices continue to rise, remaining at an average of $70 per barrel through most of 2009.

In the past, energy sources such as heavy oil and other renewable resources have been more expensive than the three traditional fossil fuels. However, the cost of producing these alternative energies has been decreasing while the costs of traditional fuels have been increasing. As a result, the world is on the threshold of a major restructuring of the global economy of energy. Many fuel sources that were uneconomical even a year ago are now cheaper than the conventional sources.

Tar Sands

In some regions, large sand deposits called **tar sands** are permeated with heavy oil and **bitumen**, a sticky, oil-like hydrocarbon. Crude oil can be obtained from both substances, but they are too thick to be pumped and require other methods of extraction.

The richest tar sands exist in Alberta (Canada), Utah, and Venezuela. In Alberta alone, tar sands contain an estimated 1 trillion barrels of petroleum. About 10 percent of this fuel is shallow enough to be surface-mined. Tar sands are dug up and heated with steam to make the heavy oil and bitumen fluid enough to separate from the sand. The oil and bitumen are then treated chemically and heated to convert them to crude oil. At present, several companies mine tar sands profitably, producing more than 13 percent of Canada's petroleum requirements. In 2008, Syncrude Canada, the world's largest producer of crude oil from oil sands, produced 105 million barrels of oil at $35 per barrel, significantly below world market prices for petroleum from conventional sources. In the first half of 2009, Syncrude Canada saw a significant increase in its operating costs, from $35 a barrel to over $50, largely due to the lower prices of conventional oil. Deeper deposits, comprising the remaining 90 percent of the reserve, can be extracted using subsurface techniques similar to those discussed for secondary and tertiary recovery.

Oil Shale

Some shales and other sedimentary rocks contain a waxy, solid organic substance called **kerogen**. Kerogen is organic material that has not yet converted to oil. Kerogen-bearing rock is called **oil shale** (Figure 5.16). If oil shale is mined, mixed with water and then heated, the kerogen converts to petroleum. In the United States, oil shales contain the energy equivalent of 2 to 5 trillion barrels of petroleum, enough to fuel the nation for 300 to 700 years at the 2006 consumption rate (Figure 5.17). However, many oil shales are of such low grade that more energy is required to mine and convert the kerogen to petroleum than is generated by burning the oil. Consequently, these low-grade shales may never be used for fuel.

FIGURE 5.16 Oil shale is an organic-rich, fine-grained sedimentary rock containing significant amounts of kerogen.

Water consumption is a serious problem in oil shale development. Approximately two barrels of water are needed to produce each barrel of oil from shale. Oil shale occurs most abundantly in the semiarid western United States. In this region, scarce water is also needed for agriculture, domestic use, and industry.

A recent study by the Rand Corporation estimates that there are 800 billion barrels of recoverable oil shale resources from the Green River Valley in Colorado, Utah, and Wyoming—enough to satisfy U.S. oil needs for 100 years at current consumption rates. However, recovering crude oil from shale is expensive, and the study predicts that oil prices will have to reach $70 to $95 per barrel before the process will be economical. As with the case of tar sands in Alberta, increasing world oil prices and technological advances in oil shale recovery techniques combine to make oil shale a likely source of energy in the future.[2]

---

2. James T. Bartis, *Oil Shale Development in the United States* (Santa Monica, CA: Rand Corporation, 2005).

Petroleum (billions of barrels)

Known and estimated petroleum reserves in the United States

Estimated petroleum in the U.S. available with secondary recovery methods

Petroleum in Canadian tar sands

Petroleum in oil shale in the United States

FIGURE 5.17 Secondary recovery, tar sands, and oil shale increase our petroleum reserves significantly.

5.7 Energy Resources: Renewable Energy

Solar, wind, geothermal, hydroelectric, wood, and other biomass fuels are renewable—natural processes replenish them as we use them. Although the amount of energy produced today by renewable sources is small compared to that provided by fossil and nuclear fuels, renewable resources have the potential to supply all of our energy needs. As the prices of conventional fossil fuels have risen, some renewables have become less expensive than conventional energy. Except for biomass fuels, renewable energy sources emit no carbon dioxide and therefore do not contribute to global warming.

Solar Energy

Current technologies allow us to use solar energy in three ways: passive solar heating, active solar heating, and electricity production by solar cells.

A passive solar house is built to absorb and store the Sun's heat directly. In active solar heating systems, solar thermal collectors absorb the Sun's energy and use it to heat water. Pumps then circulate the hot water through radiators to heat a building, or the inhabitants use the hot water directly for washing and bathing. Solar thermal collectors are becoming increasingly popular worldwide, with an estimated total capacity of more than 450 million square meters.

A **solar cell** or photovoltaic (PV) cell produces electricity directly from sunlight. A modern solar cell is a *semiconductor*, a device that can conduct electrical current under some conditions but not others. Sunlight energizes electrons in the semiconductor, producing an electric current. The sun bathes the earth with 86,000 trillion watts of energy at any given time, more than 6,600 times the amount currently used by humans each year.

Figure 5.18 shows an installation of solar panels. Although solar power still accounts for less than 1 percent of world energy demand, solar energy is our most abundant resource, and PV cell production is the fastest-growing segment of the energy industry. Photovoltaic arrays are now competitive with electricity costs during peak demand times in California, especially those installed for single-family units. PVs are also cost-effective for electricity needs far from existing power lines.

The 2008 production capacity of U.S. solar panels was 499 megawatts, amounting to much less than 1 percent of total usage. In Japan, the world leader in solar energy, the government plans to generate 10 percent of that nation's electricity with PVs by 2030.

solar cell A device that produces electricity directly from sunlight; also sometimes called a *photovoltaic (PV) cell.*

Wind Energy

In the United States, wind energy grew an average of 32 percent annually between 2004 and 2008. The total U.S. wind-generating capacity is now over 25,000 megawatts, more than 1 percent of total U.S. electricity

FIGURE 5.18 Solar panels provide clean energy.

FIGURE 5.19 Wind turbines generate electricity at Cowley Ridge, Alberta, Canada.

demand. Figure 5.19 is a wind farm in Alberta, Canada. Three of the world's largest wind farms—in California at Altamont Pass, San Gorgonio Pass, and Tehachapi Pass—are actually collections of dozens of individual wind farms with several owners and different turbine types that have been built and modified over several decades. Today, the smallest utility-scale wind turbines generate about 700 kilowatts, with some models approaching 5,000 kilowatts (5 megawatts). One megawatt of wind capacity is enough to supply 240 to 300 average American homes. Other gigantic wind farms now generate electricity in Texas, eastern Oregon, and several other states. A huge, untapped potential for wind generation exists in several midwestern and western states, where winds blow strongly and almost continuously.

Wind energy production is growing rapidly because construction of wind generators is cheaper than building new fossil fuel–fired power plants. Wind energy is also clean and virtually limitless.

Worldwide, wind is the second-fastest-growing source of energy and many countries are rapidly investing in new wind farms. The United States surpassed Germany as the world leader in wind energy production in 2008, but Germany remains a close second, with the capacity to generate just under 24,000 megawatts in 2008. This is, however, 7 percent of Germany's total energy usage, whereas the United States' larger capacity accounts for only 1 percent of total U.S. energy.

Geothermal Energy

Energy extracted from Earth's internal heat is called *geothermal energy* (Figure 5.20). Natural hot ground water can be pumped to the surface to generate electricity, or it can be used directly to heat homes and other buildings. Alternatively, cool surface water can be pumped deep into the

ground, to be heated by subterranean rock, and then circulated to the surface for use. The United States is the largest producer of geothermal electricity in the world, with a production capacity of just over 3,000 megawatts.

Hydroelectric Energy

If a river is dammed, the energy of water dropping downward through the dam can be harnessed to turn turbines that produce electricity. Hydroelectric generators supply between 15 and 20 percent of the world's electricity. They provide about 3 percent of all energy consumed in the United States, but about 8 percent of our electricity.

The United States is unlikely to increase its production of hydroelectric energy. Large dams are expensive to build, and few suitable sites remain. Environmentalists commonly oppose dam construction because the resulting reservoirs flood large areas—destroying wildlife habitats, agricultural land, towns, and migratory fish populations. For example, the dams on the Columbia River and its tributaries are largely responsible for the demise of salmon populations in the Pacific Northwest. Undammed wild rivers and their canyons are prized for their aesthetic and recreational value.

Biomass Energy

Biomass (plant) fuels currently produce about 11,000 megawatts of energy. Wood is the most productive of all biomass, followed by controlled garbage incineration and alcohol fuels. The use of biofuels—fuels

FIGURE 5.20 An example of geothermal energy. Shown here are vents emitting hot steam and other gases at Norris Geothermal Basin in Yellowstone National Park, Wyoming.

derived from crops and agricultural wastes—is growing rapidly around the world. In general, these fuels are cleaner burning than fossil fuels and can be produced domestically in most countries, thereby creating local jobs and reducing foreign oil imports. However, production of biofuels is not always a net energy gain; in some cases, more energy is used in the production and processing of these fuels than can be extracted from them. The two main types of biofuel are ethanol and biodiesel.

The Future of Renewable Energy Resources

Aside from biofuels, none of the renewable resources discussed here can be used directly to power mobile transportation systems such as cars and trucks.[3] Several methods are available, however, to convert these energy sources for use in transportation.

Perhaps the easiest way to use electricity to transport people and goods is the old-fashioned electric train. Electric streetcars, commuter trains, and subways have been used for decades. If we build more electric mass transit systems, and if people use them, we could shift away from our dependence on the internal-combustion engine and on petroleum. Eventually, the required electricity consumption could be supplied by renewable energy sources.

Another solution is the electric car. Battery-only and gasoline-electric hybrid cars have seen a recent increase in popularity and availability. If that trend continues, perhaps we can further reduce cost and dependence on petroleum.

Energy planners also envision a **hydrogen economy**, using a process in which water is dissociated into hydrogen and oxygen and the hydrogen is used as fuel. A necessary part of this process is an electrochemical device called a **fuel cell**, which uses the chemical energy of hydrogen to produce electricity cleanly and efficiently, with water and heat as by-products. One type of fuel cell separates hydrogen's negatively charged electron from the hydrogen nucleus, which then consists of a single positively charged proton. The electrons, in turn, combine with oxygen, which then reacts with the hydrogen proton to form water and heat energy. In some types of fuel cells, the electrons travel through an electrical circuit to reach the other side of the cell. This movement of electrons is an electrical current. Thus, fuel cells can produce both heat and an electrical current, which can then be used as power sources for transportation, electrical appliances, and most other energy-consuming equipment.

Fuel cells can provide energy for systems as large as a power station and as small as a laptop computer. They can also power cars, trucks, trains, and other vehicles. Fuel cells have several benefits over fossil fuel and nuclear technologies now used in power plants and vehicles. For one, fuel cells emit no pollutants that create smog and cause health problems. They emit only water vapor. Although water vapor is a greenhouse gas, it is seen as a lesser environmental threat than carbon dioxide. Hydrogen is an abundant component of water and other common materials; the supply is nearly limitless.

5.8 Energy Resources: Nuclear Fuels and Reactors

Nuclear fuels are radioactive isotopes that produce heat through nuclear reactions; the heat is used in turn to generate electricity in nuclear reactors. Uranium is the most commonly used nuclear fuel. These energy resources, like mineral resources, are nonrenewable, although uranium is abundant.

Every step in the mining, processing, and use of nuclear fuel produces radioactive wastes. The mine waste discarded during mining is radioactive. Enrichment of the ore produces additional radioactive waste. When a uranium nucleus undergoes fission in a reactor, it splits into two useless radioactive nuclei that must be discarded. After several months in a reactor, the concentration of useful uranium in the fuel rods drops until the fuel pellets are no longer useful. In some countries, these pellets are reprocessed to recover useful uranium fuel, but in the United States this process is not economical and the pellets are discarded as radioactive waste.

In the early 1970s, the nuclear industry was growing rapidly and many energy experts predicted that nuclear energy would dominate the generation of electric energy. Some experts even suggested that electricity would become "too cheap to meter." These predictions have not been realized. Four factors led to the decline of the nuclear power industry: (1) Construction of new reactors in the United States became so costly that electricity generated by nuclear power became more expensive than that generated by coal-fired power plants. (2) After major accidents at Three Mile Island in the United States and Chernobyl in Ukraine, people became concerned about safety. (3) Serious concerns remain about the safe disposal of nuclear wastes.

hydrogen economy An energy economy in which electricity is used to dissociate water into hydrogen and oxygen, and the hydrogen is used as a fuel.

fuel cell An electrochemical energy-conversion device that produces electricity from an external supply of fuel, such as hydrogen.

nuclear fuels Radioactive isotopes, such as those of uranium, used to generate electricity in nuclear reactors.

---

3. Electricity can be used to charge a battery that powers a car. This option is discussed further in the text.

(4) The demand for electricity has risen less than expected during the past three decades. As a result, growth of the nuclear power industry has nearly halted. After 1974, many planned nuclear power plants were canceled, and after 1981—almost 30 years ago—no new orders were placed for nuclear power plants in the United States. The dramatic turnaround led to serious financial altercations; in 1985 *Forbes* business magazine called the U.S. nuclear power program "the largest managerial disaster in U.S. business history, involving $1 trillion in wasted investment and $10 billion in direct losses to stockholders."

Elsewhere in the world, nuclear power production has seen a 2 percent increase in generating capacity between 2003 and 2004, the highest rate of growth ever reached. Plants are currently under construction in several countries, including India, Japan, and China. While some countries are decommissioning older plants and replacing them with other forms of power, many have plans for new nuclear reactors to be built in coming years.

Today, with rising fuel prices, many U.S. policy makers are suggesting that the nuclear option be reconsidered. However, at the same time, alternative energy resources such as wind and solar are being explored aggressively.

5.9 Conservation as an Alternative Energy Resource

The single quickest and most effective way to decrease energy consumption and to prolong the availability of fossil fuels is to conserve energy (see Figure 5.21). Policies to improve energy efficiency are more cost-effective than building new power plants. Such policies help to reduce air pollution and dependence on oil imports while saving money for consumers and industry.

Energy conservation has already produced dramatic results in the United States, where energy consumption has decreased by 47 percent per dollar of gross domestic product (GDP) in the last 30 years. However, total fuel consumption continues to rise, and there is much room for continued improvement. Some energy experts have suggested that if people in industrialized nations use more efficient equipment and develop more efficient habits, these nations could conserve as much as half of the energy that they consume.

Energy use in the United States falls under three categories: buildings, industry, and transportation. Two types of conservation strategies can be applied in each of those categories. Technical solutions involve switching to more efficient implements. Social solutions involve decisions to use existing energy systems more efficiently.

FIGURE 5.21
The end-use efficiencies of common energy-consuming systems. Home heating represents the only energy-consumption system that is even remotely efficient—and even there, 15 percent is wasted. Energy to produce incandescent lighting is 95 percent wasted; automobile transportation is 90 percent wasted.

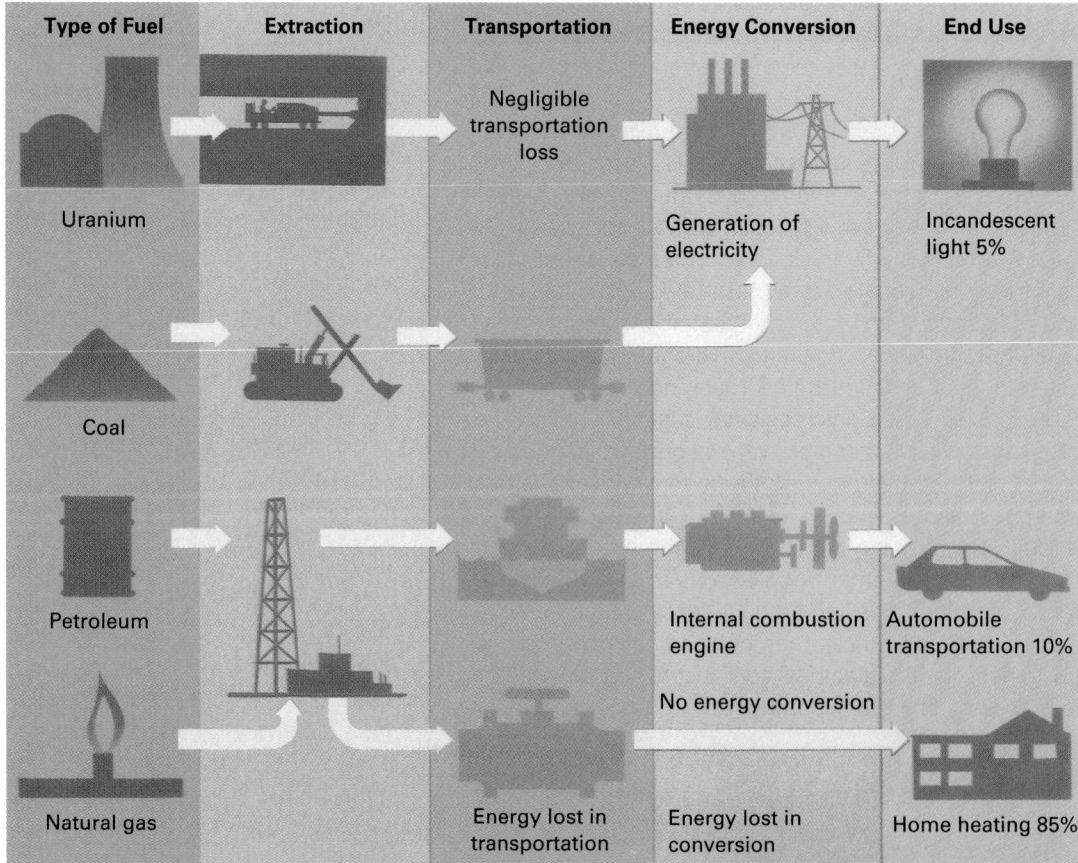

Unit 1: Earth Materials and Time

Technical Solutions

Buildings

Residential and commercial buildings consume about 39 percent of all the energy produced in the United States. Most of that energy is used for heating, air-conditioning, and lighting. Factoring energy efficiency into home and office building design or retrofitting existing structures to reduce waste can save thousands of dollars in energy costs over the lifetime of the structure.

Significant energy savings are possible in all aspects of energy consumption in buildings. As one example, about 22 percent of the electricity generated in the United States is used for lighting. Because incandescent lighting is about 95 percent inefficient in energy consumption (Figure 5.21), savings in that area alone are potentially great. A fluorescent bulb consumes one-fourth as much energy as a comparable incandescent bulb. Although fluorescent bulbs cost about $5 each, they last 10 times longer than incandescent bulbs, for an energy savings equivalent to 3 times their cost. In addition, new solid-state technology promises further advances in energy-efficient lighting, with industry experts predicting that LED lights will be 9 times more efficient than incandescent bulbs in the near future. LEDs (light-emitting diodes) do not get hot, use far less energy, and last much longer than regular incandescent bulbs, although at the present time they are too expensive for most home applications.

Switching to more efficient light sources would conserve the electricity generated by about 120 large generating plants and would save $30 billion in fuel and maintenance expenses every year. With a switch to LED lighting, these savings could be much greater.

Industry

Industry consumes 33 percent of the energy used in the United States. In general, conservation practices are cost-effective, and many companies are taking advantage of the fact that saving energy is profitable. According to the Environmental Protection Agency (EPA), average U.S. energy use per dollar of GDP declined 2.1 percent per year between 1996 and mid-2005. During that time, energy demand increased, but 78 percent of that increased demand was met by increased efficiency. However, industry still wastes great amounts of energy.

For example, about 70 percent of the electricity consumed by industry—half of all electricity produced in the United States—drives electric motors for machinery and tools. Most motors are inefficient because they run only at full speed and are slowed by brakes to operate at the proper speeds to perform their tasks. This approach is like driving your car with the gas pedal pressed to the floor and controlling your speed with the brakes. Replacing older electric motors with variable-speed motors would save an amount of electricity equal

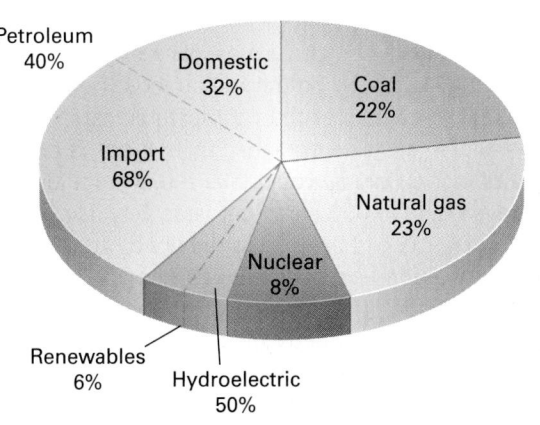

Fossil Fuels = 85%
(petroleum, coal, natural gas)

FIGURE 5.22 In the year 2008, fossil fuels supplied 85 percent of all energy used in the United States; oil alone accounted for 40 percent, natural gas for 23 percent, and coal for 22 percent.

to that generated by 150 large power plants, but such replacement has been slow.

Transportation

About 28 percent of all energy, more than 50 percent of the oil usage in the United States (Figure 5.22), and one-third of the nation's carbon emissions are consumed transporting people and goods.

The efficiency of auto and truck engines is about 10 percent (Figure 5.21). Thus, we can save much energy by using more efficient cars and trucks. Over the past few decades, automobile manufacturers have offered vehicles with increasingly efficient internal-combustion engines. Other power avenues that auto companies are exploring include electricity and hydrogen power.

A *hybrid car* uses a small, fuel-efficient gasoline engine combined with an electric motor that assists the engine when accelerating. Hybrids consume less gas and produce less pollution per mile than conventional gasoline engines. Current models of hybrid cars achieve fuel efficiencies ranging from 51/48 (traffic/highway) to

ENERGY INFORMATION ADMINISTRATION/ANNUAL ENERGY REVIEW 2004

31/27 miles per gallon, depending on make and model. The Honda Civic Hybrid gets 10 to 15 miles more per gallon than the same car with a regular gas engine. The hybrid Toyota Prius produces 90 percent fewer harmful emissions than a comparable gasoline engine.

Hybrids currently on the market cost from $3,500 to $6,000 more per car than comparable cars with conventional gasoline engines. Thus, if gas is priced at $2.50 per gallon, it would take the average driver, who averages 15,000 miles per year, 5 to 10 years to amortize the higher price for a hybrid. Additionally, the cost of hybrid batteries ranges from $1,000 to $3,000. Although the warranty may cover the hybrid battery, once the warranty expires replacement can be expensive.

Using hybrids and other energy-efficient vehicles, American motorists can achieve a 50 percent or greater increase in fuel economy, which would save 3 million barrels of oil a day—about one-third of the current oil imports.

Social Solutions

Social solutions involve altering human behavior to conserve energy. Energy-conserving actions can be used in buildings, in industry, and in transportation. Some result in inconvenience to individuals. For example, if you choose to carpool rather than drive your own car, you save fuel but inconvenience yourself by coordinating your schedule with your carpool companions. Our national fondness for large sport-utility vehicles and pickup trucks, which have poor fuel economy, has resulted in a decline in average fuel efficiency of all American cars and light trucks from 26.2 miles per gallon in 1987 to 25 miles per gallon in 2003. As explained previously, high-mileage cars are on the market, but they will make an impact only if people make the social decision to use them. People argue that this social decision comes at a cost because light vehicles make the driver and passengers more vulnerable in case of an accident—but studies have shown that lighter, more agile vehicles, with better turning capacity and more effective braking, are actually safer than heavier SUVs.

At home and in the workplace, wearing a sweater and lowering the thermostat in winter and using less air-conditioning in summer might reduce the comfort margin but can save considerable energy. Many other social solutions, however, are cost-free in terms of inconvenience.

When practiced by everyone, simply turning off the lights, the television set, and other appliances when you leave the room will conserve large amounts of energy.

© STEPHEN FINN/SHUTTERSTOCK

5.10 Energy for the 21st Century

The United States consumes 25 percent of the world's oil yet owns only 3 percent of the known reserves. In 2007, fossil fuels supplied 85 percent of all energy used in the United States; oil alone accounted for 40 percent, natural gas for 23 percent, and coal for 22 percent (Figure 5.22). Thus, oil is our major source of energy. In addition, oil is the only portable energy resource currently in popular use, and thus is the main energy resource for transportation in the United States. At current rates of consumption, we have more than 200 years of domestic coal reserves, and at least several decades of natural gas reserves. Oil, however, is another story.

In 1956, M. King Hubbert, a geologist, was working at the Shell research lab in Houston, Texas. Hubbert compared U.S. domestic oil reserves with current and predicted rates of oil consumption. He then forecast that U.S. oil production would peak in the early 1970s and would thereafter decline continuously. He predicted that Americans would have to make up an ever-increasing difference between domestic oil supply and consumption by relying on larger and larger imports, or they would have to turn to other energy resources.[4] Other experts and economists ridiculed his prediction, but in 1970 the U.S. domestic oil production reached its maximum and it has been slowly declining since then.

In 2009, the United States ranked third in global oil production. (Saudi Arabia was first; Russia second.) But because U.S. petroleum consumption was higher than it was in 1970 and production was lower, the United States imported nearly 70 percent of its petroleum (Figure 5.23). Because of this high dependence on foreign oil, the U.S. energy future is intimately linked with the global one. In 1971, Hubbert predicted that global oil production would peak between 1995 and 2000 and that the world supply of oil will be 90 percent depleted between 2020 and 2030.[5]

In 2008, oil prices rose to nearly $120 a barrel, a price that many hoped would spur producers to respond to demand by increasing production. However, some large oil producers, such as Mexico, Russia, and Saudi Arabia, actually cut back their oil production, citing cost and expense. At the same time, oil production by OPEC (Organization of the Petroleum Exporting Countries) was lower than projected because of turmoil in Iran and Iraq, further contributing to the increase in oil prices. In 2009, prices stabilized between $65 and

4. M. King Hubbert, *Nuclear Energy and the Fossil Fuels*, Publication No. 95, Shell Development Company (June 1956); available online at http://www.hubbertpeak.com/hubbert/1956/1956.pdf.
5. M. King Hubbert, "The Energy Resources of the Earth," in *Energy and Power*, a Scientific American Book (San Francisco: W. H. Freeman, 1971), 31–40; available online at <http://www.hubbertpeak.com/Hubbert/energypower.

Overview, 1949–2004

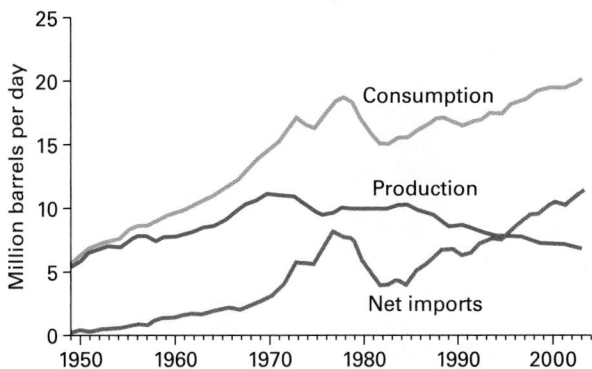

FIGURE 5.23 U.S. oil imports, production, and consumption since 1950. Our reliance on oil imports has increased steadily since 1971, as M. King Hubbert predicted.

$75 a barrel, largely due to the global economic downturn in late 2008 and 2009.

So what will happen when global oil production drops below demand? First of all, petroleum will not just "run out" one day, with all the wells suddenly going dry. As supply dwindles and demand increases, the price of fuel will rise and production of petroleum from oil shales and tar sands will increase. Some economists and geologists have suggested that the fluctuations in gasoline prices seen in recent years are the first wave of disruptions resulting from declining global oil reserves and that much greater disturbances will occur imminently.

People will not be able to afford as much fuel as they would like, so social and technical conservation strategies—previously rejected—will be implemented. As a result, demand will decrease. But if this decrease is not sufficient, petroleum prices will continue to rise. Many economists predict that the price increase could be dramatic. This is an example of an economic threshold effect. As long as potential supply is greater than demand, the price of oil reflects mainly the cost of drilling, shipping, and refining. But the moment we cross the threshold where demand is greater than supply, then an auction occurs where the rich can outbid the poor.

Is it possible to alter global energy production from a fossil fuel economy to an economy of renewable energy resources? In 2001, the UN's Intergovernmental Panel on Climate Change (IPCC) concluded that significant reduction of fossil fuel use is possible with renewable "technologies that exist in operation or pilot-plant stage today . . . without any drastic technological breakthroughs."[6] In other words, they suggested that if

we vigorously develop all the renewable energy resources listed in Section 5.7, the global economic system could absorb a drastic decline in petroleum production without massive disruptions. A year later, 18 prominent energy experts published a rebuttal in *Science*, proposing the exact opposite conclusion. Using almost the same phrases, with the simple addition of the word *not*, they argued that "energy resources that can produce 100 to 300 percent of present world power consumption without fossil fuels and greenhouse emissions do not exist operationally or as pilot plants."[7] Their basic counterargument is that global energy consumption is huge and renewable sources have low power densities. Thus we do not have the available land nor could we quickly build the required infrastructure to replace fossil fuels. This debate continues, despite continuing development and research into renewable resources.

When experts disagree, it is difficult for laypersons to evaluate the merits of the contradictory arguments. But whoever is right, it is clear that if global energy demand significantly exceeds supply, the world will fall into unprecedented economic chaos. Commerce will slip into unimaginable depression. Food supplies will diminish and food distribution will become expensive. Poor people, who are already on the edge of malnourishment, will starve. We can only hope that human ingenuity will combine with economic and political commitment to develop alternative energy resources before these catastrophes become reality.

6. Bert Metz, Ogunlade Davidson, Rob Swart, and Jiahua Pan, Eds., *Climate Change 2001: Mitigation* (New York: Cambridge University Press, 2001), 8.

7. Martin I. Hoffert, Ken Caldeira, Gregory Benford, David R. Criswell, Christopher Green, Howard Herzog, et al., "Advanced Technology Paths to Global Climate Stability: Energy for a Greenhouse Planet," *Science* 298 (November 1, 2002), 981–987.

Chapter 5: Geologic Resources

91

6

THE ACTIVE EARTH: PLATE TECTONICS

A false-color satellite image of the Sinai Peninsula and the northern part of the Red Sea shows in green the vegetation of the fertile valley of the Nile River and delta in Egypt. The city of Cairo can be seen as a gray smudge where the river widens into its broad fan-shaped delta.

visit **4ltrpress.cengage.com**

> ## There are no extra pieces in the universe. Everyone is here because he or she has a place to fill, and every piece must fit itself into the big jigsaw puzzle.
>
> *Deepak Chopra*

About 5 billion years ago, a ball of dust and gas, one among billions in the universe, collapsed into a slowly spinning disc. The inner portion of the disc then grew to form our Sun, while the outer parts coalesced to form planets orbiting the Sun. Our Earth is one of those planets. As particles of dust and gas were drawn together by gravity to form Earth, the colliding particles became hotter and hotter from impact energy. Over time, frozen crystals of carbon dioxide, methane, and ammonia melted. As the temperature continued to rise, ice melted as well. The young planet continued to heat up as asteroids, comets, and other space debris crashed into its surface. As Earth formed into a cohesive sphere, the decay of radioactive isotopes within the interior generated additional heat. About 4.6 billion years ago, the planet became hot enough that it melted. Then, as the bombardment slowed down and radioactive isotopes decayed and became less abundant, much of Earth's heat radiated into space and our planet began to cool.

Today, although Earth's surface has cooled to temperatures supportive of living organisms, the interior remains hot, both from heat left over after the early melting event and from continued decay of the remaining radioactive isotopes. Consequently, Earth becomes hotter with depth, and the temperature at Earth's center is about 6,000°C—similar to that of the Sun's surface. This heat drives Earth's internal engine, creating earthquakes, volcanic eruptions, mountain building, and continual movements of the continents and ocean basins. These effects, in turn, profoundly affect our environment—Earth's atmosphere, hydrosphere, and biosphere. The Earth engine and its effects are described in the theory of **plate tectonics**, a simple theory that provides a unifying framework for understanding the way Earth works and how Earth systems interact to create our environment. The term *tectonics* is taken from the Greek *tektonikos*, meaning "construction."

Like most great scientific revolutions, the development of plate tectonics theory developed sequentially over many years, building on earlier observations, hypotheses, and theories. The story illustrates how a scientific theory evolves and how scientists rely on the work and discoveries of earlier scientists.

6.1 Alfred Wegener and the Origin of an Idea: The Continental Drift Hypothesis

Although the theory of plate tectonics was not developed until the 1960s, it was foreshadowed early in the 20th century by a young German scientist named Alfred Wegener, who noticed that the African and South American coastlines on opposite sides of the Atlantic Ocean seemed to fit as if they were adjacent pieces of a jigsaw puzzle (Figure 6.1). He realized that the apparent fit suggested that the continents had once been joined together and had later separated by thousands of kilometers to form the Atlantic Ocean.

Wegener was not the first to make this observation, but he was the first scientist to pursue it with additional research. Studying world maps, Wegener realized that not only did the continents on both sides of the Atlantic fit together, but other continents, when moved properly, also fit like additional pieces of the same jigsaw puzzle (Figure 6.2). On his map, all the continents joined together formed one supercontinent that he called **Pangea** from the Greek root words for "all lands." The northern part of Pangea is commonly called **Laurasia** and the southern part **Gondwanaland**.

Wegener understood that the fit of the continents alone did not prove that a supercontinent

plate tectonics A theory of global tectonics stating that the lithosphere is segmented into several plates that move about relative to one another by floating on and gliding over the plastic asthenosphere. Seismic and tectonic activity occur mainly at the plate boundaries.

Pangea The supercontinent that existed when all Earth's continents were joined together, about 300 million to 200 million years ago, first identified and named by Alfred Wegener.

Laurasia The northern part of Pangea, consisting of what is now North America and Eurasia.

Gondwanaland The southern part of Pangea, consisting of what is now South America, Africa, Antarctica, India, and Australia.

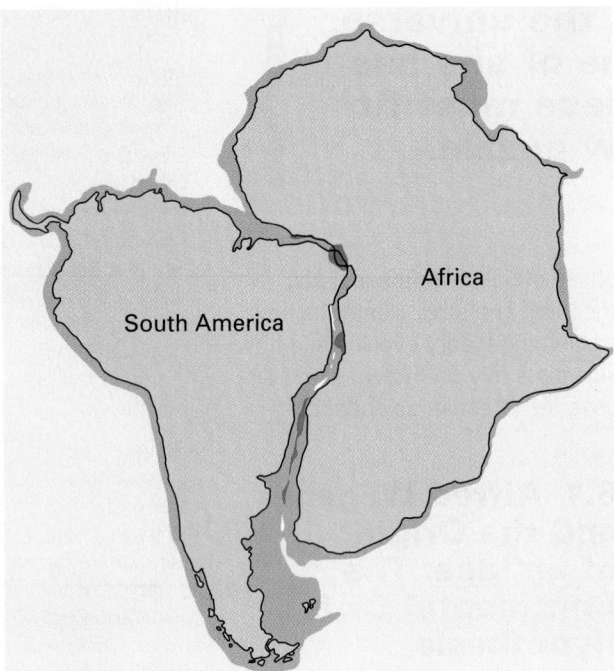

FIGURE 6.1 The African and South American coastlines appear to fit together like adjacent pieces of a jigsaw puzzle on Wegener's reconstruction map. The red areas show locations of distinctive rock types in South America and Africa. The brown regions are the continental shelves, which are the actual edges of the continents.

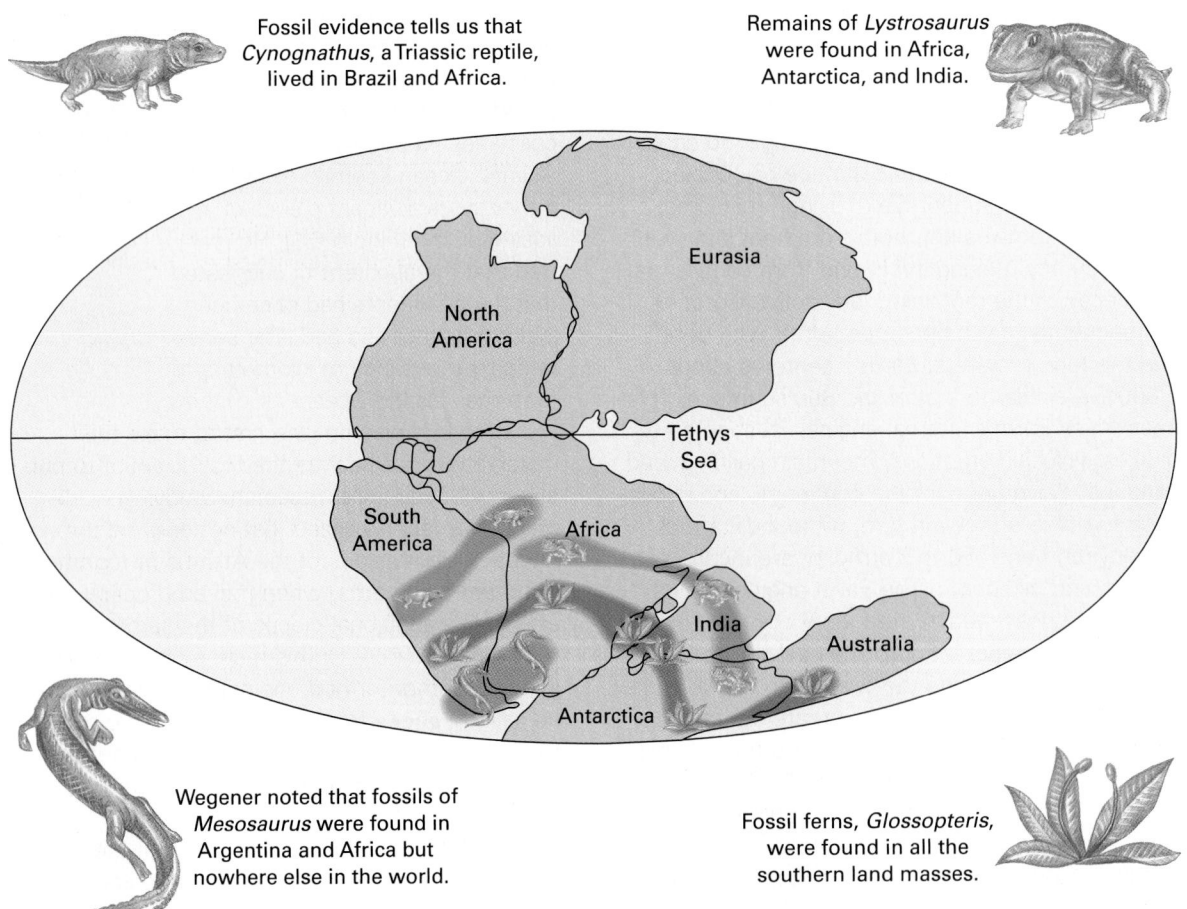

Fossil evidence tells us that *Cynognathus*, a Triassic reptile, lived in Brazil and Africa.

Remains of *Lystrosaurus* were found in Africa, Antarctica, and India.

Wegener noted that fossils of *Mesosaurus* were found in Argentina and Africa but nowhere else in the world.

Fossil ferns, *Glossopteris*, were found in all the southern land masses.

FIGURE 6.2 Geographic distributions of plant and animal fossils on Wegener's map indicate that a single supercontinent, called Pangea, existed from about 300 to 200 million years ago.

had existed. He began seeking additional evidence in 1910 and continued work on the project until his death in 1930.

He mapped the locations of fossils of several species of plants and animals that could neither swim well nor fly. Fossils of the same species are found in present-day Antarctica, Africa, Australia, South America, and India. Why would the same species be found on continents separated by thousands of kilometers of ocean? When Wegener plotted the same fossil localities on his Pangea map, he found that they were all in the same region of Pangea (Figure 6.2). He then suggested that each species, rather than migrating across the open ocean, had evolved and spread over that part of Pangea *before* the supercontinent broke apart.

Certain types of sedimentary rocks form in specific climatic zones. Glaciers and gravel deposited by glacial ice, for example, form in cold climates and are therefore found at high latitudes and high altitudes. Sandstones that preserve the structures of desert sand dunes form where deserts are common, near latitudes 30° north and south. Coral reefs and coal swamps thrive in near-equatorial tropical climates. Thus each of these rocks reflects the latitude at which it formed. Wegener plotted 250-million-year-old glacial deposits on a map showing the modern distribution of continents (Figure 6.3A). Notice that glacial deposits would have formed in tropical and subtropical zones. Figure 6.3B shows the same glacial deposits, and other climate-indicating rocks, plotted on Wegener's Pangea map. Here the glaciers cluster neatly about the South Pole. The other rocks are also found in logical locations.

Wegener also noticed several instances in which an uncommon rock type or a distinctive sequence of rocks on one side of the Atlantic Ocean is identical to rocks on the other side. When he plotted the rocks on a Pangea map, those on the east side of the Atlantic were continuous with their counterparts on the west side (Figure 6.1). For example, the deformed rocks of the Cape Fold belt of South Africa are similar to rocks found in the Buenos Aires province of Argentina. Plotted on a Pangea map, the two sequences of rocks appear as a single, continuous belt.

Wegener's concept of a single supercontinent that broke apart to form the modern continents is called the theory of **continental drift**. The theory was so revolutionary that skeptical scientists demanded an explanation of *how* continents could move. They wanted an explanation of the mechanism of continental drift. Wegener had concentrated on developing evidence that continents had drifted, not on what caused them to move. Finally, perhaps out of exasperation and as an afterthought to what he considered the important part of his theory, Wegener suggested two alternative possibilities: first, that continents plow their way through oceanic crust, shoving it aside as a

ship plows through water; or second, that continental crust slides over oceanic crust.

Physicists quickly proved that both of Wegener's mechanisms were impossible. Oceanic crust is too strong for continents to plow through it. The attempt would be like trying to push a matchstick boat through heavy tar: the boat, or the continent, would break apart. Furthermore, frictional resistance is too great for continents to slide over oceanic crust.

These conclusions were taken by most scientists as proof that Wegener's theory of continental drift was wrong. However, the physicists' calculations proved only that the mechanism proposed by Wegener was incorrect. They did not disprove, or even consider, the huge mass of evidence indicating that the continents were once joined together.

Objections to Wegener's hypothesis of continental drift continued from all sides of the scientific community. Because Wegener had been unable to defend his hypothesis, other scientists rejected it, and continental drift was largely forgotten. During the 30-year period from 1930 to 1960, a few geologists supported the continental drift hypothesis, but most ignored it.

Much of the hypothesis of continental drift is similar to modern plate tectonics theory. Modern evidence indicates that the continents were together much as Wegener had portrayed them in his map of Pangea. Today, most geologists recognize the importance of Wegener's contributions.

6.2 The Earth's Layers

The energy released by an earthquake travels through Earth as waves. After Wegener died and his theory was mostly forgotten, geologists discovered that both the speed and the direction of these waves change abruptly at certain depths as the waves pass through Earth. They soon realized that these changes reveal that Earth is a layered planet. Figure 6.4 and Table 6.1 describe the layers. It is necessary to understand Earth's layers to consider the plate tectonics theory.

The Crust

The crust is the outermost and thinnest layer. Because it is cool relative to the layers below, the crust consists of hard, strong rock (Figure 6.4). Crust beneath the oceans differs from that of continents.

continental drift The theory proposed by Alfred Wegener that Earth's continents were once joined together and later split and drifted apart. The continental drift theory has been replaced by the more complete plate tectonics theory.

© SPAULN/SHUTTERSTOCK

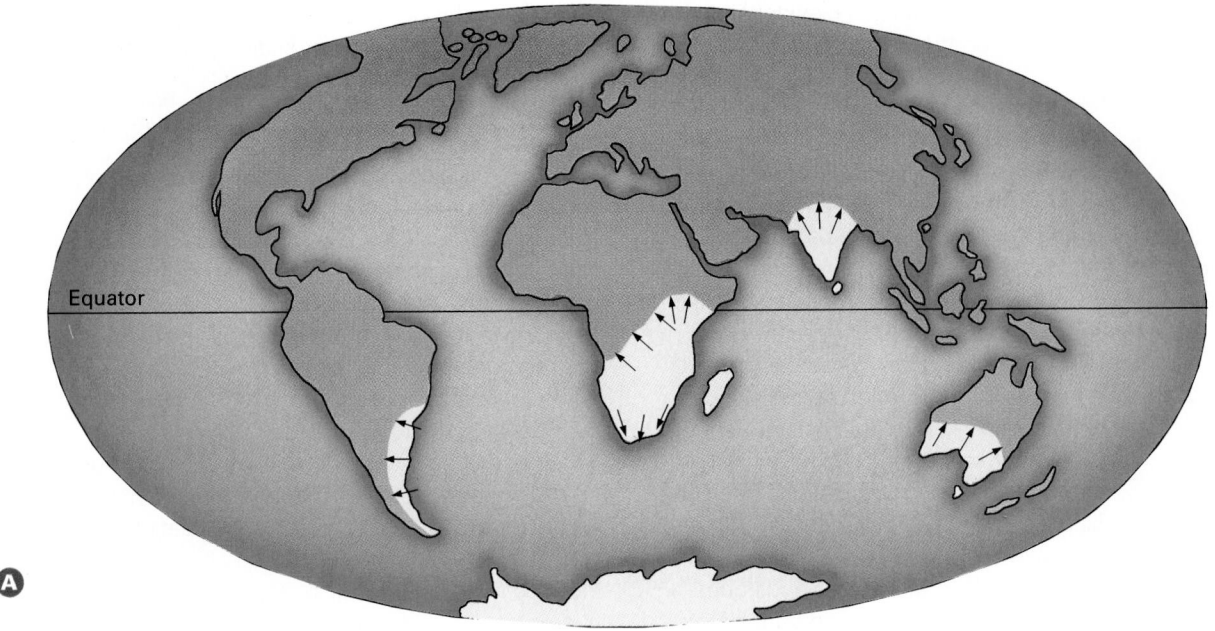

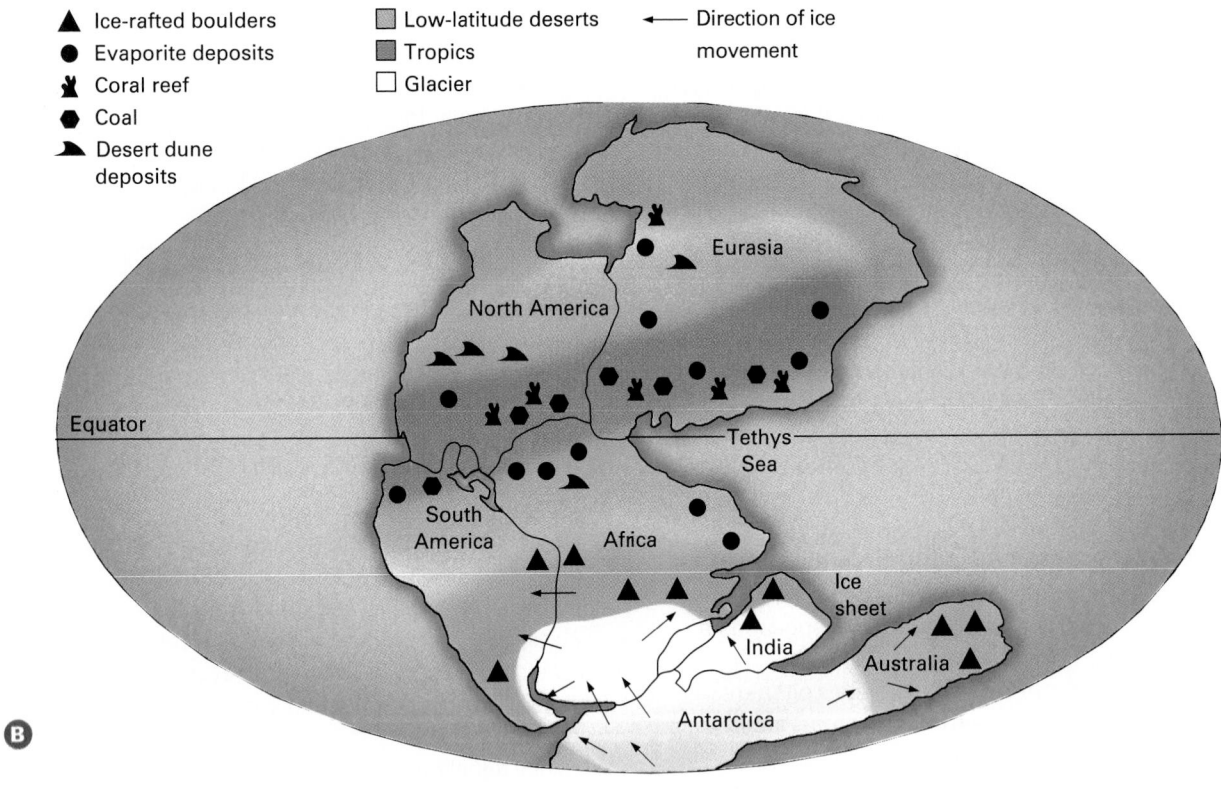

FIGURE 6.3 (A) 250-million-year-old glacial deposits are displayed in white on a map showing the modern distribution of continents. The black arrows show directions of glacial movement, indicated by glacial features described in Chapter 13. (B) 300-million-year-old glacial deposits and other climate-sensitive sedimentary rocks plotted on Wegener's map of Pangea.

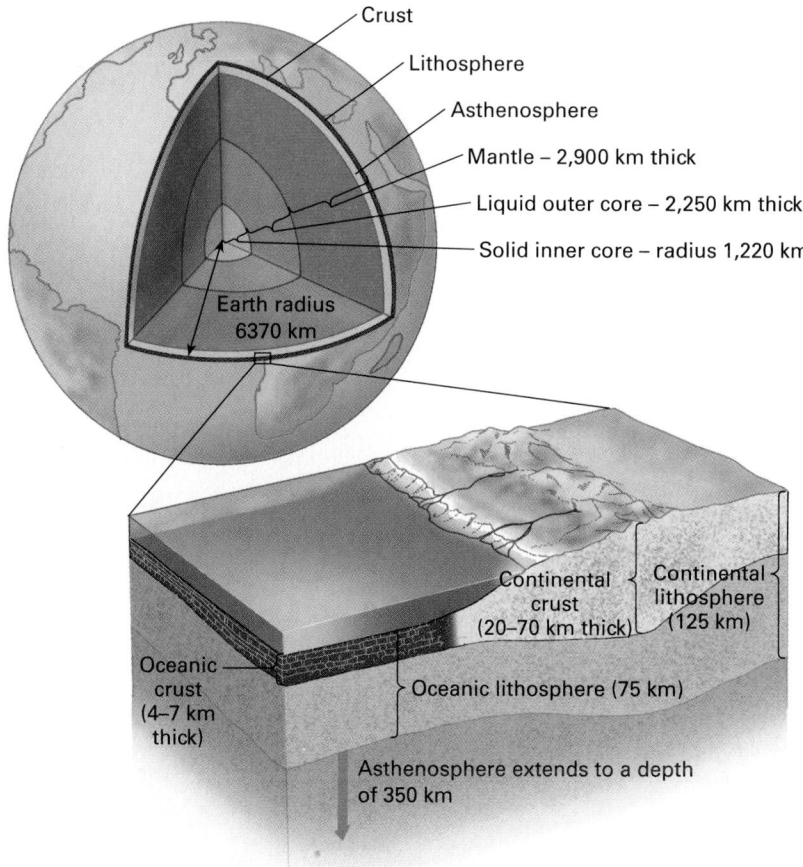

Crust

Lithosphere

Asthenosphere

Mantle – 2,900 km thick

Liquid outer core – 2,250 km thick

Solid inner core – radius 1,220 km

Earth radius 6370 km

Continental crust (20–70 km thick)

Continental lithosphere (125 km)

Oceanic crust (4–7 km thick)

Oceanic lithosphere (75 km)

Asthenosphere extends to a depth of 350 km

FIGURE 6.4 Earth is a layered planet. The insert is drawn on an expanded scale to show near-surface layering. Note that the average thickness of the lithosphere varies from about 75 kilometers beneath the oceans to about 125 kilometers beneath continents.

Oceanic crust is between 4 and 7 kilometers thick and is composed mostly of dark, dense basalt. In contrast, the average thickness of continental crust is about 20 to 40 kilometers, although under mountain ranges it can be as much as 70 kilometers thick. Continents are composed primarily of light-colored granite, less dense than basalt.

The Mantle

The mantle lies directly below the crust. It is almost 2,900 kilometers thick and makes up 80 percent of Earth's volume. The mantle is composed mainly of peridotite, a rock that is denser than the basalt and granite of the crust.

Although the chemical composition is probably essentially uniform throughout the entire mantle, temperature and pressure increase with depth. Figure 6.5A shows that the temperature at the top of the mantle is near 1,000°C and that near the mantle/core boundary, it is about 3,300°C. These changes cause the strength of mantle rock to vary with depth and create layering based on those differences in strength within the mantle. Figure 6.5B shows that internal pressure also increases with depth.

At this point in our story, it is important to understand the effects of temperature and pressure on rocks. Most

TABLE 6.1 The Layers of the Earth

	Layer	Composition	Depth	Properties
Crust	Oceanic crust	Basalt	4 to 7 km	Cool, hard, and strong
	Continental crust	Granite	20 to 70 km	Cool, hard, and strong
Lithosphere	The crust and the uppermost portion of the mantle	Varies; the crust and the mantle have different compositions	75 to 125 km	Cool, hard, and strong
Mantle (excluding the uppermost portion, which is part of the lithosphere)	Asthenosphere	Plastic, ultramafic rock, mainly peridotite, throughout entire mantle; mineralogy varies with depth	Extends to 350 km	Hot, weak, and 1% or 2% melted
	Remainder of upper mantle		Extends from 350 to 660 km	Hot, under great pressure, and mechanically strong
	Lower mantle		Extends from 660 to 2,900 km	High pressure forms minerals different from those of the upper mantle
Core	Outer core	Iron and nickel	Extends from 2,900 to 5,150 km	Liquid
	Inner core	Iron and nickel	Extends from 5,150 km to the center of Earth	Solid

asthenosphere The portion of the upper mantle just beneath the lithosphere, extending from a depth of about 100 kilometers to about 350 kilometers below the surface of Earth and consisting of weak, plastic rock where magma may form.

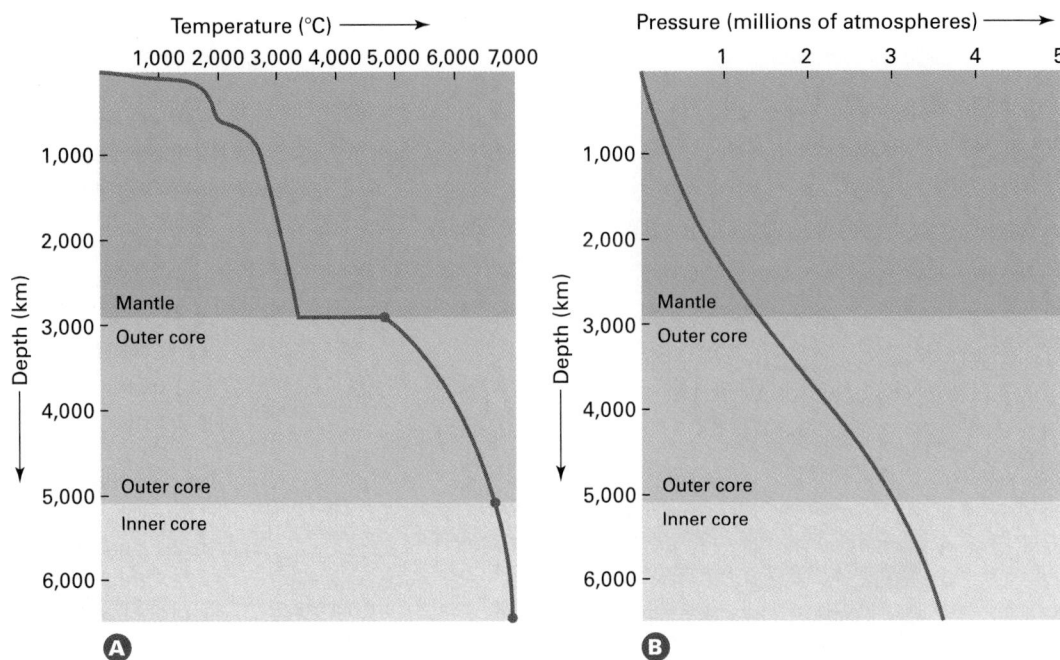

FIGURE 6.5 (A) Earth's internal temperature increases with depth. At the center of Earth, the temperature is close to 7,000°C, hotter than the Sun's surface. For reference, the temperature of an oven baking a chocolate cake is about 175°C, and a heated steel bar turns cherry red at 746°C. (B) In contrast with temperature, Earth's internal pressure increases almost linearly with depth.

people understand that increasing temperature will eventually melt a rock. Less obvious, however, is the fact that high pressure *inhibits* melting. This is because rock expands by about 10 percent when it melts. High pressure makes it more difficult for a rock to expand and therefore impedes melting. If the combined effects of temperature and pressure are close to—but just below—a rock's melting point, the rock remains solid but loses strength, so it becomes weak and plastic. Thus the rock can ooze, or flow slowly, like road tar on a hot day. At that point, if temperature rises or pressure decreases, the rock will begin to melt.

Because both temperature and pressure increase with depth in Earth, their combined effects cause changes in the physical properties of rocks with increasing depth. These changes alter the strength of upper-mantle rocks with depth, and create two distinctly different layers in the upper mantle. Although the composition of the mantle is the same in the two layers, the strength of the rocks is very different.

The Lithosphere

Figure 6.5 shows that the uppermost mantle is cool and its pressure is low, conditions similar to those in the crust. Both factors combine to produce hard, strong rock similar to that of the crust. Recall from Chapter 1 that the outer part of Earth, including both the crust and

the uppermost mantle, make up the *lithosphere*. The lithosphere averages about 100 kilometers in thickness but varies from about 75 kilometers beneath ocean basins to about 125 kilometers under the continents (Figure 6.4). The lithosphere, then, consists mostly of the cold, strong uppermost mantle; the crust is just a thin layer of buoyant rock embedded in the upper mantle. Some parts of the lithosphere have continents embedded in them, whereas other regions of lithosphere are topped by oceanic crust.

The Asthenosphere

At a depth varying from 75 to 125 kilometers beneath Earth's surface, the combined effects of rising temperature and rising pressure approach the melting point of mantle rock. As a result, the mantle abruptly loses strength and becomes weak and plastic—and 1 to 2 percent of the rock melts, although the rest remains solid. The weak, plastic, and partly molten character extends to a depth of about 350 kilometers, where increasing pressure overwhelms temperature, and the rock becomes stronger again. This layer of weak mantle rock extending from about 100 to 350 kilometers deep is the **asthenosphere** (from the Greek for "weak layer"). The average temperature in the asthenosphere is about 1,800°C, although the temperature increases with depth as it does in other Earth layers. Pressure in the asthenosphere rises from about 35 kilobars[1] near the top to about 120 kilobars at the base.

If you apply force to a plastic solid, it flows slowly. (Two familiar examples of solid but plastic materials are Silly Putty™ and hot road tar.) When building a house, you start with a strong, hard foundation. However, Earth is not constructed in this way. The strong, hard lithosphere lies on top of the soft, weak asthenosphere. Thus, the lithosphere is not rigidly supported by the rock beneath it. Instead, the lithosphere floats on the soft plastic rock of the asthenosphere. This concept of a floating lithosphere is important to our understanding of plate tectonics and Earth's internal processes.

The Mantle below the Asthenosphere

At the base of the asthenosphere, increasing pressure overwhelms the effect of rising temperature, and the strength of the mantle increases again. Although the mantle below 350 kilometers is stronger than the asthenosphere, it never regains the strength of the lithosphere. It is plastic and capable of flowing slowly, over geologic time. Recent evidence suggests that the lowermost mantle, at the core boundary, is partly molten. You can see temperature and pressure profiles in Figure 6.5.

The Core

As you learned in Chapter 1, the *core* is the innermost of Earth's layers. It is a sphere with a radius of about 3,470 kilometers, about the same size as Mars. Earth's core is composed largely of iron and nickel. The outer core is molten because of the high temperature and relatively lower pressures in that region. Near its center, the core's temperature is nearly 7,000°C, hotter than the Sun's surface. The pressure is 3.5 million times that of Earth's atmosphere at sea level. The extreme pressure compresses the inner core to a solid, despite the fact that it is even hotter than the molten outer core.

6.3 The Sea-Floor Spreading Hypothesis

Shortly after World War II, scientists began to explore the floors of Earth's oceans. Although these studies ultimately played a large role in the development of plate tectonics theory, they were initially undertaken for military and economic reasons. Defense strategists wanted a detailed knowledge of sea floor topography for submarine warfare, and the same information was needed to lay undersea telephone cables. As they mapped the sea floor, oceanographers discovered the largest mountain chain on Earth, now called the **Mid-Oceanic Ridge system** (Figure 6.6). One leg of this huge submarine mountain chain, called the **Mid-Atlantic Ridge**, lies directly in the middle of the Atlantic Ocean, halfway between North and South America to the west, and Europe and Africa to the east.

As you learned in Chapter 3, oceanic crust is composed mostly of basalt, an igneous rock rich in iron. As basaltic lava cools and becomes solid rock, the iron-rich

Mid-Oceanic Ridge system The undersea mountain chain that forms at the boundary between divergent tectonic plates within oceanic crust. It circles the planet like the seam on a baseball, forming Earth's longest mountain chain.

Mid-Atlantic Ridge The portion of the Mid-Oceanic Ridge system that lies in the middle of the Atlantic Ocean, halfway between North America and South America to the west, and Europe and Africa to the east.

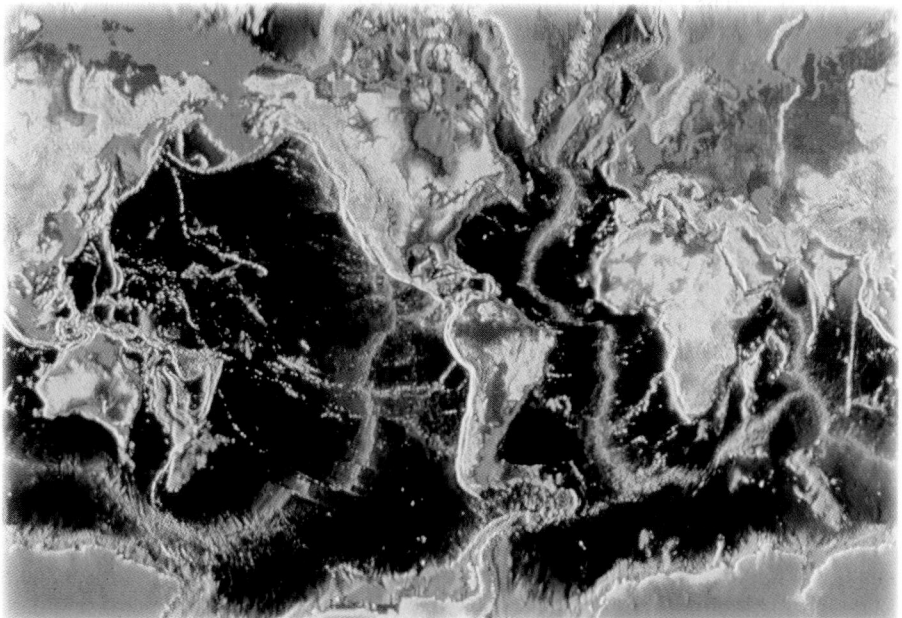

FIGURE 6.6 A chart of the sea floor. The Mid-Oceanic Ridge system is a submarine mountain chain that encircles the globe like the seam on a baseball.

OAR/NATIONAL UNDERSEA RESEARCH PROGRAM (NURP)

© LOURDU PRAKASH XAVIER/SHUTTERSTOCK

1. 1 kilobar pressure is approximately equal to 987 atmospheres, or 14,504 pounds per square inch. An *atmosphere* (abbreviated *atm*) is a unit of pressure approximately equal to the air pressure at sea level.

normal magnetic polarity A magnetic orientation the same as that of Earth's current magnetic field.

reversed magnetic polarity Magnetic orientations in rock that are opposite to the current orientation of Earth's magnetic field.

magnetic reversal A change in Earth's magnetic field in which the north magnetic pole becomes the south magnetic pole and vice versa; has occurred on average every 500,000 years over the past 65 million years.

minerals become weak magnets that align their magnetic fields parallel to the Earth's magnetic field. Thus, basalt records the orientation of Earth's magnetic field at the time the rock cools.

In addition to mapping the topography of the sea floor, oceanographers towed devices called magnetometers behind their research vessels to detect and record magnetic patterns in the deep oceans. Figure 6.7 shows the magnetic orientations of sea floor rocks near a part of the Mid-Atlantic Ridge southwest of Iceland. In this figure, green stripes represent basalt with a magnetic orientation parallel to Earth's current magnetic field, called **normal magnetic polarity**. The intervening blue stripes represent rocks with magnetic orientations that are exactly opposite to the current magnetic field, called **reversed magnetic polarity**. Notice that the stripes form a symmetric pattern of normal and reversed polarity about the axis of the ridge, and that the central stripe is green, indicating that basalt at the ridge axis has a magnetic orientation parallel to that of Earth's magnetic field today.

Why do the sea floor rocks have alternating normal and reversed polarity, and why is the pattern symmetrically distributed about the Mid-Oceanic Ridge? In the mid-1960s, three scientists—a Cambridge graduate student named Frederick Vine; his professor, Drummond Matthews; and Lawrence Morley, a Canadian working independently of the other two—proposed an explanation for these odd magnetic patterns on the sea floor. They knew that other scientists had recently discovered that Earth's magnetic field has reversed its polarity on the average of every 500,000 years during the past 65 million years. When such a **magnetic reversal** of Earth's field occurs, the north magnetic pole becomes the south magnetic pole, and vice versa. This discovery surprised most geologists, but periodic reversals of the Earth's magnetic field were well known by the early 1960s.

Vine, Matthews, and Morley suggested that the sea floor is spreading continuously away from the Mid-Oceanic Ridges, like two conveyor belts moving outward, away from each other. New basalt lava rises through cracks that form at the ridge axis as the two sides of the sea floor separate. As the lava cools and solidifies, the basalt acquires the magnetic orientation of Earth's field. Because Earth's field periodically reverses,

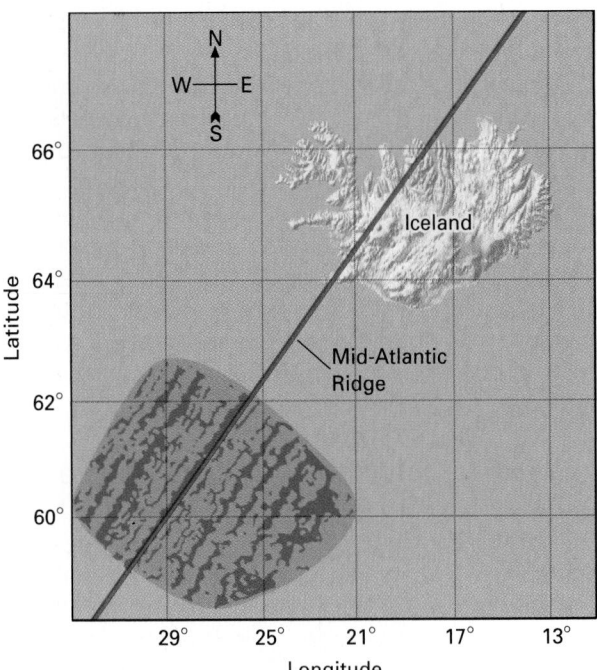

FIGURE 6.7 The Mid-Atlantic Ridge, shown in red, runs through Iceland. Magnetic orientation of sea-floor rocks near the ridge are shown in the lower-left portion of the map. The green stripes represent sea-floor rocks with normal magnetic polarity, and the blue stripes represent rocks with reversed polarity. The stripes form a symmetrical pattern of alternating normal and reversed polarity on each side of the ridge.

the sea floor acquires a striped magnetic orientation, which you can see in Figure 6.8. Thus, the sea floor and oceanic crust should become older with increasing distance from the ridge axis.

At the same time as these sea floor magnetic patterns were detected, oceanographers discovered that the thin layer of mud that overlies sea floor basalt in most parts of the oceans is thinnest at the Mid-Oceanic Ridge and becomes progressively thicker at greater distance from the ridge. They reasoned that if mud settles onto the sea floor at the same rate everywhere, and if the ridge is the newest part of the sea floor, the mud layer would be thinnest at the ridge. The mud should thicken with increasing distance from the ridge because, as stated above, oceanic crust becomes older with increasing distance from the ridge axis.

Furthermore, the oceanographers found that fossils in the deepest layers of mud overlying basalt were very young at the ridge axis but become progressively older with increasing distance from the ridge. This discovery, too, indicated that the sea floor becomes older with increasing distance from the ridge axis.

Symmetrical magnetic patterns and similar mud age and thickness trends were quickly discovered at other parts of the Mid-Oceanic Ridge system in other ocean

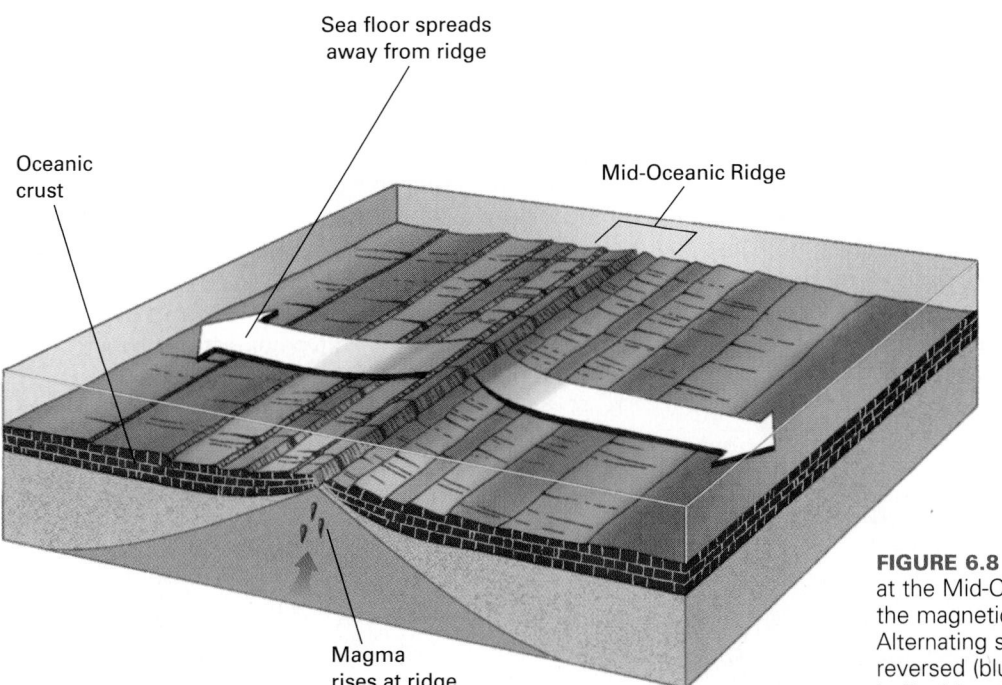

Sea floor spreads
away from ridge

Oceanic
crust

Mid-Oceanic Ridge

Magma
rises at ridge

FIGURE 6.8 As new oceanic crust cools at the Mid-Oceanic Ridge, it acquires the magnetic orientation of Earth's field. Alternating stripes of normal (green) and reversed (blue) polarity record reversals in Earth's magnetic field that occurred as the crust spread outward from the ridge.

basins, and the hypothesis of **sea-floor spreading** was proposed as a general model for the origin of all oceanic crust. In a very few years, the sea floor spreading hypothesis became the basis for development of the much broader theory of plate tectonics.

6.4 The Theory of Plate Tectonics

Like many great unifying scientific ideas, the plate tectonics theory is simple. Briefly, it states that the lithosphere is a shell of hard, strong rock about 100 kilometers thick that floats on the hot, plastic asthenosphere (Figure 6.9). As you learned in Chapter 1, the lithosphere is broken

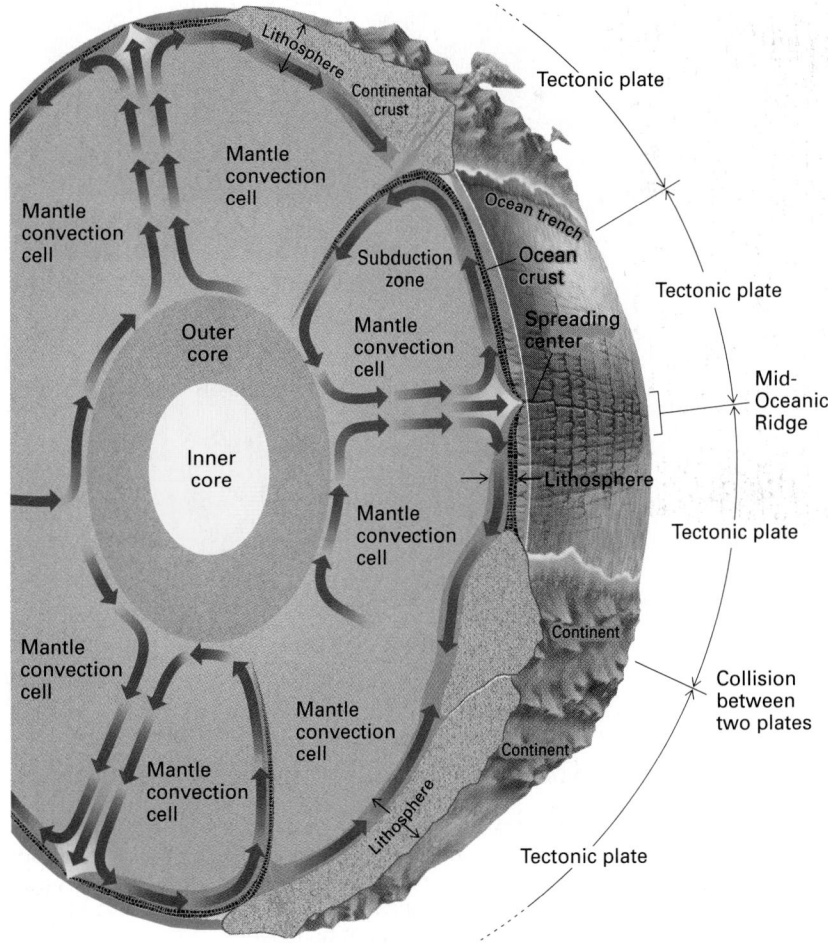

FIGURE 6.9 A cutaway view of Earth shows that the lithosphere glides horizontally across the asthenosphere. Continents and oceans ride piggyback on the lithosphere, and consequently the continents and oceans move across Earth's surface at rates of a few centimeters each year. The mantle and lithosphere circulate in huge elliptical cells that rise from the deepest mantle, then flow across Earth's surface, and finally sink back to the mantle–core boundary. In this figure, the thickness of the lithosphere is exaggerated for clarity.

into seven large (and several smaller) segments called *tectonic plates* (Figure 6.10). They are also called *lithospheric plates* or, simply, *plates*—the terms are interchangeable. The tectonic plates glide slowly over the asthenosphere at rates ranging from less than 1 to about 16 centimeters per year, about as fast as your fingernails grow. Continents and ocean basins make up the upper parts of the plates. As a tectonic

plate glides over the asthenosphere, the continents and oceans move with it.

A **plate boundary** is a fracture that separates one plate from another. Neighboring plates can move relative to one another at these boundaries in three ways, shown by the insets in Figure 6.10. At a **divergent boundary**, two plates move apart from each other. At a **convergent boundary**, two plates move toward each other, and at a **transform boundary**, they slide horizontally past each other. Table 6.2 summarizes characteristics and examples of each type of plate boundary.

The great forces generated at a plate boundary build mountain ranges and cause volcanic eruptions and

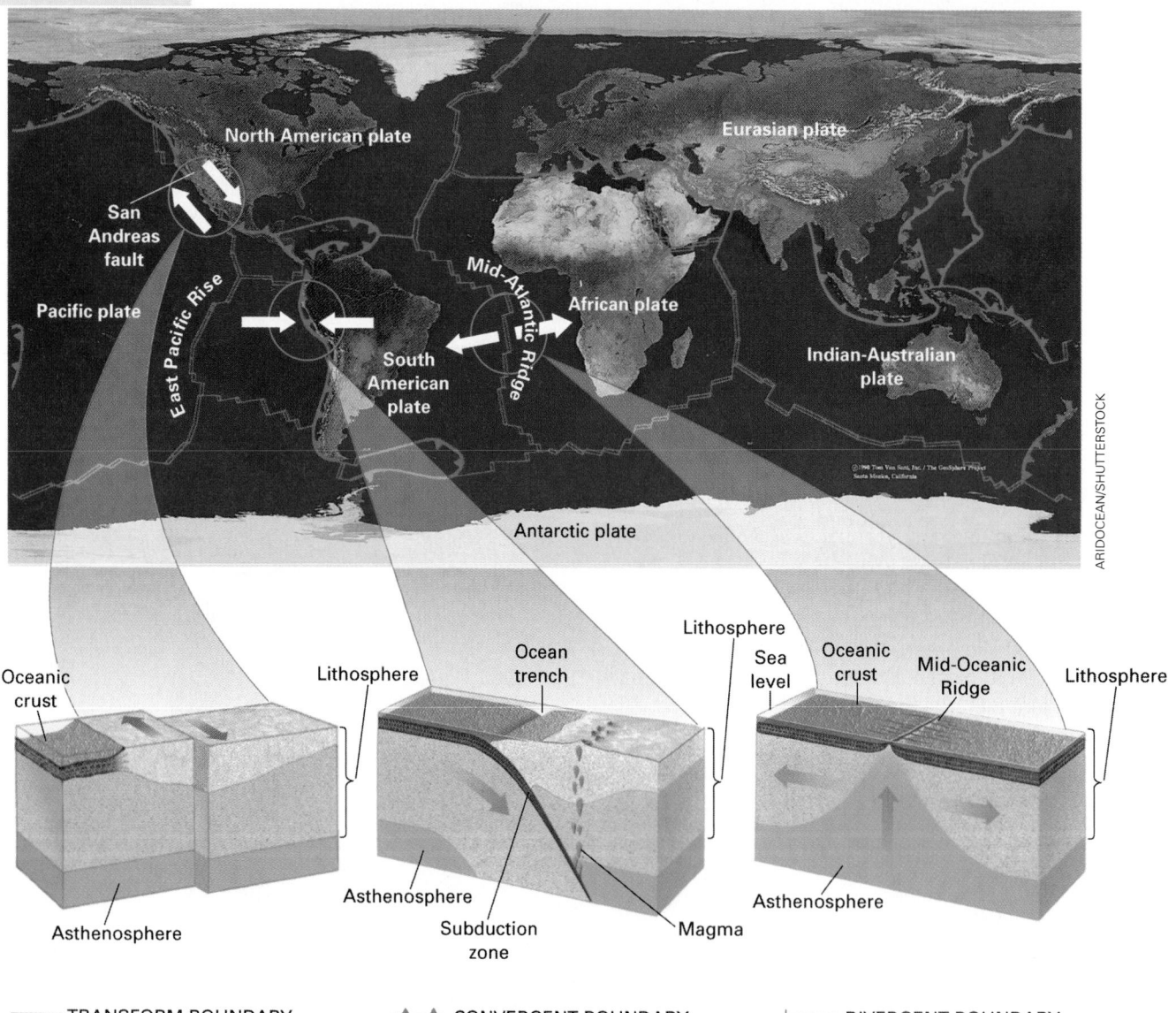

FIGURE 6.10 Earth's lithosphere is broken into seven large tectonic plates, called the African, Eurasian, Indian-Australian, Antarctic, Pacific, North American, and South American plates. The white arrows show how the plates move in different directions. The three different types of plate boundaries are shown below the map: At a transform plate boundary, rocks on opposite sides of the fracture slide horizontally past each other. Two plates move toward each other at a convergent boundary. Two plates move apart at a divergent boundary.

TABLE 6.2 Characteristics and Examples of Plate Boundaries

Type of Boundary	Types of Plates Involved	Topography	Geologic Events	Modern Examples
Divergent	Ocean–ocean	Mid-Oceanic Ridge	Sea-floor spreading, shallow earthquakes, rising magma, volcanoes	Mid-Atlantic Ridge
	Continent–continent	Rift valley	Continents torn apart, earthquakes, rising magma, volcanoes	East African rift
Convergent	Ocean–ocean	Island arcs and ocean trenches	Subduction, deep earthquakes, rising magma, volcanoes, deformation of rocks	Western Aleutians
	Ocean–continent	Mountains and ocean trenches	Subduction, deep earthquakes, rising magma, volcanoes, deformation of rocks	Andes
	Continent–continent	Mountains	Deep earthquakes, deformation of rocks	Himalayas
Transform	Ocean–ocean	Major offset of Mid-Oceanic Ridge axis	Earthquakes	Offset of East Pacific rise in South Pacific
	Continent–continent	Small deformed mountain ranges, deformations along fault	Earthquakes, deformation of rocks	San Andreas fault

earthquakes. In contrast to plate boundaries, the interior portion of a plate is usually tectonically quiet because it is far from the zones where two plates interact.

Divergent Plate Boundaries

At a divergent plate boundary (also called a *spreading center* or a *rift zone*), two plates spread apart from one another, as shown at the center of Figure 6.11. The

underlying asthenosphere then oozes upward to fill the gap between the separating plates. As the asthenosphere rises between the separating plates, some of it melts to form magma.[2] Most of the magma rises to

---

2. It seems counterintuitive that the rising, cooling asthenosphere should melt to form magma, but the melting results from the decreasing pressure, offsetting the temperature change. This process is discussed in Chapter 8.

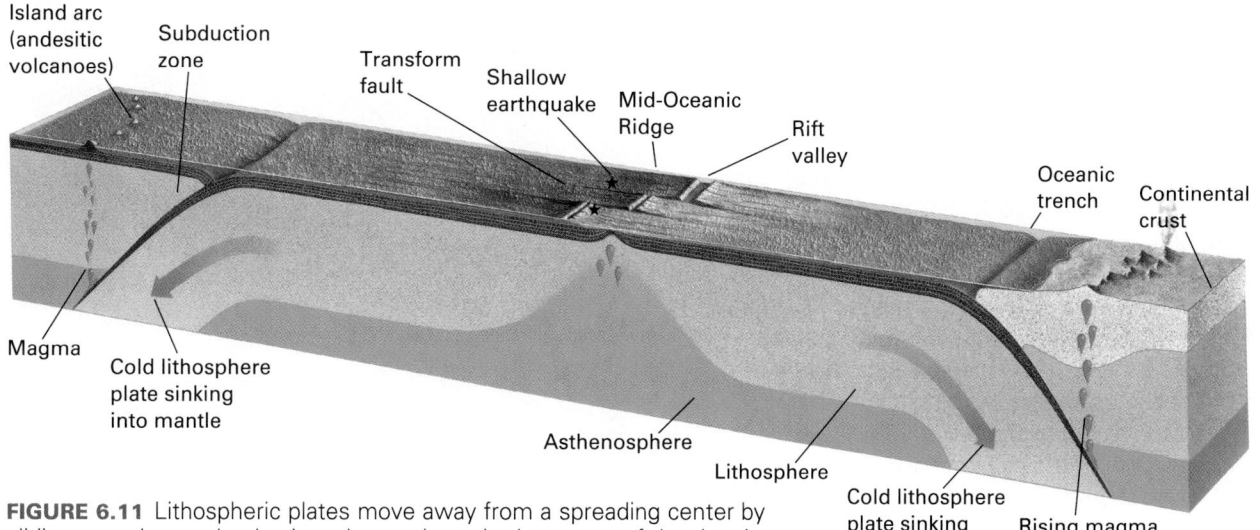

FIGURE 6.11 Lithospheric plates move away from a spreading center by gliding over the weak, plastic asthenosphere. In the center of the drawing, new lithosphere forms at a spreading center. The lithosphere beneath the spreading center is only 10 or 15 kilometers thick but it becomes thicker as the new lithosphere moves away from the spreading center and cools. At the sides of the drawing, old lithosphere sinks into the mantle at subduction zones.

Earth's surface, where it cools to form new crust. Most of this activity occurs between plates beneath the seas because most divergent plate boundaries lie in the ocean basins, but it can also occur between two continental plates.

Both the lower lithosphere (the part beneath the crust) and the asthenosphere are parts of the mantle and thus have similar chemical compositions. The main differences between the two layers are in temperature, pressure, and mechanical strength. The cool lithosphere is strong and hard, but the hot asthenosphere is weak and plastic. As the asthenosphere rises closer to Earth's surface between two separating plates, it cools, gains mechanical strength, and therefore *transforms into new lithosphere*. In this way, new lithosphere continuously forms at a divergent boundary.

At a divergent boundary, the rising asthenosphere is hot, weak, and plastic. Only the upper 10 to 15 kilometers cools enough to gain the strength and hardness of lithosphere rock. As a result, the lithosphere, including the crust and the upper few kilometers of mantle rock, can be as little as 10 or 15 kilometers thick at a spreading center. But as the lithosphere spreads, it cools from the top downward (Figure 6.11).

As it spreads outward and cools, the new lithosphere also thickens because the boundary between cool rock and hot rock migrates downward. Consequently, the thickness of the lithosphere increases as it moves away from the spreading center. Think of ice freezing on a pond. On a cold day, water under the ice freezes and the ice becomes thicker. The lithosphere continues to thicken until it attains a steady-state thickness of about 75 kilometers beneath an ocean basin, and as much as 125 kilometers beneath a continent.

The Mid-Oceanic Ridge: Rifting in the Oceans

New lithosphere at an oceanic spreading center is hotter than older lithosphere rock farther away from the divergent boundary, and therefore the new lithosphere has lower density. Consequently, it floats to a higher level, forming the undersea mountain chain called the Mid-Oceanic Ridge system (Figure 6.6). But as lithosphere migrates away from the spreading center, it cools and becomes denser. As a result, it sinks into the soft, plastic asthenosphere (Figure 6.11), making the average depth of the sea floor away from the Mid-Oceanic Ridge about 5 kilometers.

As already stated, the Mid-Oceanic Ridge system encircles the planet, forming Earth's longest mountain chain. Basaltic magma that oozes onto the sea floor at the ridge creates about 22 cubic kilometers, or 70 billion tons, of new oceanic crust each year. The Mid-Oceanic Ridge system is described further in Chapter 15.

Splitting Continents: Rifting in Continental Crust

A divergent plate boundary can divide continental crust in a process called **continental rifting**. A rift valley develops in a continental rift zone because continental crust stretches, fractures, and sinks as it is pulled apart. Continental rifting is now taking place along the East African Rift (Figure 6.12). If the rifting continues, eastern Africa will separate from the main portion of the continent, and a new ocean basin will open between the separating portions of Africa.

Convergent Plate Boundaries

At a convergent plate boundary, two lithospheric plates move toward each other. Not all lithospheric plates are made of equally dense rock. Where two plates of different densities converge, the denser one sinks into the mantle beneath the other. This sinking process is called **subduction** and is shown in both the right and left sides of Figure 6.11. A **subduction zone** is a long, narrow belt where a lithospheric plate is sinking into

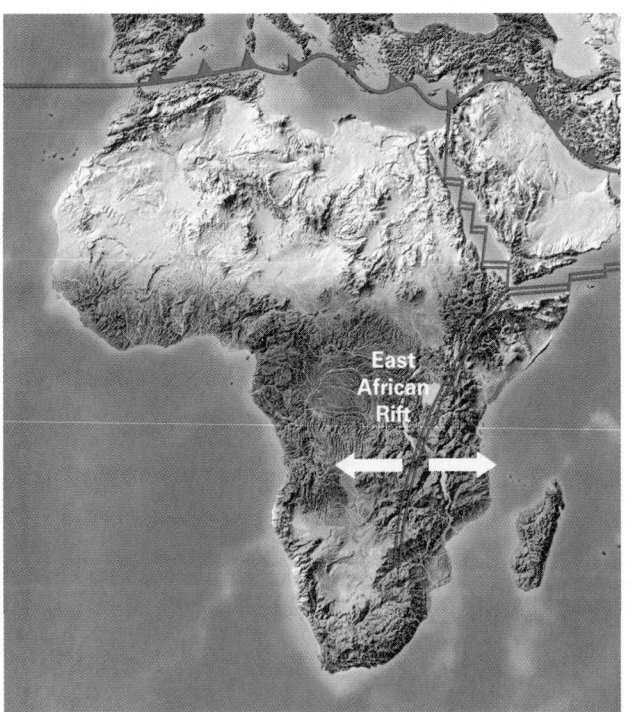

ARIDOCEAN/SHUTTERSTOCK

Divergent boundary	Convergent boundary	Transform boundary

FIGURE 6.12 The continent of Africa is splitting apart along the East African Rift.

the mantle. On a worldwide scale, the rate at which old lithosphere sinks into the mantle at subduction zones is equal to the rate at which new lithosphere forms at spreading centers. In this way, Earth maintains a global balance between the creation of new lithosphere and the destruction of old lithosphere.

Plate convergence can occur three ways: (1) between a plate carrying oceanic crust and another carrying continental crust, (2) between two plates carrying oceanic crust, and (3) between two plates carrying continental crust.

Convergence of Oceanic Crust with Continental Crust

Recall that oceanic crust is generally denser than continental crust. In fact, the entire lithosphere beneath the oceans is denser than continental lithosphere. When an oceanic plate converges with a continental plate, subduction occurs and the denser oceanic plate sinks into the mantle beneath the edge of the continent. As a result, many subduction zones are located at continental margins. Today, oceanic plates are sinking beneath the western edge of South America; along the coasts of Oregon, Washington, and British Columbia; and at several other continental margins shown in Figure 6.10. When the descending plate reaches the asthenosphere, large quantities of magma are generated by processes explained in Chapter 8. Thus volcanoes are common along subduction zones. Chapter 9 describes how the Andes—a chain of volcanic mountains—formed as a result of subduction of a Pacific oceanic plate beneath the west coast of South America.

The oldest sea floor rocks on Earth are only about 200 million years old because oceanic crust continuously recycles into the mantle at subduction zones. Rocks as old as 3.96 billion years are found on continents because subduction consumes little continental crust.

Convergence of Two Plates Carrying Oceanic Crust

Recall that newly formed oceanic lithosphere is hot, thin, and of low density, but as it spreads away from the Mid-Oceanic Ridge it becomes older, cooler, thicker, and denser. Thus, the density of oceanic lithosphere increases with its age. When two oceanic plates converge, the older, denser one sinks into the mantle. Oceanic subduction zones are common in the southwestern Pacific Ocean (Figure 6.10) and also formed the Aleutian Islands. Their effects on the geology of the sea floor are described in Chapter 15.

Convergence of Two Plates Carrying Continents

If two converging plates carry continents, neither can sink deeply into the mantle because of their low densities. A continent does not normally sink into the mantle at a subduction zone for the same reasons that a log does not sink in a lake: both are of lower density than the material beneath them. In this case, the two continents collide and crumple against each other, forming a huge mountain chain. The Himalayas, the Alps, and the Appalachians all formed as results of continental collisions. The formation of the Himalayas is described in Chapter 9.

Transform Plate Boundaries

A transform plate boundary forms where two plates slide horizontally past one another as they move in opposite directions (Figure 6.10). California's San Andreas Fault is a transform boundary between the North American plate and the Pacific plate. Frequent earthquakes occur along transform boundaries. This type of boundary can occur in both oceans and continents.

6.5 The Anatomy of a Tectonic Plate

The nature of a tectonic plate can be summarized as follows:

1. A plate is a segment of the lithosphere; thus it includes the uppermost mantle and the overlying crust.
2. A single plate can carry both oceanic crust and continental crust. The average thickness of a lithospheric plate covered by oceanic crust is 75 kilometers, whereas that of lithosphere covered by a continent is 125 kilometers. Lithosphere may be as little as 10 to 15 kilometers thick at an oceanic spreading center.
3. A plate is composed of hard, mechanically strong rock.
4. A plate floats on the underlying hot, plastic asthenosphere and glides horizontally over it.
5. A plate behaves like a slab of ice floating on a pond. It may flex slightly, as thin ice does when a skater goes by, allowing minor vertical movements. In general, however, each plate moves as a large, intact sheet of rock.
6. A plate margin is tectonically active. Earthquakes, mountain ranges, and volcanoes are common at plate boundaries. In contrast, the interior of a lithospheric plate is normally tectonically stable.
7. Tectonic plates move at rates that vary from less than 1 to 16 centimeters per year. Because continents and oceans make up the

subduction The process in which two lithospheric plates of different densities converge and the denser one sinks into the mantle beneath the other.

subduction zone A long narrow region at a convergent boundary where a lithospheric plate is sinking into the mantle during subduction; also referred to as *subduction boundary*.

convection The upward and downward flow of fluid material in response to heating and cooling. Convection occurs slowly in Earth's mantle and much more quickly in the oceans and the atmosphere.

mantle plume A relatively small rising column of mantle rock that is hotter than surrounding rock. As pressure decreases in a rising plume, magma forms at a hot spot.

upper parts of the moving lithosphere, continents and oceans migrate across Earth's surface at the same rates at which the plates move. Manhattan Island is now 9 meters farther from London than it was when the Declaration of Independence was written in 1776. Alfred Wegener was correct in saying that continents drift across Earth's surface.

6.6 Why Plates Move: The Earth as a Heat Engine

After geologists had developed the plate tectonics theory, they began to ask: "*Why* do the great slabs of lithosphere glide over Earth's surface?" Research[3] shows that subduction continues slowly all the way to the core–mantle boundary, to a depth of 2,900 kilometers. At the same time, equal volumes of hot rock rise from the deep mantle to the surface beneath a spreading center (divergent boundary), forming new lithosphere to replace that lost to subduction. The term **convection** refers to the upward and downward flow of fluid material in response to heating and cooling. The process of *mantle convection* continually stirs the entire mantle as rock that is hotter than its surroundings rises toward Earth's surface and old plates that are colder than their surroundings sink into the mantle. A single mantle convection cell may be thousands of kilometers across. In this way, the entire mantle–lithosphere system circulates in great cells, carrying rock from the core–mantle boundary to Earth's surface and then back into the deepest mantle (Figure 6.9).

A soup pot on a hot stove illustrates the process of convection. Rising temperature causes most materials, including soup (or rock), to expand. When soup at the bottom of the pot is heated by the underlying stove, it becomes warm and expands. It then rises because it is less dense than the soup at the top. When the hot soup reaches the top of the pot, it flows along the surface until it cools and sinks (Figure 6.13).

Just as the soup on the stove rests on the hot burner, the base of the mantle lies on—and is heated

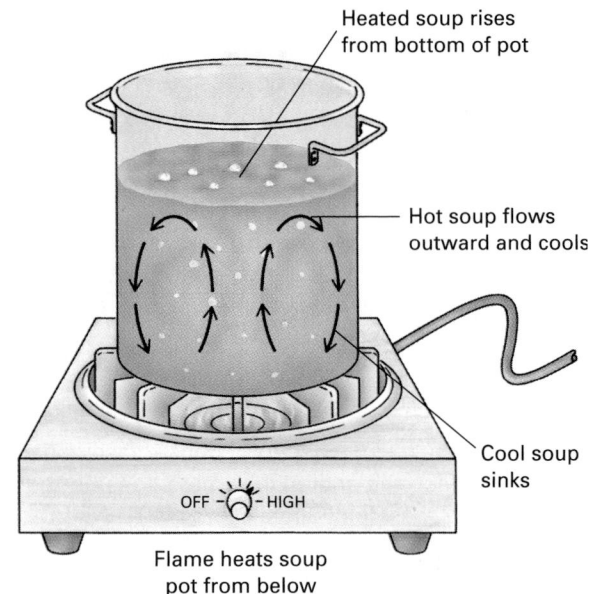

FIGURE 6.13 Soup convects when it is heated from the bottom of the pot.

by—the hotter core. Thus, heat from the core, supplemented by additional heat generated by radioactivity within the mantle, drives the entire mantle and lithosphere in huge cells of convecting rock. A tectonic plate is the upper portion of a convecting cell and thus glides over the asthenosphere as a result of the convection (Figure 6.9).

Two other processes, shown in Figure 6.14, may facilitate the movement of tectonic plates. Notice that the base of the lithosphere slopes downward from a spreading center; the grade can be as steep as 8 percent, steeper than most paved highways. Calculations show that even if the slope were less steep, gravity would cause the lithosphere to slide away from a spreading center over the soft, plastic asthenosphere at a rate of a few centimeters per year. In addition, the lithosphere becomes denser as it moves away from a spreading center and cools. Eventually, old lithosphere may become denser than the asthenosphere below. Consequently, it can no longer float on the asthenosphere and sinks into the mantle in a subduction zone, dragging the trailing portion of the plate over the asthenosphere. Both of these processes may contribute to the movement of a lithospheric plate as it glides over the asthenosphere.

Mantle Plumes and Hot Spots

In contrast to the huge, ridge-shaped mass of mantle that rises beneath a spreading center, a **mantle plume** is a relatively small rising column of plastic mantle rock that is hotter than surrounding rock. Many plumes rise from great depths in the mantle, probably because small zones of rock near the core–mantle boundary become

3. Summarized by Richard A. Kerr, "Deep-Sinking Slabs Stir the Mantle," *Science* 275 (January 31, 1997), 613–615. Also see Hans-Peter Bunge and Mark Richards, "The Origin of Large Scale Structure in Mantle Convection: Effects of Plate Motions and Viscosity Stratification," *Geophysical Research Letters* 23 (October 15, 1996), 2987–2990.

Unit 2: Internal Processes

hotter and more buoyant than surrounding regions of the deep mantle. Others may form as a result of heating in shallower portions of the mantle.

As pressure decreases in a rising plume, magma forms directly above it at a **hot spot** in the mantle—a volcanic center just beneath the lithosphere that rises to erupt from volcanoes on Earth's surface. The Hawaiian Island chain is an example of a volcanic center at a hot spot (Figure 6.15). It erupts in the middle of the Pacific tectonic plate because the plume originates deep in the mantle, far from any plate boundary and below the level of lateral plate motion.

6.7 Supercontinents

Prior to 2 billion years ago, large continents as we know them today may not have existed. Instead, many—perhaps hundreds—of small masses of continental crust and island arcs similar to Japan, New Zealand, and the modern islands of the southwest Pacific Ocean dotted a global ocean basin. Then, between 2 billion and 1.8 billion years ago, tectonic plate movements must have swept these microcontinents together to form a single landmass called a

supercontinent. After a few hundred million years, this supercontinent developed rifts and broke into fragments. The fragments then separated, each riding away from the others on its own tectonic plate. About 1 billion years ago, the fragments of continental crust reassembled, forming a second supercontinent. In turn, this continent fractured and the continental fragments reassembled into a third supercontinent about 300 million years ago, 70 million years before the appearance of dinosaurs. This third supercontinent is Alfred Wegener's Pangea, which began to break apart about 235 million years ago in late Triassic time. The tectonic plates have continued their slow movement to create the mosaic of continents and ocean basins that shape the map of the world as we know it today.

hot spot A persistent volcanic center thought to be located directly above a rising plume of hot mantle rock.

supercontinent A continent, such as Alfred Wegener's Pangea, consisting of all or most of Earth's continental crust joined together to form a single, large landmass. At least three supercontinents are thought to have existed during the past 2 billion years, and each broke apart after a few hundred million years.

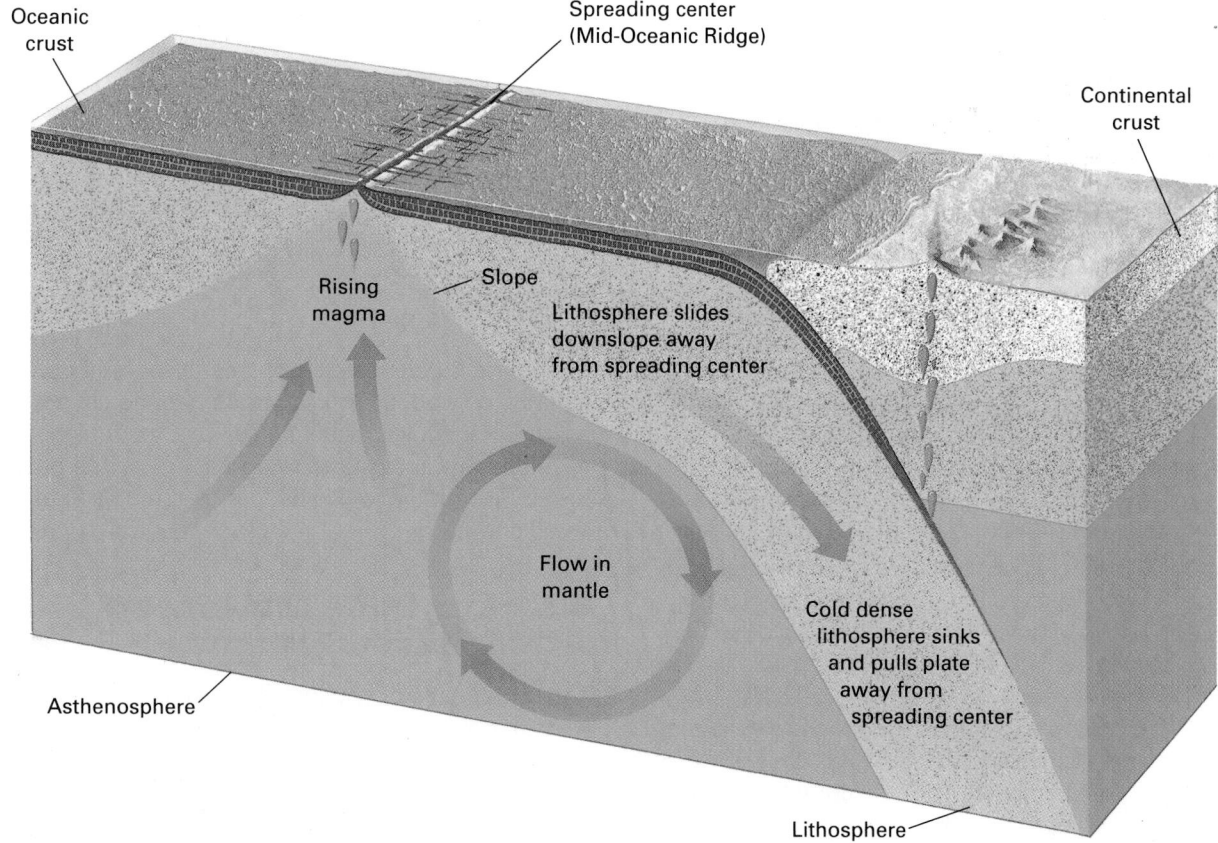

FIGURE 6.14 New lithosphere glides downslope away from a spreading center. At the same time, the old, cool part of the plate sinks into the mantle at a subduction zone, pulling the rest of the plate along with it. (The steepness of the slope at the base of the lithosphere is exaggerated in this figure.)

FIGURE 6.15 Lava flows on Hawaii provide evidence for this area being a hot spot.

6.8 Isostasy: Vertical Movement of the Lithosphere

If you have ever used a small boat, you may have noticed that the boat settles in the water as you get into it and rises as you step out. The lithosphere behaves in a similar manner. If a large mass is added to the lithosphere, the underlying asthenosphere flows laterally away from that region to make space for the settling lithosphere.

But how is mass added or subtracted from the lithosphere? One process that adds and removes mass is the growth and melting of large glaciers. When a glacier grows, the weight of ice forces the lithosphere downward. For example, in central Greenland, an ice sheet 3,000 meters thick has depressed the continental crust below sea level. Conversely, when a glacier melts, the continent rises—it rebounds. Geologists have discovered ice age beaches in Scandinavia tens of meters above modern sea level. The beaches formed when glaciers had depressed the Scandinavian crust. They now lie well above sea level because the land rose as the ice melted. The concept that the lithosphere is in floating equilibrium on the asthenosphere is called **isostasy**, and the vertical movement in response to a changing burden is called *isostatic adjustment* (Figure 6.16).

The iceberg pictured in Figure 6.17 illustrates an additional effect of isostasy. A large iceberg has a high peak, but its base extends deeply below the water surface. The lithosphere behaves in a similar manner. Continents rise high above sea level, and the lithosphere beneath a continent has a "root" that extends as much as 125 kilometers into the asthenosphere.

In contrast, most ocean crust lies approximately 5 kilometers below sea level, and oceanic lithosphere extends only about 75 kilometers into the asthenosphere. For similar reasons, high mountain ranges have deeper roots than low plains, just as the bottom of a large iceberg is deeper than the base of a small one.

6.9 How Plate Movements Affect Earth Systems

The movements of tectonic plates generate volcanic eruptions and earthquakes, which help shape Earth's surface. They also build mountain ranges and change the global distributions of continents and oceans. However, tectonic activities strongly affect our environment in other ways—impacting global and regional climate, the atmosphere, the hydrosphere, and the biosphere.

Unit 2: Internal Processes

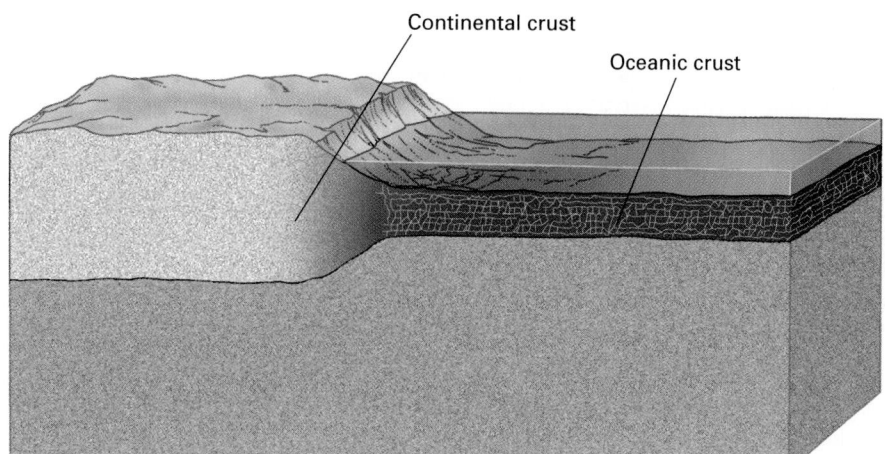

Continental crust

Oceanic crust

FIGURE 6.16 Isostatic adjustment. The weight of an ice sheet causes continental crust to sink in response to the added burden.

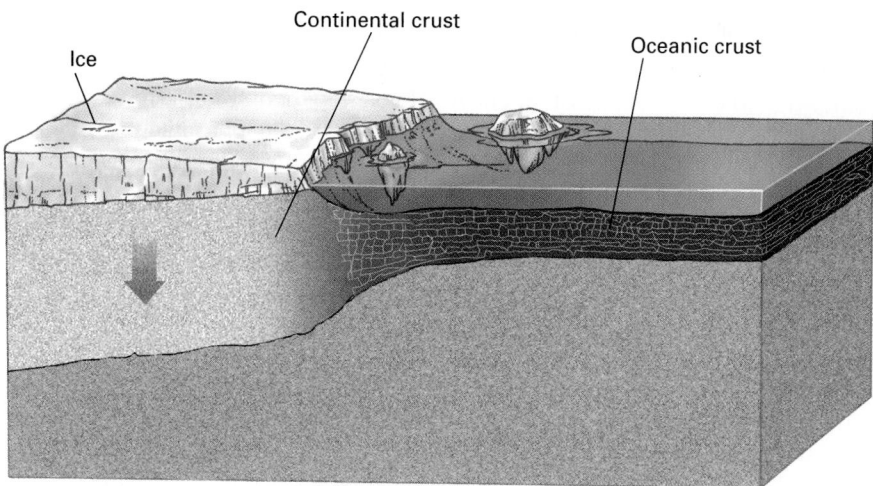

Ice

Continental crust

Oceanic crust

Volcanoes

At a divergent boundary (spreading center), hot asthenosphere oozes upward to fill the gap left between the two separating plates. Portions of the rising asthenosphere melt to form basaltic magma, which erupts onto Earth's surface. The Mid-Oceanic Ridge is a chain of submarine volcanoes and lava flows formed at divergent plate boundaries. Volcanoes are also common in continental rifts, such as the East African Rift.

Huge quantities of magma also form in the descending crust of a subduction zone and rise through the overlying lithosphere. Some solidifies within the crust, and some erupts from volcanoes on Earth's surface. Volcanoes of this type are common in the Cascade Range of Oregon, Washington, and British Columbia (Figure 6.18). They are also found in western South America, Japan, the Philippines, and near most other subduction zones.

Earthquakes

Earthquakes are common at all three types of plate boundaries but uncommon within the interior of a tec-

tonic plate. Quakes concentrate at plate boundaries simply because those boundaries are zones where one plate slips past another. The slippage is rarely smooth and continuous.

Instead, the fractures may be motionless for months to hundreds of years. Then, one plate suddenly slips a few centimeters, or even a few meters, past its neighbor. An earthquake is a vibration in rock caused by these abrupt movements. There will be more on earthquakes in the next chapter.

Mountain Building

Many of the world's great mountain chains, including the Andes and parts of the mountains of western North America, formed at subduction zones. Several processes combine to build a mountain chain at a subduction zone. The great volume of magma rising into the crust adds volume to the crust. Additional crustal thickening may occur where two plates converge for the same reason that a mound of bread dough thickens when you compress it from both sides. Finally, volcanic eruptions build chains of volcanoes. This thick crust then floats

A

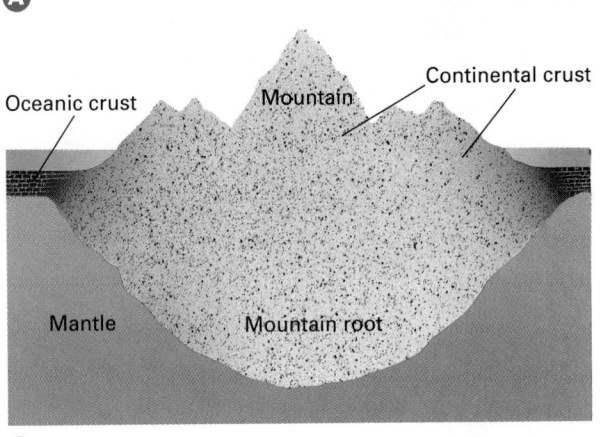

Oceanic crust / Mountain / Continental crust

Mantle / Mountain root

B

FIGURE 6.17 (A) Icebergs illustrate some of the effects of isostasy. A large iceberg has a deep root and also a high peak. (B) In an analogous manner, continental crust extends more deeply into the mantle beneath high mountains than it does under the plains. Oceanic lithosphere is thinner and denser and consequently floats at the lower level.

upward on the soft, plastic asthenosphere to form a mountain chain.

In some cases, two continents collide at a convergent plate boundary. The collision thrusts great masses of rock upward, creating huge mountain chains such as the Himalayas, the Alps, and the Appalachians.

Great chains of volcanic mountains form at rift zones because the new, hot lithosphere floats to a high level, and large amounts of magma form in these zones. The Mid-Oceanic Ridge and the volcanoes of the East African Rift and the Rio Grande Rift of the southwestern United States are examples of such mountain chains.

FIGURE 6.18 Mount St. Helens, which last erupted in 1980, is an active volcano in the Cascade Range of Washington near a convergent plate boundary.

Unit 2: Internal Processes

Migrating Continents and Oceans

Continents and ocean basins migrate over Earth's surface because they are parts of the moving lithospheric plates: they simply ride piggyback on the plates. North America is now moving away from Europe at about 2.5 centimeters per year, as the Mid-Atlantic Ridge continues to separate. Thus, during the next 10 years you will ride about 25 centimeters, about one foot, on the back of the North American plate.

In the time span of a human life, plate motion is slow. However, in the 65 million years since the extinction of the dinosaurs, North America has migrated about 1,620 kilometers, about the distance from New York City to Minneapolis. As the Atlantic Ocean widens, the Pacific is shrinking. Thus, as continents move, ocean basins open and close over geologic time.

Ocean currents carry warm water from the equator toward the poles, and cool water from polar regions toward the equator, warming polar regions and cooling the tropics. Similarly, winds transport heat and moisture over the globe. Movements of continents and ocean basins alter both ocean currents and wind patterns, and may convert a desert to a rainforest or a glacier-covered region to grassland. Streams and drainage patterns must respond to the altered rainfall distribution. Lakes may dry up, or new lakes form. Plants and animals may die or migrate away to be replaced by new species that are adapted to the new climatic conditions.

7

EARTHQUAKES AND THE EARTH'S STRUCTURE

Earthquakes occur every day. Sophisticated equipment keeps track of earthquakes as they occur. Drum-style seismographs in this exhibit at the Smithsonian Museum of Natural History show current seismic activity in Alaska, Arizona, and Japan.

visit 4ltrpress.cengage.com

As we learned in Chapter 6, the Earth's tectonic plates glide slowly over the soft asthenosphere, about as fast as your fingernail grows. At plate boundaries, friction often holds the plates locked and stationary. Rocks stretch and compress, forces build—but for decades, or even a century or more, nothing happens. Then suddenly, the plates snap free and Earth shakes. Thus, an earthquake is a classic example of a threshold effect. The interiors of plates move at a steady rate, but often motion is locked at plate boundaries. Then, when the threshold is reached, sudden and cataclysmic motion—an earthquake—occurs.

While plate tectonic motion and the resulting earthquakes may lead to death and destruction, we must appreciate that Earth's tectonic activity profoundly affects Earth systems and may cause the planet to be habitable. An earthquake, no matter how tragic, is one component of a complex system and a planet that sustains us.

The movement of tectonic plates is responsible for the following:

- Forming the oceans and the atmosphere when gases released during volcanic eruptions (a manifestation of plate movement) brought water to the surface of an otherwise dry, inhospitable Earth

- Creating the continents

- Rejuvenating soils

- Regulating global chemistry

- Concentrating metals

7.1 Anatomy of an Earthquake

Rock appears rigid, but if you apply enough stress, rock will deform. When stress is applied to a rock, the rock can deform in one of three ways: (1) elastically, (2) by fracturing, or (3) plastically.

Under small amounts of stress, the rock deforms elastically. If the stress is removed, the rock returns to its original size and shape. A rubber band deforms elastically when you stretch it. The energy used to stretch the rubber band is stored in the elongated rubber. When the stress is removed, the rubber band springs back and releases the stored energy. In the same way, an elastically deformed rock will spring back to its original shape and release its stored elastic energy when the force is removed.

However, like a rubber band, every rock has a limit beyond which it cannot deform elastically. Under certain conditions, an elastically deformed rock may suddenly fracture (Figure 7.1). When

FIGURE 7.1 Stressed rock fractured and displaced this roadway during the Loma Prieta earthquake near San Francisco in 1989.

plastic deformation Deformation that occurs without fracture after a rock's elastic limit is reached, when it continues to deform, like putty, while still solid. The rock keeps its new shape and does not store the energy used to deform it; thus, earthquakes do not occur when rocks deform plastically.

earthquake A sudden motion or trembling of Earth caused by the abrupt release of slowly accumulated elastic energy in rocks.

large masses of rock in Earth's crust deform and then fracture, the resultant rapid motion creates vibrations that travel through Earth and are felt as an earthquake.

Under other conditions, when its elastic limit is exceeded, a rock continues to deform, like putty, while still solid. This behavior is called **plastic deformation**. A rock that has deformed plastically keeps its new shape when the stress is released and, consequently, does not store the energy used to deform it. Therefore, earthquakes do not occur when rocks deform plastically. Figure 7.2 shows a rock outcrop that was deformed plastically by forces deep within Earth's crust. Parallel rock layers were bent and twisted into an S-shaped fold.

We learned in Chapter 6 that Earth's lithosphere is broken into seven large and several smaller tectonic plates. Although tectonic plates move at rates between 1 and 16 centimeters per year, friction prevents the plates from slipping past one another continuously. For decades, or even a century or two, the cool, strong rock near the plate boundary can stretch or

compress elastically, while the edges remain locked and immobile. Potential energy builds, just like the potential energy in a stretched rubber band or a compressed spring. Then, when the strain reaches a critical value, rock snaps loose or fractures. The ground rises and falls and undulates back and forth. Buildings topple, bridges fall, roadways and pipelines snap. An **earthquake** is a sudden motion or trembling of Earth caused by the abrupt release of energy that is stored in rocks.

Most movement of crustal rock occurs by slippage along established faults because the friction binding the two sides of the fault is weaker than the rock itself (Figure 7.3). But occasionally, solid rock may fracture, creating a new fault. After the earthquake dies away, the stored-up potential energy in the rock is released, but the fault remains as a weakness in the rock. When more tectonic stress builds, the rock is more likely to move along the fault than to crack again to create a new fault. Thus earthquakes occur repeatedly along established faults. For example, the San Andreas Fault in Southern California lies along a tectonic plate boundary that has moved many times in the past and will certainly move again in the future (Figure 7.4).

During an earthquake, rock moves from a few centimeters to a meter or two. Thus a single earthquake makes only small changes in the topography or geography of a region. But over tens of millions of years, the relentless motion of plates, accompanied by hundreds of thousands of earthquakes, changes the surface of the planet.

COURTESY OF GRAHAM R. THOMPSON/JONATHAN TURK

FIGURE 7.2 This rock deformed plastically when stressed.

Unit 2: Internal Processes

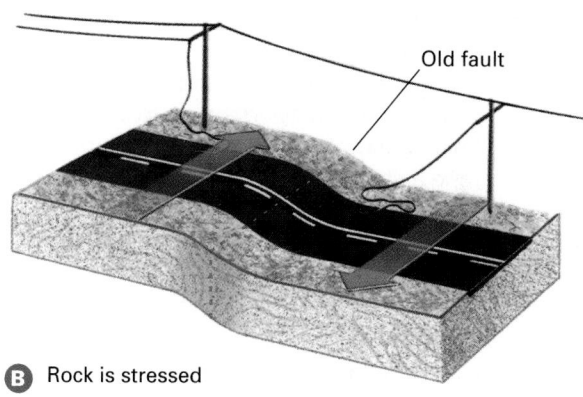

A Original position

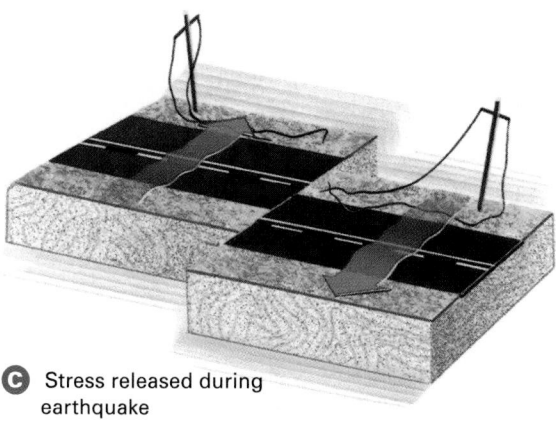

B Rock is stressed

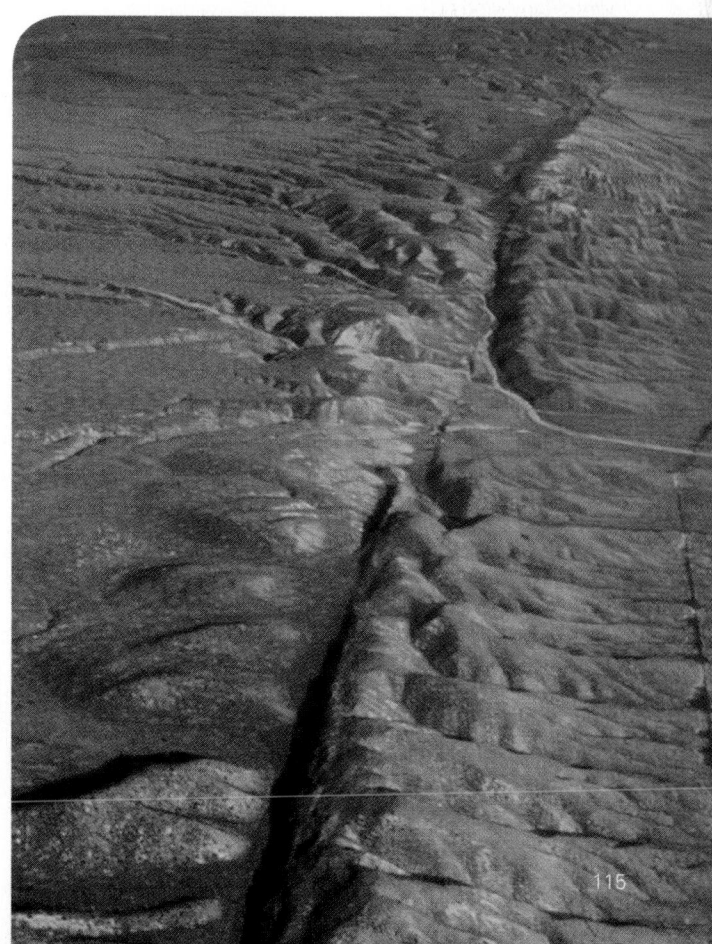

C Stress released during earthquake

FIGURE 7.3 (A) A road is built across an old fault. (B) The rock stores elastic energy when it is stressed by a tectonic force. (C) At a critical point, the rock fractures, the strain is relieved, returning the rock to its unstressed shape on either side of the fault.

7.2 Earthquake Waves

When buying a watermelon, one trick is to tap the melon gently with your knuckle. If you hear a sharp, clean sound, it is probably ripe; a dull thud indicates that it may be overripe and mushy. The watermelon illustrates two points that can be applied to Earth: (1) the energy of your tap travels through the melon, and (2) the nature of the melon's interior affects the quality of the sound.

A *wave* transmits energy from one place to another. A drumbeat travels through air as a sequence of waves; the Sun's heat travels to Earth as waves; and a tap travels through a watermelon in waves. Waves that travel through rock are called **seismic waves**. Earthquakes and explosions produce seismic waves. **Seismology** is the study of earthquakes and the nature of Earth's interior based on evidence from seismic waves.

The initial rupture point, where abrupt movement creates an earthquake, typically lies below the surface at a point called the **focus**. The point on Earth's surface

seismic wave An elastic wave that travels through rock, produced by an earthquake or explosion.

seismology The study of earthquakes and the nature of Earth's interior based on evidence from seismic waves.

focus The initial rupture point of an earthquake, typically lying below Earth's surface.

FIGURE 7.4 California's San Andreas Fault, the source of many earthquakes, is the boundary between the Pacific plate, on the left in this photo, and the North American plate on the right.

R. E. WALLACE/USGS

epicenter The point on Earth's surface directly above the initial rupture point (focus) of an earthquake.

body waves Seismic waves that travel through the interior of Earth, carrying energy from the earthquake's focus to the surface.

surface waves Seismic waves that radiate from the earthquake's epicenter and travel along the surface of Earth or along a boundary between layers within Earth.

P waves Body waves that travel faster than other seismic waves and are the first or "primary" waves to reach an observer; formed by alternate compression and expansion of rock, shaking the ground with a back-and-forth movement parallel to the direction of the wave.

S waves Seismic waves that travel slower than P waves and are the "secondary" waves to reach an observer; sometimes called *shear waves* because of the shearing motion in which they shake the ground, perpendicular to the direction of wave travel.

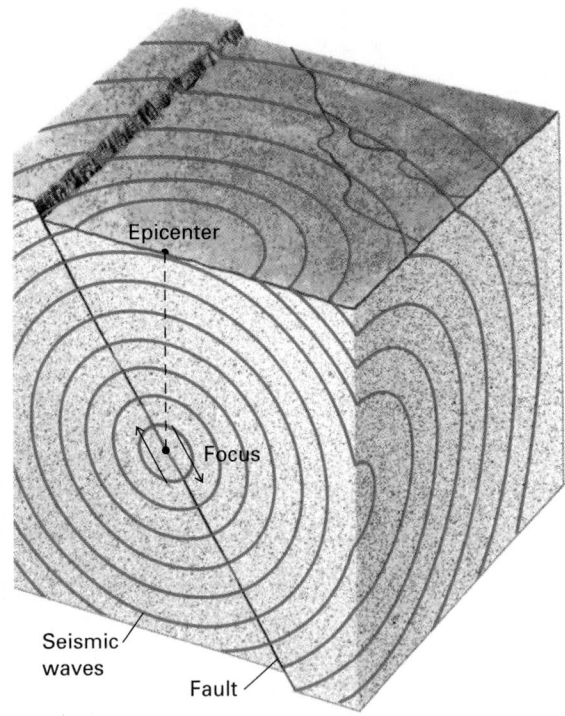

FIGURE 7.5 Body waves, generated by abrupt movement on a fault, radiate outward from the focus of an earthquake.

directly above the focus is the **epicenter**. An earthquake produces two main types of seismic waves. **Body waves** travel through Earth's interior and carry some of the energy from the focus to the surface (Figure 7.5). **Surface waves** then radiate from the epicenter and travel along Earth's surface.

Body Waves

Body waves can be two main types. **P waves** travel fast and are the first or "primary" seismic waves to reach an observer. They are compressional elastic waves that cause alternate compression and expansion of the rock (Figure 7.6). Consider a long spring such as the popular Slinky toy. If you stretch a Slinky and strike one end, a compressional wave travels back and forth along its length. P waves travel through air, liquid, and solid material. Next time you take a bath, immerse your head until your

ears are under water and listen as you tap the sides of the tub with your knuckles. You are hearing P waves.

P waves travel at speeds between 4 and 7 kilometers per second in Earth's crust and at about 8 kilometers per second in the uppermost mantle. For comparison, the speed of sound in air is only 0.34 kilometer per second, and the fastest jet fighters fly at about 0.85 kilometer per second.

S waves, the other type of body waves, are slower than P waves and thus are the "secondary" waves to reach an observer. Also called *shear waves*, S waves have a shearing motion that can be illustrated by tying a rope to a wall, holding the opposite end, and giving the rope a sharp up-and-down jerk (Figure 7.7). Although the wave travels parallel to the rope, the individual particles in the rope move at right angles to the rope length. A similar motion in an S wave produces shear stress in a rock—stress that acts in parallel but opposite directions—and gives the wave its name. S waves shake the ground at right angles to the direction in which the wave is moving. S waves travel

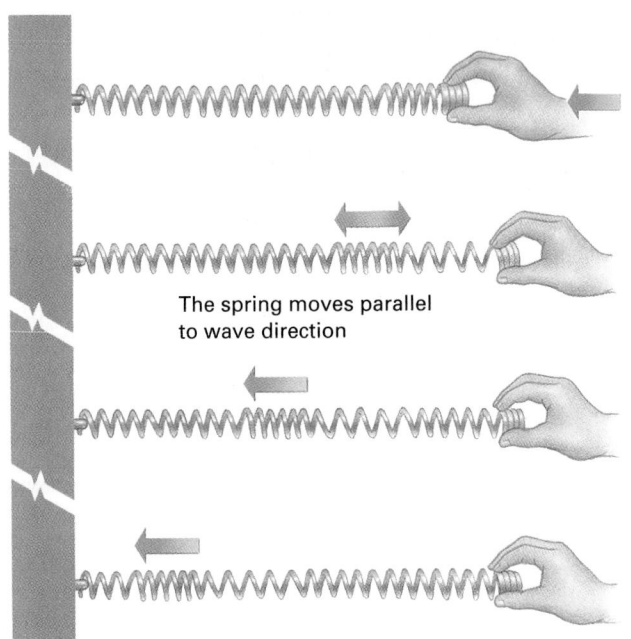

The spring moves parallel to wave direction

FIGURE 7.6 Model of a P wave (compressional wave). The spring is moving back and forth, parallel to the direction of the wave. The arrow marks the forward movement of the compressed region. In an earthquake, the P waves shake the ground parallel to the direction of the wave.

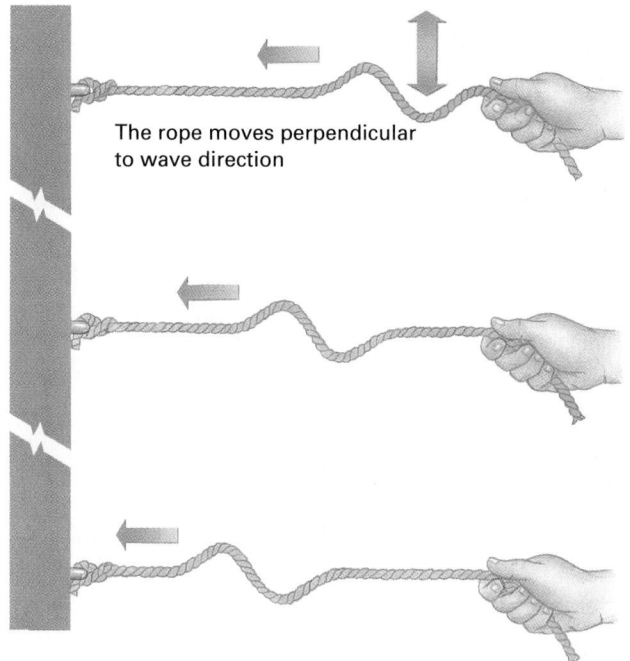

The rope moves perpendicular to wave direction

FIGURE 7.7 Model of an S wave (shear wave). The rope is moving perpendicular to the direction of wave propagation. In an earthquake, S waves shake the ground at right angles to the direction that the wave is moving.

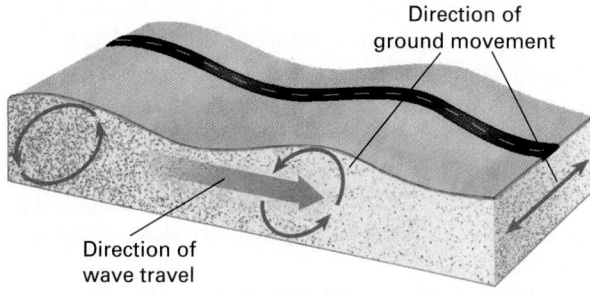

Direction of ground movement

Direction of wave travel

FIGURE 7.8 During an earthquake, surface waves move Earth's surface up and down and from side to side.

at speeds between 3 and 4 kilometers per second in the crust.

Unlike P waves, S waves move only through solids. Because molecules in liquids and gases are only weakly bound to one another, they slip past each other and thus cannot transmit a shear wave.

Surface Waves

Surface waves travel more slowly than body waves. Surface waves undulate across the ground like the waves that ripple across the water when you throw a rock into a calm lake (although the actual wave mechanism is different). Two types of surface waves occur simultaneously: an up-and-down rolling motion and a side-to-side vibration. During an earthquake, Earth's surface rolls like ocean waves and writhes from side to side like a snake (Figure 7.8). Surface waves are the principal source of movement—and damage—on the surface.

Measurement of Seismic Waves

A **seismograph** is a device that records seismic waves. In early seismographs, a heavy weight was suspended from a spring. A pen attached to the weight was aimed at the zero mark on a piece of graph paper (Figure 7.9).

seismograph An instrument that records seismic waves.

FIGURE 7.9 A seismograph records ground motion during an earthquake. In early versions of seismography, the pen draws a straight line across the rotating drum when the ground is stationary. When the ground rises abruptly during an earthquake, it carries the drum up with it. But the spring stretches, so the weight and pen hardly move. Therefore the pen marks a line lower on the drum. Conversely, when the ground sinks, the pen marks a line higher on the drum. During an earthquake the pen traces a jagged line as the drum rises and falls.

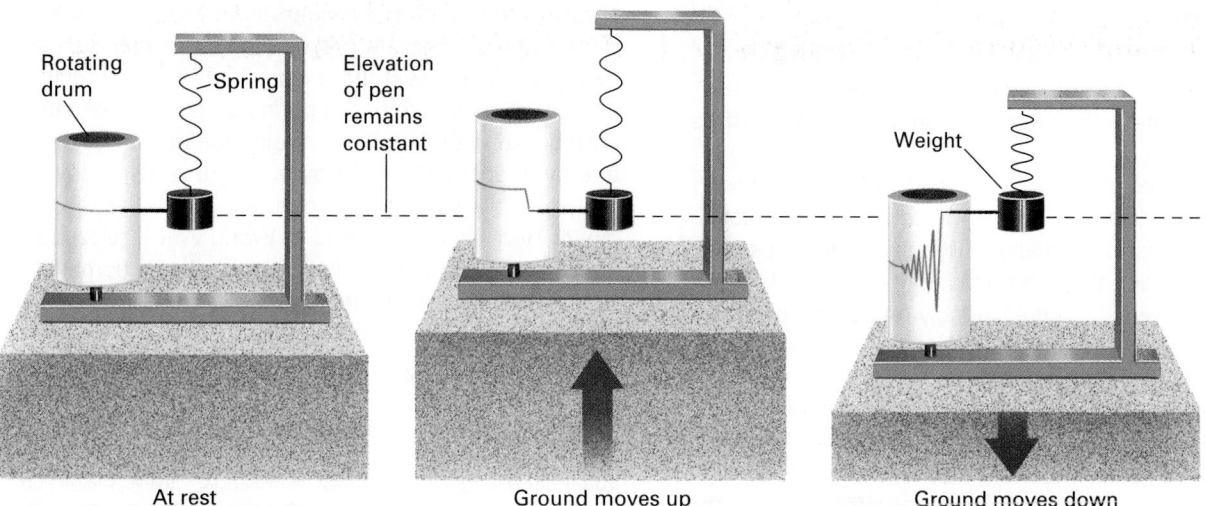

Rotating drum | Spring | Elevation of pen remains constant | Weight

At rest   Ground moves up   Ground moves down

seismogram The record of Earth vibration made by a seismograph.

Mercalli scale A scale of earthquake intensity that expresses the strength of an earthquake based on its destructive power and its effects on buildings and people, without accurately measuring the energy released by the quake.

Richter scale A scale of earthquake magnitude that expresses the amount of energy released; calculated from the amplitude of the largest body wave on a standardized seismograph, although not a precise measure of earthquake energy.

moment magnitude An earthquake scale that measures the amount of movement and the surface area of a fault during a quake, closely reflecting the total amount of energy released.

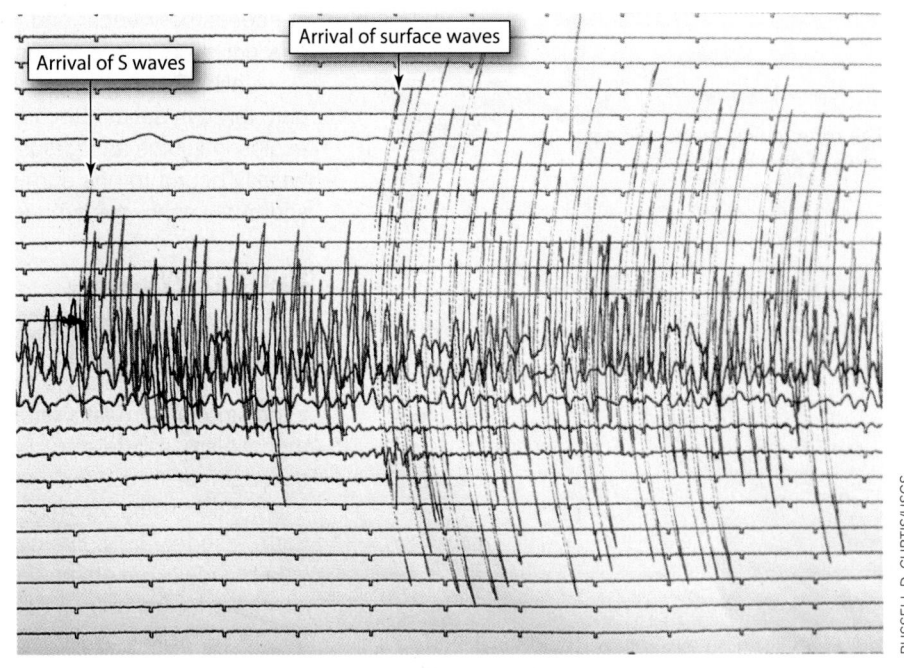

Arrival of S waves

Arrival of surface waves

RUSSELL D. CURTIS/USGS

FIGURE 7.10 This seismogram recorded north–south ground movements during the October 1989 Loma Prieta earthquake.

The graph paper was mounted on a rotating drum that was attached firmly to bedrock. During an earthquake, the graph paper jiggled up and down, but inertia kept the weight and its pen stationary. As a result, the paper moved up and down beneath the pen. The rotating drum recorded earthquake motion over time. This record of Earth vibration is called a **seismogram** (Figure 7.10). Modern seismographs use electronic motion detectors, which transmit the signal to a computer.

Measurement of Earthquake Strength

Over the past century, geologists have devised several scales to express the strength of an earthquake. Before seismographs became widespread, geologists evaluated earthquakes on the **Mercalli scale**, which was based on structural damage. On the Mercalli scale, an earthquake that destroyed many buildings was rated as more intense than one that destroyed only a few.

This system did not accurately measure the energy released by a quake, however, because structural damage also depends on distance from the focus, the rock or soil beneath the structure, and the quality of construction. In 1935 Charles Richter devised the **Richter scale** to express the amount of energy released during an earthquake. Richter magnitude is calculated from the

height of the largest earthquake body wave recorded on a specific type of seismograph. The Richter scale is more quantitative than earlier intensity scales, but it is not a precise measure of earthquake energy. A sharp, quick jolt would register as a high peak on a Richter seismograph, but a very large earthquake can shake the ground for a long time without generating extremely high peaks. Thus, a great earthquake can release a huge amount of energy that is not reflected in the height of a single peak and that is not adequately expressed by Richter magnitude.

Modern equipment and methods enable seismologists to measure the amount of movement and the surface area of a fault that moved during a quake. The product of these two values allows them to calculate the **moment magnitude**. Most seismologists now use moment magnitude rather than Richter magnitude because it more closely reflects the total amount of energy released during an earthquake. An earthquake with a moment magnitude of 6.5 has an energy of about 10^{25} (1 followed by 25 zeros) ergs.[1] The atomic bomb that was dropped on the Japanese city of Hiroshima at the end of World War II released about that much energy. (Of course, atomic bombs also release intense heat and deadly radioactivity, while earthquakes only shake the ground.)

---

1. An erg is a standard unit of energy in scientific usage. One erg is a small amount of energy; approximately 3×10^{12} ergs are needed to light a 100-watt lightbulb for 1 hour. However, 10^{25} is a very large number, and 10^{25} ergs represents a considerable amount of energy.

On both the moment magnitude and the Richter scales, the energy of the quake increases by a factor of about 30 for each increment on the scale. Thus a magnitude-6 earthquake releases roughly 30 times more energy than a magnitude-5 earthquake.

The largest possible earthquake is determined by the strength of rocks. A strong rock can store more elastic energy before it fractures than a weak rock can. The largest earthquakes ever measured had moment magnitudes of 8.5 to 8.7, about 900 times greater than the energy released by the Hiroshima bomb.

Locating the Source of an Earthquake

If you have ever watched an electrical storm, you may have used a simple technique for estimating the distance between you and the place where the lightning strikes. After the flash of a lightning bolt, count the seconds that pass before you hear the thunder. Although the electrical discharge produces thunder and lightning simultaneously, light travels much faster than sound. Light reaches you virtually instantaneously, whereas sound travels much more slowly, at 340 meters per second. If the time interval between the lightning flash and the thunder is 1 second, then the lightning struck 340 meters away and was very close.

The same principle is used to determine the distance from a recording station to both the epicenter and the focus of an earthquake. Recall that P waves travel faster than S waves and that surface waves are slower yet. If a seismograph happens to be close to an earthquake epicenter, the different waves will arrive in rapid succession for the same reason that the thunder and

lightning come close together when a storm is close. On the other hand, if a seismograph is located far from the epicenter, the S waves arrive at correspondingly later times after the P waves arrive, and the surface waves are even farther behind, as shown in Figure 7.11.

Geologists use a **time-travel curve** to calculate the distance between an earthquake epicenter and a seismograph. To plot a time-travel curve, a number of seismic stations at different locations record the time of arrival of seismic waves from an earthquake with a known epicenter and occurrence time. Using these data, a graph such as

time-travel curve
A graph that records arrival times of P and S earthquake waves, used to measure the distance from a recording station to an earthquake epicenter.

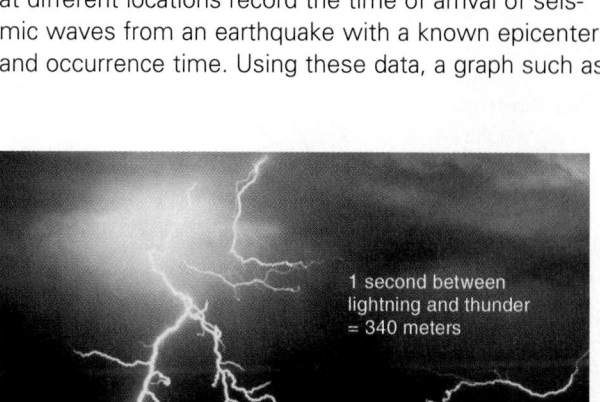

1 second between lightning and thunder = 340 meters

FIGURE 7.11 The time intervals between arrival of P, S, and surface waves at a recording station increase with distance from the focus of an earthquake.

Time of earthquake

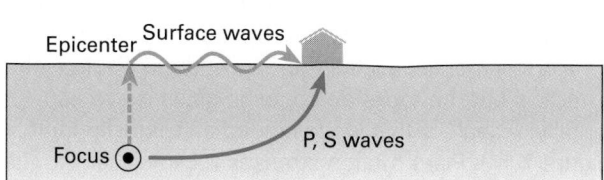

A Recording station A near focus

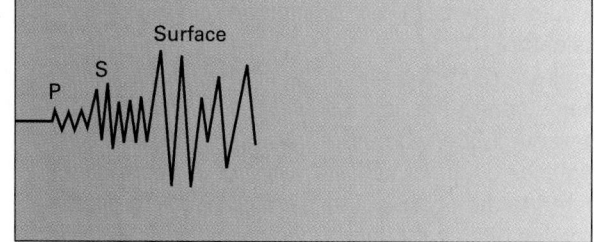

Seismogram from station A

Time

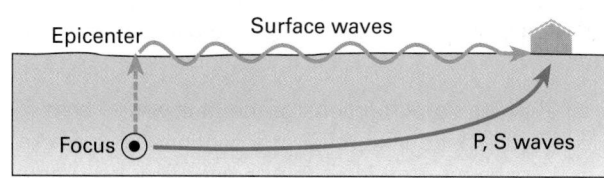

B Recording station B far from focus

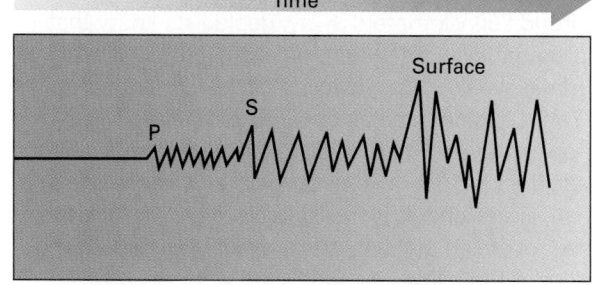

Seismogram from station B

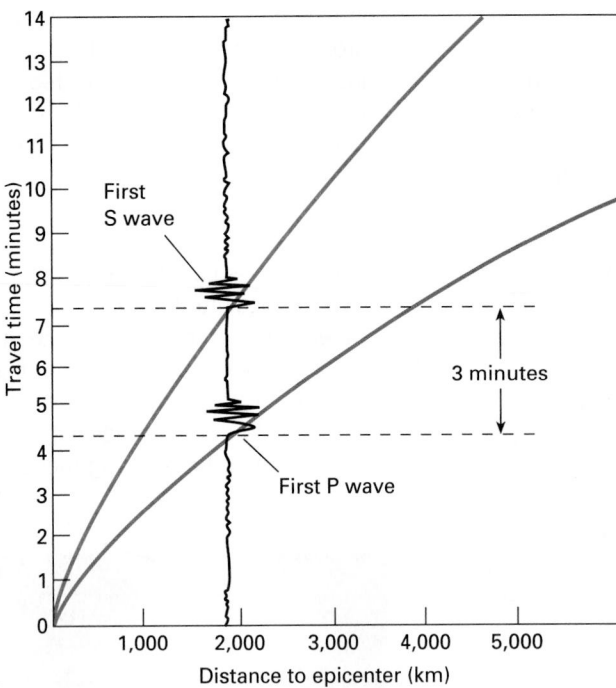

FIGURE 7.12 A time-travel curve. With this graph you can calculate the distance from a seismic station to the epicenter of an earthquake. In the example shown, a 3-minute delay between the first arrivals of P waves and S waves corresponds to an earthquake with an epicenter 1,900 kilometers from the seismic station.

San Andreas Fault zone A zone of strike-slip faults, extending from San Francisco to San Diego, which forms the transform boundary between the Pacific plate and the North American plate. Frequent earthquakes occur along this fault zone.

strike-slip fault A fault whose surface is vertical but on which the rocks on opposite sides move horizontally.

fault creep A continuous, slow movement of solid rock along a fault, resulting from a constant stress acting over a long time. Creeping faults do not usually have large earthquakes.

Figure 7.12 is drawn. This graph can then be used to measure the distance between a recording station and an earthquake whose epicenter is unknown.

Time-travel curves were first constructed using data obtained from natural earthquakes. However, scientists do not always know precisely when and where an earthquake occurs. In the 1950s and 1960s, geologists studied seismic waves from atomic bomb tests to improve the time-travel curves because they could verify both the location and timing of the explosions.

Figure 7.12 shows that if the first P wave arrives 3 minutes before the first S wave, the recording station is about 1,900 kilometers from the epicenter. But this distance does not indicate whether the earthquake originated to the north, south, east, or west. To pinpoint the location of an earthquake, geologists compare data from three or more recording stations. If a seismic station in New York City records an earthquake with an epicenter 6,750 kilometers away, geologists know that the epicenter lies somewhere on a circle 6,750 kilometers from New York City. (Draw a circle with center at New York City and a radius of 6,750 kilometers.) The same epicenter might be 2,750 kilometers from a seismic station in London and 1,700 kilometers from one in Godthab, Greenland. If a similar circle is drawn for each of these recording stations, the arcs will intersect at the epicenter of the quake.

7.3 Earthquakes and Tectonic Plate Boundaries

Before the plate tectonics theory was developed, geologists recognized that earthquakes occur frequently in some regions and infrequently in others, but they did not understand why. Modern geologists know that most earthquakes occur along plate boundaries, where tectonic plates diverge, converge, or slip past one another (Figure 7.13).

Earthquakes at a Transform Plate Boundary: The San Andreas Fault Zone

The populous region from San Francisco down to San Diego straddles the **San Andreas Fault zone**, which is a transform boundary between the Pacific plate and the North American plate. The San Andreas Fault itself is vertical but the rocks on opposite sides move horizontally. A fault of this type is called a **strike-slip fault** (Figure 7.14). Plate motion stresses rock adjacent to the fault, generating numerous smaller faults (Figure 7.15). Thus, the San Andreas Fault zone comprises the San Andreas Fault and its associated faults. In the past few centuries, hundreds of thousands of earthquakes have occurred in this zone (Figure 7.16).

The plates move past one another in three different ways along different segments of the San Andreas Fault zone:

1. Along some portions of the fault zone, rocks slip past one another at a continuous, snail-like pace referred to as **fault creep**. The movement

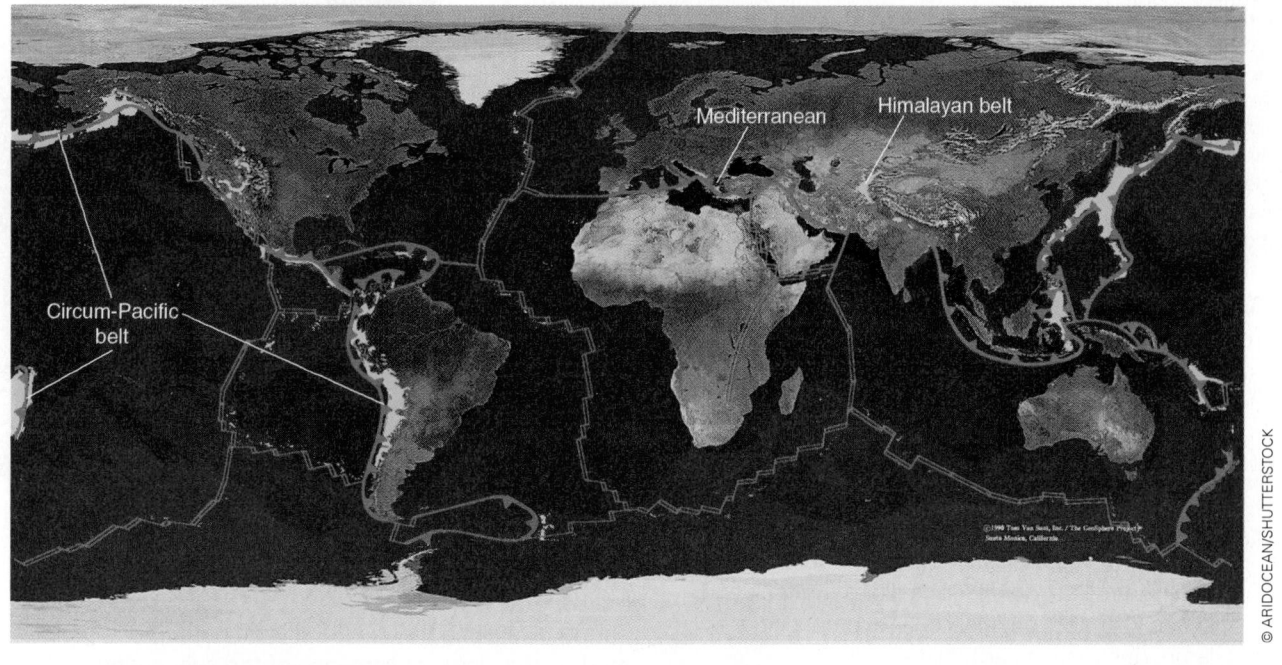

⊏⊐ Divergent boundary   ▲▲ Convergent boundary   —— Transform boundary

FIGURE 7.13 Earth's major earthquake zones coincide with tectonic plate boundaries. Each yellow dot represents a moderate or large earthquake that occurred between 1985 and 1995.

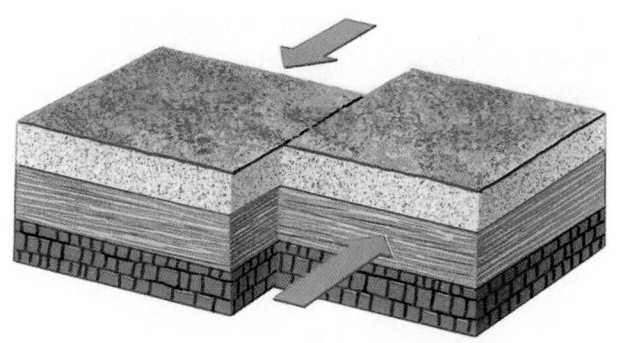

FIGURE 7.14 A strike-slip fault is vertical but the rock on opposite sides of the fracture moves horizontally.

FIGURE 7.15 An earthquake hazard map of the San Andreas Fault zone of California. (Percentages indicate the probability of an earthquake with a magnitude greater than 5 in the next 30 years.)

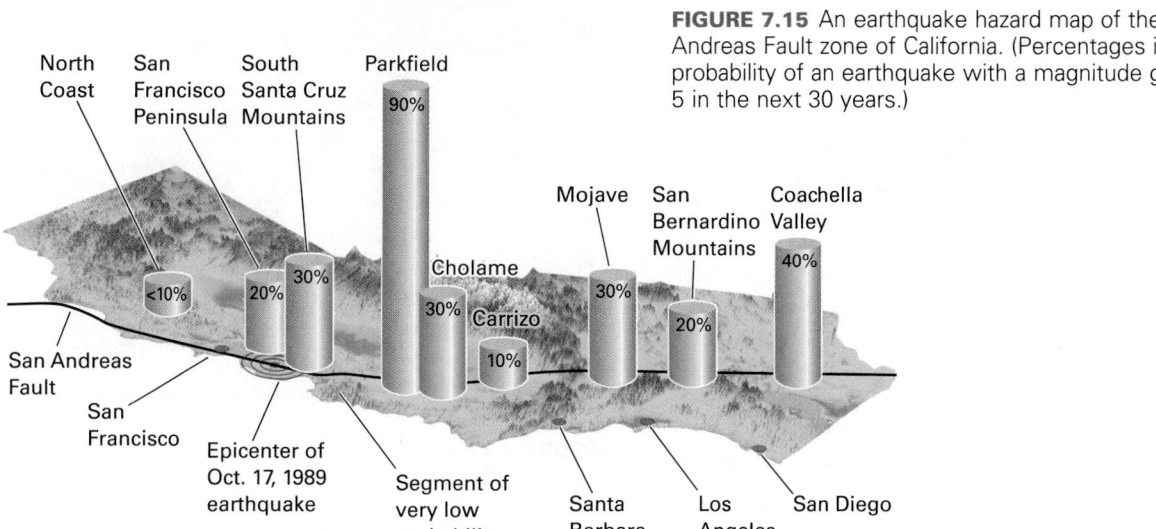

occurs without violent and destructive earthquakes because the rocks move continuously and slowly.

2. In other segments of the fault, the plates pass one another in a series of small hops, causing numerous small, nondamaging earthquakes.

3. Along the remaining portions of the fault, friction prevents slippage of the fault, although the plates continue to move past one another. In this case, rock near the fault deforms and stores elastic energy. Because the plates move past one another at an average of 3.5 centimeters per year, 3.5 meters of elastic deformation accumulate over a period of 100 years. When the accumulated elastic energy exceeds friction, the rock suddenly slips along the fault and snaps back to its original shape, producing a large, destructive earthquake.

Plates move slowly and silently every day. Rocks stretch, bend, and compress, unseen beneath our feet. Geologists understand the mechanisms of earthquakes and know for certain that more destructive quakes will occur along the San Andreas Fault. But no one knows when or where the "Big One" will strike.

Earthquakes at Convergent Plate Boundaries

Recall from Chapter 6 that in a subduction zone, a relatively cold, rigid lithospheric plate dives beneath another plate and slowly sinks into the mantle. In most places, the subducting plate slips past the plate above it with intermittent jerks, giving rise to numerous earthquakes. The earthquakes concentrate along the upper part of the sinking plate, where it scrapes past the opposing plate (Figure 7.17). This earthquake zone is called the **Benioff zone**, after Hugo Benioff, the geologist who first recognized it. Many of the world's strongest earthquakes occur in subduction zones.

The small Juan de Fuca Plate, which lies off the coasts of Oregon, Washington, and southern British Columbia, is diving beneath North America at a rate of 3 to 4 centimeters per year. Thus, the region should experience subduction zone earthquakes. Yet although small earthquakes occasionally shake the Pacific Northwest, no large ones have occurred in the past 150 to 200 years.

When two converging plates both carry continents, neither can sink deeply into the mantle. Instead the crust buckles up to form high mountains, such as the Himalayas. Rocks fracture or slip during this buckling process, generating frequent earthquakes. The 2005

USGS

FIGURE 7.16 The 1906 earthquake and fire destroyed most of San Francisco.

earthquake along the India–Pakistan border was generated by the slow convergence of the subcontinent of India with Asia.

Earthquakes at Divergent Plate Boundaries

Earthquakes frequently shake the Mid-Oceanic Ridge system as a result of faults that form as the two plates separate. Blocks of oceanic crust drop downward along most mid-oceanic ridges, forming a rift valley in the center of the ridge. Only shallow earthquakes occur along the Mid-Oceanic Ridge because here the asthenosphere rises to levels as shallow as 10 to 15 kilometers below Earth's surface and is too hot and plastic to fracture.

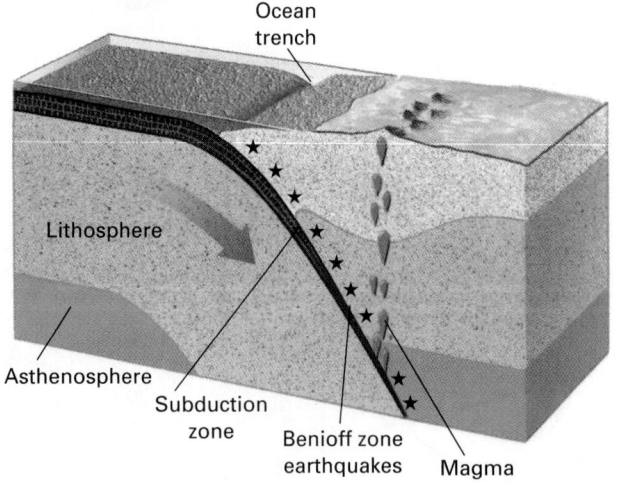

FIGURE 7.17 A descending lithospheric plate generates magma and earthquakes in a subduction zone. Earthquake foci, shown by stars, concentrate along the upper portion of the subducting plate, called the Benioff zone.

Earthquakes in Plate Interiors

No major earthquakes have occurred in the central or eastern United States in the past 100 years, and no current lithospheric plate boundaries are known in these regions. However, the largest historical earthquake sequence in the contiguous 48 states occurred near New Madrid, Missouri. In 1811 and 1812, three shocks with estimated moment magnitudes between 7.3 and 7.8 altered the course of the Mississippi River and rang church bells 1,500 kilometers away in Washington, D.C.

Earthquakes in plate interiors are not as well understood as those at plate boundaries, but modern research is revealing some clues. Tectonic forces stretched North America in Precambrian time. As the continent pulled apart, rock fractured to create two huge fault zones that crisscross the continent like a giant X. Although the fault zones failed to develop into a divergent plate boundary, they remain weaknesses in the lithosphere. New Madrid lies at a major intersection of the faults—at the center of the X. As the North American plate glides over the asthenosphere, it may pass over irregularities, or "bumps," in that plastic zone, causing slippage and earthquakes along the deep faults. In addition to the quakes recorded at New Madrid in 1811 and 1812, major earthquakes also occurred there close to the years 1600, 1300, and 900.

In 2005, researchers reported in the journal *Nature* that the rocks adjacent to the New Madrid fault are currently deforming about as fast as rocks adjacent to active tectonic plate boundaries. This deformation indicates that additional earthquakes are likely in the future.[2]

7.4 Earthquake Prediction

Long-Term Prediction

Long-term earthquake prediction recognizes that earthquakes have recurred many times along existing faults and will probably occur in these same regions again. For example, in the United States, although some faults exist in plate interiors (as in the New Madrid earthquake zone), the most active faults lie along the West Coast at plate boundaries (Figure 7.18).

Long-term predictions give only a vague idea of when the next earthquake will strike.

© STEVEN PEPPLE/SHUTTERSTOCK

2. R. Smalley Jr., M. A. Ellis, J. Paul, and R. B. Van Arsdale, "Space Geodetic Evidence for Rapid Strain Rates in the New Madrid Seismic Zone of Central USA," *Nature* 435 (June 23, 2005), 1088–1090.

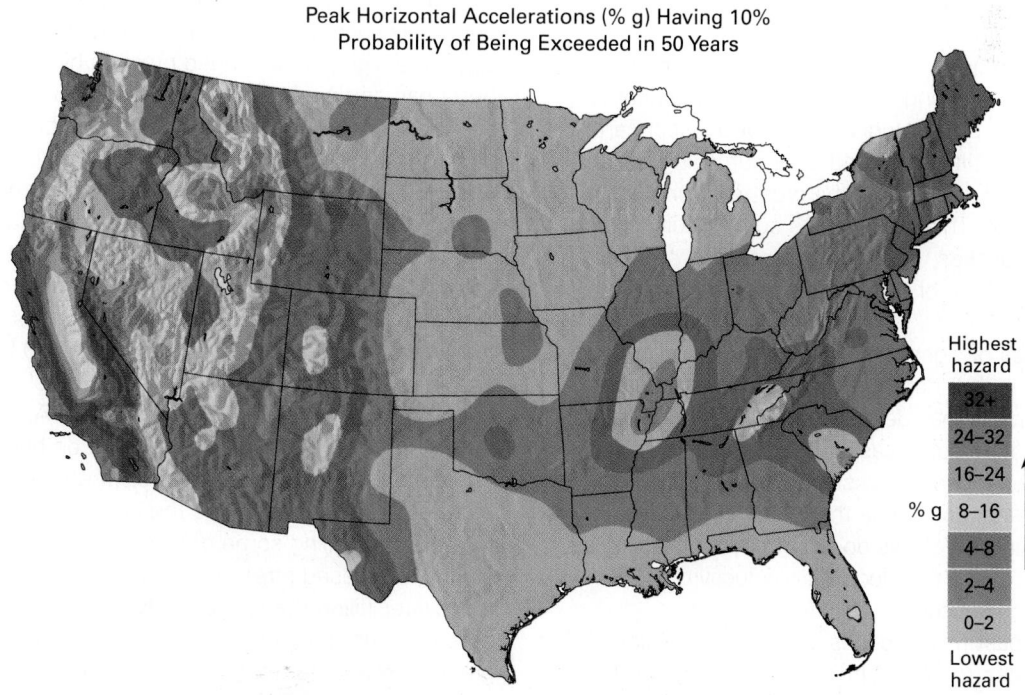

Peak Horizontal Accelerations (% g) Having 10% Probability of Being Exceeded in 50 Years

Highest hazard

%g	
32+	
24–32	
16–24	
8–16	
4–8	
2–4	
0–2	

Lowest hazard

USGS

FIGURE 7.18 This map shows earthquake hazards in the United States. The predictions are based on records of frequency and Mercalli magnitudes of historical earthquakes (% g equals percent of the acceleration of gravity).

Long-term prediction tells us *where* earthquakes are likely to occur. This information is useful because it allows engineers to establish strict building codes in earthquake-prone regions. However, long-term prediction gives only a vague idea of *when* the next earthquake will strike. Near New Madrid, earthquakes have occurred separated by 400, 300, and 200 years. The last one struck in 1812. Given an average 300-year interval, one might expect a quake in the year 2100, but such an analysis could be in error by hundreds of years.

Short-Term Prediction

Short-term predictions are forecasts that an earthquake may occur at a specific place and time. Short-term prediction depends on signals that immediately precede an earthquake.

Foreshocks are small earthquakes that precede a large quake by a few seconds to a few weeks. The cause of foreshocks can be explained by a simple analogy. If you try to break a stick by bending it slowly, you may hear a few small cracking sounds just before the final snap. Foreshocks are not a reliable tool for short-term prediction because they do not precede all earthquakes. Of those selected for study, foreshocks preceded only about half of the earthquakes. In addition, some swarms of small shocks thought to be foreshocks were not followed by a large quake.

Another approach to short-term earthquake prediction is to measure changes in the land surface near an active fault zone. Seismologists monitor unusual Earth movements with tiltmeters and laser surveying instruments because distortions of the crust may precede a major earthquake. This method has successfully predicted some earthquakes, but in other instances predicted quakes did not occur or quakes occurred that had not been predicted.

Other types of signals can be used in short-term prediction. When rock is deformed prior to an earthquake, microscopic cracks may develop as the rock approaches its rupture point. In some cases, the cracks release radon gas previously trapped in rocks and minerals. The cracks may fill with water and cause the water levels in wells to fluctuate. Air-filled cracks do not conduct electricity as well as solid rock, so the electrical conductivity of rock decreases as microscopic cracks form.

Over the past few decades, short-term prediction has not been reliable. Although some geologists continue to search for reliable indicators for short-term earthquake prediction, many other geologists believe that this goal will remain elusive.

7.5 Earthquake Damage and Hazard Mitigation

Violent ground motion may toss a person to the ground and break the person's arm, but this motion, by itself, is seldom lethal. Most earthquake fatalities and injuries occur when falling structures crush people. Structural damage, injury, and death depend on the magnitude of the quake, its proximity to population centers, rock and soil types, topography, and the quality of construction in the region.

How Rock and Soil Influence Earthquake Damage

In many regions, bedrock lies at or near Earth's surface and buildings are anchored directly to the rock. Bedrock vibrates during an earthquake and buildings may fail if the motion is violent enough. However, most bedrock returns to its original shape when the earthquake is over, so if structures can withstand the shaking, they will survive. Thus, bedrock forms a desirable foundation in earthquake-hazard areas.

In many places, structures are built on sand, silt, or clay. Sandy sediment and soil commonly settle during an earthquake. This displacement tilts buildings, breaks pipelines and roadways, and fractures dams. To avert structural failure in such soils, engineers drive steel or concrete pilings through the sand to the bedrock below. These pilings anchor and support the structures, even if the ground beneath them settles.

Mexico City provides one example of what can happen to clay-rich soils during an earthquake. The city is built on a high plateau ringed by even higher mountains. European settlers built the modern city on water-soaked, clay-rich, lake-bed sediment. On September 19, 1985, an earthquake with a moment magnitude of 8.1 struck about 500 kilometers west of the city. Seismic waves shook the wet clay beneath the city and reflected back and forth between the bedrock sides and bottom of the basin, just as waves in a bowl of Jell-O bounce off the side and bottom of the bowl. The reflections amplified the waves, which destroyed more than 500 buildings and killed between 8,000 and 10,000 people. Meanwhile, there was comparatively little damage in Acapulco, a city much closer to the epicenter but built on bedrock.

If soil is saturated with water, the sudden shock of an earthquake can cause the grains to shift closer together, expelling some of the water. When this occurs, increased stress is transferred to the pore water (the water filling the spaces between grains), and the pore pressure may rise sufficiently to suspend the grains in the water. In this case, the soil loses its shear strength—its resistance to being pulled apart—and behaves as a fluid. This process is called **liquefaction**. When soils liquefy on a hillside, the mixture of water and soil, called a slurry, flows downslope, carrying structures along with it. Landslides are common effects of earthquakes, and are discussed further in Chapter 10.

Construction Design and Earthquake Damage

Most earthquake mortality occurs when buildings fall and crush the unlucky inhabitants. Structure failure is caused by a variety of factors, including soil type and distance from the epicenter. But the tremendous differences between mortality in the poor, less-developed countries and the rich, developed countries is due to one simple factor: quality of construction, which can generate fires in addition to the increased likelihood of structural collapse.

Some common framing materials used in buildings, such as wood and steel, bend and sway during an earthquake, but they resist failure. However, brick, stone, adobe (dried mud), and other masonry products are brittle and likely to fail during an earthquake. Although masonry can be reinforced with steel, in many regions of the world people cannot afford such structural reinforcement.

Thus earthquake mortality is an example of the interface between natural systems and human ones. Forces deep within Earth, beyond human control, drive earthquake frequency and magnitude. But earthquake mortality is closely related to political and economic systems.

Tsunamis

When an earthquake occurs beneath the sea, part of the sea floor rises or falls. Water is displaced in response to the rock movement, forming a wave (Figure 7.19). Sea waves produced by an earthquake are often called tidal waves, but they have nothing to do with tides. Therefore, geologists call them by their Japanese name, **tsunami**.

In the open sea, a tsunami is so flat that it is barely detectable. Typically, the crest may be only 1 to 3 meters high, and successive crests may be more than 100 to 150 kilometers apart. However, a tsunami may travel at 750 kilometers per hour. When the wave approaches the shallow water near shore, the base of the wave drags against the bottom and the water stacks up, increasing the height of the wave. The rising wall of water then flows inland. A tsunami can flood the land for as long as 5 to 10 minutes.

Today the central part of the Indian–Australian Plate is subducting beneath the islands of Sumatra and Java. On December 26, 2004, approximately 1,200 kilometers of rock along the subducting sea floor slipped suddenly, with vertical movement up to 15 meters (Figure 7.20). The resulting earthquake of magnitude 9.0 to 9.3 lasted for almost 10 minutes and was the second-largest seis-

mic event ever recorded, second only to the 1960 Chilean quake. This tremendous displacement of rock initiated a massive tsunami that radiated in all directions, killing an estimated 283,000 people along the Indian Ocean coastlines. Survivors reported that, moments prior to the deadly wave, coastal water retreated, exposing dry mud in ocean bays. Then the wave raced inward, rearing upward as much as 30 meters, as high as a 10-story building. Although mortality was highest in Sumatra, people died in coastal areas as far away as Port Elizabeth, South Africa, 8,000 kilometers from the epicenter.

We can't prevent tsunamis, but we can, hopefully, mitigate loss of life. An early-warning system, similar to that discussed for earthquakes, could alert coastal inhabitants several minutes ahead of the deadly wave, giving people time to retreat to higher ground. In addition, barrier islands and coral reefs cause waves to break and dissipate their energy offshore; thus, preservation of these natural features protects populated regions along the coast.

tsunami A large destructive sea wave, produced by an undersea earthquake or volcano; sometimes called a *tidal wave*, although it has nothing to do with tides.

7.6 Studying the Earth's Interior

Even though the deepest well is only a 12-kilometer-deep hole in northern Russia, scientists have learned a remarkable amount about Earth's structure by studying the behavior of seismic waves. Some of the principles necessary for understanding the behavior of seismic waves are as follows:

1. In a uniform, homogeneous medium, a wave radiates outward in concentric spheres and at constant velocity.
2. The velocity of a seismic wave depends on the nature of the material that it travels through. Thus seismic waves travel at different velocities in different types of rock, varying with the rigidity and density of that rock.
3. When a wave passes from one material to another, it refracts (bends) and sometimes reflects (bounces back). Both refraction and reflection are easily observed in light waves. If you place a pencil in a glass half filled with water, the pencil appears bent. Of course the pencil does not bend; the light rays do. Light rays slow down when they pass from air to water, and as the velocity changes, the light waves *refract*. If you look in a mirror, the mirror *reflects* your image. In a similar manner, boundaries between Earth's layers refract and reflect seismic waves.

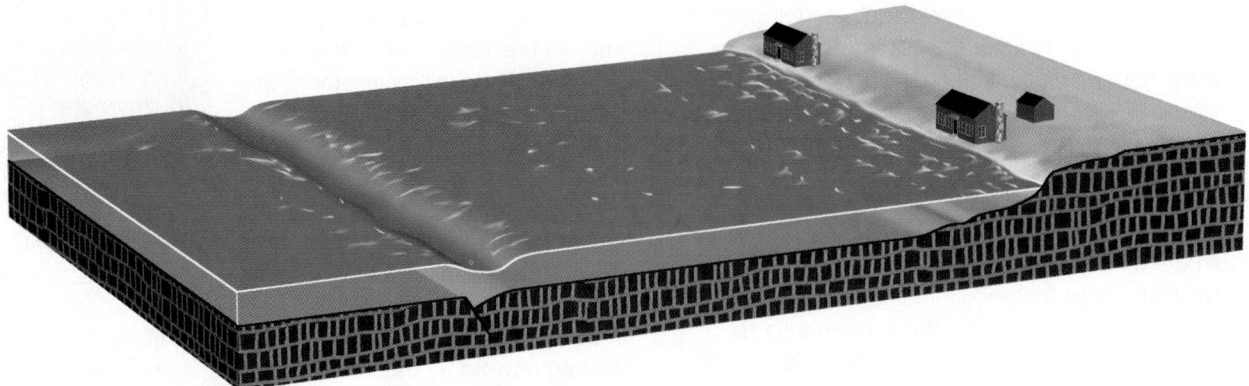

A Earthquake! Sea floor drops, sea level falls with it.

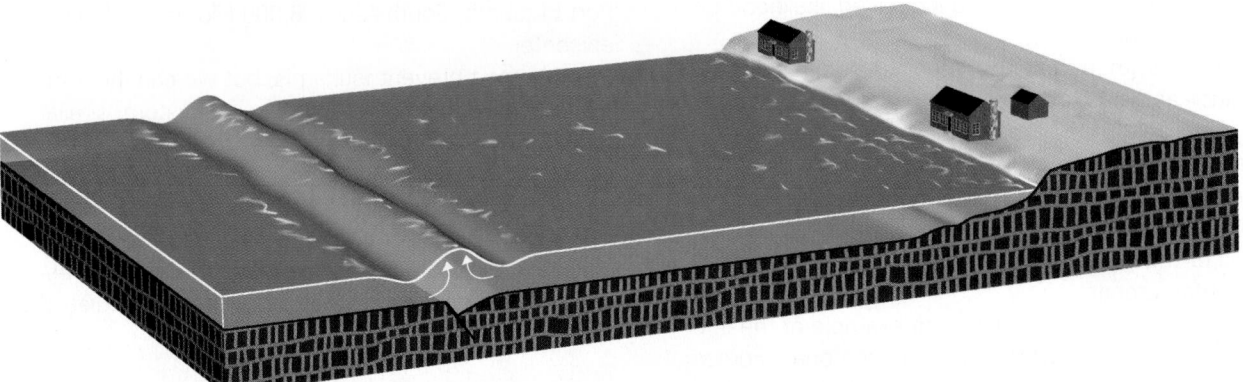

B Water rushes into low spot, and overcompensates, creating a bulge.

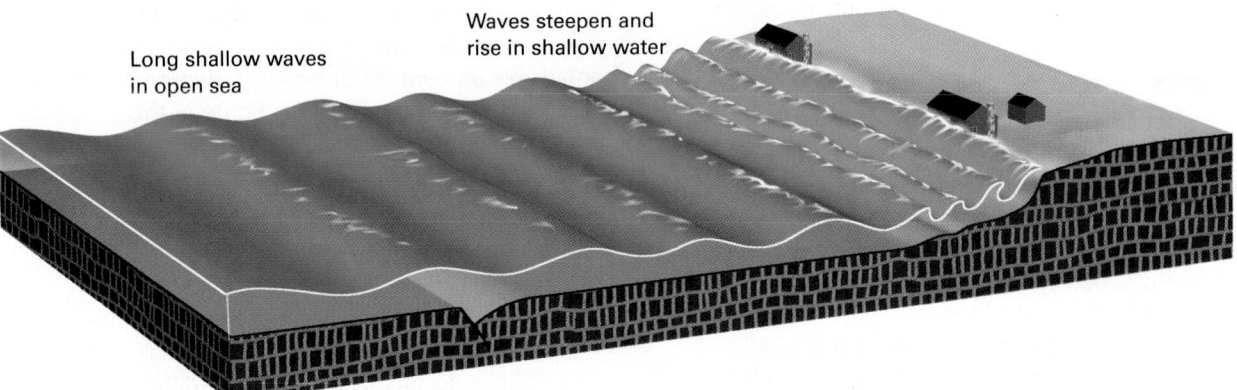

Long shallow waves in open sea

Waves steepen and rise in shallow water

C Wave rises and crashes onto land, destroying property and often causing human deaths.

FIGURE 7.19 A tsunami develops when part of the sea floor drops during an earthquake. Water rushes to fill the low spot, but the inertia of the rushing water forces too much water into the area, creating a bulge in the water surface. The long, shallow waves can build up into destructive giants when they reach shore.

4. P waves are compressional waves and can travel through gases, liquids, and solids. S waves are shear waves and travel only through solids.

Discovery of the Crust–Mantle Boundary

Figure 7.21A shows that some waves (in blue) travel directly from the earthquake focus, through the crust, to a nearby seismograph. Others (in red) travel downward into the mantle and then refract back upward to the same seismograph. The route through the mantle is longer than that through the crust. However, seismic waves travel faster in the mantle than they do in the crust. Over a short distance (less than 300 kilometers), waves traveling through the crust arrive at a seismograph before those following the longer route

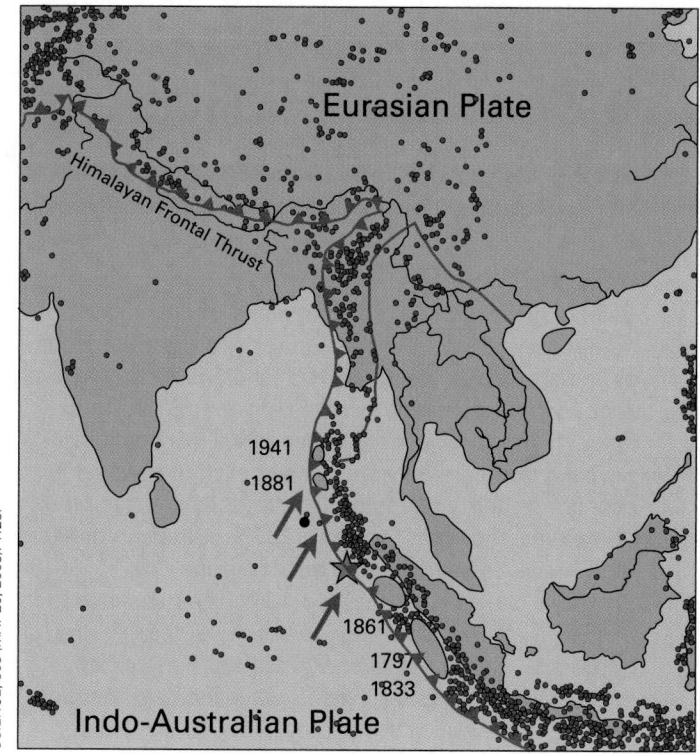

CONVERGENT BOUNDARY= ▲▲

FIGURE 7.20 Fault map of the Indo Sumatra-Andaman earthquake region. The green star shows the center of the December 2004 Sumatra-Andaman earthquake. Red dots show earthquakes greater than magnitude 5.0 from 1965 to December 2004, and the red arrows show plate motion. The frequency of past earthquakes and the continuing motion of the plates guarantee that more earthquakes will occur in the region in the near future.

SCIENCE, 308 (MAY 20, 2005): 1128.

Mohorovičić discontinuity The boundary between the crust and the mantle, identified by a change in the velocity of seismic waves; also called the *Moho*.

through the mantle. However, Figure 7.21B shows that for longer distances, the longer route through the mantle is faster because waves travel more quickly in the mantle.

The situation is analogous to the two different routes you may use to travel from your house to a friend's. The shorter route is a city street where traffic moves slowly. The longer route is an interstate highway, but you have to drive several kilometers out of your way to get to the highway and then another few kilometers from the highway to your friend's house. If your friend lives nearby, it is faster to take the city street. But if your friend lives far away, it is faster to take the longer route and make up time on the highway.

In 1909, Andrija Mohorovičić discovered that seismic waves from a distant earthquake traveled more rapidly than those from a nearby earthquake. By analyzing the arrival times of earthquake waves to many different seismographs, he identified the boundary between the crust and the mantle. Today, this boundary is called the **Mohorovičić discontinuity**, or simply the *Moho*, in honor of its discoverer.

The Moho lies at a depth ranging from 4 to 70 kilometers. As explained previously, oceanic crust is thinner than continental crust, and continental crust is thicker under mountain ranges than it is under plains.

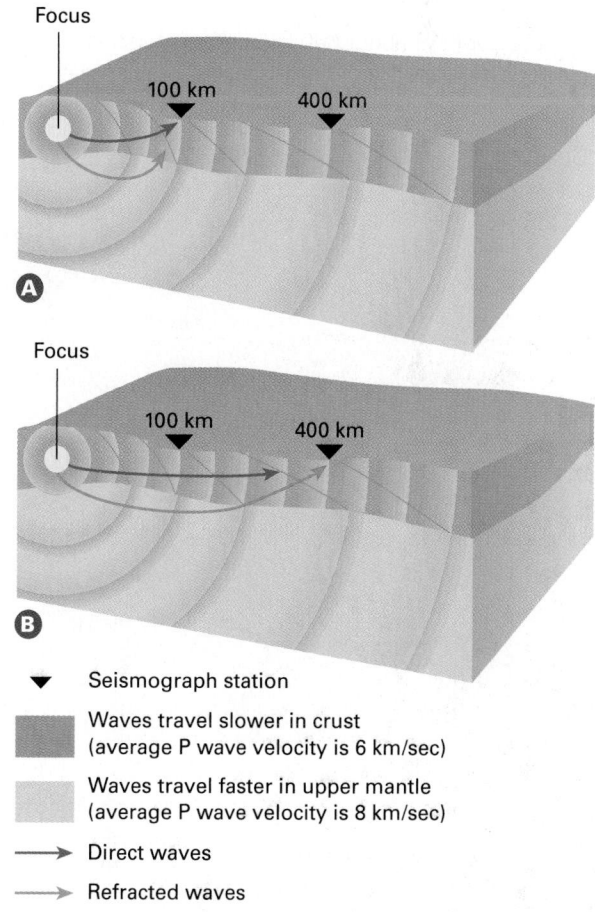

▼ Seismograph station

■ Waves travel slower in crust (average P wave velocity is 6 km/sec)

▢ Waves travel faster in upper mantle (average P wave velocity is 8 km/sec)

→ Direct waves

⟶ Refracted waves

FIGURE 7.21 The travel paths of seismic waves. (A) The direct waves reach the closer seismic station first. (B) The refracted waves reach a more distant seismic station before the direct waves. Even though the refracted waves travel a longer distance, they travel at a higher speed through the mantle.

The Structure of the Mantle

The mantle is almost 2,900 kilometers thick and composes about 80 percent of Earth's volume. Much of our knowledge of the composition and structure of the mantle comes from seismic data. As explained earlier, seismic waves speed up abruptly at the crust–mantle boundary (Figure 7.22). Seismic waves slow down again when they enter the asthenosphere at a depth between 75 and 125 kilometers. The plasticity and partially melted character of the asthenosphere slow down the seismic waves. At the base of the asthenosphere 350 kilometers below the surface, seismic waves speed up again because increasing pressure overwhelms the temperature effect and the mantle becomes stronger and less plastic.

At a depth of about 660 kilometers, seismic wave velocities increase again because pressure is great enough that the minerals in the mantle react to form denser minerals. The zone where the change occurs is called the **660-kilometer discontinuity**. The base of the mantle lies at a depth of 2,900 kilometers. Recent research has indicated that the base of the mantle, at the core–mantle boundary, may be so hot that despite the tremendous pressure, rock in this region is partially liquid.

Discovery of the Core

Using a global array of seismographs, seismologists can detect direct P and S waves up to 105° from the focus of an earthquake. Between 105° and 140° is a "shadow zone" where no direct P waves arrive at Earth's surface. This shadow zone is caused by a discontinuity, which is the mantle–core boundary. When P waves pass from the mantle into the core, they are refracted, or bent, as shown in Figure 7.23. The refraction deflects the P waves away from the shadow zone.

No S waves arrive beyond 105°. Their absence in this region shows that they do not travel through the outer core. Recall that S waves are not transmitted through liquids. The failure of S waves to pass through the outer core indicates that the outer core is liquid.

Refraction patterns of P waves, shown in Figure 7.23, indicate that another boundary exists within the core. It is the boundary between the liquid outer core and the solid inner core. Although seismic waves tell us that the outer core is liquid and the inner core is solid, other evidence tells us that the core is composed of iron and nickel.

More detailed studies, conducted in the 1980s and 1990s, show that seismic waves travel at different speeds in different directions through the inner core—thus the inner core is not homogeneous. One series of measurements suggests that the inner core, lying within its liquid sheath, is rotating at a significantly faster rate than the mantle and crust. Other researchers have proposed that the solid inner core may convect just as the solid mantle does.

Density Measurements

The overall density of Earth is 5.5 grams per cubic centimeter (g/cm^3), but both crust and mantle have average densities less than this value. The density of the crust ranges from 2.5 to 3.0 g/cm^3, and the density of the mantle varies from 3.3 to 5.5 g/cm^3. Because the mantle and crust account for slightly more than 80 percent of Earth's volume, the core must be very dense to account for the

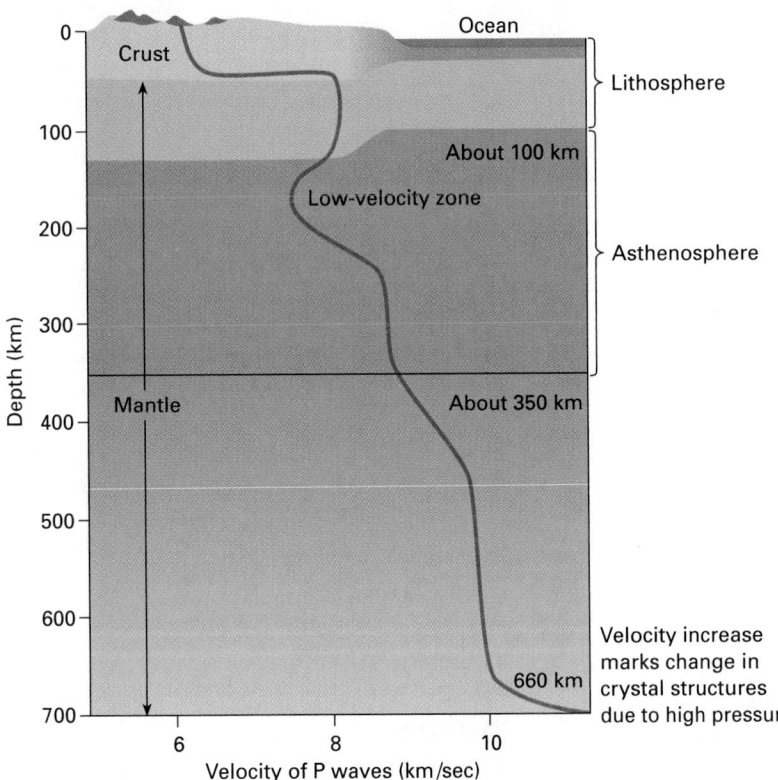

FIGURE 7.22 Velocities of P waves in the crust and the upper mantle. As a general rule, the velocity of P waves increases with depth. However, in the asthenosphere, the temperature is high enough and the pressure is low enough that rock is plastic. As a result, seismic waves slow down in this region, called the low-velocity zone. Wave velocity increases rapidly at the 660-kilometer discontinuity, the boundary between the upper and the lower mantle, probably because of a change in mineral structures due to increasing pressure.

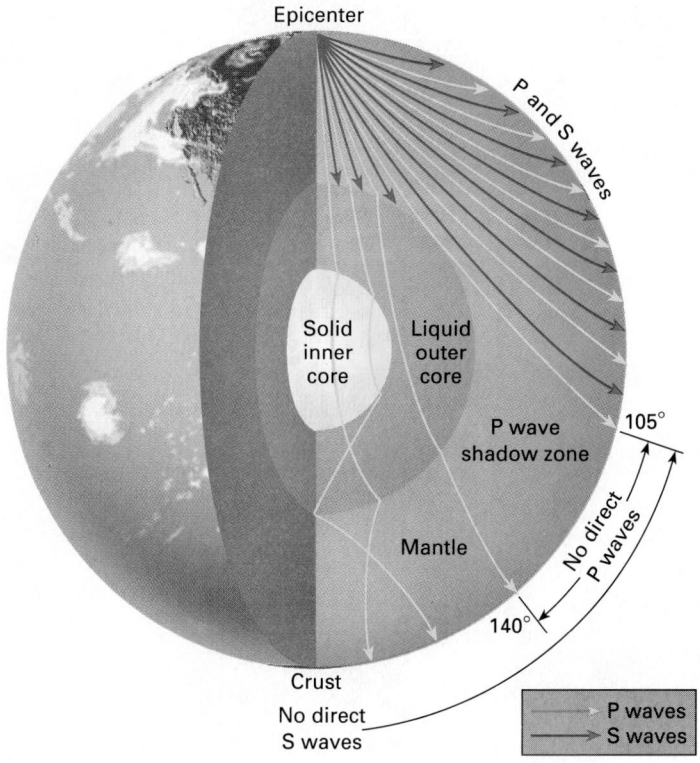

Epicenter

P and S waves

Solid inner core

Liquid outer core

P wave shadow zone

105°

Mantle

No direct P waves

140°

Crust

No direct S waves

P waves
S waves

FIGURE 7.23 Cross section of Earth showing paths of seismic waves. They bend gradually because of increasing pressure with depth. They also bend sharply where they cross major layer boundaries in Earth's interior. Note that S waves do not travel through the liquid outer core, and therefore direct S waves are observed only within an arc of 105° of the epicenter. P waves are refracted sharply at the core–mantle boundary, so there is a shadow zone of no direct P waves from 105° to 140°.

average density of Earth. Calculations show that the density of the core must be 10 to 13 g/cm³, which is the density of many metals under high pressure.

Many meteorites are composed mainly of iron and nickel. Cosmologists think that meteorites formed at about the same time that the solar system did and that they reflect the composition of the primordial solar system. Because Earth coalesced from meteorites and similar objects, scientists believe that iron and nickel must be abundant on Earth. Therefore, they conclude that the metallic core is composed of iron and nickel.

7.7 Earth's Magnetism

Early navigators learned that no matter where they sailed, a needle-shaped magnet aligned itself in an approximately north–south orientation. From this observation, they learned that Earth has a magnetic north pole and a magnetic south pole (Figure 7.24).

Earth's interior is too hot for a permanent magnet to exist. Instead, Earth's magnetic field is probably electromagnetic in origin.

Most likely, Earth's magnetic field is generated within the outer core. Metals are good conductors of electricity and the metals in the outer core are liquid and very mobile. Two types of motion occur in the liquid outer core: (1) Because the outer core is much hotter at its base than at its surface, the liquid metal convects. (2) The rising and falling metals are then deflected by Earth's spin. These convecting, spinning liquid conductors are thought to generate Earth's magnetic field.

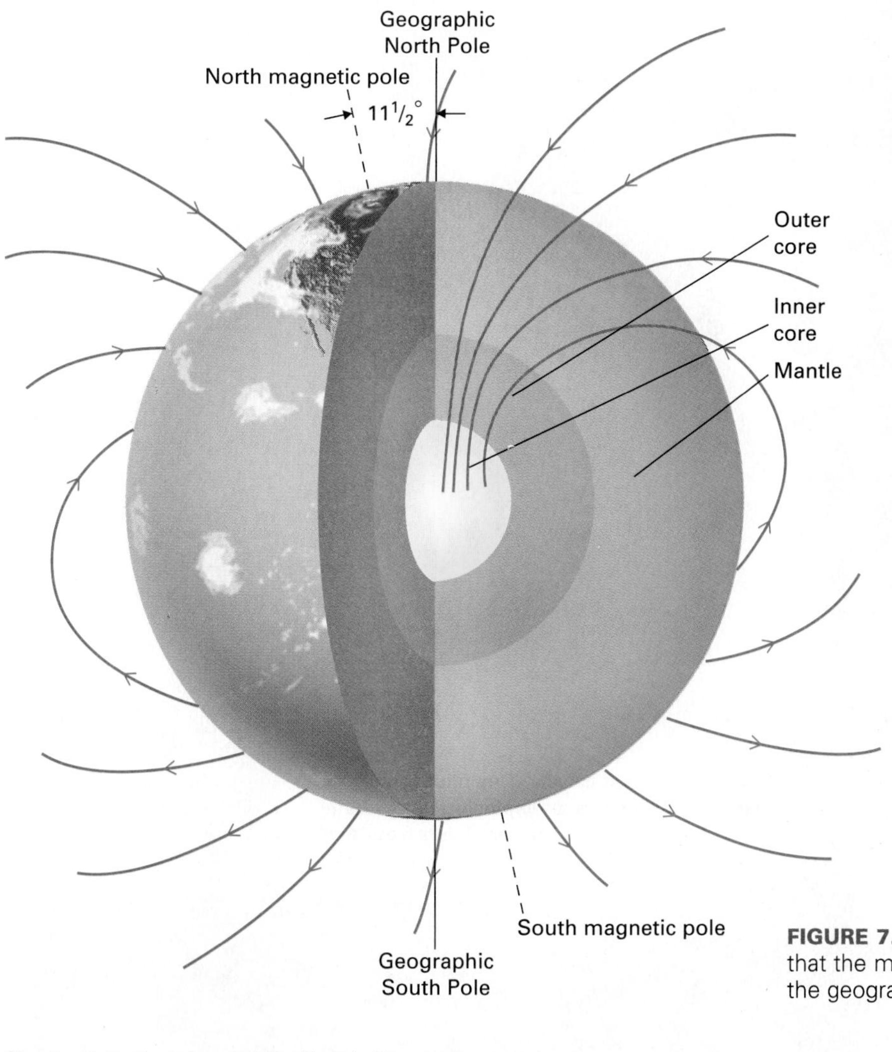

Geographic North Pole

North magnetic pole

11½°

Outer core

Inner core

Mantle

South magnetic pole

Geographic South Pole

FIGURE 7.24 The magnetic field of Earth. Note that the magnetic north pole is offset 11.5° from the geographic North Pole.

8

VOLCANOES AND PLUTONS

Eruption as dawn breaks. This photograph taken in 2009 shows an explosion of shards of still-molten lava as the lava flows from the Kilauea summit finally reach the Pacific Ocean.

visit **4ltrpress.cengage.com**

8.1 Magma

In Chapter 3 we stated that rocks melt in certain environments to form magma. This process is one example of the constantly changing nature of rocks described by the rock cycle. Why do rocks melt, and in what environments does magma form?

Processes That Form Magma

Recall that the asthenosphere is the layer in the upper mantle that extends from a depth of about 100 kilometers to 350 kilometers. In that layer the combined effects of temperature and pressure are such that 1 or 2 percent of the mantle rock is molten, as explained in Chapter 6. Although the majority of the asthenosphere is solid rock, it is so hot and so close to its melting point that relatively small changes can melt large volumes of rock. In order to understand volcanoes and plutonic rocks, we must first understand three processes that melt the asthenosphere to form magma: increasing temperature, decreasing pressure, and addition of water (Figure 8.1).

Increasing Temperature

Everyone knows that a solid melts when it becomes hot enough. Butter melts in a frying pan and snow melts under the spring Sun. For similar reasons, an increase in temperature will melt a hot rock. Oddly, however, increasing temperature is the least important cause of magma formation in the asthenosphere.

Decreasing Pressure

A mineral is a solid composed of an ordered array of atoms bonded together. When a mineral melts, the atoms become disordered and move freely, taking up more space than the solid mineral. Consequently, magma occupies about 10 percent more volume than the rock that melted to form it (Figure 8.2).

If a rock is heated to its melting point on Earth's surface, it melts readily because there is little pressure to keep it from expanding. The temperature in the asthenosphere is hot enough to melt rock, but the high pressure prevents the rock from expanding, so it does not melt. However, if the pressure were to decrease, large volumes of asthenosphere rocks would melt.

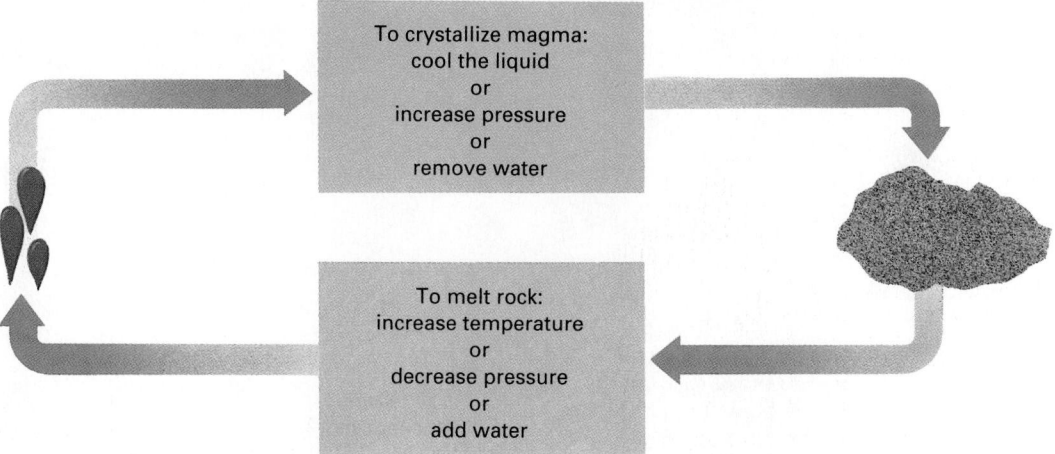

FIGURE 8.1 The lower box shows that increasing temperature, addition of water, and decreasing pressure all melt rock to form magma. The upper box shows that cooling, increasing pressure, and water loss all solidify magma to form igneous rock.

COURTESY OF GRAHAM R. THOMPSON/JONATHAN TURK

pressure-release melting Melting of asthenosphere rock that occurs when pressure in the asthenosphere decreases and the rock expands and melts.

High pressure (solid)

Low pressure (liquid)

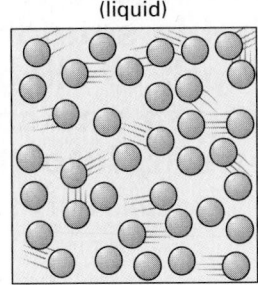

FIGURE 8.2 When most minerals melt, the volume of the resulting magma increases because there is little pressure to keep it from expanding. As a result, high pressure favors the dense, orderly arrangement of a solid mineral and low pressure favors the random, less-dense arrangement of molecules in liquid magma.

Melting caused by decreasing pressure is called **pressure-release melting**. In the section "Environments of Magma Formation," we will see how certain tectonic processes decrease pressure on asthenosphere rocks, thereby melting them.

Addition of Water

A rock containing small amounts of water generally melts at a lower temperature than an otherwise identical dry rock. Consequently, addition of water to rock that is near its melting temperature can melt the rock.

Certain tectonic processes, described shortly, add water to the hot rock of the asthenosphere to form magma.

Environments of Magma Formation

Magma forms abundantly in three tectonic environments: spreading centers, mantle plumes, and subduction zones. Let us consider each environment to see how rising temperature, decreasing pressure, and addition of water can melt the asthenosphere to create magma.

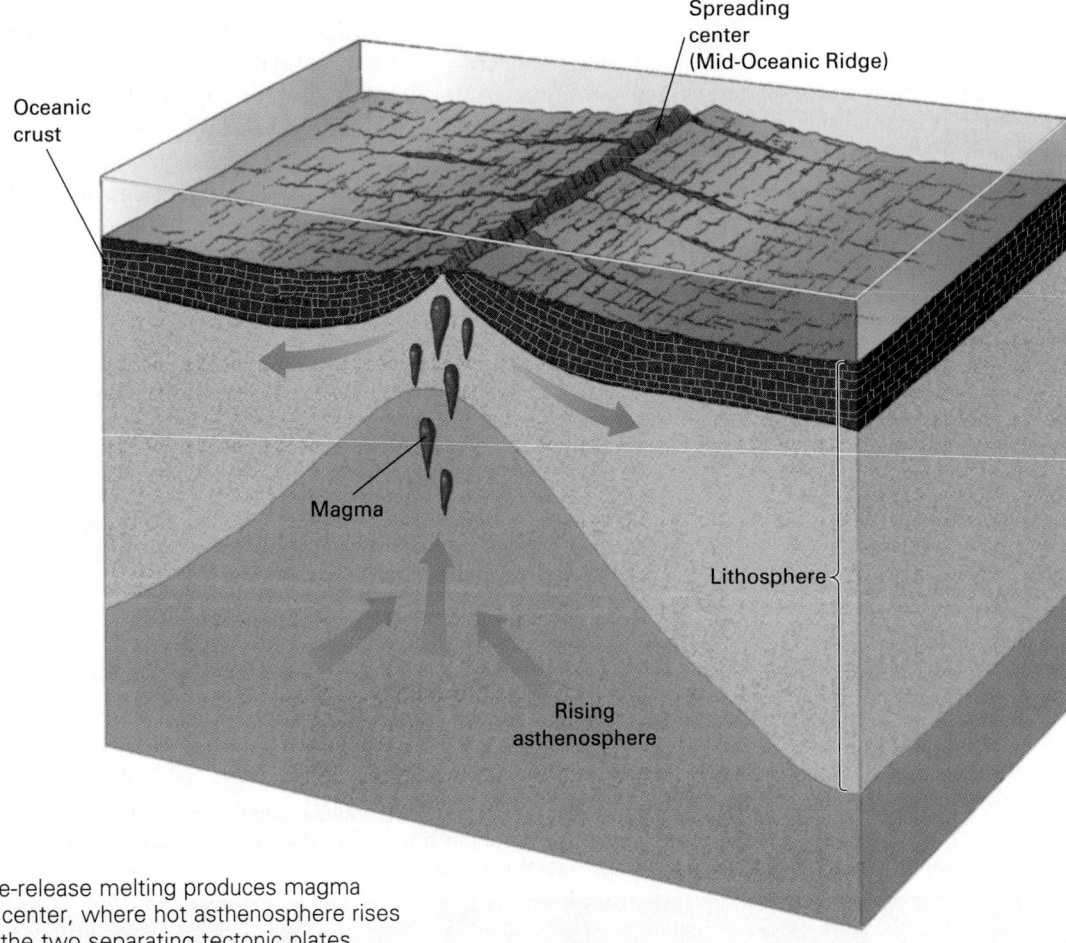

FIGURE 8.3 Pressure-release melting produces magma beneath a spreading center, where hot asthenosphere rises to fill the gap left by the two separating tectonic plates.

Unit 2: Internal Processes

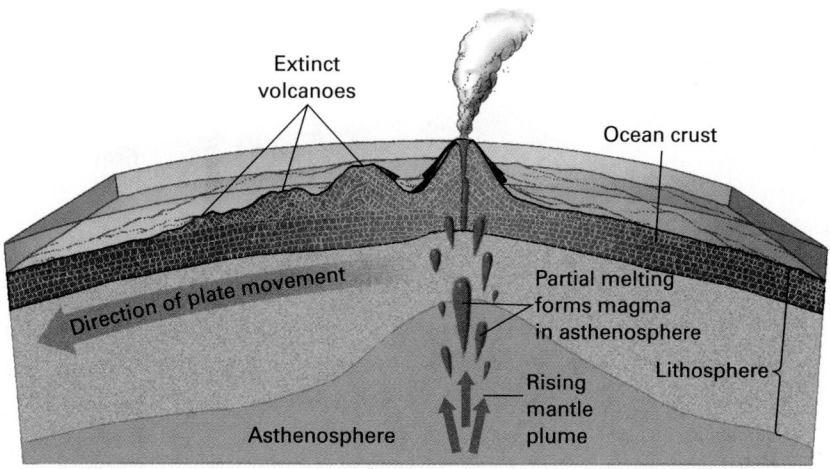

FIGURE 8.4 Pressure-release melting produces magma in a rising mantle plume. The magma rises to form a volcanic hot spot.

Magma Production in a Spreading Center

As lithospheric plates separate at a spreading center, hot, plastic asthenosphere oozes upward to fill the gap (Figure 8.3). As the asthenosphere rises, pressure drops and pressure-release melting forms basaltic magma. (The terms *basaltic* and *granitic* refer to magmas with the chemical compositions of basalt and granite, respectively.) Because the magma is of lower density than the surrounding rock, it rises buoyantly toward the surface.

Most of the world's spreading centers lie in the ocean basins, where they form the Mid-Oceanic Ridge system. The magma created at the spreading centers erupts from the ridge beneath the sea and solidifies to form new oceanic crust. The oceanic crust then spreads outward, riding atop the separating tectonic plates. Nearly all of Earth's oceanic crust is created in this way at the Mid-Oceanic Ridge system. In most places the ridge lies beneath the sea. In a few places, such as Iceland, the ridge rises above sea level and volcanic eruptions are common. Some spreading centers, like the East African Rift, occur in continents, and here, too, basaltic magma erupts onto Earth's surface.

Magma Production in a Mantle Plume

Recall from Chapter 6 that a mantle plume is a rising column of hot, plastic rock that originates deep within the mantle. The plume rises because it is hotter than the surrounding mantle and, consequently, is less dense and more buoyant. As a plume rises, pressure-release melting just beneath the lithosphere forms magma, which then rises toward Earth's surface (Figure 8.4).

Because mantle plumes and hot spots form below the lithosphere, hot spots usually occur within tectonic plates rather than at a boundary. For example, the Yellowstone hot spot, responsible for the volcanoes and hot springs in Yellowstone National Park, lies far from the nearest plate boundary. If a mantle plume rises beneath the sea, volcanic eruptions build submarine vol-

canoes and volcanic islands. The Hawaiian Islands are a chain of hot spot volcanoes that formed over a long-lived mantle plume beneath the Pacific Ocean.

Magma Production in a Subduction Zone

In a subduction zone, addition of water, decreasing pressure, and heat from friction all combine to form huge quantities of magma (Figure 8.5). As you learned in Chapter 6, a subducting plate is covered by water-saturated oceanic crust. As the wet rock dives into the hot mantle, rising temperature vaporizes the water, forming vapor that ascends into the hot asthenosphere directly above the sinking plate.

As the subducting plate descends, it drags plastic asthenosphere rock down with it, as shown by the elliptical arrows in Figure 8.5. Rock from deeper in the asthenosphere then flows upward to replace the sinking rock. Pressure decreases as this hot rock rises.

Friction generates heat in a subduction zone as one plate scrapes past the opposite plate. Figure 8.5 shows that addition of water, pressure release, and frictional heating combine to melt asthenosphere rocks in the zone where the subducting plate passes into the asthenosphere. Eventually the descending crust gets heated too.

Addition of water is probably the most important factor producing melting in a subduction zone, and frictional heating is the least important.

Volcanoes and plutonic rocks are common features of a subduction zone. The volcanoes of the Pacific Northwest, the granite cliffs of Yosemite, and the Andes Mountains are all examples of volcanic and plutonic rocks formed at subduction zones. The Ring of Fire is a zone of active volcanoes that traces the subduction zones encircling the Pacific Ocean basin. About 75 percent of Earth's active volcanoes (exclusive of the submarine volcanoes at the Mid-Oceanic Ridge) lie in the Ring of Fire (Figure 8.6).

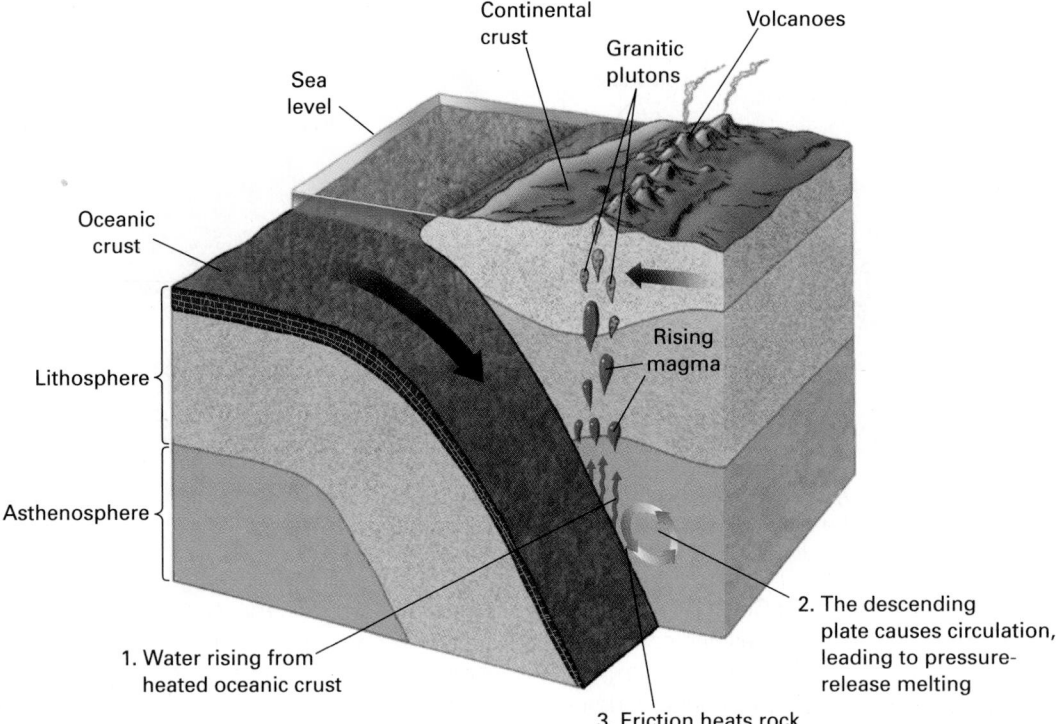

FIGURE 8.5 Three processes melt the asthenosphere to form magma at a subduction zone: (1) Steam rises from wet oceanic crust on top of the subducting plate. (2) Circulation in the asthenosphere decreases pressure on hot mantle rock. (3) Friction heats rocks in the subduction zone.

Labels on figure:
- Continental crust
- Granitic plutons
- Volcanoes
- Sea level
- Oceanic crust
- Rising magma
- Lithosphere
- Asthenosphere
- 1. Water rising from heated oceanic crust
- 2. The descending plate causes circulation, leading to pressure-release melting
- 3. Friction heats rock

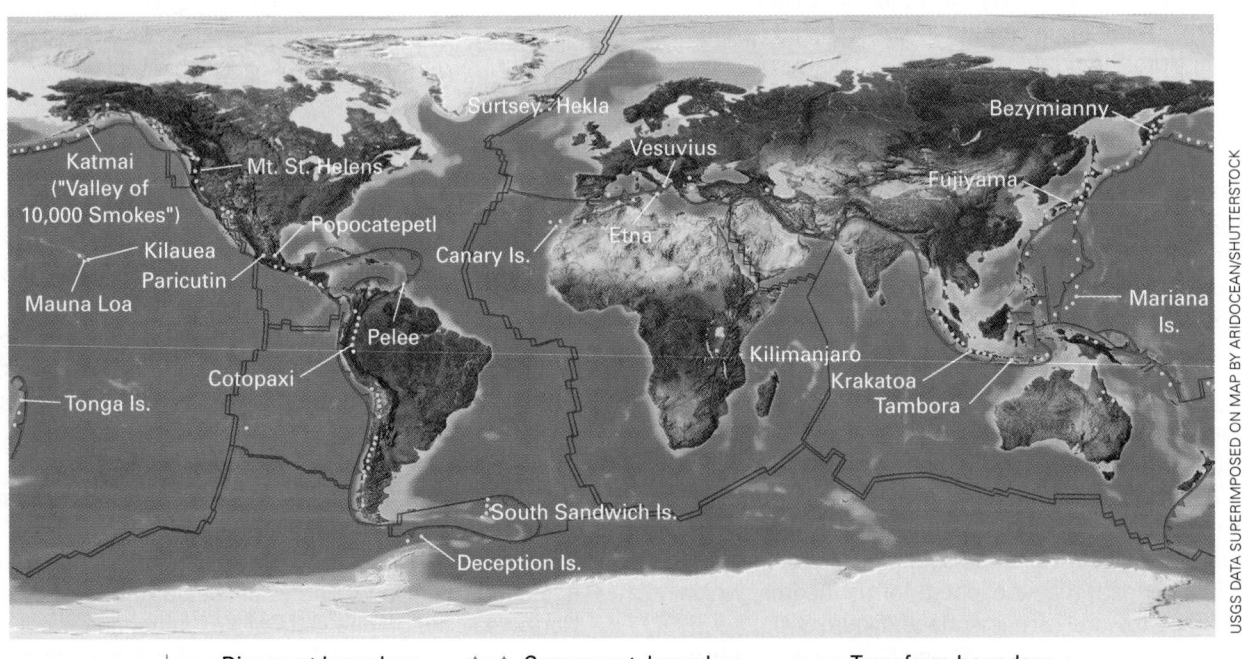

Labels on map: Katmai ("Valley of 10,000 Smokes"), Mt. St. Helens, Surtsey, Hekla, Vesuvius, Bezymianny, Fujiyama, Popocatepetl, Etna, Kilauea, Paricutin, Canary Is., Mauna Loa, Pelee, Kilimanjaro, Mariana Is., Cotopaxi, Krakatoa, Tambora, Tonga Is., South Sandwich Is., Deception Is.

⊏⊐ Divergent boundary   ▲▲ Convergent boundary   — Transform boundary

FIGURE 8.6 About 75 percent of Earth's active volcanoes (yellow dots) lie in the Ring of Fire, a chain of subduction zones at convergent plate boundaries (heavy red lines with teeth) that encircles the Pacific Ocean.

Unit 2: Internal Processes

8.2 Basalt and Granite

Basalt and granite are the most common igneous rocks. Basalt makes up most of the oceanic crust, and granite is the most abundant rock in continents. Because of their abundance, it is interesting to consider how basalt and granite form.

Recall that basaltic magma forms by melting of the asthenosphere. But the asthenosphere is peridotite. Basalt and peridotite are quite different in composition: Peridotite contains about 40 percent silica (SiO_2), but basalt contains about 50 percent. Peridotite contains considerably more iron and magnesium than basalt. How does peridotite melt to create basaltic magma? Why does the magma have a composition different from that of the rock that melted to produce it?

It is a general rule that a mixture of two or more minerals will begin to melt at a temperature lower than the melting point of any one of the minerals in its pure state. Remember that peridotite consists mainly of olivine and pyroxene, with small amounts of calcium feldspar. In one set of experiments designed to simulate melting of mantle peridotite, pure olivine melted at 1,890°C, pure pyroxene melted at 1,390°C, and pure calcium feldspar melted at 1,550°C—but peridotite rock consisting of all three minerals began to melt at 1,270°C.

Furthermore, the composition of the first bit of melt is usually richer in silica than the rock that is undergoing this process of **partial melting**. Thus, when mantle peridotite begins to melt, the magma is of basaltic composition—about 20 percent richer in silica than peridotite. Because the new basaltic magma is less dense than the peridotite rock, the magma begins to rise toward Earth's surface. This process is called partial melting because only a small amount of the original peridotite melts to form basaltic magma, leaving silica-depleted peridotite in the asthenosphere. Figures 8.3 and 8.4 show that basaltic magma forms in this way at a spreading center and in a mantle plume. Figure 8.5 shows that basaltic magma also forms in a subduction zone.

Granite and Granitic Magma

Granite contains more silica than basalt and therefore melts at a lower temperature—typically between 700°C and 900°C. Thus basaltic magma is hot enough to melt granitic continental crust. In some tectonic environments, basaltic magma forms beneath a continent and then rises into continental crust. Because the lower continental crust is hot, even a small amount of basaltic magma melts large amounts of lower continental crust to form granitic magma. Typically, the granitic magma rises a short distance and then solidifies within the crust to form plutonic rocks. Most granitic plutons solidify at depths between 5 kilometers and about 20 kilometers.

Granite forms by this process in a subduction zone at a continental margin, a continental rift zone, and a mantle plume rising beneath a continent.

Andesite and Intermediate Magma

Igneous rocks of intermediate composition, such as andesite and diorite, form by processes similar to those that generate granitic magma. Their magmas contain less silica than granite, either because they form by melting of continental crust that is lower in silica or because basaltic magma has mixed with the granitic magma.

8.3 Partial Melting and the Origin of Continents

It is believed that Earth melted shortly after its formation about 4.6 billion years ago. Magma at the surface then cooled to form the earliest crust. From the evidence of a few traces of old crust combined with computer models, geologists surmise that the first crust was lava with the composition of peridotite.

Our explanation of the formation of granitic magma by melting of granitic continental crust leaves us with two interesting questions: (1) When did granitic continents form? (2) If the early crust and mantle were composed of peridotite, how did granitic continental crust evolve at Earth's surface?

When Did Continents Form?

The 3.96-billion-year-old Acasta gneiss in Canada's Northwest Territories is among Earth's oldest known rock. It is metamorphosed granitic rock, similar to modern continental crust, and implies that at least some granitic crust had formed by early Precambrian time.

Geologists have found grains of a mineral called zircon in a sandstone in western Australia. The zircon gives radiometric dates of 4.4 billion years, although the sandstone is younger. Zircon commonly forms in granite. Geologists infer that the very old zircon initially formed in granite, which later weathered and released the zircon grains as sand. Eventually, the zircon became part of the younger sandstone. The presence of these zircon grains suggests that granitic rocks existed 4.4 billion years ago. Geologists have also found granitic rocks in Greenland and Labrador that are nearly as old as the Acasta gneiss and the Australian zircon grains.

These dates tell us that some granitic continental crust probably existed by 4.4 billion years ago.

partial melting
The process in which a silicate rock only partly melts as it is heated, forming magma that is more silica-rich than the original rock.

Partial Melting and the Origin of Granitic Continents

The differences in composition among the mantle, oceanic crust, and continental crust are reviewed in the following table:

Mantle (Peridotite)	Oceanic Crust (Basalt)	Continental Crust (Granite)
High magnesium and iron content; low silica	Less magnesium and iron and more silica compared with mantle rock	Low magnesium and iron; high silica

If the earliest crust had a composition similar to that of the mantle, we must describe how oceanic and continental crust evolved from mantle rocks. In Section 8.2 we explained that partial melting produces magma that contains a higher proportion of silica than the rock from which the melt formed. Geologists believe that the earliest crust was peridotitic lava, formed from the melted mantle as Earth cooled from the surface downward after its early pervasive melting event. Later, partial melting of this primordial crust formed a basaltic crust that was richer in silica. Then, partial melting of the basalt probably formed intermediate rocks such as andesite, which underwent another partial melting to form the silica-rich granitic continents. The process of partial melting may explain how the silica-rich continents evolved in steps from the silica-poor mantle. But what tectonic processes caused the sequence of melting steps?

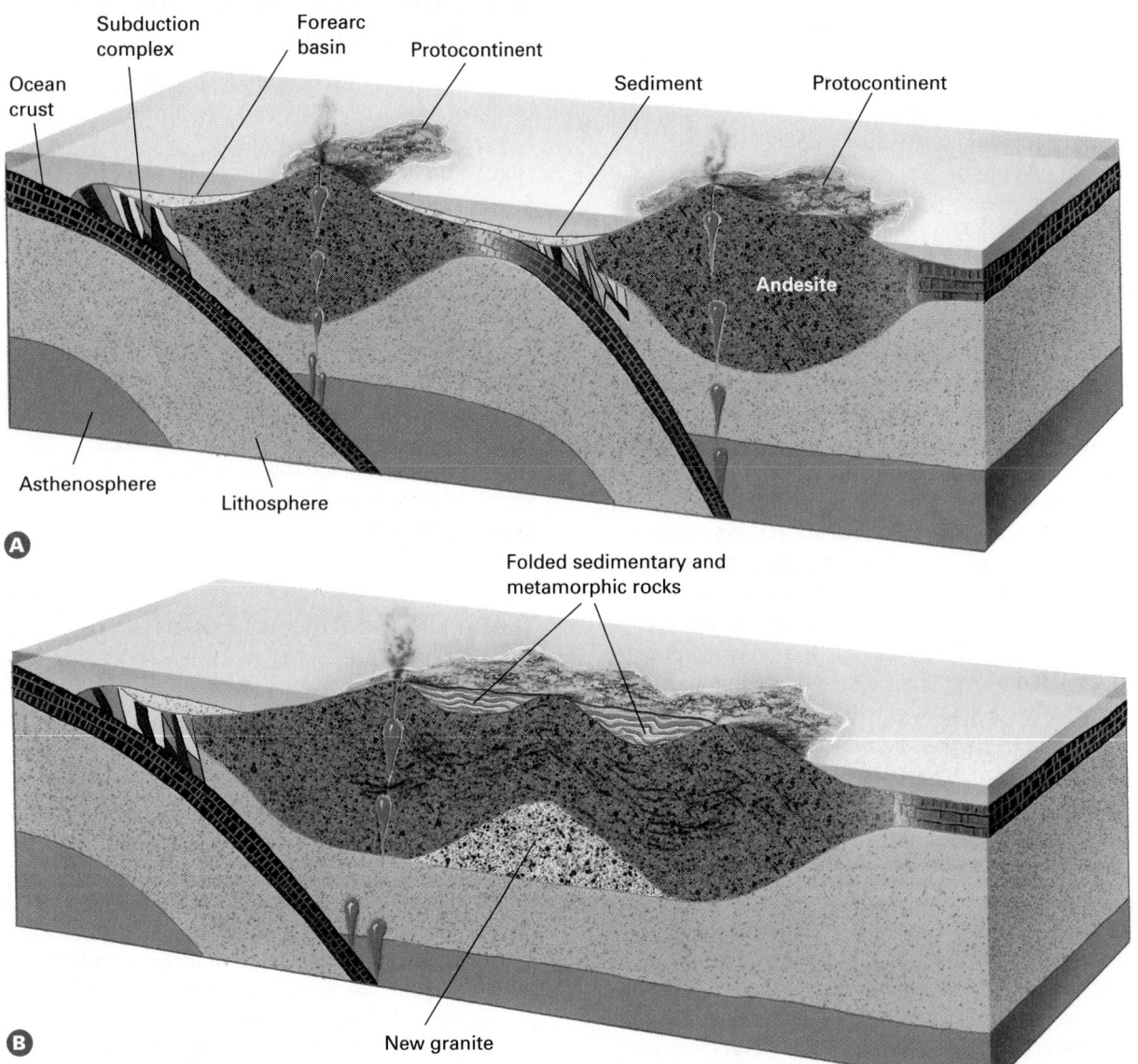

FIGURE 8.7 (A) According to one model, early continents formed in subduction zones, much like oil on the water surface of pots being heated to cook pasta. (B) As plates continued to move, small continents collided and coalesced. Sediments that had eroded from the original islands were folded, and uplifted. Some were subjected to so much heat and pressure that they became metamorphic rocks. New granite formed from partial melting of the crust at the subduction zone.

As stated earlier, in modern times magma forms in three geologic environments: spreading centers, subduction zones, and mantle plumes. Similar magma-forming environments may have existed in early Precambrian time, but geologists are uncertain which were most important.

Some observations imply that Archean tectonics was similar to modern horizontal plate movements, and most magma formed at spreading centers and subduction zones. Other evidence indicates that horizontal plate motion was minor and that vertical mantle plumes dominated early Precambrian tectonics.

Horizontal Tectonics

According to one hypothesis, heat-driven convection currents in a hot, active mantle initiated horizontal plate movement in the early Precambrian crust. The dense primordial crust dove into the mantle in subduction zones, where partial melting created basaltic magma. As a result, the crust gradually became basaltic.

At a later date, the earliest continental crust formed by partial melting of basaltic crust in a new generation of subduction zones. In Chapter 9 you will see that island arcs form today in the same manner. Therefore, the first continents probably consisted of small granitic or andesitic blobs—like island arcs—surrounded by basaltic crust (Figure 8.7A). Gradually, continued plate movement led to further subduction and isolated islands coalesced to form microcontinents. In the process, sediments accumulated and were metamorphosed during the collisions (Figure 8.7B).

Vertical Mantle Plume Tectonics

Several researchers have proposed, instead, that mantle plumes dominated early Precambrian tectonics. Upwellings of mantle rock, perhaps rising from the mantle–core boundary, led to partial melting of portions of the upper mantle. This magma then solidified to form basaltic crust. Continued partial melting eventually formed granitic continental crust.

As evidence for both models accumulates, some geologists suggest that both vertical and horizontal mechanisms were important. According to this hypothesis, mantle plumes formed thick basaltic oceanic plateaus, which then oozed outward to initiate horizontal motion. This motion caused subduction and another melting episode that generated continental crust by partial melting of the basalt plateaus.

8.4 Magma Behavior

Once magma forms, it rises toward Earth's surface because it is less dense than surrounding rock. As it rises, two changes occur: (1) magma cools as it enters shallower and cooler levels of Earth, and (2) pressure drops because the weight of overlying rock decreases.

Recall from Figure 8.1 that cooling tends to solidify magma but decreasing pressure tends to keep it liquid. So, does magma solidify or remain liquid as it rises toward Earth's surface? The answer depends on the type of magma. Basaltic magma commonly remains liquid and rises to the surface to erupt from a volcano or flow onto the sea floor at the Mid-Oceanic Ridge. In contrast, granitic magma usually solidifies within the crust.

The contrasting behavior of granitic and basaltic magmas is a consequence of their different compositions. Granitic magma contains about 70 percent silica, whereas the silica content of basaltic magma is only about 50 percent. In addition, granitic magma generally contains up to 10 percent water, whereas basaltic magma contains only 1 to 2 percent water. These differences are summarized in the following table:

Typical Granitic Magma	Typical Basaltic Magma
70% silica; up to 10% water	50% silica; 1 to 2% water

Effects of Silica on Magma Behavior

In the silicate minerals, silicate tetrahedra link together to form the chains, sheets, and framework structures described in Chapter 2. Silicate tetrahedra link together in a similar manner in magma. They form long chains and similar structures if silica is abundant in the magma, but shorter chains if less silica is present.

Because of its higher silica content, granitic magma contains longer chains than does basaltic magma. The long chains become tangled, making the magma stiffened, or viscous. It rises slowly because of its viscosity and has ample time to solidify within the crust before reaching the surface. In contrast, basaltic magma, with its shorter silicate chains, is less viscous and flows easily. Because of its fluidity, it rises rapidly to erupt at Earth's surface.

Effects of Water on Magma Behavior

A second, and more important, difference between the two magmas is that granitic magma contains more water than basaltic magma does. Water lowers the temperature at which magma solidifies. If dry granitic magma solidifies at 700°C, the same magma with 10 percent water may not become solid until the temperature drops below 600°C.

Water tends to escape as steam from hot magma. But deep in the crust where granitic magma forms, high pressure prevents the water from escaping. As the magma rises, pressure decreases and water escapes. Because the magma loses water, its solidification temperature rises, causing it to crystallize. Water loss causes rising granitic magma to solidify within the crust. Because basaltic magmas have only 1 to 2 percent

pluton A body of intrusive igneous rock.

batholith A large pluton, exposed over more than 100 square kilometers of Earth's surface.

stock A pluton exposed over less than 100 square kilometers of Earth's surface; similar to a batholith, but smaller.

dike A sheetlike igneous rock, cutting through layers of country rock, that forms when magma oozes into a fracture; as the country rock erodes, the dike is left standing on the surface.

sill A sheetlike igneous rock, parallel to the grain or layering of country rock, that forms when magma oozes between layers.

water to begin with, water loss is relatively unimportant. As a result, rising basaltic magma usually remains liquid all the way to Earth's surface, and basalt volcanoes are common.

8.5 Plutons

Recall from Chapter 3 that in most cases, granitic magma solidifies within Earth's continental crust to form a large mass of igneous rock called a **pluton** (Figure 8.8A). Many granite plutons measure tens of kilometers in diameter. How can such a large mass of viscous magma rise through solid rock?

If you place oil and water in a jar, screw the lid on, and shake the jar, oil droplets disperse throughout the water. When you set the jar down, the droplets coalesce to form larger bubbles, which rise toward the surface, easily displacing the water as they ascend. Granitic magma rises in a similar way, except that the process is slower because it rises through solid rock. Granitic magma forms near the base of continental crust, where surrounding rock is hot and plastic. As the magma rises, it pushes aside the plastic country rock, which then slowly flows back to fill in behind the rising bubble.

After a pluton forms, tectonic forces may push that part of the crust upward, and erosion may expose parts of the pluton at Earth's surface (Figure 8.8B). A **batholith** is a pluton exposed over more than 100 square kilometers of Earth's surface. An average batholith is about 10 kilometers thick, although a large one may be 20 kilometers thick. A **stock** is similar to a batholith but is exposed over less than 100 square kilometers.

Figure 8.9 shows the locations of the major batholiths of western North America. Many mountain ranges, such as California's Sierra Nevada, contain large granite batholiths (Figure 8.10). A batholith is commonly composed of numerous smaller plutons intruded sequentially over millions of years. For example, the Sierra Nevada batholith contains about 100 individual plutons, most of which were emplaced over a period of 50 million years. The formation of this complex batholith ended about 80 million years ago.

A large body of magma engulfs or pushes country rock aside as it rises. In contrast, a smaller mass of magma may flow into a fracture or between layers in country rock. A **dike** is a tabular, or sheetlike, intrusive rock that forms when magma oozes into a fracture (Figure 8.11). Dikes cut across sedimentary layers or other features in country rock and range from less than a centimeter to more than a kilometer thick (Figure 8.12). A dike is commonly more resistant to weathering than surrounding rock is. As the country rock erodes, the dike is left standing on the surface (Figure 8.13).

Magma that oozes between layers of country rock forms a sheetlike rock parallel to the layering, called a **sill** (Figure 8.11). Like dikes, sills vary in thickness from less than a centimeter to more than a kilometer and may extend for tens of kilometers in length and width (Figure 8.14).

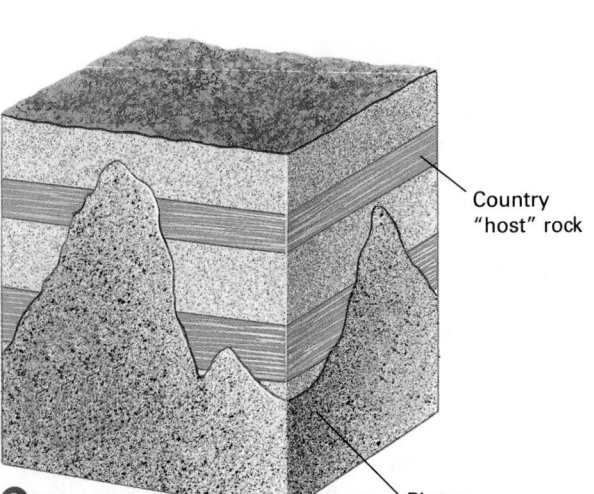

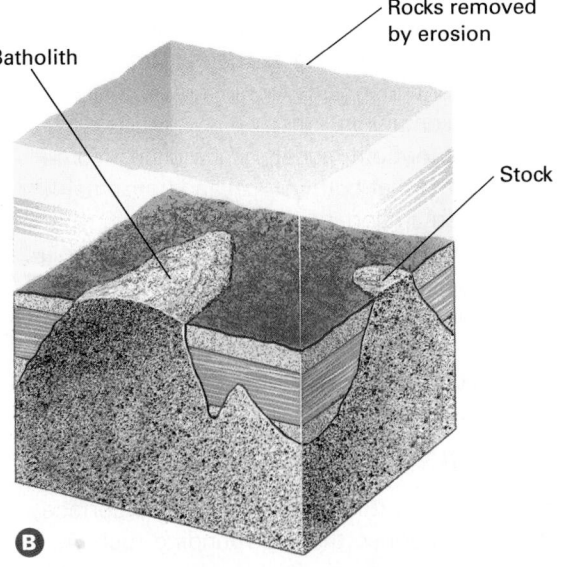

FIGURE 8.8 (A) A pluton is any intrusive igneous rock. (B) A batholith is a pluton with more than 100 square kilometers exposed at Earth's surface. A stock is similar to a batholith but has a smaller surface area.

Unit 2: Internal Processes

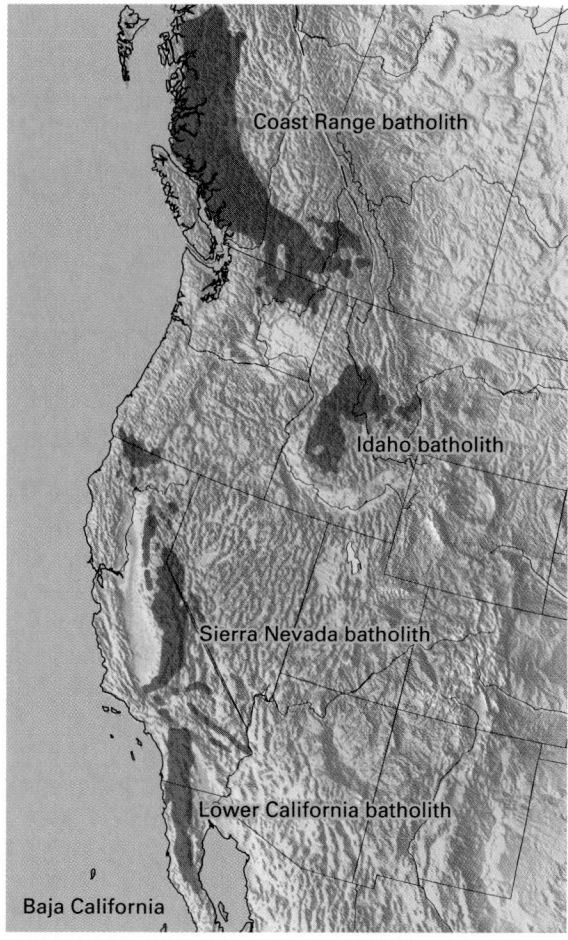

FIGURE 8.9 The large batholiths in western North America, shown here in dark gray, form high mountain ranges.

Coast Range batholith

Idaho batholith

Sierra Nevada batholith

Lower California batholith

Baja California

DON HYNDMAN

FIGURE 8.10 Granite plutons make up most of California's Sierra Nevada in Yosemite National Park.

DON HYNDMAN

FIGURE 8.12 Here, several granite dikes have been intruded into older country rock.

FIGURE 8.11 A large magma body may crystallize within the crust to form a pluton. Some of the magma may rise to the surface to form volcanoes and lava flows; some intrudes country rock to form dikes and sills.

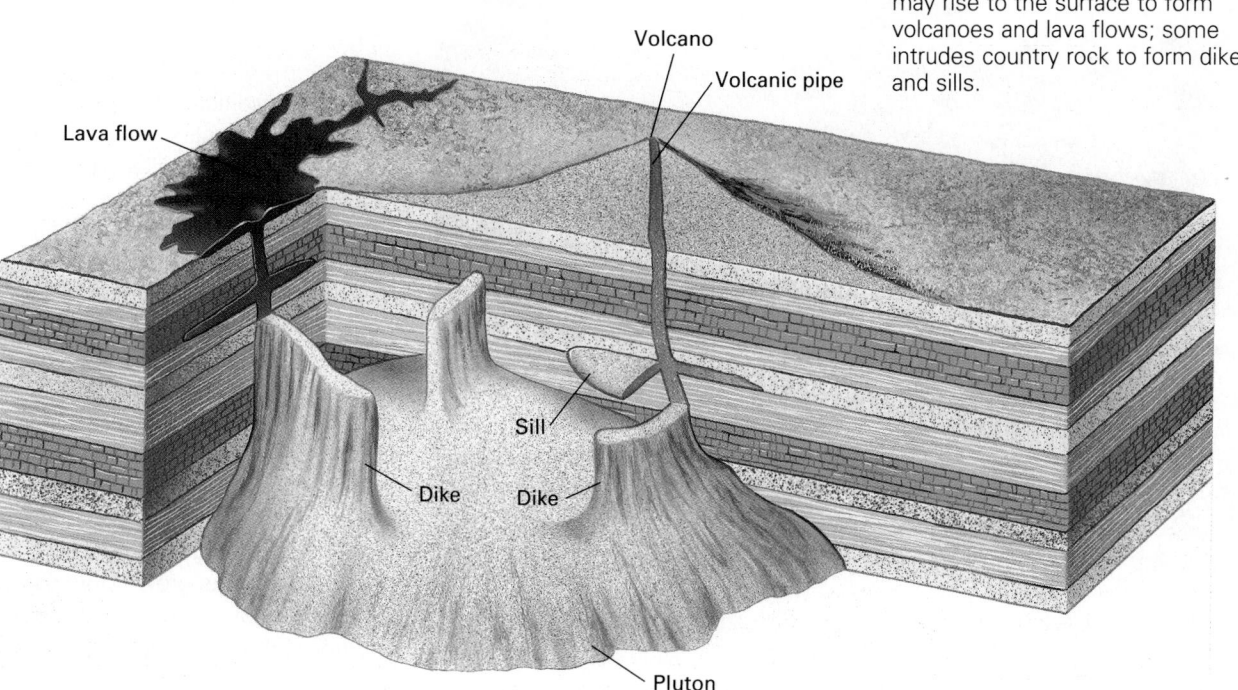

Volcano

Volcanic pipe

Lava flow

Sill

Dike

Dike

Pluton

volcano A hill or mountain formed from lava and rock fragments ejected through a volcanic vent.

pahoehoe Lava with a smooth, billowy, or ropy surface.

aa Lava that has a jagged, rubbly, broken surface.

vesicles Holes in lava rock that formed when the lava solidified before bubbles of gas or water could escape.

columnar joints Regularly spaced cracks that commonly develop in lava flows, growing downward, forming five- or six-sided columns.

FIGURE 8.13 This large dike near Shiprock, New Mexico, has been left standing after softer sandstone country rock eroded.

FIGURE 8.14 This black basalt sill in the Grand Canyon was injected between layers of sandstone.

8.6 Volcanoes

A **volcano** is a hill or mountain formed from lava and rock fragments ejected through a volcanic vent. The material erupted from volcanoes creates a wide variety of rocks and landforms, including lava plateaus and several types of volcanic outlets. Many islands, including the Hawaiian Islands, Iceland, and most islands of the southwestern Pacific Ocean, were built entirely by volcanic eruptions.

Lava and Pyroclastic Rocks

As you learned in Chapter 3, *lava* is magma that flows onto Earth's surface; the word also describes the rock that forms when the magma solidifies. Lava with low viscosity may continue to flow as it cools and stiffens, forming smooth, glassy-surfaced, wrinkled, or "ropy" ridges. This type of lava is called **pahoehoe** (pronounced "puh-HOY-hoy"), from the Hawaiian meaning "smooth" or "polished" (Figure 8.15). If the viscosity of lava is higher, its surface may partially solidify as it flows. The solid crust breaks up as the deeper, molten lava continues to move, forming **aa** (pronounced "ah-ah") lava, with a jagged, rubbled, broken surface. When lava cools, escaping gases such as water and carbon dioxide form bubbles in the lava. If the lava solidifies before the gas escapes, the bubbles are preserved as holes in the rock called **vesicles** (Figure 8.16).

As hot lava cools and solidifies, it shrinks . The shrinkage pulls the rock apart, forming cracks that grow as the rock continues to cool. In Hawaii, geologists have observed this phenomenon while watching fresh lava cool: When a solid crust measuring only 0.5 centimeter thick had formed on the surface of the glowing liquid, five- or six-sided cracks developed. As the lava continued to cool and solidify, the cracks grew downward through the flow. Such cracks, called **columnar joints**, are regularly spaced

Unit 2: Internal Processes

FIGURE 8.15 Pahoehoe lava buried this car in Hawaii.

FIGURE 8.16 Aa lava showing vesicles—gas bubbles preserved in the flow—in Shoshone, Idaho.

and intersect to form five- or six-sided columns (Figure 8.17).

If a volcano erupts explosively, it may eject both liquid magma and solid rock fragments. A rock formed from this material is called a **pyroclastic rock** (from *pyro*, meaning "fire," and *clastic*, meaning "particles"). The smallest particles, called **volcanic ash**, consist of tiny fragments of glass that formed when liquid magma exploded into the air. **Cinders** are volcanic fragments that vary in size from 4 to 32 millimeters.

Fissure Eruptions and Lava Plateaus

The gentlest type of volcanic eruption occurs when magma is so fluid that it oozes from cracks in the land surface called **fissures** and flows over the land like water. Basaltic magma commonly erupts in this manner because of its low viscosity. Fissures and fissure eruptions vary greatly in scale. In some cases, lava pours from small cracks on the flank of a volcano. Fissure flows of this type are common on Hawaiian and Icelandic volcanoes.

In other cases, however, fissures extend for tens or hundreds of kilometers and pour thousands of cubic

pyroclastic rock Rock made up of liquid magma and solid rock fragments that were ejected explosively from a volcanic vent.

volcanic ash The smallest pyroclastic particles, less than 2 millimeters in diameter.

cinders Glassy, pyroclastic volcanic fragments 4 to 32 millimeters in size.

fissures Breaks, cracks, or fractures in rocks.

FIGURE 8.17 (A) Columnar joints at Devil's Postpile National Monument in California. (B) A view from the top, where glaciers have polished the columns.

(A)

(B)

flood basalt
Basaltic lava that erupts gently in great volume from cracks at Earth's surface to cover large areas of land and form lava plateaus.

lava plateau *or* **basalt plateau** A broad plateau covering thousands of square kilometers, formed by lava deposited during a rapid sequence of fissure eruptions.

vent An opening in a volcano, typically in the crater, through which lava and rock fragments erupt.

crater A bowl-like depression at the summit of a volcano, created by volcanic activity.

FIGURE 8.18 (A) The Columbia River basalt plateau, shown here in gray, covers much of Washington, Oregon, and Idaho. (B) The Multnomah Falls cascade over individual layers of basalt flows in the Columbia River basalt plateau region in Oregon.

kilometers of basaltic lava onto Earth's surface. A fissure eruption of this type creates a **flood basalt**, which covers the landscape like a flood. It is common for many such fissure eruptions to occur in rapid succession and to create a **lava plateau**, or **basalt plateau**, covering thousands of square kilometers.

The Columbia River plateau in eastern Washington, northern Oregon, and western Idaho is a lava plateau containing 350,000 cubic kilometers of basalt (Figure 8.18). The lava is up to 3,000 meters thick and covers 200,000 square kilometers. It formed about 15 million years ago as basaltic magma oozed from long fissures in Earth's surface. The individual flows are between 15 and 100 meters thick.

Volcano Types

Volcanoes differ widely in shape, structure, and size (Table 8.1). Lava and rock fragments commonly erupt from an opening called a **vent** located in a **crater**, a bowl-like depression at the summit of the volcano that was itself created by volcanic activity (Figure 8.19). As mentioned previously, lava or pyroclastic material may also erupt from a fissure on the flanks of the volcano.

TABLE 8.1 Characteristics of Different Types of Volcanic Features

Type of Volcanic Feature	Physical Form	Size	Type of Magma	Style of Activity	Examples
Basalt plateau	Flat to gentle slope	100,000 to 1,000,000 km² in area; 1 to 3 km thick	Basalt	Formed by gentle fissure eruptions	Columbia River plateau
Shield volcano	Slightly sloped, 6° to 12°	Up to 9,000 m high	Basalt	Gentle; some fire fountains	Hawaii
Cinder cone	Moderate slope	100 to 400 m high	Basalt or andesite	Ejections of pyroclastic material	Parícutin (Mexico)
Composite volcano	Alternate layers of flows and pyroclastics	100 to 3,500 m high	Variety of types of magmas and ash	Often violent	Vesuvius (Italy); Mount St. Helens; Aconcagua (Argentina)
Caldera	Circular depression, sometimes with steep walls	Less than 40 km in diameter	Granite	Formed by a violent cataclysmic explosion; potential for violent eruption remains	Yellowstone; San Juan Mountains

FIGURE 8.19 Hot gases rise from vents in the crater of Marum volcano, in the South Pacific island nation of Vanuatu.

shield volcano A large, gently sloping volcanic mountain formed by successive flows of basaltic magma.

cinder cone A small volcano, typically less than 300 meters high, made up of loose, pyroclastic fragments blasted out of a central vent; usually active for only a short time.

Shield Volcanoes

Fluid basaltic magma often builds a gently sloping mountain called a **shield volcano** (Figure 8.20). The sides of a shield volcano generally slope away from the vent at angles between 6° and 12° from horizontal. Although their slopes are gentle, shield volcanoes can be enormous. The height of Mauna Kea volcano in Hawaii, measured from its true base on the sea floor to its top, exceeds the height of Mount Everest.

Although shield volcanoes, such as those of Hawaii and Iceland, erupt regularly, the eruptions are normally gentle and rarely life-threatening. Lava flows occasionally overrun homes and villages, but the flows advance slowly enough to give people time to evacuate.

Cinder Cones

A **cinder cone** is a small volcano composed of pyroclastic fragments. A cinder cone forms when large amounts of gas accumulate in rising magma. When the gas pressure builds sufficiently, the entire mass erupts explosively, hurling cinders, ash, and molten magma into the air. The particles then fall back around the vent to accumulate as a small mountain of pyroclastic debris. A cinder cone is usually active for only a short time because once the gas escapes, the driving force behind the eruption is gone.

As the name implies, a cinder cone is symmetrical. It also can be steep (about 30°), especially near the vent where ash and cinders pile up (Figure 8.21). Most are less than 300 meters high, although a large one can

FIGURE 8.20 Mount Skjoldbreidier in Iceland shows the typical low-angle slopes of a shield volcano.

Chapter 8: Volcanoes and Plutons

FIGURE 8.21 (A) A cinder cone volcano rises behind a field of aa in the Hawaii Volcanoes National Park. (B) Several cinder cones are visible on the flanks of the broader shield volcano in Hawaii.

composite cone *or* **stratovolcano** A steep-sided volcano formed by an alternating series of lava flows and pyroclastic eruptions and marked by repeated eruption.

be up to 700 meters high. A cinder cone erodes easily and quickly because the pyroclastic fragments are not cemented together.

Composite Cones

A **composite cone**, sometimes called a **stratovolcano**, forms over a long period of time as a sequence of lava flows and pyroclastic eruptions. The hard lava covers the loose pyroclastic material and protects it from erosion (Figure 8.22A).

Many of the highest mountains of the Andes and some of the most spectacular mountains of western North America are composite cones (Figure 8.22B). Repeated eruption is a trademark of a composite volcano. Mount St. Helens, in the state of Washington, erupted dozens of times in the 4,500 years preceding its most recent eruption in 1980. Mount Rainier, also in Washington, has been dormant in recent times but could become active again and threaten nearby populated regions.

8.7 Volcanic Explosions: Ash-Flow Tuffs and Calderas

Although granitic magma usually solidifies within the crust, under certain conditions it rises to Earth's surface, where it erupts violently. Granitic magmas that rise to the surface contain only a few percent water, like basaltic magma. But decreasing pressure allows the small amount of dissolved water in granitic magmas to form a frothy, pressurized mixture of gas and liquid magma that may be as hot as 900°C. As the mixture rises to within a few kilometers of Earth's surface, it fractures overlying rocks and explodes skyward through the fractures, as shown in panels A and B of Figure 8.23. Think of a bottle of beer or soda pop. When the cap is on and the contents are under pressure, carbon dioxide gas is dissolved in the liquid. When you remove the cap, pressure decreases and bubbles rise to the surface. If conditions are favorable, the frothy mixture erupts through the bottleneck.

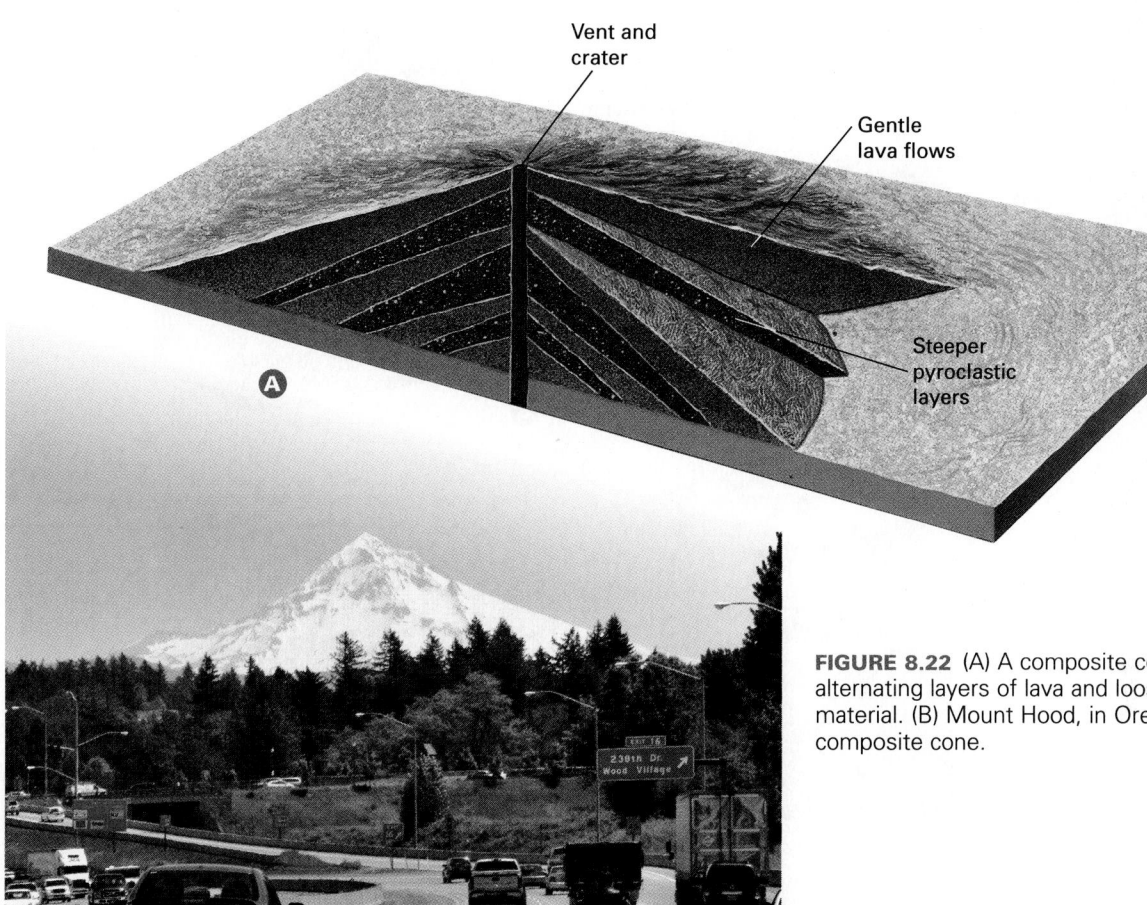

FIGURE 8.22 (A) A composite cone consists of alternating layers of lava and loose pyroclastic material. (B) Mount Hood, in Oregon, is a composite cone.

Vent and crater

Gentle lava flows

Steeper pyroclastic layers

The Destruction of Pompeii

In 79 CE, Mount Vesuvius erupted and destroyed the Roman cities of Pompeii, Herculaneum, and several neighboring villages near what is now Naples, Italy. Prior to that eruption, the volcano had been inactive for about 700 years—so long that farmers had cultivated vineyards on the sides of the mountain all the way to the summit. During the eruption, an ash flow streamed down the flanks of the volcano, burying the cities and towns under 5 to 8 meters of hot ash. When archaeologists located and excavated Pompeii 17 centuries later, they found molds of inhabitants trapped by the ash flow as they attempted to flee or find shelter. Some of the molds appear to preserve facial expressions of terror. After the 79 CE eruption, Mount Vesuvius returned to relative quiescence but became active again in 1631. It was frequently active from 1631 to 1944; in the 20th century it erupted in 1906, 1929, and 1944.

Mount Vesuvius is a stratovolcano that formed over 25,000 years by many eruptions that varied from gently flowing lava to the types of explosions that buried Pompeii. A 1996 study of seismic velocities beneath the volcano showed that seismic waves suddenly slow from 6 to 2.7 kilometers per second at a depth of 10 kilometers. The sudden velocity decrease suggests that molten magma still exists at that depth.* Because stratovolcanoes erupt frequently and remain active for long periods, and because magma underlies the volcano, geologists consider Mount Vesuvius a high risk for future eruptions.

* A. Zollo, P. Gasparini, J. Virieux, H. le Meur, G. de Natale, G. Biella, et al., "Seismic Evidence for a Low-Velocity Zone in the Upper Crust Beneath Mount Vesuvius," *Science* 274 (October 25, 1996), 592–594.

ash flow A mixture of volcanic ash, larger pyroclastic particles, and gas that flows rapidly along Earth's surface as a result of an explosive volcanic eruption.

ash-flow tuff A pyroclastic rock formed when an ash flow solidifies.

caldera A large circular depression created by the collapse of the magma chamber after an explosive volcanic eruption.

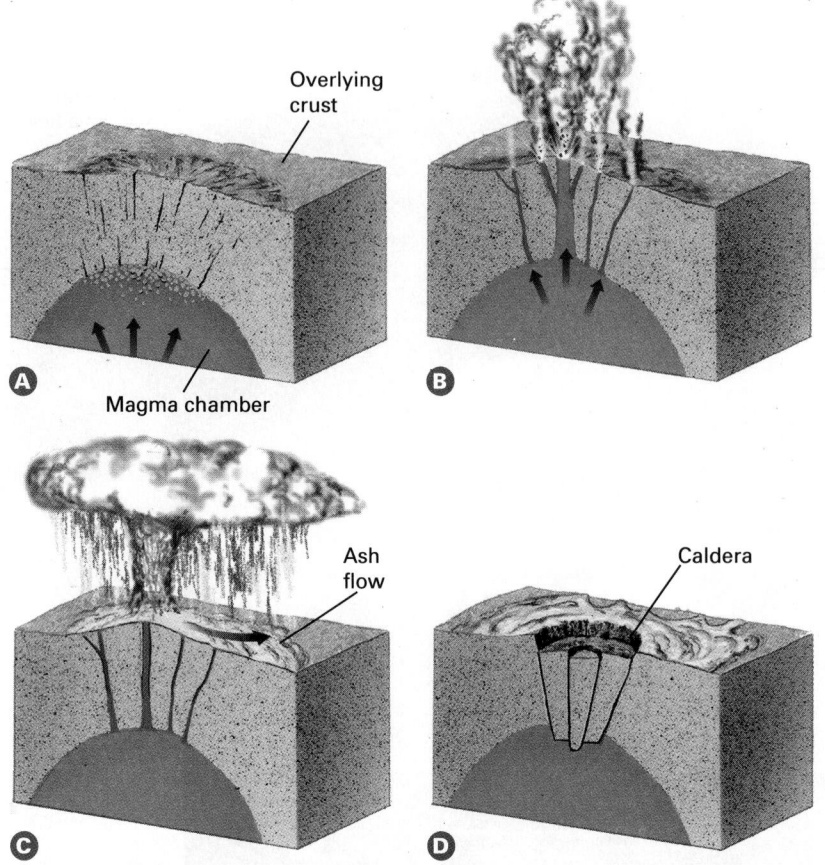

FIGURE 8.23 (A) When granitic magma rises to within a few kilometers of Earth's surface, it stretches and fractures overlying rock. Gas separates from the magma and rises to the upper part of the magma body. (B) The gas-rich magma explodes through fractures, rising as a vertical column of hot ash, rock fragments, and gas. (C) When the gas is used up, the column collapses and spreads outward as a high-speed ash flow. (D) Because so much material has erupted from the top of the magma chamber, the roof collapses to form a caldera.

A large and violent eruption can blast a column of pyroclastic material 10 or 12 kilometers into the sky, and the column might be several kilometers in diameter. A cloud of fine ash may rise even higher—into the upper atmosphere. The force of material streaming out of the magma chamber can hold the column up for hours or even days.

Ash Flows

When much of the gas has escaped from the upper layers of magma, the eruption ends. The airborne column of ash, rock, and gas then falls back to Earth's surface, spreading over the land and rushing down stream valleys (Figure 8.23C). Such a flow is called an **ash flow**.

When an ash flow stops, most of the gas escapes into the atmosphere, leaving behind a chaotic mixture of volcanic ash and rock fragments called **ash-flow tuff** (Figure 8.24).

Calderas

After the gas-charged magma erupts, the roof of the magma chamber collapses into the space that the magma had filled (Figure 8.23D). Because most magma bodies are circular when viewed from above, the collapsing roof forms a circular depression called a **caldera**. A large caldera may be 40 kilometers in diameter and have walls as much as a kilometer high. Some calderas fill up with volcanic debris as the ash column collapses; others maintain the circular depression and steep walls (Figure 8.25). We usually think of volcanic landforms as mountain peaks, but the topographic depression of a caldera is an exception. Figure 8.26 shows that calderas, ash-flow tuffs, and related rocks occur over a large part of western North America. Yellowstone National Park, Crater Lake, and Mount Vesuvius are well-known examples.

COURTESY OF GRAHAM R. THOMPSON/JONATHAN TURK

GEOFFREY SUTTON

FIGURE 8.24 (A) Ash-flow tuff forms the spectacular cliffs of Smith Rock State Park near Bend, Oregon. (B) Ash-flow tuff forms when an ash flow comes to a stop. The fragments in the tuff are pieces of rock that were carried along with the volcanic ash and gas.

COPYRIGHT AND PHOTOGRAPH BY DR. PARVINDER S. SETHI

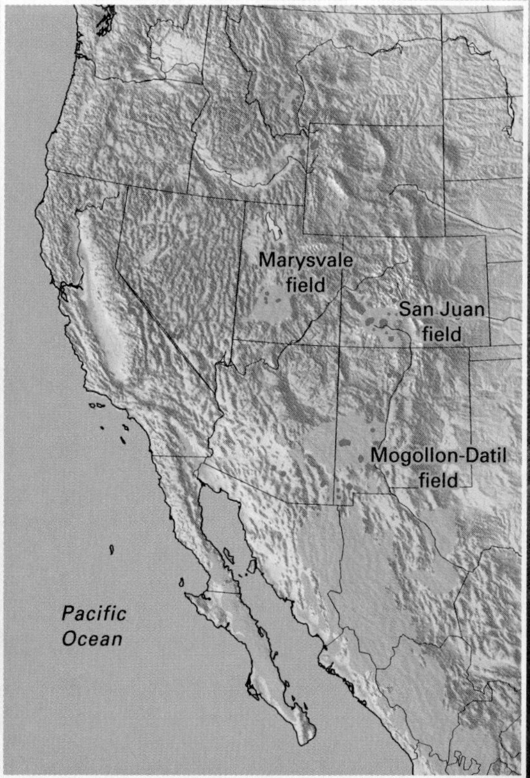

Marysvale field

San Juan field

Mogollon-Datil field

Pacific Ocean

FIGURE 8.26 Calderas (red dots) and ash-flow tuffs (orange areas) are abundant in western North America.

FIGURE 8.25 A volcanic crater at Diamond Head in Honolulu, Hawaii, is a caldera that was formed by a volcanic eruption.

Yellowstone National Park and the Long Valley Caldera

Yellowstone National Park in Wyoming and Montana is the oldest national park in the United States. Its geology consists of three large overlapping calderas and the ash-flow tuffs that erupted from them (Figure 8.27). The oldest eruption took place 1.9 million years ago and ejected 2,500 cubic kilometers of pyroclastic material. This is enough ash to bury all of Yellowstone Park to a depth of 250 meters. The next major eruption occurred 1.3 million years ago and produced about 400 cubic kilometers of ash. The most recent, 0.6 million years ago, ejected 1,000 cubic kilometers of ash and other debris and produced the Yellowstone caldera in the center of the park.

The park's geysers and hot springs are formed by hot magma beneath Yellowstone, and numerous small earthquakes indicate that the magma is moving. Geologists would not be surprised if another eruption occurred at any time.

A geologic environment similar to that of Yellowstone is found near Yosemite National Park in eastern California. Here the 170-cubic-kilometer Bishop Tuff erupted from the Long Valley caldera 0.7 million years ago. Seismic monitoring indicates that magma lies beneath Mammoth Mountain, a popular California ski area, on the southwest edge of the Long Valley caldera. Unusual amounts of carbon dioxide—a common

volcanic gas—have escaped from the caldera since 1994. Another eruption would not surprise geologists.

No event even of the magnitude of the smaller Long Valley eruption has occurred in recorded history. The 1980 Mount St. Helens eruption was 170 times smaller than the Bishop Tuff eruption. Yet the Mount St. Helens eruption killed 57 people, caused hundreds of millions of dollars in damage, and disrupted people's lives hundreds of miles from the volcano. An eruption on the scale of Yellowstone, or even of the Long Valley caldera, would probably kill many thousands of people, entomb towns and cities beneath meters of ash, and change the courses of rivers and streams. It would probably raise a dust cloud in the upper atmosphere that would darken the Sun over the entire planet for months or years, cooling the atmosphere, altering global climate, and changing global ecosystems.

FIGURE 8.27 This map shows the locations of the three Yellowstone calderas and the associated ash-flow tuff deposits.

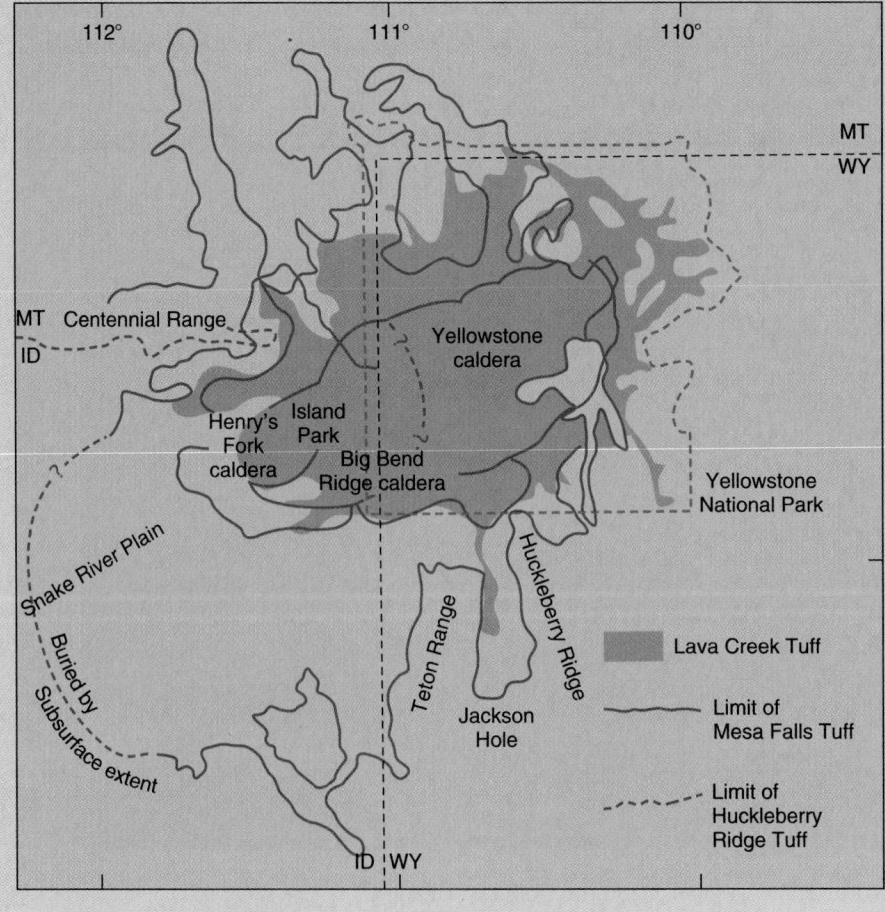

8.8 Risk Assessment: Predicting Volcanic Eruptions

Table 8.2 summarizes the major known volcanic disasters since the year 1500. The potential for such disasters in the future makes a volcanic eruption one of the greatest of geologic hazards. It also makes risk assessment and prediction of volcanic eruptions an important part of modern science.

Approximately 1,300 active volcanoes are recognized globally, and 5,564 eruptions have occurred in the past 10,000 years. These figures do not include the numerous submarine volcanoes of the Mid-Oceanic Ridge system. Many volcanoes have erupted recently, and we are certain that others will erupt soon. How can geologists predict an eruption and reduce the risk of a volcanic disaster?

Regional Prediction

Risk assessment for regional predictions is based both on the frequency of past eruptions and on potential violence. However, regional predictions based on the concentration of volcanoes in an area can only estimate *probabilities* and cannot be used to determine exactly when a particular volcano will erupt or the intensity of a particular eruption.

Short-Term Prediction

In contrast to regional predictions, short-term predictions attempt to forecast the specific time and place of an impending eruption. They are based on instruments that monitor an active volcano to detect signals that the volcano is about to erupt. The signals include changes in the shape of the mountain and surrounding land, earthquake swarms indicating movement of magma beneath the mountain, increased emissions of ash or gas, increasing temperatures of nearby hot springs, and any other signs that magma is approaching the surface.

In 1978, two U.S. Geological Survey (USGS) geologists, Dwight Crandall and Don Mullineaux, noted that Mount St. Helens had erupted more frequently and violently during the past 4,500 years than any other volcano in the contiguous 48 states. They predicted that the volcano would erupt again before the end of the 20th century.

In March 1980, about two months before the great May eruption, puffs of steam and volcanic ash rose from the crater of Mount St. Helens, and swarms of earthquakes occurred beneath the mountain. This activity convinced other USGS geologists that Crandall and Mullineaux's prediction was correct. In response, they installed networks of seismographs, tiltmeters, and surveying instruments on and around the mountain.

In the spring of 1980, the geologists warned government agencies and the public that Mount St. Helens showed signs of an impending eruption. The U.S. Forest Service and local law enforcement officers quickly evacuated the area surrounding the mountain, averting a much larger tragedy (Figure 8.28).

TABLE 8.2 Some Notable Volcanic Disasters Involving 5,000 or More Fatalities, Since the Year 1500

| Volcano | Country | Year | Primary Cause of Death and Number of Deaths | | | | |
			Pyroclastic Flow	Debris Flow	Lava Flow	Posteruption Starvation	Tsunami
Kelut	Indonesia	1586		10,000			
Vesuvius	Italy	1631	18,000				
Etna	Italy	1669			10,000		
Lakagigar	Iceland	1783				9,340	
Unzen	Japan	1792					15,190
Tambora	Indonesia	1815	12,000			80,000	
Krakatoa	Indonesia	1883					36,420
Pelée	Martinique	1902	29,000				
Santa Maria	Guatemala	1902	6,000				
Kelut	Indonesia	1919		5,110			
Nevado del Ruiz	Colombia	1985		>22,000			

FIGURE 8.28 U.S. Geological Survey geologists accurately predicted the May 1980 eruption of Mount St. Helens.

8.9 Volcanic Eruptions and Global Climate

We have learned that magma production, the first step in the development of a volcanic eruption, occurs in the upper mantle. However, as another example of a systems interaction, an eruption can profoundly affect the atmosphere, climate, and living organisms. For example, the 1991 eruptions of Mount Pinatubo in the Philippines produced the greatest ash and sulfur clouds in the latter half of the 20th century. Satellite measurements show that the total solar radiation reaching Earth's surface declined by 2 to 4 percent after the Pinatubo eruptions. The following two years, 1992 and 1993, were a few tenths of a degree Celsius cooler than the temperatures of the previous decade. Temperatures rose again in 1994, after the ash and sulfur settled out.

A plot of global temperatures before and after eight recent major volcanic eruptions shows a correla-tion between global cooling and volcanic eruptions (Figure 8.29). The correlation substantiates meteorological models showing that high-altitude dust reflects sunlight and cools the atmosphere.

Historic eruptions have been minuscule compared with some in the more distant past. About 248 million years ago, at the end of the Permian Period, 90 percent of all marine species and two-thirds of reptile and amphibian species died suddenly in the most catastrophic mass extinction in Earth history. This extinction event coincided with a massive volcanic eruption in Siberia that disgorged a million cubic kilometers of flood basalt onto Earth's surface to form a great lava plateau. The eruption must have released massive amounts of ash and sulfur compounds into the upper atmosphere, leading to cooler global climates. Many geologists think that Earth cooled enough to cause or at least contribute to the mass extinction.

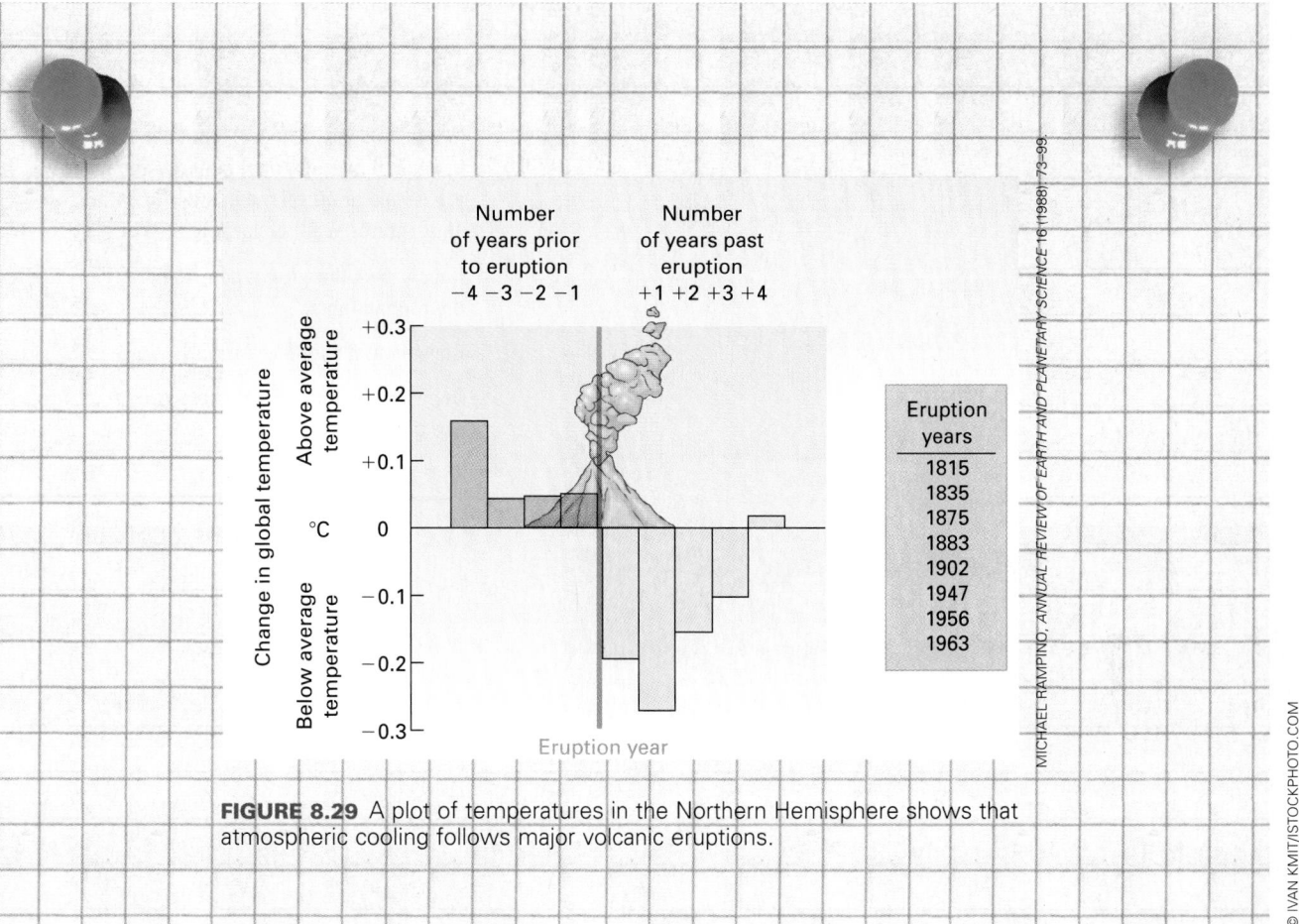

FIGURE 8.29 A plot of temperatures in the Northern Hemisphere shows that atmospheric cooling follows major volcanic eruptions.

MICHAEL RAMPINO, ANNUAL REVIEW OF EARTH AND PLANETARY SCIENCE 16 (1988) 73–99

VIRTUAL FIELD TRIP

Hydrothermal Activity

Ready to Go!

Hydrothermal is a term referring to hot water, and when geologists talk about hydrothermal activity, they are referring to the interaction between hot water and Earth. Most people associate hydrothermal activity with geysers, such as Old Faithful in Yellowstone National Park, Wyoming, or hot springs where one can go to bathe or relax. Hydrothermal activity also can result in the production of geothermal energy.

There are three areas in the world known for their hydrothermal activity. These are Iceland, New Zealand, and Yellowstone National Park, Wyoming. In this virtual field trip, we'll be visiting what is probably the best known area to view hydrothermal activity and its results: Yellowstone National Park. So, get ready to visit the crown jewel of the U.S. National Park Service.

GOALS OF THE TRIP

In this field trip, you will learn about a number of hydrothermal features and what processes are involved in their formation. The features we'll be looking at are:

1. **Hot springs, their formation, and their inhabitants.**
2. **Geysers and how they form.**
3. **Deposits associated with hot springs and geysers.**

FOLLOW-UP QUESTIONS

1. Why do you think Old Faithful erupts on such a regular basis?
2. Can the color of deposits associated with hot springs serve as an indicator of water chemistry and temperature?
3. Do you think the hydrothermal activity found in Yellowstone National Park could be tapped as a potential source of geothermal energy? What are the pros and cons of doing so?

What to See When You Go

A *hot spring* (also called a *thermal spring* or *warm spring*) is any spring in which the water temperature is higher than the temperature of the human body. Recall from the feature titled "Yellowstone National Park and the Long Valley Caldera" on page 148 that the park's hot springs are formed by hot magma beneath Yellowstone. The underground system of fractures and openings associated with hot springs is not as constrictive, nor usually as deep, as in a geyser; hence the water can bubble up and spill out onto the surface. Nevertheless, steam and volcanic gases are still associated with hot springs, as seen in this photo.

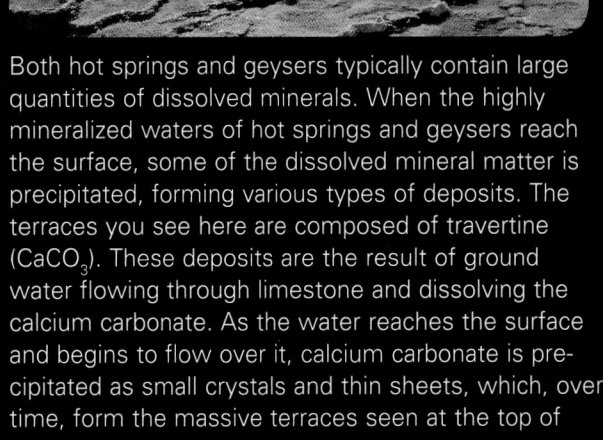

This photograph shows the vent of the Pump Geyser. Hot, acidic water containing silica and dissolved gases flows down the slope from the vent, providing an environment for *extremophiles* (organisms that live in extreme environments), in this case thermophiles (species that thrive in extremely hot water, and are typically bacteria or cyanobacteria). The reddish colors are rust-colored iron-oxide deposits resulting from the metabolism of iron by some of these bacteria. The yellow-colored deposits are sulfur, which is formed by sulfur-loving species of bacteria that reduce the hydrogen sulfide gas being emitted at the vent. The brownish bacterial mats contain bacteria that live in the cooler waters (below 140°F).

Both hot springs and geysers typically contain large quantities of dissolved minerals. When the highly mineralized waters of hot springs and geysers reach the surface, some of the dissolved mineral matter is precipitated, forming various types of deposits. The terraces you see here are composed of travertine ($CaCO_3$). These deposits are the result of ground water flowing through limestone and dissolving the calcium carbonate. As the water reaches the surface and begins to flow over it, calcium carbonate is precipitated as small crystals and thin sheets, which, over time, form the massive terraces seen at the top of this page.

Hot springs that intermittently eject hot water and steam with tremendous force are known as *geysers*. You will read more about geysers in Chapter 11, where you will learn that they are the surface expression of an extensive underground system of deep interconnected fractures within hot rocks. Ground water in these fractures is heated above the boiling point of water. Because of the pressure at depth, this water does not boil, but if there is a slight drop in pressure, such as from escaping gas, the hot water will instantly change to steam, which will then push the water upwards through the ground and into the air, producing a geyser. Old Faithful Geyser, shown here in eruption, is probably the most famous geyser in the world. Eruptions occur approximately every 90 minutes.

When hot water containing dissolved silica is erupted from a geyser, it cools and deposits silica (commonly called *sinter*) around the vent. Sinter comes in a variety of shapes and sizes. One of the largest deposits in the world is found at the Castle Geyser, so named for its resemblance to a castle ruin (see photo above). The Castle Geyser is thought to be thousands of years old and the sinter it is currently depositing is being laid down over even older and thicker deposits of sinter.

9

MOUNTAINS

Sunrise over the Grand Tetons, in Wyoming.

visit 4ltrpress.cengage.com

> { **Is not the mountain far more awe-inspiring and more clearly visible to one passing through the valley than to those who inhabit the mountain?** }
>
> *Kahlil Gibran*

stress The force exerted against an object, usually measured as force per unit area or pressure.

geologic structure Any feature produced by rock deformation, such as a fold or a fault; also refers to the combination of all such features of an area or region.

fold A bend in rock.

9.1 Folds and Faults: Geologic Structures

How Rocks Respond to Tectonic Stress

Continents move at rates between 1 to 16 centimeters per year, about as fast as your fingernail grows. Although this movement is too slow to feel or sense in any ordinary way, try to imagine the immense forces involved as continent- or ocean-sized slabs of rock slowly grind past one another or collide head-on. In Chapters 7 and 8, we studied how tectonic activity generates earthquakes and volcanic eruptions. In this chapter we will consider how rock crumples and buckles under tectonic forces to form mountains.

Stress is a force directed against an object. Stress is conventionally measured as force per unit area or pressure. When a rock is stressed, the rock may deform elastically or plastically, or it may simply break by brittle fracture, as described in Chapter 7. Several factors control how a rock responds to stress:

1. *The nature of the rock.* Think of a quartz crystal, a gold nugget, and a rubber ball. If you strike quartz with a hammer, it shatters—that is, it fails by brittle fracture. In contrast, if you strike the gold nugget, it deforms in a plastic manner—it flattens and stays flat. If you hit the rubber ball, it deforms elastically and rebounds immediately, sending the hammer flying back at you. Initially, all rocks react to stress by deforming elastically by a slight amount. Near Earth's surface, where temperature and pressure are low, different types of rocks behave differently with continuing stress. Granite and quartzite tend to fracture in a brittle manner. Other rocks, such as shale, limestone, and marble, tend to deform plastically.

2. *Temperature.* The higher the temperature, the greater the tendency of a rock to deform in a plastic manner. It is difficult to bend an iron bar at room temperature, but if the bar is heated to a red-hot temperature, it becomes plastic and bends easily.

3. *Pressure.* High nondirected (hydrostatic) pressure favors plastic behavior. Thus, because both temperature and pressure increase during burial, and because both of these factors promote plastic deformation, deeply buried rocks have a greater tendency to bend and flow under stress than do shallow rocks.

4. *Time.* Stress applied slowly, rather than suddenly, also favors plastic behavior. Marble park benches in New York City have sagged plastically under their own weight within 100 years. In contrast, rapidly applied stress, such as the blow of a hammer, causes brittle fracture in a marble bench.

Geologic Structures

Tectonic movement creates tremendous stress near plate boundaries; this stress deforms rocks. A **geologic structure** is any feature produced by rock deformation. Tectonic stress creates three types of structures: folds, faults, and joints.

Folds

A **fold** is a bend in rock. Some folded rocks display little or no fracturing, indicating that the rocks deformed in a plastic manner. In other cases, folding occurs by a combination of plastic deformation and brittle fracture (Figure 9.1). Folds formed in this manner exhibit many tiny fractures.

If you hold a sheet of clay between your hands and squeeze your hands together, the clay

anticline A fold in rock that arches upward; the oldest rocks are in the middle.

syncline A fold in rock that arches downward, and whose center contains the youngest rocks.

limbs The sides of a fold in rock.

dome A circular or elliptical anticline structure, resembling an inverted cereal bowl.

basin A bowl-shaped syncline structure, commonly filled with sediment.

deforms into a sequence of folds (Figure 9.2). This demonstration illustrates three characteristics of folds:

1. Folding usually results from compression. For example, tightly folded rocks near Palmdale, California, indicate that the region was compressed as tectonic plates converged (Figure 9.3).
2. Folding always shortens the horizontal distances in rock. Notice in Figure 9.4 that the distance between the two points A and A′ is shorter in the folded rock than it was before folding.
3. A fold usually occurs as part of a group of many similar folds.

Figure 9.5 summarizes the characteristics of five common types of folds. The figure shows

COURTESY OF GRAHAM R. THOMPSON/JONATHAN TURK

FIGURE 9.1 This rock, which is about 1 meter wide, is on the shore of the Nahanni River in the Northwest Territories of Canada. It folded plastically and then fractured.

that a fold arching upward is called an **anticline** and one arching downward is a **syncline**. The sides of a fold are called the **limbs**. Even though an anticline is structurally a high point in a fold, anticlines do not necessarily form topographic ridges. Conversely, synclines do not always form valleys. Landforms are created by combinations of tectonic and surface processes. In Figure 9.6, a syncline lies beneath the peak and an anticline forms a saddle between two peaks simply because some portions of the folded rock eroded faster than others.

A circular or elliptical anticlinal structure is called a **dome**. A dome resembles an inverted cereal bowl. Sedimentary layering dips away from the top of a dome in all directions (Figure 9.7A). A similarly shaped syncline is called a **basin** (Figure 9.7B). Domes and basins can be small structures only a few kilometers in diameter or less. Alternatively, a large basin or dome can be 500 kilometers in diameter. These huge structures form by the sinking or rising of continental crust in response to vertical movements of the underlying mantle. The

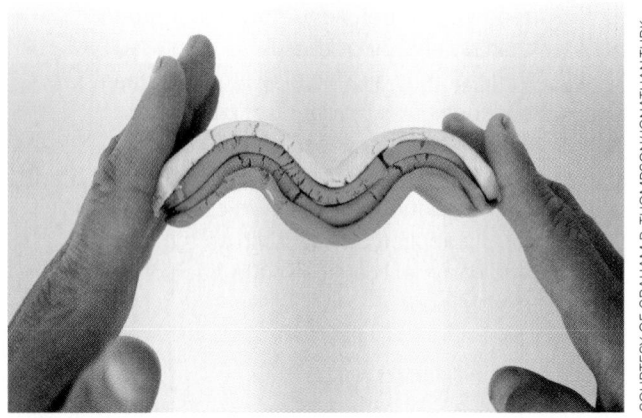

COURTESY OF GRAHAM R. THOMPSON/JONATHAN TURK

FIGURE 9.2 Clay deforms into folds when compressed.

FIGURE 9.3 Folded sedimentary rocks found near Palmdale, California, reflect tectonic compression.

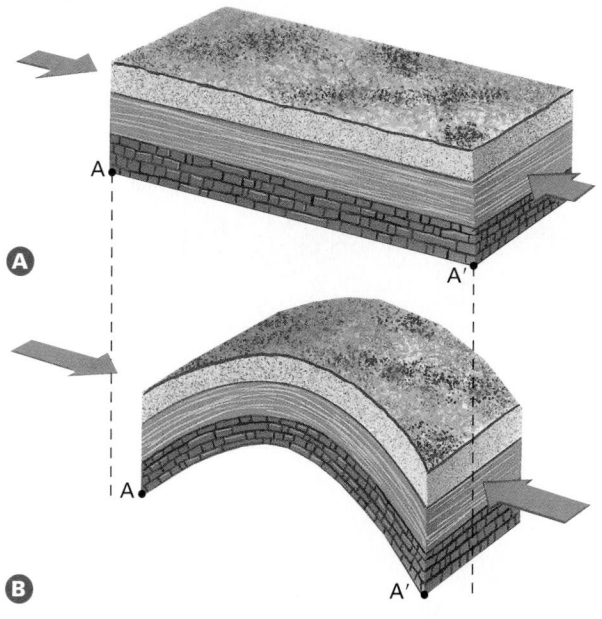

FIGURE 9.4 (A) Horizontally layered sedimentary rocks. (B) Layers fold in response to compressive stress. Notice that points A and A′ are closer after folding.

Black Hills of South Dakota are a large structural dome. The Michigan basin covers much of the state of Michigan, and the Williston basin covers much of eastern Montana, northeastern Wyoming, the western Dakotas, and southern Alberta and Saskatchewan.

Faults

A **fault** is a fracture along which rock on one side has moved relative to rock on the other side (Figure 9.8). **Slip** is the distance that rocks on opposite sides of a fault have moved. Movement along a fault may be gradual, or the rock may move suddenly, generating an earthquake. Some faults occur as single fractures in rock; most, however, consist of numerous closely spaced fractures within a **fault zone** (Figure 9.9). Rock may slide hundreds of meters or many kilometers along a large fault zone.

fault A fracture in rock along which one side has moved relative to the other side. Compare with *joint*.

slip The distance that rocks on opposite sides of a fault have moved.

fault zone An area of numerous, closely spaced faults.

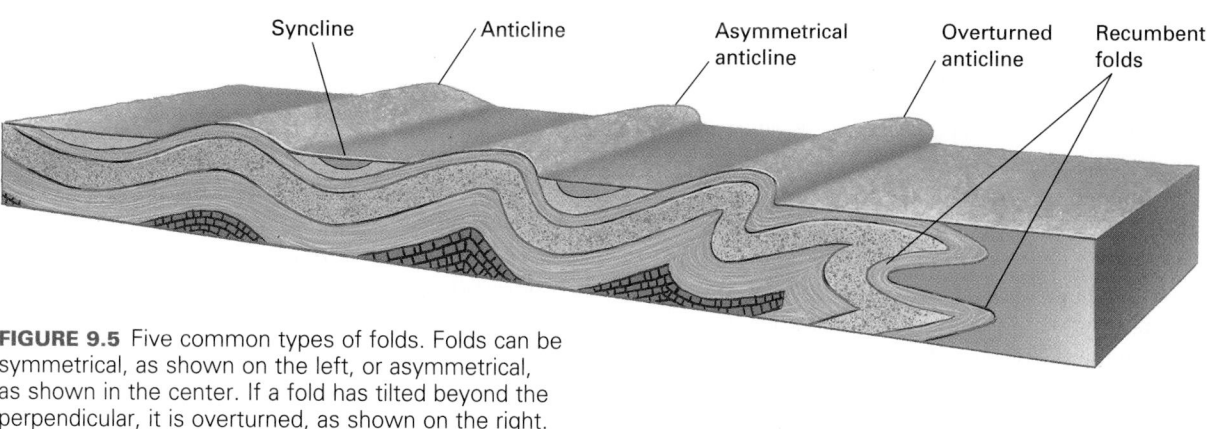

FIGURE 9.5 Five common types of folds. Folds can be symmetrical, as shown on the left, or asymmetrical, as shown in the center. If a fold has tilted beyond the perpendicular, it is overturned, as shown on the right.

FIGURE 9.6 On this peak in the Canadian Rockies, the syncline lies beneath the summit and the anticline forms the low point on the ridge.

hanging wall A term to describe the top side of an inclined fault or vein (i.e., the rock hanging above one's head).

footwall A term to describe the lower side of an inclined fault or vein (i.e., the rock one would walk on).

normal fault A fault in which the hanging wall has moved downward relative to the footwall.

graben A wedge-shaped block of rock that has dropped downward between two normal faults, forming a valley.

Rock moves repeatedly along many faults and fault zones for two reasons: (1) Tectonic forces commonly persist in the same place over long periods of time (for example, at a tectonic plate boundary). (2) Once a fault forms, it is easier for rock to move again along the same fracture than for a new fracture to develop nearby.

Hydrothermal solutions often precipitate rich ore veins in the fractures of a fault zone. Miners then dig shafts and tunnels along veins to get the ore. Many faults are not vertical but dip into Earth at an angle. Therefore, many veins have an upper and a lower side. Miners refer to the side that hangs over their heads as the **hanging wall** and the side they walk on is the **footwall** (Figure 9.10). These names are commonly used to describe both ore veins and faults.

A **normal fault** forms where tectonic movement stretches Earth's crust, pulling it apart. As the crust stretches and fractures, the hanging wall moves down relative to the footwall. Notice that the horizontal distance between points on opposite sides of the fault, such as A and A′ in Figure 9.10, is greater after normal faulting occurs.

Figure 9.11 shows a wedge-shaped block of rock called a **graben** that has dropped downward between pairs of normal faults. The word *graben* comes from the German word for "ditch or valley." (Think of a large block of rock settling downward to form a valley.)

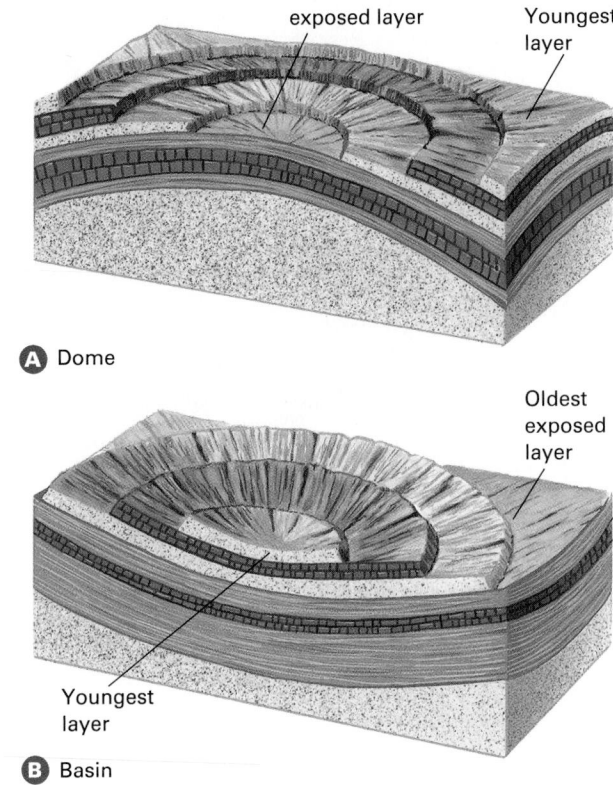

A Dome

B Basin

FIGURE 9.7 (A) Sedimentary layering dips away from a dome in all directions. (B) Layering dips toward the center of a basin.

If tectonic forces stretch the crust over a large area, many normal faults may develop, allowing numerous grabens to settle downward between the faults. The block of rock between two down-dropped grabens then appears to have moved upward relative to the grabens; it is called a **horst**.

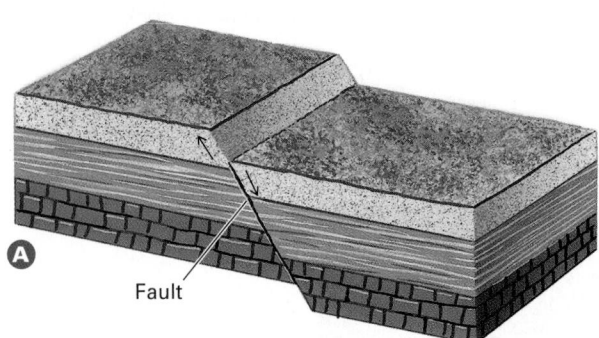

A

Fault

FIGURE 9.8 (A) A fault is a fracture along which rock on one side has moved relative to rock on the other side. (B) A small fault has dropped the right side of these volcanic ash layers downward about 60 centimeters relative to the left side.

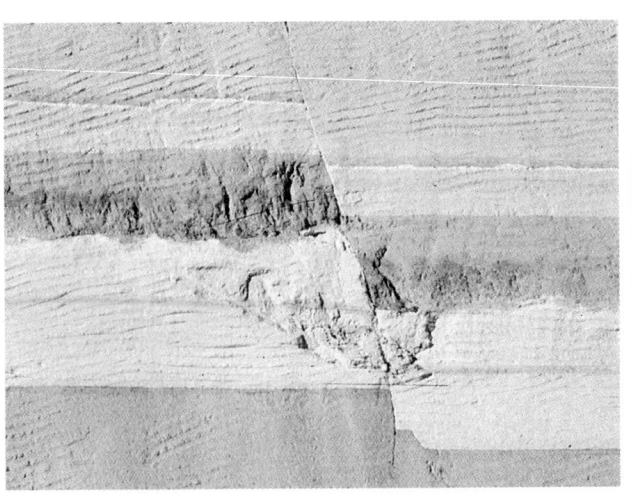

B

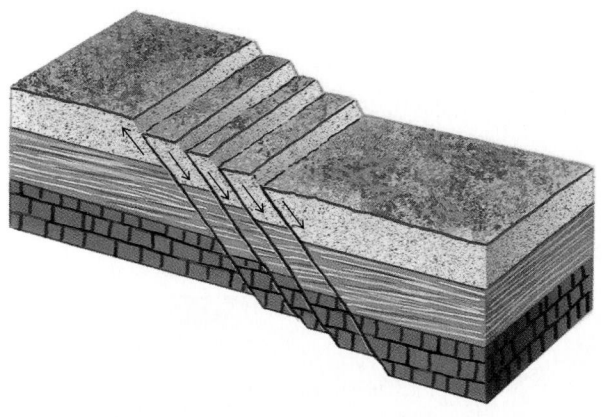

FIGURE 9.9 Faults with a large slip commonly move along numerous closely spaced fractures, forming a fault zone.

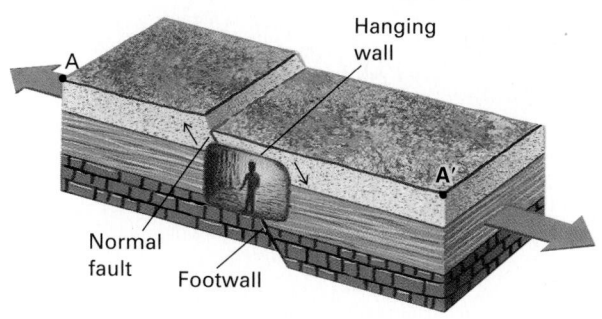

Hanging wall

Normal fault

Footwall

A

A′

FIGURE 9.10 A normal fault accommodates extension of the crust. The upper side of a fault is called the hanging wall, and the lower side is called the footwall.

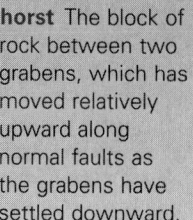

horst The block of rock between two grabens, which has moved relatively upward along normal faults as the grabens have settled downward.

reverse fault A fault in which the hanging wall has moved up relative to the footwall.

thrust fault A type of reverse fault that is nearly horizontal, with a dip of 45° or less over most of its extent.

Normal faults, grabens, and horsts are common where the crust is rifting at a spreading center, such as the Mid-Oceanic Ridge and the East African Rift zone, and where tectonic forces stretch a single plate, as in the Basin and Range Province of Utah, Nevada, and adjacent parts of western North America. They reflect extension of Earth's crust.

In a region where compressive tectonic forces squeeze the crust, geologic structures must accommodate crustal shortening. A fold accomplishes shortening.

Alternatively, compressive forces may fracture the rock to produce a **reverse fault** (Figure 9.12). Here, the hanging wall has moved upward relative to the footwall, as tectonic force squeezed the rock horizon-

tally. In Figure 9.12A, the distance between points A and A′ is shortened by the faulting.

A **thrust fault** is a special type of reverse fault that is nearly horizontal (Figure 9.13). In some thrust faults, the rocks of the hanging wall have moved many kilometers over the footwall. For example, all of the rocks of Glacier National Park in northwestern Montana slid 50 to 100 kilometers eastward along a thrust fault to their present location. This thrust is one of many that formed from about 180 to 45 million years ago as compressive tectonic forces built the mountains of western North America. Most of those thrusts moved large slabs of rock, some even larger than that of Glacier Park, from west to east in a zone extending from Alaska to Mexico.

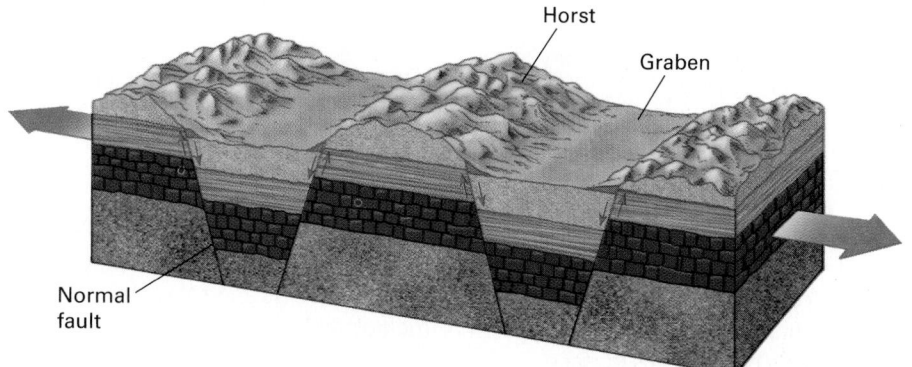

Horst

Graben

Normal fault

FIGURE 9.11 Horsts and grabens commonly form where tectonic forces stretch the crust over a broad area.

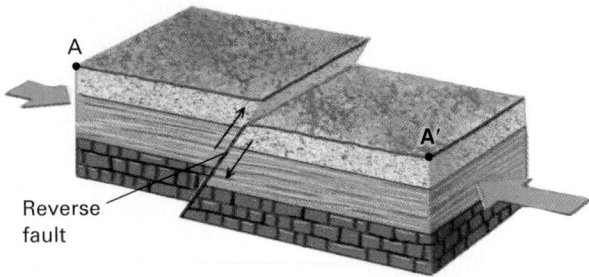

Reverse fault

A

B

FIGURE 9.12 (A) A reverse fault accommodates crustal shortening when the crust is compressed. (B) A small reverse fault in Zion National Park, Utah.

Recall from Chapter 7 that a *strike-slip fault* is one in which the fracture is vertical, or nearly so, and rocks on opposite sides of the fracture move horizontally past each other (Figure 9.14). A transform plate boundary is a strike-slip fault. As explained previously, the famous San Andreas Fault zone is a zone of strike-slip faults that form the boundary between the Pacific plate and the North American plate.

Joints

A **joint** is a fracture in which rocks on either side of the fracture have not moved. We discussed columnar joints in basalt in Chapter 8. Tectonic forces also create joints (Figure 9.15). Most rocks near Earth's surface are jointed, but joints become less abundant with depth because rocks become more plastic and less prone to fracturing at deeper levels in the crust.

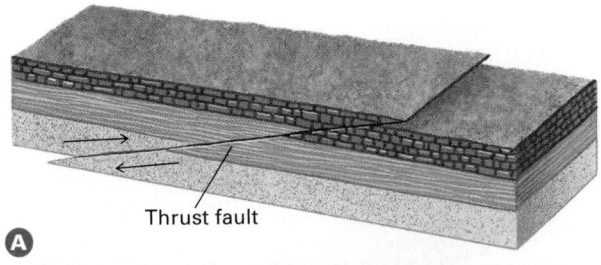

Thrust fault

A

B

FIGURE 9.13 (A) A thrust fault is a low-angle, nearly horizontal reverse fault. (B) A small thrust fault near Flagstaff, Arizona.

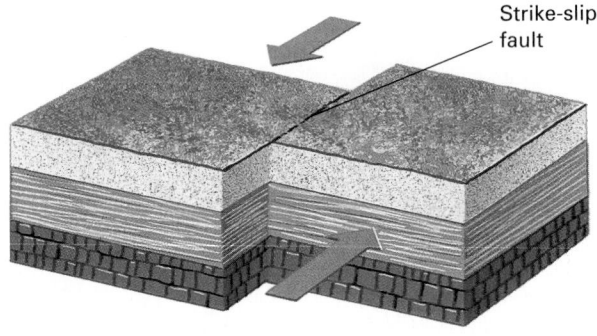

Strike-slip fault

FIGURE 9.14 A strike-slip fault is nearly vertical, but movement along the fault is horizontal.

Joints and faults are important in engineering, mining, and quarrying because they are planes of weakness in otherwise strong rock. Dams constructed in jointed rock often leak, not because the dams themselves have holes but because water seeps into the joints and flows around the dam through the fractures. You can commonly see seepage caused by such leaks in canyon walls downstream from a dam.

joint A fracture along which the rock on either side of the break does not move. Compare with *fault*.

orogeny The process of mountain building; all tectonic processes associated with mountain building.

FIGURE 9.15 Joints such as these in granite in Joshua Tree National Park in California are fractures along which the rock has not slipped.

Folds, Faults, and Plate Boundaries

Each of the three types of plate boundaries—divergent, transform, or convergent—produces different tectonic stresses and therefore different kinds of folds and faults. Tectonic plates drift apart at a divergent boundary (the Mid-Oceanic Ridge system and continental rifts), stretching adjacent rock and producing normal faults and grabens but little folding of rocks.

At a transform boundary, friction often holds rock together as the plates gradually slip past each other. The resultant stress may fold, fault, and uplift nearby rocks. Forces of this type have formed the San Gabriel Mountains along the San Andreas Fault zone, as well as mountain ranges north of the Himalayas.

Near a convergent plate boundary, compression commonly produces large regions of folds, reverse faults, and thrust faults. Folds and thrust faults are common in the mountains of western North America, the Appalachian Mountains, the Alps, and the Himalayas, all of which formed at convergent boundaries.

Although plate convergence commonly creates horizontal compression, in some instances crustal extension and normal faulting also occur at a convergent plate boundary. The Himalayas are an intensely folded mountain chain that formed by the collision between India and southern Asia. Yet, normal faults are common in parts of the Himalayas. We will describe how this occurs in Section 9.5.

9.2 Mountains and Mountain Ranges

Tectonic forces have created mountains at each of the three types of tectonic plate boundaries. As you learned in Chapter 6, the world's largest mountain chain, the Mid-Oceanic Ridge system, formed at divergent plate boundaries beneath the ocean. Mountains also rise at divergent plate boundaries on land. Mount Kilimanjaro and Mount Kenya, two volcanic peaks near the equator, lie along the East African Rift. Other ranges, such as the San Gabriel Mountains of California, formed at transform plate boundaries. However, the volcanic island arcs of the southwestern Pacific Ocean, Alaska's Aleutian Islands, and the great continental mountain chains, including the Andes, Appalachians, Alps, Himalayas, and Rockies, all rose near convergent plate boundaries. Folding and faulting of rocks, earthquakes, volcanic eruptions, intrusion of plutons, and metamorphism all occur at a convergent plate boundary. The term **orogeny** (from the Greek for "mountain producing") refers to the process of mountain building and includes all of these activities.

Because plate boundaries are nearly linear or slightly curved, mountains most commonly occur as long, linear, or slightly curved ranges and chains. For example, the Andes extend in a narrow band along the west coast of South America, and the Appalachians

underthrusting
The process by which one continent moves beneath the other during a continent–continent collision.

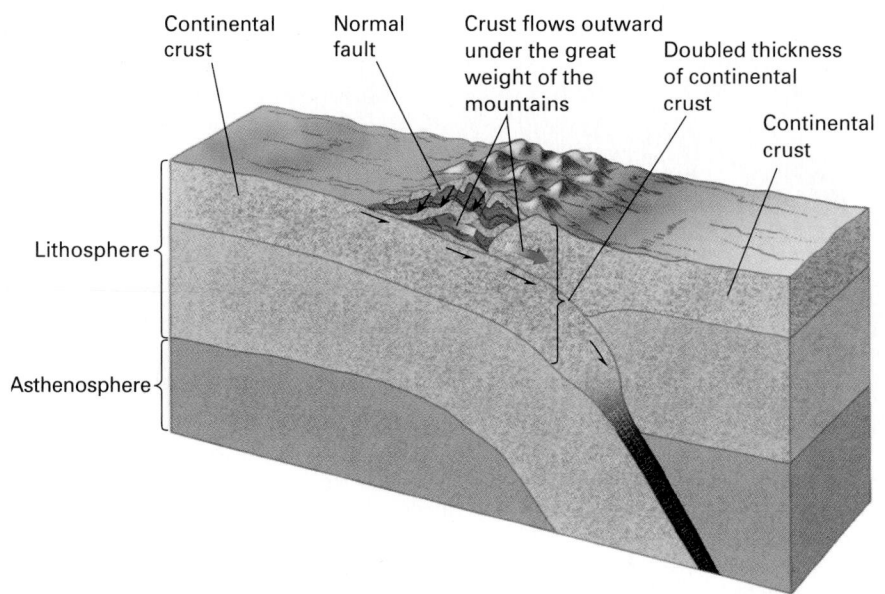

Continental crust   Normal fault   Crust flows outward under the great weight of the mountains   Doubled thickness of continental crust   Continental crust

Lithosphere

Asthenosphere

FIGURE 9.16 Several factors affect the height of a mountain range. Today the Himalayas are being uplifted by continued underthrusting. At the same time, erosion removes the tops of the peaks, and the sides of the range slip downward along normal faults. As these processes remove weight from the range, the mountains rise isostatically.

form a gently curving uplift along the east coast of North America.

Most continental mountain ranges rise isostatically because the crust becomes thicker where tectonic plates converge. Several processes thicken the crust:

1. In a subduction zone, the descending slab generates magma, which rises to cool within the overlying crust to form plutons, or erupts onto the surface to form volcanic peaks. Both the plutons and volcanic rocks thicken the continental crust by adding volumes of new material to it.

2. Magmatic activity heats the lithosphere above a subduction zone, causing it to expand and become thicker.

3. In a region where two continents collide, such as the modern Himalayas, one continent may be forced beneath the other. This process, called **underthrusting**, can double the thickness of continental crust in the collision zone (Figure 9.16).

4. Compressive forces squeeze the crust horizontally to increase its thickness. These compressive forces are important in both subduction zones and continent–continent collisions.

As the lithosphere thickens, it also rises isostatically. As the mountain chain grows higher and heavier, eventually the underlying rocks cannot support the weight of the mountains. The crust and underlying lithosphere then spread outward beneath the mountains. As an analogy, consider pouring cold honey onto a table top. At first, the honey piles up into a high, steep mound, but soon it begins to flow outward under its own weight, lowering the top of the mound.

At the same time, streams, glaciers, and landslides erode the peaks as they rise, carrying the sediment into adjacent valleys. Initially, when the mountains erode, they become lighter and rise isostatically, just as a canoe rises when you step out of it. Eventually, erosion wins over isostatic rebound. The Appalachians are an old range where erosion is now wearing away the remains of peaks that may once have been the size of the Himalayas.

With this background, let's look at mountain building in three types of convergent plate boundaries: in the ocean, at a continental margin, and between continents.

> **Initially, when mountains erode, they become lighter and rise isostatically, just as a canoe rises when you step out of it.**

9.3 Island Arcs: Subduction Where Two Oceanic Plates Converge

An **island arc** is a volcanic mountain chain that forms where two plates carrying oceanic crust converge. The convergence causes the older, colder, and denser plate to sink into the mantle beneath the other plate, creating a subduction zone and an oceanic trench (Figure 9.17). Magma forms in the subduction zone and rises to build submarine volcanoes. These volcanoes may eventually grow above sea level, creating an arc-shaped volcanic island chain next to the trench.

A layer of sediment a half-kilometer thick or more commonly covers the oldest basaltic crust of the deep seafloor. As the two plates converge, some of the sediment is scraped from the subducting slab and jammed against the inner wall (the wall toward the island arc) of the trench. Occasionally, slices of basalt from the oceanic crust, and even pieces of the upper mantle, are scraped off and mixed in with the sea-floor sediment. The process is like a bulldozer scraping soil from bedrock and occasionally knocking off a chunk of bedrock along with the soil. This scraping and compression causes folds and fractures in the sediment and rock. The rocks added to the island arc in this way are called a **subduction complex** (Figure 9.17).

Growth of the subduction complex occurs by underthrusting, thus adding the newest slices at the bottom of the complex and forcing the subduction complex upward. In addition, underthrusting thickens the crust, leading to isostatic uplift of the subduction complex. At the same time, compressive forces fold the upper crust, forming a sedimentary basin called a **forearc basin** between the subduction complex and the island arc (Figure 9.17). This process is similar to holding a flexible notebook open horizontally between your hands. If you move your hands closer together, the middle of the notebook bends downward to form a topographic depression analogous to a forearc basin. The forearc basin fills with sediment eroded from the volcanic islands.

Island arcs are abundant in the Pacific Ocean, where convergence of oceanic plates is common. The western Aleutian Islands and most of the island chains of the southwestern Pacific are island arcs.

island arc A gently curving chain of volcanic islands in the ocean formed by convergence of two plates, each bearing ocean crust, and the resulting subduction of one plate beneath the other.

subduction complex The accumulation of sea-floor sediment, basaltic oceanic crust, and slices of the upper mantle that are scraped from the upper layers of the subducting slab in a subduction zone and added to the opposite plate.

forearc basin A sedimentary basin between the oceanic trench and the magmatic arc, either in an island arc or at an Andean margin.

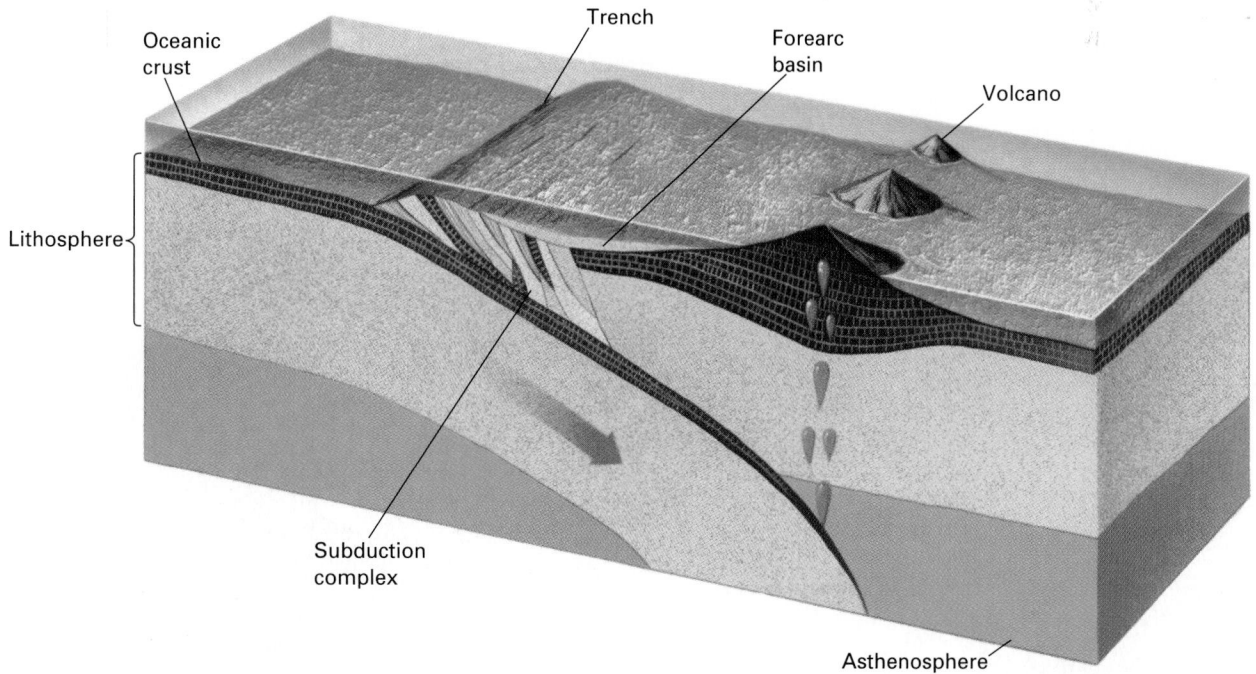

FIGURE 9.17 Formation of an island arc. A subduction complex contains slices of oceanic crust and upper mantle scraped from the top of a subducting plate. Magma forms in the subduction zone and rises to build submarine volcanoes, which eventually grow above sea level to form the volcanic mountain chain.

Andean margin
A continental margin characterized by subduction of an oceanic lithospheric plate beneath a continental plate; also called an *active continental margin*.

9.4 The Andes: Subduction at a Continental Margin

The Andes are the world's second-highest mountain chain, with 49 peaks above 6,000 meters, or nearly 20,000 feet (Figure 9.18). The highest peak is Aconcagua, at 6,962 meters. The Andes rise from the Pacific coast of South America, starting nearly at sea level. Igneous rocks make up most of the Andes, although the chain also contains folded sedimentary rocks, especially in the eastern foothills.

In early Jurassic time, about 190 million years ago, the lithospheric plate that included South America started moving westward. To accommodate the westward motion, oceanic lithosphere began to sink into the mantle beneath the west coast of South America; a subduction zone had formed by early Cretaceous time, 140 million years ago (Figure 9.19A).

By 130 million years ago, the sinking plate was generating vast amounts of basaltic magma (Figure 9.19B). Some of this magma rose to the surface to erupt from volcanoes. Most of it, however, melted portions of the lower continental crust of South America to form andesitic and granitic magma. As a result, both volcanoes and plutons formed along the entire length of western South America. As the oceanic plate sank beneath the continent, slices of seafloor mud and rock were scraped

from the subducting plate, forming a subduction complex similar to that of an island arc.

The rising magma heated and thickened the crust beneath the Andes, causing it to rise isostatically and form great peaks. At the same time, the magmatic arc shifted several kilometers eastward. When the peaks became sufficiently high and heavy, the weak, soft rock oozed outward under its own weight. This spreading formed a great belt of thrust faults and folds along the east side of the Andes (Figure 9.19C).

The Andes, then, are a mountain chain consisting predominantly of igneous rocks formed by subduction at a continental margin. The chain also contains extensive sedimentary rocks on both sides of the mountains; those rocks formed from the sediment that eroded from the rising peaks. The Andes are a good general example of subduction at a continental margin, and this type of plate margin is called an **Andean margin**.

9.5 The Himalayas: A Collision between Continents

The world's highest mountain chain, the Himalayas, separates China from India and includes the world's highest peaks (Figure 9.20). If you were to stand on the southern edge of the Tibetan Plateau and look southward, you would see the high peaks of the Himalayas. Beyond this great mountain chain lie the rainforests and hot, dry plains of India. If you had been able to stand in the same place 200 million years ago and look southward, you would have seen only ocean. At that time,

FIGURE 9.18 The Cordillera Apolobamba mountain range in Bolivia's Andes rises over 6,000 meters.

COURTESY OF GRAHAM R. THOMPSON/JONATHAN TURK

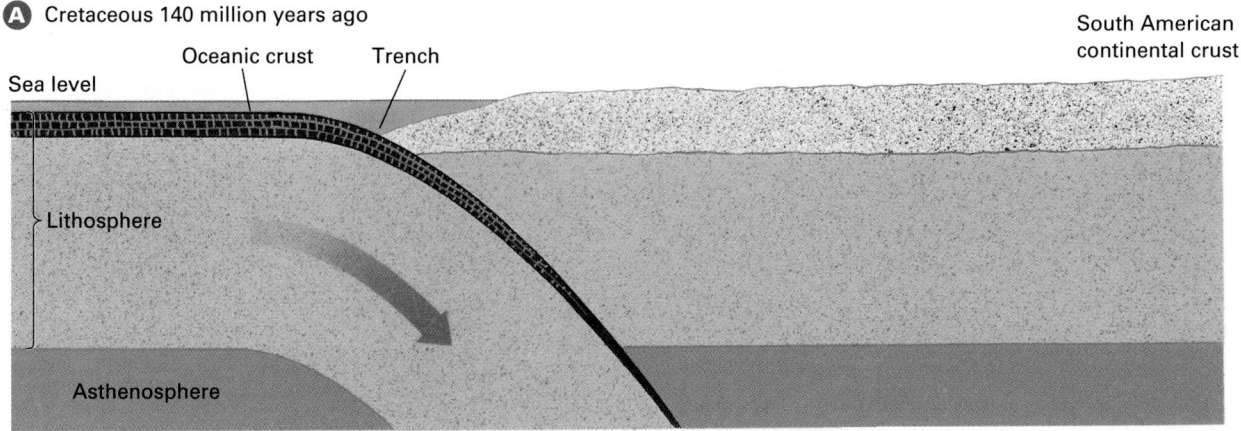

A Cretaceous 140 million years ago

Oceanic crust   Trench   South American
                                                                                 continental crust
Sea level

Lithosphere

Asthenosphere

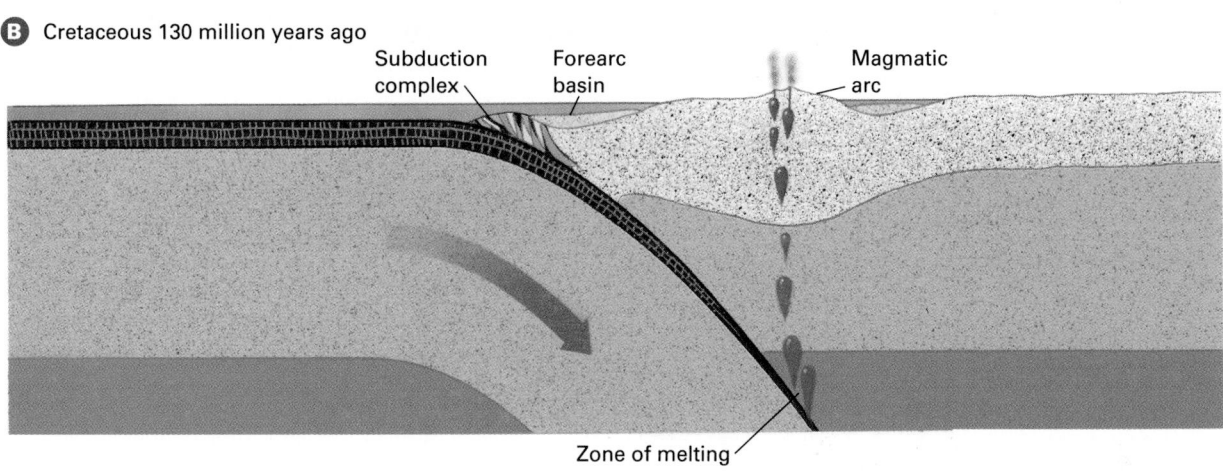

B Cretaceous 130 million years ago

Subduction   Forearc   Magmatic
complex   basin   arc

Zone of melting

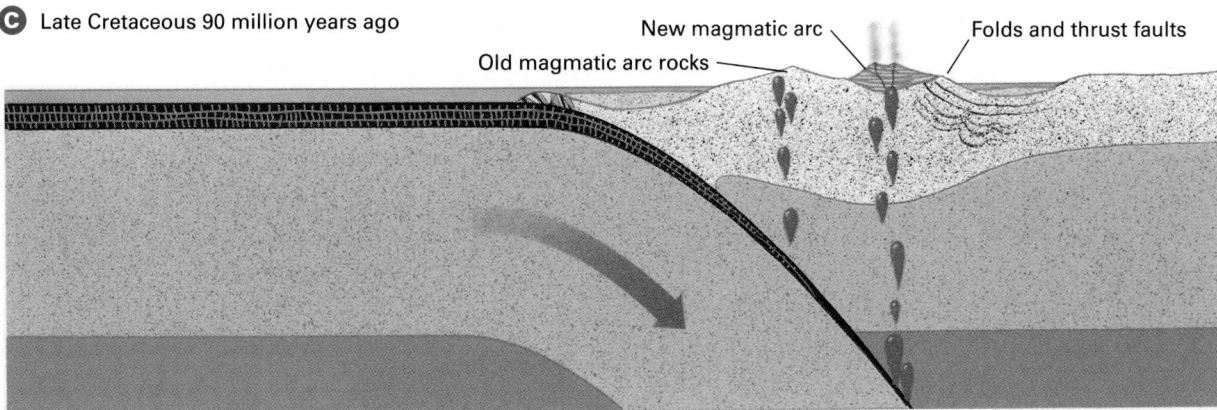

C Late Cretaceous 90 million years ago

                                     New magmatic arc   Folds and thrust faults
                        Old magmatic arc rocks

FIGURE 9.19 Development of the Andes, seen in cross section looking northward. (A) As the South American lithospheric plate moved westward in early Cretaceous time, about 140 million years ago, a subduction zone and a trench formed at the west coast of the continent. (B) By 130 million years ago, igneous activity began and a subduction complex and forearc basin formed. (C) By 90 million years ago, the trench and region of igneous activity had both migrated eastward. Old volcanoes became dormant and new ones formed to the east.

FIGURE 9.20 The Himalayas rose when a plate carrying oceanic crust converged with a plate carrying continental crust, followed by a continent–continent collision. Pictured here is Lamo-she, in China's Sichuan Province.

COURTESY OF GRAHAM R. THOMPSON/JONATHAN TURK

India was located south of the equator, separated from Tibet by thousands of kilometers of open ocean. The Himalayas had not yet begun to rise (Figure 9.21).

Formation of an Andean-Type Margin

By 80 million years ago, a triangular piece of lithosphere that included present-day India had split off from a large mass of continental crust near the South Pole (Figure 9.22A). It began drifting northward toward Asia at a high speed—geologically speaking—perhaps as fast as 20 centimeters per year. As this Indian plate started to move, its leading edge, consisting of oceanic crust,

FIGURE 9.21 (A) About 200 million years ago, India was part of a large continent located near the South Pole. (B) At that time India, southern Asia, and the intervening ocean basin were parts of the same lithospheric plate. (In this figure, the amount of oceanic crust between Indian and Asian continental crust is abbreviated to fit the diagram.)

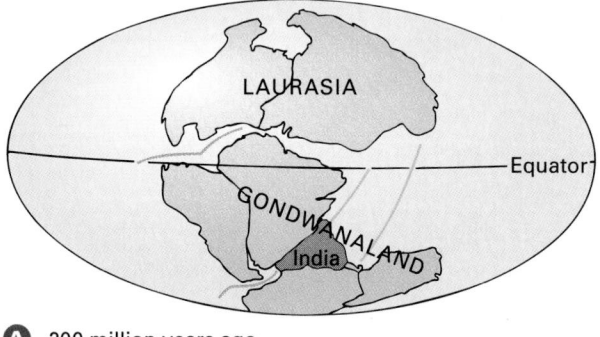

A 200 million years ago

| Spreading ridges |
| Transform/stike-slip fault |

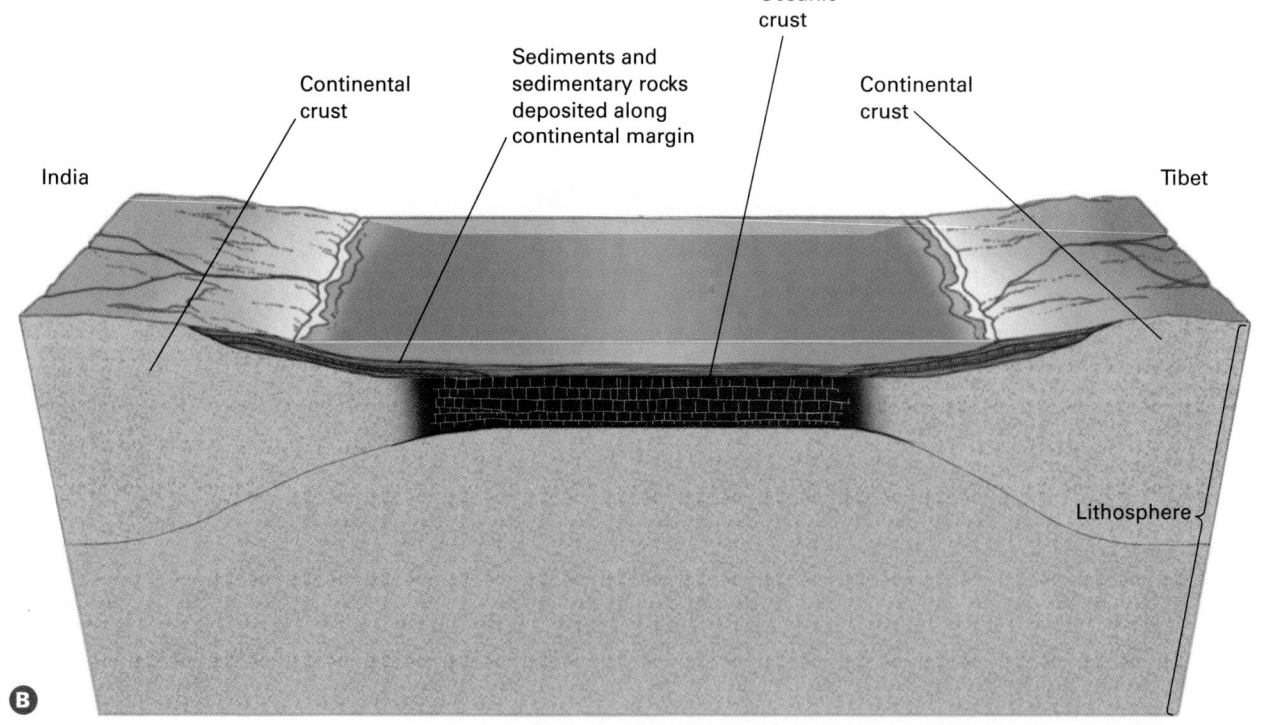

B

Unit 2: Internal Processes

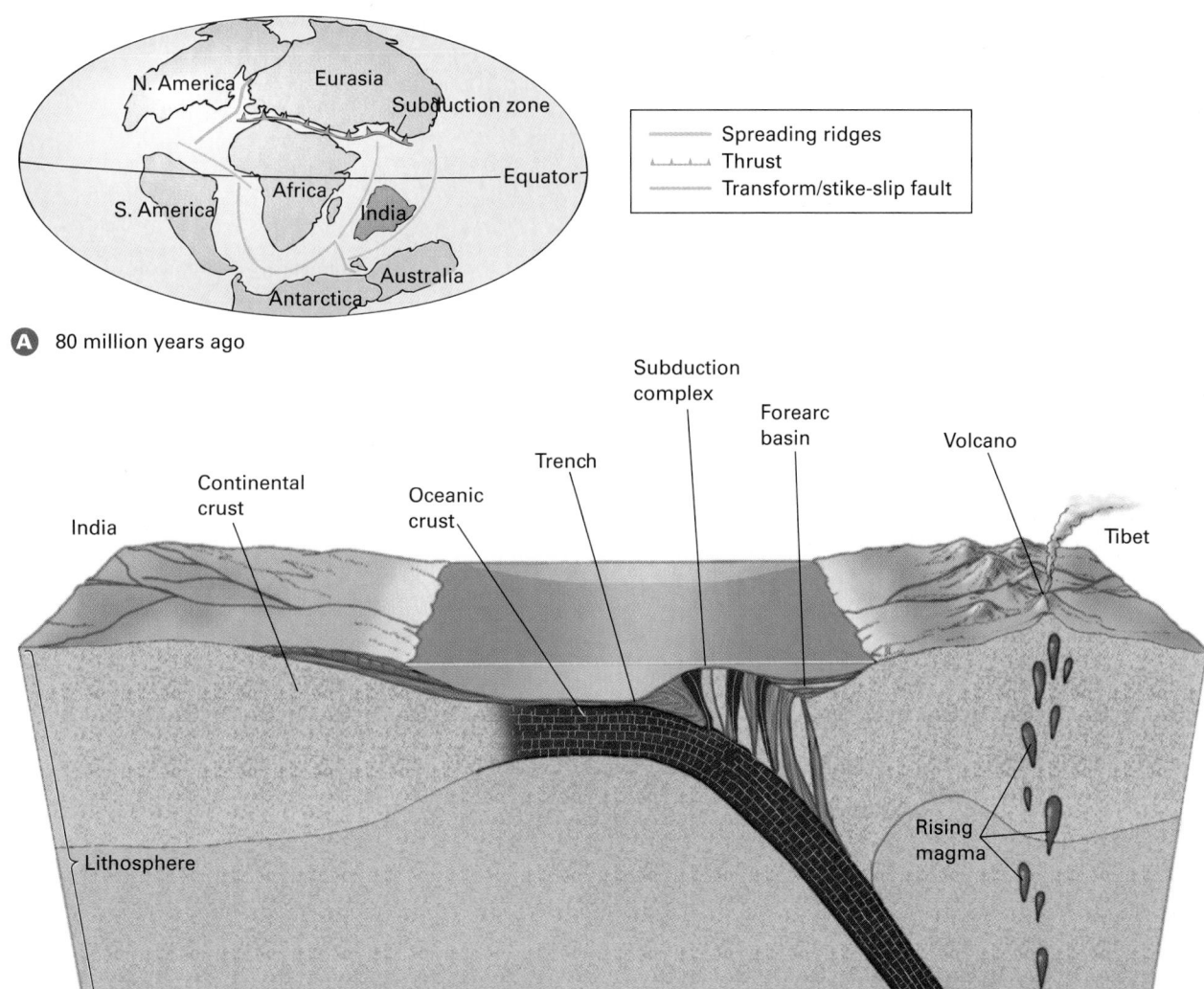

A 80 million years ago

FIGURE 9.22 (A) By 80 million years ago, India was migrating northward, approaching the equator. (B) When India began moving northward, the plate broke and subduction began at the southern margin of Asia. By 80 million years ago, an oceanic trench and subduction complex had formed. Volcanoes erupted and granite plutons formed in the region now called Tibet. (Again, the amount of oceanic crust is abbreviated.)

sank beneath Asia's southern margin. This subduction formed an Andean-type continental margin along the edge of Tibet; volcanoes erupted and granite plutons rose into the continent (Figure 9.22B).

Continent–Continent Collision

By 40 million years ago, subduction had consumed all of the oceanic lithosphere between India and Asia and the two continents collided (Figure 9.23A). Because both are continental

crust and are of comparable density, neither could sink deeply into the mantle. Igneous activity then ceased because subduction stopped (Figure 9.23B). The collision did not stop the northward movement of India, but it did slow it down to about 5 centimeters per year.

As India continued to move northward, it began to underthrust beneath Tibet, doubling the thickness of continental crust in the region (Figure 9.23C). Thick piles of sediment that had accumulated on India's northern continental shelf were scraped from harder basement rock as India slid

FIGURE 9.23 (A) By 40 million years ago, India had moved 4,000 to 5,000 kilometers northward and collided with Asia. (B) When India collided with Tibet, the leading edge of India was underthrust beneath southern Tibet. (C) Continued plate movement has doubled the thickness of continental crust, creating the high Tibetan Plateau and the Himalayas.

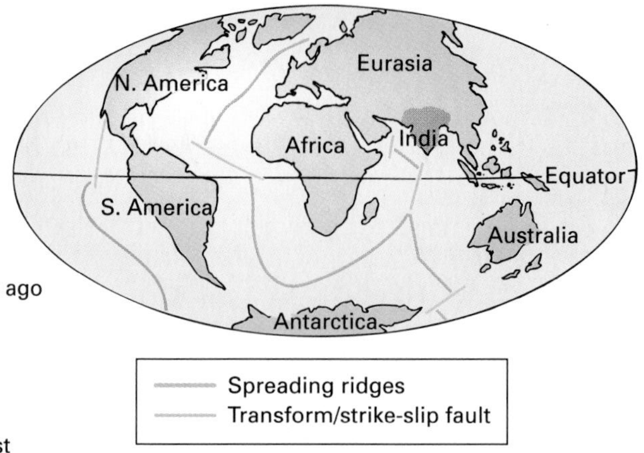

N. America
Eurasia
Africa
India
Equator
S. America
Australia
Antarctica
Spreading ridges
Transform/strike-slip fault

A 40 million years ago

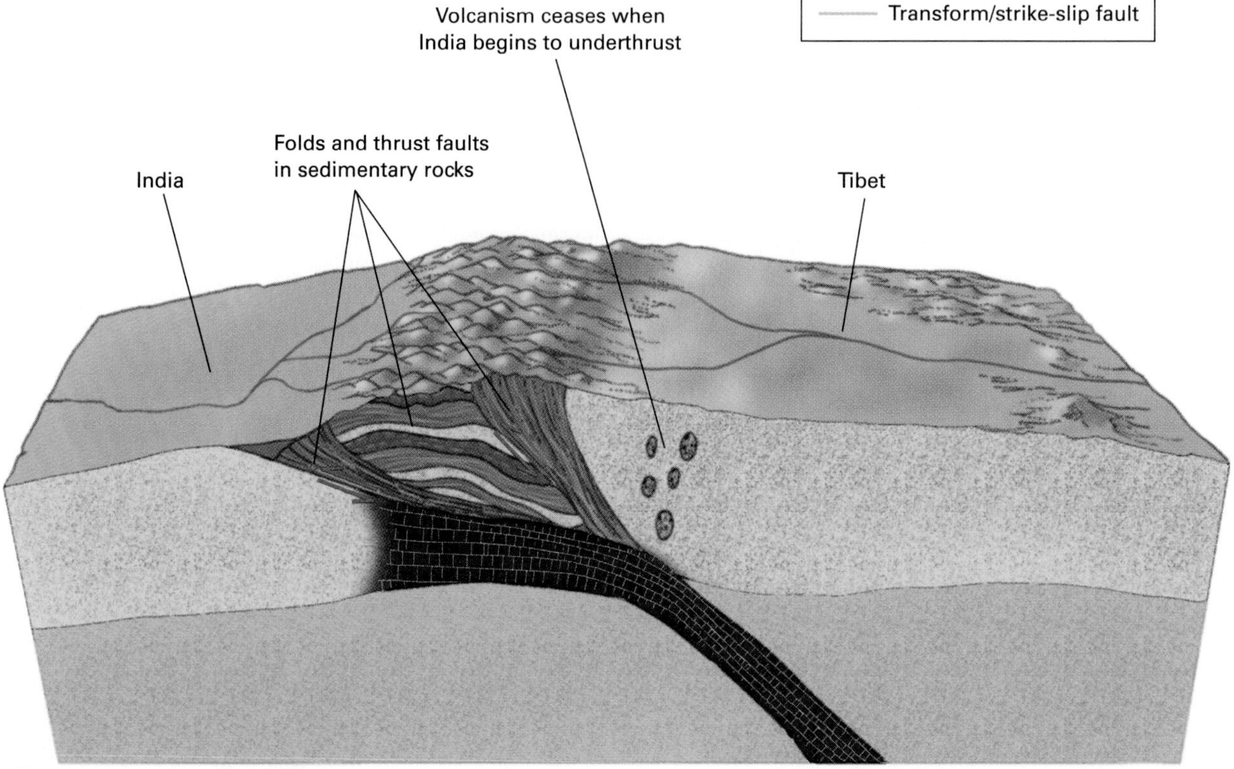

Volcanism ceases when India begins to underthrust
Folds and thrust faults in sedimentary rocks
India
Tibet

B 40 million years ago

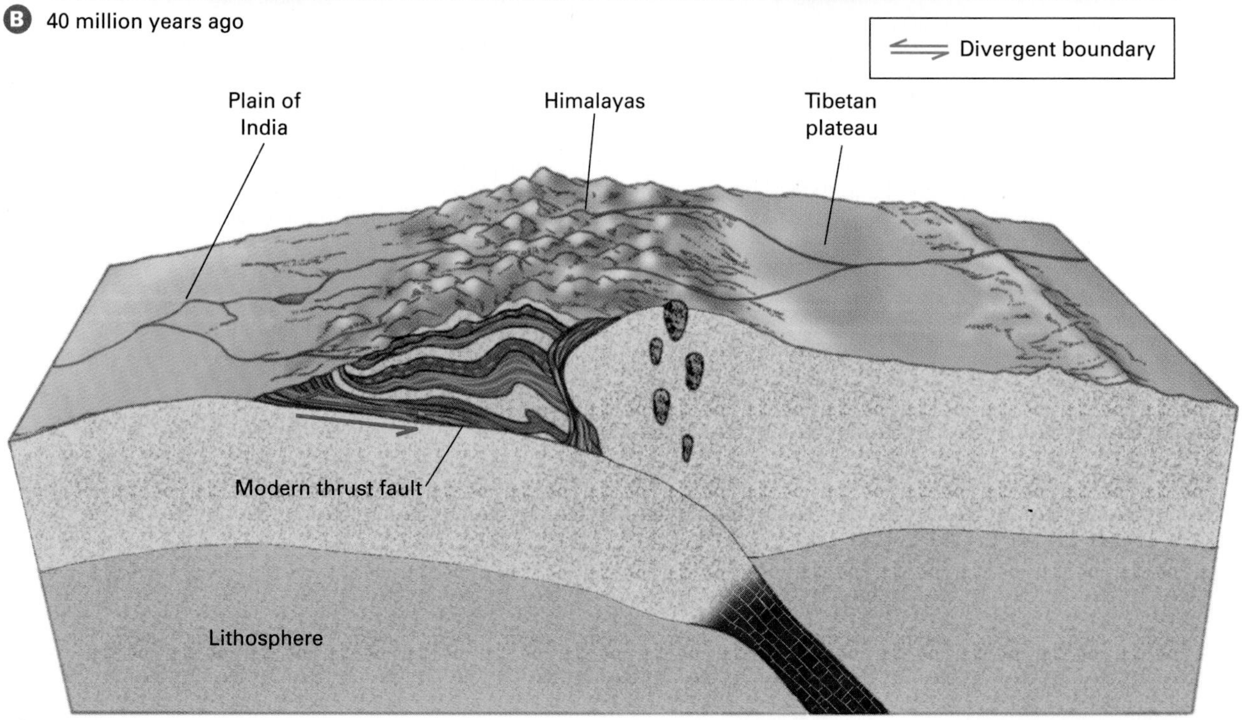

Divergent boundary
Plain of India
Himalayas
Tibetan plateau
Modern thrust fault
Lithosphere

C Today

beneath Tibet. These sediments were pushed into great folds and thrust faults (Figures 9.23B and 9.23C).

At the same time, India crushed Tibet and wedged China out of the way along huge strike-slip faults. India has pushed southern Tibet 1,500 to 2,000 kilometers northward since the beginning of the collision. These compressional forces have created major mountain ranges and basins north of the Himalayas.

The Himalayas Today

Today, the Himalayas contain igneous, sedimentary, and metamorphic rocks. Many of the sedimentary rocks contain fossils of shallow-dwelling marine organisms that lived in the shallow sea of the Indian continental shelf. Plutonic and volcanic Himalayan rocks formed when the range was an Andean margin. Rocks of all types were metamorphosed by the tremendous stresses and heat generated during subduction and continent–continent collision.

The underthrusting of India beneath Tibet and the squashing of Tibet have doubled the thickness of continental crust and lithosphere under the Himalayas and the Tibetan Plateau to the north. Consequently, the region floats isostatically at high elevation (Figure 9.23C). Even the valleys lie at elevations of 3,000 to 4,000 meters, and the Tibetan Plateau has an average elevation of 4,000 to 5,000 meters. One reason that the Himalayas contain all of Earth's highest peaks is simply that the entire plateau lies at such a high elevation. From the valley floor to the summit, Mount Everest is actually smaller than Alaska's Denali (Mount McKinley), North America's highest peak. Mount Everest rises about 3,300 meters from base to summit, whereas Denali rises about 4,200 meters. The difference in elevation of the two peaks lies in the fact that the base of Mount Everest is at about 5,500 meters, but Denali's base is at 2,000 meters.

As the Himalayas rose, they became too heavy to support their own weight. The crust beneath the range then spread outward. This stretching created normal faults in the high mountains. As the crust spread away from the highest parts of the range, it compressed rocks near the margins of the chain. As a result, extensional normal faults formed in the mountains, while compressional thrust faults developed in the foothills. At the same time, tectonic forces resulting from the continued northward movement of India pushed the mountains upward. These processes continue today and no one knows when India will stop its northward movement or how high the mountains will become. However, when the uplift ends, erosion will eventually lower the lofty peaks to rolling hills.

The Himalayan chain is only one example of a mountain chain built by a collision between two continents. The Appalachian Mountains formed when eastern North America collided with Europe, Africa, and South America between 470 and 250 million years ago. The European

Alps formed during repeated collisions between northern Africa and southern Europe beginning about 30 million years ago. The Urals, which separate Europe from Asia, formed by a similar process about 250 million years ago.

9.6 Mountains and Earth Systems

Air rises as it flows across a mountain range. Moisture condenses from rising air to produce rain or snow. Rain forms streams that race down steep hillsides, eroding gullies and canyons, while snow may accumulate to form glaciers that scour soil and bedrock. Thus mountains promote precipitation, which erodes mountains.

Recent research has shown that the rising of the Himalayas coincided with global cooling. Some scientists have suggested that the rising mountains caused this cooling trend, although many other factors may have contributed to the cooling trend.

Mountains affect plant and animal habitats. Cool climates prevail at high elevations, even at the equator. Mean annual temperature changes with each 1,000 meters of elevation as much as it does with each 1,000 kilometers of latitude. As a result, plant and animal communities change rapidly with elevation. For example, if you climb one of Colorado's 4,000-meter mountains, you reach alpine tundra near the summit. But if you traveled north from Colorado at low elevations, you would not reach similar tundra until you arrived at the Canadian Barren Lands, 3,000 kilometers away.

In many mountainous regions, human populations have increased dramatically over the past few decades. As farmland has become scarce, people have cut forests and cultivated steep slopes. Unfortunately, when the protective forest and native grass are removed, soil erosion increases dramatically. Mud washes into streams and landslides carry soil downslope (Figure 9.24). In this manner, human activity increases erosion rates.

COURTESY OF GRAHAM R. THOMPSON/JONATHAN TURK

FIGURE 9.24 As the human population in mountainous regions increases, people have cut forests on steep hillsides to plant crops. This terraced hillside in Nepal collapsed during heavy monsoon rainfall.

10

WEATHERING, SOIL, AND EROSION

Despite the best efforts of landowners along
Jenner Coast in northern California, the ocean
waves have caused severe destruction to their
properties.

visit **4ltrpress.cengage.com**

170

> ## The problem [of soil erosion], which is growing ever more critical, is being ignored because, who gets excited about dirt?
>
> *David Pimentel*

As the solar system formed, swarms of meteorites, comets, and asteroids crashed into the planets and their moons, forming huge impact craters on their surfaces. The earliest craters on Earth's surface quickly disappeared because our planet was so hot that plastic and molten rock oozed inward to fill the craters, much as soft mud will flow back in to fill a hole made by a falling rock. As the crust thickened and cooled, a second wave of bombardment formed a new set of craters. Today, however, none of these ancient craters remain on Earth's surface—but the Moon is pockmarked with craters of all ages. Why have the craters vanished from Earth but been preserved on the Moon?

Tectonic processes such as mountain building, volcanic eruptions, and earthquakes continually change the face of our planet and renew Earth's surface over geologic time. In addition, Earth is sufficiently massive to have retained its atmosphere and water. The atmosphere and hydrosphere then weather and erode the surface rocks of the geosphere. The combination of tectonic activity and erosion has eliminated all traces of early craters from Earth's surface. In contrast, the smaller Moon has lost most of its heat, so tectonic activity is nonexistent. In addition, the Moon's gravitational force is too weak to have retained an atmosphere or significant amounts of surface water to erode its surface. As a result, the Moon's ancient craters have persisted for billions of years. The combination of weathering, erosion, and tectonic activity continuously changes Earth's surface.

10.1 Weathering and Erosion

Recall from Chapter 3 that the processes that decompose rocks and convert them to loose gravel, sand, clay, and soil are called *weathering* (Figure 10.1).

Weathering involves little or no movement of the decomposed rocks and minerals. The weath-

ered material simply accumulates where it forms. However, loose soil and other weathered material offer little resistance to rain or wind and are easily eroded. **Erosion** is the removal of weathered rocks that occurs when rain, running water, wind, glaciers, or gravity transports the material to a new location. Agents of erosion may carry the weathered material great distances and finally deposit it as layers of sediment at Earth's surface.

Weathering, erosion, transport, and deposition typically occur in an orderly sequence. For example, water freezes in a crack in granite, weathering the rock and loosening a grain of quartz. A hard rain erodes the grain and washes it into a stream. The stream then transports the quartz to the seashore and deposits it as a grain of sand on the beach (Figure 10.2).

FIGURE 10.1 Weathering processes of wind, waves, and water are eroding this seaside rock near Cabo San Lucas, Mexico, creating a sculpture. The fragments from the original lie at the base.

mechanical weathering *or* **physical weathering** The disintegration of rock into smaller pieces by physical processes without altering the chemical composition of the rock.

chemical weathering The decomposition of rock when it chemically reacts with air, water, or other agents in the environment, altering its chemical composition and mineral content.

pressure-release fracturing A mechanical weathering process in which tectonic forces lift deeply buried rocks close to the surface and then erosion removes overlying rock and sediment—the net result of which is to remove the pressure from overlying material, causing the rock to expand and fracture.

Weathering occurs by both mechanical and chemical processes. **Mechanical weathering** (also called **physical weathering**) reduces solid rock to small fragments but does not alter the chemical composition of rocks and minerals. Think of crushing a rock with a hammer; the fragments are no different from the parent rock except they are smaller. In contrast, **chemical weathering** occurs when air and water chemically react with rock to alter its composition and mineral content. Chemical weathering is similar to the rusting of a steel knife blade; the final product differs both physically and chemically from the starting material.

10.2 Mechanical Weathering

Five processes cause mechanical weathering: pressure-release fracturing, frost wedging, abrasion, organic activity, and thermal expansion and contraction. Two additional processes—salt cracking and hydrolysis-expansion—result from combinations of mechanical and chemical processes.

Pressure-Release Fracturing

Many igneous and metamorphic rocks form deep below Earth's surface. Imagine, for example, that a granitic pluton solidifies from magma at a depth of 15 kilometers. At that depth, the pressure from the weight of overlying rock is about 5,000 times that at Earth's surface. Over millennia, tectonic forces may raise the granite to form a mountain range. After the granite rises and cools, the overlying rock and sediment may then erode, thus decreasing the pressure further. When the pressure diminishes, the rock expands, but because the rock is now cool and brittle, it fractures as it expands. This process is called **pressure-release fracturing**. Many igneous and metamorphic rocks that formed at depth, but now lie at Earth's surface, have fractured in this manner (Figure 10.3).

Frost Wedging

Although pressure-release fracturing occurs by processes involving only the geosphere, all other forms of weathering involve interactions among rocks of the geosphere and the hydrosphere, atmosphere, and biosphere.

Water expands by 7 to 8 percent when it freezes. If water accumulates in a crack and then freezes, the ice expands and wedges the rock apart in a process called **frost wedging**. In spring and fall in a temperate climate, water freezes at night and thaws during the day. Ice pushes rock apart but at the same time cements it together. During the day, when the ice melts, rock fragments come loose and tumble from a steep cliff. Experienced mountaineers therefore try to travel in the early morning before ice melts. A pile of loose angular rock debris, called a **talus**, will accumulate beneath many cliffs (Figure 10.4). These rocks fell from the cliffs mainly as a result of frost wedging.

TIM SCOTT/SHUTTERSTOCK

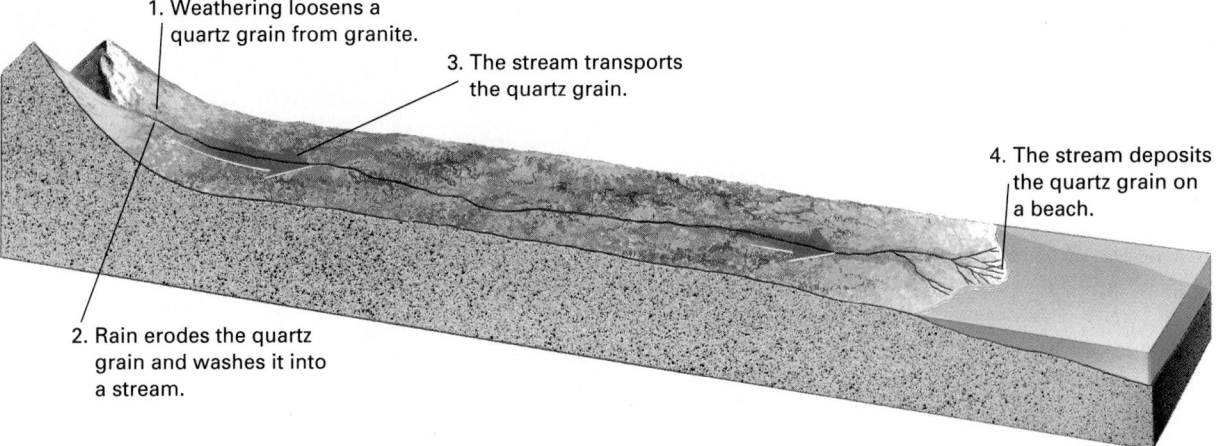

1. Weathering loosens a quartz grain from granite.

2. Rain erodes the quartz grain and washes it into a stream.

3. The stream transports the quartz grain.

4. The stream deposits the quartz grain on a beach.

FIGURE 10.2 Weathering loosens a quartz grain from granite. Rain erodes the grain, washing it into a stream. The stream transports it and finally deposits the grain on a beach.

COURTESY OF GRAHAM R. THOMPSON/JONATHAN TURK

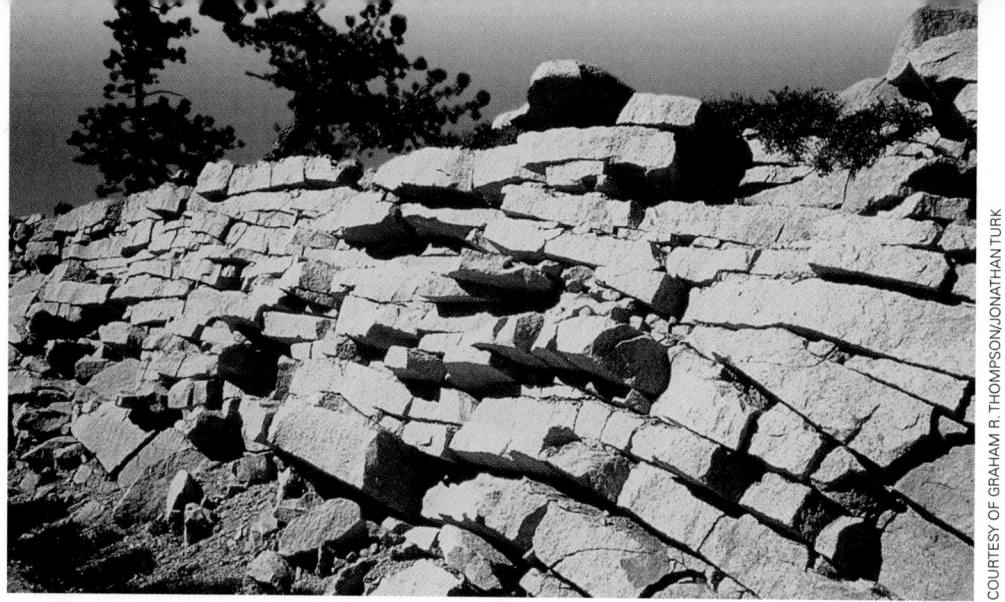

FIGURE 10.3 Pressure-release fracturing contributed to the fracturing of this granite in California's Sierra Nevada.

frost wedging A mechanical weathering process in which water freezes in a crack in rock, and then the expansion wedges the rock apart.

talus An accumulation of loose, angular rocks at the base of a cliff, created when the rocks broke off as a result of frost wedging.

abrasion A mechanical weathering process that consists of the grinding and rounding of rock surfaces by friction and impact.

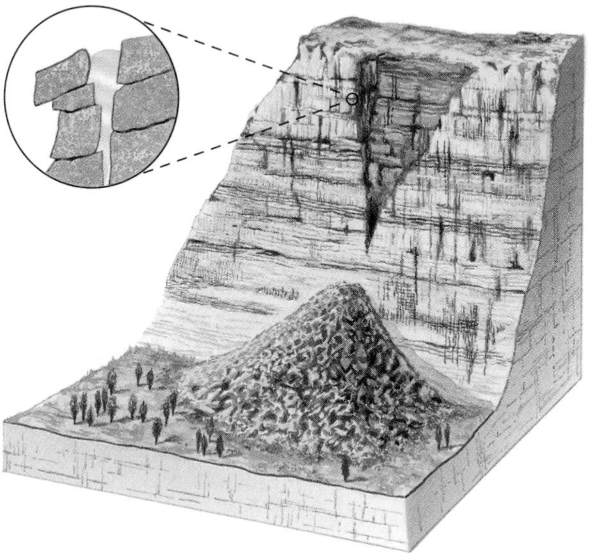

A

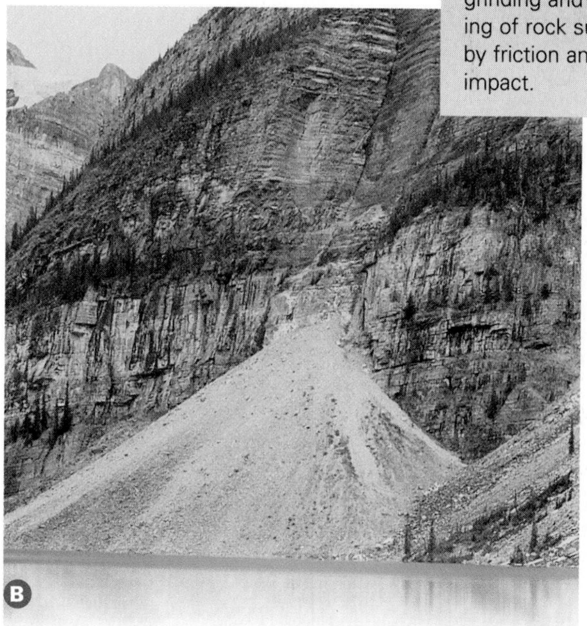

B

COURTESY OF GRAHAM R. THOMPSON/JONATHAN TURK

FIGURE 10.4 (A) Frost wedging dislodges rocks from cliffs and creates talus slopes. (B) Frost wedging produced this talus cone in the Valley of the Ten Peaks, in the Canadian Rockies.

Abrasion

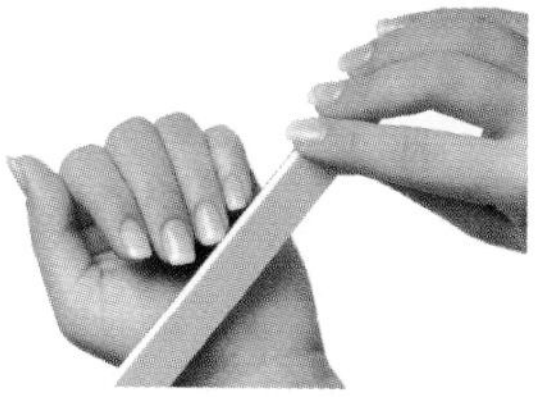

© VALUA VITALY/SHUTTERSTOCK

Rocks, grains of sand, and silt collide with one another when currents or waves carry them along a stream or beach. During these collisions, their sharp edges and corners wear away and the particles become rounded. The mechanical wearing of rocks by friction and impact is called **abrasion** (Figure 10.5). Note that

COURTESY OF GRAHAM R. THOMPSON/JONATHAN TURK

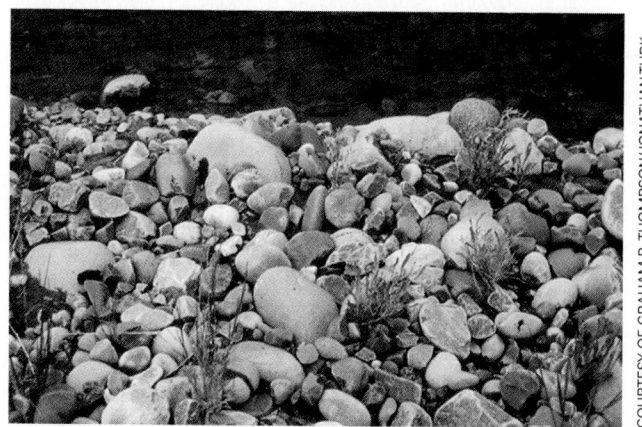

FIGURE 10.5 Abrasion rounded these rocks in a streambed in Yellowstone National Park, Wyoming.

organic activity A mechanical weathering process in which a crack in a rock is expanded by tree or plant roots growing there.

thermal expansion and contraction A mechanical weathering process that fractures rock when temperature changes rapidly, causing the surface of the rock to heat or cool faster than its interior and to expand or contract faster than the interior.

FIGURE 10.6 The powerful action of wind and sand is exhibited in this photo of sand dunes in the Mesquite Flat part of Death Valley National Park in California.

water itself is not abrasive—it is the collisions among rock, sand, and silt that round the rocks.

Wind hurls sand (Figure 10.6) and other small particles against rocks, sandblasting unusual shapes. Glaciers also abrade rock as they drag rocks, sand, and silt across bedrock.

Organic Activity

If soil collects in a crack in bedrock, a seed may fall there and sprout. The roots work their way into the crack, expand, and may eventually widen the crack (Figure 10.7). City dwellers often see the results of this type of **organic activity** when tree roots raise and crack concrete sidewalks.

FIGURE 10.7 As this tree grew in a crack in bedrock, its growing roots widened the crack.

Thermal Expansion and Contraction

Rocks at Earth's surface are exposed to daily and yearly cycles of heating and cooling. They expand when they are heated and contract when they cool. When temperature changes rapidly, the surface of a rock heats or cools faster than its interior, and as a result, the surface expands or contracts faster than the interior. The forces generated by this **thermal expansion and contraction** may fracture the rock.

In mountains or deserts at midlatitudes, temperature may fluctuate from −5°C to +25°C during a spring day. This 30° difference is probably not sufficient to fracture rocks. In contrast to small daily or annual temperature changes, fire heats rock by hundreds of degrees. If you line a campfire with granite stones, the rocks commonly break as you cook your dinner. In a similar manner, forest fires or brush fires occur commonly in many ecosystems and are an important agent of mechanical weathering.

10.3 Chemical Weathering

Rock is durable over a human lifetime. Over geologic time, however, air and water chemically attack rocks near Earth's surface. The most important processes of chemical weathering are dissolution, hydrolysis, and oxidation. Water, acids and bases, and oxygen in the atmosphere or in surface or ground water cause these processes to decompose rocks.

Dissolution

We are all familiar with the fact that some minerals dissolve readily in water while others do not. If you put a crystal of halite (rock salt, or table salt) in water, the crystals rapidly dissolve to form a solution. The process is called **dissolution**. Halite dissolves so rapidly and completely in water that the mineral is rare in natural, moist environments. On the other hand, if you drop a

Unit 3: Surface Processes

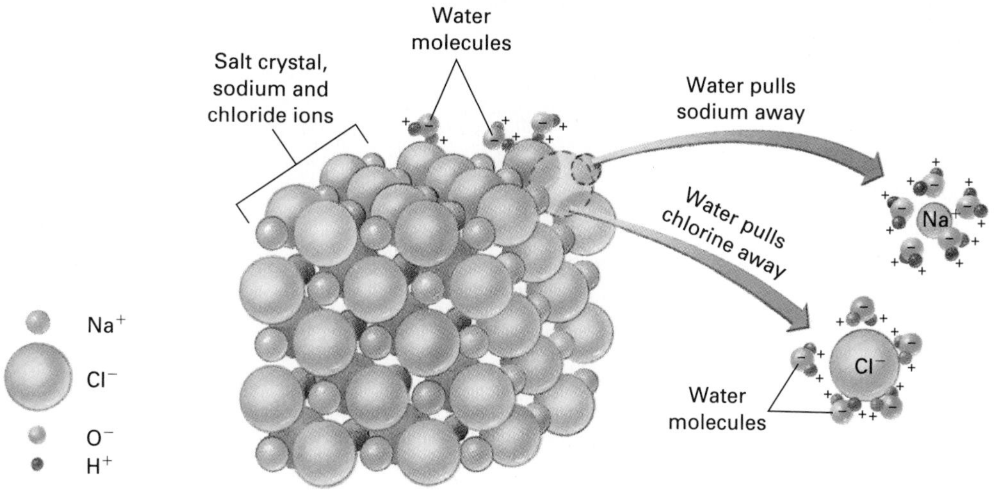

Salt crystal, sodium and chloride ions

Water molecules

Water pulls sodium away

Water pulls chlorine away

Water molecules

Na⁺

Cl⁻

O⁻

H⁺

FIGURE 10.8 Halite (NaCl) dissolves in water (H_2O) because the attractions between the water molecules and the sodium and chloride ions are greater than the strength of the chemical bonds in the crystal. Note that although the oxygen atoms are labeled positive (+) and the hydrogen atoms are labeled negative (–), these are partial charge separations only.

crystal of quartz into pure water, only a tiny amount will dissolve while nearly all of the crystal will remain intact.

To understand how water dissolves a mineral, think of an atom on the surface of a crystal. It is held in place because it is attracted to the other atoms in the crystal by the electrical forces called chemical bonds. At the same time, electrical attractions to the outside environment are pulling the atom away from the crystal. The result is like a tug-of-war. If the bonds between the atom and the crystal are stronger than the attraction of the atom to its outside environment, the crystal remains intact. If outside attractions are stronger, they pull the atom away from the crystal and the mineral dissolves.

Water (H_2O) is a polar molecule. This polarity arises because the oxygen atom in water has a great affinity for electrons, which move away from the hydrogen atoms to concentrate around the oxygen. Thus the oxygen carries a partial negative charge. The two hydrogen atoms are then left with a slight positive charge because their electrons have moved closer to the oxygen atom. When a crystal of halite (NaCl) is dropped in a glass of water, the negatively charged oxygen atoms in the water pull the positively charged sodium ions away from the halite, while the positively charged hydrogen atoms remove the negatively charged chlorine ions (Figure 10.8).

Rocks and minerals dissolve more rapidly when water is either acidic or basic. An acidic solution contains a high concentration of hydrogen ions (H^+), whereas a basic solution contains a high concentration of hydroxyl ions (OH^-). Acids and bases dissolve most minerals more effectively than pure water because they provide more electrically charged hydrogen and hydroxyl ions to pull atoms out of crystals. For example, limestone is made of the mineral calcite ($CaCO_3$). Calcite

barely dissolves in pure water but is quite soluble in acid. If you place a drop of strong acid on limestone, bubbles of carbon dioxide gas instantly form as the calcite dissolves.

Water found in nature is never pure. Atmospheric carbon dioxide dissolves in raindrops and reacts to form a weak acid called carbonic acid. As a result, even the purest rainwater falling in the Arctic or on remote mountains is slightly acidic. Air pollution can make rain even more acidic.

Thus water—especially if it is acidic or basic—dissolves ions from soil and bedrock and carries the dissolved material away. Ground water also dissolves rock to produce spectacular caverns in limestone (Figure 10.9).

COPYRIGHT AND PHOTOGRAPH BY DR. PARVINDER S. SETHI

FIGURE 10.9 Caverns form when ground water dissolves limestone. Both stalactites and stalagmites can be seen in this photo of Dixie Caverns, near Roanoke, Virginia.

COURTESY OF GRAHAM R. THOMPSON/JONATHAN TURK

FIGURE 10.10 Coarse grains of quartz and feldspar accumulate directly over weathered granite. The lens cap in the middle illustrates scale.

Hydrolysis

hydrolysis A chemical weathering process in which a mineral reacts with water to form a new mineral that has water as part of its crystal structure.

oxidation A chemical weathering process in which a mineral decomposes when it reacts with oxygen.

salt cracking A chemical weathering process in which salts that are dissolved in water in the pores of rock crystallize, widening cracks and pushing the mineral grains apart.

exfoliation A weathering process resulting in fracture when concentric plates or shells split away from a main rock mass like the layers of an onion; frequently explained as a form of pressure-release fracturing, but many geologists believe it is caused by hydrolysis-expansion.

In **hydrolysis**, water reacts with one mineral to form a new mineral that has water as part of its crystal structure. Most common minerals weather by hydrolysis. For example, feldspar, the most abundant mineral in Earth's crust, weathers by hydrolysis to form clay.

Quartz is the only rock-forming silicate mineral that does not weather to form clay. Quartz is resistant to chemical weathering because it dissolves extremely slowly and only to a small extent. It is also hard (H = 7 on the Mohs hardness scale; see Chapter 2) and has no cleavage, and so resists abrasion during erosion. When granite weathers, the feldspar and other minerals decompose to form clay but the unaltered quartz grains fall free from the rock. Hydrolysis has so deeply weathered some granites that quartz grains can be pried out with a fingernail at depths of several meters (Figure 10.10). The rock looks like granite but has the consistency of sand.

Because quartz is so tough and resistant to both mechanical and chemical weathering, it is the primary component of sand. Much of this sand is transported to streams and ultimately to the seacoast, where it concentrates on beaches and eventually forms sandstone.

Oxidation

Many elements react with atmospheric oxygen, O_2. Iron rusts when it reacts with water and oxygen. Rusting is an example of a more general weathering process called **oxidation**.[1] Iron is abundant in many minerals; if the iron in such a mineral oxidizes, the mineral decomposes.

Many valuable metals such as iron, copper, lead, and zinc occur as sulfide min-

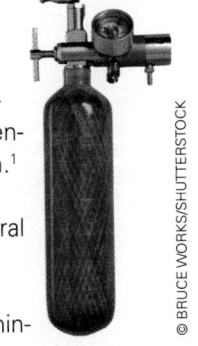
© BRUCE WORKS/SHUTTERSTOCK

erals in ore deposits. When they oxidize during weathering, the sulfur reacts to form sulfuric acid, a strong acid. The sulfuric acid washes into streams and ground water, where it may harm aquatic organisms. Many natural ore deposits generate sulfuric acid when they weather, and the reaction may be accelerated when ore is dug up and exposed at a mine site.

Chemical and Mechanical Weathering Acting Together

Chemical and mechanical weathering can work together, often on the same rock at the same time. After mechanical processes fracture a rock, water and air seep into the cracks to initiate chemical weathering.

In environments where ground water is salty, saltwater seeps through pores and cracks in bedrock. When the water evaporates, the dissolved salts crystallize. The growing crystals exert tremendous forces that loosen mineral grains and widen cracks in a process called **salt cracking**. Thus salt chemically precipitates in rock, and the growing salt crystals mechanically loosen mineral grains and break the rock apart.

Many sea cliffs show pits and depressions caused by salt cracking because spray from the breaking waves brings the salt to the rock. Salt cracking is also common in deserts, where surface water and ground water often contain dissolved salts (Figure 10.11).

Granite commonly fractures by **exfoliation**, a process in which plates or shells of weathered material split away like the layers of an onion (Figure 10.12). The plates may be only 10 or 20 centimeters thick near the surface, but they thicken with depth. Because exfoliation fractures are usually absent below a depth of 50 to 100 meters, they seem to be a result of exposure of the granite at Earth's surface.

Thus, exfoliation is frequently explained as a form of pressure-release fracturing. However, many geologists suggest that hydrolysis-expansion may be the main cause of exfoliation. During hydrolysis, feldspars and other silicate minerals react with water to form clay.

---

1. Oxidation is properly defined as the loss of electrons from a compound or element during a chemical reaction. In the weathering of common minerals, this usually occurs when the mineral reacts with molecular oxygen.

Unit 3: Surface Processes

FIGURE 10.11 Crystallizing salt loosened sand grains to form this depression in sandstone in Cedar Mesa, Utah. The white patches are salt crystals.

Old land surface before removal by weathering and erosion

Present land surface, shaped by sheeting

Uplifted granite pluton

Ⓐ

Ⓑ

FIGURE 10.12 (A) Exfoliation occurs when concentric rock layers fracture and become detached from a granite outcrop. (B) Exfoliation has fractured this granite in Pinkham Notch, New Hampshire.

As a result of the addition of water, clays have a greater volume than the original minerals have. So a chemical reaction (hydrolysis) forms clay, and the mechanical expansion that occurs as the clay forms may cause exfoliation. This explanation is compatible with the observation that exfoliation concentrates near Earth's surface because water and chemical weathering are most abundant close to the surface.

10.4 Soil

Bedrock breaks into smaller fragments as it weathers, and much of it decomposes to clay and sand. Therefore, on most land surfaces, a thin layer of loose rock fragments, clay, and sand overlies bedrock. This material is called **regolith**. Soil scientists define **soil** as the upper layers of regolith that support rooted plants (Figure 10.13), although some earth scientists and engineers use the terms *soil* and *regolith* interchangeably.

Components of Soil

Soil is a mixture of mineral grains, organic material, water, and gas. The mineral grains include clay, silt, sand, and rock fragments. Clay is so fine grained and closely packed that water and air do not flow readily through a clay-rich soil, so plants growing in clay soils suffer from lack of oxygen. In contrast, water and air flow easily through sandy soil. The most fertile soil is **loam**, a mixture especially rich in sand and silt with generous amounts of organic matter.

Minerals and organic matter contain nutrients necessary for plant growth. If you walk through a forest or prairie, you can find bits of organic **litter**—leaves, stems, and flowers that have not decomposed—on the soil surface. When this litter decomposes sufficiently that you can no longer determine the origin of individual pieces, it becomes **humus**. Humus is an essential component of most fertile soils. Humus soaks up so

regolith The thin layer of loose, unconsolidated, weathered material that overlies bedrock. Some earth scientists and engineers use the terms *regolith* and *soil* interchangeably; soil scientists include soil as only the upper layers of regolith.

soil The upper layers of regolith that support plant growth. Some earth scientists and engineers use the terms *soil* and *regolith* interchangeably.

loam The most fertile soil, a mixture especially rich in sand and silt with generous amounts of organic matter.

litter Leaves, twigs, and other plant or animal material that have fallen to the surface of the soil but have not decomposed.

humus The dark, organic component of soil consisting of litter that has decomposed enough so that the origin of the individual pieces cannot be determined.

Soil consists of the upper layers of regolith that support plants.

soil horizon A layer of soil that is distinguishable from other layers because of differences in appearance and in physical and chemical properties.

O horizon The uppermost layer of soil, named for its organic component; the combined O and A horizons are called *topsoil*.

A horizon The layer of soil below the O horizon, composed of a mixture of humus, sand, silt, and clay; combines with the O horizon as to form *topsoil*.

topsoil The fertile, dark-colored surface soil; the combined O and A soil horizons.

B horizon The soil layer just below the A horizon, containing less organic matter, and where ions leached from the A horizon accumulate; also called *subsoil*.

C horizon The lowest soil layer.

leaching The downward movement of water and dissolved ions from the O and A soil horizons into the B horizon, where they accumulate.

COURTESY OF GRAHAM R. THOMPSON/JONATHAN TURK

FIGURE 10.13 A very thin, dark-colored layer of soil overlies light-colored limestone to support this forest in Canada's Northwest Territories. Most plant life is supported by a soil layer only a few centimeters to a few meters thick.

Soil Horizons

A typical well-developed soil consists of several layers called **soil horizons**. The uppermost layer is called the **O horizon**, named for its organic component. This layer consists mostly of organic litter and humus, with a small proportion of minerals (Figure 10.14). The next layer down, called the **A horizon**, is a mixture of humus, sand, silt, and clay. The combined O and A horizons are called **topsoil**. A kilogram of average fertile topsoil contains about 30 percent by weight organic matter, including approximately 2 trillion bacteria, 400 million fungi, 50 million algae, 30 million protozoa, and thousands of larger organisms such as insects, worms, nematodes, and mites.

The third layer, the **B horizon** or **subsoil**, is a transitional zone between topsoil and weathered parent rock below. Roots and other organic material grow in the B horizon, but the total amount of organic matter is low. The lowest layer, called the **C horizon**, consists of partially weathered bedrock that grades into unweathered parent rock. This zone contains little organic matter.

When rain falls on soil, it sinks into the O and A horizons, weathering minerals, forming clay, and carrying the clay particles and dissolved ions to lower levels. This downward movement of water and dissolved ions is called **leaching**. Water, dissolved ions, and clay from the A horizon accumulate in the B horizon. This layer retains moisture because of its high clay content. Although this chemical enrichment and moisture retention are usually beneficial, too much clay accumulation on the B horizon creates a dense, waterlogged soil.

Soil-Forming Factors

Why are some soils rich and others poor, some sandy and others loamy? Six factors control soil characteristics: parent rock, climate, rates of plant growth and decay, slope aspect and steepness, time, and transport of soil materials.

Parent Rock

The texture and composition of soil depend partly on its parent rock. For example, when granite decomposes, the feldspar converts to clay and the rock releases quartz as sand grains. If the clay leaches into the B horizon, a sandy soil forms. In contrast, because basalt contains no quartz, soil formed from basalt is likely to be rich in clay and contain only small amounts of sand.

much moisture that humus-rich soil swells after a rain and shrinks during dry spells. This alternate shrinking and swelling loosens the soil, allowing roots to grow into it easily. A rich layer of humus also insulates the soil from excessive heat and cold and reduces water loss from evaporation. Humus enriches the soil in nutrients and makes them available to plants.

In intensive agriculture, farmers commonly plow the soil and leave it exposed for weeks or months. Humus oxidizes in air and decomposes; at the same time, rain dissolves soil nutrients and carries them away. Farmers replace the lost nutrients with chemical fertilizers but rarely replenish the humus. As a result, much of the soil's ability to absorb and regulate water and nutrients is lost. When rainwater flows over the surface, it transports soil particles, excess fertilizer, and pesticide residues, polluting streams and ground water.

Climate

Climate conditions such as rainfall and temperature affect soil formation largely through the chemical weathering they support. Rain seeps downward through soil, but several other factors pull the water back upward. Roots suck soil water toward the surface, and water near the surface evaporates. In addition, water molecules are electrically attracted to soil particles. This attraction initiates a process called **capillary action**, which draws water upward toward the soil surface.

During a rainstorm, water seeps through the A horizon, dissolving soluble ions such as calcium, magnesium, potassium, and sodium. In arid and semiarid regions, rainstorms typically are of short duration and little rain falls. Consequently, when the rain stops, capillary action and plant roots then draw most of the water back up toward the surface, where it evaporates or is taken up by plants. As the water escapes, many of its dissolved ions precipitate in the B horizon, encrusting the soil with salts. A soil of this type is called **pedocal** (Figure 10.15A). This process often deposits enough

calcium carbonate in the form of the mineral calcite to form a hard cement called **caliche** (pronounced "cah-LEE-chee") in the soil. The Greek word *pedon* means "soil"; *pedocal* is a composite of *pedon* and the first three letters of *calcite*. In the Imperial Valley in California, irrigation water contains high concentrations of calcium carbonate. A thick, continuous layer of caliche forms in the soil as the water evaporates. To continue growing crops, farmers must then rip this layer apart with heavy machinery.

Because nutrients concentrate when water evaporates, many pedocals are fertile if irrigation water is available. However, salts often concentrate so much that they become toxic to plants (Figure 10.16).

capillary action The process by which water is pulled upward through the soil due to the natural attraction of water molecules to soil particles.

pedocal A salt-encrusted soil in arid and semiarid climates characterized by an accumulation of calcium carbonate.

caliche A hard crust on the soil of arid and semiarid regions, formed when calcium carbonate precipitates and cements the soil particles together.

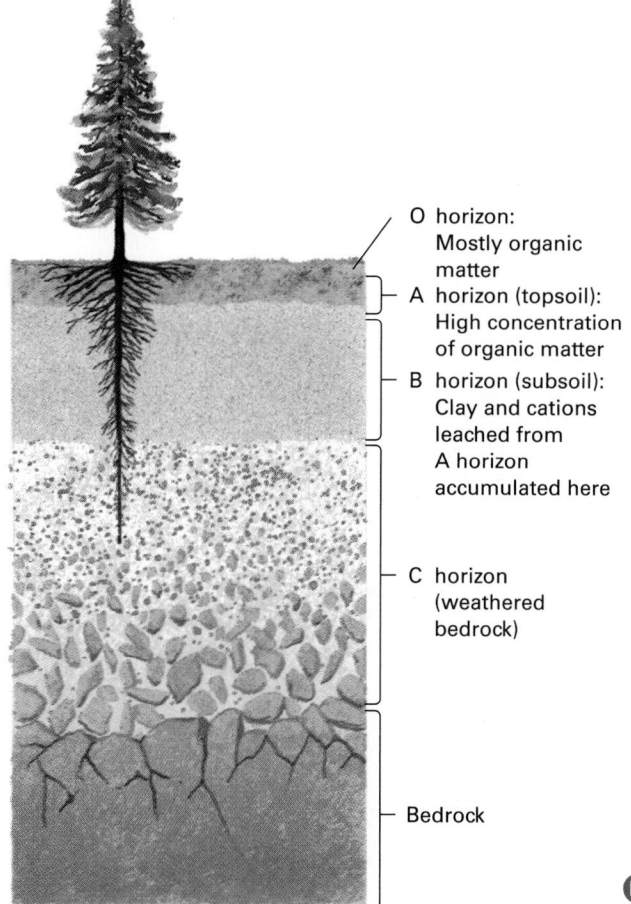

O horizon:
Mostly organic matter

A horizon (topsoil): High concentration of organic matter

B horizon (subsoil): Clay and cations leached from A horizon accumulated here

C horizon (weathered bedrock)

Bedrock

A

O

A

B

C

B

US DEPARTMENT OF AGRICULTURE

FIGURE 10.14 (A) A typical soil consists of several horizons, distinguished by color, texture, and chemistry. (B) In this soil, the darkest uppermost layer is the O horizon; the A horizon is less dark; the whiter layer is the B horizon. The B horizon grades downward into the weathered bedrock of the C horizon. The scale is in feet.

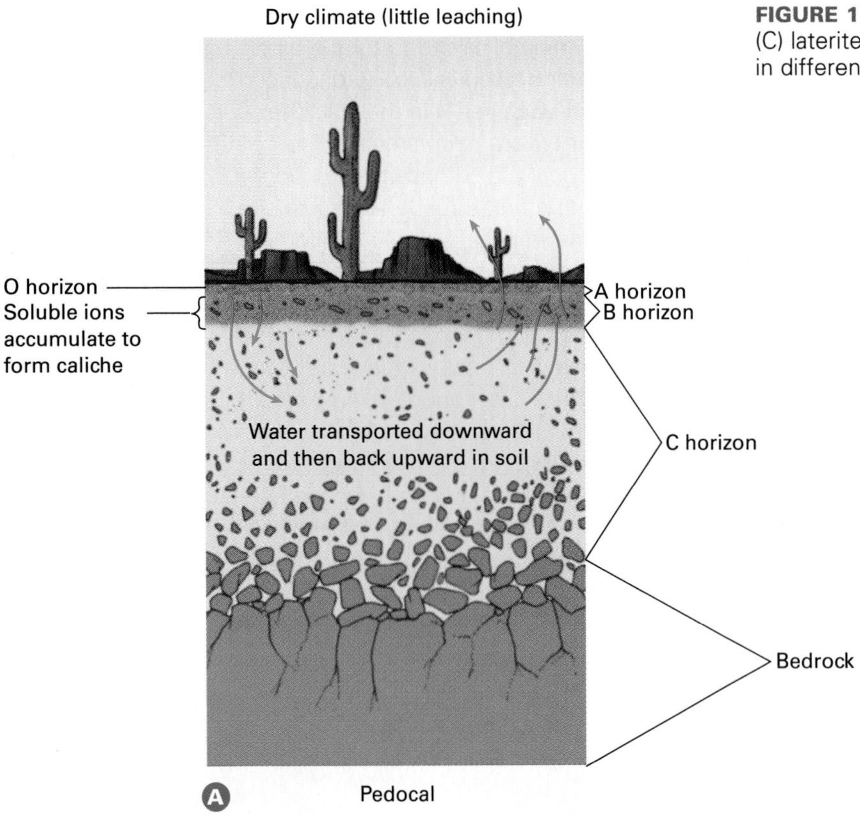

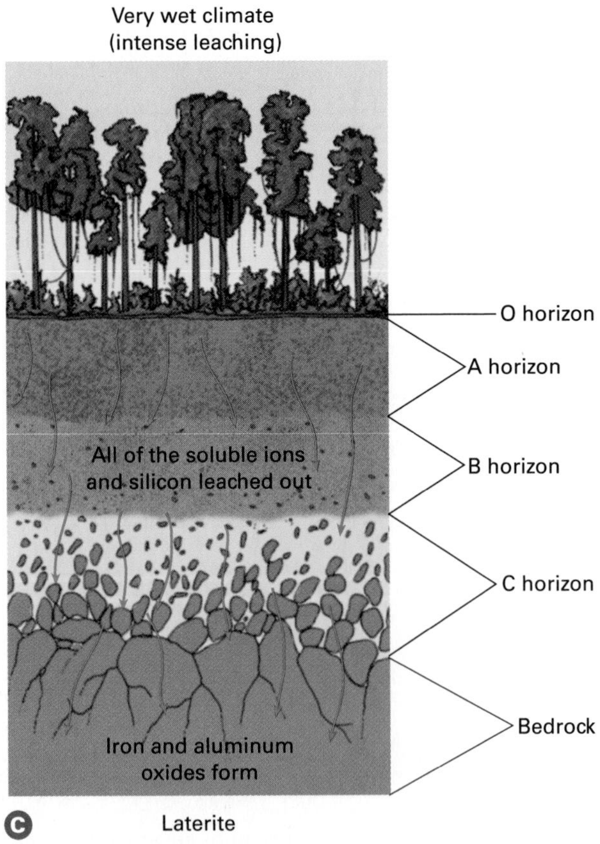

Dry climate (little leaching)

O horizon
Soluble ions accumulate to form caliche

A horizon
B horizon

Water transported downward and then back upward in soil

C horizon

Bedrock

A Pedocal

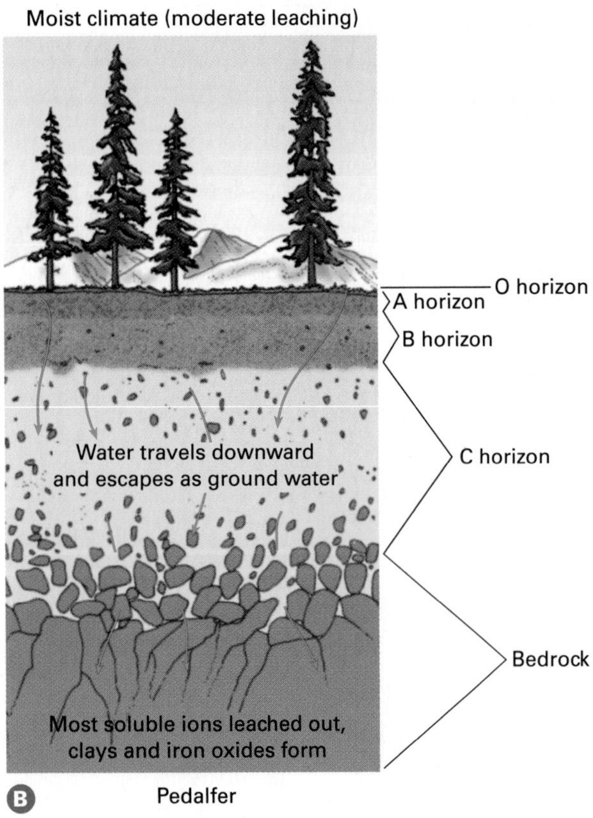

FIGURE 10.15 (A) Pedocals, (B) pedalfers, and (C) laterites are different types of soils that form in different climates.

Moist climate (moderate leaching)

O horizon
A horizon
B horizon

Water travels downward and escapes as ground water

C horizon

Bedrock

Most soluble ions leached out, clays and iron oxides form

B Pedalfer

Very wet climate (intense leaching)

O horizon

A horizon

B horizon

All of the soluble ions and silicon leached out

C horizon

Bedrock

Iron and aluminum oxides form

C Laterite

Unit 3: Surface Processes

pedalfer A common soil type that forms in moist environments, characterized by abundant iron and aluminum oxides and a concentration of clay in the B horizon.

laterite A highly weathered soil rich in oxides of iron and aluminum that usually develops in warm, moist, tropical regions.

FIGURE 10.16 Salts have poisoned this Wyoming soil. Saline water seeps into the depression and evaporates to deposit white salt crystals on the ground and on the fence posts.

In a moist climate, unlike the arid regions where pedocals form, rainstorms are of longer duration and more rain falls. As a result, water seeps downward through the soil leaching soluble ions from both the A and B horizons. The less-soluble elements, such as aluminum, iron, and some silicon, remain behind, accumulating in the B horizon to form a soil type called a **pedalfer** (Figure 10.15B). The subsoil in a pedalfer is commonly rich in clay, which is mostly aluminum and silicon and has the reddish color of iron oxide; hence the prefix *ped* (again from the Greek *pedon*) is followed by the chemical symbols *Al*, for aluminum, and *Fe*, for iron.

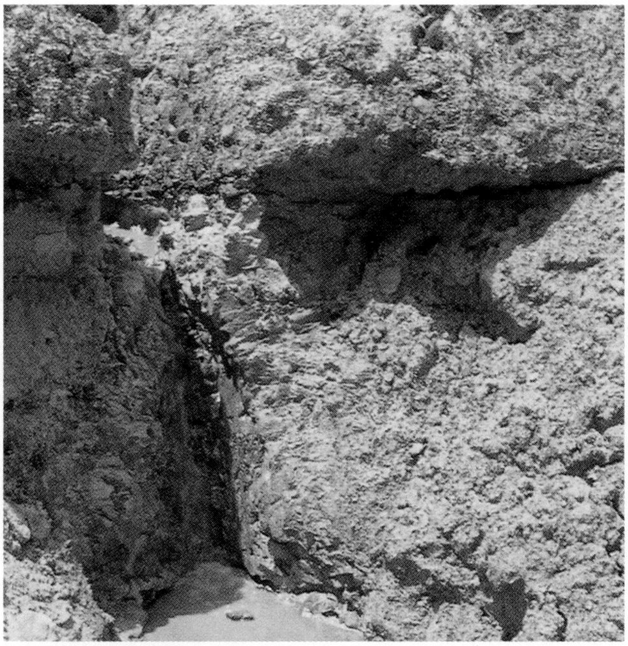

In regions of high temperature and very high rainfall, such as a tropical rainforest, so much water seeps through the soil that it leaches away nearly all the soluble cations. Only very insoluble aluminum and iron minerals remain (Figure 10.15C). Soil of this type is called a **laterite**, from the Latin *later*, meaning "brick." Laterites are often colored rust-red by iron oxide (Figure 10.17). Bauxite, the world's main type of aluminum ore (Chapter 5), is a highly aluminous laterite.

The second important component of climate, average annual temperature, affects soil formation in two ways. First, chemical reactions proceed more rapidly in warm temperatures than in cold ones. Second, plant growth and decay are temperature-dependent.

Rates of Plant Growth and Decay

In the tropics, plants grow and decay rapidly and growing plants quickly absorb the nutrients released by decaying plants. Heavy rainfall leaches nutrients from the soil, creating a laterite. As a result, little humus accumulates and few nutrients are stored in the soil. Thus even though the tropical rainforests support great populations of plants and animals, many of these forests are anchored on poor, laterite soils (Figure 10.18A). The plants depend on a rapid cycle of growth, death, and decay. Conversely, the Arctic is so cold that plant growth and decay are slow. Therefore litter and humus form slowly and Arctic soils also contain little organic matter (Figure 10.18B).

The most fertile soils are those of temperate prairies and forests. There, large amounts of plant litter

FIGURE 10.17 This aluminum-rich Georgia laterite formed when water leached away the more soluble ions.

aspect The orientation of a slope with respect to the Sun; the direction toward which it faces.

FIGURE 10.18 (A) Lateritic tropical soils, such as those on the island of Vanuatu in the South Pacific, often support lush growth but contain few nutrients and little humus. (B) The Arctic soil of Baffin Island, Canada, supports sparse vegetation and contains little organic matter.

COURTESY OF GRAHAM R. THOMPSON/JONATHAN TURK

drop to the ground in the autumn, but decay is slow during the winter, and plant growth during the growing season is not fast enough to extract all the nutrients from the soil. As a result, thick layers of nutrient-rich humus accumulate in temperate climates.

However, even regions with fertile soil are subject to soil erosion, as described in the accompanying box below.

Slope Aspect and Steepness

Aspect refers to the orientation of a slope with respect to the Sun. In other words, which direction is it facing? In the semiarid regions of the Northern Hemisphere, thick soils and dense forests cover the cool, shady north slopes of hills, but thin soils and grass dominate hot, dry, southern exposures (Figure 10.19). The reason for this difference is that in the Northern Hemisphere more water evaporates from the hot, sunny, south-

ern slopes. Therefore, fewer plants grow, weathering occurs slowly, and soil development is retarded. Plants grow more abundantly on the moister northern slopes and more rapid weathering forms thicker soils.

In general, hillsides have thin soils and valleys are covered by thicker soil, because soil erodes from hills and accumulates in valleys. When hilly regions were first settled and farmed, people naturally planted their crops in the valley bottoms, where the soil was rich and water was abundant. Recently, as population has expanded, farmers have moved to the thinner, less stable, hillside soils.

Time

Chemical weathering occurs slowly in most environments, and time is therefore an important factor in determining the extent of weathering. Recall that most minerals weather to clay. In geologically young soils,

Soil Erosion and Agriculture

In nature, soil erodes approximately as rapidly as it forms. However, improper farming, livestock grazing, and logging can accelerate erosion. Plowing removes plant cover that protects soil. Logging removes forest cover, and the machinery breaks up the protective litter layer. Similarly, intensive grazing strips away protective plants. Rain, wind, and gravity then erode the exposed soil easily and rapidly. Meanwhile, soil continues to form by weathering at its usual slow, natural pace. Thus increased rates of erosion caused by agriculture can lead to net soil loss. In addition, erosion contaminates waterways with silt as well as with herbicides and pesticides.

Soil is a natural resource that can be difficult or expensive to renew once it has become degraded. Intense pressures on the world's soil resources have resulted in loss of productivity and biodiversity and increasing desertification of once-fertile lands. When farmers use proper conservation measures, soil can be preserved indefinitely or even improved. Some regions of Europe and China have supported continuous agriculture for centuries without soil damage. However, in recent years, marginal lands on hillsides, in tropical rainforests, and along the edges of deserts have been brought under cultivation. These regions are particularly vulnerable to soil deterioration, and today soil is being lost at an alarming rate throughout the world.

SHAUNL/SHUTTERSTOCK

FIGURE 10.19 In semiarid regions, thick forests cover the cool, shady north slopes of hills (right), but grass and sparse trees dominate hot, dry southern exposures (left).

weathering may be incomplete and the soils may contain many partly weathered mineral fragments. As a result, young soils are often sandy or gravelly. As soils mature, weathering continues and the clay content increases.

Soil Transport

By studying recent lava flows, scientists have determined how quickly plants return to an area after it has been covered by hard, solid rock. In many cases, plants appear when a lava flow is only a few years old, even before weathering has formed soil. Closer scrutiny shows that the plants have rooted in tiny amounts of soil that wind or water transported from nearby areas.

In many of the world's richest agricultural areas, most of the soil was transported from elsewhere. Streams deposit sediment, wind deposits dust, and soil slides downslope from mountainsides into valleys. These foreign materials mix with locally formed soil, changing its composition and texture. The soils of river flood plains and deltas, and the rich windblown soils of central China and the American Great Plains, are examples of transported soils.

10.5 Erosion

Weathering decomposes bedrock, and plants add organic material to the regolith to create soil at Earth's surface. However, soil does not accumulate and thicken throughout geologic time. If it did, Earth would be covered by a mantle of soil hundreds or thousands of meters thick, and rocks would not exist at Earth's surface. Instead, interactions with flowing water, wind, and glaciers erode soil as it forms (Figure 10.20). In addition, some weathered material simply slides downhill under the influence of gravity. In fact, all forms of erosion combine to remove soil about as fast as it forms. For this reason, soil

is usually only a few meters thick or less in most parts of the world.

Once soil erodes, the clay, sand, and gravel begin a long journey as they are carried downhill by the same agents that eroded them: streams, glaciers, wind, and gravity. On their journey they may come to rest in a streambed, a sand dune, or a lake bed, but those environments are only temporary stops. Eventually, they erode again and are carried downhill until, finally, they are deposited where the land meets the sea. Some of the sediment accumulates on deltas, and coastal marine currents redistribute some of it along the shore. Eventually, younger sediment may bury older layers until they become lithified to form sedimentary rocks.

FIGURE 10.20 Rain and flowing water erode soil as rapidly as it forms, as seen on these slopes in Arches National Monument, Utah.

mass wasting
The downslope movement of earth material, primarily caused by gravity. *See also* landslide.

landslide A general term for mass wasting (the downslope movement of rock and regolith under the influence of gravity) and the landforms it creates.

Erosion and transport of sediment by streams, glaciers, and wind are the subjects of Chapters 11, 13, and 14. In the remainder of this chapter, we will discuss erosion by gravity: specifically, landslides.

10.6 Landslides

Mass wasting is the downslope movement of earth material, primarily caused by gravity. The word **landslide** is a general term for mass wasting and for the landforms created by mass wasting.

Although gravity acts constantly on all slopes, the strength of the rock and soil usually hold the slope in place. In some places, however, natural processes or human activity may destabilize a slope and cause mass wasting. For example, a stream can erode the base of a hillside, undercutting it until it slides. Rain, melting snow, or a leaking irrigation ditch can add weight and lubricate soil, causing it to slide downslope. Mass wasting occurs naturally in all hilly or mountainous terrain. Steep slopes are especially vulnerable, and landslide scars are common in the mountains.

Every year, small landslides destroy homes and farmland. Occasionally, an enormous landslide buries a town or city, killing thousands of people. Landslides cause billions of dollars in damage every year, a cost approximately equal to the damage caused by earthquakes over a 20-year period. In many instances, losses occur because the average person does not recognize dangers that are obvious to most geologists.

Figure 10.21 shows three examples of recent landslides that have affected humans.

Why Do Landslides Occur?
Imagine that you are a geological consultant on a construction project. The developers want to build a road at the base of a hill, and they wonder whether landslides will threaten the road. What factors should you consider?

Steepness of the Slope
Obviously, the steepness of a slope is a factor in mass wasting. If frost wedging dislodges a rock from a steep cliff, the rock tumbles to the valley below. However, a similar rock is less likely to roll down a gentle slope.

Ⓐ

J.T. GILL/USGS

FIGURE 10.21 Landslides cause billions of dollars in damage every year. (A) A few days after this photo was taken, the corner of the house hanging over the gully fell in. (B) A landslide triggered by a leaking irrigation ditch threatens a house in Darby, Montana. (C) A landslide destroyed several expensive buildings in Hong Kong.

Ⓑ

COURTESY OF GRAHAM R. THOMPSON/JONATHAN TURK

Ⓒ

HONG KONG GOVERNMENT INFORMATION SERVICES

Unit 3: Surface Processes

Type of Rock and Orientation of Rock Layers

If sedimentary rock layers dip in the same direction as a slope, the upper layers may slide over a layer of weak rock. Imagine a hill underlain by shale, sandstone, and limestone oriented so that their bedding lies parallel to the slope, as shown in Figure 10.22A. If the base of the hill is undercut (Figure 10.22B), the upper layers of sandstone and limestone may slide over the weak shale. In contrast, if the rock layers dip at an angle to the hillside, the slope may be stable even if it is undercut (Figures 10.22C and 10.22D).

Several processes can reduce the stability of a slope. A stream or ocean waves can erode its base. Road building and excavation can also destabilize it. Therefore, a geologist or engineer must consider not only a slope's stability before construction but also how the project might alter its stability.

The Nature of Unconsolidated Materials

The **angle of repose** is the maximum slope or steepness at which loose material remains stable. If the slope becomes steeper than the angle of repose, the material slides. The angle of repose varies for different types of material. Rocks commonly tumble from a cliff to collect at the base as angular blocks of talus. The angular blocks interlock and jam together. As a result, talus typically has a steep angle of repose, up to 45°. In contrast, rounded sand grains do not interlock and therefore have a lower angle of repose, about 30° to 35° (Figure 10.23).

Water and Vegetation

To understand how water affects slope stability, think of a sand castle. Even a novice sand castle builder knows that sand must be moistened to build steep walls and towers (Figure 10.24). But too much water causes the walls to collapse. Small amounts of water bind sand grains together because the electrical charges of water molecules attract the grains.

angle of repose
The maximum slope or steepness at which loose material remains stable. If the slope becomes steeper than the angle of repose, the material slides.

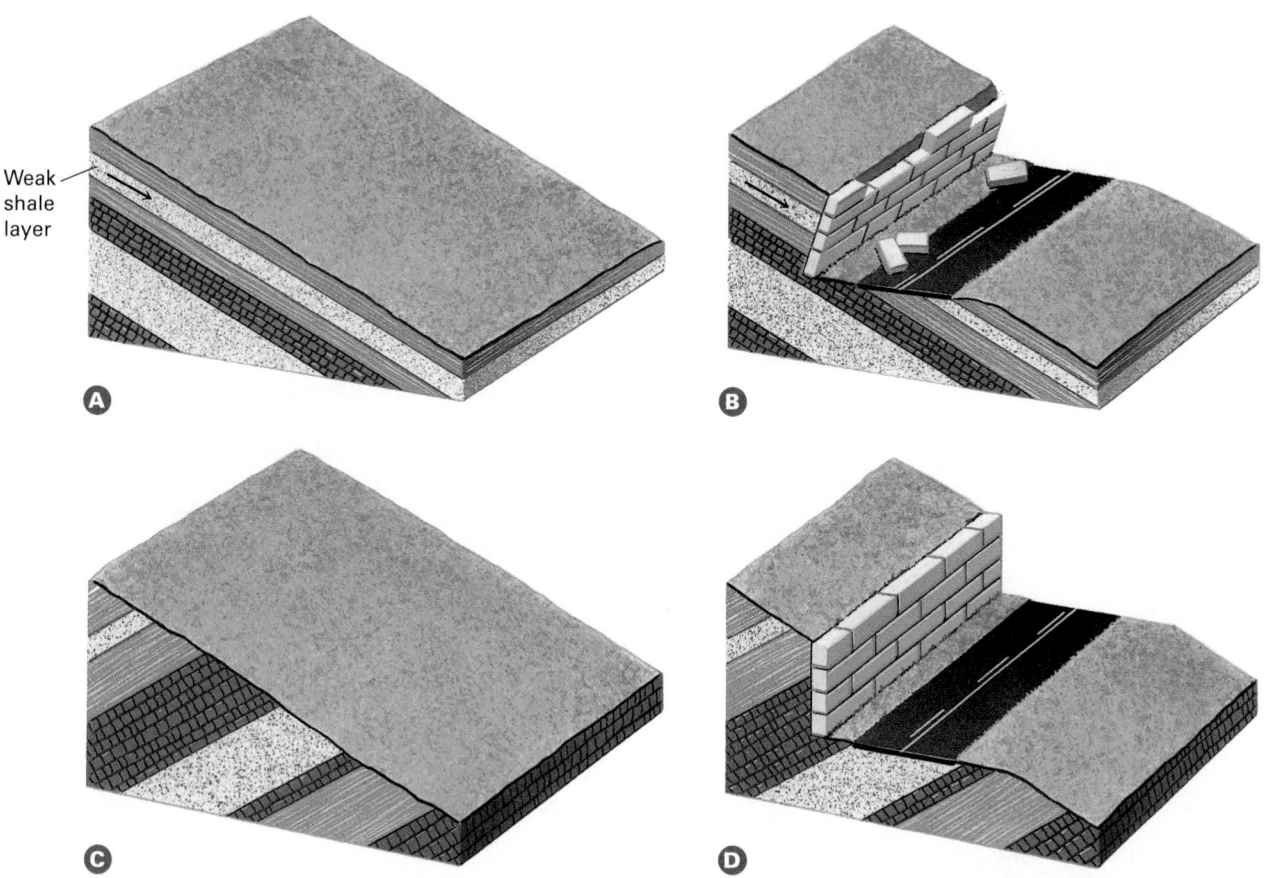

Weak shale layer

FIGURE 10.22 (A) Sedimentary rock layers dip parallel to this slope. (B) If a road cut undermines the slope, the dipping rock provides a good sliding surface, and the slope may fail. (C) Sedimentary rock layers dip at an angle to this slope. (D) The slope may remain stable even if it is undermined.

flow Mass wasting (landslide) in which loose soil or sediment moves downslope as a fluid, not as a consolidated mass; may occur slowly (less than 1 centimeter per year) or rapidly (like water over a waterfall).

slide Mass wasting in which the rock or soil initially moves as a consolidated unit along a fracture surface.

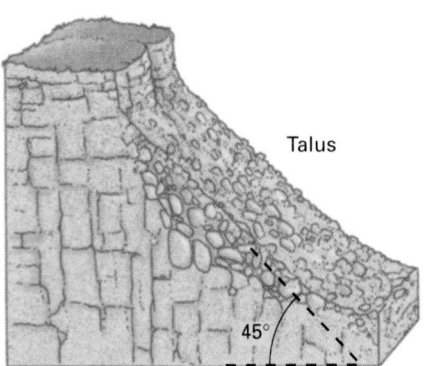

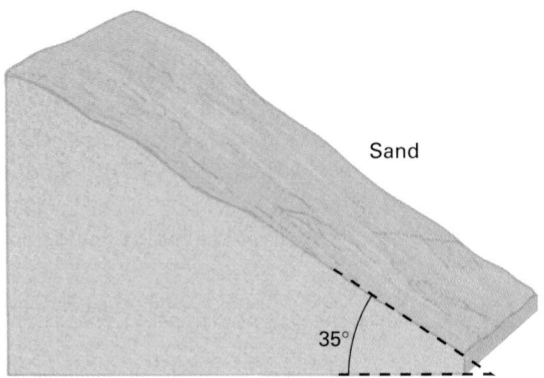

FIGURE 10.23 The angle of repose is the maximum slope at which a specific material can remain stable. Because the chunks of rock forming talus interlock and jam together, talus has a steeper angle of repose than sand, whose rounded grains do not interlock.

However, excess water lubricates the sand and adds weight to a slope.[2] When some soils become water saturated, they flow downslope, just as the sand castle collapses. In addition, if water collects on impermeable clay or shale, it may provide a weak, slippery layer so that overlying rock or soil can move easily.

Roots hold soil together and plants absorb water; therefore, a vegetated slope is more stable than a similar bare one. Many forested slopes that were stable for centuries slid when the trees were removed during logging, agriculture, or construction.

Landslides are common in deserts and regions with intermittent rainfall. For example, southern California has dry summers and occasional heavy winter rain. Vegetation is sparse because of summer drought and wildfires. When winter rains fall, bare hillsides often become saturated and slide. Mass wasting occurs for similar reasons during infrequent but intense storms in deserts.

---
2. Water fills pores in loose material and is under pressure of overlying material. Excess water raises that pressure, and it is actually the increased pore pressure that lowers slope stability.

Earthquakes and Volcanoes

An earthquake may cause a landslide by shaking an unstable slope, causing it to move. A volcanic eruption may melt the snow and ice cap at the top of the volcano; the water then flows onto the slope to release a landslide.

Thus mass wasting is common in earthquake-prone regions and in volcanically active areas. Some of the most well-known landslide disasters were caused by volcanic eruptions or earthquakes.

10.7 Types of Landslides

A landslide can occur slowly or rapidly. In some cases, rocks fall freely down the face of a steep mountain. In other instances, rock or soil creeps downslope so slowly that the movement may be unnoticed by a casual observer.

Landslides fall into three categories: flow, slide, and fall (Figure 10.25). To understand these categories, think again of a sand castle. Sand that is saturated with water flows down the face of the structure. During **flow**, loose, unconsolidated soil or sediment moves as a fluid. Some slopes flow slowly—at a speed of 1 centimeter per year or less. In contrast, mud with high water content can flow as rapidly as water falling over a waterfall.

If you undermine the base of a sand castle, the wall may fracture and a segment of the wall may slide downward. Movement of a coherent block of material along a fracture is called **slide**. Slide is usually faster than flow, but it still may take several seconds for the block to slide down the face of the castle.

FIGURE 10.24 The angle of repose depends on both the type of material and its water content. Dry sand forms low mounds, but if you moisten the sand, you can build steep, delicate towers.

Unit 3: Surface Processes

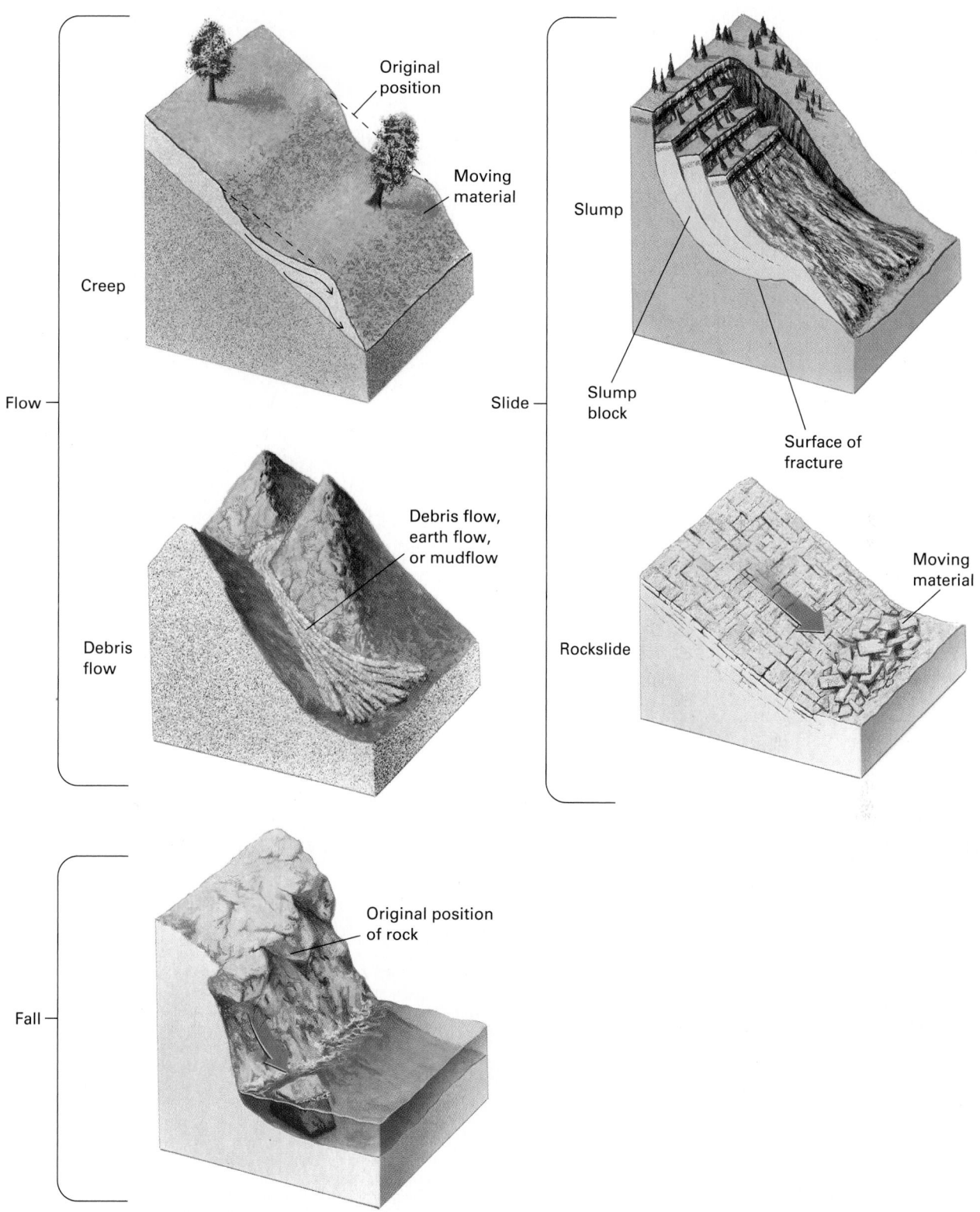

FIGURE 10.25 The three categories of mass wasting are flow, slide, and fall. During a *flow*, loose soil or sediment moves as a fluid. During a *slide*, the entire slab of rock or soil moves downslope as a unit. In a *fall*, loose material bounces freely downslope.

fall Mass wasting in which unconsolidated material falls freely or bounces down steep slopes or cliffs.

creep A subcategory of flow mass wasting; the very slow movement of loose material downslope, usually at a rate of only about 1 centimeter per year and usually on land with vegetation. Trees on a creeping block tilt downhill.

> If you ever contemplate buying hillside land for a homesite, examine the trees. If they have pistol butt bases, the slope is probably creeping, and creeping soil may tear a building apart.

© HARRIS SHIFFMAN/SHUTTERSTOCK

If you take a huge handful of sand out of the bottom of the castle, the whole tower topples. This rapid, free-falling motion of loose material is called **fall**. Fall is the most rapid type of mass wasting. In extreme cases, like the face of a steep cliff, rock can fall at a speed dictated solely by the force of gravity and air resistance.

Table 10.1 outlines the characteristics of flow, slide, and fall. Also refer back to Figure 10.25 as you read details of these three types of mass wasting, described next.

Flow

As the name implies, **creep** is the slow, downhill flow of rock or soil under the influence of gravity. A creeping slope typically moves at a rate of about 1 centimeter per year, although wet soil can creep more rapidly. During creep, the shallow soil or rock layers move more rapidly than deeper material moves. As a result, anything with roots or a foundation tilts downhill (Figure 10.26).

Trees have a natural tendency to grow straight upward. As a result, when soil creep tilts a growing

TABLE 10.1 Categories of Mass Wasting

Type of Movement	Description	Subcategory	Description	Comments
Flow	Individual particles move downslope independently of one another, not as a consolidated mass.	Creep	Rate of movement is slow, visually imperceptible.	Trees on creep slopes are tilted downhill and develop pistol butt shape.
	Typically occurs in loose, unconsolidated regolith.	Debris flow	More than half the particles are larger than sand size; rate of movement varies from less than 1 m/year to 100 km/hr or more.	Debris flows are common in arid regions with intermittent heavy rainfall, or can be triggered by volcanic eruption.
		Mudflow	Particles are fine grained, mixed with large amounts of water; rate of movement varies from less than 1 m/year to 100 km/hr or more.	Mudflows usually occur on unvegetated soil.
Slide	Material moves as consolidated blocks; can occur in regolith or bedrock.	Slump	A block of Earth material slides downslope, usually with a backward rotation on a concave surface.	Trees on slump blocks remain rooted and are tilted uphill.
		Rockslide	A newly detached segment of bedrock breaks into fragments and tumbles downslope, usually with rapid movement.	
Fall	Materials fall freely in air; typically occurs in bedrock.	—	—	Falls occur only on steep cliffs.

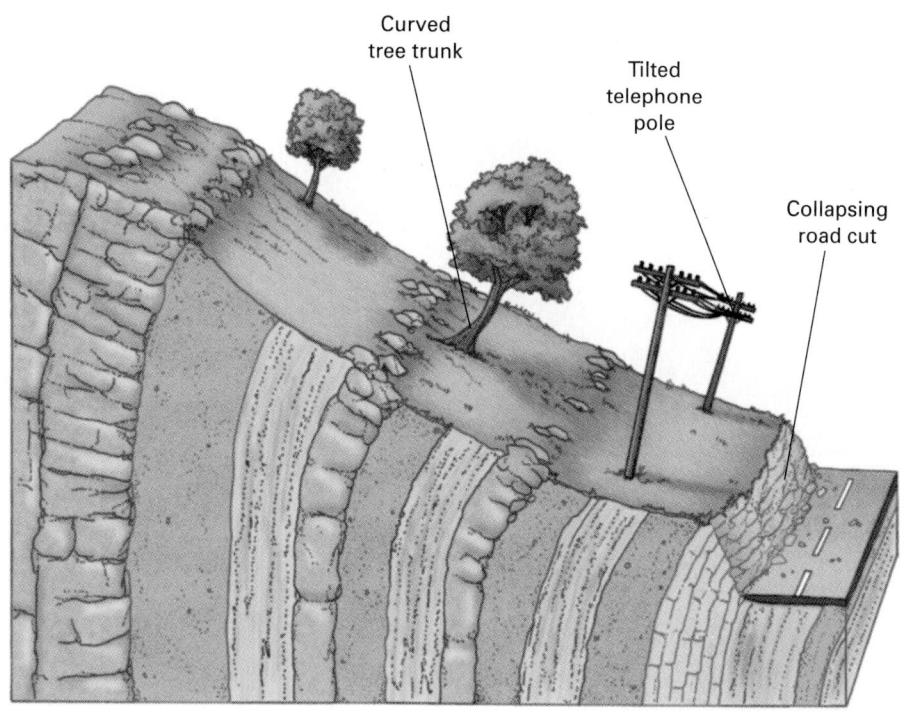

Curved
tree trunk

Tilted
telephone
pole

Collapsing
road cut

mudflow A sub-category of flow mass wasting that involves the downslope movement, usually on unvegetated land, of fine-grained soil particles mixed with a large amount of water; can be slow-moving, as slow as 1 meter per year, or as fast as a speeding car.

FIGURE 10.26 During creep, the land surface moves more rapidly than deeper layers, so objects embedded in rock or soil tilt downhill.

tree, the tree develops a J-shaped curve in its trunk, called pistol butt (Figure 10.27). If you ever contemplate buying hillside land for a homesite, examine the trees. If they have pistol butt bases, the slope is probably creeping, and creeping soil may tear a building apart.

If heavy rain falls on unvegetated soil, the water can saturate the soil to form a slurry of mud and rocks called a debris flow, earth flow, or **mudflow** depending on the size of the particles. A slurry is a mixture of water and solid particles that flows as a liquid. Wet concrete is a familiar example of a slurry. It flows easily and is routinely poured or pumped from a truck.

The advancing front of a debris flow, earth flow or mudflow often forms tongue-shaped lobes (Figure 10.28). A slow-moving mudflow travels at a rate of about 1 meter per year, but others can move as fast as a car speeding along an Interstate highway. A mudflow can pick up boulders and automobiles and destroy houses, filling them with mud or even dislodging them from their foundations.

COURTESY OF GRAHAM R. THOMPSON/JONATHAN TURK

FIGURE 10.27 If a hillside creeps as a tree grows, the tree develops pistol butt.

M. FREIDMAN, USGS

FIGURE 10.28 The 1980 eruption of Mount St. Helens melted large quantities of glacial ice near the summit of the peak. The meltwater mixed with soil to create lobe-shaped mudflows.

Slide

In some cases, a large block of rock or soil, or sometimes an entire mountainside, breaks away and slides downslope as a coherent mass or as a few intact blocks. Two types of slides occur: slump and rockslide.

A **slump** occurs when blocks of material slide downhill over a gently curved fracture in rock or rego-lith (Figure 10.29A). Trees remain rooted in the moving blocks. However, because the blocks rotate on the concave fracture, trees on the slumping blocks are tilted backward (Figure 10.29B). Thus you can distinguish slump from creep because slump tilts trees uphill, whereas creep tilts them downhill. At the lower end of a large slump, the blocks often break apart and pile up to form a jumbled, hummocky topography.

It is useful to identify slump because it often recurs in the same place or on nearby slopes. Therefore, a slope that shows evidence of past slump is not a good place to build a house.

During a **rockslide** (or rock avalanche), bedrock slides downslope over a fracture plane. Characteristically, the rock breaks up as it moves and a turbulent mass of rubble tumbles down the hillside. In a large rockslide, the falling debris traps and compresses air beneath and among the tumbling blocks. The compressed air reduces friction and allows some rockslides to attain speeds of 500 kilometers per hour. The same mechanism allows a snow or ice avalanche to cover a great distance at high speed.

FIGURE 10.29 (A) In slump, blocks of soil or rock remain intact as they move downslope. (B) Trees tilt back into the hillside on this slump along the Quesnel River in British Columbia. Slumping is facilitated by stream erosion at the base of the hill.

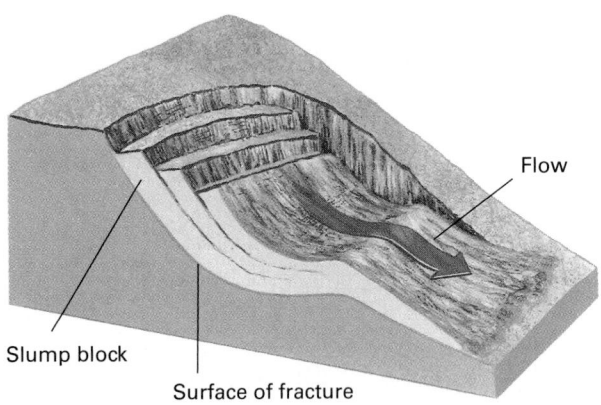

Flow

Slump block

Surface of fracture

A

COURTESY OF GRAHAM R. THOMPSON/JONATHAN TURK

B

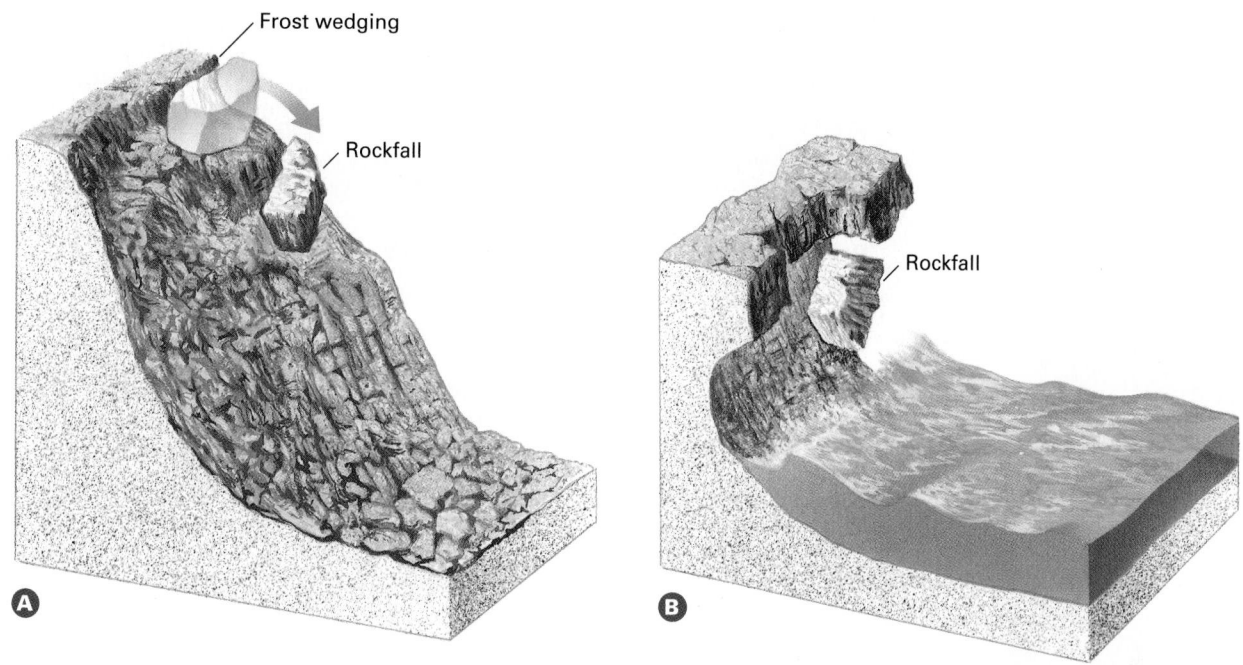

FIGURE 10.30 (A) Rockfall commonly occurs in spring or fall when freezing water dislodges rocks from cliffs. (B) Undercutting of cliffs by waves, streams, or construction can also cause rockfall.

Fall

If a rock dislodges from a steep cliff, it falls rapidly under the influence of gravity. Several processes commonly detach rocks from cliffs. Recall from our discussion of weathering that frost wedging can dislodge rocks from cliffs and cause rockfall (Figure 10.30A). Rockfall also occurs when ocean waves or a stream undercuts a cliff (Figure 10.30B).

10.8 Predicting and Avoiding Landslides

One of the most important tools in evaluating landslide hazards is to understand that landslides commonly occur in the same area as earlier landslides because the geologic conditions that cause mass wasting tend to be constant over a large area and for long periods of time. Thus, if a hillside has slumped, nearby hills may also

be vulnerable to mass wasting. In addition, landslides and mudflows commonly follow the paths of previous slides and flows. If an old mudflow lies in a stream valley, future flows may follow the same valley.

Awareness and avoidance are the most effective defenses against mass wasting. Geologists evaluate landslide probability by combining data on soil and bedrock stability, slope angle, climate, and history of slope failure in the area. They include evaluations of the probability of a triggering event, such as a volcanic eruption or earthquake. Building codes then regulate or prohibit construction in unstable areas. For example, according to the U.S. Uniform Building Code, a building cannot be constructed on a sandy slope steeper than 27°, even though the angle of repose of sand is 30° to 35°. Thus the law leaves a safety margin of 3° to 8°. Architects can obtain permission to build on more precipitous slopes if they anchor the foundation to stable rock.

FRESHWATER: STREAMS, LAKES, GROUND WATER, AND WETLANDS

Gibbon Falls in the springtime, at Yellowstone National Park.

visit **4ltrpress.cengage.com**

> ## In every glass of water we drink, some of the water has already passed through fishes, trees, bacteria, worms in the soil, and many other organisms, including people. . . . Living systems cleanse water and make it fit, among other things, for human consumption.
>
> *Elliot A. Norse*

Think of a pleasing, relaxing landscape, and you will probably envision water: A gentle surf lapping against white coral sand with palm trees in the background; a gurgling brook dropping over moss-covered rocks beneath a canopy of oak and maple; or a sunrise burning through wispy fog on a mirror-smooth lake. Oceans cover two-thirds of Earth, and water also plays a seminal role in many systems interactions on the continents. We have already learned that water is an important agent in the rock cycle. The previous chapter, "Weathering, Soil, and Erosion," was also largely about water—as a corrosive chemical, as a physical force when it freezes, and as a medium that transports rock and sediment. This chapter explores the roles played by Earth's fresh water sources: streams, lakes, ground water, and wetlands.

11.1 The Water Cycle

The **hydrologic cycle**, or the *water cycle*, describes the continuous circulation of water among the four spheres: the hydrosphere (or watery part of the planet), the geosphere (the land), the biosphere (life-forms in the sea, on land, and in the air), and the atmosphere. About 1.3 billion cubic kilometers of water exist at Earth's surface. Of this huge quantity, 97.5 percent is salty seawater, and another 1.8 percent is frozen into the great ice caps of Antarctica and Greenland. Thus, although the hydrosphere contains a great amount of water, only 0.64 percent is fresh and available in streams and rivers, ground water, lakes, and wetlands (Figure 11.1).

Water evaporates from the seas and from land to form water vapor in the atmosphere. This vapor eventually condenses and falls back to the surface as rain or snow. Most precipitation lands on the ocean, partly because the oceans cover most of the planet. The precipitation that falls on

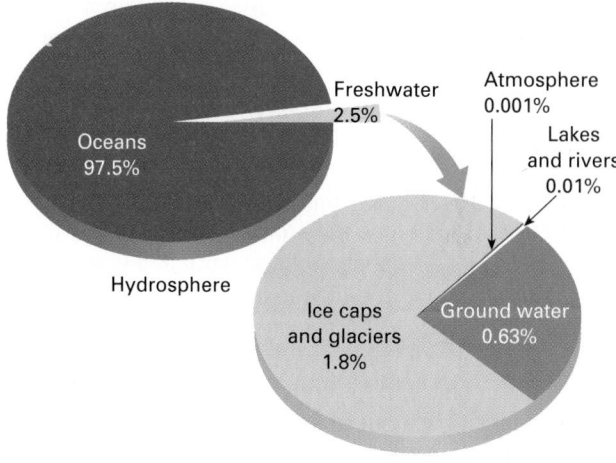

FIGURE 11.1 Of Earth's water, 97.5 percent is in the oceans, and less than 1 percent is available as freshwater in streams and rivers, lakes, ground water, and wetlands.

the continents follows four paths, as illustrated in Figure 11.2:

1. Surface water flowing to the sea in streams and rivers is called **runoff**. This water may stop temporarily in a lake or wetland, but eventually it evaporates or flows to the oceans.

2. Some water seeps into the ground—the geosphere—to become part of a vast, subterranean reservoir of ground water. Although surface water is more conspicuous, 60 times more water is stored as ground water than in all streams, lakes, and wetlands combined. Ground water seeps through bedrock and soil toward the sea, although it flows much more slowly than surface water.

transcription Direct evaporation of water into the atmosphere from the leaf surfaces of plants.

stream A moving body of water, confined in a channel and flowing downslope; a *river* is a large stream fed by smaller ones.

tributary A stream that feeds water into another stream or river.

gradient The steepness or vertical drop of a stream over a specific distance.

3. Most of the remainder of water that falls onto land evaporates back into the atmosphere. Water also evaporates directly from plants as they breathe, in a process called **transpiration**.

4. A small amount of water is also incorporated into the biosphere as plant and animal tissue when plants grow.

The hydrologic cycle not only describes the movement of water, it also describes a primary mechanism for the movement of energy from one part of the globe to another. Ocean currents transport huge quantities of heat from the equator toward the poles, thus cooling the equator and warming the higher latitudes. Evaporation is a cooling process, whereas condensation releases heat. Water vapor is a greenhouse gas, thereby warming the atmosphere, but clouds and sparkling glaciers reflect sunlight back out to space and thereby cool Earth. There are so many feedback cycles, many with opposite effects, that one scientist wrote, "Water acts as the Venetian blind of our planet, as its central heating system, and as its refrigerator, all at the same time."[1] This chapter chronicles the movement of surface water. Other components of the hydrologic cycle will be discussed throughout the remainder of the book.

11.2 Streams

Earth scientists use the term **stream** for all water flowing in a channel, regardless of the stream's size. The term *river* is commonly used for any large stream fed by smaller ones, called **tributaries**. Most streams run year-round, even during times of drought, because they are fed by ground water that seeps into the streambed.

Stream Flow and Velocity

Three factors control stream velocity: gradient, discharge, and channel characteristics.

Gradient

Gradient is the steepness, or vertical drop over a specific distance, of a stream. Obviously, if all other factors are equal, water flows more rapidly down a steep slope than a gradual one. A tumbling mountain stream may drop 40 meters or more per kilometer, whereas the lower Mississippi River has a gradient of only 0.1 meter per kilometer.

---

1. Heike Langenberg, commentary introducing the special Insight section "Climate and Water," *Nature* 419 (September 12, 2002), 187.

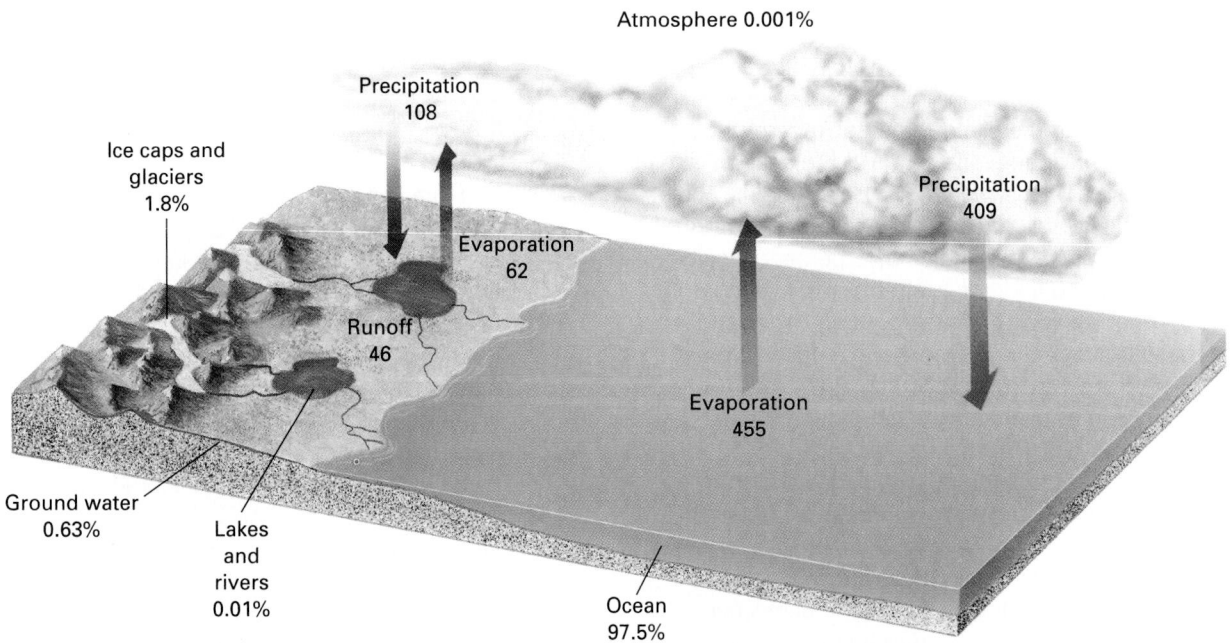

FIGURE 11.2 The hydrologic cycle shows that water circulates constantly among the sea, the atmosphere, and the land. Whole numbers indicate thousands of cubic kilometers of water transferred each year. Percentages show proportions of total global water in different portions of Earth's surface.

Discharge

Discharge is the amount of water flowing down a stream. It is expressed as the volume of water flowing past a point per unit time, usually cubic meters per second (m^3/sec). At a given point in a stream, the velocity of a stream increases when its discharge increases. Thus, a stream flows faster during flood, even though its gradient is unchanged.

When Huckleberry Finn floated down the Mississippi on a raft, he probably drifted downstream at about 7 to 9 kilometers per hour (2 to 2.5 meters per second). However, a modern rafter on one of the small mountain tributaries of the Mississippi may drift downstream at only 5.4 kilometers per hour (1.5 meters per second). Thus, although it is counterintuitive, a large, lazy-appearing river may flow more rapidly than a small, steep mountain stream because of the larger river's greater discharge.

The largest river in the world is the Amazon, with an average discharge of 150,000 m^3/sec. In contrast, the Mississippi, the largest river in North America, has an average discharge of about 17,500 m^3/sec, approximately one-ninth that of the Amazon.

A stream's discharge can change dramatically from month to month or even during a single day. For example, the Selway River, a mountain stream in Idaho, has a discharge of 100 to 130 m^3/sec during early summer, when mountain snow is melting rapidly. During the dry season in late summer, the discharge drops to about 20 m^3/sec (Figure 11.3). A desert stream may dry up completely during summer but become the site of a flash flood during a sudden thunderstorm.

Channel Characteristics

Channel characteristics refer to the shape and roughness of a stream channel. The floor of the channel is called the **bed**, and the sides of the channel are the **banks**. Friction between flowing water and the stream channel slows current velocity. Consequently, water flows more slowly near the banks than near the center of a stream. If you paddle a canoe down a straight stream channel, you move faster when you stay away from the banks. The amount of friction depends on the roughness and shape of the channel. Boulders on the banks or in the streambed increase friction and slow a stream down, whereas the water flows more rapidly if the bed and banks are smooth.

Stream Erosion and Sediment Transport

Streams shape Earth's surface by eroding soil and bedrock. The flowing water carries the eroded sediment grain by grain toward the sea. A stream may deposit some of the sediment on its flood plain, forming a level valley bottom, while it carries the remainder to the seacoast, where the sediment accumulates to form deltas and sandy beaches.

Stream erosion and sediment transport depend on a stream's energy. A rapidly flowing stream has more

discharge The volume of water flowing downstream, usually measured in units of cubic meters per second (m^3/sec).

channel characteristics Features describing the shape and roughness of a stream channel.

bed The floor of a stream channel.

banks The rising slopes bordering the sides of a stream channel.

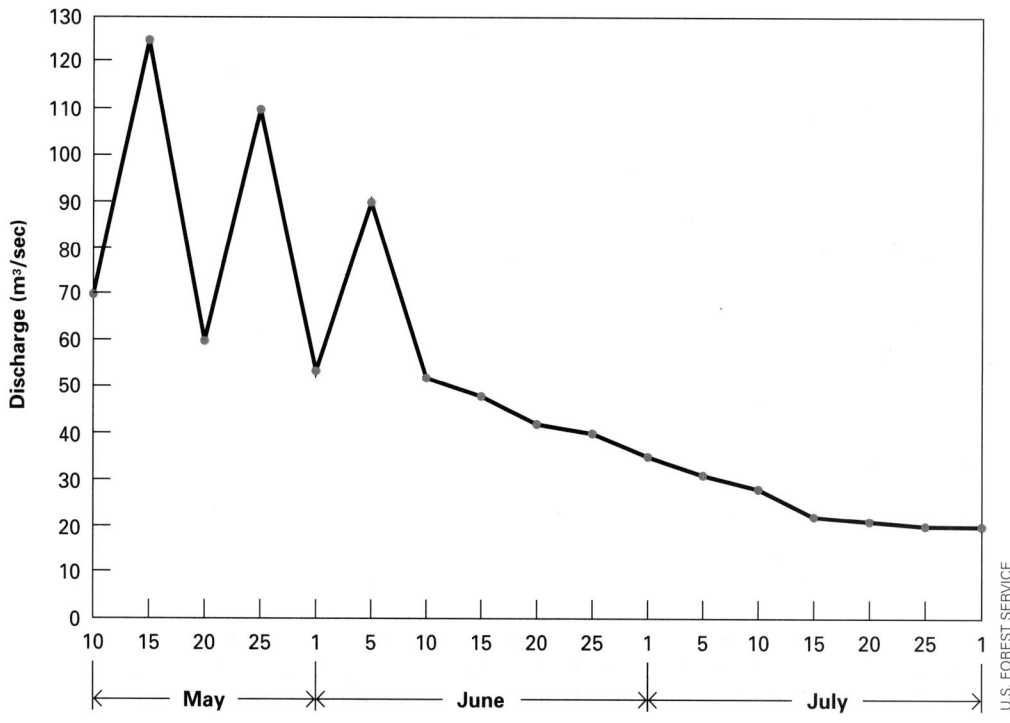

FIGURE 11.3
The 1988 hydrograph for the Selway River in Idaho shows that the discharge varied from 125 m^3/sec in the spring to 20 m^3/sec in the summer. The sharp peaks reflect high discharge during periods of rapid snowmelt.

The world's two muddiest rivers—the Yellow River in China and the Ganges River in India—each carry more than 1.5 billion tons of sediment to the ocean every year.

© JEREMY RICHARDS/SHUTTERSTOCK

competence A measure of the largest particles that a stream can transport.

capacity The maximum quantity of sediment that a stream can carry past a given point in a given amount of time.

dissolved load The portion of a stream's sediment load that consists of ions dissolved in water.

suspended load The portion of a stream's sediment load that is carried for a considerable time in suspension, free from contact with the streambed.

bed load The portion of a stream's load—boulders, cobbles, and sand—that is transported along the bottom, immediately above the streambed.

downcutting Downward erosion by a stream into its bed, usually cutting a V-shaped valley along a relatively straight path.

base level The deepest level to which a stream can erode its bed. The ultimate base level is usually sea level, but this is seldom attained.

energy to erode and transport sediment than a slow stream of the same size. The **competence** of a stream is a measure of the largest particle it can carry. A fast-flowing stream can transport cobbles and even boulders in addition to small particles. A slow stream carries only silt and clay.

The **capacity** of a stream is the total amount of sediment it can carry past a given point in a given amount of time. Capacity is proportional to both current speed and discharge. Thus a large, fast stream has a greater capacity than a small, slow one. Because the ability of a stream to erode and carry sediment is proportional to both velocity and discharge, most sediment transport and erosion occur during the few days each year when the stream is in flood. Relatively little erosion and sediment transport occur during the remainder of the year. To see this effect for yourself, look at any stream during low water. It will most likely be clear, indicating little erosion or sediment transport. Look at the same stream later, when it is flooding. It will probably be muddy and dark, indicating that the stream is eroding its bed and banks and transporting the sediment.

After a stream erodes soil or bedrock, it transports the sediment downstream in three ways. One way is in the form of ions dissolved in water, called the **dissolved load**. A stream's ability to carry dissolved ions depends mostly on its discharge and its chemistry, not its velocity. Thus even the still waters of a lake or ocean contain dissolved substances; that is why the sea and some lakes are salty. Although dissolved ions are invisible, they comprise more than half of the total sediment load carried by some rivers. More commonly, dissolved ions make up less than 20 percent of the total sediment load of streams.

If you place soil with equal amounts of sand, silt, and clay in a jar of water and shake it up, the sand grains settle quickly. But the smaller silt and clay particles remain suspended in the water as **suspended load**, giving it a cloudy appearance. Clay and silt are small enough that even the slight turbulence of a slow stream keeps them in suspension. A rapidly flowing stream can carry sand in suspension.

During a flood, when stream energy is highest, the rushing water can roll boulders and cobbles along the bottom as **bed load**. Sand also moves in this way, and if the stream velocity is sufficient, sand grains bounce and hop along the streambed like millions of tiny marbles.

Most streams carry the greatest proportion of sediment in suspension, less in solution, and the smallest proportion as bed load.

Downcutting and Base Level

A stream erodes downward into its bed and laterally against its banks. Downward erosion is called **downcutting** (Figure 11.4). The **base level** of a stream is

196

Unit 3: Surface Processes

the deepest level to which it can erode its bed. Most streams cannot erode below sea level, which is called the ultimate base level. This concept is straightforward. Water can flow only downhill. If a stream were to cut its way down to sea level, it would stop flowing and hence would no longer erode its bed.

In addition to ultimate base level, a stream may have a number of local, or temporary, base levels. For example, a stream stops flowing where it enters a lake. It then stops eroding its channel because it has reached a temporary base level (Figure 11.5). A layer of rock that resists erosion may also establish a temporary local base level because it flattens the stream gradient. Thus, the stream slows down and erosion decreases. The top of a waterfall is a temporary base level commonly established by resistant rock. For example, Niagara Falls is formed by a resistant layer of dolomite overlying

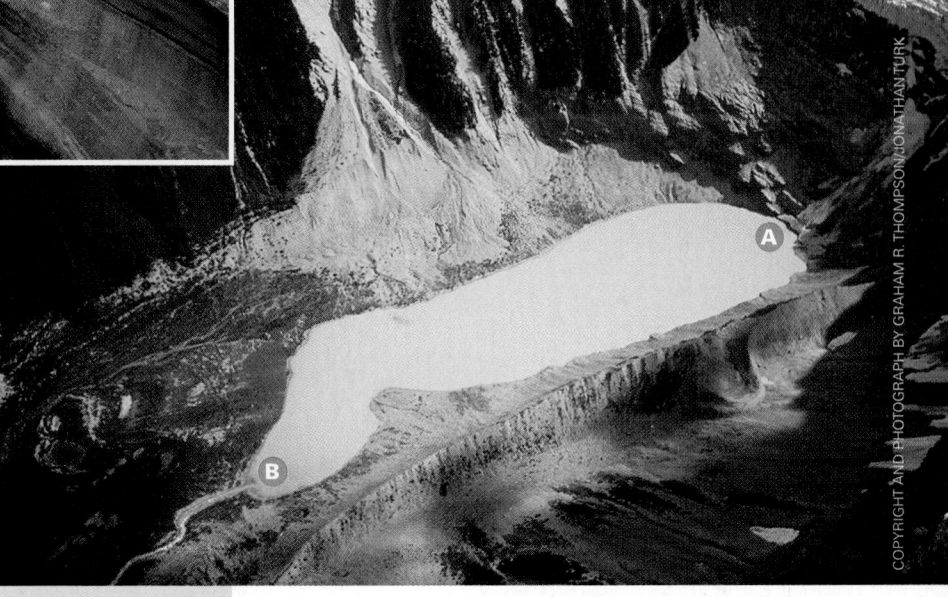

FIGURE 11.4 Deer Creek, a tributary of the Colorado River in the Grand Canyon, has downcut its channel into solid sandstone.

FIGURE 11.5 A steep mountain stream (A) flowing from a glacier in the Canadian Rockies reaches a temporary base level and stops flowing where it enters a lake. The lake drains at stream B.

softer shale. Although the dolomite resists erosion at the top of the falls, the turbulent falling water erodes the underlying shale. When its foundation is under-mined, the dolomite cap collapses and the falls migrate upstream. Niagara Falls has retreated 11 kilometers upstream since its formation about 9,000 years ago (Figure 11.6). Thus the falls retreat a little more than 1 meter per year. The erosion rate is rapid because the water generates considerable energy as it tumbles over the falls.

FIGURE 11.6 Niagara Falls has eroded 11 kilometers southward toward Lake Erie in the last 9,000 years and continues to erode today.

graded stream
A stream with a smooth, concave profile, in equilibrium with its sediment supply; it transports all the sediment supplied to it with neither erosion nor deposition in the streambed.

A stream like that in Figure 11.7A erodes rapidly in the steep places where its energy is high, and deposits sediment in the low-gradient stretches where it flows more slowly. Over time, erosion and deposition smooth out the irregularities in the gradient. The resulting **graded stream** has a smooth, concave profile (Figure 11.7B). Once a stream becomes graded, there is no net erosion or deposition and the stream profile no longer changes. An idealized graded stream such as this does not actually exist in nature, but many streams come close.

Sinuosity of a Stream Channel

A steep mountain stream usually downcuts rapidly compared with the rate of lateral erosion. As a result, it cuts a relatively straight channel with a steep-sided, V-shaped valley (Figure 11.8). The stream maintains its relatively straight path because it flows with enough energy to erode and transport any material that slumps into its channel.

In contrast, a low-gradient stream is less able to erode downward into its bed. Much of the stream

FIGURE 11.7 (A) An ungraded stream has many temporary base levels. (B) With time, the stream smoothes out the irregularities to develop a graded profile.

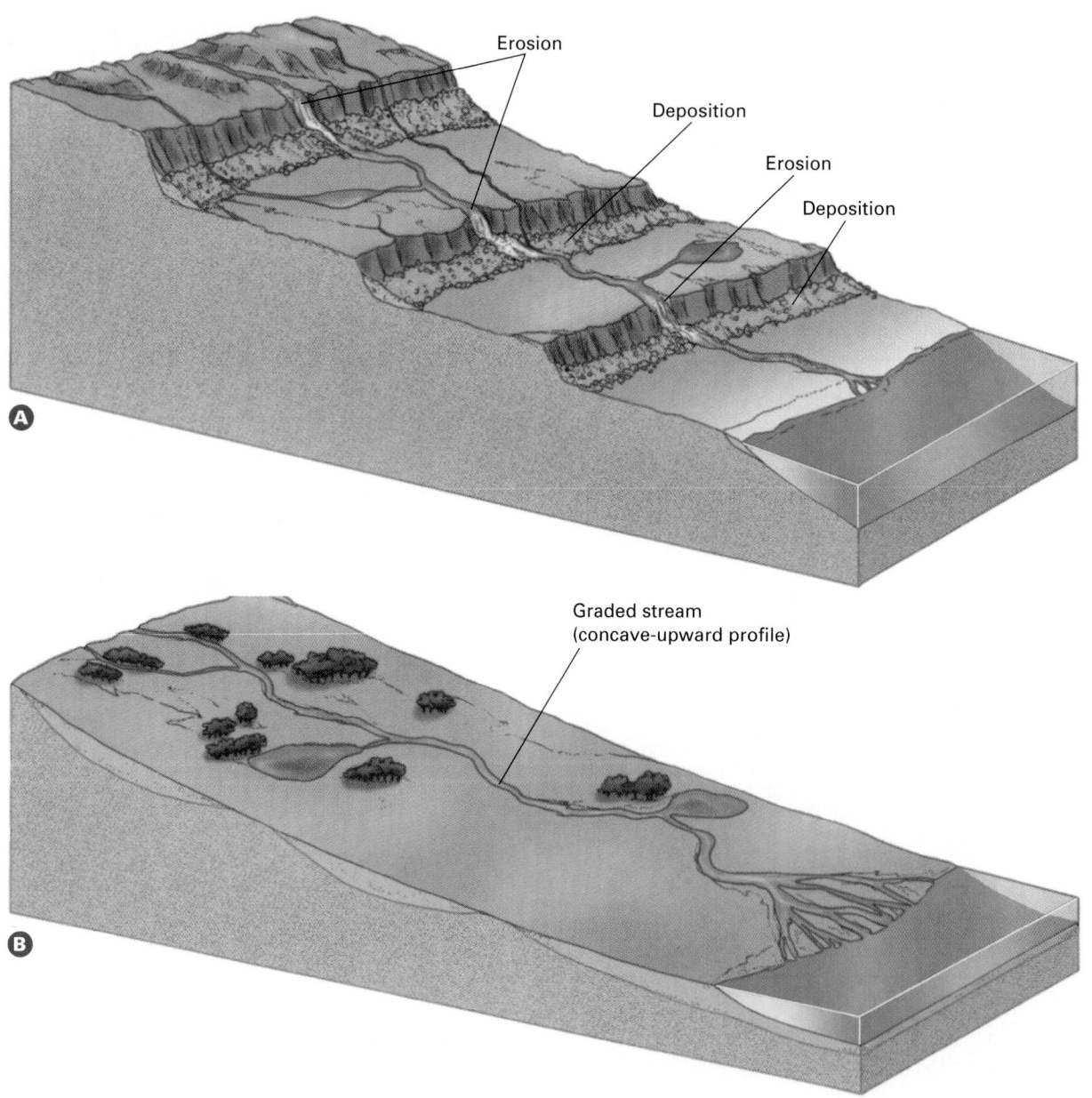

FIGURE 11.8 A steep mountain stream eroded this V-shaped valley into soft shale in the Canadian Rockies.

energy is directed against the banks, causing **lateral erosion**. Lateral erosion undercuts the valley sides and widens a stream valley. Most low-gradient streams flow in a series of bends, called **meanders** (Figure 11.9A and 11.9B). A meandering stream wanders back and forth across its flood plain, forming a wide flat valley with a flat bottom.

As a stream flows into a meander bend, the inertia of the moving water tends to keep the water moving in a straight path. Consequently, most of the current flows to the outside bank of the meander. As a result, both the velocity and channel depth are greatest near the outside of the bend, and the stream erodes its outside bank. At the same time, sediment is deposited in the slower water on the inside of the meander to form a **point bar** (Figure 11.9A). Because a meandering stream erodes the outside banks of its meanders and deposits sediment on the insides of the bends, the meanders migrate laterally, moving slowly in a sinuous course down the flood plain as the stream constantly modifies its channel. Occasionally, an **oxbow lake** forms where the stream cuts across the neck of a meander and isolates an old meander loop (Figure 11.9C and 11.9D).

FIGURE 11.9 (A) A stream erodes the outsides of meanders and deposits sand and gravel on the inside bends to form point bars. (B) Meanders and point bars form the channel of the Bitterroot River in Montana. (C) Over time, a stream may erode through the neck of a meander to form an oxbow lake. (D) This oxbow lake formed in the Flathead River in Montana.

lateral erosion The action of a low-gradient stream as it cuts into and erodes its banks, swinging from one side of its channel to the other, forming a wide flat valley.

meanders A series of twisting curves or loops in the course of a stream.

point bar A deposit of sediment in the slower water on the inside of a meander.

oxbow lake A crescent-shaped lake created where a meander loop is cut off from a stream when the ends of the meander became plugged with sediment.

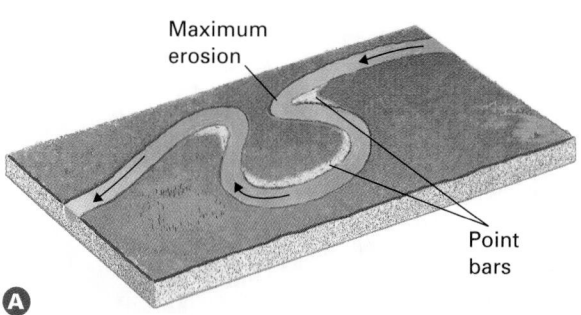

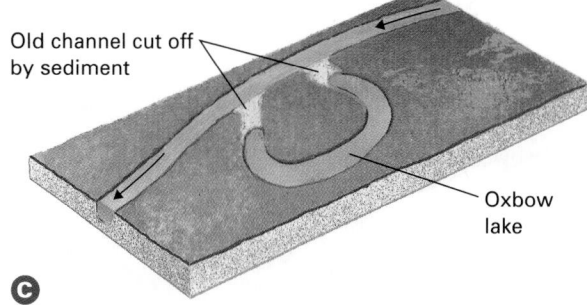

braided stream A stream that flows in many shallow, interconnecting channels; formed because more sediment was supplied to the stream than it could carry, accumulating in the channel and forcing the stream to overflow its banks and erode new channels.

drainage basin The region that is drained by a single river.

Most streams flow in a single channel. In contrast, a **braided stream** flows in many shallow, interconnecting channels (Figure 11.10). A braided stream forms where more sediment is supplied to a stream than it can carry. The excess sediment accumulates in the channel, filling it and forcing the stream to overflow its banks and erode new channels. As a result, a braided stream flows simultaneously in several channels and shifts back and forth across its flood plain.

Braided streams are common in both deserts and glacial environments because both produce abundant sediment. A desert yields large amounts of sediment because it has little or no vegetation to prevent erosion. Glaciers grind bedrock into fine sediment, which is carried by streams flowing from the melting ice.

FIGURE 11.10 Joe Creek in Canada's Yukon Territory is heavily braided because glaciers provide more sediment than the stream can carry.

which are in turn fed by smaller tributaries. Topographic highs—mountain ranges, ridges, or plateaus—separate adjacent river systems. The region drained by a single river is called a **drainage basin**. For example, the Rocky Mountains separate the Colorado and Columbia drainage basins to the west from the Mississippi and Rio Grande basins to the east. Together, those four river systems drain more than three-fourths of the United States.

Drainage Basins

Only a dozen or so major rivers flow into the sea along the coastlines of the United States (Figure 11.11). Each is fed by a number of tributaries,

FIGURE 11.11 Most of the surface water in the United States flows to the sea from approximately a dozen major rivers.

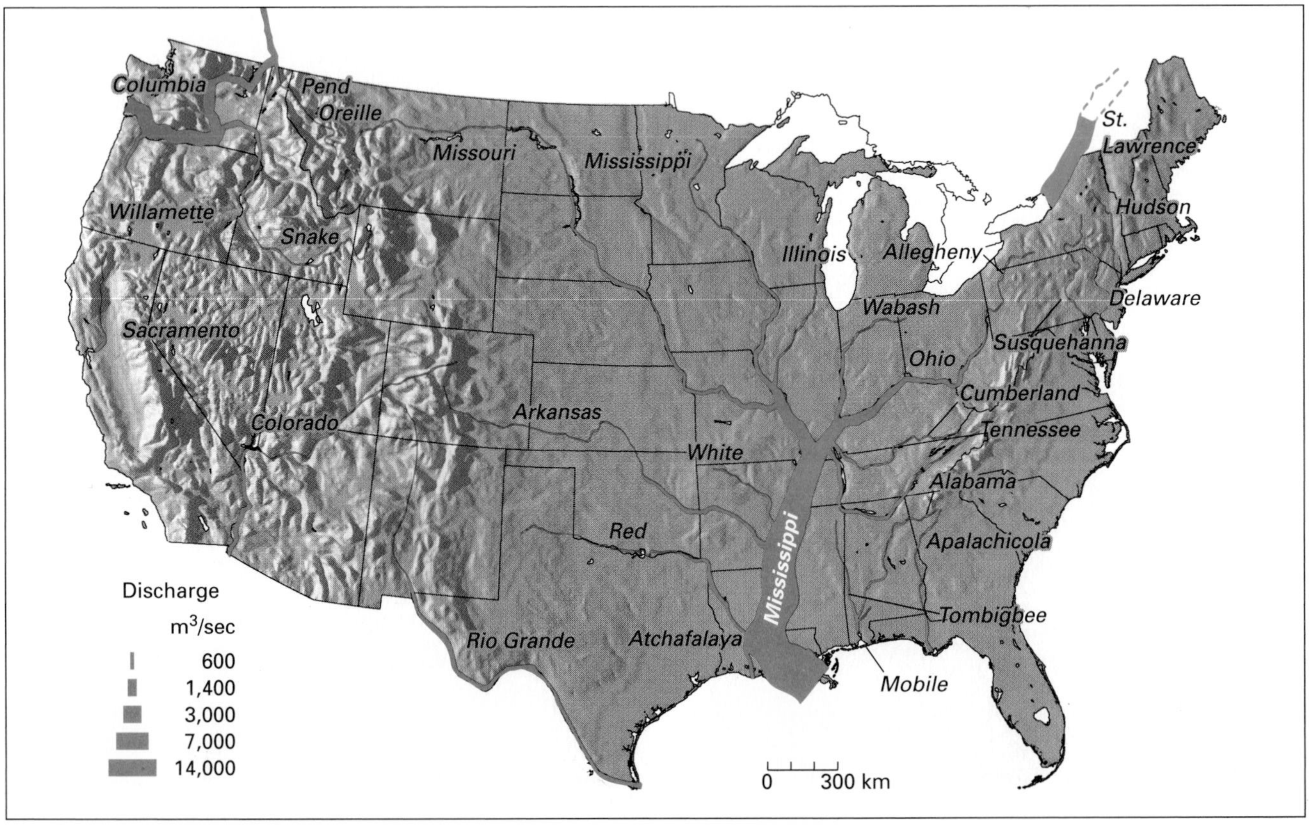

Columbia
Pend Oreille
Missouri
Mississippi
St. Lawrence
Willamette
Snake
Hudson
Illinois
Allegheny
Wabash
Delaware
Sacramento
Susquehanna
Ohio
Colorado
Arkansas
Cumberland
Tennessee
White
Mississippi
Alabama
Red
Apalachicola
Rio Grande
Atchafalaya
Tombigbee
Mobile

Discharge
m³/sec
600
1,400
3,000
7,000
14,000

0   300 km

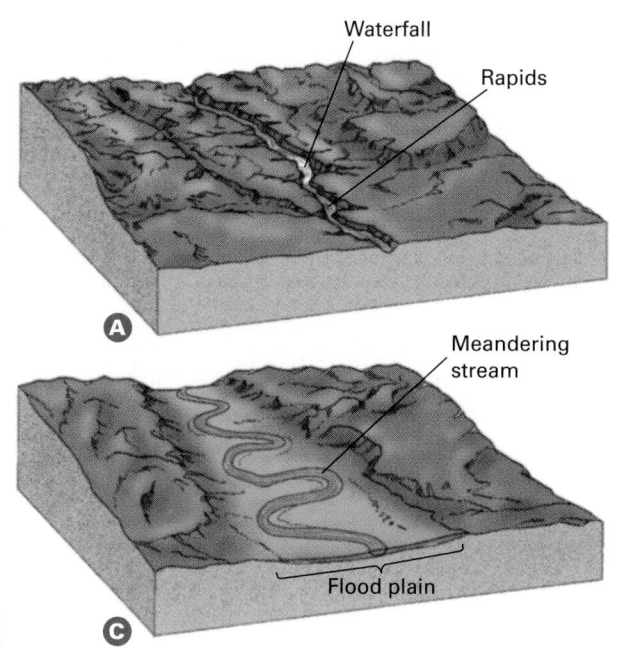

Waterfall
Rapids

A

Meandering stream

Flood plain

C

USGS

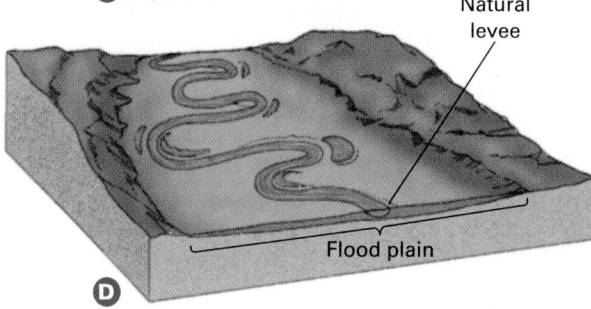

Flood plain

B

Natural levee

Flood plain

D

© APTYP KOK/SHUTTERSTOCK

alluvial fan A fan-shaped accumulation of sediment created where a steep mountain stream rapidly slows down as it reaches a relatively flat plain.

FIGURE 11.12 If tectonic activity does not uplift the land, over time streams erode the mountains away and widen the valleys into broad flood plains.

Stream Erosion and Mountains: How Landscapes Evolve

According to a model popular in the first half of the 20th century, streams erode Earth's surface and create landforms in an orderly sequence (Figure 11.12). At first, they cut steep, V-shaped valleys into mountains. Over time, the streams erode the mountains away and widen the valleys into broad flood plains. Eventually, the entire landscape flattens, forming a large, feature-less plain. However, if this were the only mechanism affecting Earth's surface during its 4.6-billion-year history, all landforms would have eroded to a flat plain. Why, then, do mountains, valleys, and high plateaus still exist?

The model tells only half the story. Streams do continuously erode the landscape, flattening mountains and widening flood plains. But at the same time, tectonic activity may uplift the land and interrupt the simple, idealized sequence. In this way, Earth's hydrosphere and atmosphere work together with tectonic processes in the geosphere to create landforms.

11.3 Stream Deposition

When streams slow down they deposit their bed load first. If water slows sufficiently, the suspended load may also fall to the bottom, but the dissolved load remains in the water until the chemical environment changes. New landforms are created when rivers deposit sediment.

If a steep mountain stream flows onto a flat plain, its gradient and velocity decrease abruptly. As a result, it deposits most of its sediment in a fan-shaped mound called an **alluvial fan**. (*Alluvium* is sediment deposited by flowing water.) Alluvial fans are common in many arid and semiarid mountainous regions (Figure 11.13).

COURTESY OF GRAHAM R. THOMPSON/JONATHAN TURK

FIGURE 11.13 This alluvial fan in Death Valley formed where a steep mountain stream deposited most of its sediment as it entered the flat valley. A road runs across the lower part of the fan.

Chapter 11: Freshwater: Streams, Lakes, Ground Water, and Wetlands

201

delta A nearly flat, fan-shaped accumulation of sediment, forming a tract of land where a stream enters a lake or ocean.

distributaries Channels that split from the main stream feeding a delta or alluvial fan, spreading out to build the area covered by the delta or fan.

submarine delta The portion of a river delta that lies underwater.

flood A relatively high stream flow that overtops the stream banks, covering land that is not usually under water.

flood plain That portion of a river valley adjacent to the channel; it is built by sediment deposited during floods and is covered by water during a flood.

A stream also slows abruptly where it enters the still water of a lake or ocean. The sediment settles out to form a nearly flat landform called a **delta**. Part of the delta lies above water level, where sediment is deposited during floods. The remainder of it lies slightly below water level. Deltas, like alluvial fans, are commonly fan shaped, resembling the Greek letter delta (Δ).

Both deltas and alluvial fans change rapidly. The stream abandons sediment-choked channels, while new channels develop, as in a braided stream. As a result, a stream feeding a delta or fan splits into many channels called **distributaries**. A large delta may spread out in this manner until it covers thousands of square kilometers (Figure 11.14). Most alluvial fans, however, are much smaller, covering just a fraction of a square kilometer to a few square kilometers. Some sediment is carried beyond the delta and deposited on the sea floor, creating a **submarine delta**—a portion of the delta that lies underwater.

Even though deltas cover only a small fraction of Earth's total land surface, they are environments that include each of Earth's four spheres. The delta is composed of solid sediment, but it is a watery zone where branched distributaries encounter the ocean. Because the land is fertile, with easy access to river and ocean, delta ecosystems are rich in natural plant and animal life—but the fertile soils and easy access to transportation systems make deltas desirable places for human habitation. Delta land is barely above sea level and river level, so it is certain to be flooded during atmospheric disturbances such as heavy rains and hurricanes.

11.4 Floods

When rainfall is heavy or snowmelt is rapid, more water flows down a stream than the channel can hold, creating a **flood**. During a flood, the stream overflows onto low-lying adjacent land called the **flood plain**. Massive floods occur somewhere in the world every year. In late August 2002, the Elbe River in eastern Europe peaked at a record 9.39 meters above flood stage, 0.6 meter higher than the previous record, recorded in 1845. One hundred people died, and the total damages exceeded $20 billion.

Although we normally think of floods as destructive events, flood plain ecosystems depend on floods. For example, cottonwood tree seeds germinate only after a flood, and waterfowl depend on flood plain wetlands. Many species of fish gradually lose out to stronger competitors during normal flows but have adapted better to floods, so their populations increase as a result of flooding. Thus species diversity is maintained by alternate periods of flooding and normal flow. Deltas are created and enlarged by floods. At normal times, when rivers are confined within their banks, the flowing water transports sediment out to sea and deposits it on the ocean floor. But during floods, river water rises above the stream banks and covers the delta land. When the flood waters slow down, they deposit sediment, thus enlarging the delta.

In some cases, flooding can directly benefit humans. Frequent small floods dredge bigger channels, which reduce the severity of large floods. Flooding streams carry large sediment loads and deposit them on flood plains to form fertile soil. So, paradoxically, the same floods that cause death and disaster create the rich soils that make flood plains and deltas so attractive for farming and human habitation.

But at the same time, floods are the costliest natural disasters on the planet—simply because people choose to live in harm's way. Many riverbank cities originally grew as ports, to take advantage of the easy transportation afforded by rivers. In addition, flood plains and deltas provide rich soils for farms, flat land for roads and buildings, and access to abundant water for industry and agriculture.

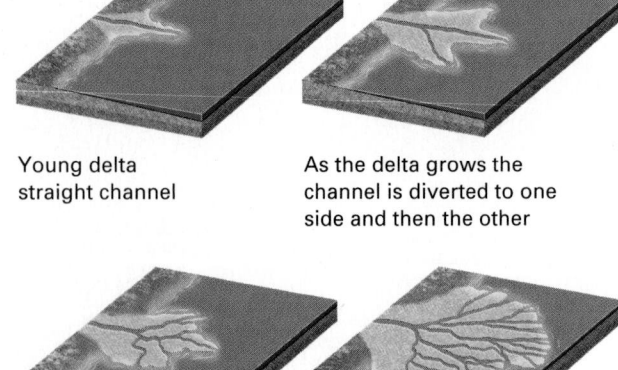

Young delta straight channel

As the delta grows the channel is diverted to one side and then the other

Distributaries form

Mature delta

FIGURE 11.14 A delta forms and grows with time where a stream deposits its sediment as it flows into a lake or the sea.

Flood Control

During the late spring and summer of 2008, heavy rain soaked the upper Midwest. Parts of eastern Iowa experienced 15 to 20 inches of rain in one month. As a result of the intense rainfall and abnormally heavy snowpack, the Mississippi River and nine Iowa rivers crested at or above record levels. In St. Louis, the Mississippi River crested 7 feet above normal. The Cedar River crested at over 32 feet, flooding 9.2 square miles of Cedar Rapids, Iowa, exceeding the highest recorded flood levels from 1929 and causing an estimated $1.5 billion in the city alone. The floods killed at least 24 people and injured hundreds more. Damages to the midwestern states exceed $6 billion.

artificial levee A wall built along the banks of a stream to prevent rising floodwater from spilling out of the channel onto the flood plain.

Artificial Levees and Channels

An **artificial levee** is a wall built along the banks of a stream to prevent rising water from spilling out of the stream channel onto the flood plain. In the past 70 years, the U.S. Army Corps of Engineers has spent billions of dollars building 11,000 kilometers of levees along the banks of the Mississippi and its tributaries. Of course, levees cannot eliminate risk of flood. During the 2008 floods in the Midwest, at least 20 levees did not hold, resulting in significant flooding and damage to large parts of Iowa and Indiana (Figure 11.15).

Unfortunately, artificial levees create conditions that may increase both flood intensity and property damage. One factor is entirely human—the protection promised by levees encourages people to build in the flood plain.

In the absence of a levee, people might decide to build on high ground, safe from floods. But when levees are built, people are more likely to construct homes and businesses in harm's way. After the 2008 floods, federal funding was used to repair the damaged levees and people returned to their homes and businesses. But these structures have proved to be unreliable in the past and may break again.

Levees also may cause much greater floods in the future, and they can cause higher floods along

FIGURE 11.15 Flooding in Iowa. When the levees along the Mississippi failed to hold the river, the results were catastrophic.

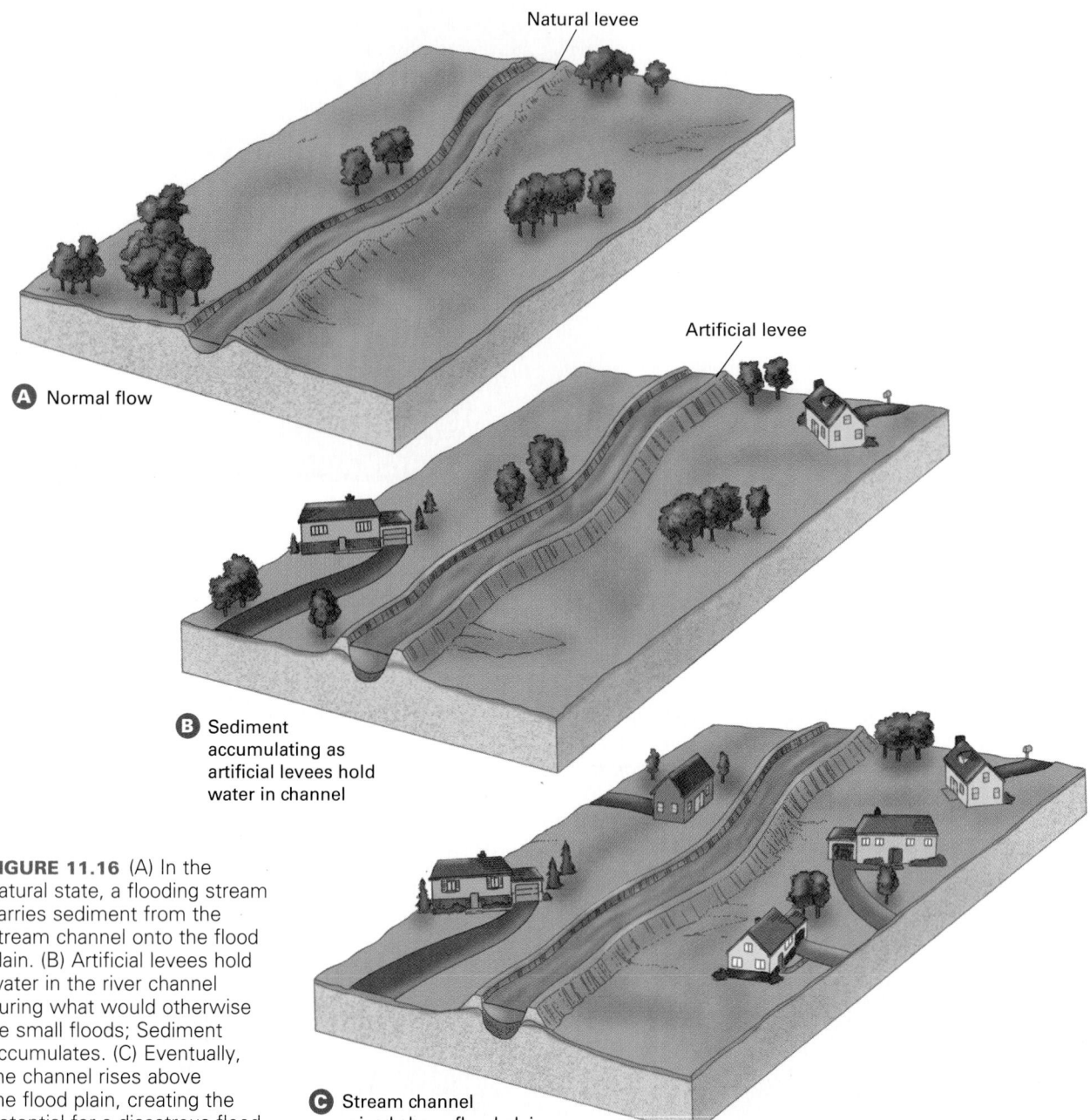

Natural levee

A Normal flow

Artificial levee

B Sediment
accumulating as
artificial levees hold
water in channel

FIGURE 11.16 (A) In the
natural state, a flooding stream
carries sediment from the
stream channel onto the flood
plain. (B) Artificial levees hold
water in the river channel
during what would otherwise
be small floods; Sediment
accumulates. (C) Eventually,
the channel rises above
the flood plain, creating the
potential for a disastrous flood.

C Stream channel
raised above flood plain

nearby reaches of a river. In the absence of levees,
when a stream floods, it deposits mud and sand on
the flood plain (Figure 11.16A). When artificial levees
are built, the stream cannot overflow during small
floods, so it deposits the sediment in its channel, raising the level of the streambed. (Figure 11.16B). After
several small floods, the entire stream may rise above
its flood plain, contained only by the levees (Figure
11.16C). This configuration creates the potential for
a truly disastrous flood because if the levee should
be breached during a large flood, the entire stream
then flows out of its channel and onto the flood plain
(Figure 11.17). As a result of artificial levees and channel sedimentation, portions of the Yellow River in

China now lie 10 meters above the flood plain. Levees
may solve flooding problems in the short term, but in
a longer time frame they can cause even larger and
more destructive floods.

Flood Control, the Mississippi River Delta, and Hurricane Katrina

In August 2005, Hurricane Katrina drove storm waters
over the levees protecting New Orleans, flooding the
city, chasing 1.3 million people from their homes, and
causing approximately $200 billion in damages. This
problem was exacerbated by decades of flood control
practices along the Mississippi. Recall that repeated

FIGURE 11.17 Levees force a flooding river into a restricted channel, forming a partial dam that raises the flood level upstream from the restriction.

small floods build a delta. If levees reduce the frequency of small floods, less sediment is deposited and the delta is not built up as rapidly.

In addition, dams trap sediment upstream and reduce the total sediment load of the river. In a natural system, a delta is built up by sediment deposition but eroded by ocean waves and currents. The Mississippi River delta grew for 200 million years because deposition was greater than erosion. But, in the last 50 years, deposition rates have decreased because of flood control and dams, while erosion rates have actually increased because urban and agricultural development has destroyed natural plant communities that normally hold the soil.

Another problem affecting the delta arises from development of abundant oil reserves beneath the delta. As the oil is removed, the surface of the delta sinks, or *subsides* (Figure 11.18). Because delta land is only marginally above sea level at best, even small amounts of subsidence can cause the land to sink below sea level.

As a result of all these processes, the Mississippi Delta has shrunk. Between 1930 and 2005, nearly 5,000

FIGURE 11.18 The city of New Orleans and the Mississippi River appear in the center of this map showing the subsidence rates—the rates at which land in the area is sinking—in millimeters per year. MRGO on the map is the Mississippi River–Gulf Outlet channel, connecting the Port of New Orleans to the Gulf of Mexico.

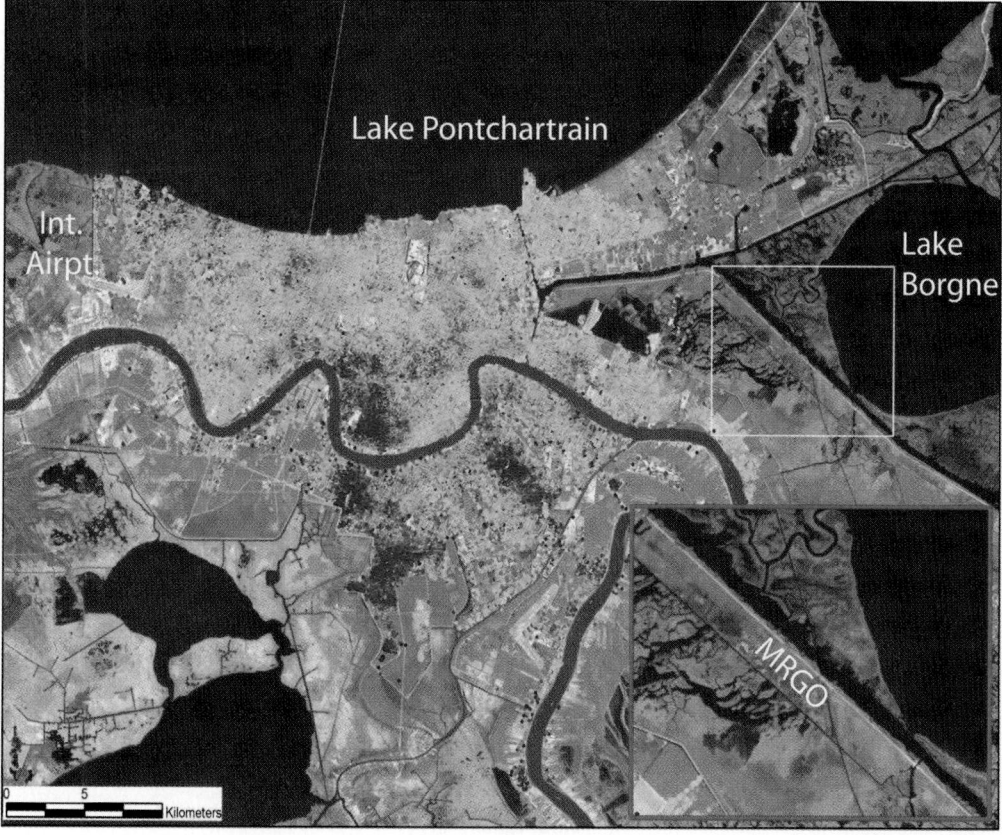

Legend

NEW ORLEANS

VEL [mm/year]

- -28.60 - -17.60
- -17.59 - -13.54
- -13.53 - -10.20
- -10.19 - -8.90
- -8.89 - -8.10
- -8.09 - -7.50
- -7.49 - -7.00
- -6.99 - -6.60
- -6.59 - -6.30
- -6.29 - -6.00
- -5.99 - -5.70
- -5.69 - -5.50
- -5.49 - -5.30
- -5.29 - -5.10
- -5.09 - -4.90
- -4.89 - -4.70
- -4.69 - -4.50
- -4.49 - -4.30
- -4.29 - -4.00
- -3.99 - -3.70
- -3.69 - -3.40
- -3.39 - -3.10
- -3.09 - -2.80
- -2.79 - -2.40
- -2.39 - -1.80
- -1.79 - 10.30

DIXON,T.H. ET AL., NEW ORLEANS SUBSIDENCE: RATES AND SPACIAL VARIATION MEASURED BY PERMANENT SCATTERER INTERFEROMETRY, NATURE, 441, 587–586, 2006. USED WITH PERMISSION.

lake A large, inland body of standing water that occupies a depression in the land surface.

square kilometers of the delta sank below sea level or washed into the sea. In 1995 the authors of this textbook wrote: "If current rates of erosion continue, the sea will ... flood New Orleans ... causing severe economic losses."[2] Tragically, this dire prediction was accurate.

While flooding is a natural event, flood frequency and severity are often augmented by logging, farming, and urbanization.

© DESHACAM/SHUTTERSTOCK

The flooding of New Orleans in 2005 is an example of both a complex systems interaction and a threshold event. While Hurricane Katrina triggered the flood, human interference and development within a complex natural river/delta system set the stage. The delta subsided and eroded slowly over 75 years. Natural buffers were removed, with no immediate ill effects. Then a catastrophic storm pushed waters inland. The levees failed, causing a huge human disaster.

Flood Plain Management

As we have shown, in many cases attempts at controlling floods either do not work or they shift the problem to a different time or place. An alternative approach to flood control is to abandon some flood control projects and let the river spill out onto its flood plain. Of course the question is: What land should be allowed to flood? Every farmer and homeowner on the river wants to maintain the levees that protect the flood plain. Currently, federal and state governments are establishing wildlife reserves in some flood plains. Because no development is allowed in these reserves, they will flood during the next high water. However, a complete river management plan involves complex political and economic considerations.

---

2. Jonathan Turk and Graham Thompson, *Environmental Geoscience* (Philadelphia: Saunders College Publishing, 1995), 428.

11.5 Lakes

Lakes and lake shores are attractive places to live and play. Clean, sparkling water, abundant wildlife, beautiful scenery, aquatic recreation, and fresh breezes all come to mind when we think of going to the lake. Despite their great value, lakes are fragile and ephemeral. Modern, post–ice age humans live in a special time in Earth's history, when Earth's surface is dotted with beautiful lakes.

The Life Cycle of a Lake

A **lake** is a large, inland body of standing water that occupies a depression in the land surface (Figure 11.19). Streams flowing into the lake carry sediment, which fills the depression in a relatively short time, geologically speaking. Soon the lake becomes a swamp, and with time the swamp fills with more sediment and vegetation and becomes a meadow or forest with a stream flowing through it.

If most lakes fill quickly with sediment, why are they so abundant today? Most lakes exist in places that were covered by glaciers during the latest Ice Age. About 18,000 years ago, great continental ice sheets

COURTESY OF GRAHAM R. THOMPSON/JONATHAN TURK

FIGURE 11.19 An alpine lake in Montana's Beartooth Mountains.

extended well south of the Canadian border, and mountain glaciers scoured the alpine valleys as far south as New Mexico and Arizona. Similar ice sheets and alpine glaciers existed in higher latitudes of the Southern Hemisphere. We are just now emerging from that glacial episode.

The glaciers created lakes in several ways. Flowing ice eroded numerous depressions in the land surface, which then filled with water. The Finger Lakes of upper New York State and the Great Lakes are examples of large lakes occupying glacially scoured depressions.

The glaciers also deposited huge amounts of sediment as they melted and retreated. Some of these great piles of glacial debris formed dams across stream valleys. When the glaciers melted, streams flowed down the valleys but were blocked by the dams. Many modern lakes occupy glacially dammed valleys (Figure 11.20). In addition, large blocks of ice may be left behind as a glacier recedes. When these ice blocks melt, they leave depressions called kettles, which then fill with water, forming a **kettle lake**.

Most of these glacial lakes formed within the past 10,000 to 20,000 years, and sediment is rapidly filling them. Many smaller lakes have already become swamps. In the next few hundred to few thousand years, many of the remaining lakes will fill with mud. The largest, such as the Great Lakes, may continue to exist for tens of thousands of years. But the life spans of lakes such as these are limited, and it will take another glacial episode to replace them.

Lakes also form by nonglacial means. A volcanic eruption can create a crater that fills with water to form a lake, such as Crater Lake in Oregon. Oxbow lakes form in abandoned river channels. Other lakes, such as Lake Okeechobee in the Florida Everglades, form in flat lands with shallow ground water. These types of lakes, too, fill with sediment and, as a result, have limited lives.

A few lakes, however, form in ways that extend their lives far beyond that of a normal lake. For example, Russia's Lake Baikal is a large, deep lake lying in a depression created by an active fault. Although rivers pour sediment into the lake, movement of the fault repeatedly deepens the basin. As a result, the lake has existed for more than a million years, so long that indigenous species of seals, other animals, and fish have evolved in its ecosystem.

Nutrient Balance in Lakes

In a deep lake, sunlight is available near the surface, but nutrients are abundant only on the bottom. Plankton (small, free-floating organisms) grow poorly on the surface due to the lack of nutrients, and bottom-rooted plants cannot grow due to lack of sunlight. Thus, the lakes contain low concentrations of nitrates, phosphates, and other critical nutrients that sustain aquatic food webs. The purity of the water gives these lakes a deep blue color that we associate with a clean, healthy lake. Such a lake is called *oligotrophic*, meaning "poorly nourished." **Oligotrophic lakes** have low productivities, meaning that they sustain relatively few living organisms, although a lake of this type is attractive for recreation and typically contains a few huge trout or similar game fish.

As a lake fills with sediment, it becomes shallower and sunlight reaches more and more of the lake bottom. The sunlight allows bottom-rooted plants to grow. As the plants die and rot, their litter adds nutrients to the lake water. Plankton increase in numbers, as do fish and other organisms. The lake becomes so productive that its surface may become covered with a green scum of plankton or a dense mat of rooted plants. The litter contributes to the sediment filling the lake, and eventually the lake becomes a swamp. A lake of this kind, with a high nutrient supply, is called a **eutrophic lake**; *eutrophic* means "well nourished." Eutrophication occurs naturally as part of the life cycle of a lake. However, addition of nutrients in the form of sewage and other kinds of pollution has greatly accelerated the eutrophication of many lakes.

kettle lake A lake that forms in a depression created by a receding glacier, filled with the water from the melting glacier.

oligotrophic lake A deep lake characterized by nearly pure water but with low concentrations of plant nutrients, thus sustaining relatively few living organisms.

eutrophic lake A relatively shallow lake characterized by abundant plant nutrients, thus sustaining multiple living organisms.

FIGURE 11.20 Gravel deposited by a glacier forms a dam to create this mountain lake in the Sierra Nevada.

thermocline The boundary between the upper warm layers and deeper cool layers of water in a lake.

turnover A process, occurring in fall and spring in temperate climates, in which a lake's surface water changes temperature in response to seasonal weather changes and convection mixes the water to equalize temperature throughout the lake.

Temperature Layering and Turnover in Lakes

If you have ever dived into a deep lake on a summer day, you probably discovered that the top meter or so of lake water can be much warmer than deeper water. This occurs because sunshine warms the upper layer of water, making it less dense than the cooler, deeper water. The warm, less-dense water floats on the cooler, denser water. The boundary between the warm and cool layers is called the **thermocline**.

In temperate climates, colder autumn weather cools the surface water to a temperature below that of deeper water, so that the surface water becomes more dense than the deeper water. Consequently it sinks, mixing the surface and deep waters and equalizing the water temperature throughout the lake. This process is called fall **turnover**. In the win-

ter, ice floats on the surface and temperature layering develops again. In spring, as ice melts on the lake, surface water again becomes more dense than deep lake waters, and spring turnover occurs. As summer comes, the lake again develops thermal layering. This seasonal process is illustrated in Figure 11.21.

Turnover in temperate lakes illustrates an important Earth systems interaction among the atmosphere, the hydrosphere, and the biosphere. During summer and winter when the lake water is layered, bottom-dwelling organisms may use up most or all of the oxygen in deep waters. At the same time, surface organisms may deplete surface waters of dissolved nutrients. However, surface water is rich in oxygen because it is in contact with the atmosphere, and deep water may be rich in nutrients because it is in contact with bottom sediment. Turnover enriches deep water in oxygen and, at

FIGURE 11.21 Lakes in temperate climates develop temperature layering in both summer and winter. As a result, bottom waters become depleted in oxygen. In fall and spring, water temperature becomes constant throughout the lake and turnover brings new supplies of oxygen to the deep waters.

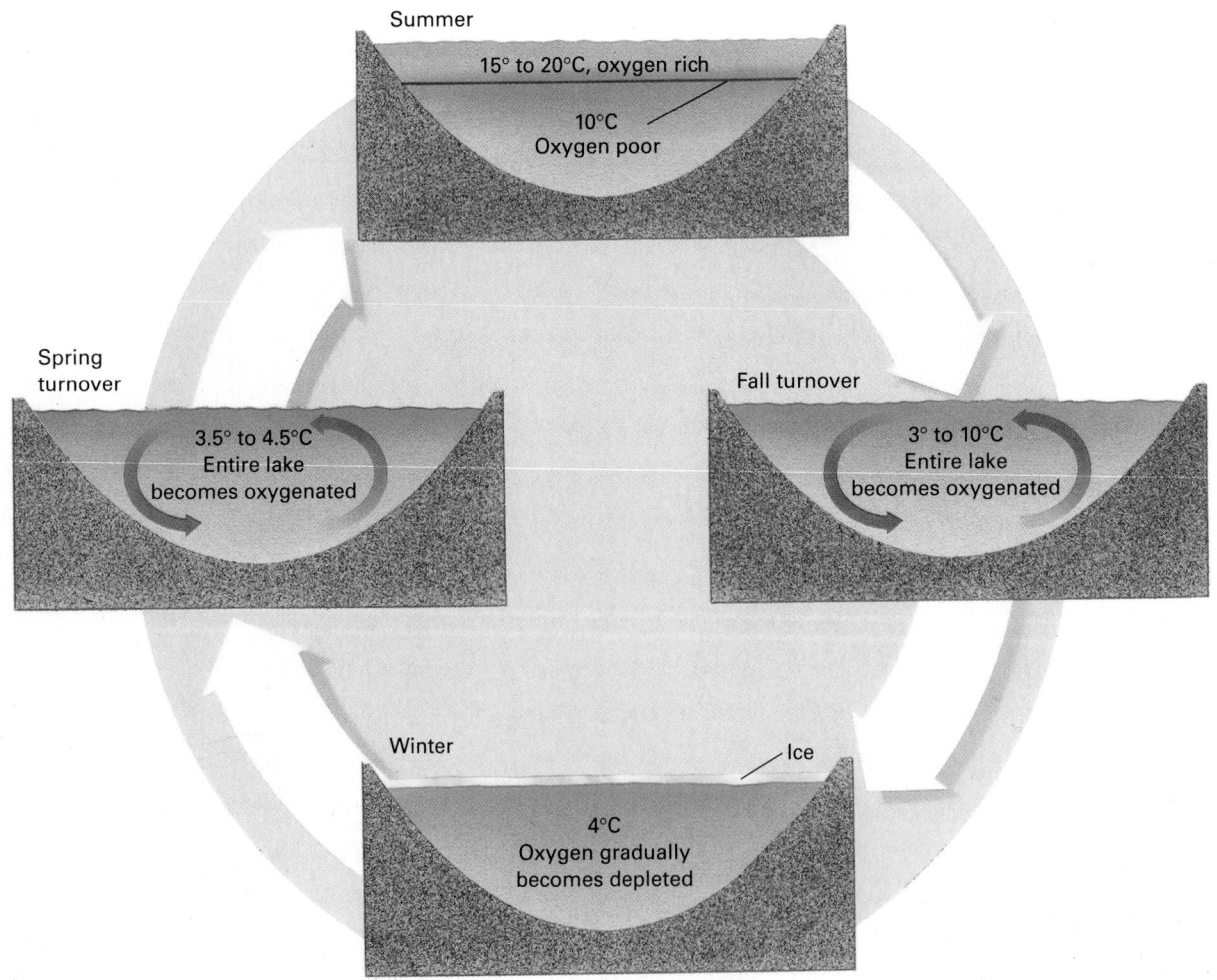

Summer

15° to 20°C, oxygen rich

10°C Oxygen poor

Spring turnover

3.5° to 4.5°C Entire lake becomes oxygenated

Fall turnover

3° to 10°C Entire lake becomes oxygenated

Winter

Ice

4°C Oxygen gradually becomes depleted

© BOB BLANCHARD/SHUTTERSTOCK

One-third of all resident bird species feed, breed, and rest in wetlands.

the same time, supplies nutrients to the surface water. The latter effect often becomes evident in the form of an *algal bloom*—a sudden and obvious increase in the amount of floating green algae on a lake's surface—in spring and fall.

11.6 Ground Water

If you drill a hole into the ground in most places, a few meters to 100 or more meters deep, its bottom fills with water after a short time, usually within a few minutes to a few days. The water appears even if no rain falls and no streams flow nearby. The water that seeps into the hole is part of our ground water reserves, which saturate Earth's crust between a few meters and a few kilometers below the surface.

Ground water is exploited by digging wells and pumping the water to the surface. It provides drinking water for more than half of the population of North America and is a major source of water for irrigation and industry. However, deep wells and high-speed pumps now extract ground water more rapidly than natural processes can replace it in many parts of the central and western United States. In addition, industrial, agricultural, and domestic contaminants seep into ground water in many parts of the world. Such pollution is often difficult to detect and expensive to clean up.

Porosity and Permeability

Ground water fills small cracks and voids in soil and bedrock. The proportional volume of these open spaces, or *pores*, is called **porosity**. Sand and gravel typically have high porosities—40 percent or more. Mud can have a porosity of 90 percent or more. It has such a high porosity because the tiny clay particles are electrically attracted to water, and consequently clay-rich mud absorbs a very high proportion of water. Most rocks

have lower porosities than loose sediment. Sandstone and conglomerate can have 5 to 30 percent porosity. Shale typically has a porosity of less than 10 percent. Igneous and metamorphic rocks have very low porosities unless they are fractured.

Porosity indicates the amount of water that rock or soil can hold. In contrast, **permeability** refers to the ability of rock or soil to transmit water (or any other fluid). Water can flow rapidly through material with high permeability. Most materials with high porosity also have high permeability. Sand and sandstone have numerous, relatively large, well-connected pores that allow the water to flow through the material. However, if the pores are very small, as in clay and shale, electrical attractions between water and soil particles slow the passage of water. Clay typically has a high porosity, but because its pores are so small and the electrical attractions slow the passage of water, it commonly has a very low permeability and transmits water slowly.

The Water Table and Aquifers

When rain falls, much of it soaks into the ground. Water does not descend into the crust indefinitely, however. Below a depth of a few kilometers, the pressure from overlying rock closes the pores, making bedrock both nonporous and impermeable. Water accumulates above this impermeable barrier, filling pores in the rock and soil. This completely wet layer of soil and bedrock above the deep, impermeable rock is called the **zone of saturation**. The **water table** is the top of the zone of saturation (Figure 11.22). The *unsaturated zone*, or **zone of aeration**, lies above the water table. In this layer, the rock or soil may be moist but is not saturated and air occupies some or all of the pore space.

If you dig into the unsaturated zone, the hole does not fill with water. However, if you dig below the water table into the zone of saturation, you have dug a **well**,

porosity The proportional volume of a material that consists of pores or open spaces, indicating the amount of water it can hold.

permeability The ability of a material to transmit water, as measured by the speed at which fluid can travel through the material.

zone of saturation A subsurface zone below the water table in which the soil and bedrock are completely saturated with water.

water table The top level of subsurface ground water, at the top of the zone of saturation and below the zone of aeration.

zone of aeration A subsurface zone above the water table where the rock or soil may be moist but not saturated, with air occupying some or all of the pore space; also called the *unsaturated zone*.

well A hole dug or drilled into Earth, generally for the production of water, petroleum, natural gas, brine, sulfur, or for exploration.

recharge To replenish an aquifer by the addition of water.

aquifer A body of rock that can yield economically significant quantities of ground water; should be both porous and permeable.

and the water level in a well is at the level of the water table (Figure 11.22). During a wet season, rain seeps into the ground to **recharge** the ground water, and the water table rises. During a dry season, the water table falls. Thus, the water level in most wells fluctuates with the seasons.

An **aquifer** is any body of rock or soil that can yield economically significant quantities of water. An aquifer must be both porous and permeable so that water flows into a well to replenish (recharge) water that is pumped out. Sand and gravel, sandstone, limestone, and highly fractured bedrock of any kind make excellent aquifers. Shale, clay, and unfractured igneous and metamorphic rocks are poor aquifers.

FIGURE 11.22 The water table is the top of the zone of saturation near Earth's surface. It intersects the land surface at lakes and streams and is the level of standing water in a well.

Ground Water Movement

Nearly all ground water seeps slowly through bedrock and soil. Ground water flows at about 4 centimeters per day (about 15 meters per year), although flow rates may be much faster or slower depending on permeability. Most aquifers are like sponges through which water seeps, rather than underground pools or streams. However, ground water can flow very rapidly through large fractures in bedrock, and in a few regions underground rivers flow through caverns.

In general, the water table mimics its surrounding topography. Refer again to Figure 11.22. Ground water flows from zones where the water table is highest toward areas where it is lowest, and so some ground water flows along the sloping surface of the water table toward the valley. But ground water also flows from zones of high pressure toward zones of low pressure. Because water pressure is greatest beneath the highest part of the water table, the pressure difference forces much of the ground water to flow downward beneath the hill, then laterally toward the valley, and finally upward beneath the lowest part of the valley where a river flows. This is how ground water feeds stream flow, and why streams flow even when no rain has fallen for weeks or months.

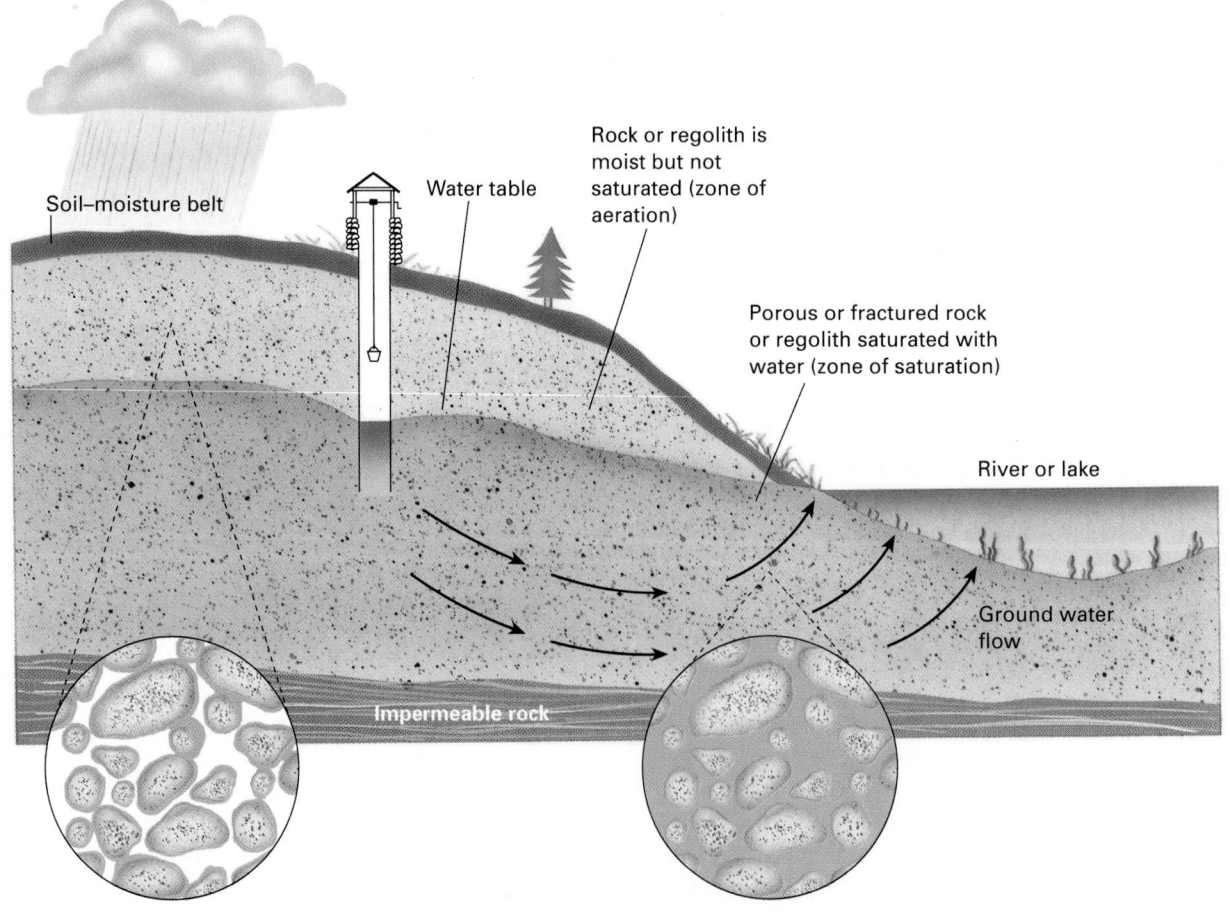

Soil–moisture belt

Water table

Rock or regolith is moist but not saturated (zone of aeration)

Porous or fractured rock or regolith saturated with water (zone of saturation)

River or lake

Ground water flow

Impermeable rock

Springs and Artesian Wells

A **spring** occurs where the water table intersects the land surface and water flows or seeps onto the surface. In some places, a layer of impermeable rock or clay lies above the main water table, creating a locally saturated zone, the top of which is called a **perched water table** (Figure 11.23). Hillside springs often flow from a perched water table. Springs also occur where fractured bedrock or cavern systems intersect the land surface.

Figure 11.24 shows a tilted layer of permeable sandstone sandwiched between two layers of imperme-able shale. An inclined aquifer such as the sandstone layer, bounded top and bottom by impermeable rock, is an **artesian aquifer**. Water in the lower part of the aquifer is under pressure from the weight of water above. Therefore, if a well is drilled through the shale and into the sandstone, water rises in the well without being pumped. A well

spring A place where the water table intersects the land surface and ground water flows or seeps onto the surface.

perched water table A localized water table above the main water table, formed where a layer of impermeable rock or clay lies above the main water table, creating a locally saturated zone.

artesian aquifer An inclined aquifer sandwiched between layers of impermeable rock, and where the water in the lower part is under pressure from the weight of water above.

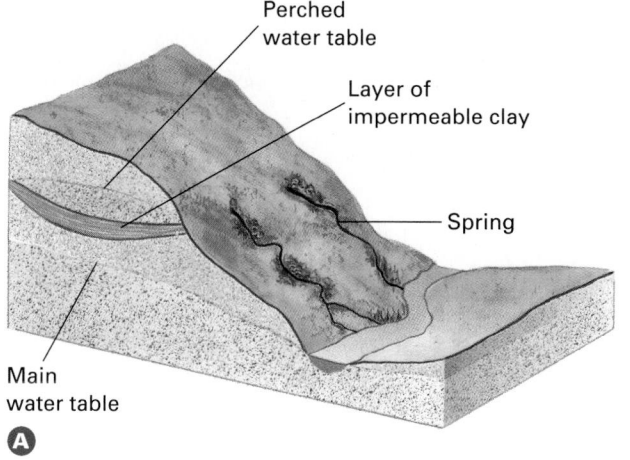

FIGURE 11.23 (A) Springs can form where a perched water table intersects a hillside. (B) Water flows from a spring on a hillside in British Columbia.

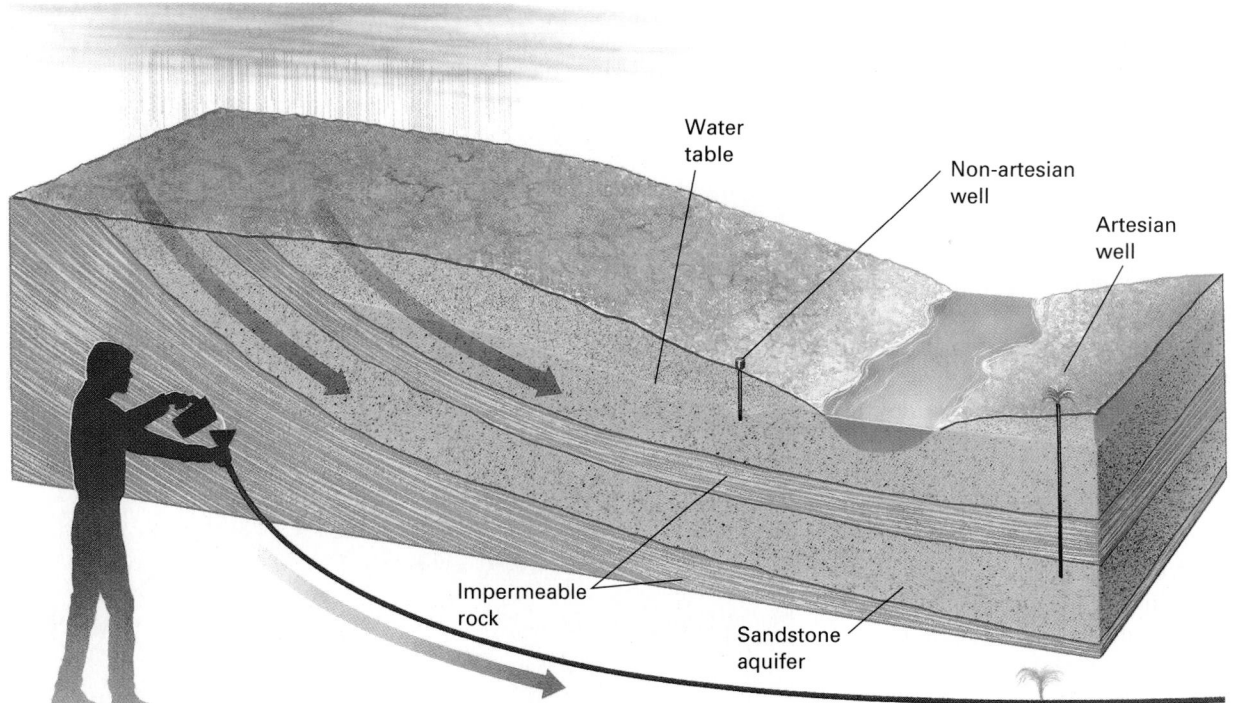

FIGURE 11.24 An artesian aquifer forms where a tilted layer of permeable rock, such as sandstone, lies sandwiched between layers of impermeable rock, such as shale. Water rises in an artesian well without being pumped. A hose with a hole (inset) shows why an artesian well flows spontaneously.

artesian well A well drilled into an artesian aquifer, in which the water rises without pumping and in some cases spurts to the surface.

cavern An underground cavity or series of chambers created when groundwater dissolves large amounts of rock, usually limestone; also called a *cave*.

stalactite An icicle-like dripstone of dissolved calcite precipitated from drops of water, which hangs from the ceiling of a cavern.

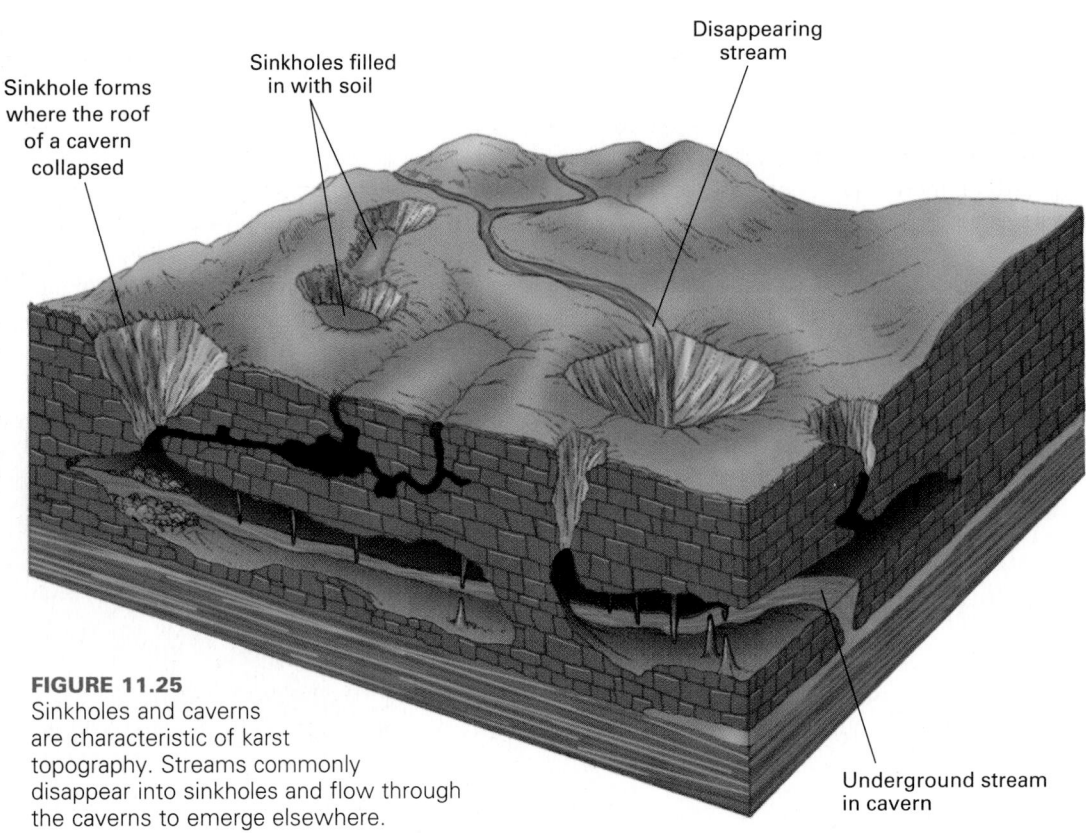

Sinkhole forms where the roof of a cavern collapsed

Sinkholes filled in with soil

Disappearing stream

Underground stream in cavern

FIGURE 11.25
Sinkholes and caverns are characteristic of karst topography. Streams commonly disappear into sinkholes and flow through the caverns to emerge elsewhere.

of this kind is called an **artesian well**. If pressure is sufficient, the water spurts out onto the land.

Caverns

Just as streams erode valleys and form flood plains, ground water also creates landforms (Figure 11.25). Recall from Chapter 10 that rainwater reacts with atmospheric carbon dioxide to become slightly acidic and capable of dissolving limestone. A **cavern** forms when acidic water seeps into cracks in limestone, dissolving the rock and enlarging the cracks. Mammoth Cave in Kentucky and Carlsbad Caverns in New Mexico are two famous caverns formed in this way.

Although caverns form when limestone dissolves, most caverns also contain features formed by deposition of calcite. When a solution of water, dissolved calcite, and carbon dioxide percolates through the ground, it is under pressure from water in the cracks above it. If a drop of this solution seeps into the ceiling of a cavern, the pressure decreases because the drop comes in contact with the air. The high humidity of the cave prevents the water from evaporating rapidly, but the lowered pressure allows some of the carbon dioxide

to escape as a gas. When the carbon dioxide escapes, the drop becomes less acidic. This decrease in acidity causes some of the dissolved calcite to precipitate as the water drips from the ceiling. Over time, a beautiful and intricate **stalactite** (from the Greek for "drip") grows to hang icicle-like from the ceiling of the cave (Figure 11.26).

FIGURE 11.26 Stalactites and stalagmites form as calcite precipitates in the limestone of Dixie Caverns, Virginia.

Only a portion of the dissolved calcite precipitates as the drop seeps from the ceiling. When the drop falls to the floor, it spatters and releases more carbon dioxide. The acidity of the drop decreases further, and another minute amount of calcite precipitates. Thus, a cone-shaped **stalagmite** (from the Greek for "drop") builds from the floor upward to complement the stalactite. Because stalagmites are formed by splashing water, they tend to be broader than stalactites. As the two features continue to grow, they may eventually meet and fuse together to form a **column**.

Sinkholes

If the roof of a cavern collapses, a **sinkhole** forms on Earth's surface (Figure 11.25). A sinkhole can also form as limestone dissolves from the surface downward. A well-documented sinkhole formed in May 1981 in Winter Park, Florida. During the initial collapse, a three-bedroom house, half a swimming pool, and six Porsches in a dealer's lot all fell into the underground cavern. Within a few days, the sinkhole was 200 meters wide and 50 meters deep, and it had devoured additional buildings and roads.

Although sinkholes form naturally, human activities can accelerate the process. The Winter Park sinkhole formed when the water table dropped, removing support for the ceiling of the cavern. The water table fell as a result of a severe drought augmented by excessive removal of ground water by humans.

Karst Topography

An irregular landscape called **karst topography** forms in regions underlain by limestone and other readily soluble rocks. Caverns and sinkholes are common features; surface streams often pour into the sinkholes and disappear into the caverns (Figure 11.25). In the area around Mammoth Cave in Kentucky, streams are given names such as Sinking Creek because of such disappearing acts. The word *karst* is derived from a region in Croatia where this type of landscape is well developed. Karst topography is found in many parts of the world.

11.7 Hot Springs, Geysers, and Geothermal Energy

At numerous locations throughout the world, hot water naturally flows to the surface to produce **hot springs**. Ground water can be heated in three ways:

1. Earth's temperature increases by about 30°C per kilometer of depth in the upper portion of the crust. Therefore, if ground water descends through cracks to depths of 2 to 3 kilometers, it is heated by 60°C to 90°C. The hot water or steam then rises because it is less dense than cold water. However, it is unusual for fissures to descend so deep into Earth, and this type of hot spring is uncommon.

2. In regions of recent volcanism, magma or hot igneous rock may remain cooling near the surface and can heat ground water at relatively shallow depths. Hot springs heated in this way are common throughout western North and South America because these regions have been magmatically active in the recent past and remain so today. Shallow magma heats the hot springs and geysers of Yellowstone National Park.

3. Many hot springs are heated by chemical reactions from small amounts of hydrogen sulfide (H_2S) dissolved in the hot water. Sulfide minerals, such as pyrite (FeS_2), react chemically with water to produce hydrogen sulfide and heat. The hydrogen sulfide rises with the heated ground water and gives it a strong odor.

Most hot springs bubble gently to the surface from cracks in bedrock. However, **geysers** violently erupt hot water and steam. Geysers generally form over open cracks and channels in hot underground rock. Before a geyser erupts, ground water seeps into the cracks and is heated by the rock. Gradually, steam bubbles form and start to rise, just as they do in a heated teakettle. If part of the channel is constricted, the bubbles accumulate and form a temporary barrier that allows the steam pressure below to increase (Figure 11.27A). The rising pressure forces some of the bubbles upward past the constriction and short bursts of steam and water spurt from the geyser (Figure 11.27B). This lowers the steam pressure at the constriction, causing the hot water to vaporize, blowing steam and hot water skyward (Figure 11.28).

The most famous geyser in North America is Old Faithful in Yellowstone National Park, which erupts on

stalagmite A cone-shaped deposit of dissolved calcite precipitated from drops of water that have fallen to the floor of a cavern.

column A cave deposit formed when a stalactite and a stalagmite meet and fuse together.

sinkhole A circular depression on Earth's surface caused by the collapse of a cavern roof or by the dissolution of surface rocks.

karst topography A type of irregular landscape that forms over limestone or other soluble rock and is characterized by caverns, sinkholes, and underground streams.

hot spring A spring formed where hot ground water flows to the surface.

geyser A type of hot spring that intermittently erupts with violent jets of hot water and steam when ground water comes in contact with hot rock.

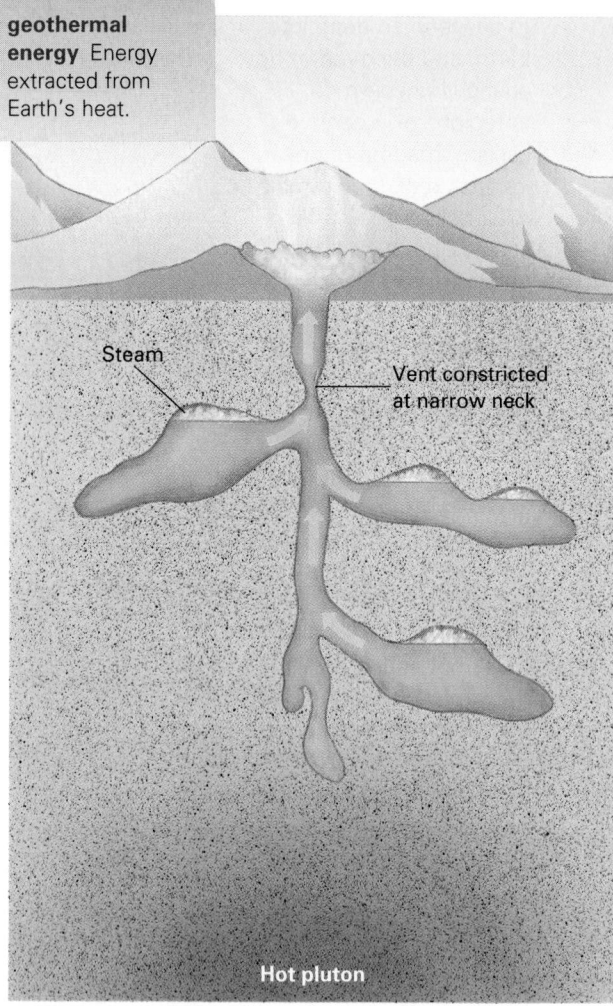

Steam

Vent constricted
at narrow neck

Hot pluton

A

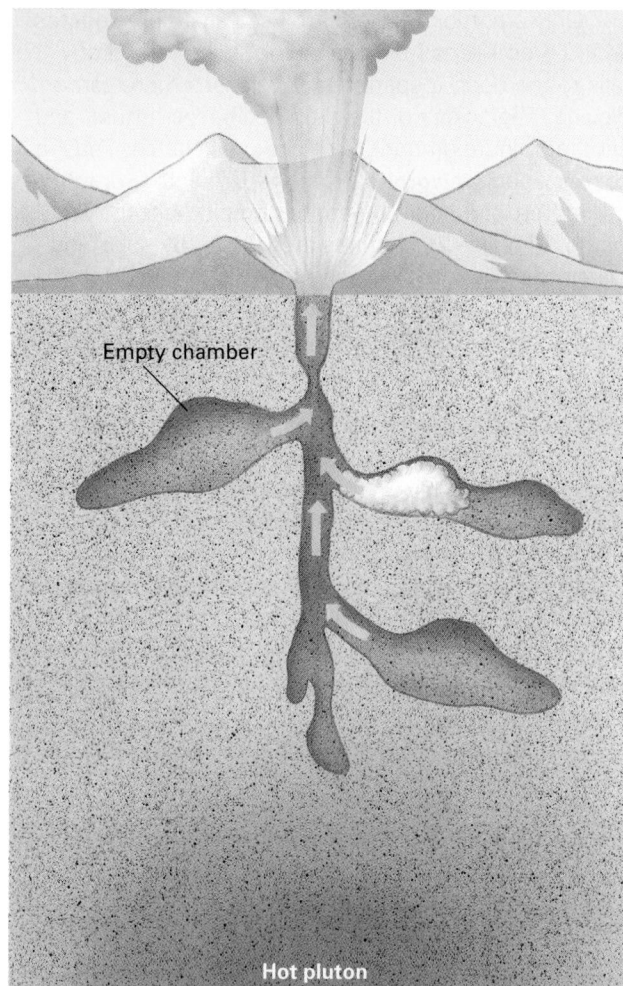

Empty chamber

Hot pluton

B

FIGURE 11.27 (A) Before a geyser erupts, ground water seeps into underground chambers and is heated by hot igneous rock. Foam constricts the geyser's neck, trapping steam and raising pressure. (B) When the pressure exceeds the strength of the blockage, the constriction blows out. Then the hot ground water flashes into vapor and the geyser erupts.

the average of once every 65 minutes. Old Faithful is not as regular as people like to believe; the intervals between eruptions vary from about 30 to 95 minutes.

Hot ground water can be used to drive turbines and generate electricity, or it can be used directly to heat homes and other buildings. Energy extracted from Earth's heat is called **geothermal energy**. In the United States, a total of 70 geothermal plants in California, Hawaii, Utah, and Nevada have a generating capacity of 2,800 megawatts, enough to supply over 2 million people with electricity and equivalent to the power out-

FIGURE 11.28 An eruption of Riverside Geyser in Upper Geyser Basin, Yellowstone National Park, Wyoming.

USGS

put of 25 million barrels of oil. However, this amount of energy is minuscule compared with the potential of geothermal energy.

11.8 Wetlands

Wetlands are known across North America as swamps, bogs, marshes, sloughs, mudflats, and flood plains. They are regions that are water soaked or flooded for all or part of the year. Some wetlands are wet only during exceptionally wet years and may be dry for several years at a time.

Wetland ecosystems vary so greatly that the concept of a wetland defies a simple definition. Wetlands share certain properties, however: The ground is wet for at least part of the time; the soils reflect anaerobic (lacking oxygen) conditions; and the vegetation consists of plants such as red maple, cattails, bulrushes, mangroves, and other species adapted to periodic flooding or water saturation. North American wetlands include all stream flood plains, frozen Arctic tundra, warm Louisiana swamps, coastal Florida mangrove swamps, boggy mountain meadows of the Rockies, and the immense swamps of interior Alaska (Figure 11.29).

Wetlands are among the most biologically productive environments on Earth. Two-thirds of the Atlantic fish and shellfish consumed by humans rely on coastal wetlands for at least part of their life cycles. One-third of the endangered species of both plants and animals in the United States also depend on wetlands for survival. More than 400 of the 800 species of protected migratory birds and one-third of all resident bird species feed, breed, and rest in wetlands.

When European settlers first arrived in North America, 87 million hectares of wetlands existed (exclusive of those in Alaska). Americans have long viewed wetlands as mosquito-infested, malarial swamps occupying land that can be farmed or otherwise developed if drained or filled. In the mid-1800s, the federal government passed legislation known as the Swamp Land Acts, which established an official policy to fill and drain wetlands to convert them to agricultural uses wherever possible.

Over the past century, farmers, ranchers, and developers drained or filled more than half of the original wetlands. California and several upper-midwestern states have lost more than 80 percent of their wetlands. Wetlands now make up between 6 and 9 percent of the lower 48 states and as much as 60 percent, or about 80 million hectares, of Alaska. Currently, between about 120,000 and 200,000 hectares of wetlands are destroyed each year.

Government policy and private practice have ignored the beneficial qualities of wetlands. Aquatic organisms consume many pollutants and degrade them to harmless by-products. Because these organisms abound in wetlands, the ecosystems are natural sewage treatment systems. Wetlands also mitigate flooding by absorbing excess water that might otherwise overrun towns and farms. Wild ducks and geese and other migratory birds depend on wetlands for breeding, food, and cover. The importance of wetlands in water purification, flood control, and wildlife habitat did not become widely recognized until the 1960s. At present, the focus of federal and state laws has changed from destruction of wetlands to their protection and preservation. But it has proved more difficult to reverse practice than policy, and wetland losses continue today.

wetlands Regions that are water soaked or flooded for all or part of the year; also known as *swamps, bogs, marshes, sloughs, mud flats,* and *flood plains.*

C. SINGLETARY, EVERGLADES NATIONAL PARK

Ⓐ

COURTESY OF GRAHAM R. THOMPSON/JONATHAN TURK

Ⓑ

FIGURE 11.29 North American wetlands extend from (A) the Everglades to (B) the immense Alaskan swamps.

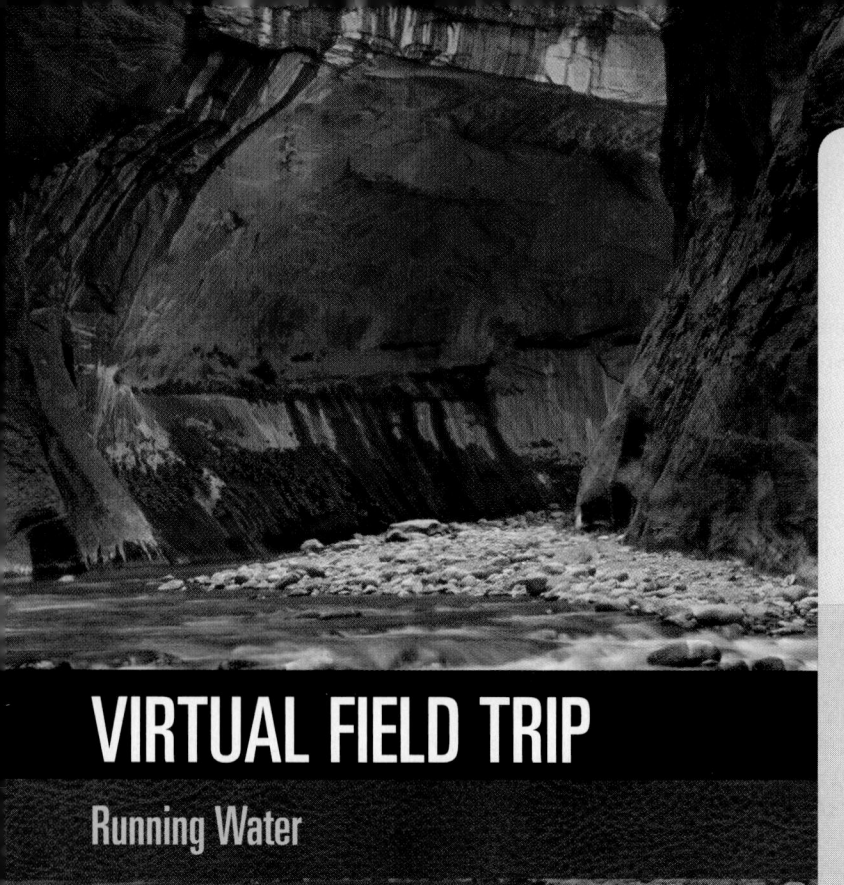

VIRTUAL FIELD TRIP

Running Water

Ready to Go!

When it comes to studying the processes and products of running water firsthand, Zion National Park in southwestern Utah is about as good as it gets. The arid climate limits vegetation cover, allowing unobstructed views of towering cliffs and monoliths, great displays of the force of running water. In this field trip, we will learn about river erosion and cut banks, and explore a world-famous slot canyon. In addition, we will study processes and products of river deposition, river abrasion, and downcutting (see "Downcutting and Base Level" on pages 196 through 198).

The Virgin River is the main waterway in the park, and it, along with its tributaries, is largely responsible for erosion of the mostly sandstone terrain. About 13 million years ago, uplift of the entire region steepened the gradient of the Virgin River, which began downcutting, a process that would eventually yield the cliffs and deep canyons now present. Let's start our trip through Zion National Park where we can make many observations relevant to our discussions of running water in this chapter.

GOALS OF THE TRIP

There are many geologic features to see in Zion National Park, but our main goals on this trip are to study:

1. Erosion
2. Sediment transport
3. Deposition by running water
4. The origin and evolution of the park's canyons

FOLLOW-UP QUESTIONS

1. Why do meandering streams form an alternating series of cut banks and point bars along their course?
2. Give a brief history of the origin and development of the canyons in Zion National Park.
3. What processes, other than down-cutting by the Virgin River, contribute to the ongoing evolution of the canyons in Zion National Park?

What to See When You Go

This image shows both erosion and deposition by a meandering stream—that is, a stream with a single, sinuous channel (see Figure 11.9 in this chapter). Notice the steep bank on the right side of the image, which is called the cut bank because the water velocity is highest here and erosion takes place. On the opposite bank, however, velocity is least and deposition occurs, forming a point bar deposit mostly of sand that slopes gently toward the cut bank. As you read under "Sinuosity of a Stream Channel," point bars are the most distinctive features of ancient meandering stream deposits. It is important to note that further downstream the maximum velocity switches from right to left, thus forming a succession of alternating point bars and cut banks.

In the photo of a point bar, notice that the water in the river is turbid (murky) because it is carrying mud (silt and clay) and probably sand along its bed. This photo shows a deposit of gravel and sand. Some of the gravel is of boulder size (greater than 25.6 cm). From observations here we can conclude that the stream was much higher when this deposit formed, and it must have been flowing very rapidly to transport such large particles. Also the gravel particles are well rounded, meaning that their sharp corners and edges were worn smooth by abrasion during transport. And finally, the deposit is poorly sorted because it contains a wide range of particle sizes.

Although this canyon (Lower Zion Canyon) and the slot canyon discussed in the previous paragraph were both eroded by running water, this one has a very different profile. As opposed to being narrow and deep, this canyon is broad and deep with alternating steep cliffs and gentle slopes. The lower, gentle slopes formed on softer rocks, mostly shale and siltstone, whereas the steep cliffs developed in the rust-colored Navajo Sandstone.

This is a great view of a slot canyon—that is, a canyon that is much deeper than it is wide, usually with vertical or near-vertical sides. Most slot canyons form where running water erodes into sandstone (as here) or limestone, although they may form in other rocks, too. Past floods are indicated by the scour marks and polished rock surface several meters above water level. Be very careful if you hike in a slot canyon, because a summer thunderstorm can cause the water level to rise several meters in just a few minutes.

As we learned under "Stream Erosion and Mountains: How Landscapes Evolve" in Section 11.2, downcutting by running water accounts for canyon deepening, but canyons also widen by a combination of processes including erosion by tributaries, weathering, and mass wasting. This blind arch formed when the softer rock below was eroded and part of the overlying sandstone collapsed. In any event, the debris from this event eventually makes it to the river only to be carried away and deposited elsewhere.

12

WATER RESOURCES

Jackson Lake Dam in Grand Tetons National Park, Wyoming.

visit 4ltrpress.cengage.com

> # When the well's dry, we know the worth of water.
>
> *Benjamin Franklin*

withdrawal Any process that uses water and then returns it to Earth locally.

More than two-thirds of Earth's surface is covered with water. But most of this water is salty, and most of the freshwater is frozen into the Antarctic and Greenland ice caps. Humans divert half of all the flowing water from rain and snow that falls on the continents, but even that is not enough for our burgeoning population and per capita consumption. Today, as we reach, and in some cases exceed, the limits of our water resources, the well is running dry. In March 2003, the United Nations published the "World Water Development Report" stating that by the middle of the 21st century, up to 7 billion people—more than the entire current population of the world—will face water scarcity. Part of the problem is that much of the Earth's rain falls in the wrong places or at the wrong times to be of much use to humans. One prominent ecologist lamented, "The wet places of the world don't need the runoff because they are wet. The dry places of the world need the runoff to irrigate the land, but they don't get much."[1]

---

1. Stuart L. Pimm, *The World According to Pimm: A Scientist Audits the Earth* (New York: McGraw-Hill, 2001), 111; online at http://www.nicholas.duke.edu/people/faculty/pimm/World%20 according%20to%20Pimm.pdf

In addition to water scarcity, much of the surface water and ground water in industrial areas is heavily polluted, destroying natural habitats and making the resource less useful for humans.

12.1 Water Supply and Demand

Rain and snow continuously replenish freshwater on land, so that the amount of available freshwater is about the same as it was two centuries ago. But the demand for water has risen dramatically until it has approached or even exceeded supply in many parts of the world (Figure 12.1). In the early 1800s, there were 1 billion people on Earth, now there are 6.5 billion. Thus as the human population has grown, the amount of available freshwater per person has diminished. In addition, our technological society uses water at rates that would have been inconceivable to Ben Franklin.

Water use falls into two categories. Any process that uses water and then returns it to Earth locally is called **withdrawal**. Most of the water used by industry and homes is returned to streams or ground water reservoirs near the

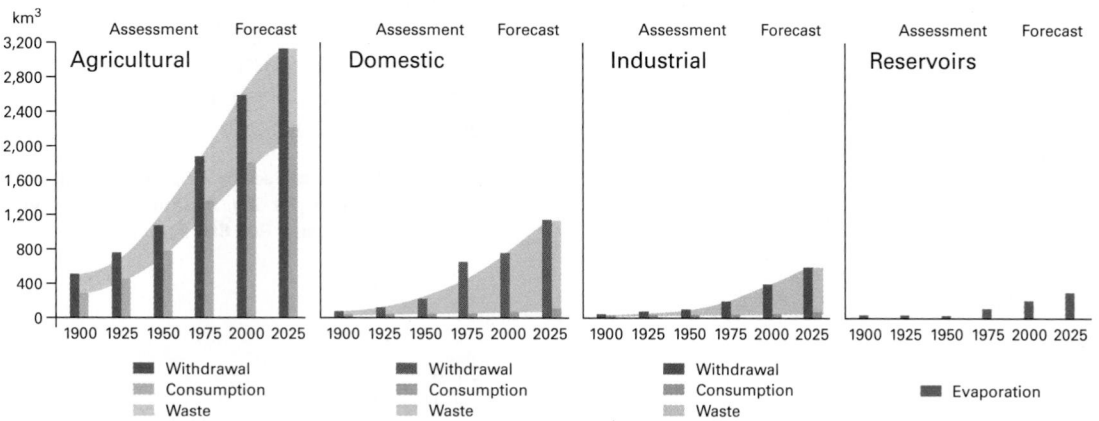

Note: Domestic water consumption in developed countries (500–800 liters per person per day) is about six times greater than in developing countries (60–150 liters per person per day).

FIGURE 12.1 Global water use, 1900 to 2025. Between 2000 and 2054, the world's population is expected to increase by 3 billion people, so water demand will continue to rise.

COPYRIGHT AND PHOTOGRAPH BY DR. PARVINDER S. SETHI

219

COURTESY OF GRAHAM R. THOMPSON/JONATHAN TURK

consumption
Any process that uses water and then returns it to Earth far from its source.

place from which it was taken. For example, river water pumped through an electric generating station to cool the exhaust is returned almost immediately to the river. Water used to flush a toilet in a city is pumped to a sewage treatment plant, purified, and discharged into a nearby stream.

In contrast, a process that uses water and then returns it to Earth far from its source is called **consumption**. Most irrigation water evaporates, disperses with the wind, and returns to Earth as precipitation hundreds or thousands of kilometers from its source (Figure 12.2). Although industry accounts for about half of all water withdrawn in the United States, agriculture accounts for most of the water that is consumed and not returned to its place of origin (Figure 12.3). Globally, about two-thirds of the total water that is used is consumed.

Water withdrawal and consumption accounts for about one-sixth (or about 17 percent) of the total run-

FIGURE 12.2 Most irrigation water evaporates and is carried away by wind.

off. Humans divert an additional 33 percent of the total runoff by damming rivers. Although this water is neither withdrawn nor consumed, dams create environmental problems, which are discussed in Section 12.2.

Water use (withdrawal and consumption) is subdivided into domestic, industrial (including power-plant cooling), and agricultural categories. As you can see from Figure 12.4, agriculture accounts for 41 percent of water use in the United States.

MICIMAKIN/SHUTTERSTOCK

Domestic Water Use

A person in an industrial urban society can live a healthy, clean, hygienic life on 50 liters of water per day. The breakdown is 5 liters for drinking, 10 liters for cooking and dishwashing, 15 liters for bathing, and 20 liters for clothes washing and toilet flushing. These last two categories are lumped together because water used to wash clothes can be recycled and used to flush the toilet. The average American actually uses about

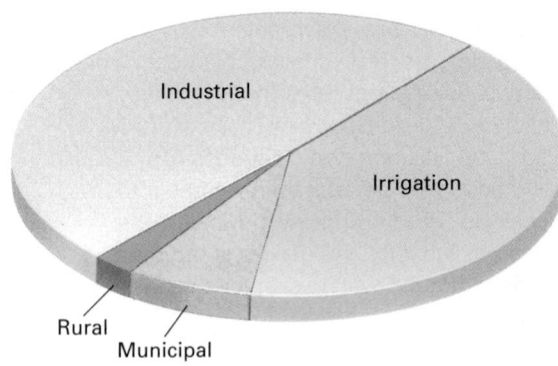

Withdrawal—1,800 billion liters per day (excluding hydropower)

A

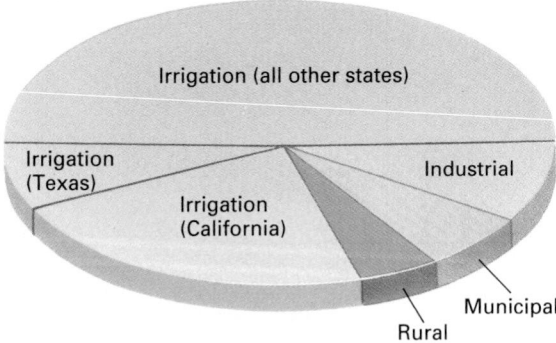

Consumption—416 billion liters per day

B

FIGURE 12.3 (A) Withdrawal: Industry accounts for about half of all water withdrawn in the United States. (B) Consumption: Agriculture accounts for more than three-fourths of the water that is consumed. Irrigation in two dry agricultural states, California and Texas, accounts for about 30 percent of all water consumed in the United States.

Water needed for a healthy, hygienic life = 50 liters per day. Amount used by typical American = 800 liters per day.

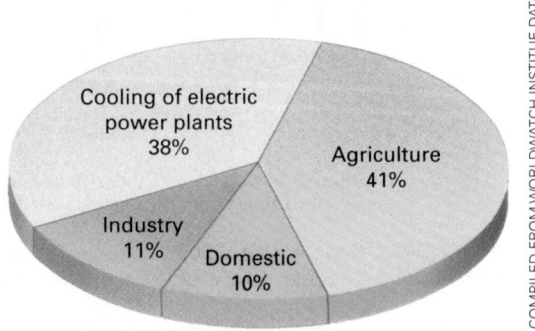

COMPILED FROM WORLDWATCH INSTITUE DATA

FIGURE 12.4 Industrial, agricultural, and domestic water use in the United States.

Unit 3: Surface Processes

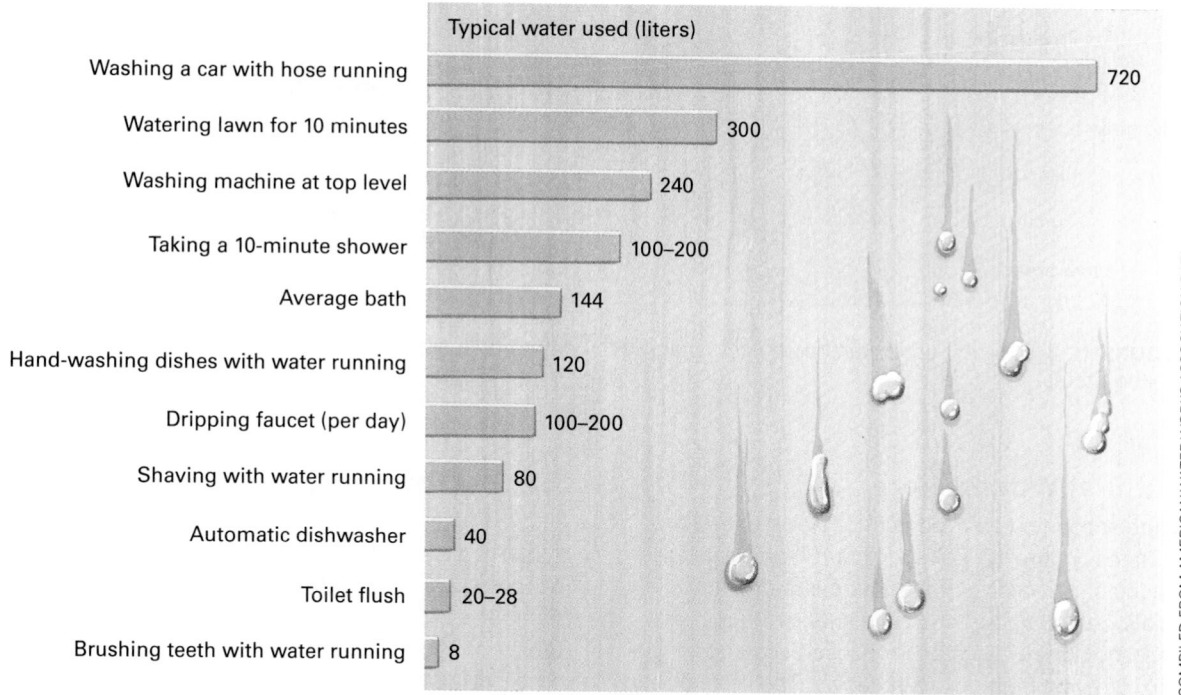

FIGURE 12.5 Domestic water use for common activities in the United States.

800 liters per day (Figure 12.5). Of this total, 200 liters is used for watering lawns. However, domestic use accounts for only 10 percent of the water used in the United States.

In contrast, the World Health Organization defines "reasonable water access" in less-developed agricultural societies as 20 liters (about 5.28 gallons) of sanitary water per person per day, assuming that the source is less than 1 kilometer from the home. Yet 1.1 billion people, about one-sixth of the global population, lack even this small amount of clean water. Water access can fall short for three reasons: either there isn't enough water, the water is unfit to drink, or the source is so far away that people must carry their water long distances from public wells or streams to their homes. The World Health Organization estimates that diarrheal illnesses caused by unsafe water account for 4.1 percent of global disease and are responsible for the deaths of 1.8 million people each year, most of them children under age five.

Industrial Water Use

The cooling systems in electric-power generating plants (run by fossil fuels or nuclear power) account for 38 percent of all water used in the United States. Although much of this water is returned to the stream from which it was taken, it is considerably warmer, which, as we shall see, affects aquatic ecosystems. All other industrial needs add an additional 11 percent to the national water demand. The quantities of water required to produce some common industrial products are shown in Figure 12.6.

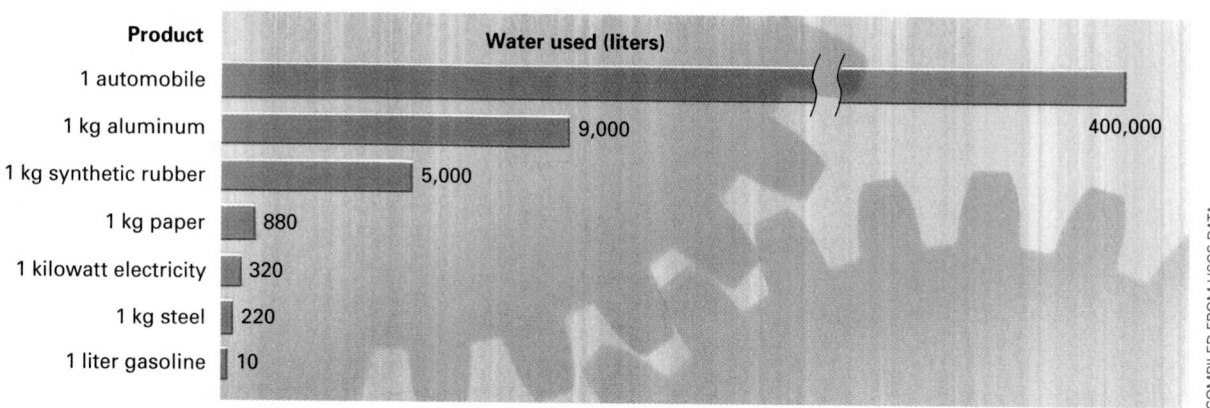

FIGURE 12.6 Large quantities of water are used to produce common industrial products in the United States.

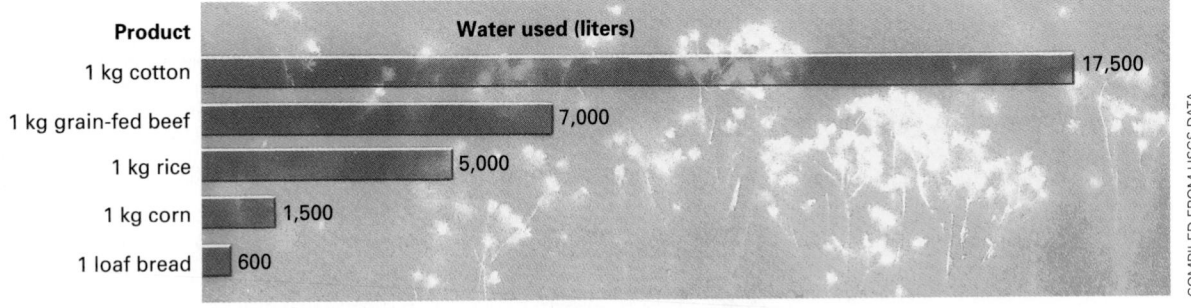

COMPILED FROM USGS DATA.

Product	Water used (liters)	
1 kg cotton		17,500
1 kg grain-fed beef		7,000
1 kg rice		5,000
1 kg corn		1,500
1 loaf bread		600

FIGURE 12.7 Different agricultural products require vastly different amounts of irrigation water in the United States.

Agricultural Water Use

Agriculture accounts for 41 percent of water use in the United States. Figure 12.7 shows that different agricultural products use vastly different amounts of water. California's Central Valley was once a desert, but irrigation has transformed it into an immensely productive region for growing fruits and vegetables. The productivity of the Great Plains in the western United States would decline by one-third to one-half if irrigation were to cease.

Globally, many nations rely on irrigation for more than half of their food production. In dry regions an even higher proportion of farmland must be irrigated, and as a result surface water resources are stressed (Figure 12.8).

12.2 Dams and Diversion

As we stated in the introduction to this chapter, much of Earth's rain falls in the wrong places or at the wrong times to be of much use to humans. For thousands of years, people realized that the problem could be ameliorated by building dams to store water that falls at the wrong time and by diverting water that falls in the wrong places. The Tigris and Euphrates rivers flow through deserts of the Middle East from distant mountains. Ancient Babylonians farmed the desert by irrigating the parched lowlands with diverted river water. In modern times, engineers build huge dams to store water. In 1950 there were 5,700 large dams in

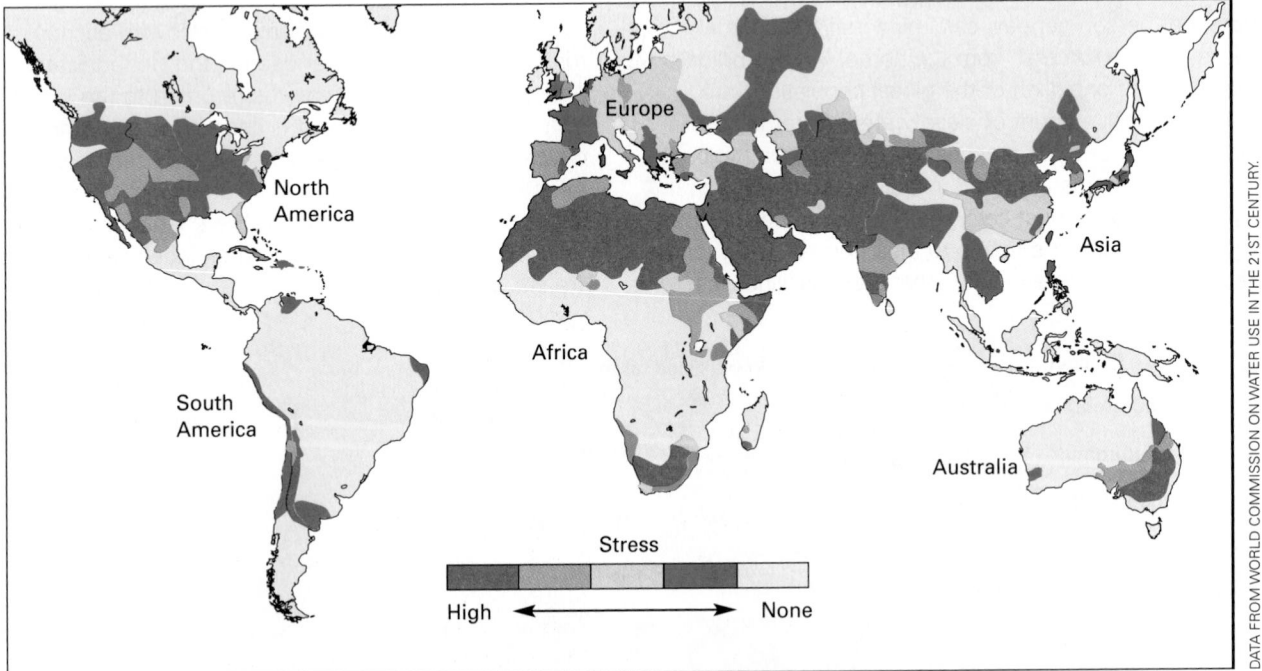

DATA FROM WORLD COMMISSION ON WATER USE IN THE 21ST CENTURY.

FIGURE 12.8 As humans use water for irrigation and other purposes, surface water resources become stressed. (Stress evaluation is based on the amount of water available compared with the amount used by people.)

Unit 3: Surface Processes

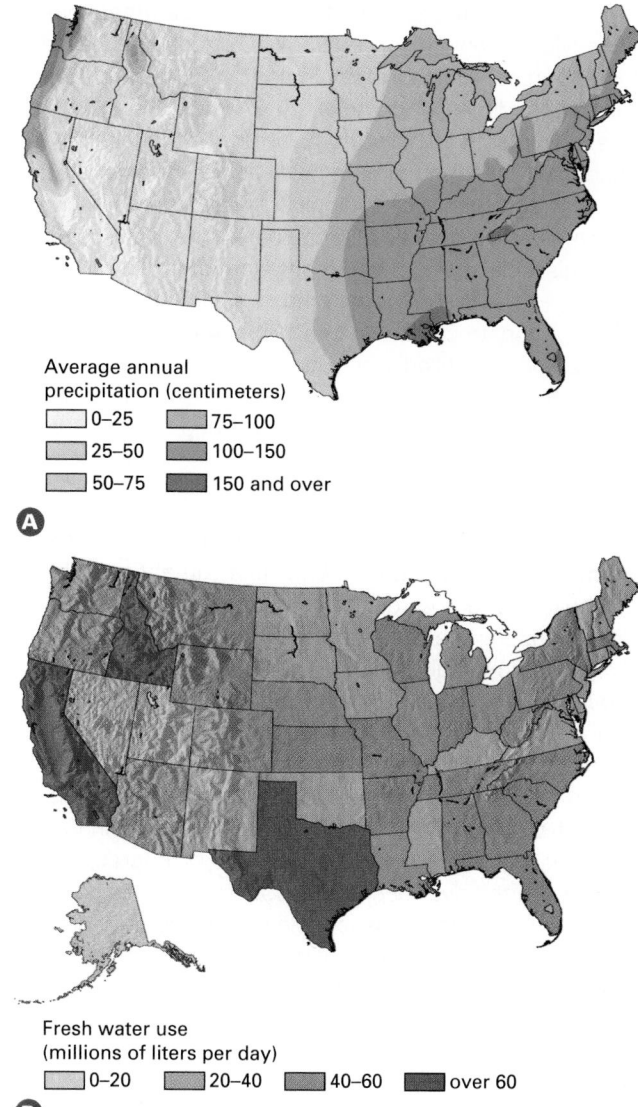

Average annual
precipitation (centimeters)
☐ 0–25   ☐ 75–100
☐ 25–50   ☐ 100–150
☐ 50–75   ☐ 150 and over

A

Fresh water use
(millions of liters per day)
☐ 0–20   ☐ 20–40   ☐ 40–60   ☐ over 60

B

FIGURE 12.9 (A) Average annual precipitation varies greatly throughout the United States. (B) Average annual water use also varies greatly. But many regions that use the greatest amounts of water receive the least rain.

the world; by 2005 there were about 50,000.[2] Almost half of the new dams were built in China. This unprecedented building boom has increased water availability in many regions and produced large amounts of hydroelectric energy, but as we will see in the discussion that follows, these benefits are offset against significant human and environmental costs.

In many regions, surface water is scarce, but huge reservoirs of ground water are accessible beneath the

surface. This ground water can be pumped to the surface for human use.

The United States receives about 3 times more water from precipitation than it uses. However, some of the driest regions are those that use the greatest amounts of water (Figure 12.9). For example, some of the most productive agricultural regions are located in the deserts of California, Texas, and Arizona, where crops are entirely supported by irrigation. Desert cities such as Phoenix, Los Angeles, Albuquerque, and Tucson support large populations that use hundreds of times more water than is locally available and renewable.

Surface Water Diversion

A **diversion system** is a pipe or canal constructed to transport water. Many use gravity to move the water, while others use pumps to lift water uphill. Diversion systems are often augmented by dams that store water. Dams are especially useful in regions of seasonal rainfall. For example, in the arid and semiarid western United States, much of the annual precipitation occurs in winter and spring. Dams store this water for the summer irrigation season, when crops need the water. In addition, the potential energy of the water in a reservoir can generate electricity. Dams supply about 5 percent of the energy used in the world today. Thus, dams are beneficial, but they can create numerous undesirable effects.

Loss of Water

The reservoir formed by a dam provides more surface area for evaporation and more bottom area for seepage into bedrock than did the stream that preceded it. For example, about 270,000 cubic meters of water per year evaporate from Lake Powell, above the Glen Canyon Dam on the Colorado River (Figure 12.10). Thus, less total water flows downstream. In addition, many canals built to carry water from a reservoir to farmland are simply ditches excavated in soil or bedrock, and they leak profusely.

Salinization

All rivers contain small concentrations of dissolved salts. When farmers use river water to irrigate cropland, the water evaporates and the salt remains in the soil. If desert or semidesert soils are irrigated for long periods of time, salt accumulates and lowers the soil fertility. In the United States, **salinization** lowers

diversion system A pipe, canal, or other infrastructure that transports water from its natural place and path in the hydrologic cycle to a new place and path to serve human needs.

salinization A process whereby salts accumulate in soil that is irrigated heavily, lowering soil fertility.

2. A large dam is defined as rising 15 meters or more from base to top and storing more than 3 million cubic meters of water.

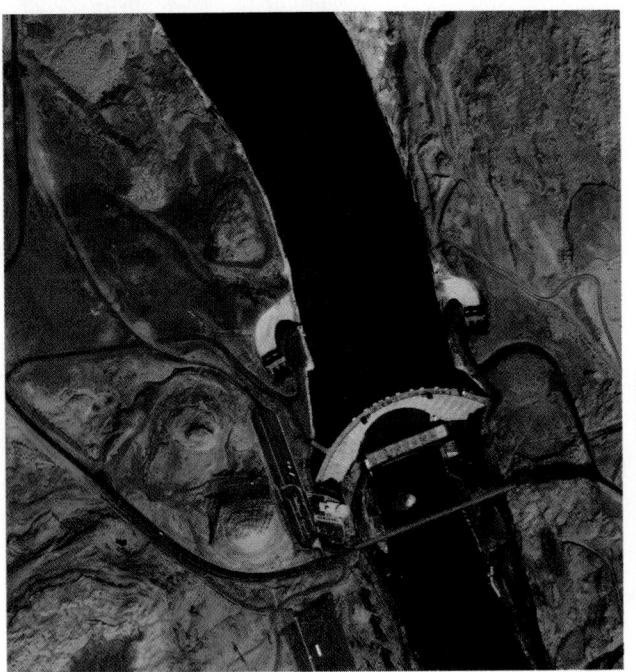

FIGURE 12.10 Construction of the Glen Canyon Dam on the Colorado River in the late 1950s created Lake Powell, shown in the upper half of this photo, between Utah and Arizona. The photo was taken from the IKONOS satellite on September 21, 2004.

crop yields on 25 to 30 percent of irrigated farmland, more than 5 million hectares (Figure 12.11). Globally, the problem affects 10 percent of irrigated land, some 25 million hectares, and it is increasing at a rate of 1 to 1.5 million hectares per year.

Silting

A stream deposits its sediment in a reservoir, where the current slows. Rates of sediment accumulation in reser-

FIGURE 12.11 White salts cover a vast expanse of Nevada desert. Salinization has reduced the productivity of millions of hectares of agricultural land globally.

voirs vary with the sediment load of the dammed river. Lake Mead, behind Hoover Dam in Arizona and Nevada, lost 6 percent of its capacity in its first 35 years as a result of sediment accumulation. At a rate such as this, a reservoir lasts for hundreds of years. Although 100 years may be a long time in political or economic terms, it is a brief moment of geological time and human civilization. In a few instances, engineers have made expensive miscalculations of sedimentation rates. The Tarbela Dam in Pakistan was completed in 1978 after nine years of construction, and by 2005 it was already more than 25 percent filled with silt. The reservoir behind the Sanmenxia Dam on the Yellow River in China filled four years after the dam was finished, making both the dam and reservoir useless.

Erosion

A stream erodes the beaches that line its banks during normal stream flow, but it deposits sediment to build the beaches back up during a flood. Recently, the beaches along the Colorado River in Grand Canyon were disappearing rapidly because management practices at the Glen Canyon Dam upstream prevented flooding in Grand Canyon. In a similar way, beaches are vanishing from many other dammed streams.

In the spring of 1996, scientists released enough water from Glen Canyon Dam to create a flood in Grand Canyon. As predicted, the floodwaters churned up sediment from the river bed and deposited it on the banks, thereby rebuilding old beaches and creating new ones. Because a reservoir accumulates silt and sand, it also interrupts the supply of sediment to the seacoast. As a result, coastal beaches erode, but natural processes do not replenish their sand. This problem has led to expensive rejuvenation projects on some popular beaches in the eastern United States.

Risk of Disaster

A dam can break, creating a disaster in the downstream flood plain. One of the greatest floods since the last ice age roared down Idaho's Teton River canyon when the Teton Dam broke on June 3, 1976. Floodwaters swept away 154 of the 155 buildings in the town of Wilford 6 miles below the dam; the town no longer exists. The flood inundated the lower half of Rexburg, a larger town farther downstream, damaging or destroying 4,000 homes and 350 businesses (Figure 12.12). Damages totaled about $2 billion, and the raging waters stripped the topsoil from tens of thousands of hectares of farmland. Only 11 people died because civil authorities warned and evacuated most of the people. But if the flood had occurred in the middle of the night when warnings and evacua-

Unit 3: Surface Processes

A

B

C

FIGURE 12.12 (A) The Teton Dam was brand new in the spring of 1976. (B) Rapidly rising water broke the dam on June 3, 1976, sending lake water pouring through the breach. (C) Flood waters from the failed Teton Dam inundated the city of Rexburg, Idaho.

tions would have been more difficult, thousands might have been killed.

Recreational and Aesthetic Losses

Dams are often built across narrow canyons to minimize engineering and construction costs. But when the canyons are flooded, unique scenery and ecosystems are destroyed. Glen Canyon Dam on the Colorado River created Lake Powell by flooding one of the most spectacular desert canyons in the American West. Today, Glen Canyon is submerged for more than 300 kilometers above the dam and can be seen only in old photos. The flooding of canyons, however, provides lakes that can be used for fishing and other water sports.

Disputes often arise among various groups that use water in dammed reservoirs. To prevent flooding, a reservoir should be nearly empty by spring so that it can store water and fill during spring runoff. Thus, a reservoir should be drawn down slowly during summer and fall. Agricultural users also prefer to have water stored during spring runoff and supplied for irrigation during summer months. However, recreational users object to a lowering of reservoir levels during summer, when they visit reservoirs most frequently. Managers of hydroelectric dams prefer to run water through their turbines during times of peak demands for electricity, which rarely correspond to flood management, irrigation, or recreational schedules.

Ecological Disruptions

A river is an integral part of the ecosystem that it flows through, and when the river is altered the ecosystem changes. Dams interrupt water flow during portions of the year, prevent flooding, change the temperature of the downstream water, and often create unnatural daily fluctuations in stream flow. All of these changes alter relative abundances of aquatic species. In addition, dams may create specific ecological problems in individual rivers.

For example, before Egypt's Aswan Dam was built, the Nile River flooded every spring, depositing nutrient-rich sediment over the flood plain and delta. The delta was in balance between erosion by the sea and addition of new silt from the yearly floods. When the dam and reservoir cut off the silt supply, erosion then dominated, and the Nile Delta is now shrinking. This loss of an annual supply of nutrients eliminated a source of free fertilizer for flood plain farms. Although the energy produced by the dam is used to manufacture commercial fertilizers, many of the poorer farmers on the flood plain cannot afford to purchase what the river once provided for free.

Human Costs

In the past 50 years, 80 million people globally, approximately equal to 27 percent of the U.S. population, have been forcibly removed from their homes and farms to make way for the reservoirs behind the dams. When these people moved into new areas, they impacted the lives of the people already living in the relocation areas. In many regions, waterborne diseases are greater surrounding stagnant water than they were near free-flowing streams. In a study of 1.5 million people affected by 50 dams across the world, only 7 percent of the affected people realized an improvement in living standards, and 70 percent were burdened by lower living standards. Twenty-three percent were unaffected.

Ground Water Diversion

Ground water provides drinking water for more than half of the population of North America and is a major source of water for irrigation and industry. It is a valuable resource because:

1. It is abundant; 60 times more freshwater exists underground than in streams and lakes combined.
2. Ground water moves very slowly; thus it is stored below Earth's surface and remains available during dry periods.

3. In some regions, ground water flows from wet environments to arid ones, making water available in dry areas.

Ground Water Depletion

If ground water is pumped to the surface faster than it can flow through the aquifer to the well, a **cone of depression** forms near the well (Figure 12.13). When the pump is turned off, ground water flows back toward the well in a matter of days or weeks if the aquifer has good permeability, and the cone of depression disappears. Conversely, if water is continuously removed more rapidly than it can flow to the well through the aquifer, the water table drops.

Before the development of advanced drilling and pumping technologies, human impact on ground water was minimal. Today, however, deep wells and high-speed pumps can extract ground water more rapidly than the hydrologic cycle recharges it. Where such excessive pumping is practiced, the water table falls as the ground water reservoir becomes depleted. In some cases the aquifer is no longer able to supply enough water to support the farms or cities that have overexploited it. This situation is common in the central, western, and southwestern United States (Figure 12.14).

The Ogallala aquifer extends almost 900 kilometers from the Rocky Mountains eastward across the prairie, and from Texas to South Dakota. It consists of water-saturated porous sandstone and conglomerate within 350 meters of the surface (Figure 12.15A). The aquifer averages about 65 meters in thickness and is the world's largest known aquifer. Between 1930 and 2000, about 200,000 wells were drilled into the Ogallala aquifer, and extensive irrigation systems were installed throughout the region.

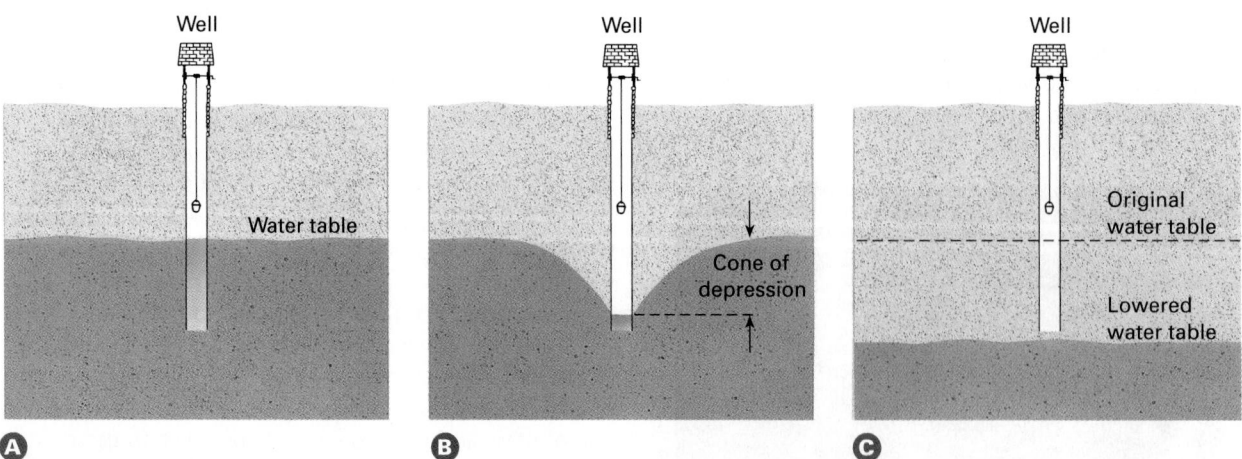

FIGURE 12.13 (A) A well is drilled into an aquifer. (B) A cone of depression forms because a pump draws water faster than the aquifer can recharge the well. (C) If the pump continues to extract water at the same rate, the water table falls.

Unit 3: Surface Processes

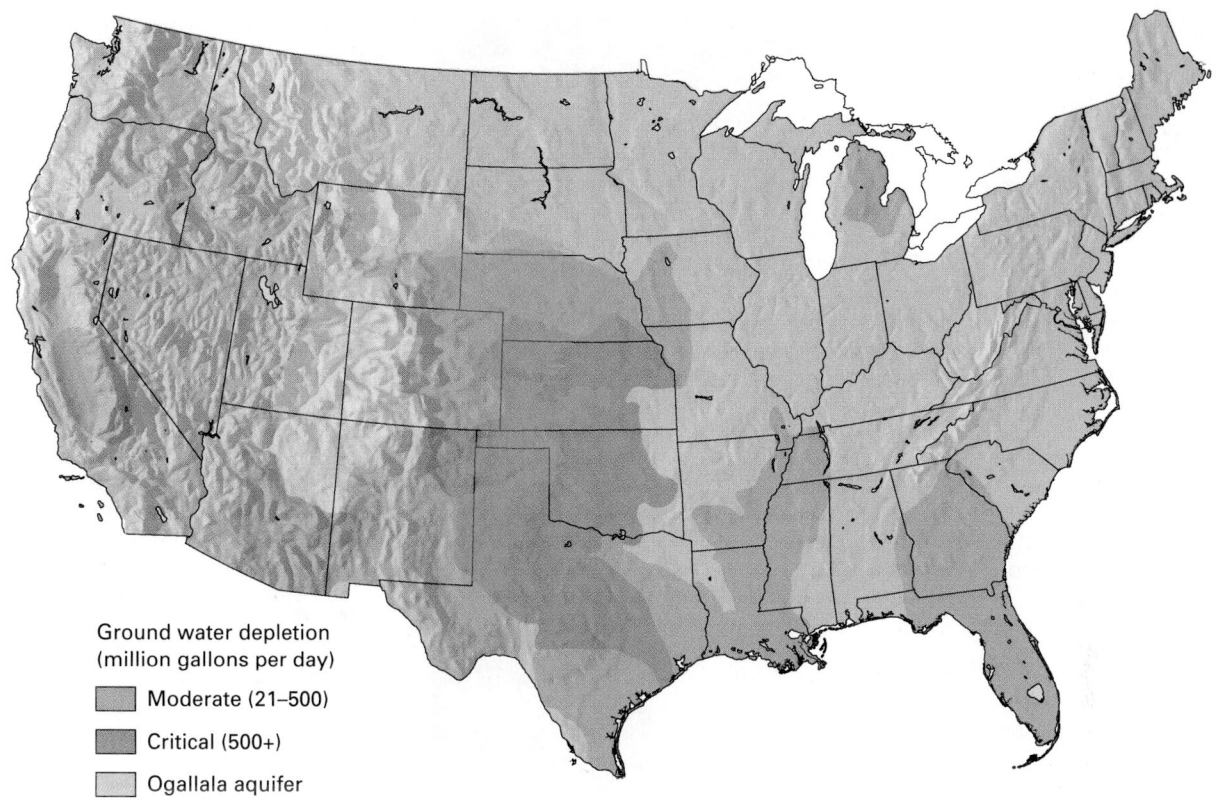

Ground water depletion
(million gallons per day)

■ Moderate (21–500)

■ Critical (500+)

□ Ogallala aquifer

FIGURE 12.14 Aquifer depletion is a common problem in the United States. The numbers refer to the amount of water used in excess of the recharge rate.

Most of the water in the Ogallala aquifer accumulated when the last Pleistocene ice sheet melted, about 15,000 years ago. But because the High Plains receive little rain, the aquifer is now mostly recharged by rain and snowmelt in the Rocky Mountains, hundreds of kilometers to the west. The problem is that water moves very slowly through the aquifer, at an average rate of only about 15 meters per year. At this rate,

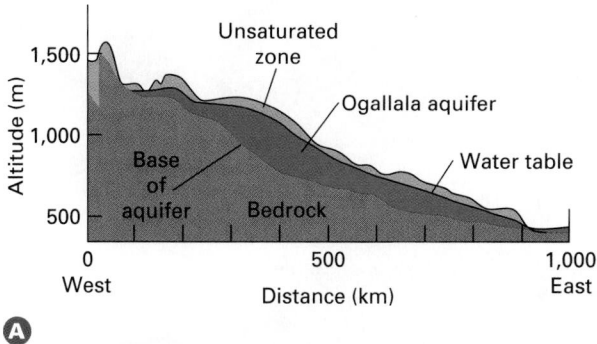

A

FIGURE 12.15 The Ogallala aquifer supplies water to much of the High Plains. (A) A cross-sectional view of the aquifer shows that much of its water originates in the Rocky Mountains and flows slowly as ground water beneath the High Plains. (B) This map shows the extent and water level changes in the Ogallala aquifer from 1980 and 1997.

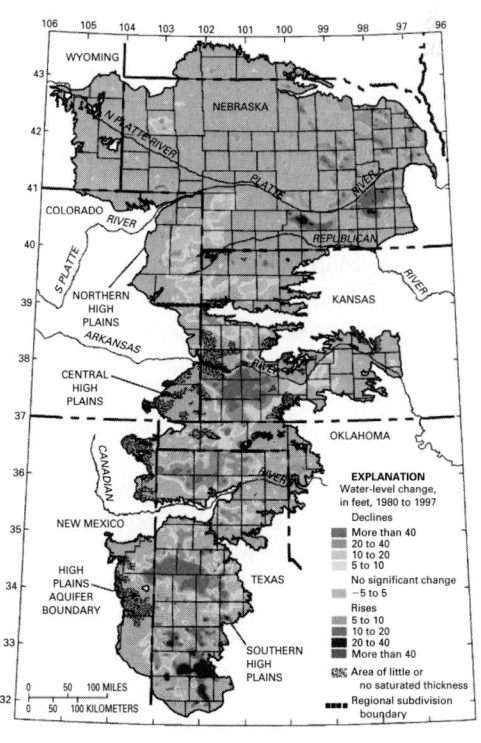

B

subsidence The irreversible sinking or settling of Earth's surface.

saltwater intrusion A condition in coastal regions in which excessive pumping of fresh ground water causes salty ground water to invade an aquifer.

ground water takes 60,000 years to travel from the mountains to the eastern edge of the aquifer. Presently, approximately 9.6 billion liters are returned to the aquifer every year through rainfall and ground water flow. However, farmers remove 90 billion liters annually. Under such conditions, the deep ground water is, for all practical purposes, nonrenewable.

If the present pattern of water use continues, one-quarter of the original water in the Ogallala aquifer will be consumed by 2020. However, in some of the drier regions of northwest Texas and west central Kansas, the wells are already depleted (Figure 12.15B). The aquifer irrigates about 5 million hectares of farms and ranches (an area about the size of the states of Massachusetts, Vermont, and Connecticut combined). About 20 percent of the total U.S. agricultural output, including 40 percent of the feedlot beef, is produced in this region, with a total value of $32 billion. If the aquifer is depleted and another source of irrigation water is not found, productivity in the central High Plains is expected to decline by 80 percent. Farmers will go bankrupt, and food prices will rise throughout the nation.

Subsidence

Excessive removal of ground water can cause **subsidence**, the sinking or settling of Earth's surface. When water is withdrawn from an aquifer, rock or soil particles shift to fill the space left by the lost water. As a result, the volume of the aquifer decreases and the overlying ground subsides.

Subsidence rates can reach 5 to 10 centimeters per year, depending on the rate of pumping and the nature of the aquifer. Some areas in the San Joaquin Valley of California have sunk nearly 10 meters (Figure 12.16). The land surface has subsided by as much as 3 meters in the Houston–Galveston area of Texas. The

US GEOLOGICAL SURVEY

FIGURE 12.16 The land surface at this point in San Joaquin, California, was at the height of the 1925 marker in the year 1925. Subsidence resulting from ground water extraction for irrigation lowered the surface by about 10 meters in 52 years.

problem is particularly severe in cities. For example, Mexico City is built on an old marsh. Over the years, as the weight of buildings and roadways has increased and much of the ground water has been removed, parts of the city have settled as much as 8.5 meters. Many millions of dollars have been spent to maintain the city on its unstable base. Similar problems are affecting Phoenix, Arizona, and other U.S. cities.

Unfortunately, subsidence is irreversible. When rock and soil contract, their porosity is permanently reduced so that ground water reserves cannot be completely recharged, even if water becomes abundant again.

Saltwater Intrusion

Two types of ground water occur in coastal areas: fresh ground water and salty ground water that seeps in from the sea. Freshwater floats on top of salty water because it is less dense. If too much freshwater is removed from the aquifer, salty ground water rises to the level of wells (Figure 12.17). It is unfit for drinking, irrigation, or industrial use. **Saltwater intrusion** has affected much of south Florida's coastal ground water reservoirs. Once an area is contaminated by saltwater intrusion, it dramatically reduces freshwater storage in that aquifer.

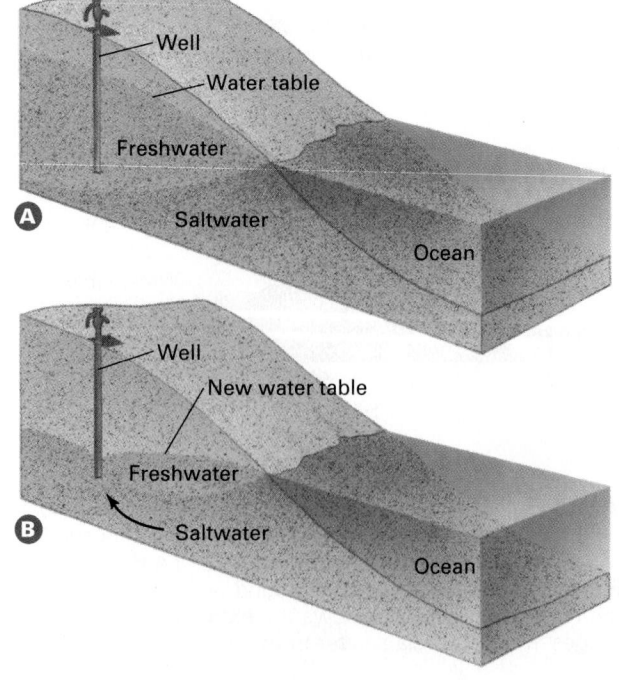

FIGURE 12.17 Saltwater intrusion can pollute coastal aquifers. (A) Freshwater lies above saltwater, and the water in the well is fit to drink. (B) If too much freshwater is removed, the water table falls. The level of saltwater rises and contaminates the well.

12.3 The Great American Desert

In the years following the Civil War, an American geologist, John Wesley Powell, explored the area between the western mountains (the Sierra Nevada–Cascade crest) and the 100° meridian (the line of longitude running through the Dakotas; Nebraska; Kansas; and Abilene, Texas). He recognized that most of this region is arid or semiarid, and he called it the Great American Desert. Despite Powell's warning about inadequate water, Americans flocked to settle there. As described earlier, several parts of this region now use huge amounts of water for cities, industry, and agriculture, although they receive little precipitation. For example, arid California, Nevada, Arizona, Texas, and Idaho use the greatest quantities of water in the United States. Some desert cities, including Phoenix, El Paso, Reno, and Las Vegas, rely on diversion of surface water and/or overpumping of ground water for nearly all of their water.

Billions of dollars have been spent on heroic projects to divert rivers and pump ground water to serve this area of the country. For example, 1,200 dams and the two largest irrigation projects in the world have been built in California alone. More than 80 percent of the water diverted to southern California is used to irrigate desert and near-desert cropland. The water supplied to these farms commonly irrigates crops such as cotton, rice, and alfalfa. Cotton requires 17,000 liters per kilogram; rice and alfalfa also use great amounts of water. The U.S. government pays large subsidies to farmers in naturally wet southeastern and south central states not to grow those same crops. The irrigation later used to support cattle ranching and other livestock in California is enough to supply the needs of the entire human population of the state. Controversy over water diversion projects is not limited to the American West. New York City obtains much of its water from reservoirs in upstate New York, and the New York City Board of Water Supply controls those reservoirs and watersheds hundreds of miles from the city. Opposing interests of local land-use groups and city water users have provoked heated conflicts in the state of New York.

The Colorado River

The Colorado River runs through the southwestern portion of the Great American Desert. Starting from the snowy mountains of Colorado, Wyoming, Utah, and New Mexico, it flows across the arid Colorado Plateau and southward into Mexico, where it empties into the Gulf of California.

Because the river flows through a desert, farmers, ranchers, cities, and industrial users along the entire length of the river compete for rights to use the water. In the 1920s, the Colorado River discharged 18 billion cubic meters of water per year into the Gulf of California. In 1922 the U.S. government apportioned 9 billion cubic meters for the river's Upper Basin (Colorado, Utah, Wyoming, and New Mexico) and the remaining 9 billion cubic meters for Lower Basin users in Arizona, California, and Nevada. Thus, all the water was allocated and none of it was set aside for maintenance of the river ecosystem or for Mexican users south of the border. No water flowed into the sea. Twenty years later, an international treaty awarded Mexico 1.8 billion cubic meters to be taken equally from Upper Basin and Lower Basin users.

The Colorado and many of its tributaries flow across sedimentary rocks, many of which contain soluble salt deposits. As a result, the Colorado is naturally a salty river. In addition, the U.S. government built 10 large dams on the Colorado, and evaporation from the reservoirs has further concentrated the salty water. In 1961, the water flowing into Mexico contained 27,000 ppm (parts per million) salts, compared with an average salinity of about 100 ppm for large rivers. For comparison, 27,000 ppm is 77 percent as salty as seawater.

Mexican farmers used this water for irrigation, and their crops died. In 1973, the U.S. government built a desalinization plant to reduce the salinity of Mexico's share of the water.

There are also significant questions about the possibility of a long-term drought in the Southwest. Tree ring studies have shown that more extreme and extended droughts have occurred in the past several centuries than have been experienced in modern times. With current demands on the Colorado River increasing, even a minor drought can have severe consequences. The end of 2004 marked five consecutive years of lower-than-average rainfall and subsequent decline in reservoir levels. In January 2005, Lake Powell was at 35 percent of its capacity, the lowest level since 1969 when the reservoir was being filled. Maintaining a reliable supply of water for the growing population of the Colorado River basin is one of the most challenging public policy issues of the Southwest.

12.4 Water and International Politics

The Jordan River flows along the border between Israel and its enemies, Jordan and Syria. In the north it forms the boundary between Israel and the disputed Golan Heights, and downstream it forms the boundary between Israel and the disputed West Bank. Population in this water-parched desert is expected to skyrocket from 32 million at the end of 2001 to 52 million by 2025.

The Los Angeles Water Project

In the mid-1800s Los Angeles was a tiny, neglected settlement on the California coast. The Los Angeles River frequently flooded in winter and diminished to a trickle in summer. The average annual precipitation of the Los Angeles area is 65 centimeters, most of which falls during a few winter weeks.

Among the early settlers were Mormons, who had become experts at irrigating dry farmland from their experience in the Utah desert. By the late 1800s, irrigated farms in the Los Angeles basin were producing a wealth of fruits and vegetables. Suddenly, Los Angeles was an attractive, growing town. In 1848 the town had a population of 1,600. The population passed 100,000 by 1900 and 200,000 by 1904. Then the city ran out of water. Wells dried up and the river was inadequate to meet demand. Los Angeles was surrounded on three sides by deserts and on the fourth by the Pacific Ocean. There was no nearby source of freshwater.

However, 250 miles to the northeast, the Owens River flowed out of the Sierra Nevada. The city of Los Angeles bought water rights, bit by bit, from farmers and ranchers in the Owens Valley and at the headwaters of the river. The city then spent millions of dollars building the Los Angeles Aqueduct across some of the most difficult, earthquake-prone terrain in North America. When finished, the aqueduct was 357 kilometers long, including 85 kilometers of tunnel. Siphons and pumps carried the water over hills and mountains too treacherous for tunneling (Figure 12.18). On November 5, 1913, the first water poured from the aqueduct into the San Fernando Valley.

The following 10 to 15 years were unusually rainy in the Los Angeles area and in the Owens Valley. The rain recharged the ground water under the Los Angeles basin, and the Los Angeles River flowed freely again. As a result, irrigated farmland increased from 1,200 hectares in 1913, when the aqueduct opened, to more than 30,000 hectares in 1918.

In the 1920s, normal dryness and drought returned. Los Angeles and the farms of the San Fernando Valley demanded more water from the Owens River, and the Owens Valley dried up.

By the mid-1930s, Los Angeles owned 95 percent of the farmland and 85 percent of the residential and commercial property in the Owens Valley. Although the city leased some of the farmland back to the farmers and ranchers, the water supply became so unpredictable that agriculture slowly dried up with the land. As Los Angeles continued to grow, the water from the Owens River was not sufficient, so the Los Angeles Department of Water and Power drilled wells into the Owens Valley and began pumping its ground water aquifer dry.

Today, the Owens Valley is parched. Its remaining citizens pump gas, sell beer, and make up motel beds for tourists driving through on their way to somewhere else (Figure 12.19). Ironically, the water that once irrigated farms and ranches in the naturally arid Owens Valley was diverted through the Los Angeles Aqueduct, at great cost to taxpayers, to irrigate farms in the naturally arid San Fernando Valley. The aqueduct converted a farmer's paradise to a desert and a desert to a farmer's paradise.

FIGURE 12.18 (A) In this region upstream from the town of Bishop, California, most of the Owens River flows through a pipe. (B) Only a small trickle remains in the river's old streambed.

FIGURE 12.19 As most of the Owens River is diverted to Los Angeles, farms and ranches have dried up, and so have local businesses.

Unit 3: Surface Processes

MICHAEL KLENETSKY/SHUTTERSTOCK

Compounding the region's political tensions are the following issues regarding the water supply:

- Syria has planned to build dams that will divert one of the major tributaries of the Jordan; Israel has threatened to bomb the dam before it is completed.

- Jordan's late King Hussein (father of the current king) said in 1990 that water was the only issue that could take him to war with Israel.

- While the Palestinians struggle for an independent homeland, the Israelis are pumping 35 percent of their sustainable freshwater supply from an aquifer that lies beneath the disputed West Bank. At the present time, the Israelis are overpumping the aquifer and at the same time rationing water use by the West Bank Arabs. If they cede the land to their enemies, they lose the water.

Current international water law offers little help in resolving conflicts over water. In general, downstream nations argue that a river or aquifer is a communal resource to be shared equitably among all adjacent nations. Predictably, many upstream nations maintain that they have absolute control over the fate of water within their borders and have no responsibility to downstream neighbors.

Any international code of water use must be based on three fundamental principles:

1. Water users in one country must not cause major harm to water users in other countries downstream.
2. Water users in one country must inform neighbors of actions that may affect them before the actions are taken. (For example, if a dam is built, engineers must inform downstream users that they plan to stop or reduce river flow to fill a reservoir.)
3. People must distribute water equitably from a shared river basin.

Unfortunately, these principles, especially the last one, are open to such subjective and self-interested interpretations that their intent is easily and often subverted.

12.5 Water Pollution

While consumption reduces the *quantity* of water in a stream, lake, or ground water reserve, **pollution** is the reduction of the *quality* of the water by the introduction of impurities (Figure 12.20). A polluted stream may run full to the banks, yet it may be toxic to aquatic wildlife and unfit for human use.

In the early days of the Industrial Revolution, factories and sewage lines dumped untreated wastes into rivers. As the population grew and industry expanded, many of our waterways became fetid, foul-smelling, and unhealthy. The first sewage treatment plant in the United States was built in Washington, D.C., in 1889, more than 100 years after the Revolutionary War. Soon other cities followed suit, but few laws regulated industrial waste discharge.

In 1952 and again in 1969, the Cuyahoga River in Cleveland, Ohio, was so heavily polluted with oil and industrial chemicals that it caught fire, spreading flame and smoke across the water. While the highly publicized photographs of the flaming river flowing through downtown Cleveland had a powerful visual impact, other water pollution disasters were equally insidious, such as Love Canal in New York State.

These and similar incidents made it clear that water pollution was endangering health and reducing the quality of life for people throughout the United States. As a result, Congress passed the **Clean Water Act** in 1972. The legislation states that:

pollution The reduction of the quality of a resource by the introduction of impurities.

Clean Water Act A federal law passed in 1972, mandating the cleaning of the nation's rivers, lakes, and wetlands and forbidding the discharge of pollutants into waterways.

COURTESY OF GRAHAM R. THOMPSON/JONATHAN TURK

FIGURE 12.20 Pollution is the reduction of the quality of a resource by the introduction of impurities.

biodegradable pollutants
Pollutants that decay naturally in a reasonable amount of time, being consumed or destroyed by organisms that live naturally in soil or water.

persistent bioaccumulative toxic chemicals (PBTs)
Nonbiodegradable toxins that are released into and accumulate in the environment.

1. It is the national goal that the discharge of pollutants into the navigable waters be eliminated by 1985.
2. It is the national goal that, wherever attainable, an interim goal of water quality which provides for the protection and propagation of fish, shellfish, and wildlife and provides for recreation in and on the water be achieved by July 1, 1983.
3. It is the national policy that the discharge of toxic pollutants in toxic amounts be prohibited.

Types of Pollutants

Biodegradable Pollutants

A biodegradable material is one that decays naturally, being consumed or destroyed in a reasonable amount of time by organisms that live naturally in soil and water. Several types of contaminants are **biodegradable pollutants**:

1. Sewage is wastewater from toilets and other household drains. It includes biodegradable organic material such as human and food wastes, soaps, and detergents. Sewage also includes some nonbiodegradable chemicals because people flush paints, solvents, pesticides, and other chemicals down the drains.
2. Disease organisms, such as typhoid and cholera, are carried into waterways in the sewage of infected people.
3. Phosphates and nitrate fertilizers flow into surface water and ground water mainly from agricultural runoff. Phosphate detergents and phosphates and nitrates from feedlots also fall into this category.

Nonbiodegradable Pollutants

Many materials are not decomposed naturally by environmental chemicals or consumed by decay organisms, and therefore they are *nonbiodegradable*. Not all of these are poisons, but some are. In modern parlance, nonbiodegradable poisons are called **persistent bioaccumulative toxic chemicals (PBTs)**. The Environmental Protection Agency (EPA) recognizes tens of thousands

Love Canal

Love Canal in Niagara Falls, New York, was excavated to provide water to an industrial park that was never built. In the 1940s, the Hooker Chemical Company purchased part of the old canal as a site to dispose of toxic manufacturing wastes. The engineers at Hooker considered the relatively impermeable ground surrounding the canal to be a reasonably safe disposal area. During the following years, the company disposed of approximately 19,000 tons of chemical wastes by loading them into 55-gallon steel drums and dumping the drums in the canal. In 1953, the company covered one of the sites with dirt and sold the land to the Board of Education of Niagara Falls for $1, after warning the city of the buried toxic wastes. The city then built a school and playground on the site.

In the process of installing underground water and sewer systems to serve the school and growing neighborhood, the relatively impermeable soil surrounding the old canal and waste burial site was breached. Gravel used to backfill the water and sewer trenches then provided a permeable path connecting the toxic waste dump to surrounding parts of the city.

During the following decades, the buried drums rusted through, and the chemical wastes seeped into the ground water. In the spring of 1977, heavy rains raised the water table to the surface, and the area around Love Canal became a muddy swamp. But it was no ordinary swamp—the leaking drums had contami-

nated the ground water with toxic and carcinogenic compounds. The poisonous fluids soaked the playground, seeped into basements of nearby homes, and saturated gardens and lawns. Children who attended the school and adults who lived nearby developed epilepsy, liver malfunctions, skin sores, rectal bleeding, and severe headaches. In the years that followed, an abnormal number of pregnant women suffered miscarriages, and large numbers of babies were born with birth defects.

Thus a human tragedy at Love Canal was caused by deliberate, but legal, burial of industrial poisons, coupled with flawed hydrogeology, and complete disregard of public health by both the Hooker Chemical Company and local officials.

In the decades following this public health tragedy, environmental laws have become much stricter and public awareness has increased dramatically. It is almost unimaginable that anyone would build a school over a toxic waste site today. Yet, less flagrant releases of toxic and radioactive materials continue to this day, and the Love Canal disaster is a reminder to maintain vigilance.

of PBTs. Yet, even nontoxic materials can pollute and adversely affect ecosystems. **Nonbiodegradable pollutants** can be roughly subdivided into three broad categories:

1. Many organic industrial compounds are particularly troublesome because they are toxic in high doses, are suspected carcinogens in low doses, and can survive for decades in aquatic systems. Examples include some pesticides, dioxin, and PCBs.
2. Toxic inorganic compounds include mine wastes, road salt, and mineral ions such as those of cadmium, arsenic, and lead.
3. When sediment enters surface waters, it neither fertilizes nor poisons the aquatic system. However, the sediment muddies streams and buries aquatic habitats, thus degrading the quality of an ecosystem.

Radioactive Materials

Radioactive materials include wastes from the mining of radioactive ores, nuclear power plants, nuclear weapons, and medical and scientific applications.

Heat

Heat can also pollute water. In a fossil fuel or nuclear electric generator, an energy source (coal, oil, gas, or nuclear fuel) boils water to form steam. The steam runs a generator to produce electricity. The exhaust steam must then be cooled to maintain efficient operation.

The cheapest cooling agent is often river or ocean water. A 1,000-megawatt power plant heats 10 million liters of water by 35°C every hour. The warm water can kill fish directly, affect their reproductive cycles, and change the aquatic ecosystems to eliminate some native species and favor the introduction of new species. For example, if water is heated, indigenous cold-water species like trout will die out and be replaced by carp or other warm-water fish.

Any of these categories of pollutants can be released in either of two ways. **Point source pollution** arises from a specific site such as a septic tank, a gasoline spill, or a factory. In contrast, **nonpoint source pollution** is generated over a broad area. Fertilizer and pesticide runoff from lawns and farms fall into this latter category.

12.6 How Sewage, Detergents, and Fertilizers Pollute Waterways

Recall from Chapter 11 that an oligotrophic stream or lake contains few nutrients and clear, nearly pure water (Figure 12.21A). Sewage, detergents, and agricultural fertilizers are plant nutrients. They do not poison aquatic systems; they nourish them. If humans dump one or more of these nutrients into an oligotrophic waterway, aquatic organisms multiply so quickly that they consume most of the dissolved oxygen. Many fish, such as trout and salmon, die. Other organisms, such as algae, carp, and water worms, proliferate. Thus an increased supply of nutrients transforms a clear, sparkling stream or lake into a slime-covered, eutrophic waterway (Figure 12.21B).

If humans release even more sewage, detergent, or fertilizer, decay organisms multiply so rapidly that they consume all the oxygen. As a result, most aquatic life dies. Then, **anaerobic** bacteria (bacteria that live without oxygen) take over, releasing noxious hydrogen sulfide gas as they feast on the remaining organic matter (Figure 12.21C). Even carp and worms cannot survive in such an environment. As they die and rot, the hydrogen sulfide that bubbles to the surface smells like rotten eggs.

When a discharge of raw sewage flows into a stream, decay organisms immediately begin feasting, and at the same time the current carries the pollutants and the decay organisms downstream. Over time, the decay organisms consume most of the pollutants. When this food supply is consumed, the decay organisms die off. Natural turbulence replenishes oxygen in the water and aerobic organisms become reestablished. Thus, the river cleanses itself. However, if the original source of pollution is too large, or if every town along the river dumps pollutants into the same waterway, then these natural cleaning mechanisms become overwhelmed, and the river remains polluted all the way to the ocean.

12.7 Toxic Pollutants, Risk Assessment, and Cost–Benefit Analysis

Although natural decay organisms decompose and decontaminate human feces over a period of days or weeks, many organic compounds, such as the banned pesticide DDT, persist for decades, and heavy metal contaminants such as arsenic or lead never decompose. (These heavy metals may change chemical form and therefore become benign, but the atoms never decompose.)

nonbiodegradable pollutants Pollutants that do not decay naturally in a reasonable amount of time, including some industrial compounds, toxic inorganic compounds, and nontoxic sediment that muddies streams and habitats.

point source pollution Pollution that arises from a specific site such as a septic tank or a factory.

nonpoint source pollution Pollution that is generated over a broad area, such as fertilizers and pesticides spread over agricultural fields.

anaerobic Without oxygen; anaerobic bacteria are bacteria that live without oxygen.

A

B

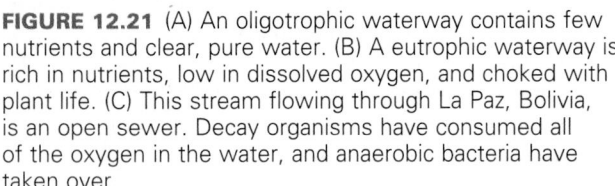

C

COURTESY OF GRAHAM R. THOMPSON/JONATHAN TURK

FIGURE 12.21 (A) An oligotrophic waterway contains few nutrients and clear, pure water. (B) A eutrophic waterway is rich in nutrients, low in dissolved oxygen, and choked with plant life. (C) This stream flowing through La Paz, Bolivia, is an open sewer. Decay organisms have consumed all of the oxygen in the water, and anaerobic bacteria have taken over.

Environmental laws in developed nations regulate the discharge of pollutants into waterways. Yet pollutants continue to enter streams and rivers. Even in situations where waterways are usually contaminated with only small or even minuscule amounts of these toxic nonbiodegradable compounds, a great many different pollutants may be found in a single river, and even small concentrations may cause cancer or birth defects.

Scientists do not know the doses at which many compounds become harmful to human health. The uncertainty results, in part, from delayed effects of some chemicals. For example, a contaminant might increase the risk of a certain type of cancer, but the cancer may not develop until 10 to 20 years after exposure. Because scientists cannot perform direct toxicity experiments on humans, and because they cannot wait decades to observe results, they frequently attempt to assess the carcinogenic properties of a contaminant by feeding it to laboratory rats. If rats are fed very high doses, sometimes hundreds of thousands of

times more concentrated than environmental pollutants, they may then contract cancer in weeks or months—rather than in years or decades. Suppose that a rat gets cancer after drinking the equivalent of 100,000 glasses of polluted well water each day for a few weeks. Can we say that the same contaminant will cause cancer in humans who drink 10 glasses a day for 20 years? No one knows. It may or may not be legitimate to extrapolate from high doses to low doses and from one species to another.

Scientists also use epidemiological studies to assess the risk of a pollutant. For example, if the drinking water in a city is contaminated with a pesticide and a high proportion of people in the city develop an otherwise rare disease, then the scientists may infer that the pesticide caused the disease and that its presence in drinking water constitutes a high level of risk to human health.

Because neither laboratory nor epidemiological studies can prove that low doses of a pollutant are harmful to humans, scientists are

PAKHNYUSHCHA/SHUTTERSTOCK

Unit 3: Surface Processes

faced with a question: Should businesses and governments spend money to clean up the pollutant? Some argue that such expenditure is unnecessary until we can prove that the contaminant is harmful. Others invoke the precautionary principle that it's better to be safe than sorry. Proponents of the latter approach argue that people commonly act on the basis of incomplete proof. For example, if a mechanic told you that your brakes were likely to fail within the next 1,000 miles, you would recognize this as an opinion, not a fact. Yet would you wait for the brakes to fail or replace them now?

Pollution control is expensive. However, pollution is also expensive. If a contaminant causes people to sicken, the cost to society can be measured in medical bills and loss of income resulting from missed work. Many contaminants damage structures, crops, and livestock. People in polluted areas also bear expense because tourism diminishes and land values are reduced when people no longer want to visit or live in a contaminated area. All of these costs are called **externalities**—additional or unintended consequences of the original activity or condition.

12.8 Ground Water Pollution

Water in a sponge saturates tiny pores and passages. To clean a contaminant from a sponge, it would be necessary to clean every pore of the sponge that came into contact with contaminated water. Because this is nearly impossible to achieve completely, no one would wash dishes with a sponge that had been used to clean a toilet.

Many different types of sources contaminate ground water (Figure 12.22). Because these contaminants permeate an aquifer in a manner analogous to the way in which contaminated water saturates a sponge, the removal of a contaminant from an aquifer is extremely difficult and expensive. Once a pollutant enters an aquifer, the natural flow of ground water disperses it as a growing **plume of contamination**. Because ground water flows slowly, usually at a few centimeters per day, the plume spreads slowly (Figure 12.23).

Most contaminants persist in a ground water aquifer for much longer than in a stream or lake. The rapid flow of water through streams and lakes replenishes their water quickly, but ground water flushes much more slowly (Table 12.1). In addition, oxygen, which reacts to decompose many contaminants, is less abundant in ground water than in surface water.

As a result, despite nearly 40 years of government concern and action, the following situations exist:

externalities Additional or unintended consequences of an activity or condition, such as the indirect costs of environmental degradation, including reduction in tourism and lowered land values.

plume of contamination The slow-growing, three-dimensional zone within an aquifer or of surface water affected by a dispersing pollutant.

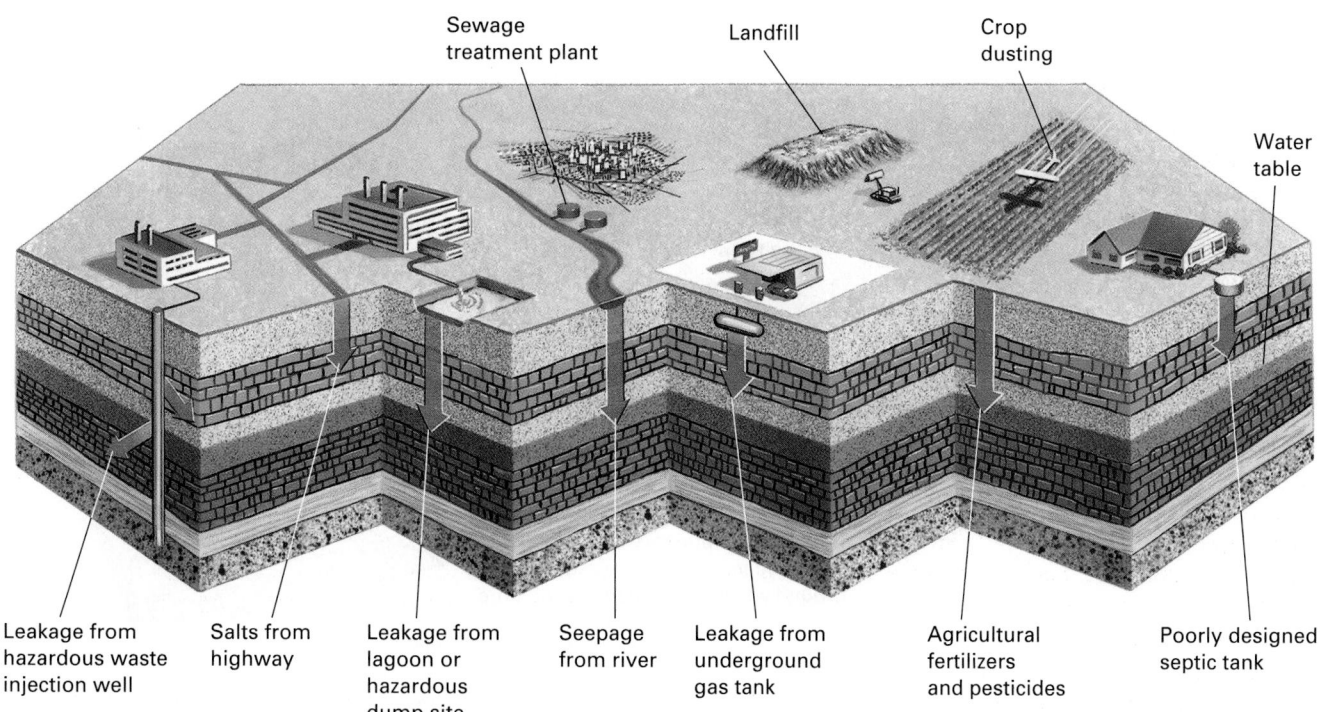

Sewage treatment plant

Landfill

Crop dusting

Water table

Leakage from hazardous waste injection well

Salts from highway

Leakage from lagoon or hazardous dump site

Seepage from river

Leakage from underground gas tank

Agricultural fertilizers and pesticides

Poorly designed septic tank

FIGURE 12.22 Many different sources contaminate ground water.

- Up to 25 percent of the usable ground water in the United States is contaminated. About 45 percent of municipal ground water supplies in the United States are contaminated with organic chemicals.

- The gasoline additive MTBE is a frequent and widespread contaminant of underground drinking water. According to the EPA, 5 to 10 percent of drinking water in areas using reformulated gasoline shows MTBE contamination.

- Wells in 38 states contain pesticide levels high enough to threaten human health. Every major aquifer in New Jersey is contaminated.

- In Florida, where 92 percent of the population drinks ground water, more than 1,000 wells have been closed because of contamination and over 90 percent of the remaining wells have detectable levels of industrial or agricultural chemicals.

- In 2003 the EPA reported 436,494 confirmed releases of dangerous volatile organic compounds leaking from underground fuel storage tanks in the United States.

remediation The treatment of a contaminated area, such as an aquifer, to remove or decompose a pollutant.

TABLE 12.1 Average Residence Times for Water in Various Reservoirs

Atmosphere	8 days
Rivers	16 days
Soil moisture	75 days
Seasonal snow cover	145 days
Glaciers	40 years
Large lakes	100 years
Shallow ground water	200 years
Deep ground water	10,000 years

Treating a Contaminated Aquifer

The treatment, or **remediation**, of a contaminated aquifer commonly occurs in a series of steps.

Eliminating the Source

The first step in treating an aquifer is to eliminate the pollution source so additional contaminants do not escape. If an underground tank is leaking, the remaining liquid in the tank can be pumped out and the tank dug from the ground. If a factory is discharging toxic

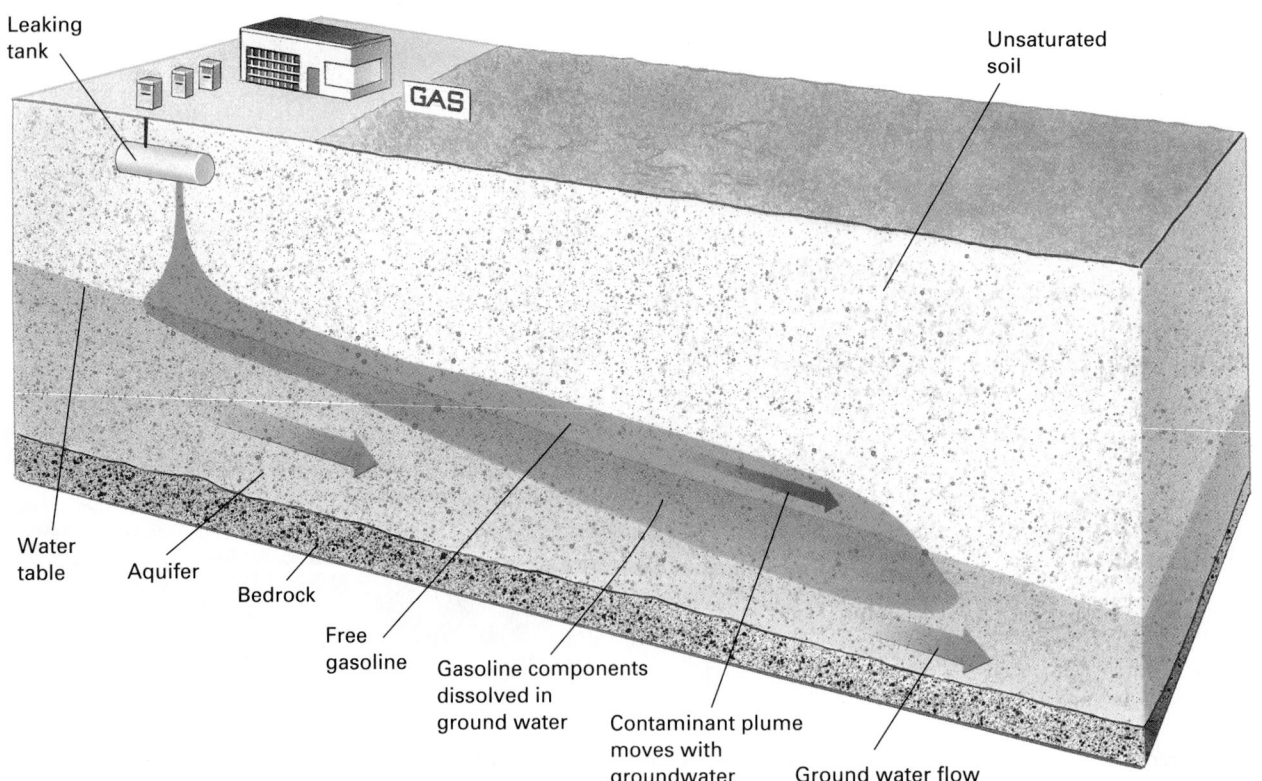

FIGURE 12.23 Pollutants disperse as a plume of contamination. Gasoline and many other contaminants are lighter than water. As a result, they float and spread on top of the water table. Soluble components may dissolve and migrate with ground water.

FIGURE 12.24 An oil refinery in New Jersey. New Jersey suffers from some of the worst ground water pollution in North America as a result of heavy industry.

COURTESY OF GRAHAM R. THOMPSON/JONATHAN TURK

bioremediation The use of micro-organisms to decompose an environmental contaminant.

chemical remediation Treatment of a contaminated area by injecting it with a chemical compound that reacts with the pollutant to produce harmless products.

chemicals into the ground water, courts may issue an injunction ordering the factory to stop the discharge (Figure 12.24).

Elimination of the source prevents additional pollutants from entering the ground water, but it does not solve the problem posed by the pollutants that have already escaped. For example, if a buried gasoline tank has leaked slowly for years, many thousands of gallons of gas may have contaminated the underlying aquifer. Once the tank has been dug up and the source eliminated, people must deal with the gasoline in the aquifer.

Monitoring

When aquifer contamination is discovered, a hydrogeologist monitors the contaminants to determine how far, in what direction, and how rapidly the plume is moving and whether the contaminant is becoming diluted. The hydrogeologist may take samples from domestic wells. If too few wells surround a pollution source, the hydrogeologist may drill additional wells to monitor the movement of the plume through the aquifer.

Modeling

After measuring the rate at which the contaminant plume is spreading, the hydrogeologist develops a computer model to predict future dispersion of the contaminant through the aquifer. The model considers the local geologic structure, the permeability of the aquifer, directions of ground water flow, and mixing rates of ground water to predict dilution effects.

Remediation

Several processes are currently used to clean up a contaminated aquifer. Contaminated ground water can be contained by building an underground barrier to isolate it from other parts of the aquifer. If the trapped contaminant does not decompose by natural processes, hydrogeologists drill wells into the contaminant plume and pump the water to the surface, where it is treated to destroy the pollutant.

Bioremediation uses microorganisms to decompose a contaminant. Specialized microorganisms can be fine-tuned by genetic engineers to destroy a particular contaminant without damaging the ecosystem. Once a specialized microorganism is developed, it is relatively inexpensive to breed it in large quantities. The microorganisms are then pumped into the contaminant plume, where they attack the pollutant. When the contaminant is destroyed, the microorganisms run out of food and die, leaving a clean aquifer. Bioremediation can be among the cheapest of all cleanup procedures.

Chemical remediation is similar to bioremediation. If a chemical compound reacts with a pollutant to produce harmless products, the compound can be injected into an aquifer to destroy contaminants. Common reagents used in chemical remediation include oxygen and dilute acids and bases. Oxygen may react with a pollutant directly or provide an environment favorable for microorganisms, which then degrade the pollutant. Thus, we can sometimes reduce contamination simply by pumping air into the ground. Acids or bases neutralize certain contaminants or precipitate dissolved pollutants.

Reclamation teams can also dig up the entire contaminated portion of an aquifer. The contaminated soil is treated by incineration or with chemical processes to destroy the pollutant. The treated soil is then returned to fill the hole. This process is prohibitively expensive, however, and is used only in extreme cases.

Yucca Mountain Controversy

In December 1987, the U.S. Congress chose a site near Yucca Mountain, Nevada, about 175 kilometers from Las Vegas, as the national burial ground for all spent reactor fuel unless sound environmental objections were found. Since that time, scientists have spent $4 billion testing and analyzing the area, making Yucca Mountain probably the most closely studied piece of real estate in the history of geology.

The Yucca Mountain site is located in the Basin and Range Province, a region noted for faulting and volcanism related to ongoing tectonic extension of that part of the western United States. Bedrock at the Yucca Mountain site is welded tuff, a hard volcanic rock. The tuffs erupted from several volcanoes that were active from 16 to 6 million years ago. The last eruption near Yucca Mountain occurred 15,000 to 25,000 years ago. In addition, geologists have mapped 32 faults that have moved during the past 2 million years adjacent to the Yucca Mountain site. The site itself is located within a structural block bounded by parallel faults (Figure 12.25). On July 29, 1992, a magnitude 5.6 earthquake occurred about 20 kilometers from the Yucca Mountain site and damaged the Department of Energy (DOE) facilities on the site. Critics of the Yucca Mountain site argue that recent earthquakes and volcanoes prove that the area is geologically active.

The site's environment is desert dry, and the water table lies 550 meters beneath the surface. The repository will consist of a series of tunnels and caverns dug into the tuff 300 meters beneath the surface and 250 meters above the water table. Thus, it is designed to isolate the waste from both surface water and ground water. However, it is impossible to predict the climatological changes that may occur over the next 10,000 years. If the climate becomes appreciably wetter, the water table may rise or surface water could percolate downward. In addition, if an earthquake fractures rocks beneath the site, then contaminated ground water might disperse more rapidly than predicted. Furthermore, critics point out that construction of the repository will involve blasting and drilling, and these activities could fracture underlying rock, opening conduits for flowing water.

After decades of political wrangling, Congress approved the Yucca Mountain site in the summer of 2002. In July 2004 the EPA was forced to change its standard for the time period necessary to safeguard the nuclear material from leaking into the environment from 10,000 to 1 million years. The site could open as early as 2010. But the debate continues. Supporters of the Yucca Mountain repository argue that we need nuclear power and that therefore as a society we must accept a certain level of risk. They argue further that at present, 45,000 tons of high-level radioactive waste lie in 126 unstable temporary storage facilities across the United States and that the Yucca Mountain site is safer than the temporary storage sites now being used. Opponents argue that the site is geologically unsound and that new alternatives must be found.

In November, 2009, the Obama administration announced plans to stop the pursuit of a license for the planned Yucca Mountain nuclear waste repository. Whether this is the end of the project or yet another twist in a long saga, remains to be seen. For timely updates on the politics of this contentious issue, see: http://www.yuccamountain.org/new.htm

12.9 Nuclear Waste Disposal

In a nuclear reactor, radioactive uranium nuclei split into smaller nuclei, many of which are also radioactive. Most of these radioactive waste products are useless and must be disposed of without exposing people to the radioactivity. In the United States, military processing plants, 104 commercial nuclear reactors, and numerous laboratories and hospitals generate more than 3,000 tons of high-level radioactive wastes every year.

Chemical reactions cannot destroy radioactive waste because radioactivity is a nuclear process, and atomic nuclei are unaffected by chemical reactions. Therefore, the only feasible method for disposing of radioactive wastes is to store them in a place safe from geologic hazards and human intervention and to allow them to decay naturally. The U.S. Department of Energy defines a permanent repository as one that will isolate radioactive wastes for 10,000 years. However, this number is totally arbitrary, based on the fact that 10,000 years is so far away that most people won't worry about conditions that far in the future. Actually, radioactive wastes will remain harmful for 1 million years because of the long half-lives of some radioactive isotopes. As a result, radioactive wastes could become harmful to human or nonhuman ecosystems over the course of geologic time.

For a repository to keep radioactive waste safely isolated for long periods of time, it must meet at least three geologic criteria:

1. It must be safe from disruption by earthquakes and volcanic eruptions.
2. It must be safe from landslides, soil creep, and other forms of mass wasting.
3. It must be free from floods and seeping ground water that might corrode containers and carry wastes into aquifers.

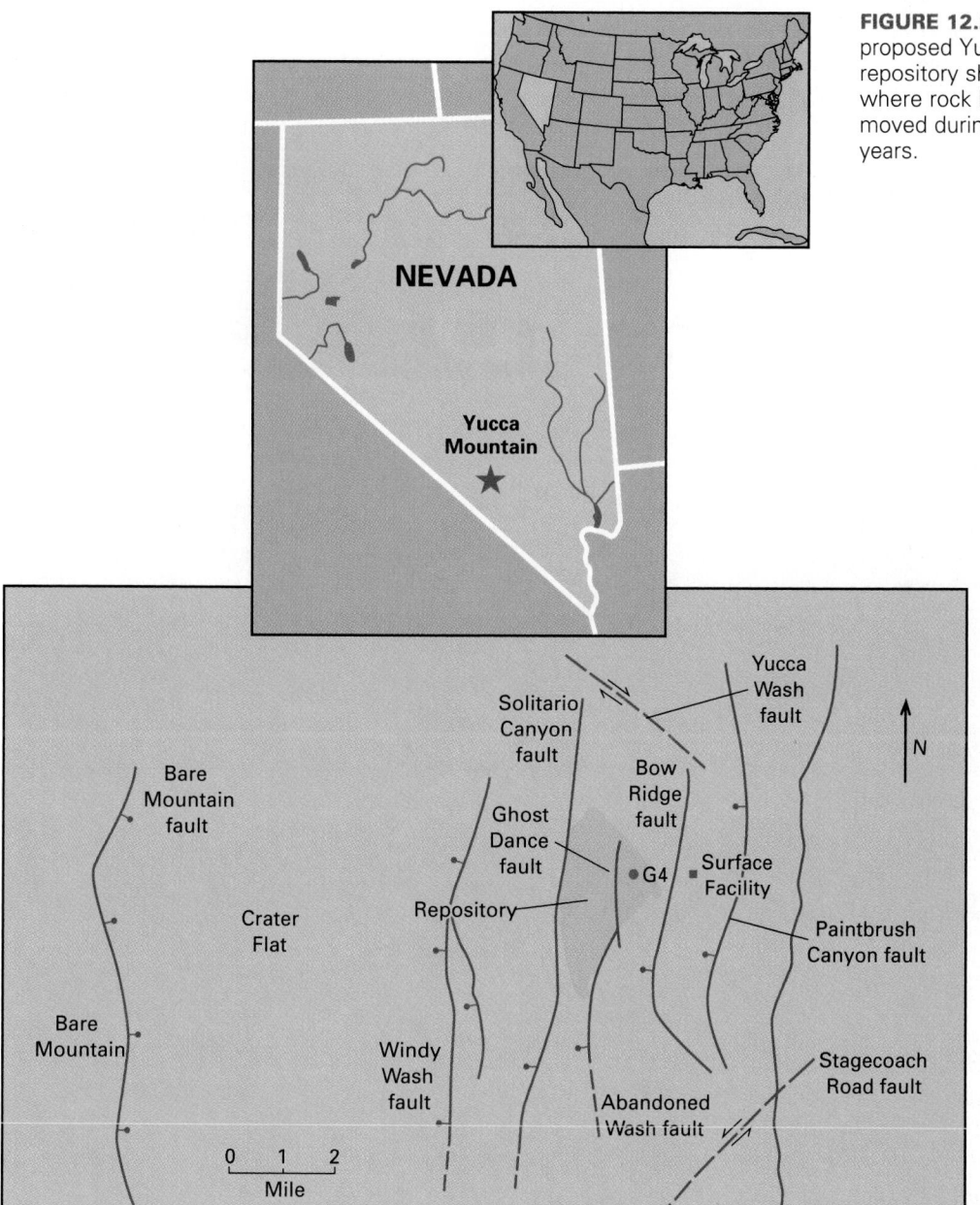

FIGURE 12.25 A map of the proposed Yucca Mountain repository shows numerous faults where rock has fractured and moved during the past 2 million years.

12.10 The Clean Water Act: A Modern Perspective

The Clean Water Act set an ambitious agenda for cleaning the nation's rivers, lakes, and wetlands. Have we achieved our goals? The good news is that municipalities no longer dump raw sewage and factories no longer discharge untreated waste directly into our waterways. The bad news is that many of our streams remain polluted.

In 2000, the EPA reported that 40 percent of the nation's streams, 45 percent of lakes, and 50 percent of estuaries (where a stream or river meets the sea) do not meet the standards of the Clean Water Act. Forty percent of the country's freshwater remained unsafe for human use. According to a 2004 EPA study, one-third of lakes and one-fourth of the nation's rivers contain fish that are contaminated with mercury, dioxin, and PCBs at levels considered dangerous for human consumption.

13

GLACIERS AND ICE AGES

Matanuska Glacier in Alaska is a prime example of an alpine glacier. The glacial front is 2 miles long.

visit **4ltrpress.cengage.com**

{ **Glaciers are delicate and individual things, like humans. Instability is built into them.** }

Will Harrison

> **glacier** A massive, long-lasting accumulation of compacted snow and ice that forms on land and moves downslope or spreads outward under its own weight.

We often think of glaciers as features of high mountains and the frozen polar regions, yet anyone living in the northern third of the United States is familiar with landscapes created by a vast ice sheet that covered this region 18,000 years ago. The low, rounded hills of upper New York State, Wisconsin, and Minnesota are composed of gravel deposited by the ice. In addition, people in this region swim and fish in lakes created by those glaciers.

Numerous times during Earth's history, glaciers grew to cover large parts of Earth and then melted away. Before the most recent major glacial advance, beginning about 100,000 years ago, the world was free of ice except for high mountains and the polar ice caps of Antarctica and Greenland. Then, in a relatively short time—perhaps only a few thousand years—Earth's climate cooled. As winter snow failed to melt completely in summer, the polar ice caps spread into lower latitudes. At the same time, glaciers formed near mountain summits, even near the equator. They flowed down mountain valleys into nearby lowlands. When the glaciers reached their maximum size 18,000 years ago, they covered one-third of Earth's continents.

Then about 15,000 years ago, Earth's climate warmed again and the glaciers began to melt rapidly. Although 18,000 years is a long time when compared with a single human lifetime, it is but an eyeblink in geologic time. In fact, humans, who have been on this planet for about 100,000 years, lived through the most recent glaciation. In south-

west France and northern Spain, humans developed sophisticated spearheads and carved body ornaments between 40,000 and 30,000 years ago. People first began experimenting with agriculture about 12,000 to 10,000 years ago.

13.1 Formation of Glaciers

In most temperate regions, winter snow melts completely in spring and summer. However, in certain cold, wet environments, some of the winter snow remains unmelted during the summer and accumulates year after year. During summer, the snow crystals become rounded and denser as the snowpack is compressed and alternately warmed during daytime and cooled at night. If snow survives through one summer, it converts to rounded ice grains called *firn*. Mountaineers like firn because the sharp points of their ice axes and crampons sink into it easily and hold firmly. If firn is buried deeper in the snowpack, it converts to glacial ice, which consists of closely packed ice crystals (Figure 13.1).

A **glacier** is a massive, long-lasting, moving mass of compacted snow and ice. Glaciers form only on land, wherever the amount of snow that falls in winter exceeds the amount that melts in summer. Because ice under pressure is plastic and deforms easily, mountain glaciers flow downhill. Glaciers on level land flow outward under their own weight.

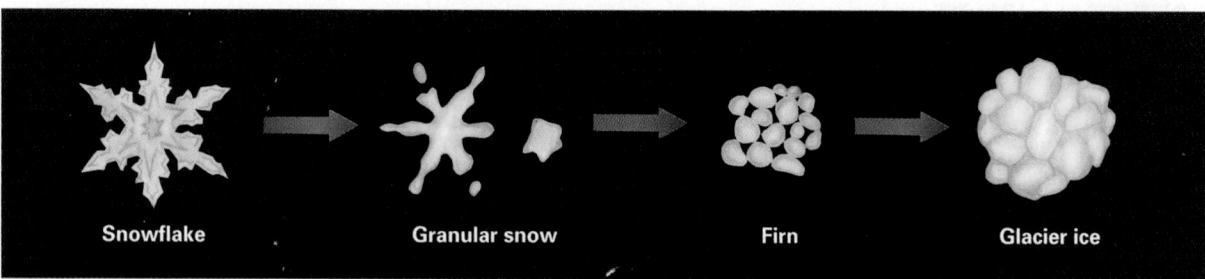

FIGURE 13.1 Newly fallen snow changes through several stages to form glacier ice.

Glaciers form in two environments. Alpine glaciers form at all latitudes on high, snowy mountains. Continental ice sheets form at all elevations in the cold polar regions.

Alpine Glaciers

Mountains are generally colder and wetter than adjacent lowlands. Near the summits, winter snowfall is deep and summers are short and cool. These conditions create **alpine glaciers** (Figure 13.2). Alpine glaciers exist on every continent—in the Arctic and Antarctica, in temperate regions, and in the tropics. Glaciers cover the summits of Mount Kenya in Africa and Mount Cayambe in South America, even though both peaks are near the equator.

Some alpine glaciers flow great distances from the peaks into lowland valleys. For example, the Kahiltna Glacier, which flows down the southwest side of Denali (Mount McKinley) in Alaska, is about 65 kilometers long, 12 kilometers across at its widest point, and about 700 meters thick. Although most alpine glaciers are smaller than the Kahiltna, some are larger.

The growth of an alpine glacier depends on both temperature and precipitation. The average annual temperature in the state of Washington is warmer than in Montana, yet alpine glaciers in Washington are larger and flow to lower elevations than those in Montana. Winter storms buffet Washington from the moisture-laden Pacific. Consequently, Washington's mountains receive such heavy winter snowfall that even though summer melting is rapid, snow accumulates every year. In much drier Montana, snowfall is light enough that most of it melts in the summer, and thus Montana's mountains have only a few small glaciers.

Continental Glaciers

In polar regions, winters are so long and cold and summers so short and cool that glaciers cover most of the land regardless of its elevation. An **ice sheet**, or **continental glacier**, covers an area of 50,000 square kilometers or more. The ice spreads outward in all directions under its own weight.

Today, Earth has only two ice sheets, one in Greenland and the other in Antarctica. These two ice sheets contain 99 percent of the world's ice and about three-fourths of Earth's freshwater. The Greenland sheet is more than 2.7 kilometers thick in places and covers 1.8 million square kilometers. Yet it is small compared with the Antarctic ice sheet, which blankets

FIGURE 13.2 This alpine glacier flows around granite peaks in British Columbia, Canada.

COURTESY OF GRAHAM R. THOMPSON/JONATHAN TURK

about 13 million square kilometers, almost 1.5 times the size of the United States. The Antarctic ice sheet covers entire mountain ranges, and the mountains that rise above its surface are islands of rock in a sea of ice.

Whereas the South Pole lies in the interior of the Antarctic continent, the North Pole is situated in the Arctic Ocean. At the North Pole, only a few meters of ice freeze on the relatively warm sea surface, and the ice fractures and drifts with the currents. As a result, no ice sheet exists there.

13.2 Glacial Movement

The rate of glacier movement varies with slope steepness, precipitation, and air temperature. In the coastal ranges of southeast Alaska, where annual precipitation is high and average temperature is relatively warm (for glaciers), some glaciers move 15 centimeters to 1 meter per day. In contrast, in the interior of Alaska, where conditions are colder and drier, glaciers move only a few centimeters per day. At these rates, ice can flow the length of an alpine glacier in a few hundred to a few thousand years. In some instances, a glacier may surge at a speed of 10 to 100 meters per day.

Glaciers move by two mechanisms: *basal slip* and *plastic flow*. In **basal slip**, the entire glacier slides over bedrock in the same way that a bar of soap slides down a tilted board. Just as wet soap slides more easily than dry soap, water between bedrock and the base of a glacier accelerates basal slip.

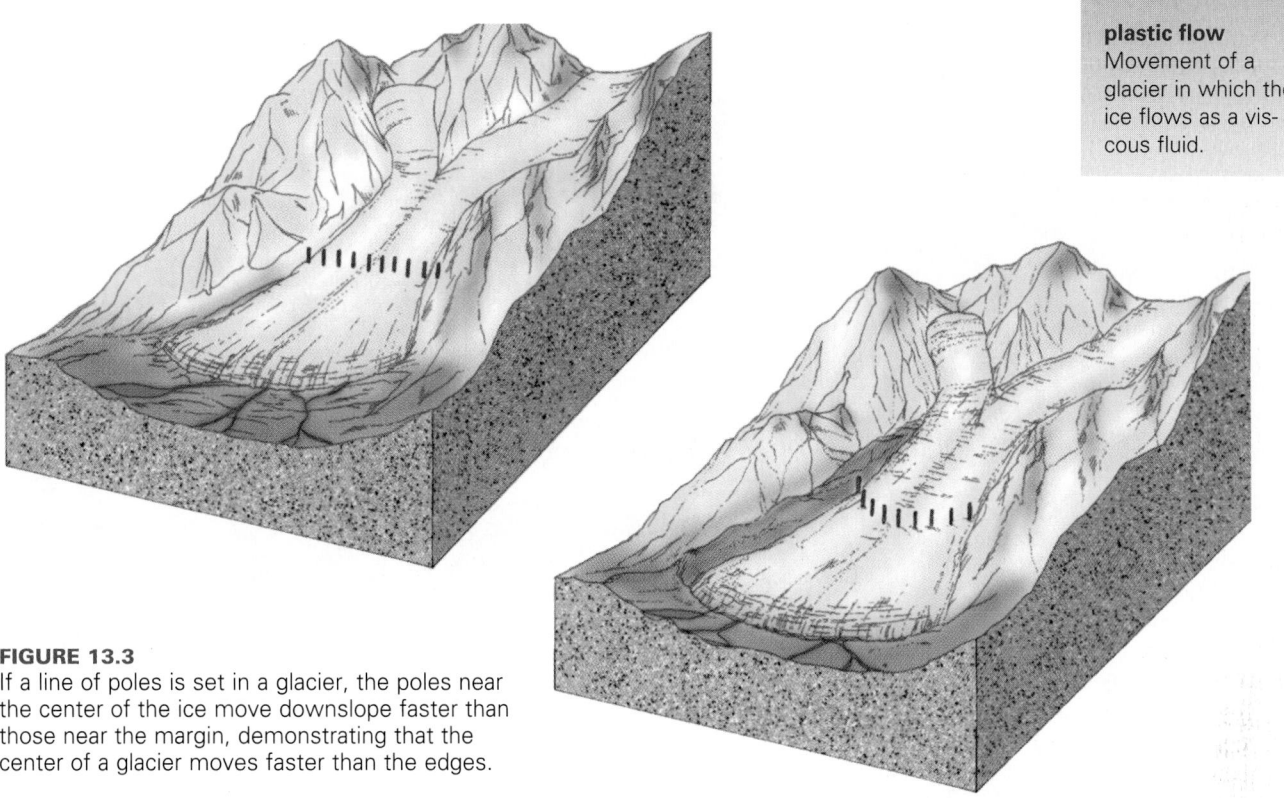

FIGURE 13.3
If a line of poles is set in a glacier, the poles near the center of the ice move downslope faster than those near the margin, demonstrating that the center of a glacier moves faster than the edges.

Several factors cause water to accumulate near the base of a glacier. Earth's heat melts ice near bedrock. Friction from glacial movement also generates heat. Water occupies less volume than an equal amount of ice. As a result, pressure from the weight of overlying ice favors melting. Finally, during the summer, water melted from the surface of a glacier may seep downward to its base.

A glacier also moves by **plastic flow**, in which the ice flows as a viscous fluid. Plastic flow is demonstrated by two experiments. In one, scientists set a line of poles in the ice (Figure 13.3). After a few years, the ice moved downslope so that the poles formed a U-shaped array. This experiment shows that the center of the glacier moves faster than the edges. Frictional resistance with the valley walls slows movement along the edges and glacial ice flows plastically, allowing the center to move faster than the sides.

In another experiment, scientists drove a straight, flexible pipe downward into a glacier to study the flow of ice at depth (Figure 13.4). At a later date, they discovered that the entire pipe moved downslope and also became bent. At the surface of a glacier, the ice is brittle, like an ice cube or the ice found on the surface of a lake. In contrast, at depths greater than about 40 meters, the pressure is sufficient to allow ice to deform plastically. The curvature in the pipe shows that the ice moved plastically and that middle levels of the glacier moved faster than the lower part. The base of

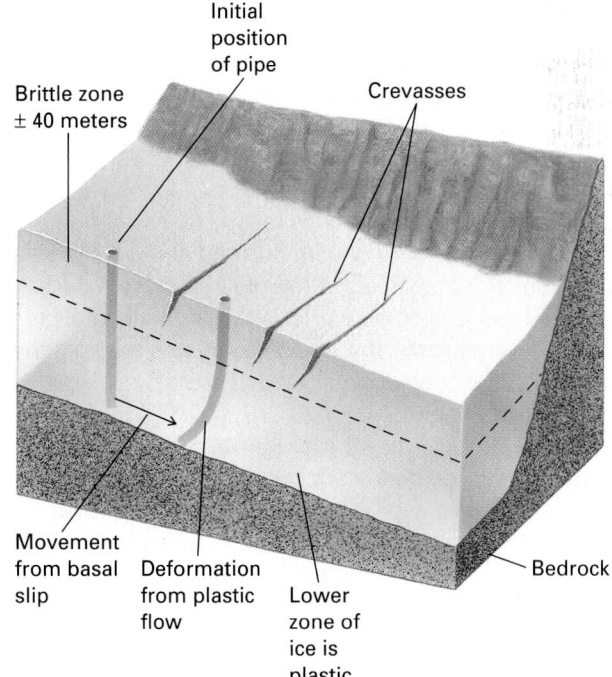

FIGURE 13.4 In this experiment, a pipe was driven through a glacier until it reached bedrock. The entire pipe moved downslope with the ice but also became curved. The pipe became curved because friction with bedrock slowed movement of the bottom of the glacier. Middle layers of ice flowed more rapidly because there the ice is plastic. At a depth shallower than 40 meters, the ice does not flow plastically, and the pipe in that zone remained straight.

crevasse A fracture or crack in the brittle upper 40 meters of a glacier, formed when the glacier flows over uneven bedrock.

ice fall A section of a glacier consisting of numerous crevasses and towering ice pinnacles.

zone of accumulation The higher-elevation upper end of an alpine glacier, where more snow falls in winter than melts in summer and snow accumulates from year to year, and where the glacier's surface is covered by snow year-round.

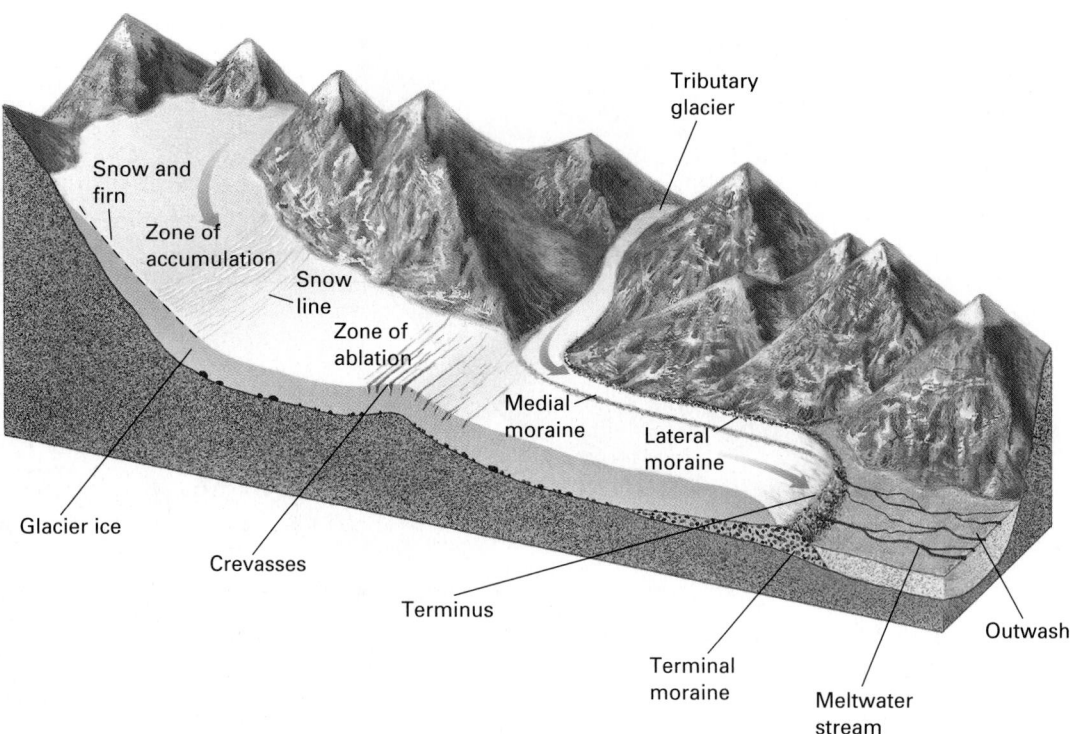

FIGURE 13.5 A schematic view of an alpine glacier. Crevasses form in the upper, brittle zone of a glacier where the ice flows over uneven bedrock.

the glacier is slowed by friction against bedrock, so it moved more slowly than the plastic portion above it.

The relative rates of basal slip and plastic flow depend on the steepness of the bedrock underlying the glacier and the thickness of the ice. A small alpine glacier on steep terrain moves mostly by basal slip. In contrast, the bedrock beneath portions of the Antarctica and Greenland ice sheets is relatively level, so the ice has no slope to slide down. Thus these continental glaciers are huge plastic masses of ice (with a thin, brittle cap) that ooze outward, mainly under the forces created by their own weight.

When a glacier flows over uneven bedrock, the deeper plastic ice bends and flows over bumps, while the brittle, upper layer stretches and cracks, forming a **crevasse** (Figures 13.5 and 13.6). Crevasses form only in the brittle, upper 40 meters of a glacier, not in the lower plastic zone. Crevasses open and close slowly as a glacier moves. An **ice fall** is a section of a glacier consisting of crevasses and towering ice pinnacles. The pinnacles form where ice blocks break away from the crevasse walls and rotate as the glacier moves. With crampons, ropes, and ice axes, a skilled mountaineer might climb into a crevasse. The walls are a pastel blue, and sunlight filters through the narrow opening above. The ice shifts and cracks, making creaking sounds as the glacier advances. Many mountaineers have been crushed by falling ice while traveling through ice falls.

COURTESY OF GRAHAM R. THOMPSON/JONATHAN TURK

FIGURE 13.6 Crevasses in the Bugaboo Mountains of British Columbia.

The Mass Balance of a Glacier

Consider an alpine glacier flowing from the mountains into a valley. At the upper end of the glacier, snowfall is heavy, temperatures are below freezing for much of the year, and avalanches carry large quantities of snow from the surrounding peaks onto the ice. There, more snow falls in winter than melts in summer, and snow accumulates from year to year. This higher-elevation part of the glacier is called the **zone of accumulation**. There the glacier's surface is covered by snow year-round.

Unit 3: Surface Processes

Lower in the valley, the temperature is higher throughout the year, and less snow falls. This lower part of a glacier, where more snow melts in summer than accumulates in winter, is called the **zone of ablation**. When the snow melts, a surface of old, hard, glacial ice is left behind. The **snow line** is the boundary between permanent snow and seasonal snow—in other words, between the zone of accumulation and the zone of ablation. The snow line shifts up and down the glacier from year to year, depending on weather. Ice exists in the zone of ablation because the glacier flows downward from the accumulation area. Even farther down the valley, the rate of glacial flow cannot keep pace with melting, so the glacier ends at its **terminus**.

Glaciers grow and shrink. If annual snowfall increases or average temperature drops, more snow accumulates; then the snow line of an alpine glacier descends to a lower elevation, and the glacier grows thicker. At first the terminus may remain stable, but eventually it advances farther down the valley. The lag time between a change in climate and a glacial advance may range from a few years to several decades, depending on the size of the glacier, its rate of motion, and the magnitude of the climate change. If annual snowfall decreases or the climate warms, the zone of accumulation shrinks and the glacier retreats.

When a glacier retreats, its ice continues to flow downhill, but the terminus melts back faster than the glacier flows downslope. In Glacier Bay, Alaska, glaciers have retreated 60 kilometers in the past 125 years. Near the terminus, newly exposed rock is bare and lifeless.

A few kilometers from the glacier, where rock has been exposed for a few decades, scattered lichens grow on otherwise bare rock. Even farther away, seabird droppings have mixed with windblown silt and weathered rock to form thin soil that supports mosses in sheltered cracks. Near the head of the bay, 60 kilometers from the terminus, tidal currents and ocean storms have washed enough sediment over the glaciated land to create soil and support stunted trees.

In equatorial and temperate regions, glaciers commonly terminate at an elevation of 3,000 meters or higher. However, in a cold, wet climate, a glacier may extend into the sea (Figure 13.7). Giant chunks of ice break off, forming **icebergs**.

The largest icebergs in the world are those that break away from the Antarctic ice shelf. Between the years 2000 and 2002, two plates of ice the size of Connecticut and at least one the size of Rhode Island broke free from the West Antarctic Ice Sheet and floated into the Antarctic Ocean. The tallest icebergs in the world break away from tidewater glaciers in Greenland and are about 300 to 400 meters thick.

zone of ablation The lower-altitude part of an alpine glacier, where more snow melts in summer than accumulates in winter, and where the melting snow leaves behind a surface of old, hard, glacial ice.

snow line The boundary between permanent snow and seasonal snow.

terminus The end, or foot, of a glacier.

iceberg A large chunk of ice that breaks from a glacier into a body of water.

FIGURE 13.7 A kayaker paddles among small icebergs that calved from the LeConte Glacier in Alaska.

13.3 Glacial Erosion

Rock at the base and sides of a glacier may have been fractured by tectonic forces, frost wedging, or pressure-release fracturing. The moving ice dislodges the loosened rock (Figure 13.8). Ice is viscous enough to pick up and carry particles of all sizes, from silt-sized grains to house-sized boulders.

FIGURE 13.8 (A) A glacier plucks rocks from bedrock and then drags them along, abrading both the loose rocks and the bedrock. (B) These crescent-shaped depressions in granite at LeConte Bay, Alaska, were formed by glacial plucking. They indicate that ice flow along this steep wall was from right to left.

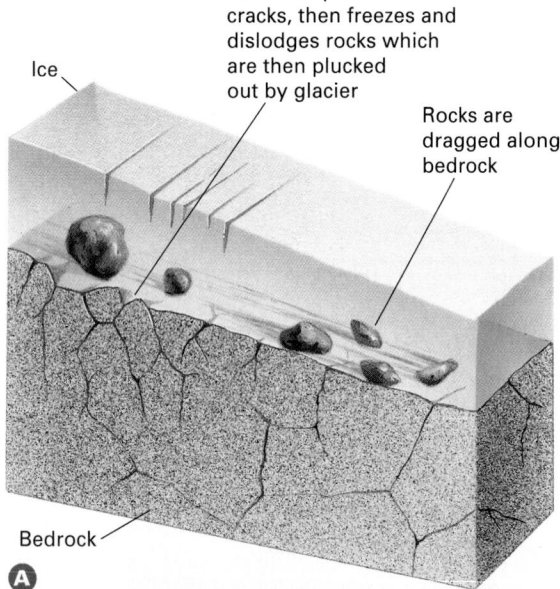

Water seeps into cracks, then freezes and dislodges rocks which are then plucked out by glacier

Ice

Rocks are dragged along bedrock

Bedrock

Ⓐ

COURTESY OF GRAHAM R. THOMPSON/JONATHAN TURK

Ⓑ

COURTESY OF GRAHAM R. THOMPSON/JONATHAN TURK

FIGURE 13.9 Stones embedded in the base of a glacier gouged these striations in bedrock in British Columbia.

Thus glaciers erode and transport huge quantities of rock and sediment.

Ice itself is not abrasive to bedrock because it is too soft. However, rocks embedded in the ice scrape across bedrock, cutting deep, parallel grooves and scratches called **glacial striations** (Figure 13.9). When glaciers melt and striated bedrock is exposed, the markings show the direction of ice movement. Glacial striations are used to map the flow directions of glaciers.

Erosional Landforms Created by Alpine Glaciers

Let's take an imaginary journey through a mountain range that was glaciated in the past but is now mostly ice free (Figure 13.10). We start with a helicopter ride to the summit of a high, rocky peak. Our first view from the helicopter is of sharp, jagged mountains rising steeply above smooth, rounded valleys carved by glaciers.

We've already seen that mountain streams in unglaciated regions commonly erode downward into their beds, cutting steep-sided, V-shaped valleys (Chapter 11). A glacier, however, is not confined to a narrow streambed but instead fills its entire valley. As a result, it scours the sides of the valley as well as the bottom, carving a broad, rounded, **U-shaped valley** (Figures 13.10 and 13.11).

We land on one of the peaks and step out of the helicopter. Beneath us, a steep cliff drops off into a spoon-shaped depression in the mountainside called a **cirque**. A small glacier at the head of the cirque reminds

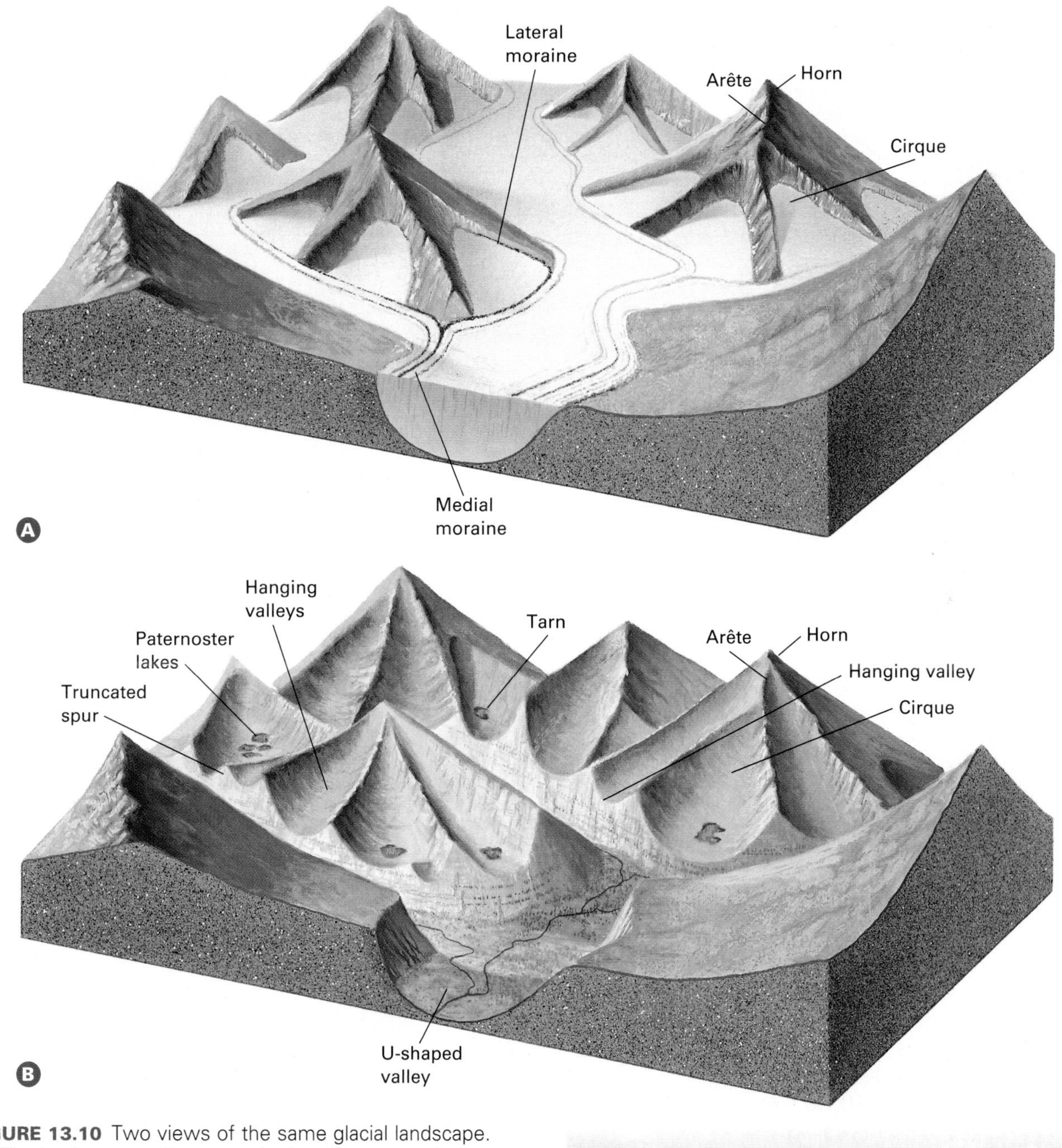

Lateral
moraine

Arête Horn

Cirque

Medial
moraine

A

Hanging
valleys

Tarn

Paternoster
lakes

Arête Horn

Hanging valley

Truncated
spur

Cirque

U-shaped
valley

B

FIGURE 13.10 Two views of the same glacial landscape.
(A) The landscape as it appeared when it was mostly
covered by glaciers. (B) The same landscape as it appears
now, after the glaciers have melted.

us of the larger mass of ice that existed in a colder,
wetter time (Figure 13.12A).

To understand how a glacier creates a cirque, imagine a gently rounded mountain. As snow accumulates
and a glacier forms, the ice flows down the mountainside (Figure 13.12B). The ice erodes a small depression
that grows slowly as the glacier flows (Figure 13.12C).
With time, the cirque walls become steeper and higher.
The glacier carries the eroded rock from the cirque to
lower parts of the valley (Figure 13.12D). When the

FIGURE 13.11 A U-shaped valley in the Purcell Mountains
of British Columbia.

tarn A small lake at the base of a cirque.

paternoster lakes A series of lakes in a glacial valley, strung out like beads and connected by short streams and waterfalls.

horn A sharp, pyramid-shaped rock summit where three or more cirques intersect near the summit.

arête A sharp, narrow rib of rock forming a border between adjacent valleys or between two cirques, created when two alpine glaciers moved along opposite sides of the mountain ridge and eroded both sides.

hanging valley A small glacial valley lying high above the floor of the main valley.

Basin formed by glacial weathering and erosion

Exposed rocks dislodged and transported by weathering and erosion

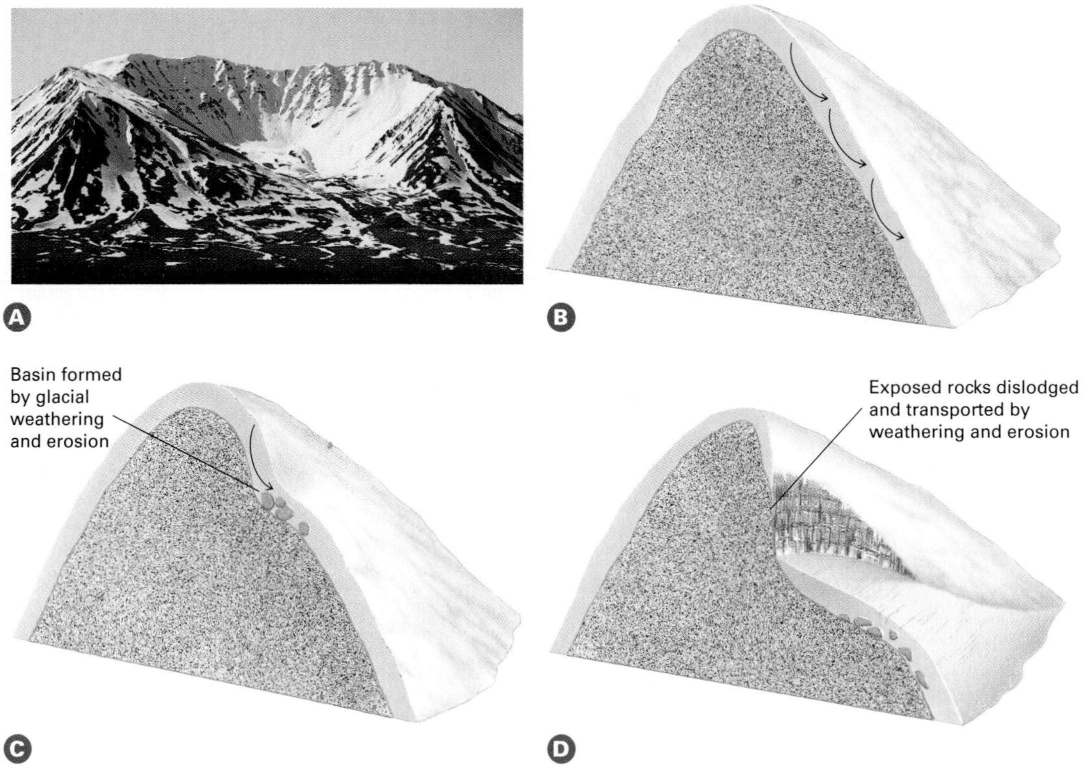

FIGURE 13.12 (A) A glacier eroded this concave cirque into a mountainside in the Mat-Su Valley, just south of the Alaska Range near Anchorage, Alaska. (B) To form a cirque, snow accumulates, and a glacier begins to flow from the summit of a peak, shown here in cross section. (C) Glacial weathering and erosion form a small depression in the mountainside. (D) Continued glacial movement enlarges the depression. When the glacier melts, it leaves a cirque carved in the side of the peak, as in the photograph.

glacier finally melts, it leaves a steep-walled, rounded cirque.

Streams and lakes are common in glaciated mountain valleys. As a cirque forms, the glacier may erode a depression into the bedrock beneath it. When the glacier melts, this depression fills with water, forming a small lake, or **tarn**, nestled at the base of the cirque. If we hike down the valley below the high cirques, we may encounter a series of lakes called **paternoster lakes**, which are commonly connected by rapids and waterfalls (Figures 13.10 and 13.13). Paternoster lakes are a sequence of small basins plucked out by a glacier. The term *paternoster* refers to a string of rosary beads and evokes the image of a string of lakes in a glacial valley. When the glacier recedes, the basins fill with water.

If glaciers erode three or more cirques into different sides of a peak, they may create a steep, pyramid-shaped rock summit called a **horn**. The Matterhorn in the Swiss Alps is a famous horn (Figures 13.10 and 13.14). Two alpine glaciers flowing along opposite sides of a mountain ridge may erode both sides of the ridge, forming a sharp, narrow rib of rock that lies approxi-

mately perpendicular to the main ridge. This feature, called an **arête** often forms a border between adjacent valleys or cirques.

Looking downward from our peak, we may see a waterfall pouring from a small, high valley into a larger, deeper one. A small glacial valley lying high above the floor of the main valley is called a **hanging valley** (Figures 13.10 and 13.15). The famous waterfalls of Yosemite Valley in California cascade from hanging valleys. A hanging valley forms where a small tributary glacier joined a much larger one. The tributary glacier eroded a shallow valley while the massive main glacier gouged a deeper one. When the glaciers

FIGURE 13.13 Glaciers eroded bedrock to form this string of paternoster lakes in the Sierra Nevada.

FIGURE 13.14 The Matterhorn in Switzerland formed as three alpine glaciers eroded cirques into the peak from three different sides. The ridge between two cirques is called an arête.

melted, they exposed an abrupt drop where the small valley joins the main valley.

Deep, narrow inlets called **fjords** extend far inland on many high-latitude seacoasts. Most fjords are glacially carved valleys that were later flooded by encroaching seas as the glaciers melted (Figure 13.16).

fjord A deep, narrow, glacially carved valley on a high-latitude seacoast that was later flooded by encroaching seas as the glaciers melted.

FIGURE 13.15 Yosemite Falls cascades from a hanging valley in Yosemite National Park.

FIGURE 13.16 A steep-sided fjord bounded by 1,000-meter-high cliffs on Baffin Island, Canada.

Erosional Landforms Created by a Continental Glacier

A continental glacier erodes the landscape just as an alpine glacier does. However, a continental glacier is considerably larger and thicker and is not confined to a valley. As a result, it covers vast regions, often including entire mountain ranges. The recent ice sheets have scoured deep basins that have filled with water to form the Great Lakes and the Finger Lakes in New York (Figure 13.17).

13.4 Glacial Deposits

In the 1800s, geologists recognized that the large deposits of sand and gravel found in the Alps and other places had been transported from distant sources. A popular hypothesis at the time explained that this material had drifted in on icebergs during catastrophic floods. The deposits were called *drift* after this inferred mode of transport.

Today we know that continental glaciers covered vast parts of the land only 10,000 to 20,000 years ago and that these glaciers carried and deposited drift. Although the term *drift* is a misnomer, it remains in common use. Now geologists define **drift** as all rock or sediment transported and deposited by a glacier. Glacial drift averages 6 meters thick over the rocky hills and pastures of New England and 30 meters thick over the plains of Illinois.

Drift is divided into two categories. **Till** was deposited directly by glacial ice. **Stratified drift** was first carried by a glacier and then transported and deposited by a stream.

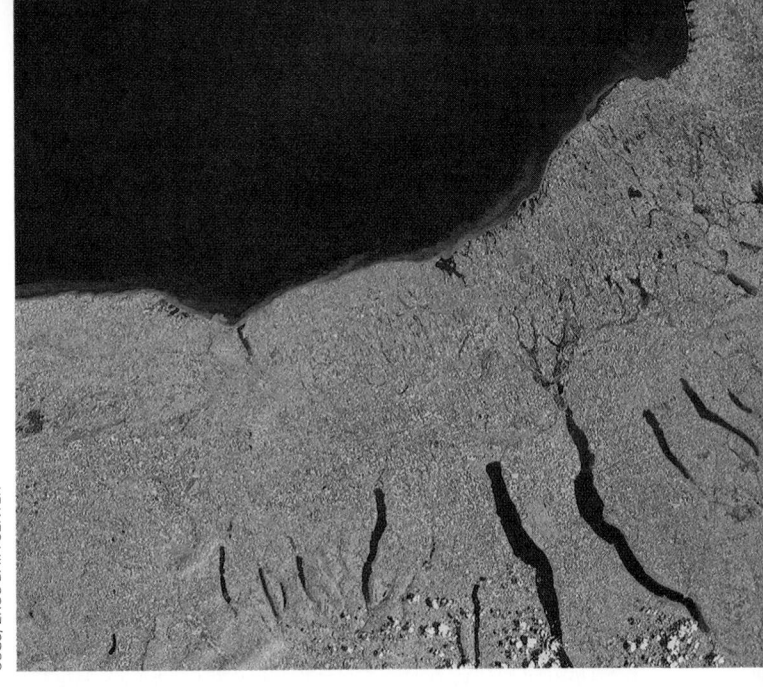

USGS, EROS DATA CENTER

FIGURE 13.17 This satellite view shows the Finger Lakes (bottom right), which formed when Pleistocene glaciers scoured deep valleys from preexisting stream valleys and the deepened valleys filled with water. Pleistocene glaciers also formed Lake Ontario (top).

Landforms Composed of Till

Ice is so much more viscous than water that it carries a wide range of particle sizes. When a glacier melts, it deposits particles of all sizes—from fine clay to huge boulders—in an unsorted, unstratified mass (Figure 13.18). Within a glacier, each rock or grain of sediment is protected by the ice that surrounds it. Therefore, the pieces do not rub against one another, and glacial transport does not round sediment as a stream does. If you find rounded gravel in till, it became rounded by a stream before the glacier picked it up.

Occasionally, large boulders lie on the surface in country that was once glaciated. In many cases the boulders are of a rock type different from the bedrock in the immediate vicinity. Boulders of this type are called **erratics** and were transported to their present locations by a glacier. The origins of erratics can be determined

© CARLOS AMEGLIO/SHUTTERSTOCK

FIGURE 13.18 Unsorted glacial till of all sizes. The large cobbles are mixed in with smaller sediments. The cobbles were rounded by stream action before being transported by the Matanuska Glacier in Alaska.

by exploring the terrain in the direction the glacier came from until the parent rock is found. Some erratics were carried 500 or even 1,000 kilometers from their points of origin and provide clues to the movement of glaciers.

Moraines

A **moraine** is a mound or a ridge of till. Think of a glacier as a giant conveyor belt. An old-fashioned airport conveyor belt simply carries suitcases to the end of the belt and dumps them in a pile. Similarly, a glacier carries sediment and deposits it at its terminus. If a glacier is neither advancing nor retreating, its terminus may remain in the same place for years. During that time, sediment accumulates at the terminus to form a ridge called an **end moraine** (Figure 13.19). An end moraine that forms when a glacier is at its greatest advance, before beginning to retreat, is called a **terminal moraine** (refer again to Figure 13.5).

If warmer conditions prevail, the glacier recedes. If the glacier then stabilizes again during its retreat and the terminus remains in the same place for a sufficient amount of time, a new end moraine, called a **recessional moraine**, forms.

When a glacier recedes steadily, till is deposited in a relatively thin layer over a broad area, forming a **ground moraine**. Ground moraines fill old stream channels and other low spots. Often this leveling process disrupts drainage patterns. Many of the swamps in the northern Great Lakes region and northern New England lie on ground moraines formed when the most recent continental glaciers receded.

End moraines and ground moraines are characteristic of both alpine and continental glaciers. An end moraine deposited by a large alpine glacier may extend for several kilometers and be so high that even a person in good physical condition would have to climb for an hour to reach the top. Moraines may be dangerous to hike over if their sides are steep and the till is

FIGURE 13.19 The end moraine of an alpine glacier on Baffin Island, Canada, in midsummer. Dirty, old ice forms the lower part of the glacier below the snow line, and clean snow lies higher up on the ice in the zone of accumulation.

moraine A mound or ridge of till deposited directly by glacial ice.

end moraine A ridge of till that forms at the end, or terminus, of a glacier that is neither advancing nor retreating and whose terminus has remained in the same place for years.

terminal moraine An end moraine that forms when a glacier is at its greatest advance before beginning to retreat.

recessional moraine A moraine that forms at the new terminus of a glacier as the glacier stabilizes temporarily during retreat.

ground moraine The moraine formed when a glacier recedes steadily and deposits till in a relatively thin layer over a broad area.

lateral moraine A ridge-like moraine that forms from sediment on or adjacent to the sides of a mountain glacier.

medial moraine A moraine formed in or on the middle of a glacier by the merging of lateral moraines as two glaciers flow together.

drumlins Elongate hills, usually occurring in clusters, formed when a glacier flows over and reshapes a mound of till or stratified drift.

loose. Large boulders are mixed randomly with rocks, cobbles, sand, and clay. A careless hiker can dislodge boulders and send them tumbling to the base.

The most recent Pleistocene continental glaciers reached their maximum extent about 18,000 years ago. Their terminal moraines record the southernmost extent of those glaciers. In North America, the terminal moraines lie in a broad, undulating front extending across the northern United States. Enough time has passed since the glaciers retreated that soil and vegetation have stabilized the till and most of the hills are now covered by vegetation (Figure 13.20).

When an alpine glacier moves downslope, it erodes the valley walls as well as the valley floor. Therefore, the edges of the glacier carry large loads of sediment. Additional debris falls from the valley walls and accumulates on and near the sides of mountain glaciers. Sediment near the glacial margins forms a **lateral moraine** (Figure 13.21).

If two alpine glaciers converge, their lateral moraines merge into the middle of the resulting larger glacier. This till forms a visible dark stripe on the surface of the ice called a **medial moraine** (Figure 13.22).

FIGURE 13.21 A lateral moraine lies against the valley wall in the Bugaboo Mountains of British Columbia.

Drumlins

Elongate hills, called **drumlins**, cover parts of the northern United States and are best exposed across the rolling farmland in upstate New York. *Drumlin* is from the Old Irish for "back" or "ridge," and each one looks like a whale swimming through the ground with its back in the air. They usually occur in clusters. An individual drumlin is typically about 1 to 2 kilometers long and about 15 to 50 meters high. Drumlins are created when a glacier flows over and reshapes a mound of till; some consist partly of bedrock. In either case, the glacier generally erodes a steep-sided face as it advances. It then deposits some sediment on the downslope side to form a long, pointed slope. Thus a geologist can determine the direction of motion of an ancient glacier by studying drumlins.

FIGURE 13.22 Merging lateral moraines from coalescing glaciers formed three separate medial moraines (the dark stripes) on Baffin Island, Canada.

FIGURE 13.20 This wooded terminal moraine in New York State marks the southernmost extent of glaciers in that region.

Unit 3: Surface Processes

FIGURE 13.23 Streams flowing from the terminus of a glacier filled this valley on Baffin Island with outwash.

Landforms Composed of Stratified Drift

Because a glacier erodes great amounts of sediment, a stream flowing from a glacier is commonly laden with silt, sand, and gravel. The stream deposits this sediment beyond the glacier terminus as **outwash** (Figures 13.5 and 13.23). Glacial streams carry such a heavy load of sediment that they often become braided, flowing in multiple channels. Outwash deposited in a narrow valley is called a **valley train**. If the sediment spreads out from the confines of the valley into a larger valley or plain, it forms an **outwash plain** (Figure 13.24). Outwash plains are also characteristic of continental glaciers.

During summer, when snow and ice melt rapidly, streams form on the surface of a glacier. Some may be quite large. Many of these streams flow off the front or sides of the glacier, while others plunge into crevasses and run beneath the glacier over bedrock or drift. These streams commonly deposit small mounds of sediment, called **kames**, at the margin of a receding

outwash
Sediment deposited by streams flowing from the terminus of a melting glacier.

valley train
Outwash deposited in a narrow mountain valley by the streams flowing from an alpine glacier.

outwash plain
A broad, level surface formed when outwash spreads onto a wide valley or plain beyond a glacier.

kame A small mound or ridge of stratified drift deposited by a stream that flows on top of, within, or beneath a glacier.

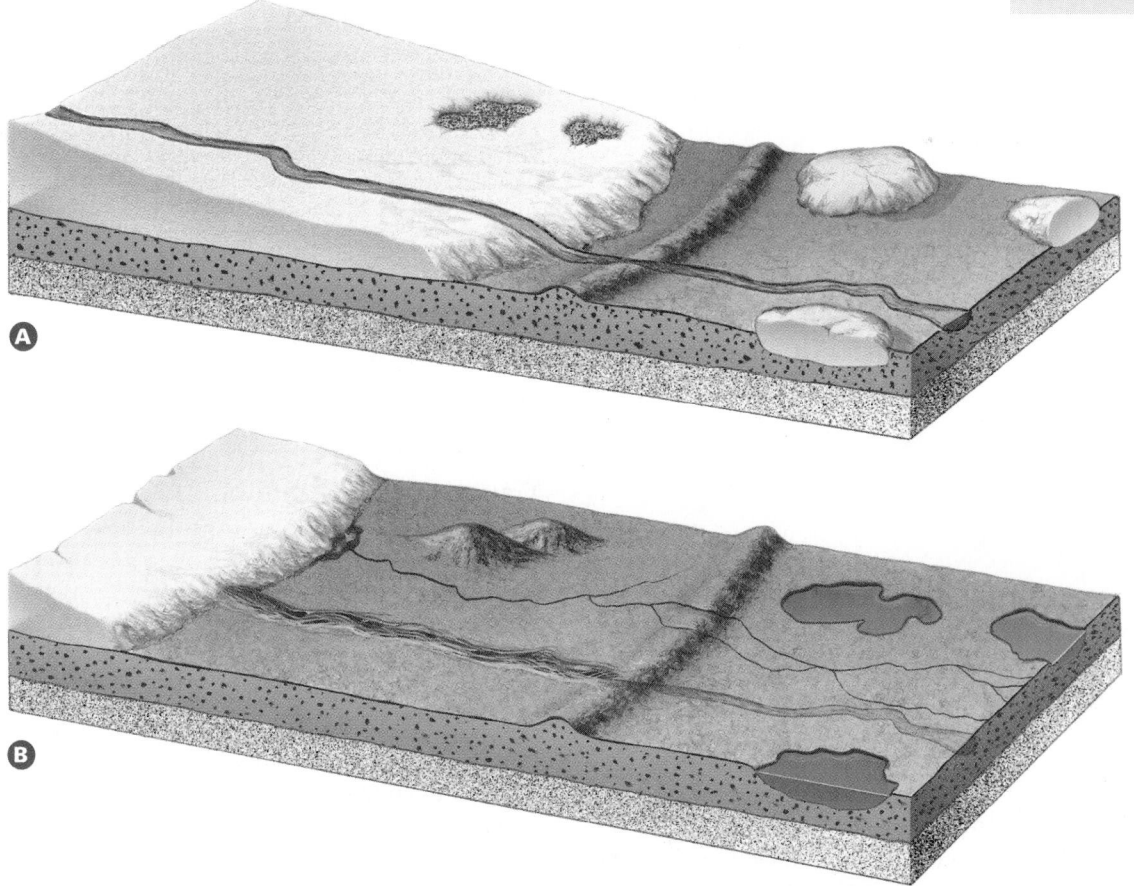

FIGURE 13.24 (A) A melting glacier exposes several different glacial landforms that were created beneath the ice. (B) Sediment flowing from the melting ice creates an outwash plain beyond the glacial terminus.

esker A long, snake-like ridge formed as the channel deposit of a stream that flowed within or beneath a melting glacier.

ice age A time of extensive glacial activity, when alpine glaciers descended into lowland valleys and continental glaciers spread over the higher latitudes.

Pleistocene Ice Age The most recent ice age, which began roughly 2 or 3 million years ago, characterized by several advances and retreats of glaciers. Most climate models indicate that Earth is still in the Pleistocene Ice Age.

glacier or where sediment collects in a crevasse or other depression in the ice. An **esker** is a long, sinuous ridge that forms as the channel deposit of a stream that flowed within or beneath a melting glacier (Figure 13.24).

Because kames, eskers, and other forms of stratified drift are stream deposits and were not deposited directly by ice, they show sorting and sedimentary bedding, which distinguishes them from unsorted and unstratified till. In addition, the individual cobbles or grains deposited from these streams are usually rounded.

13.5 The Pleistocene Ice Age

Geologists have found terminal moraines extending across all high-latitude continents. By studying those moraines, as well as lakes, eskers, outwash plains, and other glacial landforms, geologists have determined that massive glaciers once covered large portions of the continents, altering Earth systems. A time when alpine glaciers descend into lowland valleys and continental glaciers spread over land in high latitudes is called an **ice age**. During an ice age, glaciers several kilometers thick spread across the landscape. Beneath the burden of ice, the continents sink deeper into the asthenosphere. The ice weathers rock and erodes soil, altering the landscape.

Geologic evidence shows that Earth has been warm and relatively ice free for at least 90 percent of the past 1 billion years. However, at least six major ice ages occurred during that time. Each one lasted from 2 to 10 million years.

The most recent ice age took place mainly during the Pleistocene Epoch and is called the **Pleistocene Ice Age**. It began about 2 million years ago in the Northern Hemisphere, although evidence of an earlier beginning has been found in the Southern Hemisphere. However, Earth was not glaciated continuously during the Pleistocene Ice Age; instead, climate fluctuated and continental glaciers grew and then melted away several times (Figure 13.25). During the most recent interglacial period, the average temperature was about the same as it is today, or perhaps a little warmer. Then, high-latitude temperature dropped at least 15°C, causing the ice to

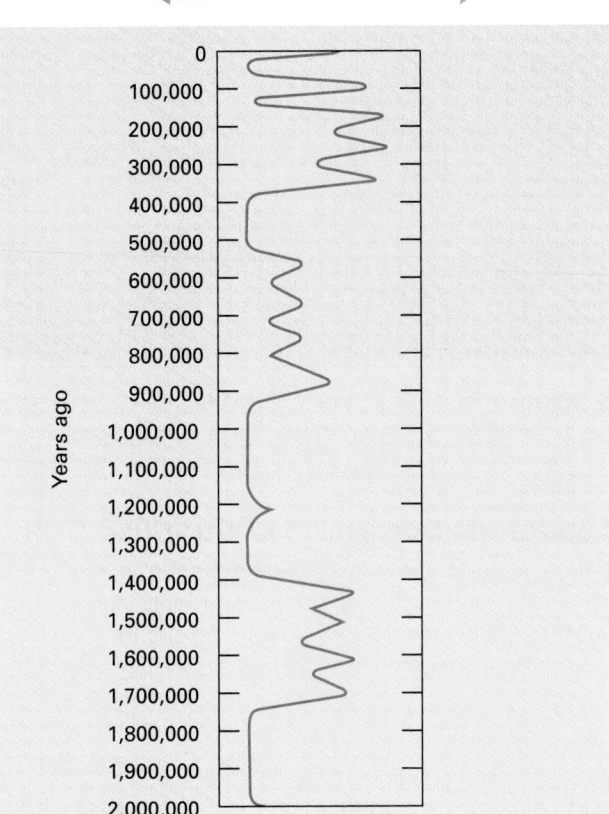

FIGURE 13.25 Glacial cycles during the Pleistocene Ice Age. Cold temperatures coincided with Pleistocene glacial advances, and warm intervals coincided with glacial melting. We are probably living within a warm period of the Pleistocene Ice Age, and continental ice sheets may advance once again.

advance.[1] Most climate models indicate that we are still in the Pleistocene Ice Age and the continental ice sheets may advance again.

Causes of the Pleistocene Ice Age and Glacial Cycles

For reasons that are poorly understood, prior to 2 million years ago Earth's climate had been cooling for tens of millions of years—and then something happened to push the planet over a climate threshold and plunge it into an ice age.

An increase in volcanic ash 2 million years ago coincided closely with a dramatic increase in sediment grains that showed glacial markings. Thus an increase in volcanic activity occurred at about the same time that

1. Mark Chandler, "Trees Retreat and Ice Advances," Glacial Cycles, *Nature* 381 (June 6, 1996), 477–478.

glaciers began to spread. A possible period of intense volcanic activity injected enough dust into the atmosphere to reflect appreciable quantities of sunlight. This slight additional cooling event may have initiated the Pleistocene Ice Age.

Although volcanic dust may have triggered the *onset* of the Pleistocene glacial epoch, no such events seem to be associated with the repeated growth and melting of glaciers that characterize the Pleistocene Epoch. Instead, scientists have found that slight, periodic variations in Earth's orbit and orientation relative to the Sun coincided with Pleistocene glacial expansion and shrinking.

In the 19th century, astronomers detected three periodic variations in Earth's orbit and spin axis (Figure 13.26):

1. Earth's orbit around the Sun is elliptical rather than circular. The shape of the ellipse is called **eccentricity**. The more elliptical the orbit, the more *eccentric* it is said to be. Earth's orbit eccentricity varies in a regular cycle lasting about 100,000 years.

eccentricity A term referring to the elliptical shape of Earth's orbit around the Sun; the more elliptical the orbit, the more eccentric it is said to be.

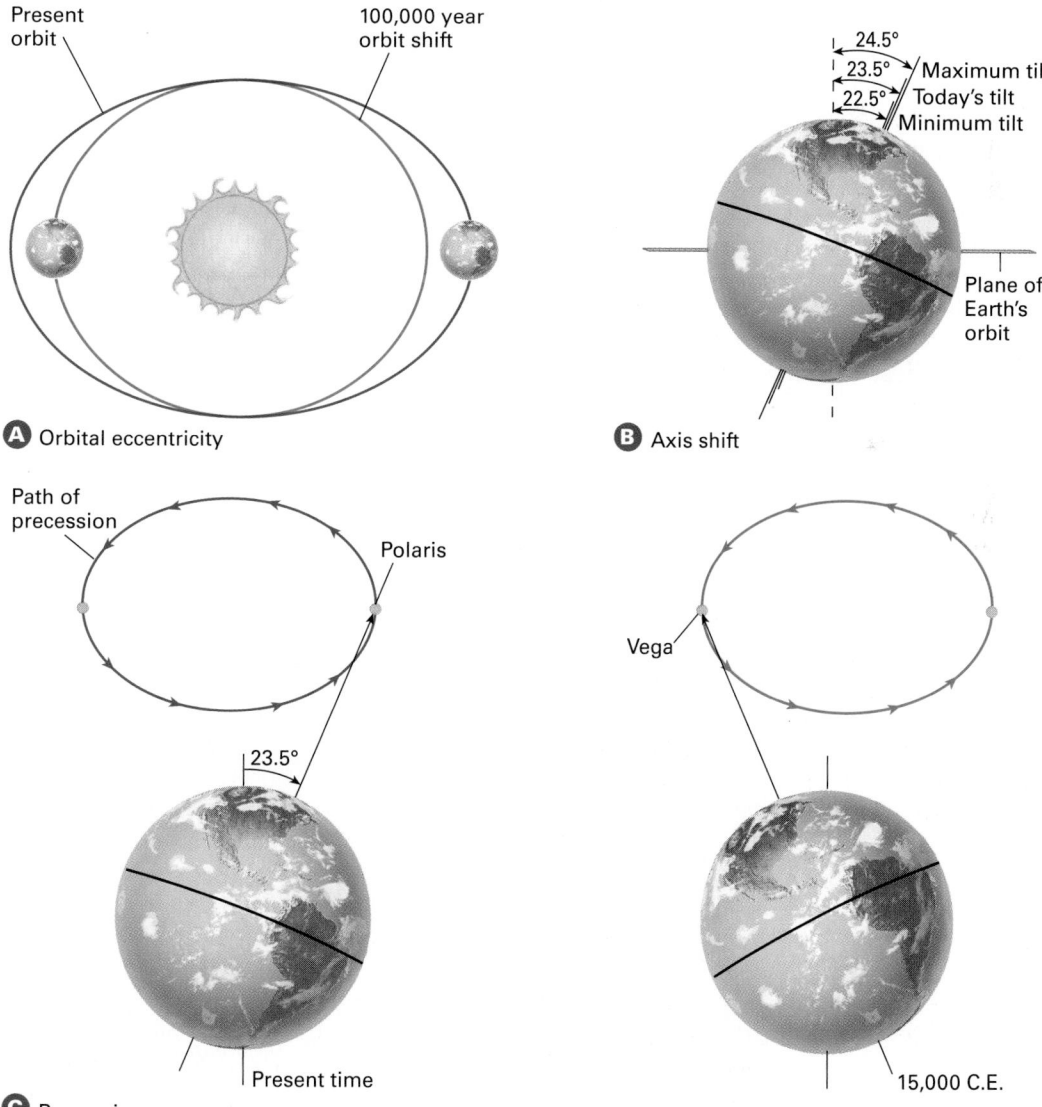

A Orbital eccentricity

B Axis shift

C Precession

FIGURE 13.26 Earth's orbital variations may explain the temperature oscillations and glacial advances and retreats during the Pleistocene Epoch. Earth's orbit and spin axis vary in three ways: (A) the elliptical shape of the orbit changes over a cycle of about 100,000 years; (B) the tilt of Earth's axis of rotation oscillates by about 2° over a cycle of about 41,000 years; (C) Earth's axis completes a full cycle of precession about every 26,000 years.

tilt The angle of Earth's axis with respect to a line perpendicular to the plane of its orbit around the Sun. Earth's axis is tilted by about 23.5°.

precession The circling or wobbling of Earth's axis as the planet travels in its orbit, like that of a wobbling top.

2. The **tilt**, or angle of Earth's axis, is currently about 23.5° with respect to a line perpendicular to the plane of its orbit around the Sun. The tilt oscillates by about 2° on about a 41,000-year cycle.

3. Earth's axis, which now points directly toward the North Star (Polaris), circles like that of a wobbling top. This circling, called **precession**, completes a full cycle every 26,000 years.

These changes affect both the total solar radiation received by Earth and the distribution of solar energy with respect to latitude and season. Seasonal changes in sunlight reaching higher latitudes can reduce summer temperature. If summers are cool and short, winter snow and ice persist, leading to growth of glaciers.

Early in the 20th century, a Yugoslavian astronomer, Milutin Milankovitch, calculated that the orbital variations generate alternating cool and warm climates in the midlatitudes and higher latitudes. Moreover, the timing of the calculated cooling coincided with that of Pleistocene glacial advances. Therefore, he concluded that orbital variations caused Pleistocene glacial cycles.

Modern calculations indicate that orbital cycles by themselves are not sufficient to cause glaciers to advance and retreat. Instead, orbital cycles disturb other Earth systems, which in turn cause additional cooling. Thus a relatively small initial disturbance is amplified to cause a major climate change. In one recent study, researchers calculated that orbital variations probably caused high-latitude climate to cool enough to kill vast regions of Pleistocene northern forest. Forests control climate by absorbing solar energy and warming the atmosphere. When the forests died, more solar energy reflected back out to space. This loss of solar energy caused Earth to cool even more, and the glaciers advanced.[2] But ice reflects more solar radiation back into space, so the feedback process then became further amplified as the growth of glaciers amplified global cooling. Thus variations in Earth's orbit could have altered the biosphere and the altered biosphere triggered additional cooling that led to a major glacial advance.

Effects of Pleistocene Continental Glaciers

At its maximum extent about 18,000 years ago, the most recent North American ice sheet covered 10 million square kilometers—most of Alaska, Canada, and parts of the northern United States (Figure 13.27). At the same time, alpine glaciers flowed from the mountains into the lowland valleys.

2. R. G. Gallimore and J. E. Kutzbach, "Role of Orbitally Induced Changes in Tundra Area in the Onset of Glaciation," *Nature* 381 (June 6, 1996), 503–505.

FIGURE 13.27 Maximum extent of the continental glaciers in North America during the latest glacial advance, approximately 18,000 years ago. The arrows show directions of ice flow.

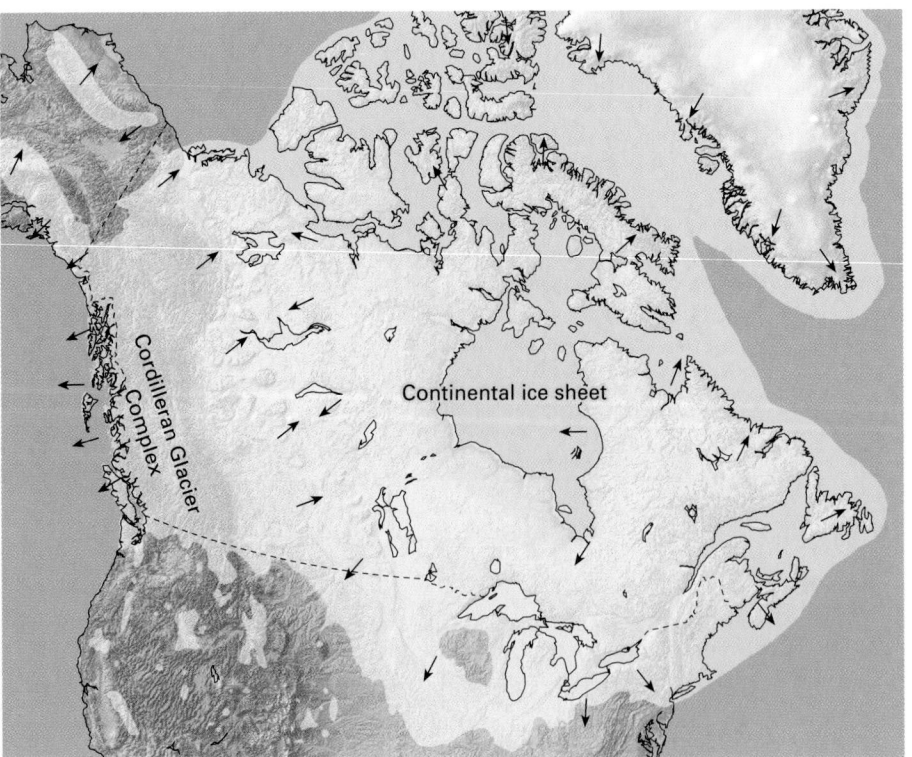

Cordilleran Glacier Complex

Continental ice sheet

© JAN MARTIN WILL/SHUTTERSTOCK

The erosional features and deposits created by these glaciers dominate much of the landscape of the northern states. Today, terminal moraines form a broad band of rolling hills from Montana across the Midwest and eastward to the Atlantic Ocean. Long Island, New York, and Cape Cod, Massachusetts, are composed largely of terminal moraines. Kettle lakes or lakes dammed by moraines are abundant in northern Minnesota, Wisconsin, and Michigan. Drumlins dot the landscape in the northern states. Ground moraines, outwash, and loess (windblown glacial silt) cover much of the northern Great Plains. These deposits have weathered to form the fertile soil of North America's breadbasket.

Pleistocene glaciers advanced when midlatitude and high-latitude climates were colder and wetter than today. When the glaciers melted, the rain and meltwater flowed through streams and collected in numerous lakes. Later, as the ice sheets retreated and the climate became drier, many of these streams and lakes dried up.

The basin that is now Death Valley was once filled with water to a depth of 100 meters or more. Most of western Utah was covered by Lake Bonneville. As drier conditions returned, Lake Bonneville shrank to become Great Salt Lake, west of Salt Lake City.

Sea-Level Changes with Glaciation

When glaciers grow, they accumulate water that would otherwise be in the oceans, and sea level falls. When glaciers melt, sea level rises again. When the Pleistocene glaciers reached their maximum extent 18,000 years ago, global sea level fell to about 130 meters below its present elevation. As submerged continental shelves became exposed, the global land area increased by 8 percent (although about one-third of the land was ice covered).

When the ice sheets melted, most of the water returned to the oceans, raising sea level again. At the same time, portions of continents rebounded isostati-

cally as the weight of the ice was removed. The effect along any specific coast depends upon the relative amounts of sea-level rise and isostatic rebound. The rising seas submerged some coastlines. Others rebounded more than sea level rose. Today, beaches in the Canadian Arctic lie tens to a few hundred meters above the sea. Portions of the shoreline of Hudson Bay have risen isostatically 300 meters.

tillite Till that was deposited by glaciers so long ago that it became lithified into solid rock.

13.6 Snowball Earth: The Greatest Ice Age in Earth's History

Research suggests that at least twice, and perhaps as many as five times, in late-Precambrian time between 800 and 550 million years ago, massive ice sheets completely covered all continents and the world's oceans froze over—even at the equator—entombing the entire globe in a 1-kilometer-thick shell of ice. This ice age, called Snowball Earth by the researchers who discovered it, contrasts sharply with the Pleistocene Ice Age, when ice covered just a third of the continents and only the polar seas froze over.[3]

The main evidence for these Precambrian global glaciations is based on a unique rock called **tillite**. Recall that till consists of an unsorted mixture of boulders, silt, and clay that was deposited by a glacier. Pleistocene tills are loose gravel; you can dig them up with a shovel. Tillite, however, is hard, solid rock that in every other respect resembles the Pleistocene tills. It is till that was deposited by glaciers so long ago that it has become cemented into hard rock. Researchers have found at least two thick layers of tillite between 750 and 580 million years old on almost every continent. Recall that continents have moved around Earth through geologic time. Other types of evidence show that some of the continents lay at the equator when the tillites formed. Other continents were nearer to the poles at the same time, showing that the glaciations were global in scope. In some localities, the glacial deposits lie on top of thick limestone layers. The meaning of these limestone deposits is explained later.

© INFOMAGES/SHUTTERSTOCK

3. Paul F. Hoffman, Alan J. Kaufman, Galen P. Halverson, and Daniel P. Schrag, "A Neoproterozoic Snowball Earth," *Science* 281 (August 28, 1998), 1342–1346.

13.7 The Earth's Disappearing Glaciers

Glaciers are now shrinking in more places and at more rapid rates than at any other time since scientists began keeping records (Table 13.1). In 2003, during a single, hot summer in Europe, 10 percent of the glacial ice in the Alps melted. Scientists point out that this accelerated loss of glacial ice coincides with an increase in atmospheric carbon dioxide levels of about 50 percent in the past century, and they suggest that the melting may be one of the first symptoms of human-caused global warming.

TABLE 13.1 Selected Examples of Ice Melt around the World

Name	Location	Measured Loss
Arctic sea ice	Arctic Ocean	Has shrunk by 7% since 1978, with a 14% loss of thicker, year-round ice. Has thinned by 40% in less than 30 years.
Greenland Continental Ice Sheet	Greenland	Has thinned by more than 1 meter a year on its southern and eastern edges since 1993.
Columbia Glacier	Alaska	Has retreated nearly 13 kilometers since 1982. In 1999, retreat rate increased from 25 meters per day to 35 meters per day.
Glacier National Park	Rocky Mountains, U.S.	Since 1850, the number of glaciers has dropped from 150 to fewer than 50. Remaining glaciers could disappear completely in 30 years.
Antarctic sea ice	Southern Ocean	Ice to the west of the Antarctic Peninsula decreased by some 20% between 1973 and 1993, and continues to decline.
Pine Island Glacier	West Antarctica	Grounding line (where glacier hits ocean and floats) retreated 1.2 kilometers a year between 1992 and 1996. Ice thinned at a rate of 3.5 meters per year.
Larsen B Ice Shelf	Antarctic Peninsula	Calved a 200 km² iceberg in early 1998. Lost an additional 1,714 km² during the 1998–1999 season, and at least 300 km² during the 1999–2000 season.
Tasman Glacier	New Zealand	Terminus has retreated 3 kilometers since 1971, and main front has retreated 1.5 kilometers since 1982. Has thinned by up to 200 meters on average since the 1971–1982 period. Icebergs began to break off in 1991, accelerating the collapse.
Meren Glacier, Carstenz Glacier, and Northwall Firn	Irian Jaya, Indonesia	Rate of retreat increased to 45 meters a year in 1995, up from only 30 meters a year in 1936. Glacial area shrank by some 84% between 1936 and 1995. Meren Glacier is now close to disappearing altogether.
Dokriani Bamak Glacier	Himalayas, India	Retreated by 20 meters in 1998, compared with an average retreat of 16.5 meters over the previous 5 years. Has retreated a total of 805 meters since 1990.
Duosuogang Peak	Ulan Ula Mountains, China	Glaciers have shrunk by some 60% since the early 1970s.
Tien Shan Mountains	Central Asia	Some 22% of glacial ice volume has disappeared in the past 40 years.
Caucasus Mountains	Russia	Glacial volume has declined by 50% in the past century.
Alps	Western Europe	Glacial area has shrunk by 35 to 40% and volume has declined by more than 50% since 1850. Glaciers could be reduced to only a small fraction of their present mass within decades.
Mount Kenya	Kenya	Largest glacier has lost 92% of its mass since the late 1800s.
Speka Glacier	Uganda	Retreated by more than 150 meters between 1977 and 1990, compared with only 35 to 45 meters between 1958 and 1977.
Upsala Glacier	Argentina	Has retreated 60 meters a year on average over the last 60 years, and rate is accelerating.
Quelccaya Glacier	Andes, Peru	Rate of retreat increased to 30 meters a year in the 1990s, up from only 3 meters a year between the 1970s and 1990.

Source: From "Melting of Earth's Ice Cover Reaches New High," *Worldwatch News Brief,* by Lisa Mastny, Worldwatch Institute, March 6, 2000 (http://www.worldwatch.org/alerts/000306.html).

US GEOLOGICAL SURVEY, GLACIER FIELD STATION

A

B

US GEOLOGICAL SURVEY, GLACIER FIELD STATION

FIGURE 13.28 (A) Boulder Glacier in Glacier National Park, Montana, in July 1932. (B) The glacier in July 1988, 56 years later, photographed from the same point. The glacier had disappeared completely by 1988.

Alpine glaciers reflect these changes in a highly visible way. The larger glaciers in Glacier National Park, Montana, have shrunk to a third of their size since 1850, and they continue to melt away today. Many of the smaller glaciers have disappeared completely (Figure 13.28). Since 1850, the number of glaciers in the park has decreased from 150 to fewer than 50. One computer model predicts that all of the glaciers will be gone from the park by 2030 if global temperatures continue to rise as predicted. Even if temperatures remain constant, the glaciers will disappear by 2100.

On a global average, "small" (relative to the Greenland and Antarctic ice sheets) glaciers similar to those of Glacier National Park lost about 7 meters in thickness between 1961 and 1998. In some instances, disintegration has been even more dramatic. For example, half of Alaska's Columbia Glacier has melted in the past 20 years and the glacier currently releases 5 cubic kilometers of meltwater into Prince William Sound every year. Scientists predict that up to a quarter of the total global mass of alpine ice could disappear by 2050, and as much as half by 2100.

Global climate changes do not occur uniformly around the globe. The North Polar regions are warming faster than the average rate for all of Earth. Arctic sea ice covers an area about the size of the United States. Between 1978 and 2009, the ice pack decreased by 7 percent of its area. The sea ice has also thinned from an average thickness of 3.1 meters to an average of 1.8 meters since the 1960s. The Greenland Continental Ice Sheet has thinned by more than a meter per year on its southern and eastern edges since 1993. These rates are accelerating. For example, in 2005, the Kangerdlugssuaq Glacier in Greenland was retreating at a rate of 14 kilometers a year, compared with 4.8 kilometers a year in 2001.

The Antarctic ice cap, which contains 91 percent of Earth's ice and averages 2.3 kilometers thick, has been shrinking for the past 10,000 years as the most recent of the Pleistocene glaciers melted away.

Ice shelves are thick masses of ice that are floating in the ocean but are connected to glaciers on land. They gain ice by flow from the land glaciers and lose ice by melting and when large chunks break off and float into the ocean. Most of Earth's ice shelves surround Antarctica. Ice shelves respond to rising temperature more sensitively than glaciers do. Since 1974, seven Antarctic ice shelves have shrunk by a total of 13,500 square kilometers. These ice shelves are described further in Chapter 21.

The melting of the Arctic and Antarctic ice has profound effects for the planet. This melting contributes to the rise in sea level, and the freshwater flowing into the ocean may ultimately alter ocean currents and global climate.

DESERTS AND WIND

Sand dunes encroach upon the Mesquite Flats, Death Valley, California.

visit **4ltrpress.cengage.com**

260

desert Any region that receives less than 25 centimeters (10 inches) of rain per year and consequently supports little or no vegetation.

Earth's continents are categorized into climate zones based primarily on precipitation and temperature. In turn, climate determines the communities of plants and animals that live in a region. Thus the nature of Earth's surface is determined by complex interactions among the four spheres. A **desert** is any region that receives less than 25 centimeters (10 inches) of rain per year and consequently supports little or no vegetation.[1] Most deserts are surrounded by *semiarid zones* that receive 25 to 50 centimeters of annual rainfall—more moisture than a true desert but less than adjacent regions.

Deserts cover 25 percent of Earth's land surface outside of the polar regions and make up a significant part of every continent. If you were to visit the great deserts of Earth, you might be surprised by their geologic and topographic variety. You would see coastal deserts along the beaches of Chile; shifting dunes in the Sahara; deep, red sandstone canyons in southern Utah; stark granite mountains in Arizona; and bitter-cold polar deserts with a few lichens hanging tenaciously to the otherwise barren rock. The world's deserts are similar only in that they all receive scant rainfall.

14.1 Why Do Deserts Exist?

Rain and snow are unevenly distributed over Earth's surface. The wettest place on Earth, Mount Waialeale, Hawaii, receives an average of 1,168 centimeters (38 feet) of rain annually. In contrast, 10 years or more may pass between rains or snowfalls in the Atacama Desert of Peru and Chile. Several factors—including latitude, mountains, and overall climate—control rainfall patterns and therefore the global distribution of deserts and semiarid lands.

Latitude

The Sun shines most directly near the equator, warming air near Earth's surface. The air absorbs moisture from the equatorial oceans and rises because it is warmer, and therefore less dense, than surrounding air. Rising air cools as pressure decreases. But cool air cannot hold as much water as warm air, so the water vapor condenses and falls as rain (Figure 14.1). For this reason, vast tropical rainforests grow near the equator.

This rising equatorial air, which is now drier because of the loss of moisture, flows northward and southward at high altitudes. The air cools, becomes denser, and sinks back toward Earth's surface at about 30° north and south latitudes. As the air falls, it is compressed and becomes

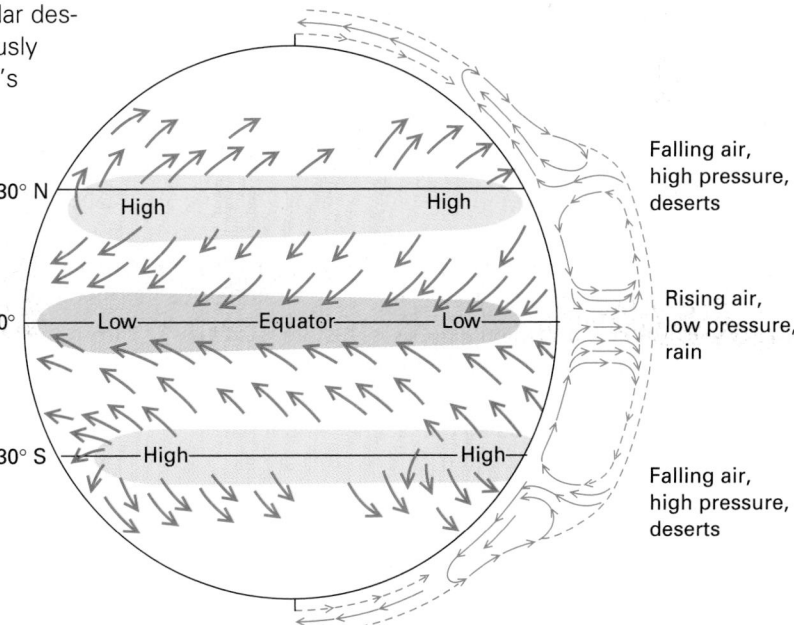

FIGURE 14.1 Falling air creates deserts at 30° north and south latitudes. The red arrows inside the globe indicate surface winds. The blue arrows on the right show airflow on the surface and at higher elevations.

---

1. The definition of *desert* is linked to soil moisture and depends on temperature and amount of sunlight in addition to rainfall. Therefore, the 25-centimeter criterion is approximate.

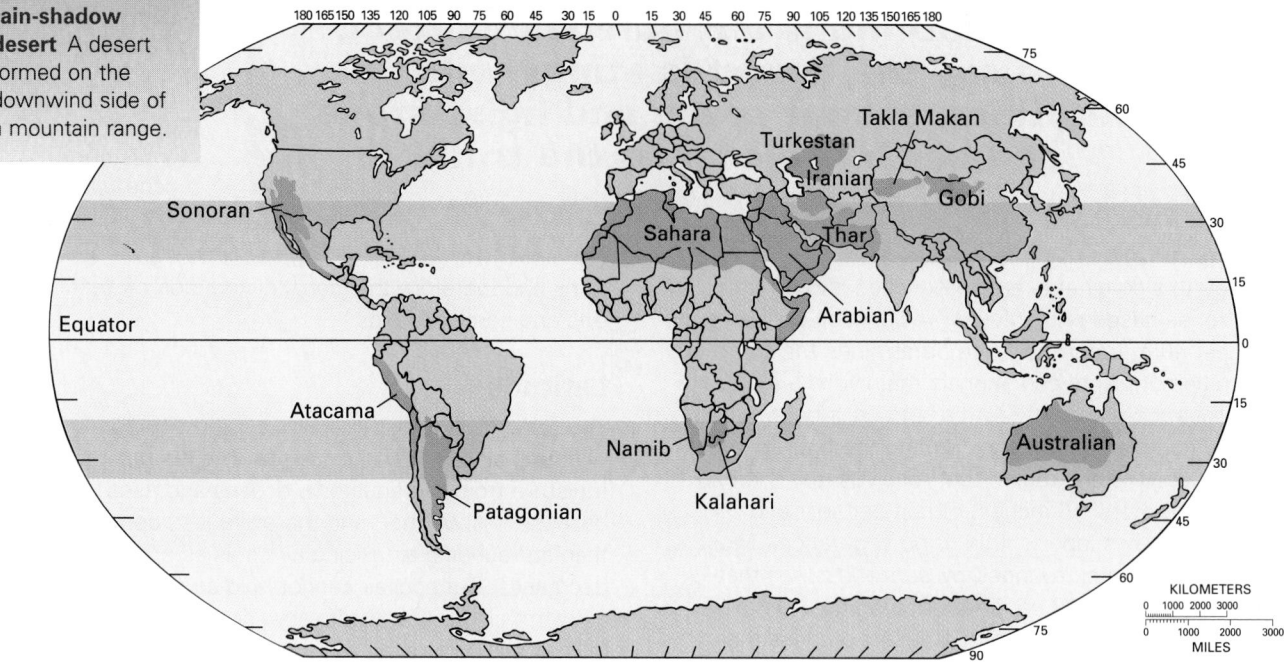

FIGURE 14.2 The major deserts of the world are concentrated at approximately 30° north and south latitudes.

warmer, which enables it to hold more water vapor. As a result, water evaporates from the land surface into the air. Because the sinking air absorbs water, the ground surface is dry and rainfall is infrequent. Thus, many of the world's largest deserts lie at about 30° north and south latitudes (Figure 14.2).

Mountains: Rain-Shadow Deserts

When moisture-laden air flows over a mountain range, it rises. As the air rises, it cools and its ability to hold water decreases. As a result, the water vapor condenses into rain or snow, which falls as precipitation on the windward side and on the crest of the range (Figure 14.3). This cool air flows down the leeward (or downwind) side and sinks. As in the case of sinking air at 30° latitude, the air is compressed and warmed as it descends and it has already lost much of its moisture. This warm, dry air creates an arid zone called a **rain-shadow desert** on the leeward side of the range.

In this way, tectonic forces, which create mountains, affect very long-term rainfall patterns. In turn, as we learned in Chapter 10, falling and flowing water will weather and erode rocks and define living conditions for plants and animals. Thus we encounter yet another example of Earth systems interactions as the building of a mountain range (a tectonic process) alters rainfall (an atmospheric process) and ultimately defines the types of ecosystems that exist in a region (the biosphere).

Coastal and Interior Deserts

Because most evaporation occurs over the oceans, one might expect that coastal areas would be moist and climates would become drier with increasing distance from the sea. This is generally true, but a few notable exceptions exist.

The Atacama Desert along the west coast of South America is so dry that portions of Peru and Chile often receive no rainfall for a decade or more. Cool ocean currents flow along the west coast of South America. When the cool marine air encounters warm land, the

FIGURE 14.3 A rain-shadow desert forms where warm, moist air from the ocean rises as it flows over mountains. As it rises, it cools and water vapor condenses to form rain. The dry, descending air on the lee side absorbs moisture, forming a desert.

air is heated. The warm, expanding air absorbs moisture from the ground, creating a coastal desert.

The Gobi Desert is a broad, arid region in central Asia. The center of the Gobi lies at about 40°N latitude, and its eastern edge is a little more than 400 kilometers from the Yellow Sea. As a comparison, Pittsburgh, Pennsylvania, lies at about the same latitude and is 400 kilometers from the Atlantic Ocean. If latitude and distance from the ocean were the only factors, both regions would have similar climates. However, the Gobi is a barren desert and western Pennsylvania receives enough rainfall to support forests and rich farmland. The Gobi is bounded by the Himalayas to the south and the Altai and Tien Shan mountain ranges to the west, which shadow it from the prevailing winds. In contrast, winds carry abundant moisture from the Gulf of Mexico, the Great Lakes, and the Atlantic Ocean to western Pennsylvania.

Thus, in some regions, deserts extend to the seashore and in other regions the interior of a continent is humid. The climate at any particular place on Earth results from a combination of many factors. Latitude and proximity to the ocean are important, but complex interactions involving the direction of prevailing winds, the direction and temperature of ocean currents, and the positions of mountain ranges also control climate.

14.2 Water and Deserts

Although rain and snow rarely fall in deserts, water plays an important role in these dry environments. Thus the hydrosphere affects the geosphere in even the driest places on Earth. Water can reach a desert from three sources. Streams flow from adjacent mountains or other wetter regions, bringing surface water to some

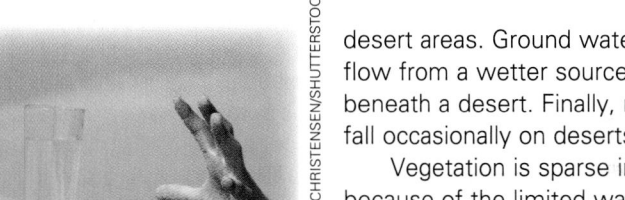

desert areas. Ground water may also flow from a wetter source to an aquifer beneath a desert. Finally, rain and snow fall occasionally on deserts.

Vegetation is sparse in most deserts because of the limited water supply. Thus much bare soil is exposed, unprotected from erosion. As a result, rain easily erodes desert soils and flowing water is an important factor in the evolution of desert landscapes.

Desert Streams

Large rivers flow through some deserts. For example, the Nile River flows through North African deserts and the Colorado River crosses the arid southwestern United States (Figure 14.4). Desert rivers receive most of their water from wetter, mountainous regions bordering the arid lands.

FIGURE 14.4 The Colorado River flows from the Rocky Mountains through the arid southwestern United States.

In a desert, the water table is commonly deep below a streambed, so that water seeps downward from the stream into the ground. As a result, many of the smaller desert streams flow only for a short time after a rainstorm, or during the spring when winter snows are melting. A streambed that is dry for most of the year is called a **wash** (Figure 14.5).

Desert Lakes

While most lakes in wetter environments lie at the level of the water table and are fed, in part, by ground water, many desert lakes lie above the water table. During the wet season, rain and streams fill a desert lake. Some desert lakes are drained by outflowing streams, while many lose water only by evaporation and seepage. During the dry season, inflowing streams may dry up and evaporation and seepage may be so great that the lake dries up com-

pletely. An intermittent desert lake like this is called a **playa lake**, and the dry lake bed is called a **playa** (Figure 14.6).

Recall from Chapter 11 that streams and ground water contain dissolved salts. When this slightly salty water fills a desert lake and then evaporates, the ions precipitate to deposit the salts on the playa. Over many years, economically valuable mineral deposits, such as those of Death Valley, may accumulate (Figure 14.7).

Flash Floods

Bedrock or tightly compacted soil covers the surface of many deserts, and little vegetation is present to help absorb moisture. As a result, rainwater runs over the surface to collect in gullies and washes. During a rainstorm, a dry streambed may fill with water so rapidly that a **flash flood**—a brief, intense local flood—occurs. Occasionally, novices to desert camping pitch their tents in a wash, where they find soft, flat sand to sleep on and shelter from the wind. However, if a thunderstorm occurs upstream during the night, a flash flood may fill the wash with a wall of water mixed with rocks and boulders, creating disaster for the campers. By midmorning

A

B

COURTESY OF GRAHAM R. THOMPSON/JONATHAN TURK

FIGURE 14.5 Courthouse Wash, Utah (A) in the spring, when rain and melting snow fill the channel with water, and (B) in midsummer, when the creek bed is a dry wash.

FIGURE 14.6 Mud cracks pattern the floor of a playa in Utah.

FIGURE 14.7 Borax and other valuable minerals are abundant in the evaporite deposits of Death Valley. Mule teams hauled the ore from the valley in the 1800s.

of the next day, the wash may contain only a tiny trickle, and within 24 hours it may be completely dry again.

When rainfall is unusually heavy and prolonged, the desert soil itself may become saturated enough to create a mudflow that carries boulders and anything else in its path downslope. Some of the most expensive homes in Phoenix, Arizona, and other desert cities are built on alluvial fans and steep mountainsides, where they afford good views but are prone to mudflows during wet years.

Pediments and Bajadas

Recall from Chapter 11 that when a steep, flooding mountain stream empties into a flat valley, the water slows abruptly and deposits most of its sediment at the mountain front, forming an alluvial fan. Although

fans form in all climates, they are particularly conspicuous in deserts (Figure 14.8). A large fan may be several kilometers across and rise a few hundred meters above the surrounding valley floor.

If the mouths of several canyons are spaced only a few kilometers apart, the alluvial fans extending from each canyon may merge. A **bajada** is a broad, gently sloping depositional surface formed by merging alluvial fans and extending into the center of a desert valley. Typically, the fans merge incompletely, forming an undulating surface that follows the mountain front for tens of kilometers. The sediment that forms the bajada may fill the valley to a depth of several thousand meters.

A **pediment** is a broad, gently sloping surface eroded into bedrock. Pediments commonly form along the front of desert mountains. The bedrock surface of a pediment is covered with a thin veneer of gravel that is in the process of being transported from the mountains, across the pediment, to the bajada.

Together, a pediment and bajada form a smooth surface from the mountain front to the valley center (Figure 14.9). The surface steepens slightly near the mountains, so it is concave. To distinguish a pediment from a bajada, you would have to dig or drill a hole. If you were on a pediment, you would strike bedrock after only a few meters, but on a bajada, bedrock would be buried beneath hundreds or even thousands of meters of gravel.

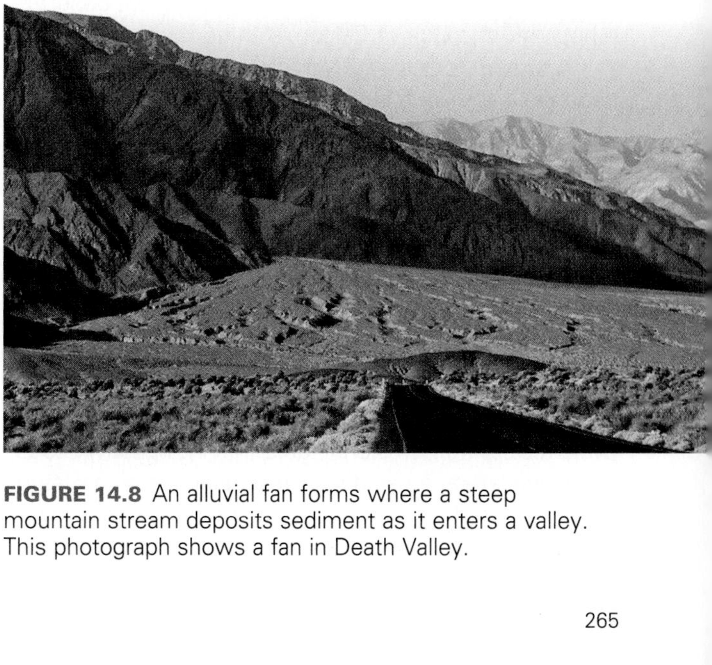

bajada A broad, gently sloping depositional surface formed by the merging of alluvial fans from closely spaced canyons, extending outward into a desert valley.

pediment A broad, gently sloping erosional surface that forms along the front of desert mountains uphill from a bajada, usually covered by a patchy veneer of gravel only a few meters thick.

FIGURE 14.8 An alluvial fan forms where a steep mountain stream deposits sediment as it enters a valley. This photograph shows a fan in Death Valley.

plateau A large, elevated area of relatively flat land.

mesa A flat-topped mountain, shaped like a table, that is smaller than a plateau and larger than a butte.

butte A flat-topped mountain, smaller and more tower-like than a mesa, characterized by steep cliff faces.

FIGURE 14.9 The bajada in the foreground merges with a gently sloping pediment to form a continuous surface in front of these mountains in Mongolia. This basin is filling with sediment from the surrounding mountains because it has no external drainage.

14.3 Two American Deserts

The Colorado Plateau

The Colorado Plateau covers a broad region encompassing portions of Utah, Colorado, Arizona, and New Mexico (Figure 14.10). During the past billion years of Earth's history, this region has been alternately covered by shallow seas, lakes, and deserts. Sediment accumulated, sedimentary rocks formed, and tectonic forces later uplifted the land to form the plateau. The Colorado River cut through the bedrock as the plateau rose, to form the 1.6-kilometer-deep Grand Canyon and its tributary canyons. The modern Colorado River receives most of its water from snowmelt and rains in the high Rocky Mountains east and north of the plateau, and the river then flows through the heart of the great desert to empty into the Gulf of California, in northern Mexico.

A stream forms a canyon by eroding downward into bedrock. If the downcutting stream reaches a resistant rock layer, it may erode laterally, widening the canyon. In some places on the Colorado Plateau, downcutting predominates and streams erode deep, narrow canyons. In other places, lateral erosion undercuts canyon walls, and the rock collapses along vertical joints to form flat-topped mesas and buttes that rise above a relatively flat plain. The river eventually carries the sediment from the plateau to the Gulf of California.

The flat tops of mesas and buttes usually form on a bed of sandstone or other sedimentary rock that is relatively resistant to erosion (Figure 14.11). In many cases, extensive lateral erosion leaves spectacular pinnacles or spires, isolated remnants of once-continuous rock layers that the streams have all but completely eroded away.

A **plateau** is a large elevated area of fairly flat land. The term *plateau* is used for regions as large as the Colorado Plateau as well as for smaller, elevated flat surfaces. A **mesa** is smaller than a plateau and is a flat-topped mountain shaped like a table; in fact, *mesa* is Spanish for "table." A **butte** is also a flat-topped mountain, smaller and more tower-like than a mesa, and characterized by steep cliff faces. These landforms are common features of the Colorado Plateau.

Death Valley and the Great Basin

Death Valley lies in the rain shadow of the High Sierras in California. The deepest part of the valley is 82 meters below sea level. It is a classic rain-shadow desert, receiving a scant 5 centimeters of rainfall per year. The mountains to the west receive abundant moisture, and during the winter rainy season and spring snowmelt, streams flow from the mountains into the valley.

In the Colorado Plateau, the Colorado River and its tributaries carry sediment away from the desert to the Gulf of California, leaving deep canyons. In contrast, streams flow into Death Valley from the surrounding mountains, but no streams flow out. Because Death Valley has no external drainage, the valley is filling with sediment eroded from the surrounding mountains. The sediment collects to form vast alluvial fans and bajadas (Figure 14.12). Stream water collects in broad playa lakes that dry up under the hot summer sun.

Death Valley is just one small part of a vast desert region in the American West that has no external drainage (Figure 14.10). The Great Basin—which includes most of Nevada; the western half of Utah; and parts of California, Oregon, and Idaho—is a large desert region

Great Basin
Colorado Plateau

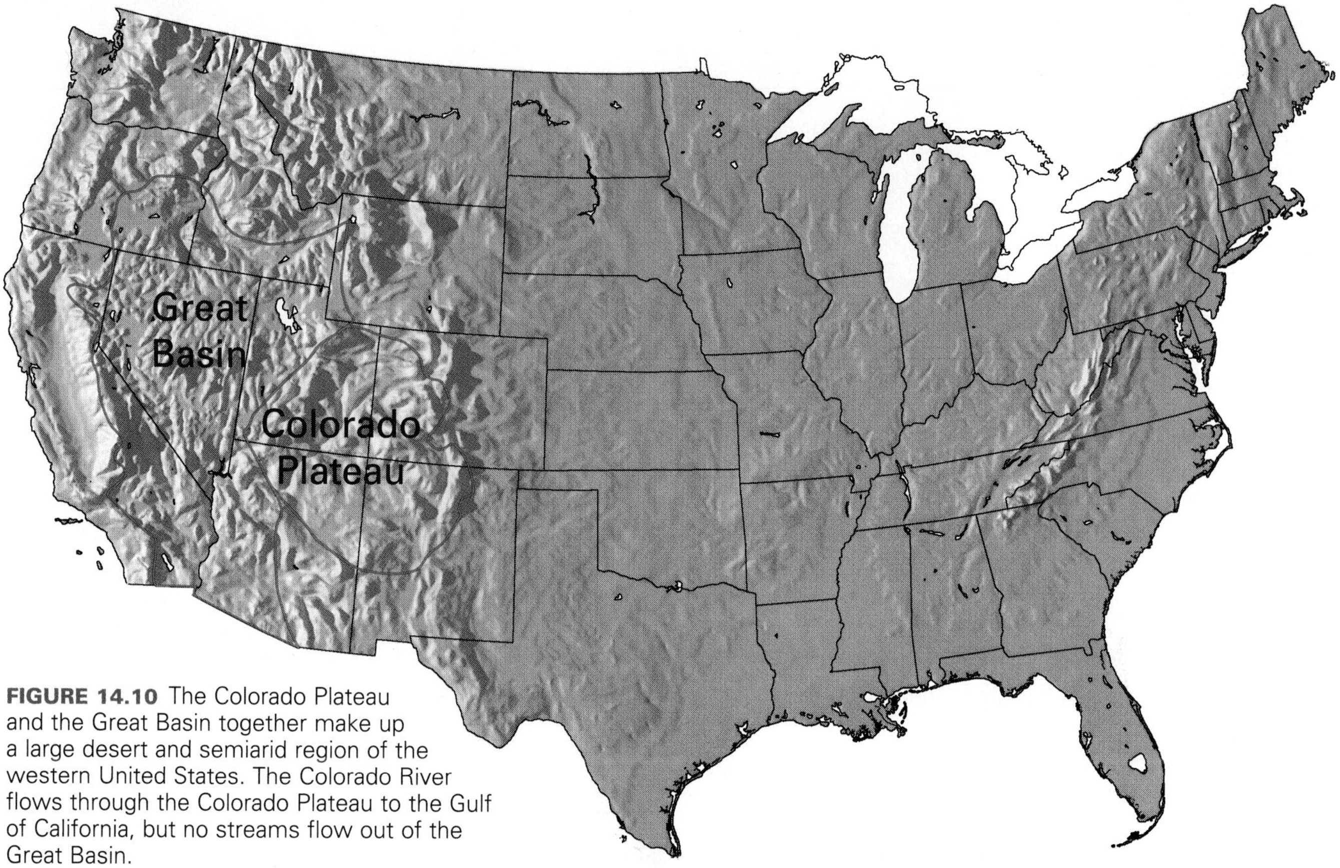

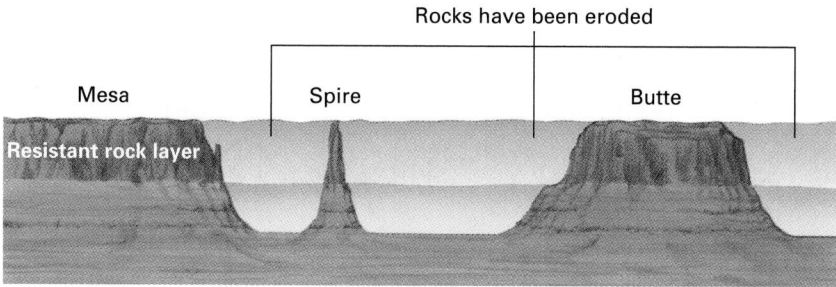

FIGURE 14.10 The Colorado Plateau and the Great Basin together make up a large desert and semiarid region of the western United States. The Colorado River flows through the Colorado Plateau to the Gulf of California, but no streams flow out of the Great Basin.

Rocks have been eroded
Mesa
Spire
Butte
Resistant rock layer
A

FIGURE 14.11 (A) Spires and buttes form when streams reach a temporary base level and erode laterally. The streams transport the eroded sediment away from the region. (B) Spires and buttes in Monument Valley, Arizona.

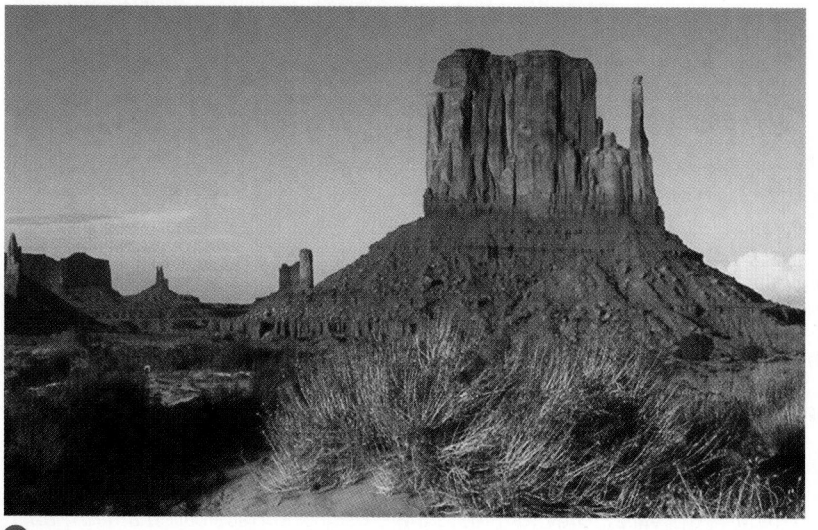

B

COURTESY OF GRAHAM R. THOMPSON/JONATHAN TURK

FIGURE 14.12 Sediment eroded from surrounding mountains is slowly filling this valley in Nevada.

between the Sierra Nevada and the Rocky Mountains. Because there are no streams flowing out of the Great Basin, sediment is not flushed out and has accumulated to become thousands of meters thick.

Features common to the Great Basin (such as bajadas and pediments) result from a combination of tectonic, erosional, and depositional processes. The mountains and valleys initially formed by block faulting

(Figure 14.13A). The valleys are downdropped blocks of rock called *grabens*, and the mountain ranges are uplifted blocks called *horsts* (as discussed in Chapter 9). Desert streams partially eroded the mountains and deposited sediment to form pediments, alluvial fans, and bajadas (Figure 14.13B). Today, with no streams carrying deposits out of the Great Basin, the mountains are drowning in their own sediment (Figure 14.13C).

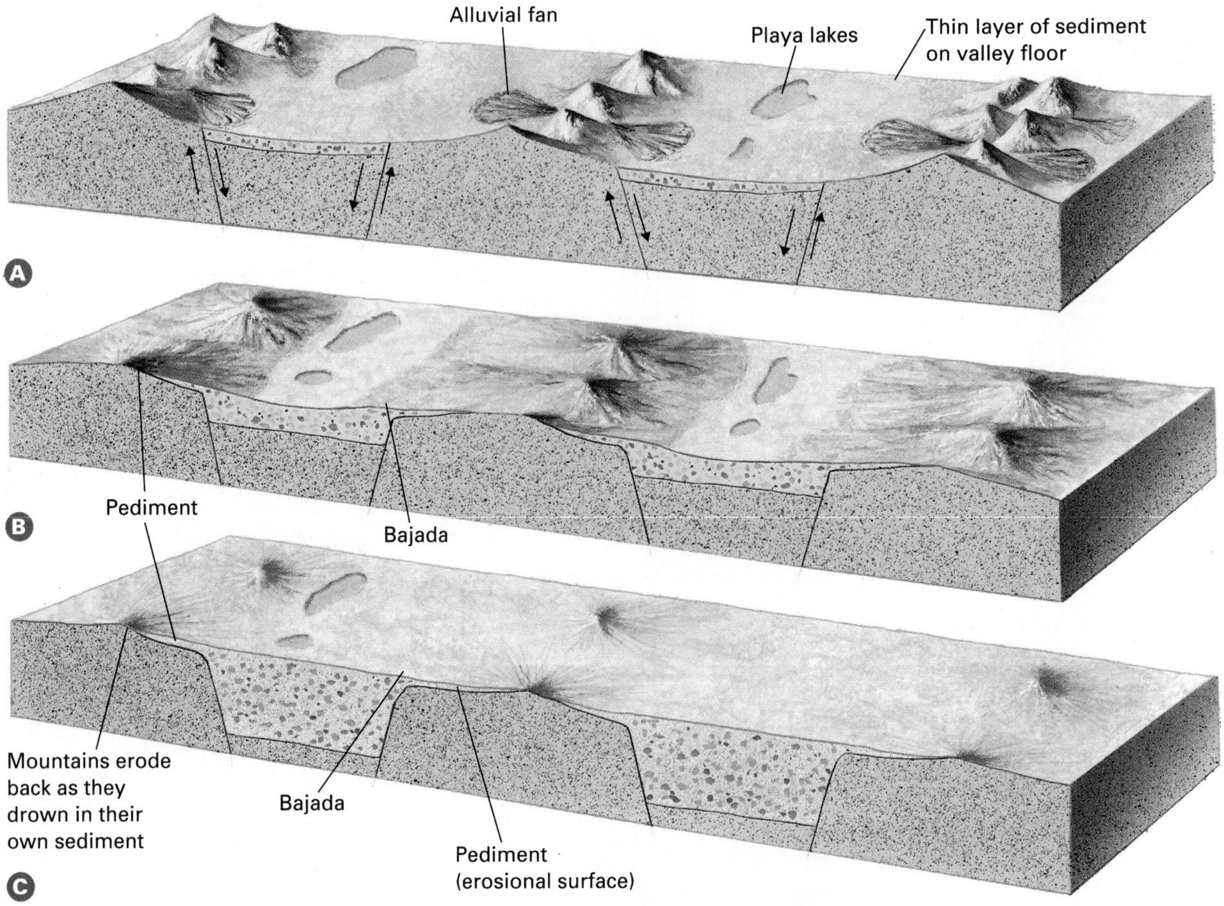

Alluvial fan · Playa lakes · Thin layer of sediment on valley floor

A

Pediment · Bajada

B

Mountains erode back as they drown in their own sediment · Bajada · Pediment (erosional surface)

C

FIGURE 14.13 A scenario for the formation of bajadas and pediments. (A) The mountains and valleys form by block faulting. Desert streams deposit sediment to form alluvial fans. (B) As the streams erode the mountains and deposit sediment in the valley, they form both the erosional surface called a *pediment*, and the depositional surface called a *bajada*. (C) Eventually the mountains drown in their own sediment.

Unit 3: Surface Processes

14.4 Wind

Just as water from the hydrosphere plays an important role in the desert environment, moving air of the atmosphere also shapes and sculpts the desert landscape. When wind blows through a forest or across a prairie, the trees or grasses protect the soil from wind erosion. In wet climates, rain accompanies most windstorms; the water dampens the soil and binds particles together, and little wind erosion occurs. In contrast, a desert commonly has little or no vegetation and rainfall, so wind erodes bare, unprotected desert soil.

Wind erosion is not limited to deserts. Wind is an important agent of erosion wherever the wind blows over unvegetated soil. Windblown dunes and other features created by wind are common along seacoasts, where salty sea spray limits plant growth, and in regions recently abandoned and left bare by receding glaciers.

Wind Erosion

Wind erosion, called **deflation**, is a selective process. Because air is much less dense than water, wind moves only small particles, mainly silt and sand. (Clay particles usually stick together, and, consequently, wind does not erode clay effectively.) Imagine bare soil containing silt, sand, pebbles, and cobbles. When wind blows, it removes only the silt and sand, leaving the pebbles and cobbles as a continuous cover of stones called **desert pavement** (Figure 14.14). Desert pavement prevents the wind from eroding additional sand and silt, even though this finer sediment may be abundant beneath the layer of stones. As a result of this process, approximately 80 percent of the world's desert area is rocky and only 20 percent is covered by sand (Figure 14.15).

Transport and Abrasion

Because sand grains are relatively heavy, wind rarely lifts sand more than 1 meter above the ground and carries it only a short distance. In a windstorm, the sand grains bounce and hop over the ground in a process called **saltation** (from the Latin for "dance"). In contrast, wind carries fine silt in suspension. Skiers in the Alps commonly encounter a silty surface on the snow, blown from the Sahara Desert across the Mediterranean Sea.

Windblown sand is abrasive and erodes bedrock. Because wind carries sand close to the surface, wind erosion occurs near ground level. If the base of a

deflation Erosion by wind.

desert pavement A continuous cover of closely packed stones left behind when wind erodes smaller particles such as silt and sand.

saltation The bouncing, hopping movement of sand grains as the wind blows them along the ground or as they are carried in suspension by water or some other fluid.

Wind removes surface sand

Formation of desert pavement complete—no further wind erosion

Ⓐ

Ⓑ

COURTESY OF GRAHAM R. THOMPSON/JONATHAN TURK

FIGURE 14.14 (A) Wind erodes silt and sand but leaves larger rocks behind to form desert pavement. (B) Desert pavement is a continuous cover of stones left behind when wind blows silt and sand away.

COPYRIGHT AND PHOTOGRAPH BY DR. PARVINDER S. SETHI

FIGURE 14.15 About 80 percent of Earth's deserts are covered by stony desert pavement.

FIGURE 14.16 Wind abrasion near Grand Canyon, Arizona, selectively eroded the base of this rock because windblown sand moves mostly near the surface.

pinnacle is sculpted as in Figure 14.16, wind may be the responsible agent, although salt cracking (see Chapter 10) at ground level can also erode the base of a desert pinnacle.

Dunes

A **dune** is a mound or ridge of wind-deposited sand (Figure 14.17). As explained earlier, wind removes sand from the surface in many deserts, leaving behind a rocky, desert pavement. The wind then deposits the sand in a topographic depression or other place where the wind slows down. Dunes commonly grow to heights of 30 to 100 meters, and some giants exceed 500 meters. In some places they are tens or even hundreds of kilometers long. Although some desert dune fields cover only a few square kilometers, the largest is the Rub al-Khali ("Empty Quarter") in Arabia, which covers 560,000 square kilometers, larger than the state of California.

Dunes also form where glaciers have recently melted and along sandy coastlines. Glacier deposits consist of large quantities of bare, unvegetated sediment. A sandy beach is commonly unvegetated because sea salt prevents plant growth. As a result, both of these environments contain the essentials for dune formation: an abundant supply of sand and a windy environment with sparse vegetation.

Most dunes are asymmetrical. Wind erodes sand from the windward side of a dune, carries it up to the dune crest, and then the sand slides down the sheltered leeward side. In this way, dunes migrate in the downwind direction (Figure 14.18). The leeward face

of a dune is called the **slip face**. Typically, the slip face dips about 35° from the horizontal, which is the angle of repose for dry sand. That is about twice as steep as the windward face. In addition, the sand on a slip face is usually very loose, whereas the sand on the windward face is packed by the wind.

Migrating dunes overrun buildings and highways. For example, near the town of Winnemucca, Nevada, dunes advance across U.S. Highway 95 several times a year. Highway crews must remove as much as 4,000 cubic meters of sand to reopen the road. Engineers often attempt to stabilize dunes in inhabited areas. One method is to plant vegetation to reduce deflation and stop dune migration. The main problem with this approach is that desert dunes commonly form in regions that are too dry to support vegetation. Another solution is to build artificial wind-breaks to create dunes in places where they do the least harm. For example, a fence traps blowing sand and forms a dune, thereby protecting areas downwind. Fencing is a temporary solution, however, because eventually the dune covers the fence and resumes its migration. In Saudi Arabia, dunes are sometimes stabilized by covering them with tarry wastes from petroleum refining.

Fossil Dunes

When dunes are buried by younger sediment and lithified over geologic time, the resulting sandstone retains the original sedimentary structures of the dunes. Figure 14.19 shows a rock face in Zion National Park in Utah. The steep layering is not evidence of tectonic

FIGURE 14.17 Windblown sand formed these dunes near Lago Poopo, Bolivia.

Unit 3: Surface Processes

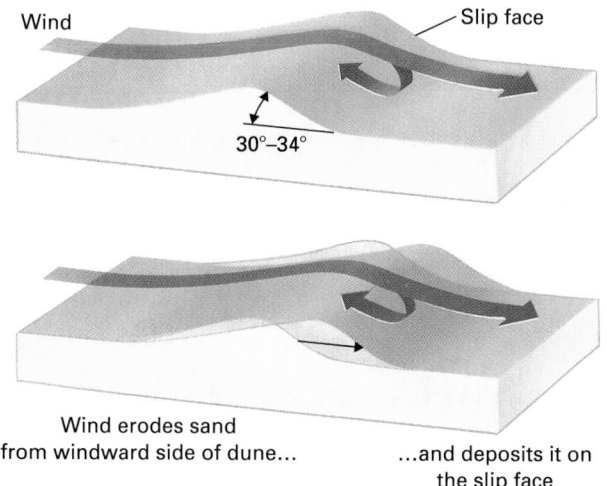

FIGURE 14.18 Sand dunes migrate in the downwind direction.

Wind

Slip face

30°–34°

Wind erodes sand from windward side of dune...

...and deposits it on the slip face

tilting but is the original, steeply dipping layering of the dune slip face. The beds dip in the direction that the wind was blowing when it deposited the sand. Notice that the planes dip in different directions, indicating shifting wind direction, generating an example of cross-bedding (described in Chapter 3).

Types of Sand Dunes

Wind speed and sand supply control the shapes and orientation of dunes. A **barchan dune** forms in rocky deserts where there is little sand. The center of the dune grows higher than the edges (Figure 14.20A). When the dune

migrates, the edges move faster because there is less sand to transport. The resulting barchan dune is crescent shaped with its tips pointing downwind (Figure 14.20B). Barchan dunes are not connected to one another but instead migrate independently. In a rocky desert, barchan dunes cover only a small portion of the land; the remainder is bedrock or desert pavement (Figure 14.20C).

If sand is plentiful and evenly dispersed, it accumulates in long ridges called **transverse dunes** aligned perpendicular to the prevailing wind (Figure 14.21). If sparse desert vegetation is present, the wind may form a small depression called a *blowout* in a bare area among the desert plants. As sand is carried out of the blowout, it accumulates in a **parabolic dune**, the tips of which are

A saucer or trough-shaped hollow formed by wind erosion is called a blowout. In the 1930s, intense, dry winds eroded large areas of the Great Plains and created the Dust Bowl.

barchan dune A crescent-shaped dune, highest in the center, with the tips facing downwind; generally forms in rocky deserts where there is little sand.

transverse dune A relatively long, straight dune with a gently sloping windward side and a steep lee face that is perpendicular to the prevailing wind; forms where sand is plentiful and evenly dispersed.

parabolic dune A crescent-shaped dune with tips pointing into the wind; forms in moist semidesert regions and along seacoasts where sparse vegetation is present to anchor the tips of the dune.

FIGURE 14.19 This cross-bedded sandstone in Zion National Park preserves the sedimentary bedding of ancient sand dunes.

longitudinal dune
A long, symmetrical dune that is parallel with the direction of the prevailing wind; forms where sand is limited and wind direction is erratic but from the same general direction.

loess A homogenous, porous deposit of wind-blown silt, typically unlayered, that forms vertical bluffs and cliffs.

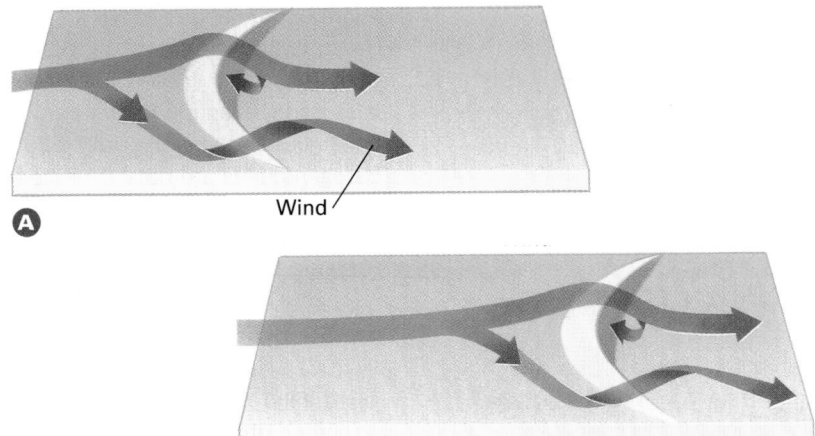

Wind

A

B

C

COURTESY OF GRAHAM R. THOMPSON/JONATHAN TURK

FIGURE 14.20 (A and B) When sand supply is limited, the tips of a barchan dune travel faster than the center and point downwind. (C) A barchan dune in Coral Pinks, Utah.

anchored by plants on each side of the blowout (Figures 14.22A and 14.22B). A parabolic dune is similar in shape to a barchan dune, except that the tips of the parabolic dune point into the wind. Parabolic dunes are common in moist semidesert regions and along seacoasts, where sparse vegetation grows in the sand.

If the wind direction is erratic but prevails from the same general quadrant of the compass and the supply of sand is limited, then long, straight **longitudinal**

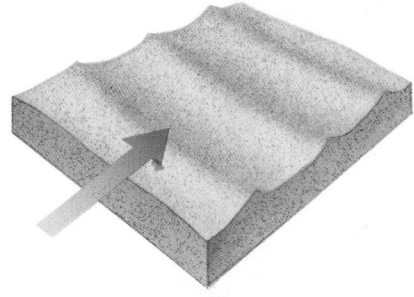

Transverse

FIGURE 14.21 Transverse dunes form perpendicular to the prevailing wind direction in regions with abundant sand.

dunes form parallel to the prevailing wind direction (Figure 14.23). In portions of the Sahara Desert, longitudinal dunes reach 100 to 200 meters in height and are as much as 100 kilometers long.

Loess

Wind can carry silt for hundreds or even thousands of kilometers and then deposit it as **loess** (pronounced "luss"). Loess is porous, uniform, and typically lacks layering. Often the angular silt particles interlock. As a result, even though the loess is not cemented, it typically forms vertical cliffs and bluffs (Figure 14.24).

The largest loess deposits in the world, found in central China, cover 800,000 square kilometers and are more than 300 meters thick. The silt was blown from the Gobi and the Taklimakan deserts of central Asia. The particles interlock so effectively that people have dug caves into the loess cliffs to make their homes. However, in 1920 a great earthquake caused the cave system to collapse, burying and killing an estimated 100,000 people.

Large loess deposits accumulated in North America during the Pleistocene Ice Age, when continental ice

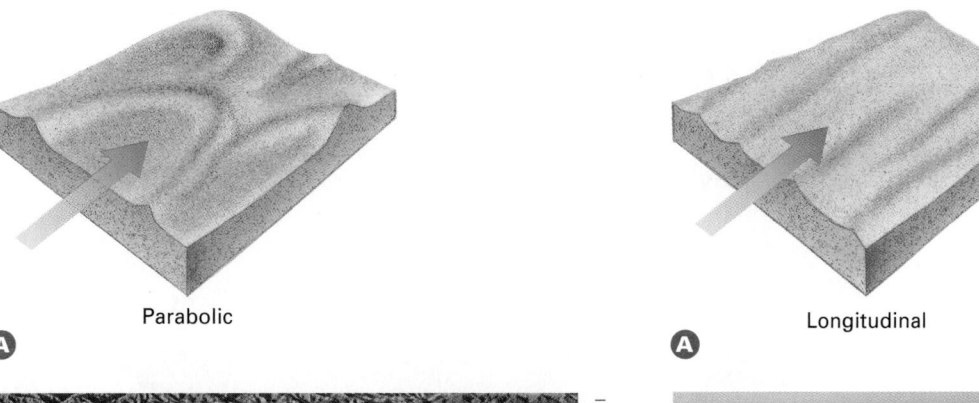

Parabolic

Ⓐ

Longitudinal

Ⓐ

Ⓑ

COPYRIGHT AND PHOTOGRAPH BY DR. PARVINDER S. SETHI

Ⓑ

ALBERT COPLEY/VISUALS UNLIMITED

FIGURE 14.22 (A) A parabolic dune is crescent shaped with its tips pointing upwind. It forms where wind blows sand from a blowout, and grass or shrubs anchor the dune tips. (B) Grass and shrubs anchor the tip of this parabolic dune in the southern California desert.

FIGURE 14.23 Longitudinal dunes are long, straight dunes that form where the wind is erratic and sand supply is limited.

COURTESY OF GRAHAM R. THOMPSON/JONATHAN TURK

FIGURE 14.24 Villagers in Askole, Pakistan, have dug caves in these vertical loess cliffs.

desertification A process by which semiarid land is converted to desert, by human mismanagement or by climate change.

FIGURE 14.25 Loess deposits cover large areas of the United States.

Thick, continuous deposits (8–30 m thick)

Thinner, intermittent deposits (1.5–8 m thick)

sheets ground bedrock into silt. Streams carried this fine sediment from the melting glaciers and deposited it in vast plains. These zones were cold, windy, and devoid of vegetation, and wind easily picked up and transported the silt, depositing thick layers of loess as far south as Vicksburg, Mississippi. Residents of Vicksburg took shelter from federal bombardment in nearby loess caves during the Civil War's Battle of Vicksburg.

Loess deposits in the United States range from about 1.5 meters to 30 meters thick (Figure 14.25). Soils formed on loess are generally fertile and make good farmland. Much of the rich soil of the central plains of the United States and eastern Washington State formed on loess.

14.5 Desertification

The Sahara is the largest desert on the planet. South of the Sahara lies the semiarid Sahel (Figure 14.26). During the 1960s, unusually heavy rains caused the Sahel to bloom. People expanded their flocks to take advantage of the additional forage. Rich countries contributed foreign aid. As a result, medical attention and sanitation improved, and the human population grew dramatically. Many people predicted a new era of prosperity for the Sahel, but the favorable rains were an anomaly. In the late 1960s and early 1970s, drought destroyed the range. During this period, governments in North Africa began to enforce national borders more strictly, curtailing nomadism. When people settled in specific regions,

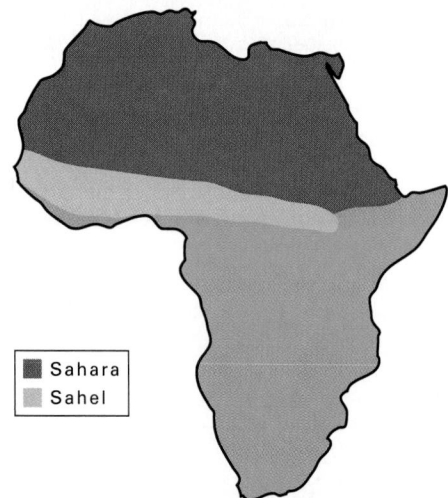

FIGURE 14.26 The Sahara Desert and the semiarid Sahel region dominate the ecosystems of northern Africa.

■ Sahara
▨ Sahel

their flocks grazed the same area throughout the year. Plants did not have time to regenerate, and the hungry animals chewed the grasses down to the roots. Civil and international war brought instability to the region, and famine struck.

Reports issued in the 1970s and 1980s claimed that the Sahara was expanding southward into the Sahel at a rate of 5 kilometers per year. Scientists argued that overgrazing, farming, and firewood gathering had caused the desert expansion. This growth of the desert caused by human mismanagement has been called **desertification**.

© ERIC ISSELEE/SHUTTERSTOCK

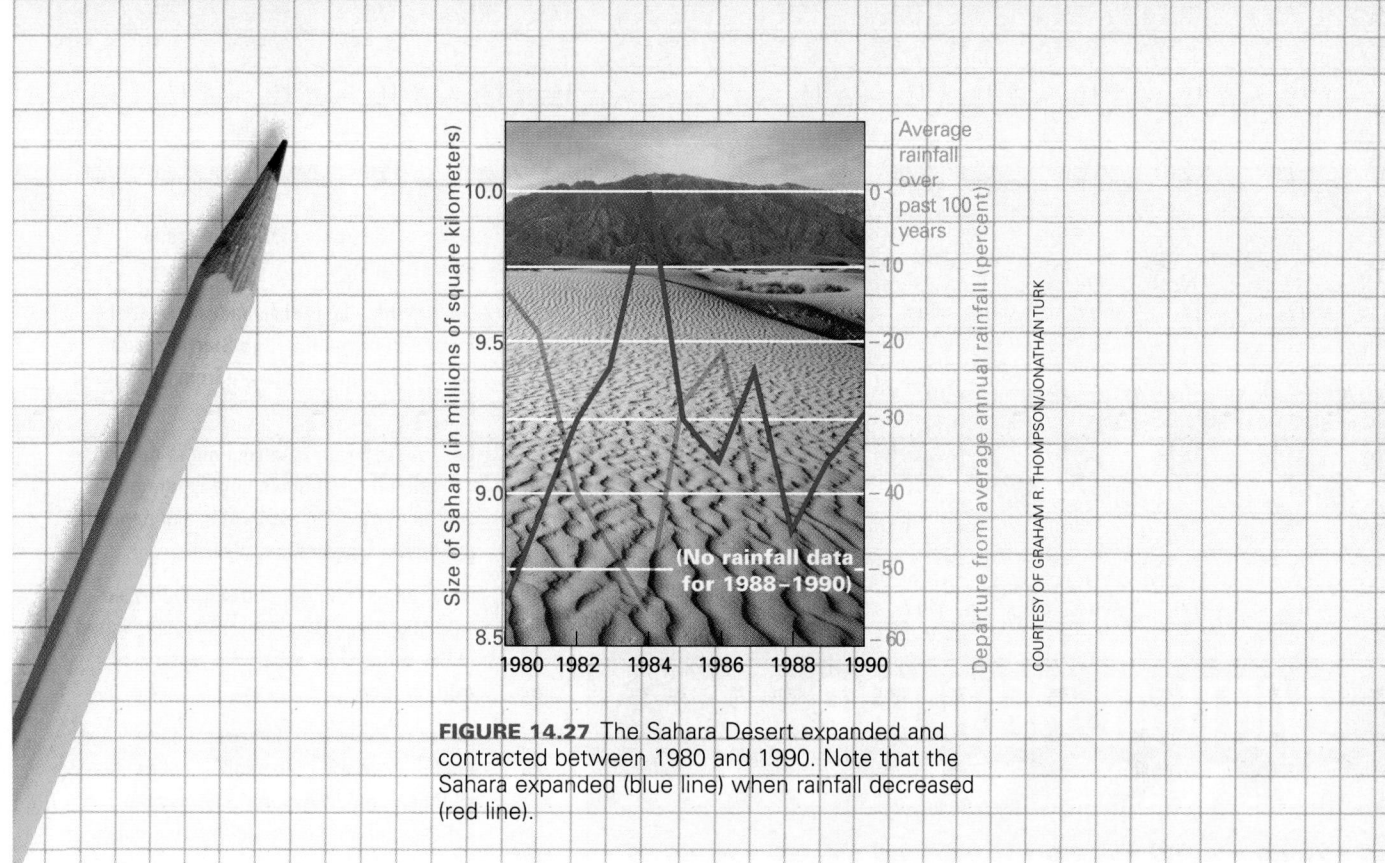

FIGURE 14.27 The Sahara Desert expanded and contracted between 1980 and 1990. Note that the Sahara expanded (blue line) when rainfall decreased (red line).

More recent research has shown that overgrazing a semiarid region causes land degradation but does not cause a desert to expand. The Sahel–Sahara desert boundary is clearly visible on satellite photographs as a boundary separating a region of sparse vegetation from one with almost no plants. Researchers plotted changes in the size of the desert by studying satellite photographs taken between 1980 and 1990. As shown in Figure 14.27, the desert expands (blue line) when rainfall declines (red line).[2] Thus, decreasing rainfall—not overgrazing—may have been responsible for expansion of the Sahara.

Since 1990, rainfall has moisturized the soil and caused the desert to retreat again. But, in addition, massive agricultural aid to the region has led to improved farming and herding practices in some regions. Farmers have learned land management techniques specific to desert environments, which has helped keep the land usable despite fluctuations in rainfall. As a result, thousands of acres of land have been rehabilitated, and millet and sorghum yields have increased by 50 to 75 percent. This story reminds us once again that while a desert ecosystem is created by low rainfall, human management can significantly alter the productivity of the land.

2. William H. Schlesinger, James F. Reynolds, Gary L. Cunningham, Laura F. Huenneke, Wesley M. Jarrell, Ross A. Virginia, and Walter G. Whitford, "Biological Feedbacks in Global Desertification," *Science* 247 (March 1990), 1043–1048.

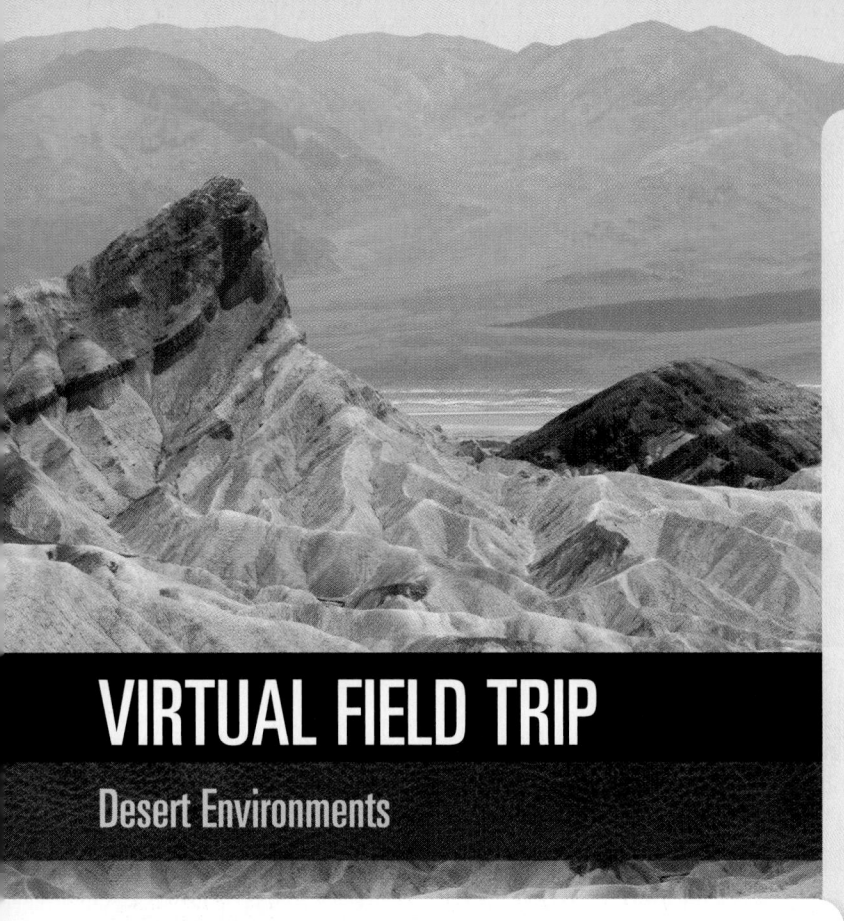

VIRTUAL FIELD TRIP

Desert Environments

Ready to Go!

On this virtual field trip we will visit Death Valley in California, one of the two examples of desert environments described in the text in Chapter 14. In these scenic views, stark rock surfaces and undulating sandy dunes are plainly visible because they are not obscured by vegetation. The landscapes are primarily sculpted and shaped through the combined actions of wind and intermittent but often violent rainstorms, which may result in flash floods. Wind transport produces desert pavement and dunes, while at the same time, abrading and sculpting rocks, as discussed in Chapter 14 and this virtual field trip. Water erodes and transports sediment from the surrounding mountains and deposits it to form alluvial fans and bajadas, two dominant features of Death Valley, discussed in Chapter 14 under "Death Valley and the Great Basin." Because there are no streams flowing out of the valley, various salts, including commercial borax deposits, accumulate from evaporating water.

GOALS OF THE TRIP

Some of the goals of this trip are to understand that:

1. Death Valley is a rain-shadow desert, formed because the high Sierra Nevada Mountains block moisture from the Pacific Ocean to the west.

2. Even though rain is infrequent in deserts, occasional but intense downpours erode sediment from the mountains and deposit it in alluvial fans and bajadas.

3. No streams flow out of Death Valley, so salts dissolved in rainwater are deposited in the valley floor when the water evaporates.

4. Because there is little vegetation to bind the surface of the soil, wind erodes sediment and transports it across the valley, sculpting rock and forming dunes.

5. Mountains both east and west of Death Valley show presence of sedimentary, metamorphic, and volcanic rocks (discussed in Chapter 3).

FOLLOW-UP QUESTIONS

1. Compare the cracking patterns of evaporite deposits with the mud-cracks illustrated in the virtual field trip "Sedimentary Rocks: Formation and Correlation (pages 44–45)."

2. Both the sedimentary rocks virtual field trip and this one on Death Valley take you into desert environments. Use the principle of superposition and the principle of original horizontality, introduced in Chapter 4 and the virtual field trip in Chapter 3 to explain the colorful and tilted layering of these sedimentary rocks near Zabriskie Point (photo left).

3. Speculate on whether a conglomerate is more likely to form in a dune environment or an alluvial fan. Explain your reasoning.

What to See When You Go

If you were to walk through Death Valley on a typical July afternoon, the air might be stiflingly still and unbearably hot. No water would be visible anywhere. Yet, if you look around you with the eye of a geologist, you will see abundant evidence of the effects of both wind and flowing water. Wind transport and deposition forms beautiful sand dunes on the valley floor.

Movement of sediment by streams is an important process in the evolution of landscapes, as discussed in the text in Chapter 10, 11, and 14. In the photo above, streams carried dark gravel from higher up on the mountain across the lower foothills, to create a visual landscape of alternating colors.

Huge alluvial fans along the mountains were formed by water from occasional heavy rains that rushed down the slopes, unimpeded by vegetation. This is a good time to reflect on the concepts of uniformitarianism and catastrophism, introduced in Chapter 1, stating that geologic change can occur slowly and steadily, bit by bit, day in and day out, or through more occasional but more violent change.

Review again each of the features illustrated in this virtual field trip: The alluvial fans, the salt flats, the wind sculpted rocks, the dunes, and the various types of rocks in the mountains. Now review the processes involved in the formation of each of these features and reflect on the time scales and the relative importance of uniform and catastrophic change.

Chemical weathering is relatively less important in desert environments than in humid ones because water is less abundant. This type of weathering, however, still occurs; the rock shown above was coated with desert varnish as rainwater poured over a cliff face.

15

OCEAN BASINS

This image shows the Earth's ocean basins mapped with satellite altimetry. Sea-floor features larger than 10 kilometers are detected by gravitational distortion of the sea surface.

visit **4ltrpress.cengage.com**

NOAA/NATIONAL GEOPHYSICAL DATA CENTER

> **Perhaps, to man, the most important interdependence of the sea and the land is through their interaction with the overlying atmosphere.**
>
> *H. B. Stewart*

15.1 The Origin of Oceans

The primordial Earth, heated by the impacts of colliding planetesimals and the decay of radioactive isotopes, was molten, or near molten. The sky, without an atmosphere, was black. There were no oceans—no life. Today, we consider Earth in terms of four spheres: the geosphere, hydrosphere, atmosphere, and biosphere. Each sphere is as different from the others as a rock is different from a flowing stream, a breath of air, or a butterfly.

For the moment, let's abandon our view of Earth's four spheres and think of only two kinds of Earth materials: volatile substances and nonvolatile ones. (A *volatile* substance is a compound that evaporates rapidly and therefore easily escapes into the atmosphere.) Most scientists agree that the surface of primordial Earth contained few volatiles. How, then, did enough of these compounds collect to form a thick atmosphere, vast oceans, and a global biosphere of living organisms?

For many years, geologists hypothesized that abundant volatiles, including water and carbon dioxide, were trapped within early Earth's interior. This reasoning was based on three observations and inferences. First, our cosmogenic models show that volatiles were evenly dispersed in the cloud of dust, gas, and planetesimals that coalesced to form the planets. It seemed likely that some of those volatiles would have become trapped within Earth as it formed. Second, scientists have detected volatiles in modern comets, meteoroids, and asteroids. If volatiles were trapped within the small objects that passed through our neighborhood in space, it seemed logical to infer that they also accumulated in Earth's interior as the original cloud of dust and gas coalesced. Finally, modern volcanic eruptions eject

gases and water vapor into the air. Geologists inferred that these gases originate in the mantle and are remnants of the original volatiles trapped during Earth's formation. Geologists concluded that some of these volatiles escaped during volcanic eruptions early in Earth's history and that they formed the atmosphere, the oceans, and living organisms.

Today, many scientists question this conclusion. To understand their questions, let's return to the cloud of dust and gas that coalesced to form the planets. Recall that our region of space heated up as dust, gas, and planetesimals collided to become planets. At the same time, hydrogen fusion began within the Sun and solar energy radiated outward to heat the inner Solar System. The newly born Sun also emitted a stream of ions and electrons, called the *solar wind*, that swept across the inner planets, blowing their volatile compounds into outer regions of the Solar System. As a result, most of Earth's volatile compounds boiled off and were swept into the cold outer regions of the Solar System (Figure 15.1).

According to a currently popular hypothesis, shortly after our planet formed and lost its volatiles, a Mars-sized object smashed into Earth. The cataclysmic impact blasted through the crust and deep into the mantle, ejecting huge quantities of pulverized rock into orbit. The fragments eventually coalesced to form the Moon. The impact also ejected most of Earth's remaining volatiles with enough velocity that they escaped Earth's gravity and disappeared into space. According to this hypothesis, Earth's surface then was left barren and rocky, with few volatiles either on the surface or in the deep mantle. Thus it had neither water nor an atmosphere. The hot mantle churned and volcanic eruptions repaved the surface with lava, but these events added few volatiles to Earth's

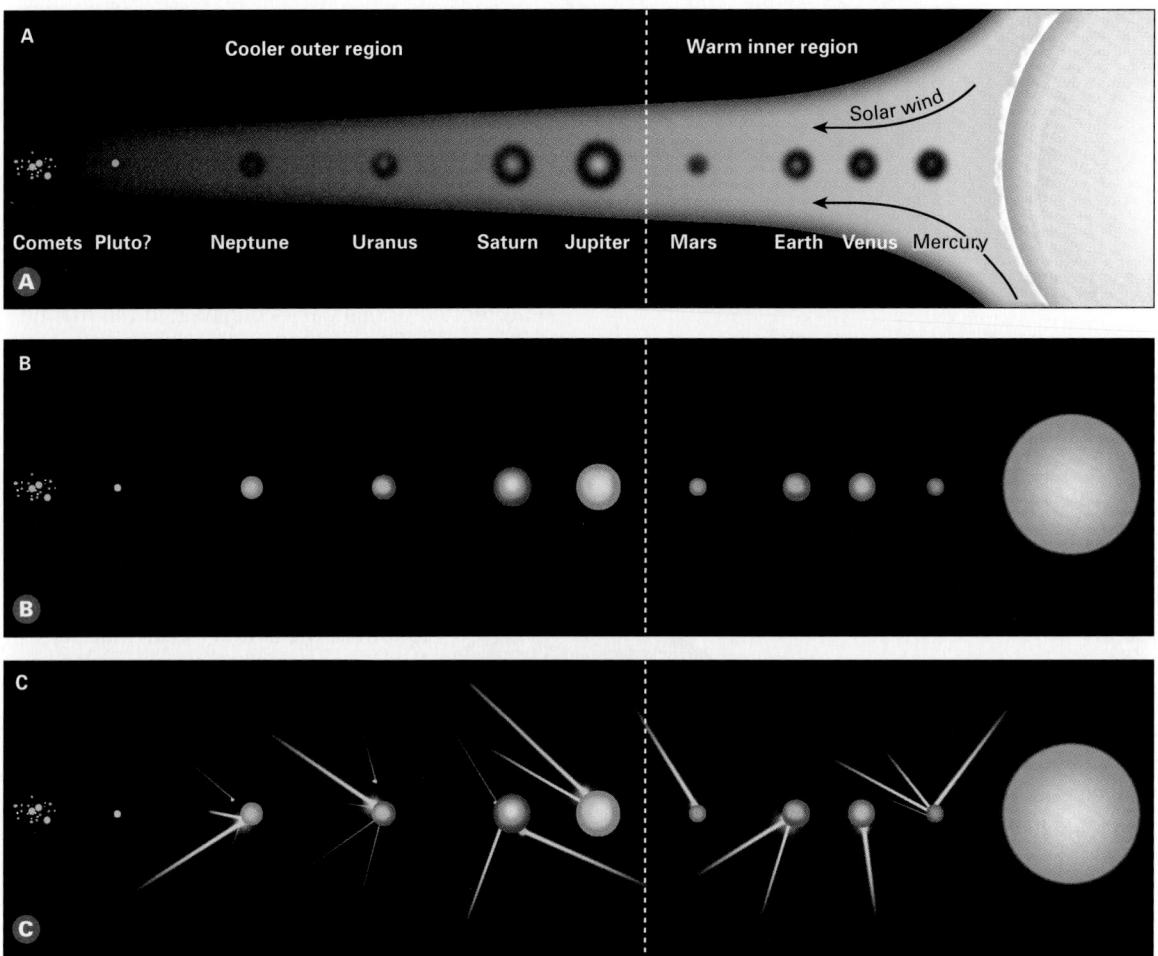

FIGURE 15.1 (A) As the Solar System formed, volatiles boiled away from the inner planets. The solar wind then blew them into the cool region beyond Mars. (B) The outer planets captured some of the volatiles, and some condensed to form comets in the frigid zone beyond Neptune. As a result, the inner four planets were left with no atmospheres or oceans. The outer planets grew to become giants. (Pluto is an anomalous planet, and its origin is discussed in Chapter 23) (C) Comets and meteorites crashed into Earth and other planets, returning some of the volatiles to their surfaces. The volatiles accumulated to form the atmosphere, oceans, and the foundation of life on Earth. (Sizes and distances in this drawing are not to scale.)

surface. According to one estimate, outgassing of the deep mantle accounted for no more than 10 percent of Earth's hydrosphere, atmosphere, and biosphere.[1]

If this scenario is correct, why do modern volcanoes emit volatiles? According to one hypothesis, most of the gases given off by modern volcanoes are recycled from the surface. Water, carbon (in the form of carbonate rocks such as limestone), and other light compounds are carried into shallow parts of the mantle by

subducting slabs. These volatiles return to the surface during volcanic eruptions. Therefore, modern volcanic eruptions, like their primordial ancestors, do not outgas appreciable quantities of volatiles from the deep mantle.

Now let's return to the volatiles that streamed away from the hot, inner Solar System. As they flew away from the Sun, volatiles entered a cooler region beyond Mars. Most of the volatiles were captured by the outer planets—Jupiter, Saturn, Uranus, and Neptune—but some continued their journey toward the outer fringe of the Solar System (Figure 15.1B). Here, beyond the orbits of the known planets, volatiles from the inner Solar System combined with residual

1. Paul J. Thomas, Christopher F. Chyba, and Christopher P. McKay, *Comets and the Origin and Evolution of Life* (New York: Springer-Verlag, 1997).

Unit 4: The Oceans

FIGURE 15.2 Comet Hale–Bopp. The early Solar System was crowded with comets, meteoroids, and asteroids; these bodies are composed of rock and condensed volatile materials such as ice and solid carbon dioxide.

© ARTSHOTS/SHUTTERSTOCK

dust and gas to form comets (Figure 15.2). A comet's nucleus has been compared to a dirty snowball because it is composed mainly of ice and rock. Other compounds not common in snowballs exist in comets as well. These include frozen carbon dioxide, ammonia, and simple organic molecules. Volatiles are also abundant in certain types of meteoroids and asteroids in the region between Mars and Jupiter.

Astronomers calculate that the early Solar System was crowded with comets, meteoroids, and asteroids—space debris left over from planetary formation. Many contained volatile compounds. When a large piece of space debris crashes into a planet, it is called a **bolide**. A large number of bolides crashed into Earth, nearby planets, and moons (Figure 15.1C). While falling space debris added only 0.001 percent to Earth's total mass, it imported 90 percent of its modern reservoir of volatiles.

Upon entry and impact, the frozen volatiles in the bolides vaporized, releasing water vapor, carbon dioxide, ammonia, simple organic molecules, and other volatiles. As the planet cooled and atmospheric pressure increased, the water vapor condensed to liquid, forming the first oceans. The light molecules transported to Earth in bolides also provided gases that formed the atmosphere and the raw materials for life. (The formation and evolution of the atmosphere is discussed in Chapter 17.)

Thus, at least some, and probably most, of the compounds necessary to produce the hydrosphere, the atmosphere, and the biosphere traveled to Earth from outer regions of the Solar System. The water that fills Earth's oceans came from interplanetary space.

bolide A large piece of space debris, such as an asteroid, that crashes into a planet.

Later in Earth's history, impacts from outer space blasted rock and dust into the sky, causing mass extinctions and killing large portions of life on Earth. Thus extraterrestrial impacts may have provided the raw materials for the oceans, the atmosphere, and for life—and later caused mass extinctions.

15.2 The Earth's Oceans

If you were to ask most people to describe the difference between a continent and an ocean, they would almost certainly reply, "Why, obviously, a continent is land and an ocean is water!" This observation is true, of course, but to a geologist another distinction is more important. The geologist would explain that rocks beneath the oceans are different from those of a continent. The accumulation of seawater in the world's ocean basins is a *result* of that difference.

Modern oceanic crust is dense basalt and varies from 4 to 7 kilometers thick. Continental crust is made of lower-density granite and averages 20 to 40 kilometers in thickness. In addition, the entire continental lithosphere is both thicker and less dense than oceanic lithosphere. As a result of these differences, the thick, lower-density continental lithosphere floats isostatically at high elevations, whereas oceanic lithosphere sinks to low elevations. Most of Earth's water flows downhill to collect in the depressions formed by oceanic lithosphere. Even if no water existed on Earth's surface, oceanic crust would form deep basins and continental crust would rise to higher elevations.

Oceans cover about 71 percent of Earth's surface. The sea floor is about 5 kilometers deep in the central parts of the ocean basins, although it is only 2 to 3 kilometers deep above the Mid-Oceanic Ridge and plunges to 11 kilometers in the Mariana Trench (Figure 15.3).

The ocean basins contain 1.4 billion cubic kilometers of water—18 times more than the volume of all land above sea level. So much water exists at Earth's surface that if Earth were a perfectly smooth sphere, it would be covered by a global ocean 2,000 meters deep.

The size and shape of Earth's ocean basins change over geologic time. At present, the Atlantic Ocean is growing wider at a rate of a few centimeters each year as the sea floor spreads apart at the Mid-Atlantic Ridge and as the Americas move away from Europe and Africa. At the same time, the Pacific is shrinking at a similar rate, as oceanic crust sinks into subduction

rock dredge An open-mouthed steel net dragged along the sea floor behind a research ship for the purpose of sampling rocks from submarine outcrops.

sea-floor drilling A process in which drill rigs mounted on offshore platforms or on research vessels cut cylindrical cores from both sediment and rock of the sea floor, which are then brought to the surface for study.

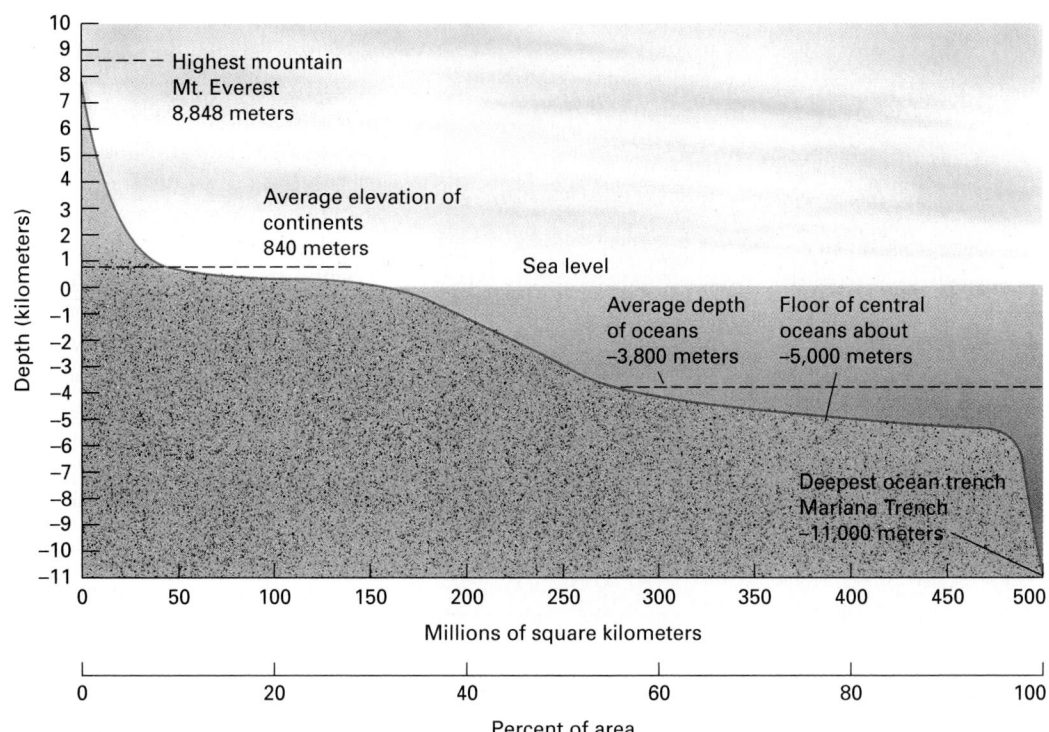

FIGURE 15.3 A schematic cross section of the continents and ocean basins. The vertical axis shows elevations relative to sea level. The horizontal axis shows the relative areas of the types of topography. Thus, 30 percent, or about 150 million square kilometers, of Earth's surface lies above sea level.

zones around its edges. In short, the Atlantic Ocean basin is now expanding at the expense of the Pacific.

The oceans (hydrosphere) affect global climate (atmosphere) and the biosphere in many ways. The seas absorb and store solar heat more efficiently than do rocks and soil. As a result, oceans are generally warmer in winter and cooler in summer than adjacent land is. Most of the water that falls as rain or snow is water that evaporated from the seas. In addition, ocean currents transport heat from the equator toward the poles, cooling equatorial climates and warming polar environments. Because plate tectonic activities alter the sizes and shapes of ocean basins, they also alter oceanic currents and profoundly affect regional climates over geologic time. In these and other ways, the oceans play a large role in Earth systems interactions.

15.3 Studying the Sea Floor

Seventy-five years ago, scientists had better maps of the Moon than of the sea floor. The Moon is clearly visible in the night sky, and we can view its surface with a telescope. The sea floor, however, is deep, dark, and inhospitable to humans. Modern oceanographers use a variety of techniques to study the sea floor, including several types of sampling and remote sensing.

Sampling

Several devices collect sediment and rock directly from the ocean floor. A **rock dredge** is an open-mouthed steel net dragged along the sea floor behind a research ship. The dredge breaks rocks from submarine outcrops and hauls them to the surface. Oceanographers sample sea-floor mud by lowering a weighted, hollow steel pipe from a research vessel. The weight drives the pipe into the soft sediment, which is forced into the pipe. The sediment core is retrieved from the pipe after it is winched back to the surface. If the core is removed from the pipe carefully, even the most delicate sedimentary layering is preserved.

Sea-floor drilling methods developed for oil exploration also take core samples from oceanic crust. Large drill rigs are mounted on offshore platforms and on research vessels. The drill cuts cylindrical cores from both sediment and rock, which are then brought to the surface for study. Although this type of sampling is expensive, cores can be taken from depths of several kilometers into oceanic crust.

Remote Sensing

Remote sensing methods do not require direct physical contact with the ocean floor, and for some studies this approach is both effective and economical (Figure

282

Unit 4: The Oceans

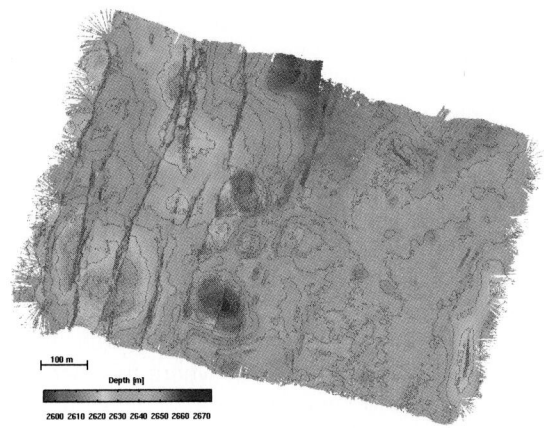

FIGURE 15.4 The deep-sea remote ABE sensor can image the sea floor in several ways on a variety of scales. This bathymetric map (measuring the water's depth) from the Lau Basin near Fiji covers an area of 1 by 0.6 kilometer, showing mounds, fissures, and hydrothermal vent spires with a resolution of about 1 meter.

15.4). The **echo sounder** is an instrument commonly used to map sea-floor topography. It emits a sound signal from a research ship and then records the signal after it bounces off the sea floor and travels back up to the ship. The water depth is calculated from the time required for the sound to make the round trip. A topographic map of the sea floor is constructed as the ship steers a carefully navigated course with the echo sounder operating continuously. Modern echo sounders, called SONAR, transmit 1,000 signals at a time to create more complete and accurate maps.

A **seismic profiler** works in the same way but uses a higher-energy signal that penetrates and reflects from layers in the sediment and rock. This gives a picture of the layering and structure of oceanic crust, as well as the sea-floor topography (Figure 15.5).

A **magnetometer** is an instrument that measures a magnetic field. Magnetometers towed behind research ships measure the magnetism of sea-floor rocks. Data collected by ship-borne magnetometers

echo sounder An instrument that emits sound waves and then records them as they reflect off the sea floor; the data are then used to record the topography of the sea floor.

seismic profiler A device that emits a high-energy signal that penetrates and reflects from layers in sediment and rock beneath the sea floor; the data are used to construct a topographic profile of the sea floor.

magnetometer An instrument that measures a magnetic field.

Recently, scientists have used deep-diving robots and laser imagers to sample and photograph the sea floor. A robot is cheaper and safer than a submarine, and a laser imager penetrates up to eight times farther through water than a conventional camera does.

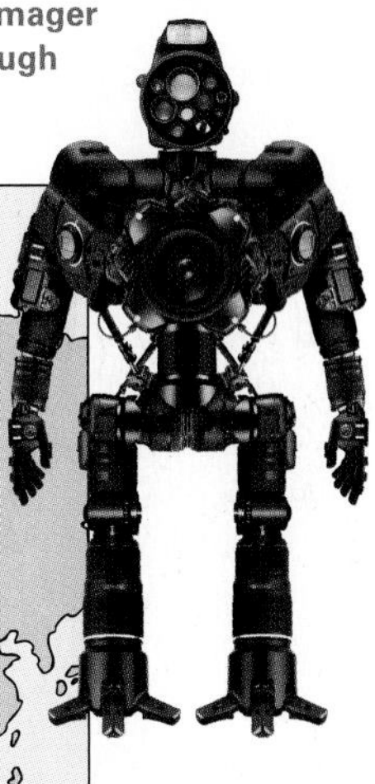

© VOLODYMYR KRASYUK/SHUTTERSTOCK

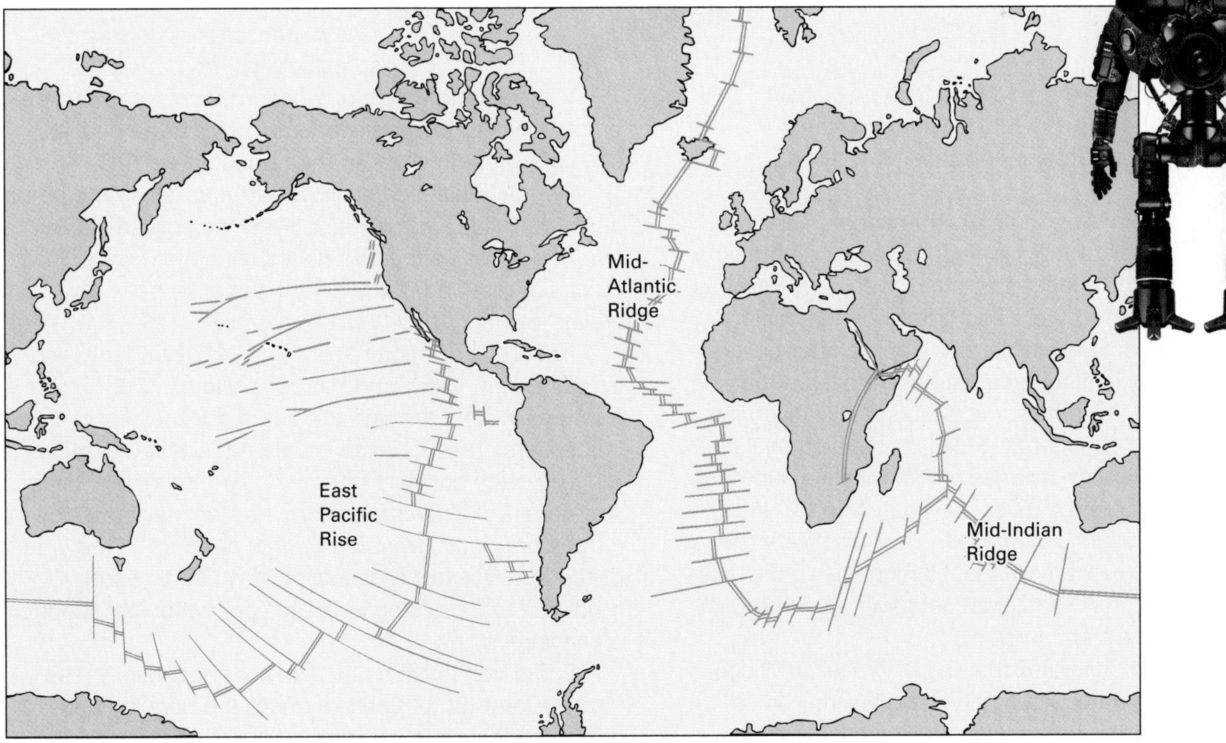

FIGURE 15.5 This map shows that divergent plate boundaries, or spreading centers, coincide exactly with the Mid-Oceanic Ridge system in the world's oceans. The spreading centers are shown in double red lines; the single red lines are transform faults.

resulted in the now-famous discovery of symmetric magnetic stripes on the sea floor. That discovery rapidly led to the development of the sea-floor spreading hypothesis and of the theory of plate tectonics shortly thereafter, as described in Chapter 6.

Satellite-based **microwave radar** instruments measure the echo of microwave pulses to detect subtle swells and depressions on the sea surface. These features reflect sea-floor topography. For example, the mass of a sea-floor mountain 4,000 meters high creates sufficient gravitational attraction to produce a gentle, 6-meter-high swell on the sea surface directly above it. The data are used to make sea-floor maps.

15.4 Features of the Sea Floor

The Mid-Oceanic Ridge System

Following World War I (1914–1918), oceanographers began using early versions of echo-sounding devices to measure ocean depths. Those surveys showed that the sea floor was much more rugged than previously thought, and the surveys further identified the continuity and size of Middle Ground, which is now called the Mid-Atlantic Ridge.

During World War II, naval commanders needed topographic maps of the sea floor to support submarine warfare. Those detailed maps, made with early versions of the echo sounder, were kept secret by the military. When they became available to the public after peace was restored, scientists were surprised to learn that the ocean floor has at least as much topographic diversity and relief as the continents. Broad plains, high peaks, and deep valleys form a varied and fascinating submarine landscape. In the 1950s, oceanographic surveys conducted by several nations led to the discovery that Middle Ground, or the Mid-Atlantic Ridge, is just part of a great submarine mountain range, now called the *Mid-Oceanic Ridge system*.

Recall from Chapter 6 that the Mid-Oceanic Ridge system is a continuous submarine mountain chain that encircles the globe. Its total length exceeds 80,000 kilometers, and it is more than 1,500 kilometers wide in places. The ridge rises an average of 2 to 3 kilometers above the surrounding deep sea floor. Although it lies almost exclusively beneath the seas, it is Earth's largest mountain chain, covering more than 20 percent of Earth's surface, about two-thirds as much as all continents combined. Even the Himalayas, Earth's largest continental mountain chain, occupy only a small fraction of that area.

A **rift valley** is an elongate depression that develops at a divergent plate boundary. In the Mid-Oceanic Ridge system, a rift valley 1 to 2 kilometers deep and several kilometers wide splits many segments of the ridge crest. Oceanographers in small research submarines can dive into the rift valley, where they see gaping vertical cracks up to 3 meters wide on the valley floor. Recall that the Mid-Oceanic Ridge system is a spreading center, where two lithospheric plates are spreading apart from each other. The cracks form as brittle oceanic crust separates at the ridge axis. Basaltic magma then rises through the cracks and flows onto the floor of the rift valley. This basalt becomes new oceanic crust as two lithospheric plates spread outward from the ridge axis.

The new crust (and the underlying lithosphere) at the ridge axis is warmer, and therefore of relatively lower density, than older crust and lithosphere located farther from the ridge axis. Its buoyancy causes it to float high above the surrounding sea floor, elevating the Mid-Oceanic Ridge system 2 to 3 kilometers above the deep sea floor. The new lithosphere cools as it spreads away from the ridge. As a result of cooling, it becomes thicker and denser and sinks to lower elevations, forming the deeper sea floor on both sides of the ridge (Figure 15.6).

Normal faults and shallow earthquakes are common along the Mid-Oceanic Ridge system because oceanic crust fractures as the two plates separate (Figure 15.7). Blocks of crust drop downward along the sea-floor cracks, forming the rift valley.

Hundreds of fractures, called **transform faults**, cut across the rift valley and the ridge (Figure 15.8). These fractures extend through the entire thickness of the lithosphere. They develop because the Mid-Oceanic Ridge system consists of many short segments. Each segment is slightly offset from adjacent segments by a transform fault. Transform faults are original features of the Mid-Oceanic Ridge; they form, as an accommodation to Earth's spherical shape, when lithospheric spreading begins.

Some transform faults displace the ridge by less than a kilometer, but others offset the ridge by hundreds of kilometers. In some cases, a transform fault can grow so large that it forms a transform plate boundary. The San Andreas Fault in California is a transform plate boundary.

© FENG YU/SHUTTERSTOCK

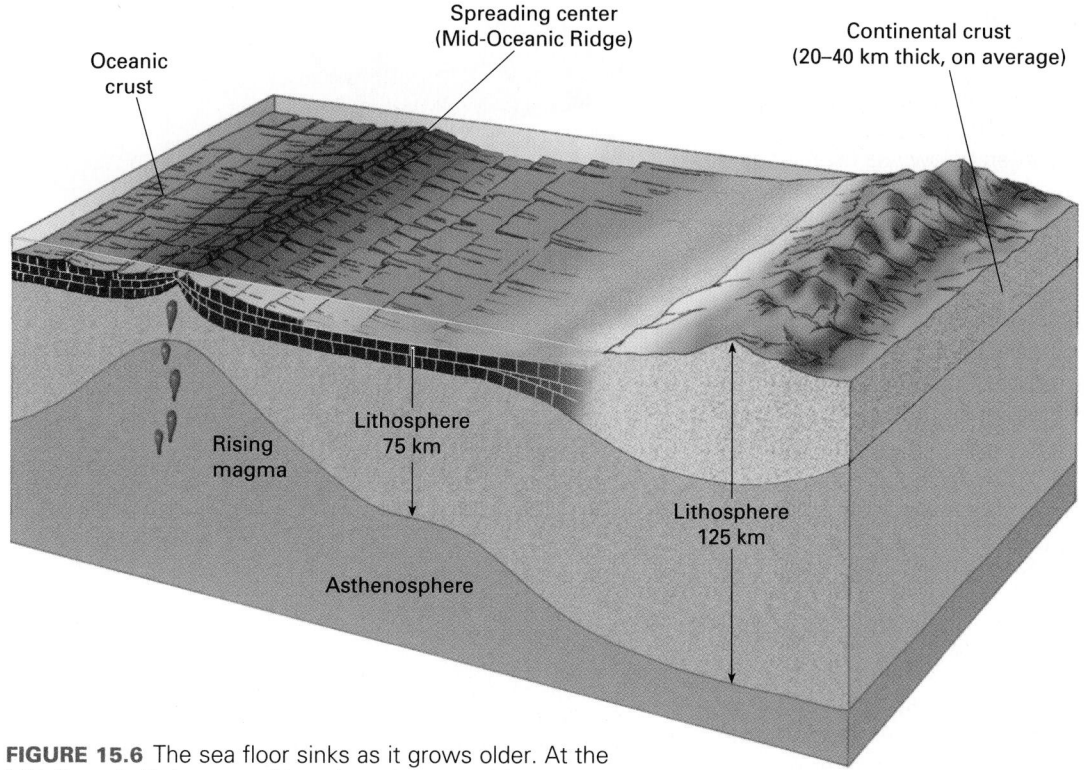

FIGURE 15.6 The sea floor sinks as it grows older. At the Mid-Oceanic Ridge, new lithosphere is buoyant because it is hot and of low density. It ages, cools, thickens, and becomes denser as it moves away from the ridge and consequently sinks. The central portion of the sea floor lies at a depth of about 5 kilometers.

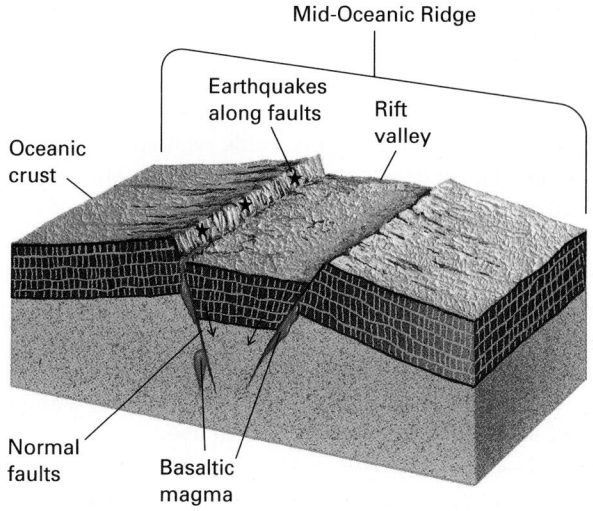

FIGURE 15.7 A cross-sectional view of the central rift valley in the Mid-Oceanic Ridge. As the plates separate, blocks of rock drop down along the fractures to form the rift valley. The moving blocks cause earthquakes.

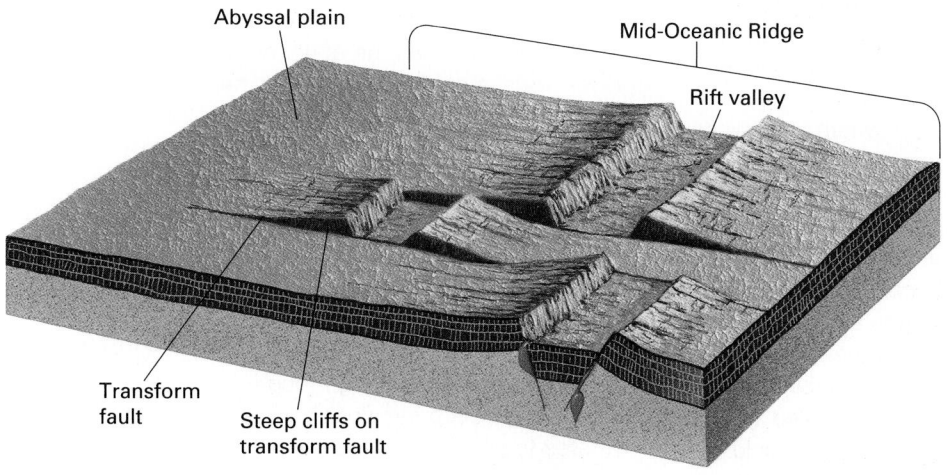

FIGURE 15.8 Transform faults offset segments of the Mid-Oceanic Ridge. Adjacent segments of the ridge may be separated by steep cliffs 3 kilometers high. Note the flat abyssal plain far from the ridge.

Global Sea-Level Changes and the Mid-Oceanic Ridge System

A thin layer of marine sedimentary rocks blankets large areas of Earth's continents. These rocks tell us that those places must have been below sea level when the sediment accumulated.

Tectonic activity can cause a continent to sink, allowing the sea to flood a large area. However, at particular times in the past (most notably during the Cambrian, Carboniferous, and Cretaceous Periods), marine sediments accumulated on low-lying portions of all continents simultaneously, indicating simultaneous global flooding of low parts of all continents. Although our plate tectonics model explains the sinking of individual continents, or parts of continents, it does not explain why all continents should sink at the same time. Therefore, we need to explain how sea level could rise globally by hundreds of meters to flood all continents simultaneously.

Continental glaciers have advanced and melted numerous times in Earth's history. During the growth of continental glaciers, seawater evaporates and is frozen into the ice that rests on land. As a result, sea level drops. When glaciers melt, the water runs back into the oceans and sea level rises. The alternating growth and melting of glaciers during the Pleistocene Epoch caused sea level to fluctuate by as much as 200 meters. However, the ages of most marine sedimentary rocks on continents do not coincide with times of glacial melting. Therefore, we must look for a different cause to explain continental flooding.

Recall that the new, hot lithosphere at a spreading center is buoyant, causing the Mid-Oceanic Ridge system to rise above the surrounding sea floor. This submarine mountain chain displaces a huge volume of seawater. If the Mid-Oceanic Ridge system were smaller, it would displace less seawater and sea level would fall. If it were larger, sea level would rise.

The Mid-Oceanic Ridge rises highest at the spreading center, where new lithosphere rock is hottest and has the lowest density. The elevation of the ridge decreases on both sides of the spreading center because the lithosphere cools and shrinks as it moves outward.

Now consider a spreading center where spreading is very slow (perhaps 1 to 2 centimeters per year). At such a slow rate, the newly formed lithosphere would cool before it migrated far from the spreading center. As a result, the ridge would be narrow and of low volume, as shown in Figure 15.9A. In contrast, rapid sea-floor spreading of 10 to 20 centimeters per year would create a high-volume ridge because the newly formed, hot lithosphere would be carried a considerable distance away from the spreading center before it cooled and shrank (Figure 15.9B). This high-volume ridge would

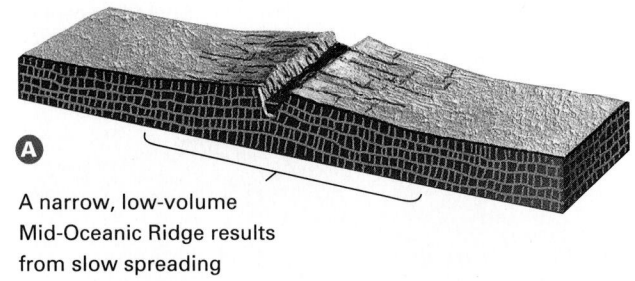

A narrow, low-volume Mid-Oceanic Ridge results from slow spreading

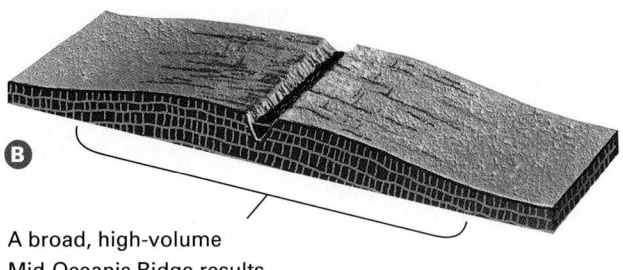

A broad, high-volume Mid-Oceanic Ridge results from rapid spreading

FIGURE 15.9 (A) Slow sea-floor spreading creates a narrow, low-volume Mid-Oceanic Ridge that displaces less seawater and lowers sea level. (B) Rapid sea-floor spreading creates a wide, high-volume ridge that displaces more seawater and raises sea level.

displace considerably more seawater than a low-volume ridge would displace, and the high-volume ridge would cause a global sea-level rise. If the modern Mid-Oceanic Ridge system were to disappear completely, sea level would fall by about 400 meters.

Sea-floor age data indicate that the rate of sea-floor spreading has varied from about 2 to 16 centimeters per year since Jurassic time, about 200 million years ago. Sea-floor spreading was unusually rapid during Late Cretaceous time, between 110 and 85 million years ago. That rapid spreading should have formed an unusually high-volumed Mid-Oceanic Ridge and resulted in flooding of low-lying portions of continents. Geologists have found marine sedimentary rocks of Late Cretaceous age on nearly all continents, indicating that Late Cretaceous time was, in fact, a time of abnormally high global sea level. Thus, a process initiated by heat transfer deep within the geosphere profoundly affected sea level of the hydrosphere and life throughout the biosphere. Unfortunately, because no oceanic crust is older than about 200 million years, the hypothesis cannot be tested for earlier times when extensive marine sedimentary rocks accumulated on continents.

Life on the Mid-Oceanic Ridge System

Oceanographers had long thought that little life could exist on the deep sea floor because no sunlight penetrates to those depths to support photosynthesis.

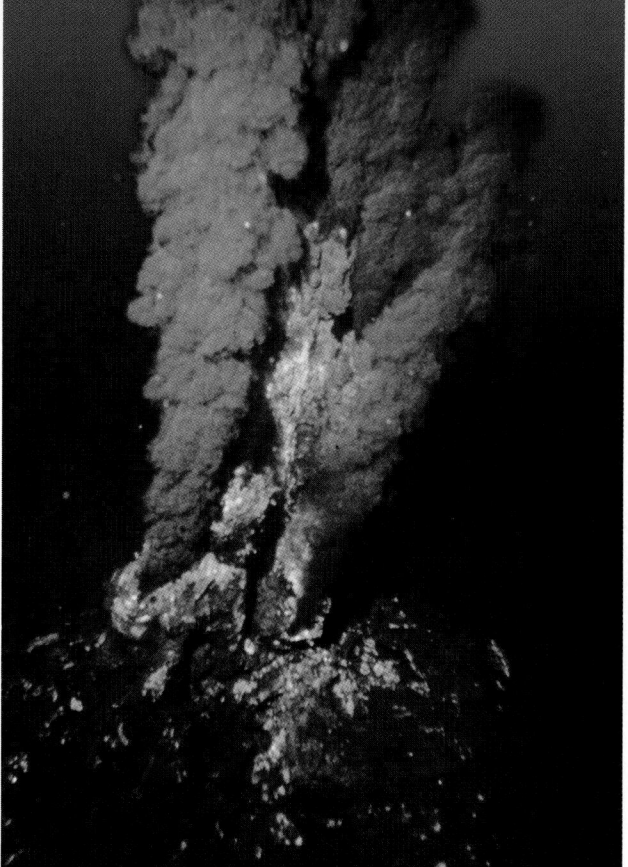

FIGURE 15.10 A black smoker spouts from the East Pacific rise. Seawater is heated as it circulates through hot sea-floor rocks, and it dissolves metals and sulfur from the rocks. The ions precipitate as "smoke," consisting of tiny mineral grains, when the hot solution spouts into cold ocean water.

However, scientists using modern diving and sampling techniques have discovered thriving communities and a unique food chain in isolated parts of the deep sea floor.

On the volcanically active Mid-Oceanic Ridge system, the hot rocks heat seawater as it circulates through fractures in oceanic crust. The hot water dissolves metals and sulfur from the rocks. Eventually, the hot, metal-and-sulfur-laden water rises back to the sea-floor surface, spouting from fractures as a jet of black water called a *black smoker* (described in Chapter 5). The black color is caused by precipitation of fine-grained metal sulfide minerals as the solutions cool on contact with seawater (Figure 15.10).

These scalding, sulfurous waters are as hot as 400°C and would be toxic to life on land. Yet the deep sea floor around a black smoker teems with life. At the vents, bacteria produce energy from hydrogen sulfide in a process called **chemosynthesis**. Thus, the bacteria release energy from chemicals and are not dependent on photosynthesis. The chemosynthetic bacteria are the foundation of a deep-sea food chain—either larger vent organisms eat them or the larger organisms live symbiotically with the bacteria. For example, instead of a digestive tract, the red-tipped tube worm (Figure

15.11) has a special organ that hosts the chemosynthetic bacteria. It provides a home for the bacteria and in return receives nutrition from the bacteria's wastes. Other vent organisms in this unique food chain include giant clams and mussels, eyeless shrimp, crabs, and fish.

Oceanic Trenches and Island Arcs

In many parts of the Pacific Ocean and in some other ocean basins, two oceanic plates converge. One dives beneath the other, forming a subduction zone. The sinking plate drags the sea floor downward, forming a long, narrow depression called an **oceanic trench**. The deepest place on Earth is in the Mariana

chemosynthesis
A process in which bacteria produce energy from hydrogen sulfide and thus are not dependent on photosynthesis.

oceanic trench
A long, narrow, steep-sided depression of the sea floor formed where a subducting oceanic plate sinks into the mantle, dragging the sea floor downward.

FIGURE 15.11 These red tubeworms are part of a thriving plant and animal community living near a black smoker in the Guaymas Basin in the Gulf of California.

accreted terrane
A landmass that originated as an island arc or a microcontinent and was later added onto a continent.

Trench, north of New Guinea in the southwestern Pacific, where the ocean floor sinks to nearly 11 kilometers below sea level. Depths of 8 to 10 kilometers are common in other trenches.

Huge amounts of magma are generated in the subduction zone. The magma rises and erupts on the sea floor to form submarine volcanoes next to the trench. The volcanoes eventually grow to become a chain of islands called an *island arc* (Figure 15.12), as we learned in Chapter 9. The western Aleutian Islands are an example of an island arc. Many others occur at the numerous convergent plate boundaries in the western Pacific (Figure 15.13).

If subduction stops after an island arc forms, volcanic activity also ends. The island arc may then ride quietly on a tectonic plate until it arrives at another subduction zone at a continental margin. However, the density of island arc rocks is relatively low, making them too buoyant to sink into the mantle (Figures 15.14A and 15.14B). Instead, the island arc collides with the continent. When this happens, the subducting plate commonly fractures on the seaward side of the island arc to form a new subduction zone. In this way, the island arc breaks away from the ocean plate and becomes part of the continent (Figure 15.14C). Much of western California, Oregon, Washington, and western

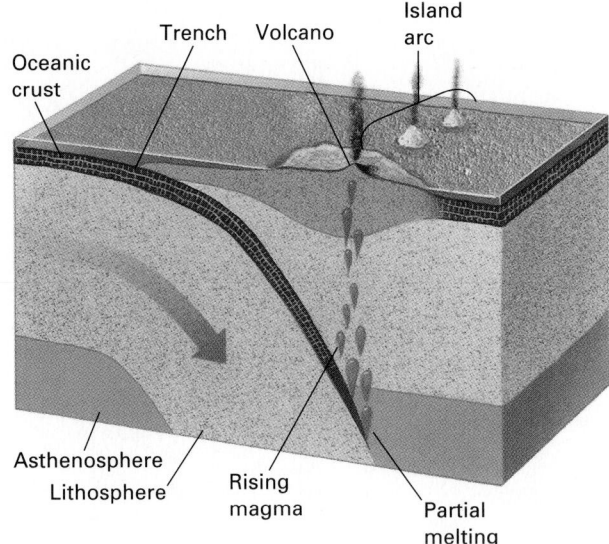

FIGURE 15.12 An oceanic trench forms at a convergent boundary between two oceanic plates. One of the plates sinks, generating magma that rises to form a chain of volcanic islands called an island arc.

British Columbia were added to North America in this way from 180 million to about 50 million years ago. These late additions to our continent, called **accreted terranes**, are shown in Figure 15.15.

FIGURE 15.13 The Lesser Sunda Islands are part of the Indonesian island arc.

Unit 4: The Oceans

Note the following points:

1. An island arc forms as magma rises from the mantle at an oceanic subduction zone.
2. The island arc eventually migrates toward the edge of a continent and becomes part of it.
3. The continent with the added island arc cannot sink into the mantle at a subduction zone because of its buoyancy.
4. Thus, material is transferred from the mantle to a continent. This aspect of the plate tectonics model suggests that the amount of continental crust has increased throughout geologic time. However, some geologists feel that small amounts of continental crust can also

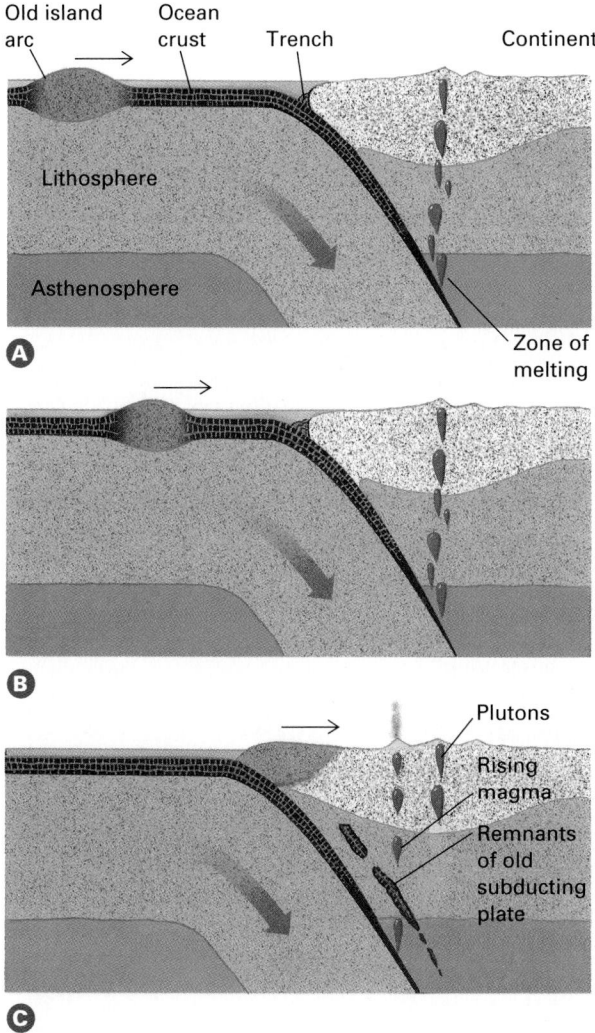

FIGURE 15.14 (A) An island arc is part of a lithospheric plate that is sinking into a subduction zone beneath a continent. (B) The island arc reaches the subduction zone but cannot sink into the mantle because of its low density. (C) The island arc is jammed onto the continental margin and becomes part of the continent. The subduction zone and trench step back to the seaward side of the island arc.

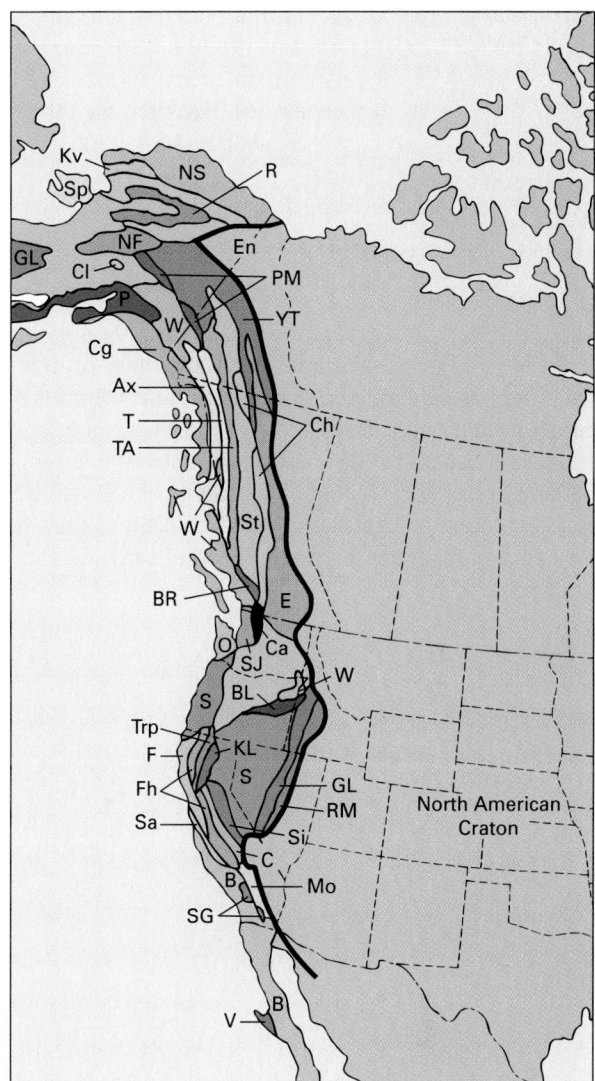

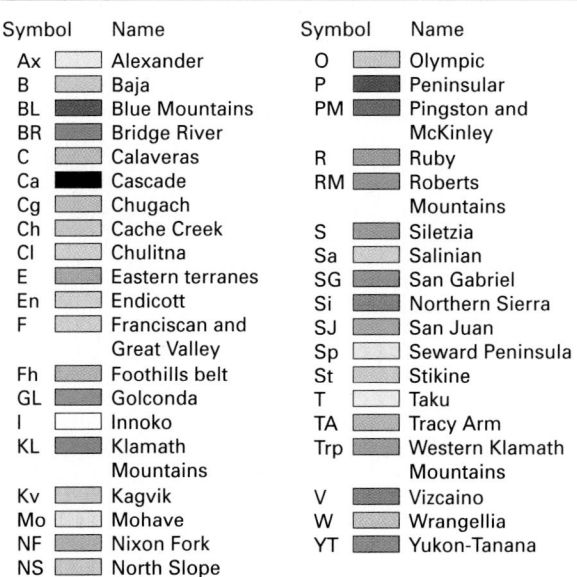

Symbol	Name	Symbol	Name
Ax	Alexander	O	Olympic
B	Baja	P	Peninsular
BL	Blue Mountains	PM	Pingston and McKinley
BR	Bridge River		
C	Calaveras	R	Ruby
Ca	Cascade	RM	Roberts Mountains
Cg	Chugach		
Ch	Cache Creek	S	Siletzia
Cl	Chulitna	Sa	Salinian
E	Eastern terranes	SG	San Gabriel
En	Endicott	Si	Northern Sierra
F	Franciscan and Great Valley	SJ	San Juan
		Sp	Seward Peninsula
Fh	Foothills belt	St	Stikine
GL	Golconda	T	Taku
I	Innoko	TA	Tracy Arm
KL	Klamath Mountains	Trp	Western Klamath Mountains
Kv	Kagvik	V	Vizcaino
Mo	Mohave	W	Wrangellia
NF	Nixon Fork	YT	Yukon-Tanana
NS	North Slope		

FIGURE 15.15 The accreted terranes of western North America are microcontinents and island arcs from the Pacific Ocean that were added to the continent.

seamount A sub-
marine mountain,
usually of volcanic
origin, that rises 1
kilometer or more
above the sur-
rounding sea floor.

oceanic island A
submarine moun-
tain (seamount)
that rises above
sea level.

return to the mantle in subduction zones, and that the total amount of continental crust has been approximately constant for the past 2.5 billion years.

Seamounts, Oceanic Islands, and Atolls

A **seamount** is a submarine mountain that rises 1 kilometer or more above the surrounding sea floor. An **oceanic island** is a seamount that rises above sea level. Both are common in all ocean basins but are particularly abundant in the southwestern Pacific Ocean. Seamounts and oceanic islands sometimes occur as isolated peaks, but they are more commonly found in chains. Dredge samples show

that seamounts, oceanic islands, and the ocean floor itself are all made of basalt.

Most seamounts and oceanic islands are volcanoes that formed at a hot spot above a mantle plume, and most form within a tectonic plate rather than at a plate boundary. An isolated seamount or short chain of small seamounts probably formed over a plume that lasted for only a short time. In contrast, a long chain of large islands, such as the Hawaiian Island–Emperor Seamount chain, formed over a long-lasting plume. In this case the lithospheric plate migrated over the plume as the magma continued to rise from a source beneath the lithosphere. Each volcano formed directly over the plume and then became extinct as the moving plate carried it away from the plume. As a result, the seamounts and oceanic islands become progressively younger toward the end of the chain that is volcanically active today (Figure 15.16).

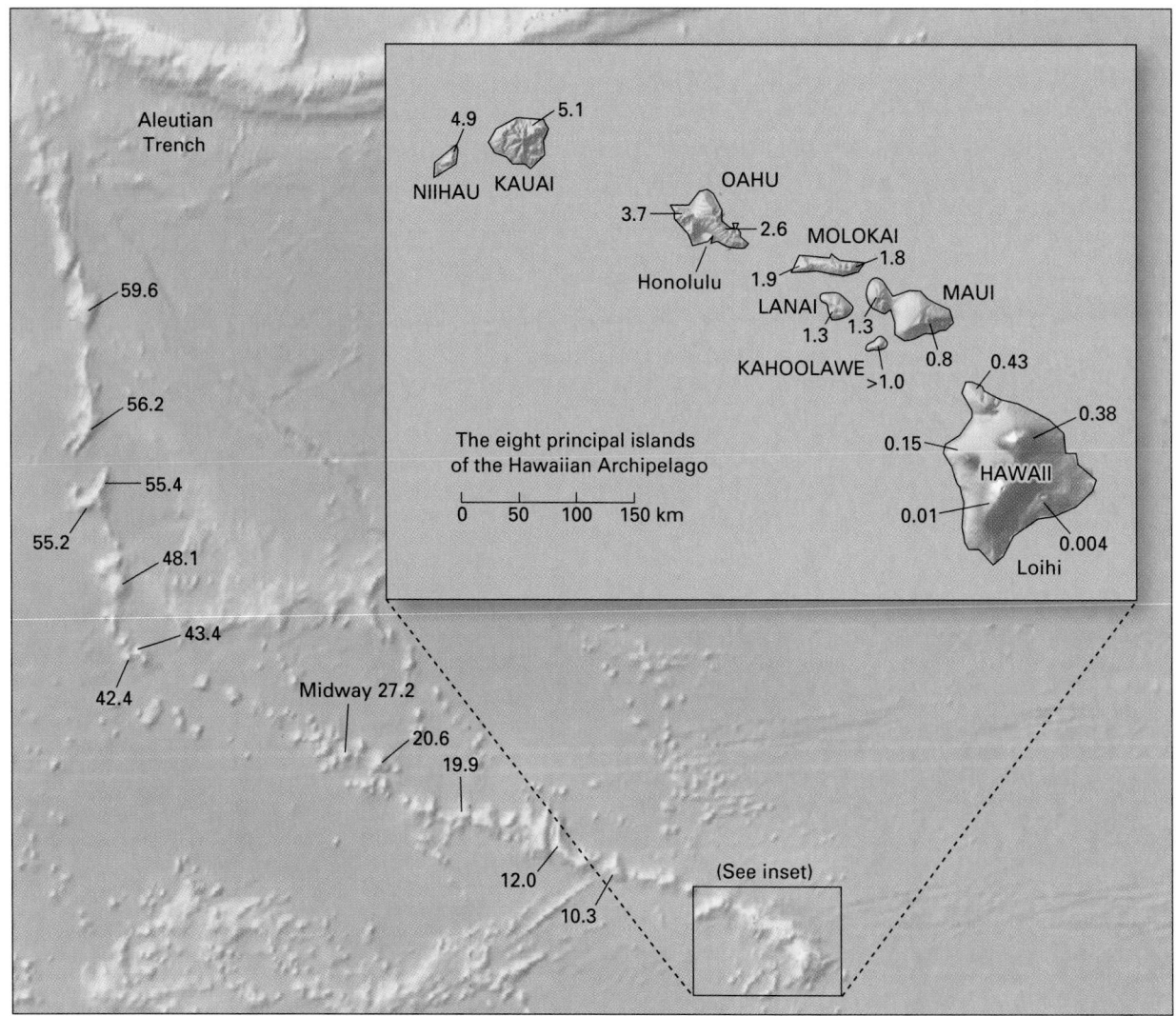

FIGURE 15.16 The Hawaiian Island–Emperor Seamount chain becomes older in a direction going away from the island of Hawaii. The numbers represent ages, in millions of years, of the oldest volcanic rocks of each island or seamount.

Unit 4: The Oceans

After a volcanic island forms, it begins to sink. Three factors contribute to the sinking:

1. If the mantle plume stops rising, it stops producing magma. Then the lithosphere beneath the island cools and becomes denser, and the island sinks. Alternatively, a moving plate may carry the island away from the hot spot. This also results in cooling, contraction, and sinking of the island.
2. The weight of the newly formed volcano causes isostatic sinking.
3. Erosion lowers the top of the volcano.

These three factors gradually transform a volcanic island to a seamount (Figure 15.17). If the Pacific Ocean plate continues to move at its present rate, the island of Hawaii may sink beneath the sea within 10 to 15 million years. Sea waves may erode a flat top on the sinking island, forming a flat-topped seamount called a **guyot** (pronounced "gee-o," after Swiss-born American geologist Arnold Henri Guyot), as illustrated in Figure 15.18.

The South Pacific and portions of the Indian Ocean are dotted with numerous other islands called *atolls*. An **atoll** is a circular coral reef that forms a ring of islands around a central lagoon (Figure 15.19). Atolls vary from 1 to 130 kilometers in diameter and are surrounded by deep water of the open sea. If corals live only in shallow water, how did atolls form in the deep sea? Charles Darwin studied this question during his famous voyage on the *Beagle* from 1831 to 1836. He reasoned that a coral reef must have formed in shallow water on the flanks of a volcanic island. Eventually the island sank, but the reef continued to grow upward, so that the living portion always remained in shallow water (Figure 15.20). This proposal was not accepted at first because scientists could not explain how a

guyot A flat-topped seamount, formed when the top of a sinking island is eroded by sea waves.

atoll A circular coral reef that forms a ring of islands around a central lagoon and is bounded on the outside by deep water of the open sea.

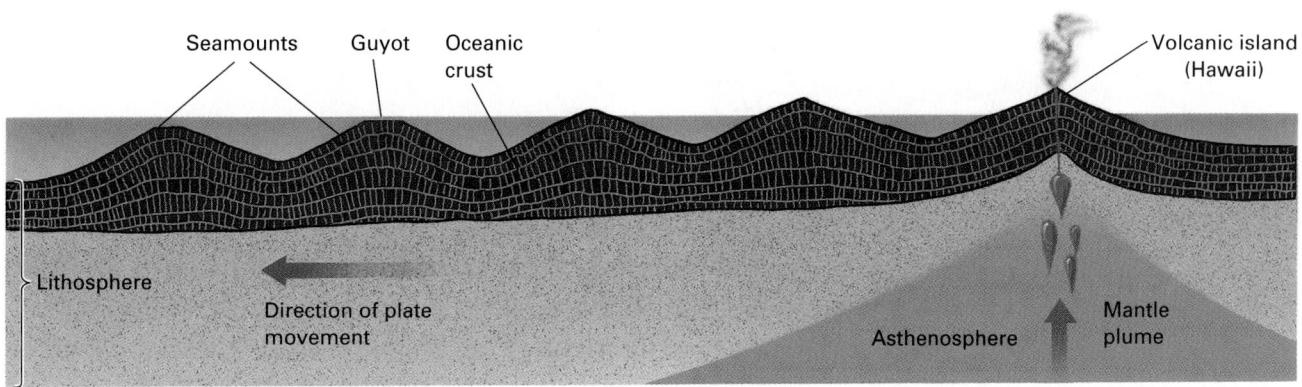

FIGURE 15.17 The Hawaiian Islands and Emperor Seamounts sink as they move away from the mantle plume.

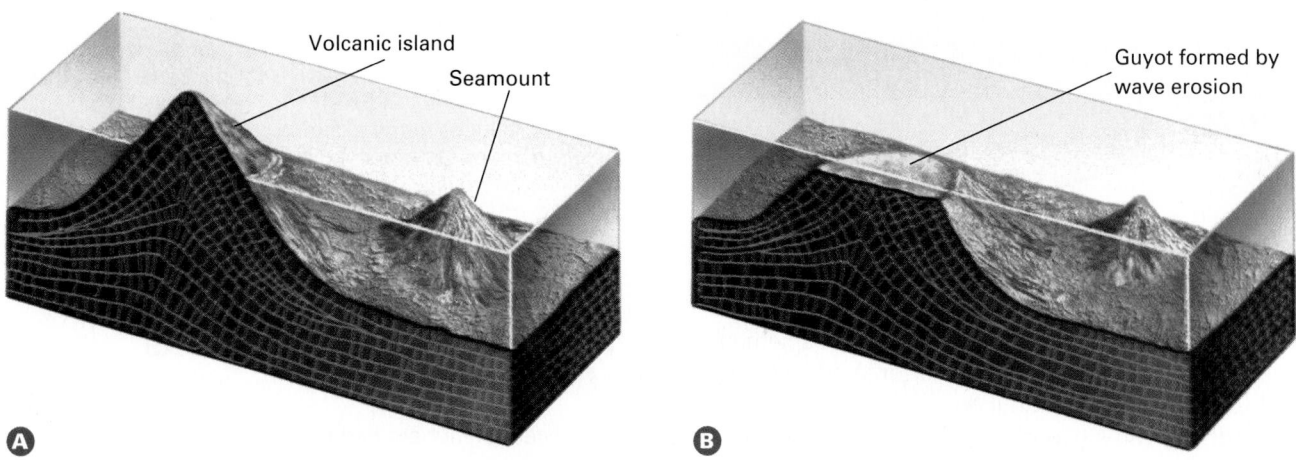

FIGURE 15.18 (A) A volcanic island rises above sea level. (B) Waves erode a flat top on a sinking island to form a guyot.

FIGURE 15.19 The Tetiaroa Atoll in French Polynesia formed by the process described in Figure 15.20. Over time, storm waves wash coral sands on top of the reef and vegetation grows on the sand, forming the individual islands in the atoll.

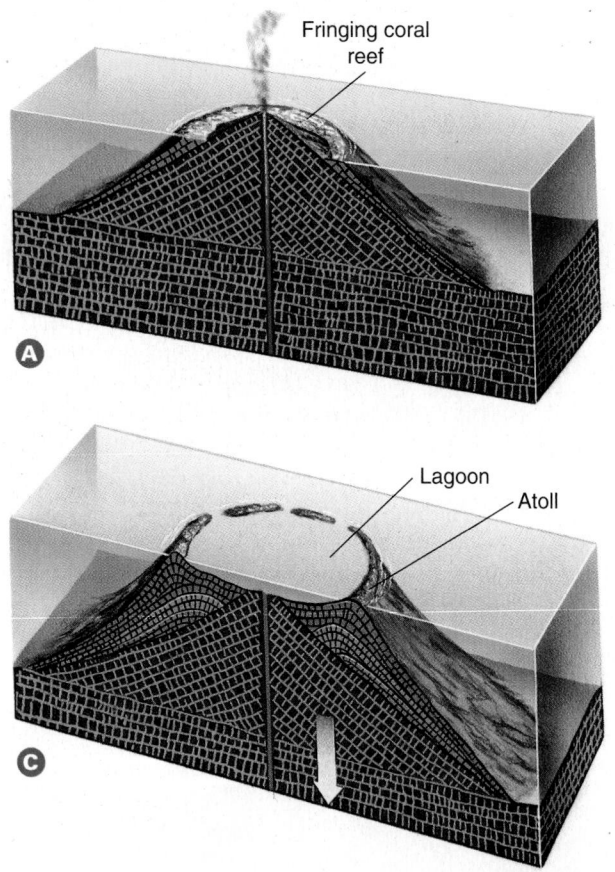

Fringing coral reef

A

Barrier reef

B

Lagoon

Atoll

C

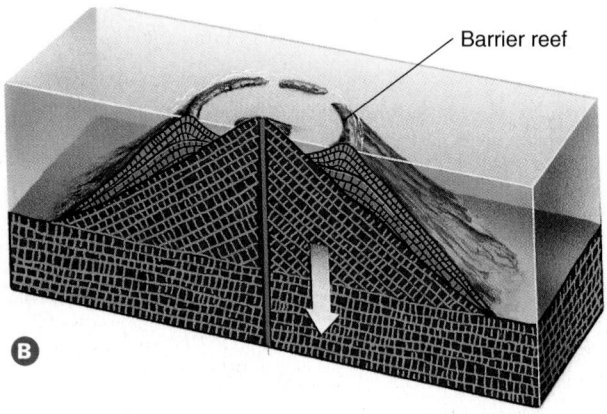

FIGURE 15.20 (A) A fringing reef grows along the shore of a young volcanic island. (B) As the island sinks, the reef continues to grow upward to form a barrier reef that encircles the island. (C) Finally the island sinks below sea level and the reef forms a circular atoll. This model of atoll formation was proposed by Charles Darwin in 1842, following his voyage on the *Beagle*.

volcanic island could sink. However, when scientists drilled into a Pacific atoll shortly after World War II and found volcanic rock hundreds of meters beneath the reef, Darwin's hypothesis was revived. It is considered accurate today, in light of our ability to explain why volcanic islands sink.

15.5 Sediment and Rocks of the Sea Floor

Early oceanographers had believed that the oceans are 4 billion years old, so mud on the ancient sea floor should have been very thick, having had so much time to accumulate. In 1947, however, scientists on the U.S. research ship *Atlantis* discovered that the mud layer on the bottom of the Atlantic Ocean is much thinner than they expected. Why is there so little mud on the sea floor? The answer to this question would be a crucial piece of evidence in development of the theory of plate tectonics.

Earth is 4.6 billion years old, and rocks as old as 3.96 billion years have been found on continents. Once formed, most continental crust remains near Earth's surface because of its buoyancy. In contrast, no parts of the sea floor are older than about 200 million years because oceanic crust forms continuously at the Mid-Oceanic Ridge and then recycles into the mantle at subduction zones. As the theory of plate tectonics was emerging in the early 1960s, it became apparent that the fact that only a relatively thin layer of mud covers the sea floor could be easily explained if the oldest sea floor is less than 200 million years old, not 4 billion

years old as oceanographers had once believed. This evidence supported the fledgling plate tectonics theory in its early days.

Seismic profiling and sea-floor drilling show that oceanic crust consists of three layers. The uppermost layer consists of sediment, and the lower two are basalt (Figure 15.21).

Ocean-Floor Sediment

The uppermost layer of oceanic crust, called **layer 1**, consists of two types of sediment. **Terrigenous sediment** is sand, silt, and clay eroded from the continents and carried to the deep sea floor by gravity and submarine currents. Most of this sediment is found close to the continents. **Pelagic sediment**, however, collects even on the deep sea floor far from continents. It is a gray and red-brown mixture of clay that was mostly carried from continents by wind, and the remains of tiny plants and animals that live in the surface waters of the oceans (Figure 15.22). When these organisms die, their remains slowly settle to the ocean floor.

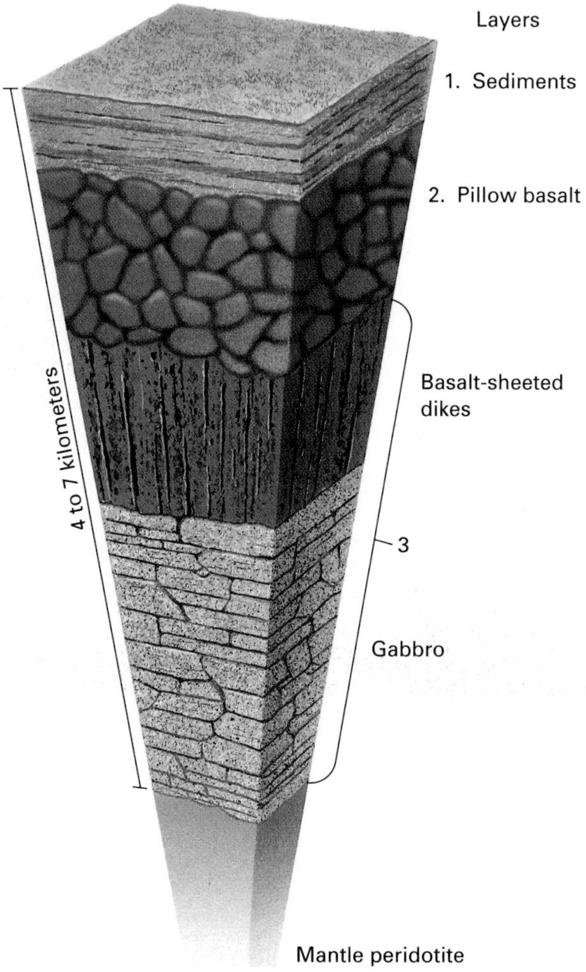

layer 1 The uppermost layer of oceanic crust, composed of terrigenous and pelagic sediment.

terrigenous sediment Sea-floor sediment composed of sand, silt, and clay eroded from the continents.

pelagic sediment Muddy ocean sediment that consists of a mixture of clay carried from continents and the skeletons of tiny marine organisms.

Layers

1. Sediments

2. Pillow basalt

Basalt-sheeted dikes

3

Gabbro

4 to 7 kilometers

Mantle peridotite

FIGURE 15.21 The three layers of oceanic crust. Layer 1 consists of sediment. Layer 2 is pillow basalt. Layer 3 consists of vertical dikes overlying gabbro. Below layer 3 is the upper mantle.

INTEGRATED OCEAN DRILLING PROGRAM

FIGURE 15.22 This scanning electron microscope photo shows foraminifera, tiny organisms that float near the surface of the seas. When these organisms die, their remains sink to the sea floor to become part of the pelagic mud layer. Each of the fossils is the size of a fine sand grain.

FIGURE 15.23 Underwater photo of pillow basalt off the island of Hawaii.

Pelagic sediment accumulates at a rate of about 2 to 10 millimeters per 1,000 years. Near the Mid-Oceanic Ridge system there is virtually no sediment because the sea floor is so young. The sediment thickness increases with distance from the ridge because the sea floor becomes older as it spreads away from the ridge. Close to shore, pelagic sediment gradually merges with the much-thicker layers of terrigenous sediment, which can be 3 kilometers or more thick. The observation of increasing thickness of sea-floor mud away from the ridge also supported the plate tectonics theory in its early days.

Parts of the ocean floor beyond the Mid-Oceanic Ridge system are flat, level, featureless surfaces called the **abyssal plains**. They are the flattest surfaces on Earth. Seismic profiling shows that the basaltic crust is rough and jagged throughout the ocean. On the abyssal plains, however, pelagic sediment buries this rugged profile, forming the smooth abyssal plains. If you were to remove all of the sediment, you would see rugged topography similar to that of the Mid-Oceanic Ridge.

Basaltic Oceanic Crust

Layer 2 lies below layer 1 and is about 1 to 2 kilometers thick. It consists mostly of **pillow basalt**, which forms as hot magma oozes onto the sea floor. Contact with cold sea water causes the molten lava to contract into pillow-shaped spheroids (Figure 15.23), much as molten candy "balls" in cold water.

Layer 3, 3 to 5 kilometers thick, is the deepest and thickest layer of oceanic crust. It directly overlies the mantle. The upper part consists of vertical basalt dikes, which formed as the magma oozing toward the surface froze in the cracks of the rift valley. The lower portion of layer 3 consists of gabbro, the coarse-grained equivalent of basalt. The gabbro forms as pools of magma cool slowly (because they are insulated by the basalt dikes above them).

The basaltic crust of layers 2 and 3 forms at the Mid-Oceanic Ridge. However, these rocks make up the foundation of all oceanic crust because all oceanic crust forms at the ridge axis and then spreads outward. In some places, chemical reactions with seawater have altered the basalt of layers 2 and 3 to a soft, green rock that contains up to 13 percent water.

15.6 Continental Margins

A continental margin is a place where continental crust meets oceanic crust. Two types of continental margins exist. A **passive continental margin** occurs where continental and oceanic crust are firmly joined together. Because it is not a plate boundary, little tectonic activity occurs at a passive margin. Continental margins on both sides of the Atlantic Ocean are passive margins.

In contrast, an **active continental margin** occurs at a convergent plate boundary, where oceanic lithosphere sinks beneath the continent in a subduction zone. The west coast of South America is an active continental margin, also called an *Andean margin*, as you learned in Chapter 9.

Passive Continental Margins

Recall from Chapter 6 that about 250 million years ago all of Earth's continents were joined into the supercontinent called Pangea. Shortly thereafter, Pangea began to rift apart into the continents as we know them today. The Atlantic Ocean opened as the east coast of North

America separated from Europe and Africa. As Pangea broke up, the continental crust fractured and thinned near the fractures (Figure 15.24A). Basaltic magma rose at the new spreading center, forming oceanic crust between North America and Africa (Figure 15.24B). All tectonic activity then focused on the spreading Mid-Atlantic Ridge, and no further tectonic activity occurred at the continental margins; hence the term *passive continental margin* (Figure 15.24C).

active continental margin A continental margin that occurs at a convergent plate boundary, characterized by subduction of an oceanic lithospheric plate beneath a continental plate; also called an *Andean margin*.

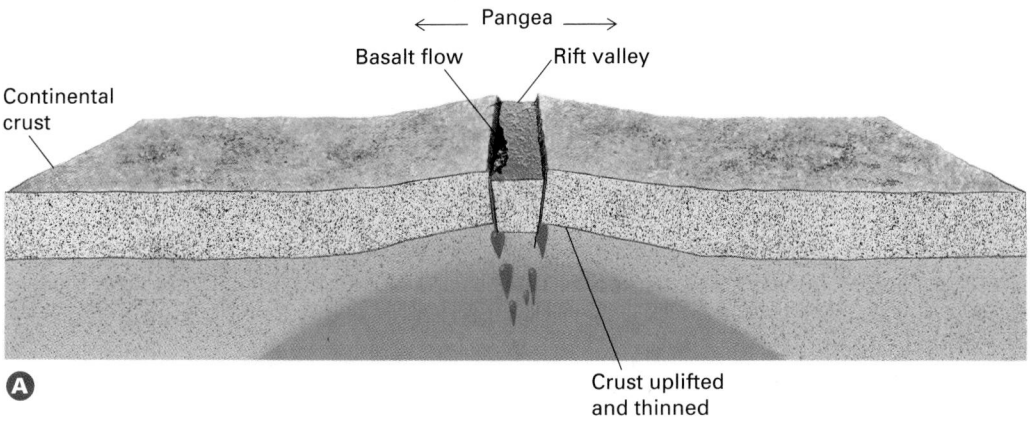

Ⓐ

FIGURE 15.24
(A) Continental crust fractured as Pangea began to rift. (B) Faulting and erosion thinned the crust as it separated. Rising basaltic magma formed new oceanic crust in the rift zone. (C) Sediment eroded from the continents formed broad continental shelves on the passive margins of North America and Africa.

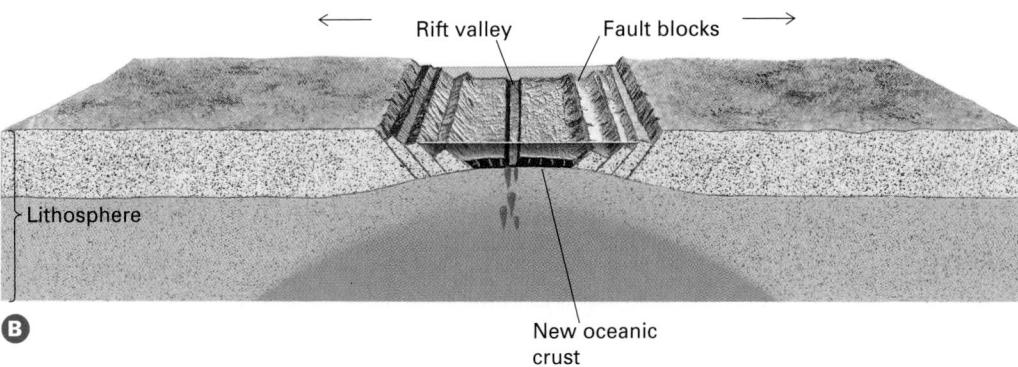

Ⓑ

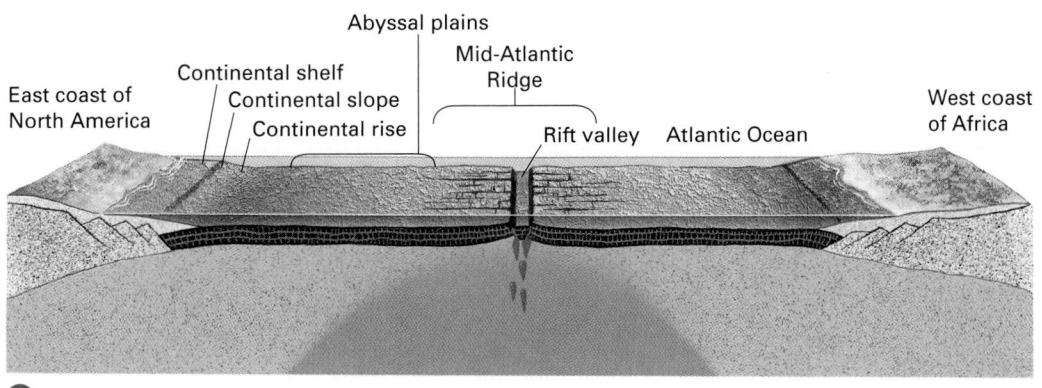

Ⓒ

continental shelf
A shallow, nearly
level area of con-
tinental crust cov-
ered by sediment
and sedimentary
rocks, submerged
below sea level
at the edge of a
continent between
the shoreline
and the continen-
tal slope.

**carbonate plat-
forms** Extensive
accumulations of
limestone, such
as the Florida
Keys and the
Bahamas, formed
on a continental
shelf in warm
regions where
sediment does not
muddy the water
and reef-building
organisms thrive.

The Continental Shelf

On all continents, streams and rivers deposit sediment on coastal deltas, like the Mississippi Delta. Then, ocean currents redistribute the sediment along the coast, depositing it both on the thin margin of continental crust and on oceanic crust close to the continent. The sediment forms a shallow, gently sloping, submarine surface on the edge of the continent called a **continental shelf** (Figure 15.25). As sediment accumulates on a continental shelf, the edge of the continent sinks isostatically because of the added weight. This effect keeps the shelf slightly below sea level.

Over millions of years, thick layers of sediment accumulated on the passive east coast of North America, forming a broad continental shelf along the entire coast. The depth of the shelf increases gradually from the shore to about 200 meters at the outer shelf edge. The average inclination of the continental shelf is about 0.1°. A continental shelf on a passive margin can be a large feature. The shelf off the coast of southeastern Canada is about 500 kilometers wide, and parts of the shelves of Siberia and northwestern Europe are even wider.

In some places, a supply of sediment may be lacking, either because no rivers bring sand, silt, or clay to the shelf or because ocean currents bypass that area. In warm regions where sediment does not muddy the water, reef-building organisms thrive. As a result, in tropical and subtropical latitudes where clastic sediment is lacking, thick beds of limestone accumulate. Limestone accumulations of this type may be hundreds of meters thick and hundreds of kilometers across and are called **carbonate platforms**. The Florida Keys and the Bahamas are modern-day examples of carbonate platforms on continental shelves (Figure 15.26).

Some of the world's richest petroleum reserves occur on the continental shelves of the North Sea between England and Scandinavia, in the Gulf of Mexico, and in the Beaufort Sea on the northern coast of Alaska and western Canada. In recent years, oil companies have explored and developed these offshore reserves. Deep drilling has revealed that granitic continental crust lies beneath the sedimentary rocks, confirming that the continental shelves are truly parts of the continents, despite the fact that they are covered by seawater.

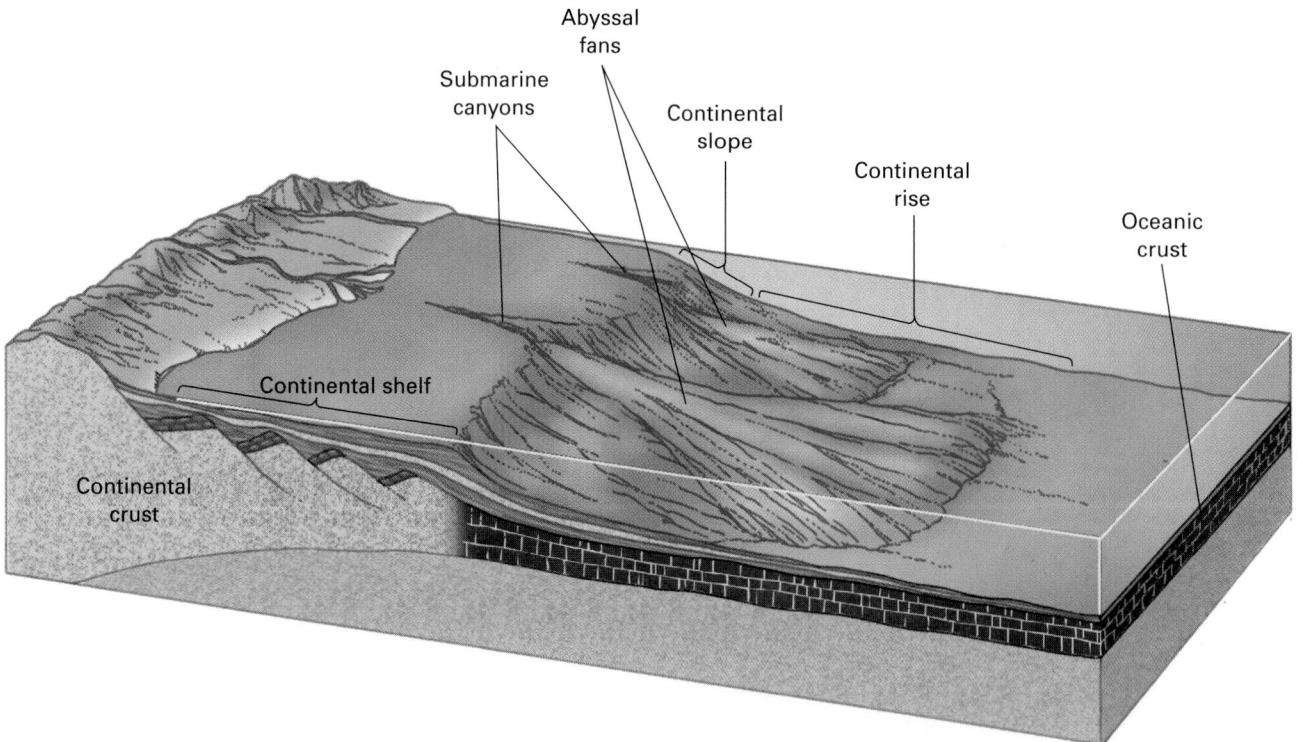

FIGURE 15.25 A passive continental margin consists of a broad continental shelf, slope, and rise formed by accumulation of sediment eroded from the continent.

continental slope The relatively steep (averaging 3° but varying between 1° and 10°) underwater slope between the continental shelf and the continental rise.

continental rise An apron of sediment at the foot of the continental slope where it merges with the deep sea floor.

submarine canyon A deep, V-shaped, steep-walled trough eroded into a continental shelf and slope.

abyssal fan A large, fan-shaped accumulation of sediment deposited at the base of many submarine canyons adjacent to the deep sea floor; also called a *submarine fan*.

FIGURE 15.26 Clear skies and calm waters allowed the MODIS satellite to capture this stunning image of southern Florida, the Bahamas, and Cuba. In the center, Andros Island is surrounded by the bright blue halo of Great Bahama Bank, a carbonate platform that was inundated by a rising sea level between 10,000 and 2,500 years ago, as the last ice age glaciers were melting. In most places the water above the platform doesn't exceed 6 meters.

The Continental Slope and Rise

At the outer edge of a shelf, the sea floor suddenly steepens to an average slope of about 3° (though it can vary from 1° to 10°) as it falls away from 200 meters to about 5 kilometers in depth. This steep region of the sea floor averages about 50 kilometers wide and is called the **continental slope**. It is a surface formed by sediment accumulation, much like the shelf. Its steeper angle is due primarily to thinning of continental crust where it nears the junction with oceanic crust. Seismic profiler exploration shows that the sedimentary layering is commonly disrupted where sediment has slumped and slid down the steep incline.

A continental slope becomes less steep as it gradually merges with the deep ocean floor. This region, called the **continental rise**, consists of an apron of terrigenous sediment that was transported across the continental shelf and deposited on the deep ocean floor at the foot of the slope. The continental rise averages a few hundred kilometers wide. Typically, it joins the deep sea floor at a depth of about 5 kilometers.

In essence, then, the shelf–slope–rise complex is a smoothly sloping, submarine surface on the edge of a continent, formed by accumulation of sediment eroded from the continent.

Submarine Canyons and Abyssal Fans

In many places, sea-floor maps show deep valleys, called **submarine canyons**, eroded into the continental shelf and slope (Figure 15.25). They look like submarine stream valleys. A canyon typically starts on the outer edge of a continental shelf and continues across the slope to the rise. At its lower end, a submarine canyon commonly leads into an **abyssal fan** (also called a *submarine fan*), a large, fan-shaped pile of sediment lying on the continental rise.

Most submarine canyons occur where large rivers enter the sea. When they were first discovered, geologists thought the canyons had been eroded by rivers during the Pleistocene Epoch, when accumulation of glacial ice on land lowered sea level by as much as 150 meters. However, this explanation cannot account for the deeper portions of submarine canyons cut into the

turbidity current
A rapidly flowing submarine current laden with suspended sediment, which develops when loose, wet sediment tumbles down the continental slope in a submarine landslide.

lower continental slopes at depths of a kilometer or more. Therefore, the deeper parts of the submarine canyons must have formed underwater, and a submarine mechanism must be found to explain them.

Geologists subsequently discovered that **turbidity currents** erode the continental shelf and slope to create or deepen the submarine canyons. A turbidity current develops when loose, wet sediment tumbles down the slope in a submarine landslide. The movement may be triggered by an earthquake or simply by oversteepening of the slope as sediment accumulates. When the sediment starts to move, it mixes with water. Because the mixture of sediment and water is denser than water

alone, it flows down the shelf and slope as a turbulent, chaotic avalanche. A turbidity current can travel at speeds greater than 100 kilometers per hour and for distances up to 700 kilometers.

Sediment-laden water traveling at such speed has tremendous erosive power. Once a turbidity stream or current cuts a small channel into the shelf and slope, subsequent currents follow the same channel, just as a stream uses the same channel year after year. Over time, the currents erode a deep submarine canyon into the shelf and slope. Turbidity currents slow down when they reach the deep sea floor. The sediment accumulates there to form an abyssal fan. Most submarine canyons and fans form near the mouths of large rivers because the rivers supply the great amount of sediment needed to create turbidity currents.

Large abyssal fans form only on passive continental margins. They are uncommon at active margins because

FIGURE 15.27 At an active continental margin, an oceanic plate sinks beneath a continent, forming an oceanic trench. The continental shelf is narrow, the slope is steep, and the continental rise is nonexistent.

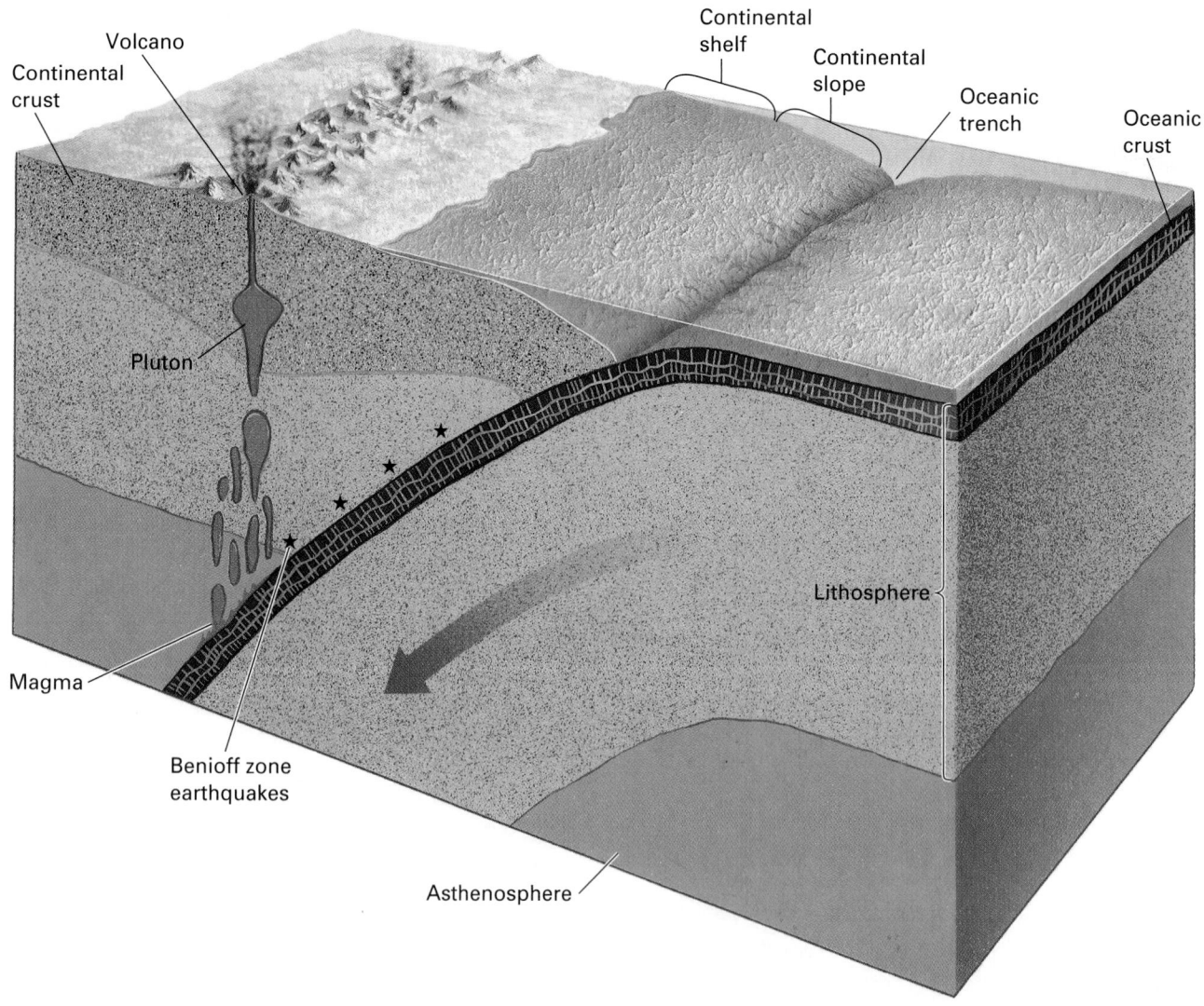

in that environment, the trench swallows the sediment. Furthermore, most of the world's largest rivers drain toward passive margins. The largest known fan is the Bengal Fan, which covers about 4 million square kilometers beyond the mouth of the Ganges River in the Indian Ocean east of India. More than half of the sediment eroded from the rapidly rising Himalayas ends up in this fan. Interestingly, the Bengal Fan has no associated submarine canyon, perhaps because the sediment supply is so great that the rapid accumulation of sediment prevents erosion of a canyon.

Active Continental Margins

An active continental margin forms at a subduction zone, where an oceanic plate converges with a continent. The oceanic plate sinks into the mantle, forming an oceanic trench similar to the trenches associated with island arcs described in Section 15.4 (Figure 15.27). A trench can form wherever subduction occurs—where oceanic crust sinks beneath the edge of a continent, or where it sinks beneath another oceanic plate.

Recall that at a passive margin, the continental crust merges gradually with oceanic crust and a wide continental shelf–slope–rise complex develops as thick deposits of sediment accumulate on the passive margin. At an active margin, however, the thick continental crust meets the thin oceanic crust at a subduction zone. Most of the sediment transported from a continent to an active margin is swallowed up in the trench. As a result, an active margin commonly has a narrow continental shelf or none at all. The landward wall of the trench (the side toward the continent) is the continental slope of an active margin. It typically inclines at 4° or 5° in its upper part and steepens to 15° or more near the bottom of the trench. The continental rise is absent because sediment flows into the trench instead of accumulating on the ocean floor.

16

OCEANS AND COASTLINES

Cliffs along the Oregon coast.

visit **4ltrpress.cengage.com**

> ## Blue, green, grey, white, or black; smooth, ruffled, or mountainous; that ocean is not silent.
>
> *H. P. Lovecraft*

About 60 percent of the world's human population lives within 100 kilometers of the coast. One million people live on low coral islands, and many millions more live on low-lying coastal land vulnerable to coastal flooding. At risk are not just individuals and villages but unique human cultures. Faced with changing coastal environments, these people are threatened with forced abandonment of their nations.

Popular media coverage has linked damage along the American Gulf Coast to sea-level rise induced by global warming. However, many geologists and oceanographers argue that other factors such as crustal subsidence, described in Chapter 15, may be responsible for the submergence of Pacific island nations and coastal regions of the United States.

Regardless of the cause, the image of an entire country disappearing beneath the sea, or a great seaport city being ravaged by a hurricane, is receiving considerable media attention. For many people, it is disturbing that solid land should sink beneath the waves or that a modern city could be defenseless against a sea storm. Yet geologists know that continents and islands have risen from the sea and sunk back beneath the waves throughout Earth's history. In this chapter, we will continue studying Earth's oceans and the processes that form, shape, or destroy islands and coastlines.

Coastlines are among the most geologically active environments on Earth. Sea level rises and falls, flooding shallow parts of continents and stranding beaches high above sea level. Regions of continental crust also rise and sink, with results similar to those caused by fluctuating sea level. Rivers deposit great quantities of sand and mud on coastal deltas. Waves and currents erode beaches and transport sand along hundreds or even thousands of kilometers of shoreline. Converging tectonic plates buckle coastal regions, creating mountain ranges, earthquakes, and volcanic eruptions.

16.1 Geography of the Oceans

All of Earth's oceans are connected, and water flows from one to another, so in one sense Earth has just one global ocean. However, several distinct ocean basins exist within the global ocean (Figure 16.1). The largest and deepest is the Pacific. It covers one-third of Earth's surface, more than all land combined, and contains more than half of the world's water. The Atlantic Ocean has about half the surface area of the Pacific. The Indian Ocean is slightly smaller than the Atlantic. The Arctic Ocean surrounds the North Pole and extends southward to the shores of North America, Europe, and Asia. Therefore, it is bounded by land, with only a few straits and channels connecting it to the Atlantic and Pacific Oceans. The surface of the Arctic Ocean freezes in winter, and parts of it melt for a few months during summer and early fall. The Antarctic Ocean, feared by sailors for its cold and ferocious winds, has no sharp northern boundary. The northernmost limit of the Antarctic Ocean is the zone where warm currents from the north converge with cold Antarctic water.

16.2 Seawater

Salts and Trace Elements

The **salinity** of seawater is the total quantity of dissolved salts, expressed as a percentage. In this case, the term *salts* refers to all dissolved ions, not only sodium and chloride.

Dissolved ions make up about 3.5 percent of the weight of ocean water. The six ions listed in Figure 16.2 make up 99 percent of the ocean's dissolved material. However, almost every other element found on land is also found dissolved in seawater, albeit in trace amounts. For example, seawater contains about 0.000000004 (4×10^{-9}) percent gold. Although the concentration is small, the oceans are large and therefore contain a lot

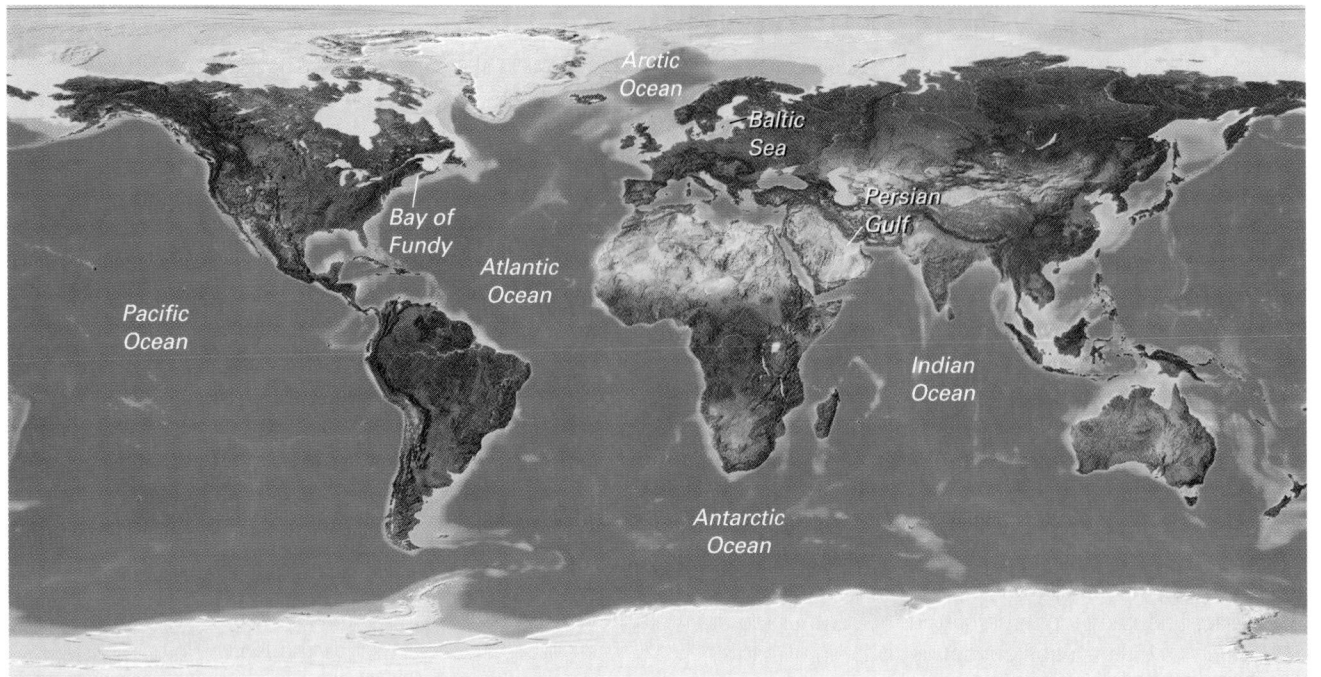

FIGURE 16.1 The oceans of the world. The "seven seas" are the North Atlantic, South Atlantic, North Pacific, South Pacific, Indian, Arctic, and Antarctic. However, these designations are related more closely to commerce than to the geology and oceanography of the ocean basins. Geologists and oceanographers recognize four major ocean basins: the Atlantic, Pacific, Indian, and Arctic. The map also designates a few smaller seas and bays that are mentioned in the text.

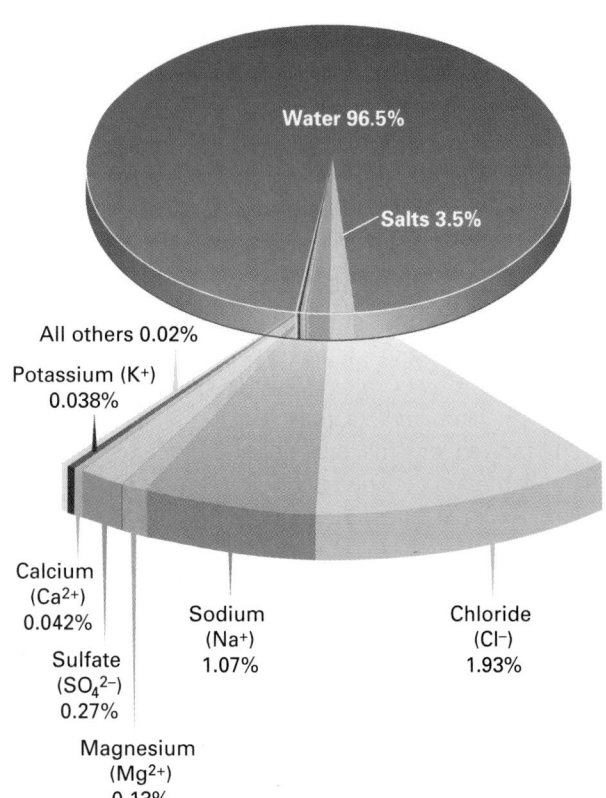

FIGURE 16.2 Six common ions form most of the salts in seawater.

of gold. About 4.4 kilograms of gold are dissolved in each cubic kilometer of seawater. Because the oceans contain about 1.3 billion cubic kilometers of water, about 5.7 billion kilograms of gold exist in the oceans. Unfortunately, it would be hopelessly expensive to extract even a small portion of this amount.

The world's rivers carry more than 2.5 billion tons of dissolved salts to the oceans every year. Underwater volcanoes contribute additional dissolved ions. However, the salinity of the oceans has been relatively constant throughout much of geologic time because salt has been removed from seawater at the same rate at which it has been added. When a portion of a marine basin becomes cut off from the open oceans, the water evaporates, precipitating thick sedimentary beds of salt. Additionally, large amounts of salt become incorporated into shale and other sedimentary rocks.

Dissolved Gases

In addition to trace elements and salts, seawater also contains dissolved gases, especially carbon dioxide and oxygen. These dissolved gases exchange freely and continuously with the atmosphere. If the atmospheric concentration of carbon dioxide or oxygen rises, much of the gas quickly dissolves into seawater; if the atmospheric concentration of a gas falls, that gas *exsolves* (comes out of solution) from the sea and enters the atmosphere. Thus, the seas buffer atmospheric concentrations of gases.

At the same time, the atmosphere buffers the chemistry of seawater. When carbon dioxide dissolves in seawater, the seawater becomes more acidic. In recent years, the acidification of seawater from the increase in atmospheric carbon dioxide concentration has seriously affected marine ecosystems. Corals, many kinds of algae, shellfish, foraminifera, and other organisms that make hard body parts out of calcium are adversely affected, thus adversely affecting oceanic ecosystems.

The capacity of seawater to absorb carbon dioxide from the atmosphere has a profound effect on the concentration of this greenhouse gas in the atmosphere, and therefore is an important factor in global climate and climate change. These processes are discussed in detail in Chapter 21. Oxygen dissolved in seawater is critical to most forms of marine life, which require oxygen for survival.

Temperature

Recall from Chapter 11 that lakes are comfortably warm for swimming during the summer because warm water floats on the surface and does not readily mix with the deep, cooler water. Oceans develop a similar temperature layering, but because sea waves and currents stir the surface water, the warm layer in an ocean extends from the surface to a depth as great as 450 meters (Figure 16.3). Below this layer is the *thermocline*, where the temperature drops rapidly with depth. The thermocline extends to a depth of 2 kilometers. Beneath the thermocline, the temperature of ocean water varies from about 1°C to 2.5°C. The cold, dense water in the ocean depths mixes very little with the surface. Thus, there are three distinct temperature zones in the ocean, as shown in Figure 16.3. This layered structure does not exist in the polar seas because cold surface water sinks, causing vertical mixing.

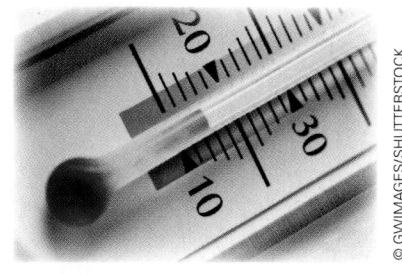

tides The cyclic rise and fall of ocean water caused by the gravitational force of the Moon and, to a lesser extent, of the Sun.

16.3 Tides

Even the most casual observer will notice that on any beach the level of the ocean rises and falls on a cyclical basis. If the water level is low at noon, it will reach its maximum height at about 6:13 p.m. and be low again at about 12:26 a.m. These vertical displacements are called **tides**. Most coastlines experience two high tides and two low tides during an interval of about 24 hours and 53 minutes.

Tides are caused by the gravitational pull of the Moon and Sun on the sea surface. Although the Moon is much smaller than the Sun, it is so much closer to Earth that its influence predominates. At any given time, one region of Earth (point A in Figure 16.4) lies directly under the Moon. Because gravitational force is greater for objects that are closer together, the part of the ocean nearest to the Moon is attracted with the strongest force. The water rises, resulting in a high tide in that region.

But so far our explanation is incomplete. As Earth spins on its axis, a given point on Earth passes directly under the Moon approximately once every 24 hours and 53 minutes, but the period between successive high tides is only 12 hours and 26 minutes. Why are there ordinarily two high tides in a day? The tide is high not only when a point on Earth is directly under the Moon but also when it is 180° away. To understand this effect, we must consider the Earth–Moon

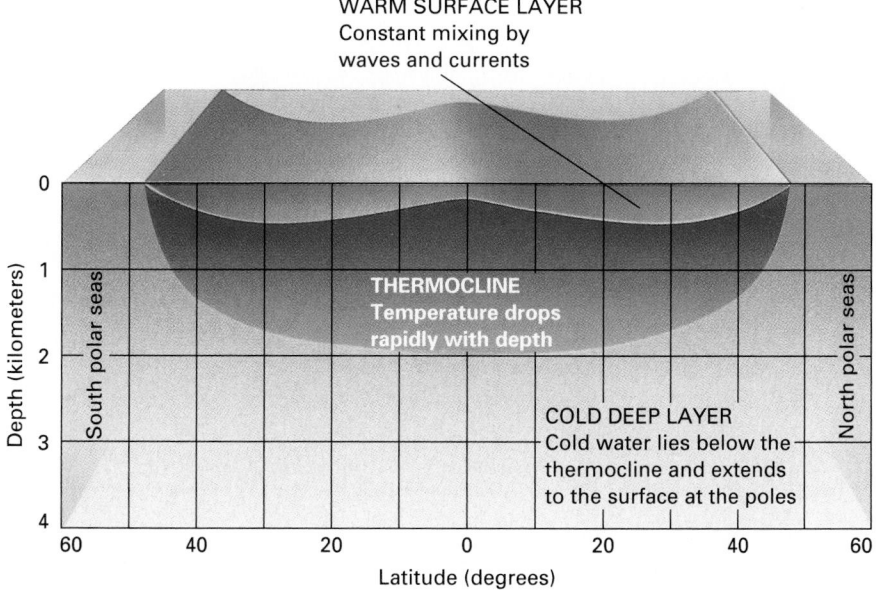

WARM SURFACE LAYER
Constant mixing by waves and currents

THERMOCLINE
Temperature drops rapidly with depth

COLD DEEP LAYER
Cold water lies below the thermocline and extends to the surface at the poles

South polar seas

North polar seas

Depth (kilometers)

Latitude (degrees)

FIGURE 16.3 There are three temperature layers in the ocean. The surface depth to about 450 meters is warm; temperature cools rapidly with depth in the thermocline, and the ocean depths are cold.

spring tides The very high and very low tides that occur when the Sun, Moon, and Earth are aligned and their gravitational fields combine to create a strong tidal bulge.

FIGURE 16.4 The Moon's gravity causes a high tide at point A, directly under the Moon. The motion of the Earth–Moon system causes a high tide at point B, opposite point A.

FIGURE 16.5 The Moon moves 13.2° every day. (A) The Moon is directly above an observer on Earth. (B) One day later, Earth has completed one complete rotation, but the Moon has traveled 13.2°. Earth must now travel for another 53 minutes before the observer is directly under it.

orbital system. Most people visualize the Moon orbiting around Earth, but it is more accurate to say that Earth and the Moon orbit around a common center of gravity. The two celestial partners are locked together like dancers spinning around in each other's arms. Just as the back of a dancer's dress flies outward as she twirls, the oceans on the opposite side of Earth from the Moon bulge outward. This bulge is the high tide 180° away from the Moon (point B in Figure 16.4). Thus, the tides rise and fall twice daily.

High and low tides do not occur at the same time each day but are delayed by approximately 53 minutes every 24 hours. Earth makes one complete rotation on its axis in 24 hours, but at the same time, the Moon is orbiting Earth in the same direction. After a point on Earth makes one complete rotation in 24 hours, that point must spin for an additional 53 minutes to catch up with the orbiting Moon. This is why the Moon rises approximately 53 minutes later each day. In the same manner, the tides are approximately 53 minutes later each day (Figure 16.5).

Although the Sun's gravitational pull on the oceans is smaller than the Moon's, it does affect tides. When the Sun and Moon are directly in line with Earth, their gravitational fields combine to create a strong tidal bulge. During these times, the variation between high and low tides is large, producing **spring tides** (Figure 16.6A). When the Moon is 90° out of alignment with the Sun and Earth, each partially

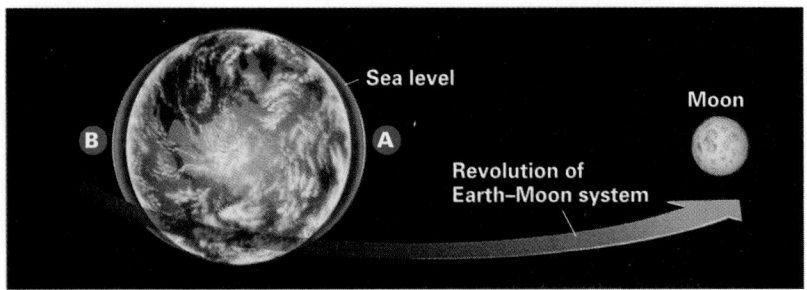

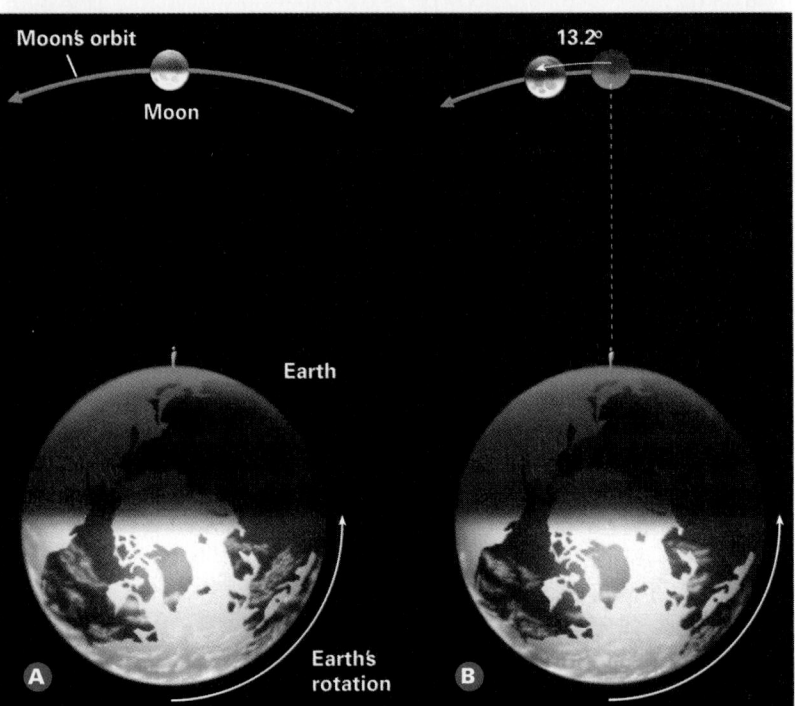

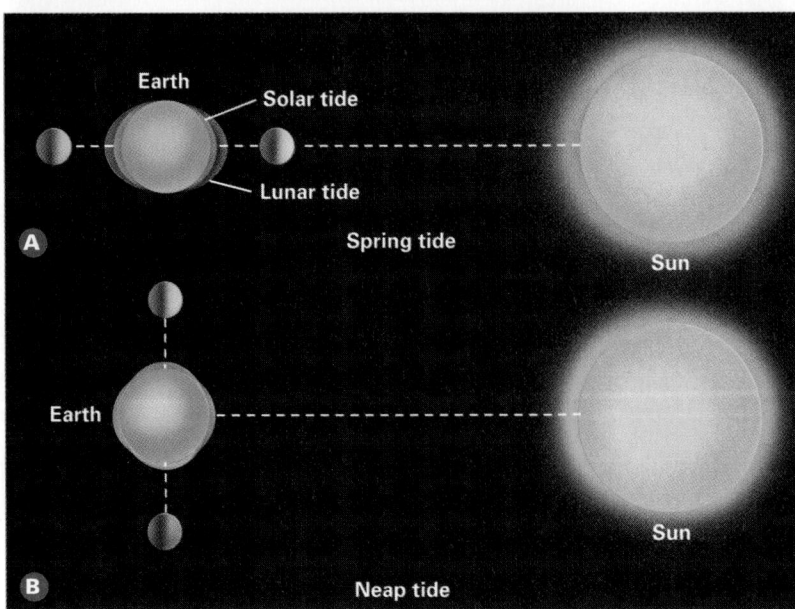

FIGURE 16.6 (A) Spring tides occur when Earth, Sun, and Moon are lined up, with their gravitational fields combining to create a strong tidal bulge. (B) Neap tides occur when the moon lies at right angles to a line drawn between the Sun and Earth.

offsets the effect of the other and the differences between the levels of high and low tide are smaller. These relatively small tides are called **neap tides** (Figure 16.6B).

Tidal variations differ from place to place. For example, the Bay of Fundy is shaped like a giant funnel (Figure 16.1). The shape of the bay concentrates the rising and falling tides, and as a result the tidal variation is as much as 15 meters during a spring tide. In contrast, tides vary by less than 2 meters along a straight stretch of coast such as that in Santa Barbara, California. In the central oceans, the tides average about 1 meter. Mariners consult tide tables that give the time and height of the tides in any area on any day.

© EPIC STOCK/SHUTTERSTOCK

neap tides The relatively small tides that occur when the Moon is 90° out of alignment with the Sun and Earth.

crest The highest part of a wave.

trough The lowest part of a wave.

wavelength The distance between successive wave crests (or troughs).

wave height The vertical distance from the crest to the trough of a wave.

16.4 Sea Waves

Most ocean waves develop when wind blows across water. Waves vary from gentle ripples to destructive giants that can erode coastlines, topple beach houses, and sink ships. In deep water, the size of a wave depends on (1) the wind speed, (2) the length of time that the wind has blown, and (3) the distance that the wind has traveled (sailors call this last factor *fetch*). A 25-kilometer-per-hour wind blowing for 2 to 3 hours across a 15-kilometer-wide bay generates waves about 0.5 meter high. But if a storm blows at 90 kilometers per hour for several days over a fetch of 3,500 kilometers, it can generate 30-meter-high waves, as tall as a ship's mast.

The highest part of a wave is called the **crest**; the lowest is the **trough** (Figure 16.7). The **wavelength** is the distance between successive crests. The **wave height** is the vertical distance from the crest to the trough.

Recall from our discussion of earthquakes in Chapter 7 that when a wave travels through rock, the energy of the wave is transmitted rapidly over large distances, but the rock itself moves only slightly. In a similar manner, a single water molecule in a water wave does not travel with the wave. While the wave moves horizontally, the water molecules move in small circles, as shown in Figure 16.8. That is why a ball on the ocean bobs up and down and sways back and forth as the waves pass, but it does not travel along with the waves. In addition, the circles of water movement become smaller with depth. At a depth equal to about half the wavelength, the disturbance becomes negligible. Thus if you dive deep enough, you escape wave motion. No one gets seasick in a submarine.

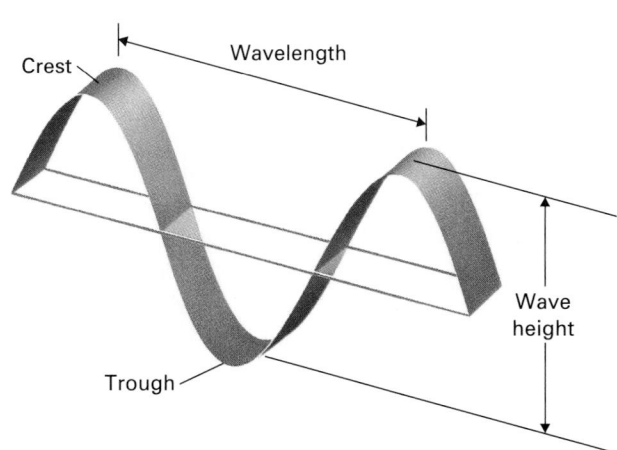

FIGURE 16.7 Terminology used to describe a wave.

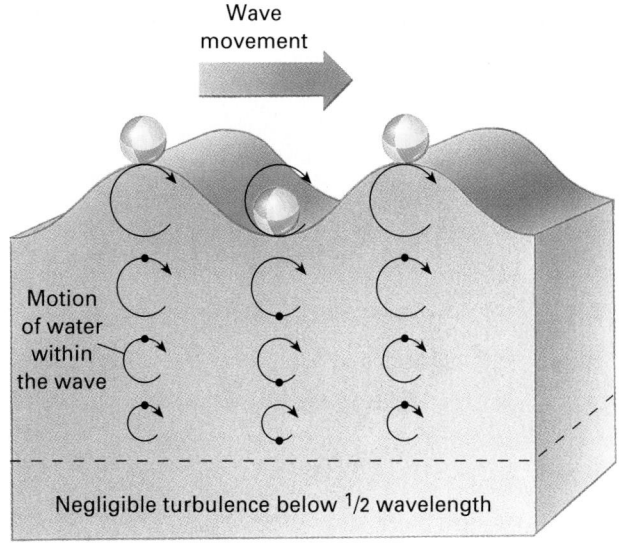

FIGURE 16.8 While a wave moves horizontally across the sea surface, the water itself moves only in small circles.

current A continuous flow of water in a particular direction.

surface current Horizontal flow of water in the upper 400 meters of the oceans, caused by wind blowing over the sea surface.

deep-sea current Vertical and horizontal flow of water below a depth of 400 meters in the oceans, caused mainly by gravity.

16.5 Ocean Currents

The water in an ocean wave oscillates in circles, ending up where it started. In contrast, a **current** is a continuous flow of water in a particular direction. Although river currents are familiar and easily observed, ocean currents were not recognized by early mariners. **Surface currents**, caused by wind blowing over the sea surface, flow in the upper 400 meters of the

seas and involve about 10 percent of the water in the world's oceans. In contrast, **deep-sea currents** transport seawater both vertically and horizontally below a depth of 400 meters and are driven by gravity as denser water sinks and less-dense water rises. Furthermore, in certain places several processes (discussed below under the headings "Deep-Sea Currents" and "Upwelling") cause surface water to sink to great depths, and in other places deep water rises to the sea surface. Thus, the entire global ocean circulates continuously. Figure 16.9 shows both major surface currents and deep-sea currents in the Atlantic Ocean.

FIGURE 16.9 Surface and deep-sea currents of the Atlantic Ocean. Both types of currents profoundly affect climate and other components of Earth systems. This artist's rendition has simplified the currents to emphasize the major flow patterns.

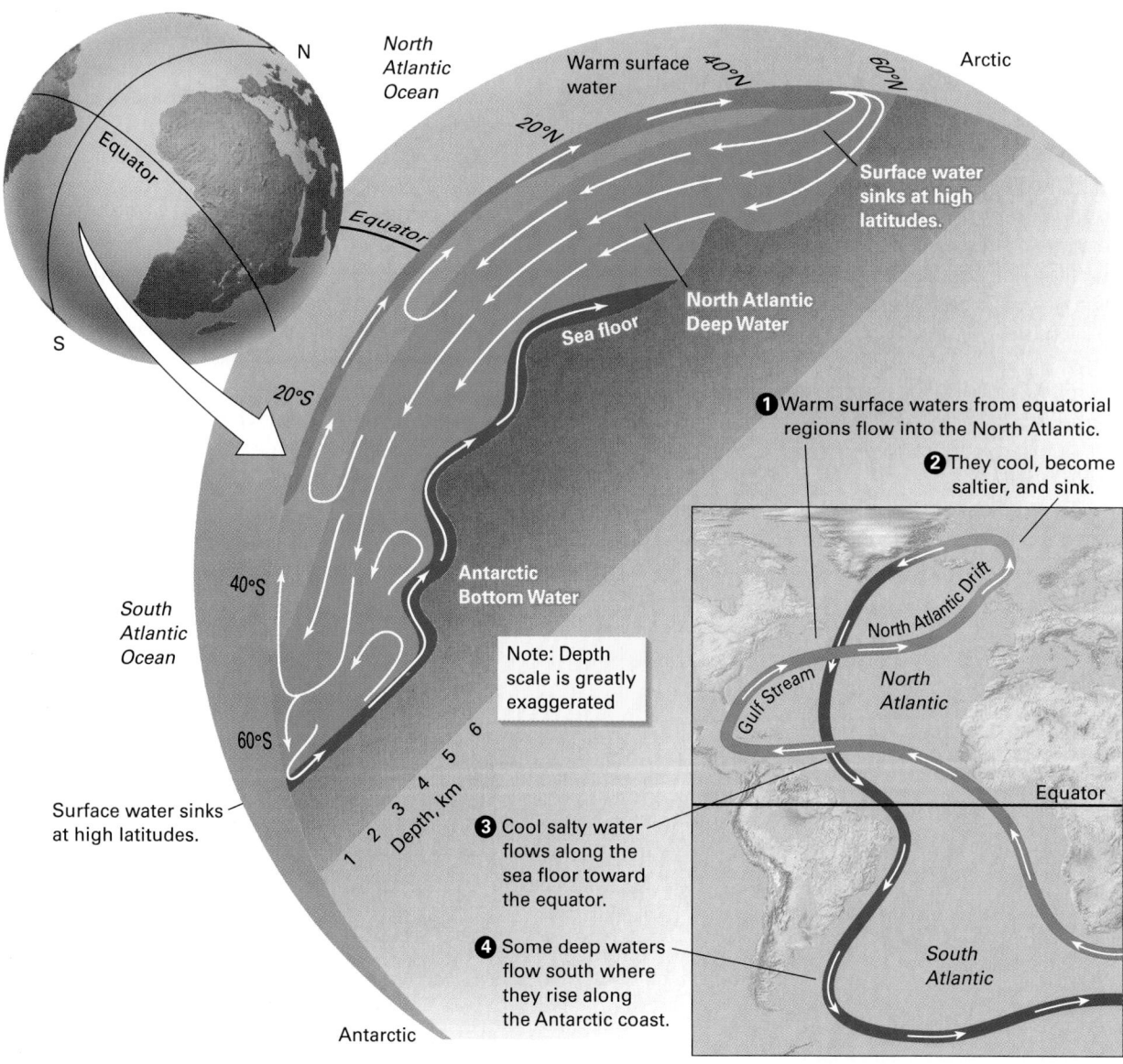

A storm surge is an onshore flood of water created by a low-pressure weather system such as a hurricane or tropical cyclone. The surge is caused primarily by strong winds of the hurricane or cyclone blowing over the sea surface and causing seawater to pile up above sea level.

Surface Currents

In the mid-1700s, Benjamin Franklin lived in London, where he was deputy postmaster general for the American colonies. He noticed that mail ships took 2 weeks longer to sail from England to North America than did merchant ships. Franklin learned that the captains of the merchant ships had discovered a current flowing northward along the east coast of North America and then across the Atlantic to England. When sailing from Europe to North America, the merchant ships saved time by avoiding the current. The captains of the mail ships were unaware of this current and lost time sailing against it on their westward journeys. In 1769, Franklin and his cousin, Timothy Folger, a merchant captain, charted the current and named it the **Gulf Stream**.

As ship traffic increased and navigators searched for the quickest routes around the globe, they discovered other ocean currents (Figure 16.10). Ocean currents have been described as rivers in the sea. The analogy is only partially correct; ocean currents have no well-defined banks, and they carry much more water than even the largest river. The Gulf Stream is 80 kilometers wide and 650 meters deep near the east coast of Florida and moves at approximately 5 kilometers per hour, a moderate walking speed. As it moves northward and eastward, the current widens and slows; east of New York it is more than 500 kilometers wide and travels at less than 8 kilometers per day.

Another important difference between rivers and surface currents is that rivers flow in response to gravity, whereas ocean surface currents are driven primarily by wind. When wind blows across water in a constant direction for a long time, it drags surface water along with it, forming a current. In many regions, the wind blows in the same direction throughout the year, forming currents that vary little from season to season.

Gulf Stream One of many oceanic surface currents. It begins near the Gulf of Mexico and travels northward up the Atlantic coast, growing wider and slower as it moves to the northeast.

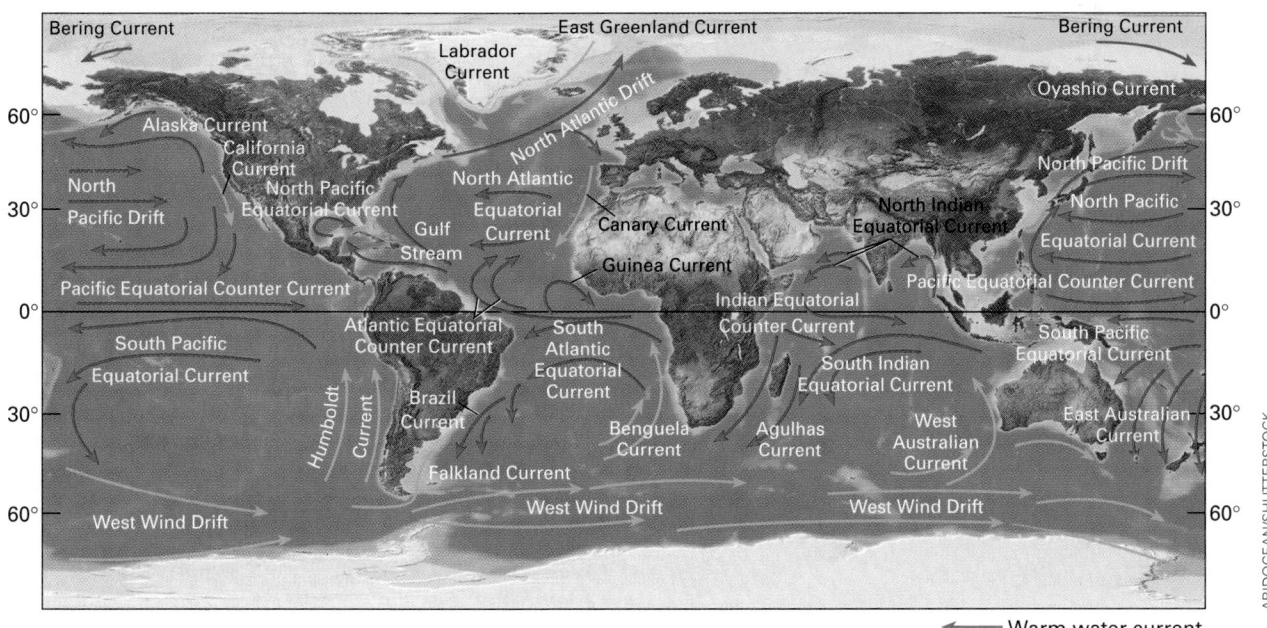

FIGURE 16.10 Major oceanic surface currents of the world.

gyre A circular or elliptical current in either water or air.

Coriolis effect A deflection of air or water currents, caused by the rotation of Earth.

However, in other places, changing winds cause currents to change direction. For example, in the Indian Ocean prevailing winds shift on a seasonal basis. When the winds shift, the ocean currents follow.

Ocean currents profoundly affect Earth's climate. The Gulf Stream transports 1 million cubic meters of warm water northward past any point every second, warming North America and Europe. For example, Churchill, Manitoba, is a village in the interior of Canada. It is frigid and icebound for much of the year. Polar bears regularly migrate through town. Yet it is at about the same latitude as Glasgow, Scotland, which is warmed by the Gulf Stream and therefore experiences relatively mild winters. In general, the climate in western Europe is warmer than that at similar latitudes in other regions not heated by tropical ocean currents.

Why Surface Currents Flow in the Oceans

Surface currents are driven primarily by friction between wind blowing over the sea surface and surface water. The winds simply drag the sea surface along in the same direction that the winds are blowing. In Figure 16.11, the green arrows show the major prevailing winds over the North and South Atlantic oceans. The orange arrows show the elliptical surface currents, called **gyres**, in the same regions, simplified from Figure 16.9. The North Atlantic gyre circulates in a clockwise direction, and the southern gyre circulates counterclock-

wise. But notice that the currents do not flow in exactly the same directions as the prevailing winds. Instead, the east–west surface currents in the Northern Hemisphere are deflected to the right of the winds. Where their flow is blocked by a continent, the currents veer clockwise to continue their circuit. In the Southern Hemisphere the currents are deflected to the left of the winds and turn counterclockwise when they encounter land. These differences between prevailing wind directions and the flow directions of the surface currents suggest that forces other than wind also affect the great gyres.

One of those forces that deflect the gyres away from the prevailing winds is the **Coriolis effect**, named for Gaspard-Gustave de Coriolis, the 19th century French scientist who described it. To understand this effect, consider the rotating Earth: The circumference of Earth is greatest at the equator and decreases to zero at the poles. But all parts of the planet make one complete rotation every day. Therefore, a point on the equator must travel farther and faster than any point closer to the poles. At the equator, all objects move eastward with a velocity of about 1,600 kilometers per hour; at the poles there is no eastward movement at all and the velocity is 0 kilometers per hour.

Now imagine a rocket fired from the equator toward the North Pole. Before it was launched it was traveling eastward at 1,600 kilometers per hour with the

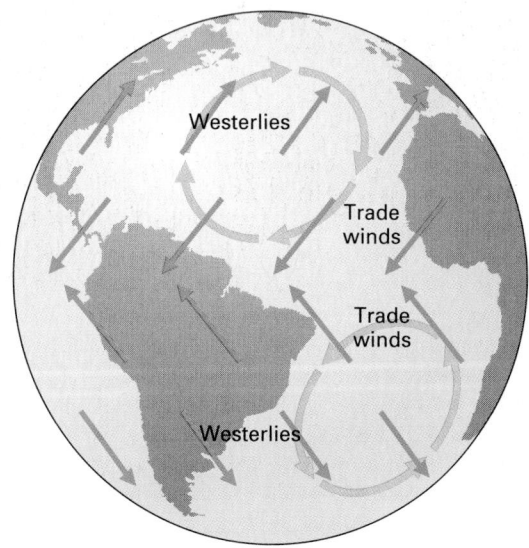

FIGURE 16.11 The green arrows show major prevailing winds over the North and South Atlantic oceans—the Trade Winds (easterlies) and the westerlies. The orange arrows show the surface oceanic currents in the same regions.

rotating Earth. As it takes off, it is moving eastward at 1,600 kilometers per hour and northward at its launch speed. As it moves north from the equator, it is still traveling eastward at 1,600 kilometers per hour, but points on Earth beneath it move eastward at a slower and slower speed as the rocket approaches the pole. As a result, the rocket curves toward the east, or the right. In a similar manner, a mass of water or air deflects in an easterly direction as it moves poleward from the equator, as shown in Figure 16.12A.

Conversely, consider an ocean current flowing southward from the Arctic Ocean toward the equator. Since it started near the North Pole, this water moves more slowly than Earth's surface near the equator, and therefore the current veers toward the west, or to the right, as shown in Figure 16.12B. Thus, north–south currents always veer to the right in the Northern Hemisphere. In the Southern Hemisphere currents turn toward the left for the same reason.

At the sea surface, current directions are affected by both the prevailing winds and the Coriolis effect. However, below the surface, the water doesn't "feel" the wind; it "feels" only the movement of water directly above. The Coriolis effect is as strong at depth as it is at the sea surface. As a result, each successively deeper layer is less affected by the wind and more affected by the Coriolis effect. Consequently, deeper layers of water in the surface currents are deflected even more to the right in the Northern Hemisphere, and to the left in the Southern Hemisphere, than is the shallowest water. The net effect of this process on the flow directions of the gyres extends down to a depth of about 100 meters and is called **Ekman transport**, after Vagn Walfrid Ekman, the Swedish scientist who developed the mathematics that describes the process.

Deep-Sea Currents

Wind does not affect the ocean depths, and oceanographers once thought that deep ocean water was almost motionless. In her 1951 book *The Sea around Us*, Rachael Carson wrote that the ocean depths are "a place where change comes slowly, if at all." However, in 1962 ripples and small dunes were photographed on the floor of the North Atlantic. Because flowing water forms these features, the photographs suggested that water was moving in the ocean depths. More recently, oceanographers have measured deep-sea currents directly with flow meters, and photographed moving sand and mud with underwater television cameras.

Wind drives surface currents, but deep-sea currents are driven by differences in water density. Dense water sinks and flows horizontally along the sea floor to form a deep-sea current. Two factors cause water to become dense and sink: cooling temperature and rising salinity. The global deep-sea circulation shown in Figures 16.9 and 16.13 is caused by these two factors and is called **thermohaline circulation** (*thermo* for temperature, and *haline* for salinity).

Recall that water is most dense when it is cold, close to freezing. Therefore, as tropical surface water moves poleward and cools, it becomes denser and sinks. In addition, water density increases as salinity

Ekman transport
The natural process by which wind causes movement of water near the ocean surface. Each layer of water in the ocean drags with it the layer beneath; thus the movement of each layer of water is affected by the movement of the layer above.

thermohaline circulation The force behind deep-sea currents, created by differences in water temperature and salinity (and therefore differences in water density).

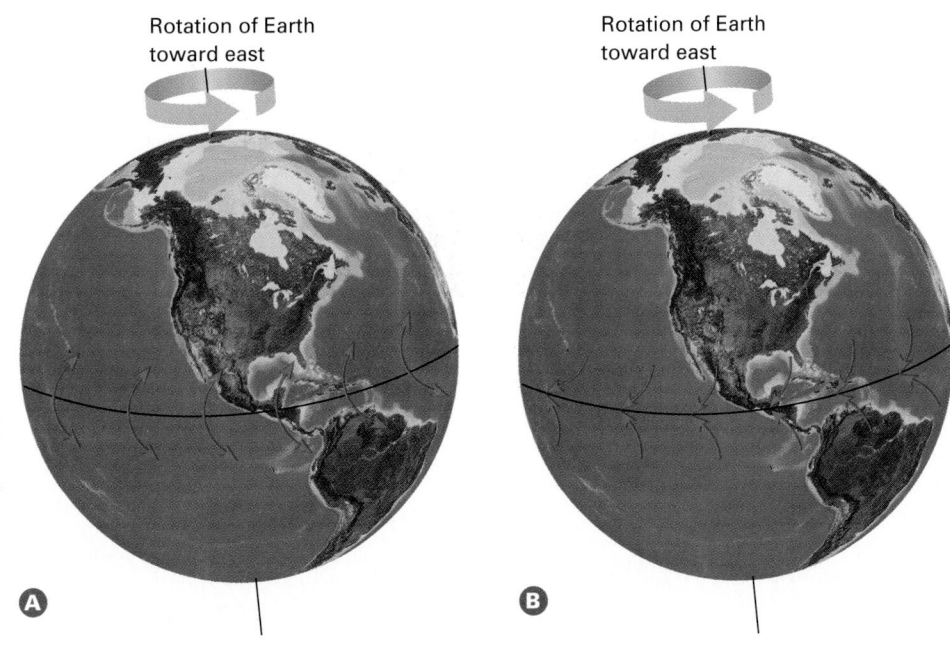

ARIDOCEAN/SHUTTERSTOCK

FIGURE 16.12 The Coriolis effect deflects water and wind currents. (A) Water or air moving poleward from the equator is traveling east faster than the land beneath it and veers to the east (turns right in the Northern Hemisphere and left in the Southern Hemisphere). (B) Water or air moving toward the equator is traveling east slower than the land beneath it and veers to the west (turns right in the Northern Hemisphere and left in the Southern Hemisphere).

upwelling A rising ocean current that transports cold water (and nutrients) from the depths to the surface.

increases. So water sinks when it becomes saltier. Seawater can become saltier if surface water evaporates. Polar seas also become saltier when the surface freezes, because salt does not become incorporated in the ice. Arctic and Antarctic water is dense because the water is both cold and salty. In contrast, addition of freshwater makes seawater less salty and less dense. This effect is pronounced in enclosed bays with abundant, inflowing rivers and also in the polar regions when icebergs float into the ocean and melt. As we will see in Chapter 21, recent rapid melting of Greenland glaciers introduces enough freshwater into the North Atlantic to reduce the density of surface water and alter its buoyancy.

Figure 16.10 shows that the Gulf Stream originates in the subtropics. When this water reaches the northern part of the Atlantic Ocean near the tip of Greenland, it cools and sinks. When the sinking water reaches the sea floor, it is deflected southward to form the North Atlantic Deep Water (Figure 16.13), which flows along the sea floor all the way to Antarctica. An individual water molecule that sinks near Greenland may travel for 500 to 2,000 years before resurfacing half a world away in the south polar sea.

Upwelling

If water sinks in some places, it must rise in others to maintain mass balance. This upward flow of water is called **upwelling**. Upwelling carries cold water from the ocean depths to the surface. Upwelling also brings nutrients from the deep ocean to the surface, creating rich fisheries along the coasts of California and Peru. Several processes can cause upwelling, both in the open oceans and along coastlines.

In Figure 16.10, note that the California Current flows southward along the coast of California. In the Southern Hemisphere, the Humboldt Current moves northward along the west coast of South America. Both currents are deflected westward—away from shore—by the Coriolis effect modified by Ekman transport. As these surface currents veer away from shore, water from the ocean depths rises toward the surface along the California coast and the west coast of South America. In August, water on the east coast of the United States is warmed by the Gulf Stream and may be a comfortable 21°C. However, on the central California coast, the cool California Current combines with the upwelling deep water to produce water that is only 15°C, and surfers and swimmers must wear wetsuits to stay in the water for long.

The frictional drag of a prevailing offshore wind can pull surface water away from a coast. Deep water then upwells along the continental shelf to replace the surface water flowing away from shore. Winds blowing parallel to shore can also create upwelling. The wind causes water to flow parallel to shore, but the Coriolis effect and Ekman transport deflect the current, driving it away from the coast. Deep water then rises to replace the surface water.

In most years, an offshore wind drives surface water away from the coast of Peru and adjacent portions of western South America, creating a strong,

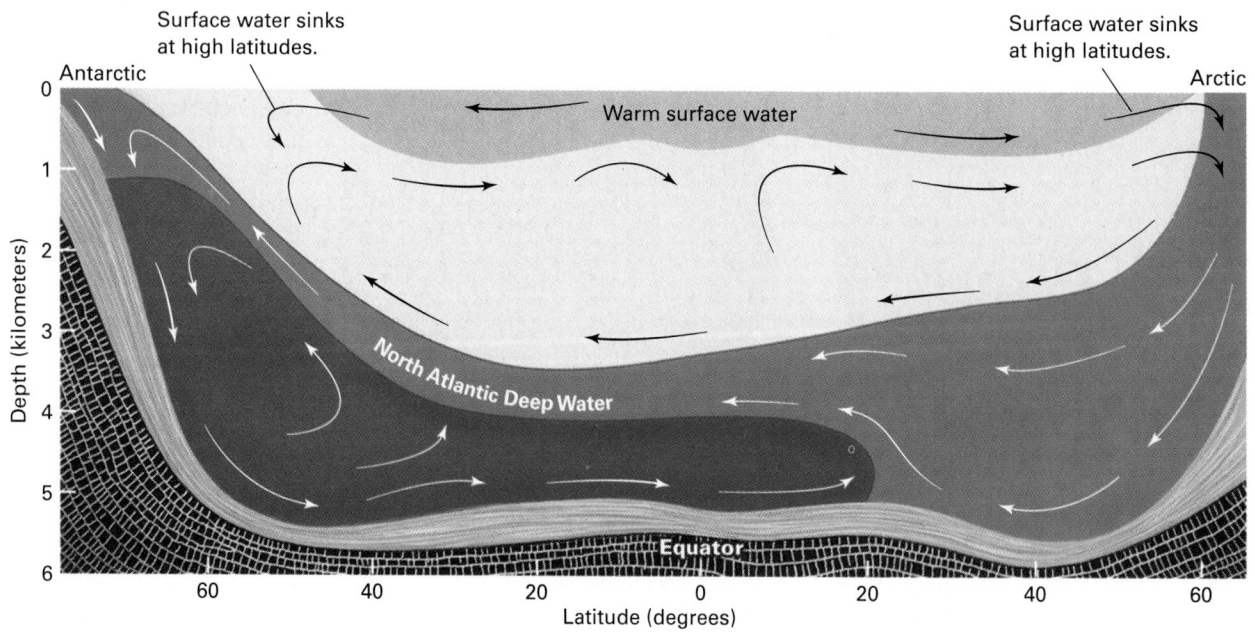

FIGURE 16.13 A profile of the Atlantic Ocean shows surface and subsurface currents.

Unit 4: The Oceans

nutrient-rich upwelling current that produces rich fisheries along that coast. However, about every 3 to 7 years—in *El Niño* years—the offshore wind weakens and the upwelling does not occur. As a result unusually warm, nutrient-poor water accumulates along the west coast of South America, displacing the cold Humboldt Current. El Niño's effects last for about a year before conditions return to normal. Many meteorologists now think that El Niño affects weather patterns for nearly three-quarters of Earth. The causes and effects of El Niño are described in Chapter 19.

Figure 16.10 shows that the south equatorial currents of both the Atlantic and Pacific flow near the equator. Although the Coriolis effect is weak near the equator, water in those currents is deflected poleward, and deep, cold water rises from the ocean depths to replace the surface water in **equatorial upwelling**, which occurs in the open oceans.

16.6 The Seacoast

As we mentioned in the introduction to this chapter, coastlines are among the most geologically active zones on Earth, where atmosphere, geosphere, hydrosphere, and biosphere all affect the local environments. Subduction occurs along many continental coasts. Waves and currents weather, erode, transport, and deposit sediment continuously on all coastlines. In addition, the shallow waters of continental shelves are among the most productive biological ecosystems on Earth. Many organisms that live in these regions build hard shells or

skeletons of calcium carbonate. Eventually, the remains of these shells and skeletons become lithified to form limestone. Thus living organisms become part of the rock cycle.

Weathering and Erosion on the Seacoast

Waves batter most coastlines. If you walk down to the shore, you can watch turbulent water carry sand grains or even small cobbles along the beach. Even on a calm day, waves steepen as they approach shore and then crash against the beach. When a wave enters shallow water, the bottom of the wave drags against the sea floor. This drag compresses the circular motion of the wave into ellipses. This deformation slows the lower part of the wave, so that the upper part moves more rapidly than the lower. As the front of the wave rises over the base, the wave steepens until it collapses forward, or **breaks** (Figure 16.14). Chaotic, turbulent waves breaking along a shore are called **surf**.

Most coastal erosion occurs during intense storms because storm waves are much larger and more energetic than normal waves (Figure 16.15). A 6-meter-

equatorial upwelling Oceanic upwelling in which surface currents flowing westward on both sides of the equator are deflected poleward and are replaced by upward flow of deep, nutrient-rich waters.

break To collapse or crash, as when a wave approaches a beach and the front of the wave rises over its base, growing steeper until it collapses forward.

surf The chaotic, turbulent waves breaking along the shore.

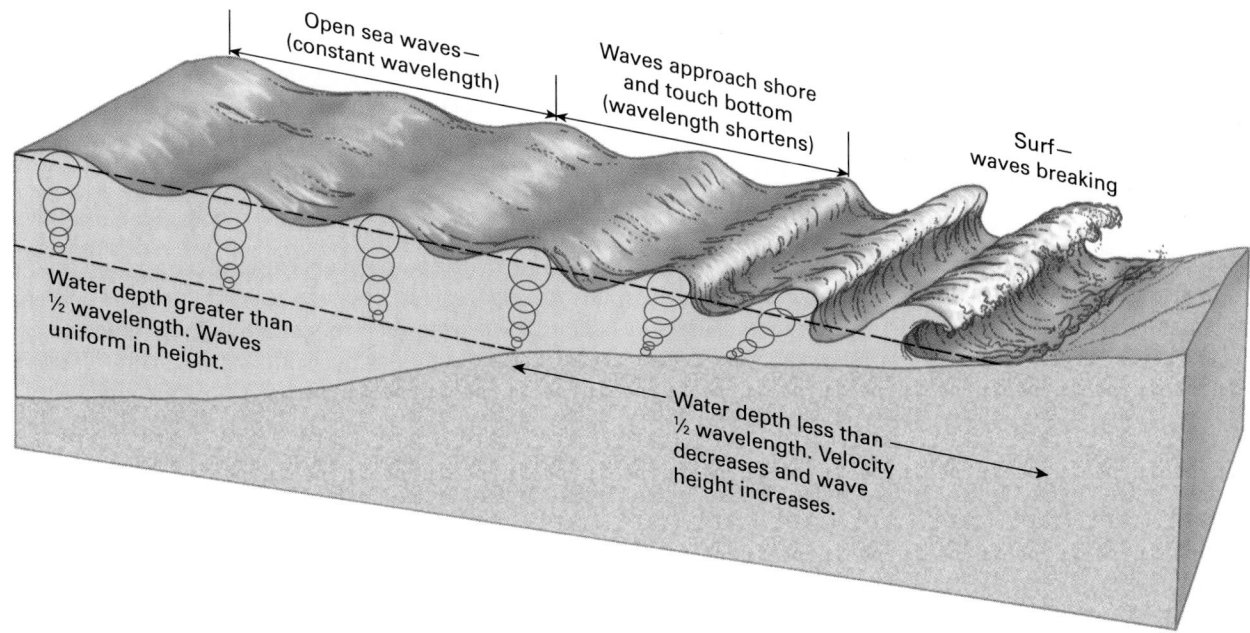

FIGURE 16.14 When a wave approaches the shore, the circular motion at the bottom of the wave flattens out and becomes elliptical. The bottom of the wave drags against the sea floor. As a result, the wavelength shortens and the wave steepens until it finally breaks, creating surf.

refraction The bending of a wave that occurs when it approaches the shore at an angle; the end of the wave in shallow water slows down while the end in deeper water continues at a faster speed.

© PETE NIESEN/SHUTTERSTOCK

FIGURE 16.15 Expensive seashore houses like these along the California coast are often threatened by coastal erosion that occurs during intense storms.

high wave strikes shore with 40 times the force of a 1.5-meter-high wave. A giant, 10-meter-high storm wave strikes a 10-meter-wide sea wall with four times the thrust energy of the three main orbiter engines of a space shuttle.

Seawater weathers and erodes coastlines by hydraulic action, abrasion, solution, and salt cracking, processes that are familiar from our earlier discussions of weathering and streams.

A wave striking a rocky cliff drives water into cracks or crevices in the rock, compressing air in the cracks. The air and water combine to create hydraulic forces strong enough to dislodge rock fragments or even huge boulders. Storm waves create forces as great as 25 to 30 tons per square meter. Engineers built a breakwater in Wick Bay, Scotland, of car-sized rocks weighing 80 to 100 tons each. The rocks were bound together with steel rods set in concrete, and the sea wall was topped by a steel-and-concrete cap weighing more than 800 tons. A large storm broke the cap and scattered the rocks about the beach. The breakwater was rebuilt, reinforced, and strengthened, but a second storm destroyed this wall as well. On the Oregon coast, the impact of a storm wave tossed a 60-kilogram rock over a 25-meter-high lighthouse. After sailing over the lighthouse, it crashed through the roof of the keeper's cottage, startling the inhabitants.

While images of flying boulders are spectacular, most wave erosion occurs gradually, by abrasion. Water is too soft to abrade rock, but waves carry large quantities of silt, sand, and gravel. Breaking waves roll this sediment back and forth over bedrock, acting like liquid sandpaper, eroding the rock. At the same time, smaller cobbles are abraded as they roll back and forth in the surf zone.

Seawater slowly dissolves rock and carries ions in solution. Saltwater also soaks into bedrock; when the water evaporates, the growing salt crystals pry the rock apart.

Sediment Transport along Coastlines

Most waves approach the shore at an angle rather than head-on. When this happens, one end of the wave encounters shallow water and slows down, while the rest of the wave is still in deeper water and continues to advance at a relatively faster speed. As a result, the wave bends (Figures 16.16A and 16.16B). This effect is called **refraction**. Consider the analogy of a sled gliding down a snowy hill onto a cleared pathway. If the sled hits the surface at an angle, one runner will reach it before the other. The runner that hits the pavement first slows down, while the other, which is still on the snow, continues to travel rapidly (Figure 16.16C). As a result, the sled turns abruptly.

As waves approach an irregular coast, they reach the headlands first, breaking against the point and eroding it. The waves then refract around the headland and travel parallel to its sides. Surfers seek such refracted waves because these waves move nearly parallel to the coast and therefore travel for a long distance before breaking. For example, the classic big-wave mecca at Waimea Bay of Oahu, Hawaii, forms along a point of land that juts into the sea. Waves build when they strike an offshore coral reef, then refract along the point and charge into the bay.

Refracted waves transport sand and other sediment toward the interior of a bay. As the headlands erode and the interiors of bays fill with sand, an irregular coastline eventually straightens (Figure 16.17).

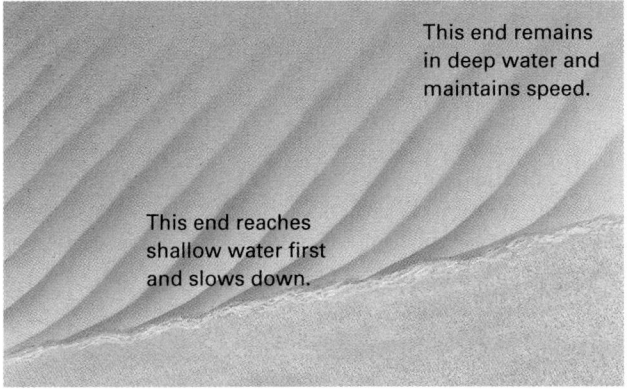

This end remains in deep water and maintains speed.

This end reaches shallow water first and slows down.

A

B

COURTESY OF GRAHAM R. THOMPSON/JONATHAN TURK

Sled analogy

Snowy hill

Road

C

FIGURE 16.16 (A) When a water wave strikes the shore at an angle, the end in shallow water slows down, causing the wave to bend, or refract. (B) Wave refraction on a lake shore. (C) A sled turns when it strikes a paved surface at an angle because one of the sled's runners hits the roadway and slows down before the other does.

FIGURE 16.17 (A) When a wave strikes a headland, the shallow water causes that portion of the wave to slow down. (B) Part of the wave breaks against the headland, weathering the rock. A portion of the wave refracts, transporting sediment and depositing it on the beach inside the bay. (C) Eventually, this selective weathering, erosion, and deposition will straighten an irregular coastline.

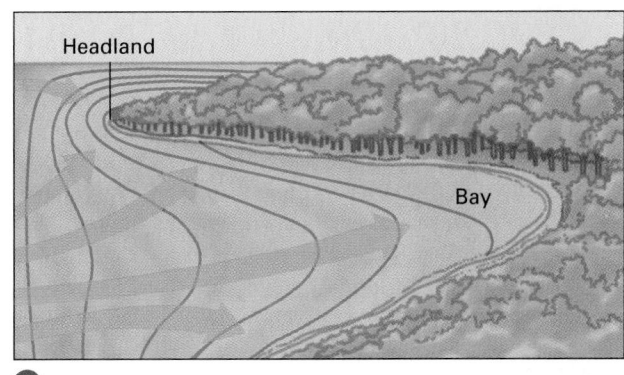

Headland

Bay

A

Cliff formed by wave erosion

Deposition

B

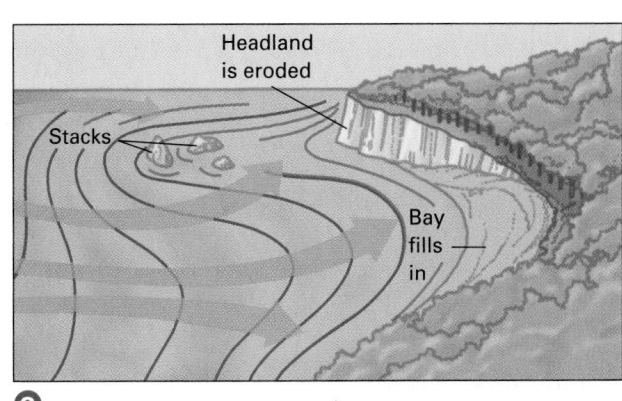

Headland is eroded

Stacks

Bay fills in

C

When waves strike shore at an angle, they form a **longshore current** that flows parallel to the shore (Figure 16.18). Longshore currents flow in the surf zone and a little farther out to sea and may travel for tens or even hundreds of kilometers. They transport sand for great distances along coastlines. Sediment transport also occurs by **beach drift**. If a wave strikes the beach obliquely, it pushes sand up and along the beach in the direction that the wave is traveling. When water recedes, the sand flows straight down the beach as shown in Figure 16.18. Thus, at the end of one complete wave cycle, the sand has moved a short distance parallel to the coast. The next wave transports the sand a little further, until, over time, sediment moves long distances.

Longshore currents and beach drift work together to transport and deposit huge amounts of sand along a coast. Much of the sand found at Cape Hatteras, North Carolina, originated hundreds of kilometers away, from the mouth of the Hudson River in New York and from glacial deposits on Long Island and southern New England. Midway along this coast, at Sandy Hook, New Jersey, an average of 2,000 tons a day will move past any point on the beach. As a result of this process, beaches have been called rivers of sand.

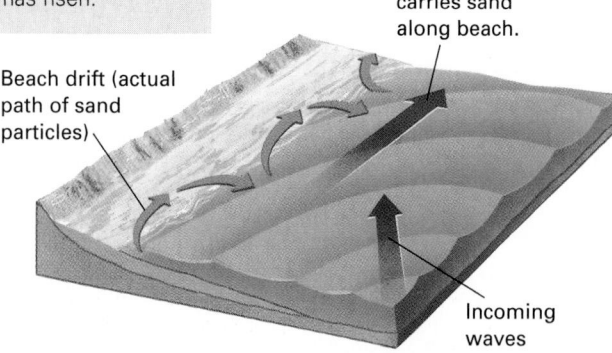

Beach drift (actual path of sand particles)

Longshore current carries sand along beach.

Incoming waves

FIGURE 16.18 Longshore currents and beach drift transport sediment along a coast.

Tidal Currents

When tides rise and fall along an open coastline, water moves in and out from the shore as a broad sheet. If the flow is channeled by a bay with a narrow entrance or by islands, the moving water funnels into a **tidal current**, which is a flow of ocean water caused by tides. Tidal currents can be intense where large differences exist between high and low tides and narrow constrictions occur in the shoreline. On parts of the west coast of British Columbia, a diesel-powered fishing boat cannot make headway against tidal currents flowing between closely spaced islands. Fishermen must wait until the tide, and hence the tidal current, reverses direction before proceeding.

16.7 Emergent and Submergent Coastlines

Geologists have found drowned river valleys and fossils of land animals on continental shelves beneath the sea. They have also found sedimentary rocks containing fossils of fish and other marine organisms in continental interiors. As a result, we infer that sea level has changed, sometimes dramatically, throughout geologic time. An **emergent coastline** forms when a portion of a continent that was previously under water becomes exposed as dry land (Figure 16.19). Falling sea level or rising land can cause emergence. As explained in Section 16.8, many emergent coastlines are sandy. In contrast, a **submergent coastline** develops when the sea floods low-lying land and the shoreline moves inland (Figure 16.19). Submergence occurs when sea level rises or coastal land sinks. A submergent coast is commonly irregular, with many bays and headlands. The coast of Maine, with its numerous fjords, inlets, and rocky bluffs, is a submergent coastline (Figure 16.20). Small, sandy beaches form in protected coves, but most of the headlands are rocky and steep.

Factors That Cause Coastal Emergence and Submergence

Tectonic processes can cause a coastline to rise or sink. Isostatic adjustment can also depress or elevate a portion of a coastline. About 18,000 years ago, a huge continental glacier covered most of Scandinavia, causing it to sink isostatically. As the lithosphere settled, the displaced asthenosphere flowed southward, causing the Netherlands to rise. When the ice melted, the process reversed as the asthenosphere flowed back from below the Netherlands to Scandinavia. Today, Scandinavia is rebounding and the Netherlands is sinking. During the Pleistocene Ice Age, Canada was depressed by the ice, and asthenosphere rock flowed southward. Today, the asthenosphere is flowing back north, much of Canada is rebounding, and much of the United States is sinking.

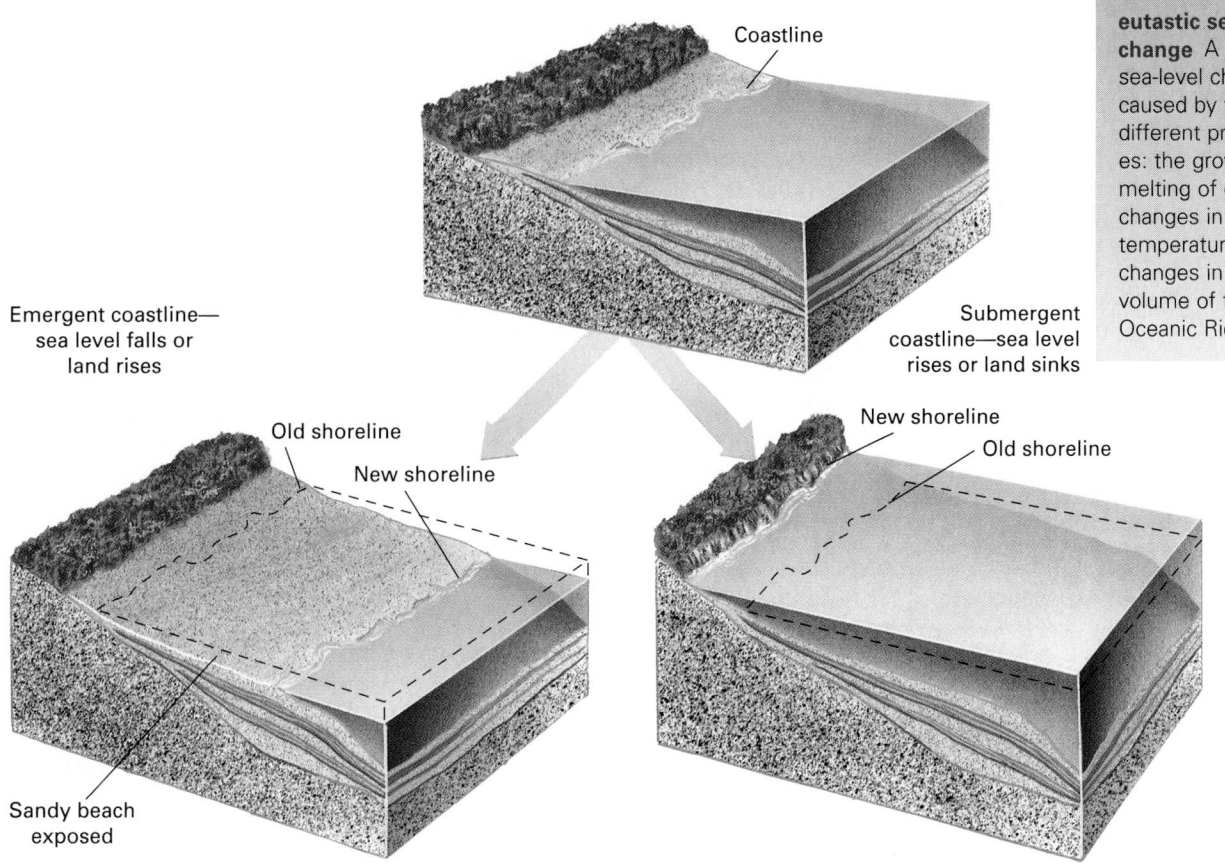

Coastline

Emergent coastline—
sea level falls or
land rises

Submergent
coastline—sea level
rises or land sinks

eutastic sea-level change A global sea-level change, caused by three different process-es: the growth or melting of glaciers, changes in water temperature, and changes in the volume of the Mid-Oceanic Ridge.

Old shoreline

New shoreline

New shoreline

Old shoreline

Sandy beach
exposed

FIGURE 16.19 If sea level falls or if the land rises, the new coastline is emergent. Offshore sand is exposed to form a sandy beach. If coastal land sinks or sea level rises, the new coastline is submergent. Areas that were once land are flooded. Irregular shorelines develop and beaches are commonly rocky.

FIGURE 16.20 The Maine coast is a rocky, irregular, submergent coastline.

COURTESY OF GRAHAM R. THOMPSON/JONATHAN TURK

Sea level can also change globally. A global sea-level change, called **eustatic sea-level change**, occurs by three mechanisms: the growth or melting of glaciers, changes in water temperature, and changes in the volume of the Mid-Oceanic Ridge.

During an ice age, vast amounts of water move from the sea to form continental glaciers, and sea level falls, resulting in global emergence. Similarly, when glaciers melt, sea level rises globally, causing submergence.

Seawater expands when it is heated and contracts when it is cooled. Although this change is not notice-able in a glass of water, the volume of the oceans is so great that a small temperature change can alter sea level measurably. As a result, global warming causes sea-level rise, and cooling leads to falling sea level.

Temperature changes and glaciation are linked. When global temperature rises, seawater expands and glaciers melt. Thus, even minor global warming can lead to a large sea-level rise. The opposite effect is also true. When temperature falls, seawater contracts, glaciers grow, and sea level falls.

As explained in Chapter 15, changes in the volume of the Mid-Oceanic Ridge can also affect sea level.

beach Any strip of shoreline washed by waves and tides.

foreshore *or* **intertidal zone** The part of a beach that lies between the high-tide and low-tide lines, and is exposed to the air at low tide but is covered by water at high tide.

backshore The upper zone of a beach that is usually dry but is washed by waves during storms.

FIGURE 16.21 (A) Sea lions doze on a sandy California beach. (B) Rocky beaches are found along the Oregon coast to the north.

The Mid-Oceanic Ridge displaces seawater. When lithospheric plates spread slowly from the Mid-Oceanic Ridge, they create a narrow ridge that displaces relatively little seawater, resulting in low sea level. In contrast, rapidly spreading plates produce a high-volume ridge that displaces more water, causing a global sea-level rise. At times in Earth's history, spreading has been relatively rapid, and as a result, global sea level has been high.

16.8 Beaches

When most people think about going to the **beach**, they think of gently sloping expanses of sand. However, a *beach* is any strip of shoreline that is washed by waves and tides. Although many beaches are sandy, others are swampy or rocky (Figure 16.21).

A beach is divided into two zones, the **foreshore**, also called the **intertidal zone**, and the **backshore**. The foreshore lies between the high- and low-tide lines and is alternately exposed to the air at low tide and covered by water at high tide. The backshore is usually dry but is washed by waves during storms. Many terrestrial plants cannot survive in saltwater, so specialized, salt-resistant plants live in the backshore. The backshore can be wide or narrow, depending on its topography, the local tidal difference, and the frequency and intensity of storms. In a region where the land rises steeply, the backshore may be a narrow strip. In contrast, if the coast consists of low-lying plains and if coastal storms occur regularly, the backshore may extend several kilometers inland.

If weathering and erosion occur along all coastlines, why are some beaches sandy and others rocky? The answer lies partly in the fact that most sand is not formed by weathering and erosion at the beach itself. Instead, several processes transport sand to a seacoast. Rivers carry large quantities of sand, silt, and clay to the sea and deposit it on deltas that may cover thousands of square kilometers. In some coastal regions, glaciers deposited large quantities of sandy till along coastlines during the Pleistocene Ice Age. In tropical and subtropical latitudes, eroding reefs supply carbonate sand to nearby beaches. A sandy coastline is one with abundant sediment from any of these sources.

Longshore currents transport and deposit the sand along the coast. Much of the sand carried by these currents accumulates on underwater offshore bars. Thus, a great deal of sand may be stored offshore from a beach. If such a coastline emerges, this vast supply of sand becomes exposed as dry land. Thus, sandy beaches are abundant on emergent coastlines.

In contrast, rocky coastlines occur where sediment from any of these sources is scarce. With no abundant sources of sand, small sandy beaches may form in protected bays but most of the coast will be rocky. On a submergent coastline, rising sea level puts the stored offshore sand even farther out to sea and below the depth of waves. As a result, submergent coastlines commonly have rocky beaches.

Sandy Coastlines

A **spit** is a small, finger-like ridge of sand or gravel that extends outward from a beach (Figure 16.22). As sediment migrates along a coast, the spit may continue to grow. A well-developed spit may rise several meters above high-tide level and may be tens of kilometers long. A spit may block the entrance to a bay, forming a **baymouth bar**. A spit may also extend outward into the sea, creating a trap for other moving sediment.

A **barrier island** is a long, low-lying island that extends parallel to the shoreline. It looks like a beach or spit and is separated from the mainland by a sheltered body of water called a **lagoon**. Barrier islands extend along the east coast of the United States from New York to Florida. They are so nearly continuous that a sailor in a small boat can navigate the entire coast inside the barrier island system and remain protected from the open ocean most of the time. Barrier islands also line the Texas Gulf Coast.

Barrier islands form in several ways. The two essential ingredients are a large supply of sand and the

spit A small ridge of sand or gravel extending from a beach into a body of water.

baymouth bar A spit that extends partially or completely across the entrance to a bay.

barrier island A long, narrow, low-lying island that extends parallel to the shoreline.

lagoon A sheltered body of water separated from the sea by a reef or barrier island.

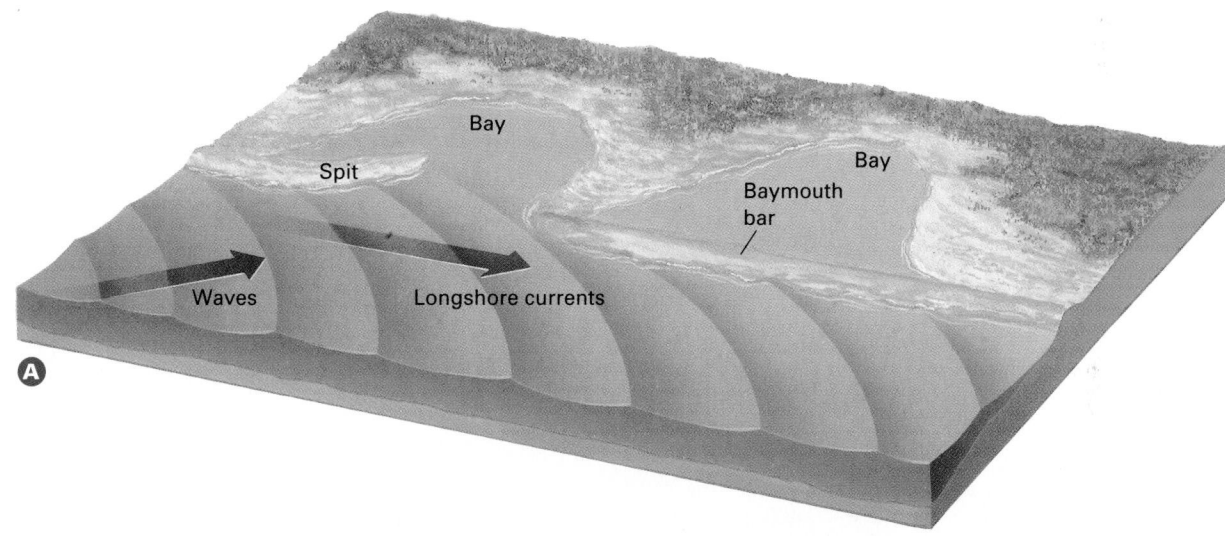

COURTESY OF GRAHAM R. THOMPSON/JONATHAN TURK

FIGURE 16.22 (A) Spits and baymouth bars are common features of sandy emergent coastlines. (B) Aerial photograph of a spit that formed along a low-lying coast in northern Siberia.

waves or currents to transport it. If a coast is shallow for several kilometers outward from shore, breaking storm waves may carry sand toward shore and deposit it just offshore as a barrier island. Alternatively, if a longshore current veers out to sea, it slows down and deposits sand where it reaches deeper water. Waves may then pile up the sand to form a barrier island.

Other mechanisms that create barrier islands involve sea-level change. Underwater sand bars may be exposed as a coastline emerges. Alternatively, sand dunes or beaches may form barrier islands if a coastline sinks.

Development on Sandy Coastlines

The Atlantic coast of the United States is fringed with the longest chain of barrier islands in the world. Many seaside resorts are built on these islands, and developers often ignore the fact that they are transient and changing landforms (Figure 16.23). If the rate of erosion exceeds that of deposition for a few years in a row, a barrier island can shrink or disappear completely, leading to destruction of beach homes and resorts. In addition, barrier islands are especially vulnerable to hurricanes, which can wash over low-lying islands and move enormous amounts of sediment in a very brief

time. In September 1996, Hurricane Fran flattened much of Topsail Island, a low-lying barrier island in North Carolina. Geologists were not surprised because the homes were not only built on sand—they were built on sand that was virtually guaranteed to move.

As a second example, Long Island extends eastward from New York City and is separated from Connecticut by Long Island Sound. Longshore currents flow westward, eroding sand from glacial deposits at the eastern end of the island and depositing it to form beaches and barrier islands on the south side of the island (Figure 16.24). At any point along the beach, the currents erode and deposit sand at approximately the same rates (Figure 16.25A). Geologists calculate that supply of sand at the eastern end of the island is large enough to last for a few hundred years. When the glacial deposits at the eastern end of the island become exhausted, the flow of sand will cease. Then the entire coastline will erode and the barrier islands and beaches will disappear.

However, if we narrow our time perspective and look at a Long Island beach over a season or during a single storm, the rates of erosion and deposition are not equal. Thus beaches shrink and expand with the seasons or the passage of violent gales. In the winter,

© FLORIDA STOCK/SHUTTERSTOCK

FIGURE 16.23 Many resorts, condos, and homes are built on transient and changing barrier islands, such as these condos on Hutchinson Island, Florida.

Unit 4: The Oceans

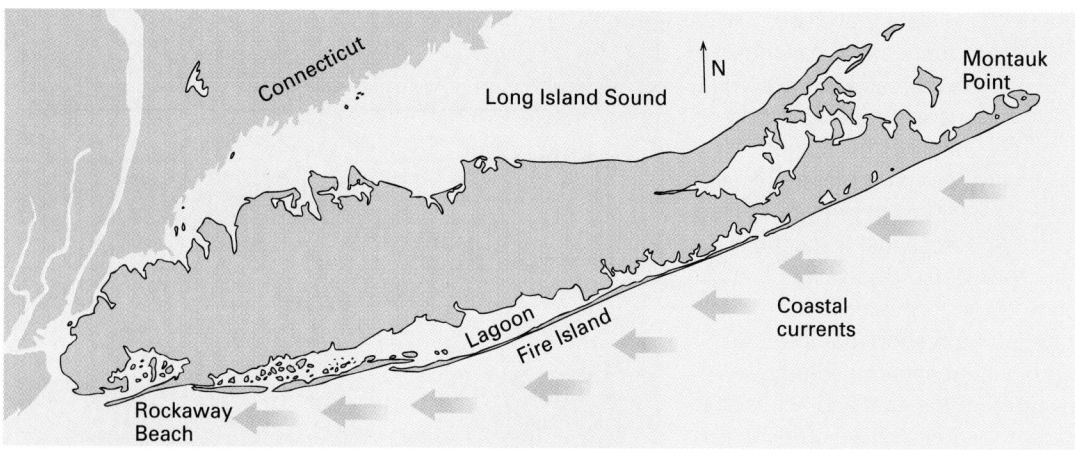

FIGURE 16.24 Longshore currents carry sand westward along the south shore of Long Island to create a series of barrier islands.

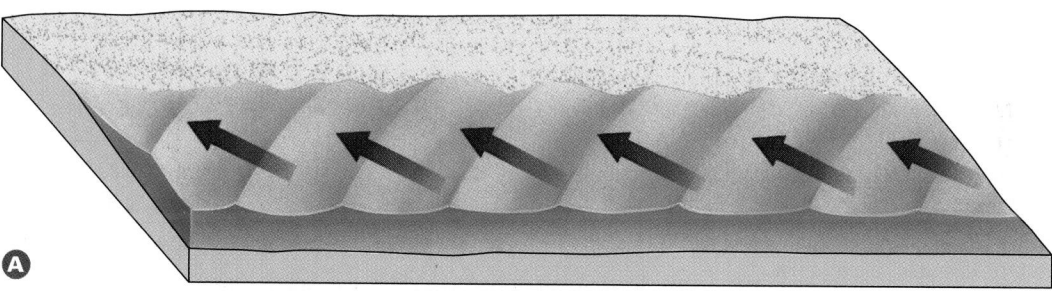

Undeveloped beach; ocean currents (arrows) carry sand along the shore, simultaneously eroding and building the beach.

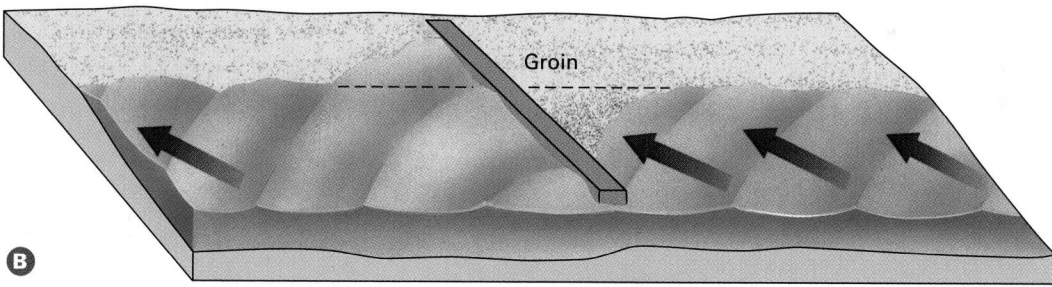

A single groin or breakwater; sand accumulates on upstream side and is eroded downstream.

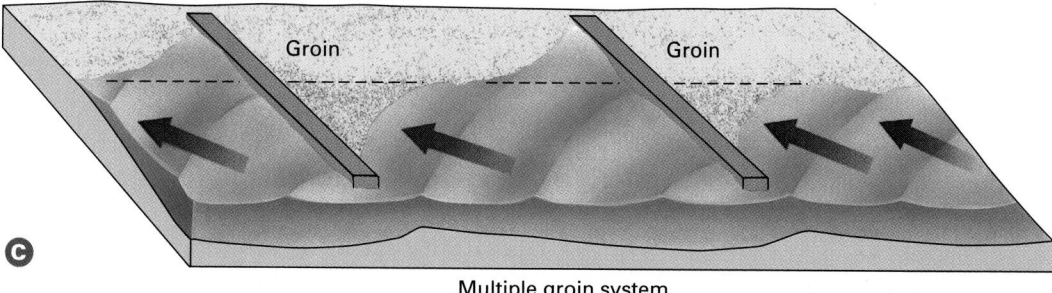

Multiple groin system

FIGURE 16.25 (A) Longshore currents simultaneously erode and deposit sand along an undeveloped beach. (B) A single groin or breakwater traps sand on the upstream side, resulting in erosion on the downstream side. (C) A multiple groin system propagates the uneven distribution of sand along the entire beach.

violent waves and currents erode beaches, whereas sand accumulates on the beaches during the calmer summer months. In an effort to prevent these seasonal fluctuations and to protect their personal beaches, Long Island property owners have built stone barriers called **groins** from shore out into the water. The groin intercepts the steady flow of sand moving from the east and keeps that particular part of the beach from eroding. But the groin impedes the overall flow of sand. West of the groin the beach erodes as usual, but the sand is not replenished because the upstream groin traps it. As a result, beaches downcurrent from the groin erode away (Figure 16.25B). The landowners living downcurrent from groins may then decide to build other groins to protect their beaches (Figure 16.25C). The situation has a domino effect, with the net result that millions of dollars are spent in ultimately futile attempts to stabilize a system that was naturally stable in its own dynamic manner (Figure 16.26).

Storms pose another dilemma. Hurricanes commonly strike Long Island in the late summer and fall, generating storm waves that completely overrun the barrier islands, flattening dunes and eroding beaches. When the storms are over, gentler waves and longshore currents carry sediment back to the beaches and rebuild them. As the sand accumulates again, salt marshes rejuvenate and the dune grasses grow back within a few months.

These short-term fluctuations are incompatible with human ambitions. People build houses, resorts, and hotels on or near the shifting sands. The owner of a home or resort hotel cannot allow the buildings to be flooded or washed away. Therefore, property owners construct large sea walls along the beach. When a storm wave rolls across an undeveloped low-lying beach, it dissipates its energy gradually as it flows over the dunes and transports sand. The beach is like a judo master who defeats an opponent by yielding with the attack, not countering it head-on. A sea wall interrupts this gradual absorption of wave energy. The waves crash violently against the barrier and erode sediment at its base until the wall collapses. It may seem surprising that a reinforced concrete sea wall is more likely to be

FIGURE 16.26 (A) This aerial photograph of a Long Island beach shows sand accumulating on the upstream side of a groin and erosion on the downstream side. (B) A close-up of one house on that Long Island beach shows waves lapping against the foundation.

permanently destroyed than a beach of grasses and sand dunes, yet this is often the case (Figure 16.27).

Rocky Coastlines

A rocky coastline is one without any of the abundant sediment sources described previously. In many areas on land, bedrock is exposed or covered by only a thin layer of soil. If this type of sediment-poor terrain is sub-

Unit 4: The Oceans

Winter

Dunes

Beach

Sand carried out to sea;
shoreline recedes .

A

Summer

Dunes   Shore
         in winter   Shore
                      in summer

Sand replaced

B

Sea
wall

C

D

FIGURE 16.27 (A) In a natural beach, the violent winter waves often move sand out to sea. (B) The gentler summer waves push sand toward shore and rebuild the beach. (C) Wave energy concentrates against a sea wall and (D) may eventually destroy it.

merged, and if there are no other sources of sand, the coastline is rocky.

A **wave-cut cliff** forms when waves erode the headland into a steep profile. As the cliff erodes, it leaves a flat or gently sloping **wave-cut platform** (Figure 16.28). If waves cut a cave into a narrow headland, the cave may eventually erode all the way through the headland, forming a scenic **sea arch**. When an arch collapses or when the inshore part of a headland erodes faster than the tip, it leaves behind a pillar of rock called a **sea stack**

plankton Small marine organisms that live mostly within a few meters of the sea surface, where sunlight is available, and that conduct most of the photosynthesis and nutrient consumption in the ocean and form the base of the marine food web.

phytoplankton Plankton that conduct photosynthesis like land-based plants and that are the base of the food chain for aquatic animals.

zooplankton Tiny marine animals that live mostly within a few meters of the sea surface and feed on phytoplankton.

(Figure 16.29). As waves continue to batter the rock, eventually the sea stack crumbles.

If the sea floods a long, narrow, steep-sided coastal valley, a sinuous bay called a *fjord* is formed (Chapter 13). Fjords are common at high latitudes, where rising sea level has flooded coastal valleys scoured by Pleistocene glaciers. Fjords may be hundreds of meters deep, and often the cliffs drop straight into the sea.

16.9 Life in the Sea

On land, most photosynthesis is conducted by multicellular plants such as mosses, ferns, grasses, and trees. Large animals such as cows, deer, elephants, and bison consume the plants. In contrast, most of the photosynthesis and consumption in the ocean is carried out by small organisms called **plankton**. Many plankton are single-celled and microscopic; others are more complex and are up to a few centimeters long. One major difference between terrestrial and aquatic ecosystems is that on land, soil nutrients are abundant on the surface, where light is also abundant. However, in the oceans, light is available only on the surface, whereas nutrients tend to settle to the dark depths. Plankton live mostly within a few meters of the sea surface, where light is available. However, the growth and productiv-

FIGURE 16.29 Massive waves of the Antarctic Ocean eroded cliffs to form these sea stacks near Cape Horn.

ity of these organisms is limited by the fact that most of the nutrients such as nitrates, iron, and phosphates, are pulled downward toward the sea floor by gravity. There are two types of plankton: phytoplankton and zooplankton.

Phytoplankton conduct photosynthesis like land-based plants do. Therefore, they are the base of the food chain for aquatic animals. Although phytoplankton are not readily visible, they are so abundant that they supply about 50 percent of the oxygen in our atmosphere. **Zooplankton** are tiny animals that feed on the phytoplankton (Figure 16.30). The larger and more familiar marine plants (such as seaweed) and animals (such as fish, sharks, and whales) play a relatively small role in oceanic photosynthesis and consumption. However these organisms are important to humans, because people depend on fish for a vital source of protein.

World Fisheries

The shallow water of a continental shelf supports large populations of marine organisms. In addition, many deep-sea fish spawn in shallow water within 1 or 2 kilometers of shore. Shallow zones in bays, lagoons, and estuaries are especially hospitable to life because they have (1) easy access to the deep sea, (2) lower salinity than the open ocean, (3) a high concentration of nutrients originating from land and sea, (4) shelter, and (5) abundant plant life rooted to the sea floor in addi-

FIGURE 16.28 Waves hurl sand and gravel against solid rock to erode cliffs and create a wave-cut platform along the Oregon coast.

.56–1.0 mm (0.03937 inch.)

FIGURE 16.30 Zooplankton are tiny animals that live near the sea surface and feed on phytoplankton.

tion to the phytoplankton floating on the surface. As a result, about 99 percent of the marine fish caught every year are harvested from the shallow waters adjacent to shore.

In 1970, fishermen harvested 3 million tons of cod. But this catch was biologically unsustainable and declined to 1 million tons by 1993. In 1995, biologists suspended cod fishing in many areas to allow the fish populations to recover. In 1978, the herring and mackerel stocks began to decline, so boats switched nets to trawl for squid. In the 1980s the squid population began to decline. Thus fishing pressure has worked through the food chain, disrupting the entire ecosystem.

In one study, oceanographers documented that industrial fishing fleets have caused a 90 percent reduction of large predatory fish such as tuna, marlin, swordfish, cod, halibut, and flounder. At the advent of intensive commercial fishing 100 years ago, most fleets caught 6 to 12 fish for every 100 baited hooks. Today, despite sophisticated electronic and aerial fish-finding techniques, the catch rate has plummeted to 1 fish per 100 hooks. The Worldwatch Institute has pointed out that as industrial fishing depletes "large, long-lived predatory species ... that occupy the highest levels of the food chain, they move down to the next level—to species that tend to be smaller, shorter-lived, and less valuable. As a result, fishers worldwide now fill their nets with plankton-eating species such as squid, jacks, mackerel, sardines, and invertebrates including oysters, mussels, and shrimp." Commercial fishermen now work harder, spend more time, and consume more fuel to capture smaller quantities of "less valuable species—they are essentially fishing down the marine food web." The authors of the study continue:

> But the cycle of fishing down the marine food web can't go on forever. (At lower trophic levels, the species are so small and diluted that it is no longer economically feasible to fish.) At the current rate of descent, it will take only 30 to 40 years to fish down to the level of plankton.[1]

The UN's Food and Agricultural Organization predicts that world fish harvests will remain fairly constant at around 90 million tons per year until the year 2010. After 2010, ecosystem destruction and overfishing will lead to a sharp decline in populations and harvests.

Reefs

A **reef** is a wave-resistant ridge or mound built by corals, oysters, algae, or other marine organisms. Because corals need sunlight and warm, clear water to thrive, coral reefs develop in shallow, tropical seas where little

---

1. From Anne Platt McGinn, "Freefall in Global Fish Stocks," *World Watch Magazine* 11 (May–June 1998), 10.

FIGURE 16.31 (A) Coral reefs form abundantly in clear, shallow, tropical water. An aerial view of a fringing reef adjacent to a high volcanic island. (B) Reefs grow in the clear, shallow water near Vanuatu and many other South Pacific Islands.

DR. JAMES P. MCVEY, NOAA SEA GRANT PROGRAM

COURTESY OF GRAHAM R. THOMPSON/JONATHAN TURK

suspended clay or silt muddies the water (Figure 16.31). As the corals die, their offspring grow on their remains. Oyster reefs, on the other hand, form in temperate estuaries and can grow in more turbid water.

Globally, coral reefs cover about 600,000 square kilometers—about the area of France—but they spread out in long, thin lines. They are extraordinarily productive ecosystems because the corals provide shelter for many fish and other marine species. Within the past 50 years, 10 percent of the world's coral reefs have been destroyed and an additional 30 percent are in critical condition. Several factors contribute to this destruction:

- Silt from cities, urban roadways, farms, and improper logging smother the delicate reef organisms.

- Fertilizer runoff from farms and sewage runoff from cities have added nutrients to coastal waters, feeding coral predators. For example, starfish thrive in nutrient-rich water and starfish eat corals. Microorganisms fertilized by agricultural runoff and sewage have caused massive disease epidemics that kill the coral. In addition, toxic chemicals poison corals in polluted, industrial areas.

- Overfishing or improper fishing can kill reefs. Parrot fish and sea urchins eat algae that smother corals. If fishermen harvest too many parrot fish and sea urchins, the algae grow unchecked and kill the reefs. Also, in many parts of the world, fishermen dynamite coral reefs so their nets do not get tangled. Unfortunately, when the reefs are destroyed, fish populations decline. Therefore, while dynamiting reefs improves short-term gain, the practice diminishes long-term, sustainable harvests.

- Corals thrive best in a narrow temperature range. In recent years, oceanographers have compiled considerable evidence that sea surface temperatures have become warmer in recent decades, and the warm water is leading to massive deaths of the corals.

16.10 Global Warming and Sea-Level Rise

Sea level has risen and fallen repeatedly in the geologic past, and coastlines have emerged and submerged throughout Earth's history. During the past 40,000 years, sea level has fluctuated by 150 meters, primarily in response to growth and melting of glaciers (Figure 16.32). The rapid sea-level rise that started about 18,000 years ago began to level off about 7,000 years ago. By coincidence, humans began to build cities about 7,000 years ago. Thus, civilization has developed during a short time when sea level has been relatively constant.

Shore-based gauging stations and satellite radar studies agree that global sea level is presently rising at about 3 millimeters (about the thickness of a nickel) per year. Thus if present rates continue, sea level will rise 20 centimeters in 100 years. Records from the last century indicate an average rise in sea level of 1 to 2 millimeters per year. Such a rise would be significant along very low-lying areas such as the Netherlands, Bangladesh, and many Pacific islands. As explained previously, global warming causes sea level rise by two mechanisms. First, water expands when it is heated. Second, warm air temperatures cause glaciers to melt, which adds freshwater to the oceans. The expansion of seawater with increasing temperature is gradual because a small temperature increase causes a small sea-level rise. However, melting of glaciers can be caused by threshold mechanisms and therefore can occur rapidly.

Consequences of rising sea level vary with location and economics. Some villages on South Pacific islands have already been impacted. The wealthy, developed nations could build massive barriers to protect cities and harbors from a small sea-level rise. In regions where global sea-level rise is compounded by local tectonic sinking, dikes are already in place or planned. Portions of Holland lie below sea level, and the land is protected by a massive system of dikes. In London, where the high-tide level has risen by 1 meter in the past century, multimillion-dollar storm gates have been built on the Thames River. Venice, Italy, which is built over sea-level canals, has flooded frequently in recent years, and here, too, expensive engineering projects are under way. However, it is unlikely that people could protect against a dramatic sea-level rise. Coastal cities worldwide would be inundated.

Many poor countries cannot afford coastal protection even for a small sea-level rise. A 1-meter rise in sea level would flood 17 percent of the land area of Bangladesh, displacing 38 million inhabitants.

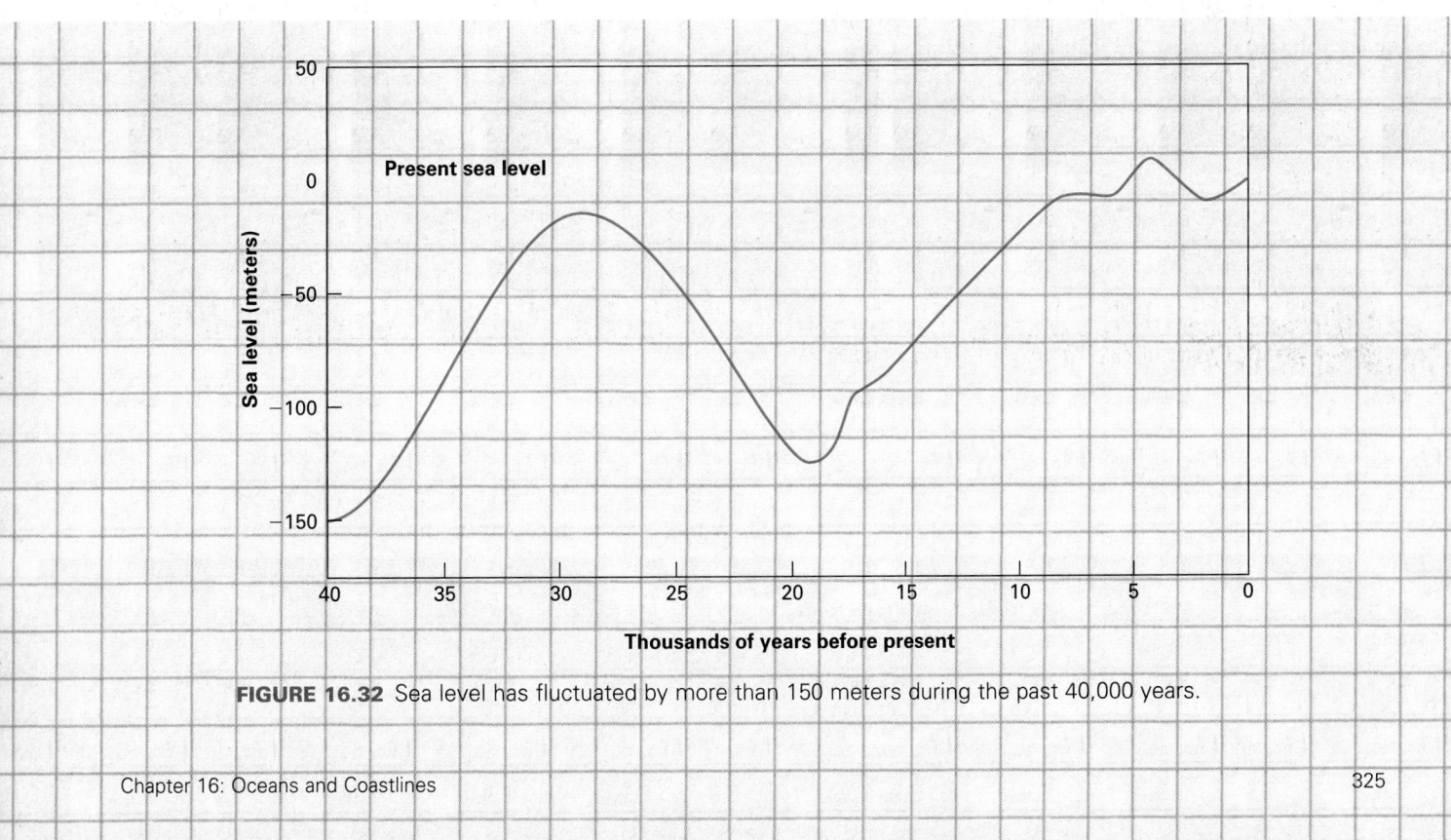

FIGURE 16.32 Sea level has fluctuated by more than 150 meters during the past 40,000 years.

17

THE ATMOSPHERE

1999 Hurricane Dennis.

visit **4ltrpress.cengage.com**

326

> # { Among the notable things about fire is that it also requires oxygen to burn—exactly like its enemy, life. }
>
> *Otto Weininger*

Every multicellular organism needs oxygen to survive. If the oxygen abundance in the atmosphere were to drop below 44 percent of its current value, life on Earth as we know it would perish. If oxygen is essential to life, would we be better off if we had an even greater supply? The answer is yes, to a limit. Even at sea level, athletes can enhance their performance by breathing a small amount of bottled oxygen. But, paradoxically, too much oxygen is poisonous. If you breathe air that has 55 percent or more oxygen than is found at sea level, your body metabolism is so rapid that essential molecules and enzymes decompose. In addition, fires burn more rapidly with increased oxygen concentration. If the oxygen level in the atmosphere were to rise significantly, fires would burn uncontrollably across the planet, altering ecosystems as we know them.

Earth is the near-perfect size and distance from a stable and medium-temperature star to permit optimal atmospheric conditions for life. But in addition, atmospheric composition and temperature are not determined solely by planetary size and distance from the Sun. Our planet's environment is finely regulated by Earth systems interactions. Over the past 4 billion years, the Sun's output has slowly increased, although there have been numerous fluctuations during this period. However, Earth's temperature has remained remarkably constant.

17.1 Earth's Early Atmospheres

Scientists have a nearly continuous record of rocks of different ages, from 3.96 billion years ago to the present. Therefore when they study the history of the crust, they can analyze the chemical composition of ancient rocks. But there are no samples of very old atmospheres. So how can we determine atmospheric composition millions to billions of years ago? There are many gaps in our understanding, but the history chronicled next comes from two sources: modeling and the study of

rocks. Modeling involves calculations about how atmospheric gases would have behaved under the presumed environment of early Earth. To test these models, scientists study the geochemistry of ancient rocks. Rocks react with water, rock, and air. By studying the rocks that existed at a specific time period, scientists deduce the other components of the environment that would have produced those reactions. For example, as we will discuss, iron reacts with oxygen to produce iron oxides. Thus, if we find iron oxides in certain types of sedimentary rocks that formed 2.6 billion years ago, we deduce that oxygen must have been present in the air and water.

The First Atmospheres: 4.6 to 4.0 Billion Years Ago

Our Solar System formed from a cold, diffuse cloud of interstellar gas and dust. About 99.8 percent of this cloud was composed of the two lightest elements: hydrogen and helium. Consequently, when Earth formed, its primordial atmosphere was composed almost entirely of these two light elements. But because Earth is relatively close to the Sun and its gravitational force is relatively weak, its primordial hydrogen and helium atmosphere rapidly boiled off into space and escaped. Table 17.1 shows this and the subsequent atmospheres of Earth described here.

In Chapter 15, we learned that most of the volatile compounds that form Earth's hydrosphere, atmosphere, and biosphere originated from outer parts of the Solar System. Recall that, in its infancy, the Solar System was crowded with bits of rock, comets, ice chunks, and other debris left over from the initial coalescence of the planets. These bolides crashed into the planet in a near-continuous rain that lasted almost 800 million years. Carbonate compounds and carbon-rich rocks reacted under the heat and pressure of impact to form carbon dioxide. Ice quickly melted into water. Ammonia, common in the icy tail of comets, reacted to form nitrogen (Figure 17.1).

outgassing The release of volatiles from Earth's mantle to the surface in volcanic eruptions.

TABLE 17.1 The Earth's Atmosphere through Time

Events That Formed the Atmosphere	Age of Atmosphere	Composition of Atmosphere
Primordial atmosphere: From initial accretion of planets	4.6 billion years ago	Hydrogen (H_2) and helium (He)
Secondary atmosphere: Bolide impact from outer space	4.5 billion years ago	Carbon dioxide (CO_2), water (H_2O), and nitrogen (N_2)
Atmosphere formed by outgassing and modified by reactions of gases with geosphere	4.5 to 2.7 billion years ago	Predominantly hydrogen (H_2) and carbon dioxide (CO_2); some water (H_2O) and nitrogen (N_2)
Evolution of cyanobacteria, which begin producing oxygen	2.7 billion years ago	Reactions with the environment causing oxygen (O_2) produced by cyanobacteria to be removed as quickly as it is produced
First Great Oxidation Event	2.4 billion years ago	Oxygen (O_2) accumulating in the atmosphere; hydrogen (H_2) becoming a trace gas
Second Great Oxidation Event	600 million years ago	Buildup of oxygen concentration in the atmosphere as biological and geological processes slow decay
Modern atmosphere: High oxygen concentration maintained by biological photosynthesis	Today	Primarily nitrogen (N_2) and oxygen (O_2), with smaller concentrations of other gases

When Life Began: 4.0 to 2.6 Billion Years Ago

The carbon dioxide, water, and nitrogen atmosphere changed as volatiles escaped from Earth's mantle to the surface in volcanic eruptions, in a process called **outgassing**. In 1953, Stanley Miller and Harold Urey hypothesized—on the basis of little direct modeling—that by 4 billion years ago, Earth's atmosphere consisted primarily of methane (CH_4), ammonia (NH_3), hydrogen (H_2), and water (H_2O). They mixed these gases in a glass tube and fired sparks across the tube to simulate Earth's early atmosphere, beset by lightning storms.

Amino acids, the building blocks of proteins, formed in the tube. The Miller–Urey model immediately became popular because scientists speculated that the first living organisms formed by accretion of these abiotic (non-living) amino acids.

In the early 1970s, the idea of a methane–ammonia–hydrogen atmosphere was largely discredited, and most scientists postulated that the Hadean atmosphere was composed mainly of carbon dioxide (CO_2), with smaller amounts of nitrogen (N_2), water (H_2O), and other gases. However, in 2005 scientists used models of Earth's Hadean geosphere to calculate how gases trapped in

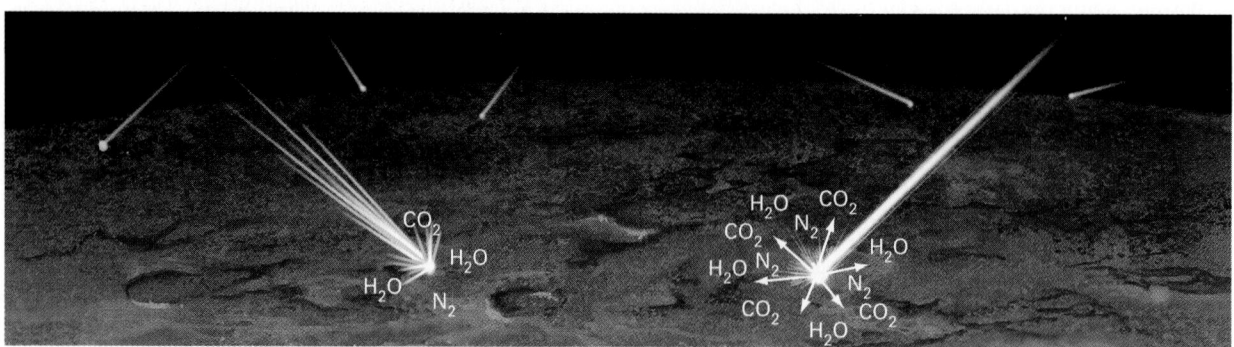

FIGURE 17.1 Comets, meteoroids, and asteroids imported Earth's volatiles from outer parts of the Solar System.

the interior would react with rocks and minerals in the planet's interior and surface. One pivotal question is: When did Earth's core form? Earth's core is the deepest layer of the geosphere. The atmosphere is a thin veneer surrounding the crust. How could one affect the other? Recall that the core is composed primarily of iron and nickel. Before Earth's interior dissociated into a layered core and mantle, the mantle contained more iron than it does today. Then, when iron settled into the core, the mantle became relatively iron-poor.

Mantle rocks rise to the surface through volcanic eruptions. In turn, gases in the atmosphere react with surface rocks. When gases react with rocks, the reactions not only change the chemical composition of the rocks, but they also alter the composition of the atmosphere. Thus, the formation of the core affects the evolution of the atmosphere. As we have said many times before, Earth is a system. Rocks affect the air. Air affects rocks.

Modern hypotheses state that Earth was hot at the time of its formation and most of the planet's iron was sequestered in its core shortly after the planet evolved. By modeling the reactions of volatiles in this iron-poor mantle, researchers concluded in 2005 that Earth's early atmosphere contained large amounts of hydrogen, as Miller and Urey postulated, but also high concentrations of carbon dioxide, as later models had proposed.[1] The argument is ongoing, and we can expect further discussion and modifications.

Thus our models of Earth and its evolution are constantly changing. Scientists propose hypotheses and theories, consider new ideas—and often other scientists disagree. This is the nature and the joy of science. In a review article about Earth's early atmospheres, Christopher Chyba of SETI Institute and Stanford University wrote that the argument about the composition of the Hadean atmosphere "makes it a great time for young scientists to enter the field, but it also reminds us that some humility regarding our favorite models is in order."[2]

17.2 Life, Iron, and the Evolution of the Modern Atmosphere

As explained previously, the first living organisms may have formed by accretion of complex abiotic (nonliving) organic molecules. But these complex organic molecules are oxidized and destroyed in an oxygen-rich environment. (This oxidation is analogous to slow burn-ing.) If large amounts of oxygen were present in Earth's early atmosphere, the abiotic precursors to living organisms could not have formed.

By studying the minerals in a rock, a geochemist can determine whether it formed in an oxygen-rich or an oxygen-poor environment. Recent studies of Earth's oldest rocks indicate that the atmospheric oxygen concentration in Hadean time was extremely low. Thus, the molecules necessary for the emergence of living organisms would have been preserved in the primordial atmosphere.

Although life could not have emerged in an oxygen-rich environment, complex multicellular life requires an oxygen-rich atmosphere to survive. How did oxygen become abundant in our atmosphere?

The world's earliest organisms probably obtained their energy from reactions with minerals such as iron and sulfur in an extremely inefficient process. Later, organisms subsisted, in part, by eating each other. But these food chains were limited because there were only a few organisms on Earth. A crucial step in evolution occurred when primitive bacteria evolved the ability to harness the energy in sunlight and produce organic tissue. This process, known as **photosynthesis**, is the foundation for virtually all modern life. During photosynthesis, organisms convert carbon dioxide and water to organic sugars. They release oxygen as a by-product. In 1972, an English chemist named James Lovelock hypothesized that the oxygen produced by primitive organisms gradually accumulated to create the modern atmosphere. When the oxygen concentration reached a critical level to sustain efficient metabolism in late-Precambrian time, multicellular organisms evolved and the biosphere as we know it was born. Lovelock was so overwhelmed by the intimate connection between living and nonliving components of Earth's systems that he likened our planet to a living creature, which he called **Gaia** (Greek for "Earth").

The Lovelock hypothesis, now nearly 40 years old, remains generally accepted, but scientists are still investigating many of the details. For example, blue-green algae called **cyanobacteria** began producing oxygen 2.7 billion years ago, but appreciable quantities of oxygen didn't appear in the atmosphere until 2.4 billion years ago (Figure 17.2). The production of oxygen by cyanobacteria was occurring slowly and steadily. But

photosynthesis
The process by which chlorophyll-bearing plant cells convert carbon dioxide and water to organic sugars, using sunlight as an energy source; oxygen is released in the process.

Gaia The term (Greek for "Earth") used by James Lovelock to refer to our planet, which he likened to a living creature due to the interconnectivity of all of Earth's systems.

cyanobacteria
Blue-green bacteria that were among the earliest photosynthetic life-forms on Earth.

1. Feng Tian, Owen B. Toon, Alexander A. Pavlov, and H. De Sterck, "A Hydrogen-Rich Early Earth Atmosphere," *Science* 308 (May 13, 2005), 1014–1017.
2. Christopher F. Chyba, "Rethinking Earth's Early Atmosphere," *Science* 308 (May 13, 2005), 962–963.

First Great Oxidation Event The process, occurring approximately 2.4 billion years ago, when the oxygen concentration in Earth's atmosphere increased suddenly from trace amounts to appreciable quantities, probably because of a combination of biological and geochemical processes.

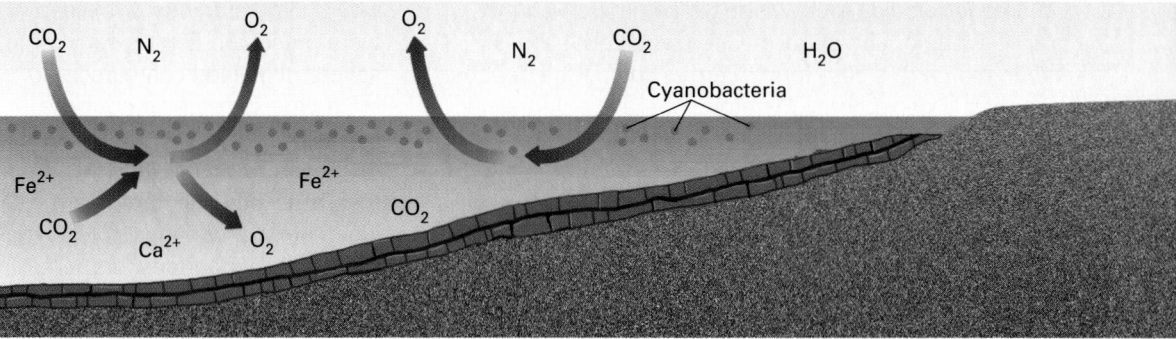

FIGURE 17.2 The oxygen content of the atmosphere slowly and steadily increased after cyanobacteria plants began to release oxygen 2.7 billion years ago.

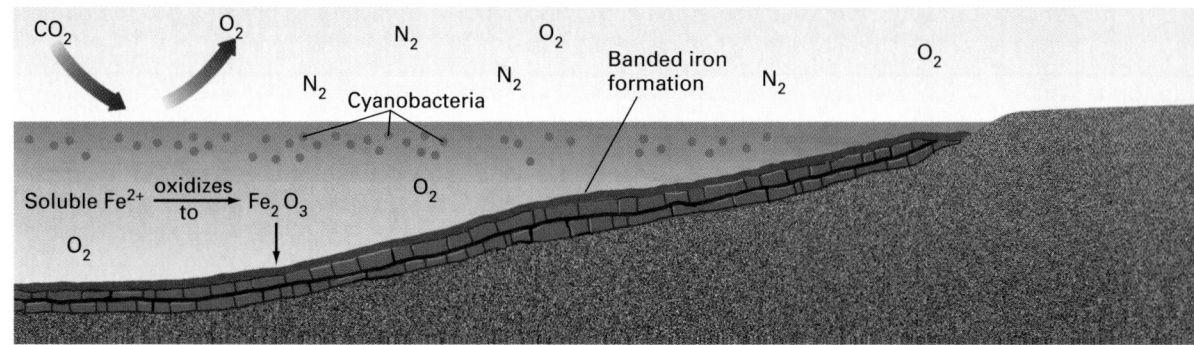

FIGURE 17.3 When the oxygen concentration in the atmosphere and seawater reached a threshold, the oxygen combined with dissolved iron to form iron minerals. The iron minerals precipitated to the sea floor to form the first layer of a banded iron formation.

for 300 million years, the concentration of oxygen in the atmosphere didn't rise. Then, 2.4 billion years ago, in a process called the **First Great Oxidation Event**, the concentration of atmospheric oxygen rose abruptly.

To understand the mechanism of this threshold reaction, we must study atmosphere–geosphere and atmosphere–biosphere systems interactions.

Systems Interactions That Affected Oxygen Concentration in the Hadean Atmosphere

Geosphere–Atmosphere Interactions

Even though large quantities of iron coalesced into Earth's core, appreciable quantities remained in the mantle and crust. Today, about 5 percent of the weight of the crust is iron.

Oxygen dissolves in water. As a result, when early cyanobacteria released oxygen, some of this gas dissolved in seawater. In an oxygen-poor environment, iron also dissolves in water. However, when oxygen is abundant, the dissolved oxygen reacts with dissolved iron and causes the iron to precipitate rapidly. Thus, if air contains little or no oxygen, there is also little oxygen

in seawater, and large amounts of iron dissolve in the seas. If the oxygen concentration of the atmosphere and the ocean rises to a threshold level, iron precipitates rapidly, forming a layer of iron oxide minerals on the sea floor. Thus, when enough oxygen had accumulated in seawater to react with the dissolved iron, vast quantities of iron minerals precipitated onto the sea floor, producing layers of iron-rich minerals (Figure 17.3). This process also removed oxygen, explaining why the oxygen concentration in the atmosphere didn't rise, even though cyanobacteria were releasing this gas. Iron, the main ingredient in steel, is the world's most commonly used metal. About 1 billion tons of iron are mined every year, 90 percent from banded iron formations, sedimentary layers of iron-rich minerals sandwiched between beds of clay and other silicate minerals. Most of Earth's banded iron deposits formed from 2.6 to 1.9 billion years ago, although a few are older and some are younger. The alternating layers are a few centimeters thick and give the rocks their banded appearance (Figure 17.4). A single iron formation of this type may be hundreds of meters thick and cover tens of square kilometers.

The alternating layers of iron minerals and other minerals may have developed because, for a long time,

FIGURE 17.4 In this banded iron formation from Michigan, the red bands are iron oxide minerals and the dark layers are chert (silica).

the oxygen level in the seas hovered near the threshold at which soluble iron converts to the insoluble variety. When the dissolved oxygen concentration increased, the oxygen reacted with the dissolved iron to precipitate iron-oxide minerals on the sea floor. But the formation of those minerals extracted oxygen as well as iron from the seawater, and lowered its oxygen concentration below the threshold. Then, dissolved iron accumulated again in the seas while the oxygen was slowly replenished. During that time, clay and other minerals washed from the continents and accumulated on the sea floor as they do today, forming the thin layers of silicate minerals that lie between the iron-rich layers. When the oxygen concentration rose above the threshold again, another layer of iron minerals formed.

Banded iron formations contain thousands of alternating layers of iron minerals and silicates. The great thickness of the iron formations, coupled with the fact that they continued to form from 2.6 to 1.9 billion years ago, suggests that these reactions must have kept the levels of dissolved oxygen close to the threshold for 700 million years. Thus the iron-rich rocks that support our industrial society were formed by interactions among early photosynthetic organisms, sunlight, air, and the oceans.

Biosphere–Atmosphere Interactions
In the primordial atmosphere, free oxygen also reacted with hydrogen to form water. This process helped keep the oxygen concentration in the atmosphere low. However, after life evolved, bacteria in the oceans removed atmospheric hydrogen in a process that produced methane. When the hydrogen concentration decreased sufficiently, free oxygen became chemically stable in the atmosphere and the oxygen concentration could rise.

Whatever the exact combination of biological and geochemical processes, evidence in the rocks indicates that the oxygen concentration in the atmosphere remained low and then jumped suddenly from trace to appreciable quantities approximately 2.4 billion years ago.

Evolution of the Modern Atmosphere

Several deposits of banded iron formed after the First Great Oxidation Event, and this process continued to remove oxygen that was released during photosynthesis. The last, major, banded iron layer was deposited about 1.9 billion years ago, but the oxygen concentration in the atmosphere did not increase dramatically. At least two more critical steps were required before efficient multicellular organisms could evolve.

The Sun emits energy largely in the form of high-energy ultraviolet light. These rays are energetic enough to break complex molecules apart and kill evolving multicellular organisms. But high-altitude oxygen absorbs ultraviolet radiation in a process that forms ozone (O_3). Thus the oxygen concentration couldn't increase in the lower atmosphere until appreciable concentrations accumulated in the upper atmosphere. To summarize: oxygen, largely produced by the earliest photosynthetic organisms, was not only necessary for life as we know it today, but as ozone it also protected multicellular life by filtering out harmful solar rays.

Multicellular plants and animals emerged in late-Precambrian time, between 1 billion and 543 million years ago. About 600 million years ago, the oxygen level in the atmosphere increased rapidly a second time, in a process called the **Second Great Oxidation Event**. What changed abruptly 1.3 billion years after the last banded iron layers were deposited to allow oxygen to accumulate? Scientists propose that prior to 600 million years ago, biological decay was almost as rapid as photosynthesis. Therefore, the oxygen that was released into the atmosphere was immediately consumed during respiration, according to the following reactions:

During photosynthesis: Carbon dioxide + Water → Sugars + Oxygen

During respiration and decay: Sugars + Oxygen → Carbon dioxide + Water

Then beginning abruptly, 600 million years ago, several processes locked up organic matter in sediments before it could decay. When decay and respiration slowed down, excess oxygen accumulated in the atmosphere. All the proposed processes for sequestering organic matter involve complex reactions among Earth's four spheres:

Second Great Oxidation Event
The process, occurring about 600 million years ago, when the oxygen concentration in Earth's atmosphere increased abruptly a second time in response to interactions among chemical, physical, and biological processes.

- Geochemical processes produced an abundance of clays 600 million years ago. These clays buried and preserved organic matter on the sea floor.

- Zooplankton evolved in the seas. These organisms produced dense, organic-laden feces that fell to the sea floor and accumulated in the clays mentioned above.

- Simple lichens evolved on land. The lichens accelerated weathering, and the weathered ions washed into the sea and provided nutrients for phytoplankton. In turn, the phytoplankton fed the zooplankton, which sequestered nutrients as described previously.

Thus numerous complex chemical, physical, and biological processes combined to set the stage for the Second Great Oxidation Event.

If all life on Earth were to cease, the atmosphere would revert to an oxygen-poor composition and become poisonous to modern plants and animals. But at the same time, geochemical mechanisms also contribute to the atmosphere that sustains us. Fires burn rapidly if oxygen is abundant; if its concentration were to increase even by a few percent, fires would burn uncontrollably across the planet. If atmospheric oxygen levels were to decrease appreciably, most modern plants and animals would not survive. If the carbon dioxide concentration were to increase by a small amount, atmospheric temperature would rise as a result of greenhouse warming. Earth's atmosphere not only sustains us, it insulates Earth's surface as winds distribute the Sun's heat around the globe, so the surface is neither too hot nor too cold for life to exist. Clouds form from water vapor in the atmosphere and rain falls from clouds. In addition, the atmosphere filters out much of the Sun's ultraviolet radiation, which can destroy living tissue and cause cancer. The atmosphere carries sound; without air we would live in silence. Without an atmosphere, airplanes and birds could not fly, wind would not transport pollen and seeds, the sky would be black rather than blue, and no reds, purples, and pinks would color the sunset. When we understand this complex web of interacting processes, and realize that we are the only planet in the Solar System to be so fortunate, we can only wonder at the fragility of Earth's atmosphere.

17.3 The Modern Atmosphere

The modern atmosphere is mostly gas, with small quantities of water droplets and dust. The gaseous composition of dry air is roughly 78 percent nitrogen, 21 percent oxygen, and 1 percent other gases (Figure 17.5). Nitrogen, the most abundant gas, does not react readily with other substances. Oxygen, though, reacts chemically as fires burn, iron rusts, and plants and animals respire. Carbon dioxide, which formed 80 percent

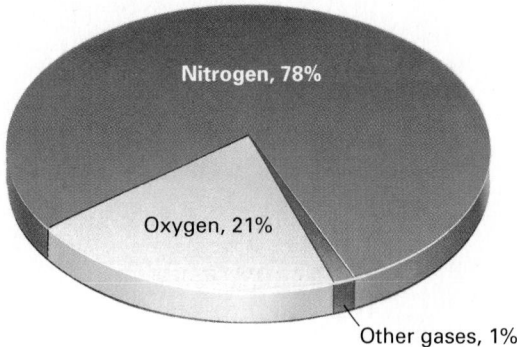

Composition of the Modern Atmosphere

FIGURE 17.5 Composition of the modern atmosphere.

of the secondary Hadean atmosphere, is a trace gas in the modern atmosphere, with a concentration of only 0.035 percent.

In addition to the gases listed above, air contains water vapor, water droplets, and dust. The types and quantities of these components vary with both location and altitude. In a hot, steamy jungle, air may contain 5 percent water vapor by weight, whereas in a desert or cold polar region, only a small fraction of a percent may be present.

If you sit in a house on a sunny day, you may see a sunbeam passing through a window. The visible beam is light reflected from tiny specks of suspended dust. Clay, salt, pollen, bacteria, viruses, bits of cloth, hair, and skin are all components of dust. People travel to the seaside to enjoy the "salt air." Visitors to the Great Smoky Mountains in Tennessee view the bluish, hazy air formed by sunlight reflecting from pollen and other dust particles.

Within the past century, humans have altered the chemical composition of the atmosphere in many different ways. We have increased the carbon dioxide concentration by burning fuels and igniting wildfires. Factories release chemicals into the air—some are benign, others are poisonous. Smoke and soot change the clarity of the atmosphere. These changes are discussed in Section 17.6.

17.4 Atmospheric Pressure

The molecules in a gas zoom about in a random manner. For example, at 20°C an average oxygen molecule is traveling at 425 meters per second (950 miles per hour). In the absence of gravity, or where there are temperature differences or other perturbations, a gas will fill a space homogeneously.

Thus, if you floated a cylinder of gas in space, the gas would disperse until there was an equal density of molecules and an equal pressure throughout the cylinder. But gases that surround Earth are perturbed by many

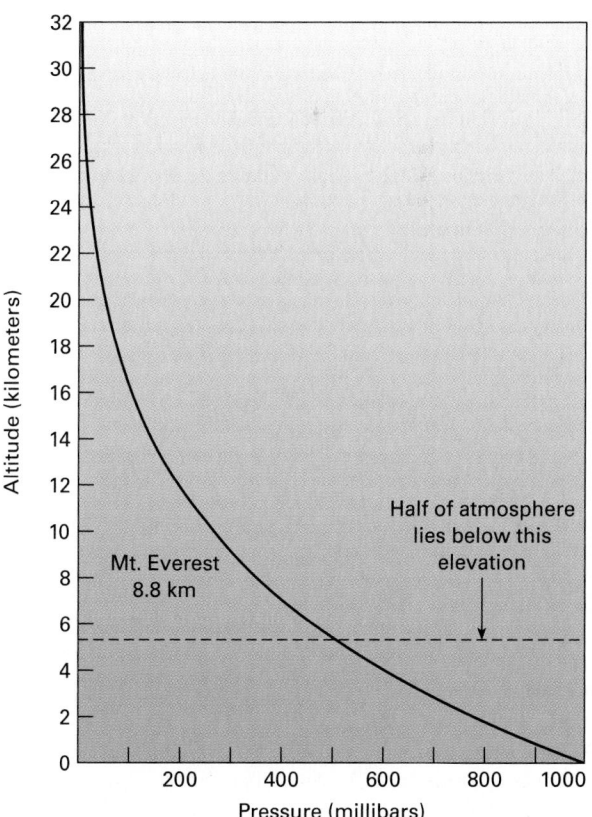

Near Mount Everest base camp, at about 5,000 meters, there is half as much oxygen in the air as there is at sea level.

© MY:SUMMIT/SHUTTERSTOCK

atmospheric pressure or **barometric pressure** The pressure of the atmosphere at any given location and time.

barometer A device used to measure barometric pressure.

bar Unit of measurement for atmospheric pressure. One bar is equal to sea-level atmospheric pressure.

influences, which ultimately create the complex and turbulent atmosphere that helps shape the world we live in.

Within our atmosphere, gas molecules zoom about, as in the imaginary cylinder, but in addition, gravity pulls them downward. As a result of this downward force, more molecules concentrate near the surface of Earth than at higher elevations. Therefore, the atmosphere is denser at sea level than it is at higher elevations—and the pressure is higher. Density and pressure then decrease exponentially with elevation (Figure 17.6). At an elevation of about 5,000 meters, the atmosphere contains about half as much oxygen as it does at sea level. If you ascended in a balloon to 16,000 meters (16 kilometers) above sea level, you would be above 90 percent of the atmosphere and would need an oxygen mask to survive. At an elevation of 100 kilometers, pressure is only 0.00003 that of sea level, approaching the vacuum of outer space. There is no absolute upper boundary to the atmosphere.

Atmospheric pressure, often called **barometric pressure**, is measured with a **barometer**. A simple but accurate barometer is constructed from a glass tube that is sealed at one end. The tube is evacuated and the unsealed end placed in a dish of a liquid such as mercury. The mercury rises in the tube because atmospheric pressure depresses the level of mercury in the dish but there is no air in the tube (Figure 17.7). At sea level mercury rises approximately 76 centimeters, or 760 millimeters (about 30 inches), into an evacuated tube.

Meteorologists express pressure in inches or millimeters of mercury, referring to the height of the column of mercury in a barometer. They also express pressure in bars and millibars. A **bar** is approximately equal to sea-level atmospheric pressure. A millibar is 0.001 of a bar.

A mercury barometer is a cumbersome device nearly a meter tall, and mercury vapor is poisonous. A safer and more portable instrument for measuring pressure, called an *aneroid barometer*, consists of a partially evacuated metal chamber connected to a pointer. When atmospheric pressure increases, it compresses the chamber and the pointer moves in one direction. When pressure

decreases, the chamber expands, directing the pointer the other way (Figure 17.8).

Changing weather can also affect barometric pressure. On a stormy day at sea level, pressure may be 980 millibars (28.94 inches), although it has been known to drop to 900 millibars (26.58 inches) or less during a hurricane. In contrast, during a period of clear, dry weather, a typical high-pressure reading may be 1,025 millibars (30.27 inches). These changes are discussed in Chapter 19.

FIGURE 17.6 Atmospheric pressure decreases with altitude. One-half of the atmosphere lies below an altitude of 5,600 meters.

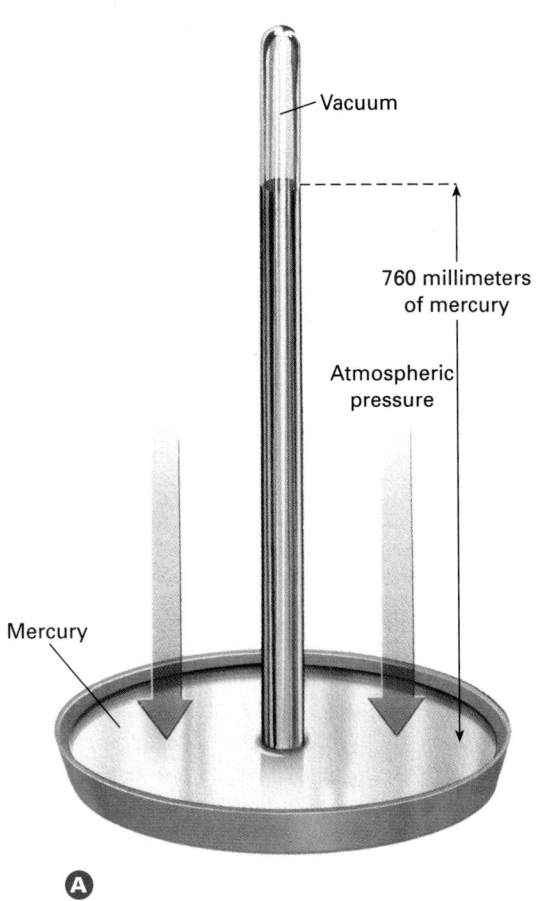

FIGURE 17.7 (A) Atmospheric pressure forces mercury upward in an evacuated glass tube. The height of the mercury in the tube is a measure of air pressure. (B) Three common scales for reporting atmospheric pressure and the conversion among them.

Vacuum

760 millimeters of mercury

Atmospheric pressure

Mercury

A

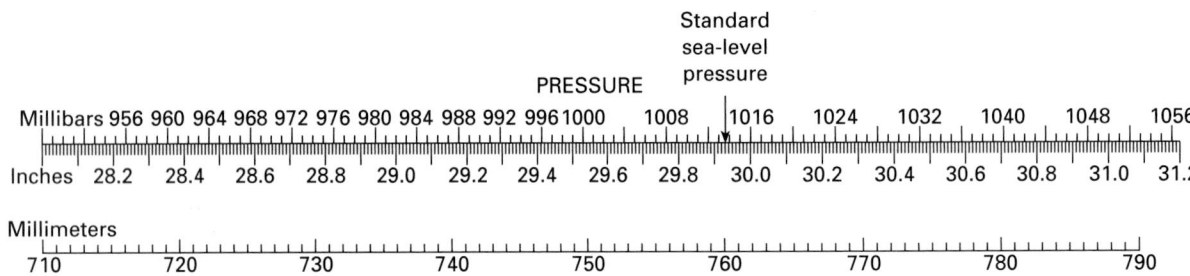

Standard sea-level pressure

PRESSURE

Millibars 956 960 964 968 972 976 980 984 988 992 996 1000 1008 1016 1024 1032 1040 1048 1056

Inches 28.2 28.4 28.6 28.8 29.0 29.2 29.4 29.6 29.8 30.0 30.2 30.4 30.6 30.8 31.0 31.2

Millimeters
710 720 730 740 750 760 770 780 790

B

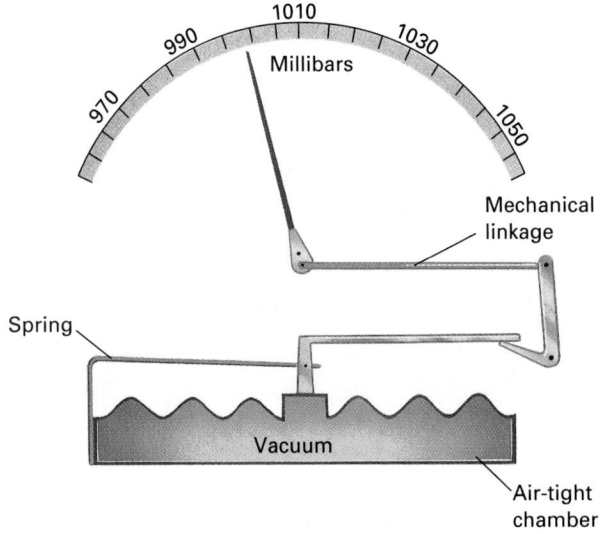

1010
990
1030
970
Millibars
1050
Mechanical linkage
Spring
Vacuum
Air-tight chamber

FIGURE 17.8 In an aneroid barometer, increasing air pressure compresses the air-tight chamber and causes the connected pointer to move in one direction. When the pressure decreases, the chamber expands, deflecting the pointer the other way.

Unit 5: The Atmosphere: Evolution and Composition

17.5 Atmospheric Temperature

The temperature of the atmosphere changes with altitude (Figure 17.9). The layer of air closest to Earth, the layer we live in, is the **troposphere**. Virtually all of the water vapor and clouds exist in this layer, and almost all weather occurs here. Earth's surface absorbs solar energy, and thus the surface of the planet is warm. But, as explained earlier, continents and oceans also radiate heat, and some of this energy is absorbed in the troposphere. At higher elevations in the troposphere, the atmosphere is thinner and absorbs less energy; in addition, lower parts of the troposphere have absorbed much of the heat radiating from Earth's surface. Consequently, temperature decreases at higher levels in the troposphere; mountaintops are generally colder than valley floors, and pilots flying at high altitudes must heat their cabins.

The top of the troposphere is the **tropopause**, which lies at an altitude of about 17 kilometers at the equator, although it is lower at the poles. At the tropopause the steady decline in temperature with altitude ceases abruptly. Cold air from the upper troposphere is too dense to rise above the tropopause. As a result, little mixing occurs between the troposphere and the layer above it, called the **stratosphere**.

In the stratosphere, temperature remains constant to 35 kilometers and then increases with altitude until, at about 50 kilometers, it is as warm as that at Earth's surface. This reversal in the temperature profile occurs because the troposphere and stratosphere are heated by different mechanisms. As already explained, the troposphere is heated primarily from below, by Earth. The stratosphere, however, is heated primarily from above, by solar radiation.

Oxygen molecules (O_2) in the stratosphere absorb energetic ultraviolet rays from the Sun. The radiant energy breaks the oxygen molecules apart, releasing free oxygen atoms. The oxygen atoms then recombine to form ozone (O_3). Ozone absorbs ultraviolet energy more efficiently than oxygen does, warming the upper stratosphere. Ultraviolet radiation is energetic enough to affect organisms. Small quantities give us a suntan, but large doses cause skin cancer and cataracts of the eye, inhibit the growth of many plants, and otherwise

troposphere The layer of air that lies closest to Earth's surface and extends upward to about 17 kilometers.

tropopause The top of the troposphere; the boundary between the troposphere and the stratosphere.

stratosphere The layer of air above the tropopause, extending upward to about 55 kilometers.

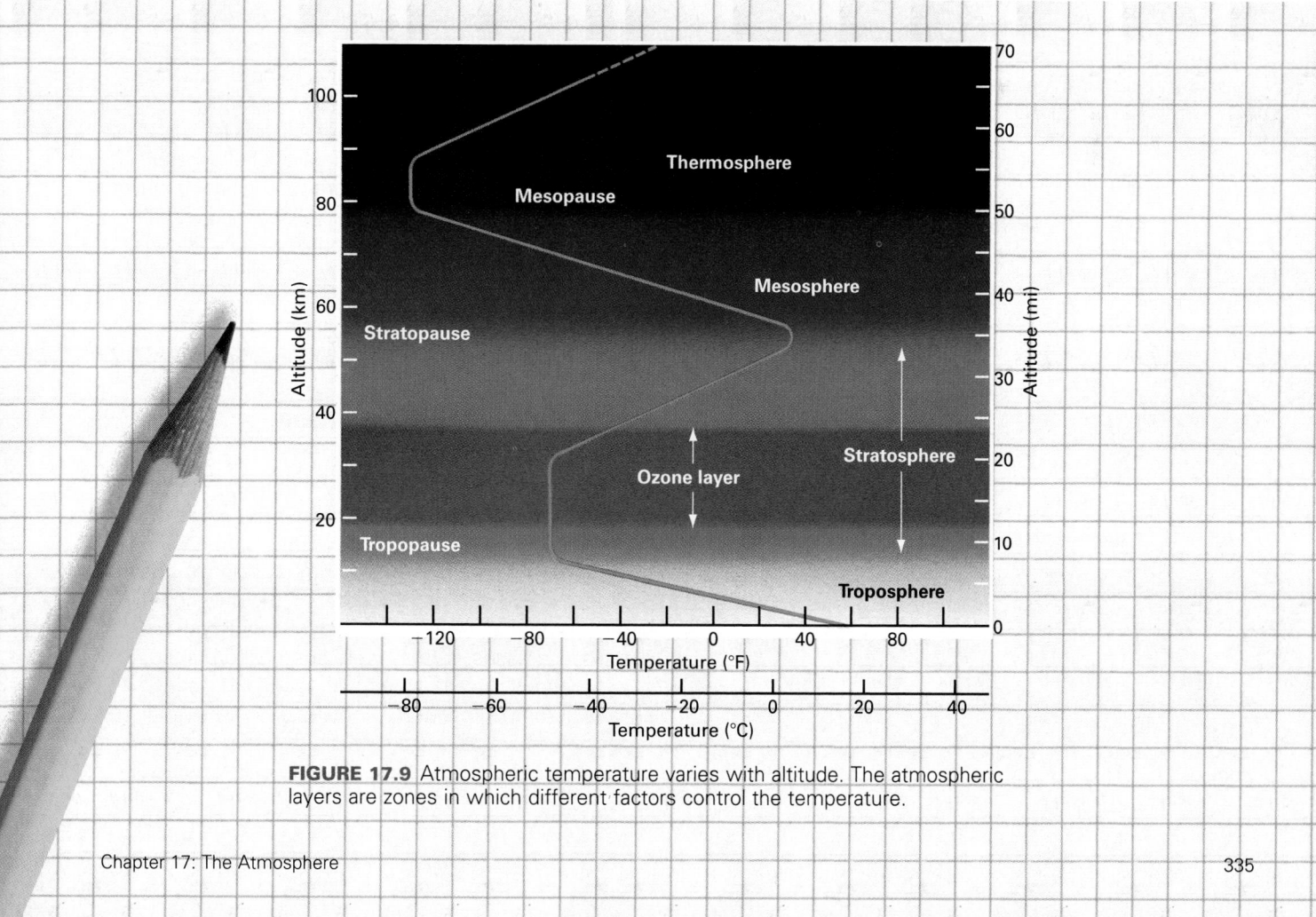

FIGURE 17.9 Atmospheric temperature varies with altitude. The atmospheric layers are zones in which different factors control the temperature.

stratopause
The ceiling of the stratosphere; the boundary between the stratosphere and the mesosphere.

mesosphere The layer of air that lies above the stratopause, extending upward from about 55 kilometers to about 80 kilometers above Earth's surface.

mesopause The ceiling of the mesosphere; the boundary between the mesosphere and the thermosphere.

thermosphere An extremely high and diffuse region of the atmosphere lying above the mesosphere, from about 80 kilometers upward.

harm living tissue. The ozone in the upper atmosphere protects life on Earth by absorbing much of this high-energy radiation before it reaches Earth's surface.

Ozone concentration declines in the upper portion of the stratosphere, and therefore at about 55 kilometers above Earth temperature once more begins to decline rapidly with elevation. This boundary between rising and falling temperature is the **stratopause**, the ceiling of the stratosphere. The second zone of declining temperature is the **mesosphere**. Little radiation is absorbed in the mesosphere, and the thin air is extremely cold. The ceiling of the mesosphere is the **mesopause**. Starting at about 80 kilometers above Earth, the temperature again remains constant and then rises rapidly in the **thermosphere**. Here the atmosphere absorbs high-energy X-rays and ultraviolet radiation from the Sun. High-energy reactions strip electrons from atoms and molecules to produce ions. The temperature in the upper portion of the thermosphere is just below freezing, not extremely cold by surface standards.

17.6 Air Pollution

Ever since the first cave dwellers huddled around a smoky fire, people have introduced impurities into the air. The total quantity of these impurities is minuscule compared with the great mass of our atmosphere and with the monumental changes that occurred during the evolution of the planet. Yet air pollution remains a significant health, ecological, and climatological problem for modern industrial society.

In 1948, Donora was an industrial town of about 14,000 located 50 kilometers south of Pittsburgh, Pennsylvania. One large factory in town manufactured structural steel and wire and another produced zinc and sulfuric acid. During the last week of October 1948, dense fog settled over the town. But it was no ordinary fog; the moisture contained pollutants from the two factories. After four days, visibility became so poor that people could not see well enough to drive, even at noon with their headlights on. Gradually at first, and then in increasing numbers, residents sought medical attention for nausea, shortness of breath, and constrictions in the throat and chest. Within a week, 20 people had died and about half of the town was seriously ill.

Other incidents similar to what happened in Donora occurred worldwide. In response to the growing problem, the United States enacted the Clean Air Act in 1963. As a result of the Clean Air Act and its amendments, total emissions of air pollutants have decreased and air quality across the country has improved (Figure 17.10). It is even more encouraging to note that this decrease in emissions has occurred at a time when population, energy consumption, vehicle miles traveled, and gross domestic product (GDP) have increased dramatically (Figure 17.11). Donora-type incidents have not been repeated. Smog has decreased, and rain has become less acidic. Yet some people believe that we have not gone far enough and that air pollution regulations should be strengthened further.

Sources and types of air pollution are listed in Figure 17.12 and discussed in the following section.

Gases Released When Fossil Fuels Are Burned

Coal is largely carbon, which, when burned completely, produces carbon dioxide. Petroleum is a mixture of *hydrocarbons*, compounds composed of carbon and hydrogen. When hydrocarbons burn completely, they produce carbon dioxide and water as the only combustion products. Neither is poisonous, but both are greenhouse gases. If fuels were composed purely of compounds of carbon and hydrogen, and if they always burned completely, air pollution from burning of fossil fuels would pose little direct threat to our health (although combustion of fossil fuels would still

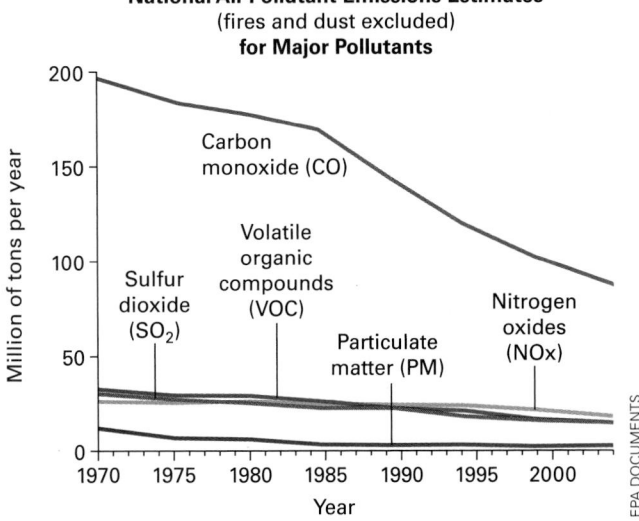

FIGURE 17.10 Emission of five air pollutants in the United States from 1970 to 2004. The Clean Air Act was first enacted in 1963. (This graph does not show local concentrations in heavily congested areas such as Los Angeles.)

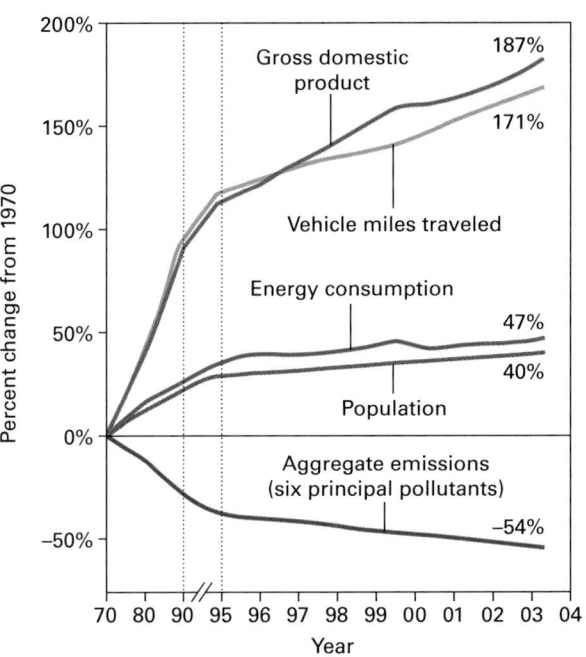

Comparison of Growth Areas and Emissions

FIGURE 17.11 Air pollution emissions declined by 54 percent between 1970 and 2004, despite the fact that gross domestic product, vehicle miles traveled, energy consumption, and population have all increased.

contribute to global warming). However, fossil fuels contain impurities, and combustion is usually incomplete. As a result, other products form—most of which are harmful.

Products of incomplete combustion include hydrocarbons such as benzene and methane. Benzene is a carcinogen (a compound that causes cancer), and methane is another greenhouse gas. Incomplete combustion of fossil fuels releases many other pollutants, including carbon monoxide (CO), which is colorless and odorless yet very toxic.

Additional problems arise because coal and petroleum contain impurities that generate other kinds of pollution when they are burned. Small amounts of sulfur are present in coal and, to a lesser extent, in petroleum. When these fuels burn, the sulfur forms oxides, mainly sulfur dioxide (SO_2) and sulfur trioxide (SO_3). High sulfur dioxide concentrations have been associated with major air pollution disasters of the type that occurred in Donora. Today the primary global source of sulfur dioxide pollution is coal-fired electric generators.

Nitrogen, like sulfur, is common in living tissue and therefore is found in all fossil fuels. This nitrogen, together with a small amount of atmospheric nitrogen, reacts when coal or petroleum is burned. The products are mostly nitrogen oxide (NO), and nitrogen dioxide (NO_2). Nitrogen dioxide is a reddish-brown gas with a strong odor. It therefore contributes to the "browning" and odor of some polluted urban atmospheres. Automobile exhaust is the primary source of nitrogen oxide pollution.

Acid Rain

As explained earlier, sulfur and nitrogen oxides are released when coal and petroleum burn. These oxides are also released when metal ores are refined. In moist air, sulfur dioxide reacts to produce sulfuric acid and nitrogen oxides react to form nitric and nitrous acid. These strong atmospheric acids dissolve in water droplets and fall as **acid precipitation**, also called **acid rain** (Figure 17.13).

Acidity is expressed on the **pH scale**. A solution with a pH of 7 is neutral, neither acidic nor basic. On a

acid precipitation *or* **acid rain** Rain, snow, fog, or mist that has become acidic after reacting with air pollutants.

pH scale A logarithmic scale that measures the acidity of a solution. A pH of 7 is neutral; numbers lower than 7 represent acidic solutions, and numbers higher than 7 represent basic ones.

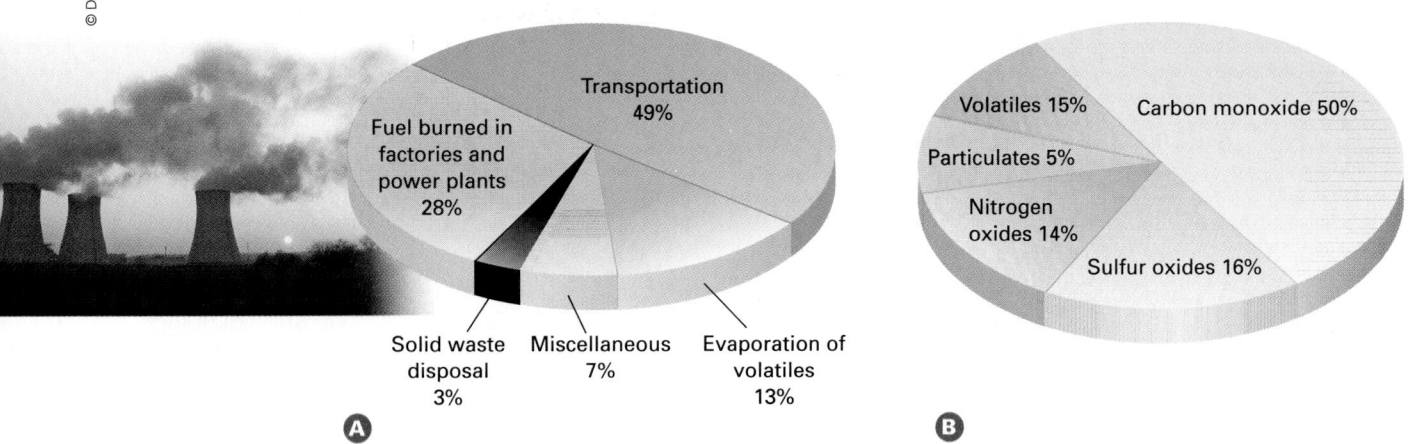

FIGURE 17.12 (A) Sources of air pollution in the United States. (B) Types of air pollutants in the United States. (Although carbon dioxide is a greenhouse gas, it is not listed as a pollutant because it is not toxic.)

pH scale, numbers lower than 7 represent acidic solutions, and numbers higher than 7 represent basic ones. For example, soapy water is basic and has a pH of about 10, whereas vinegar is an acid with a pH of 2.4.

Rain reacts with carbon dioxide in the atmosphere to produce a weak acid. As a result, natural rainfall has a pH of about 5.7. However, in the "bad old days" before the Clean Air Act was properly enforced, rain was much more acidic. A fog in southern California in 1986 reached a pH of 1.7, which approaches the acidity of toilet bowl cleaners.

Consequences of Acid Rain

Sulfur and nitrogen oxides impair lung function, aggravating diseases such as asthma and emphysema. They also affect the heart and liver and have been shown to increase vulnerability to viral infections such as influenza.

Acid rain corrodes metal and rock. Limestone and marble are especially susceptible because they dissolve rapidly in mild acid. In the United States the cost of deterioration of buildings and materials from acid precipitation is estimated at several billion dollars per year.

Acid rain also affects plants (Figure 7.14).

Smog and Ozone in the Troposphere

Imagine that your great-grandfather had entered the exciting new business of making moving pictures. Old-

FIGURE 17.14 Acid rain has killed these trees in the mountains of North Carolina, near Mount Mitchell.

time photographic film was "slow" and required lots of sunlight, so many filmmakers left the polluted, overcast, industrial Northeast. Southern California, with its warm, sunny climate and little need for coal, was preferable. Thus, a district of Los Angeles called Hollywood became the center of the movie industry. Its population boomed, and after World War II automobiles became about as numerous as people. Then the quality of the air deteriorated in a strange way. People noted four different kinds of changes: (1) A brownish haze called **smog** settled over the city (Figure 17.15); (2) people felt irritation in their eyes and throats; (3) vegetable crops became damaged; and (4) the sidewalls of rubber tires developed cracks.

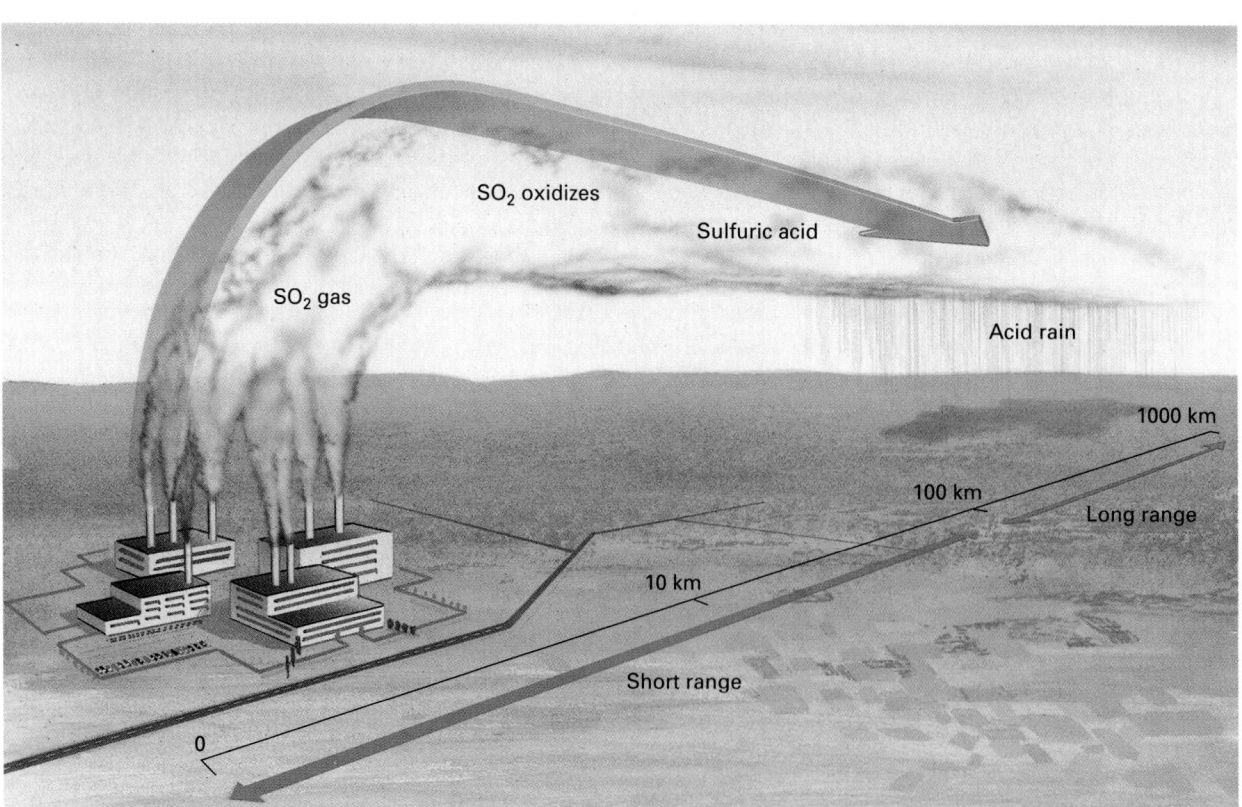

FIGURE 17.13 Acid rain develops from the addition of sulfur compounds to the atmosphere by industrial smokestacks.

FIGURE 17.15 A brownish haze of smog settles in the valley. Note that the smog lies beneath a distinct line caused by an atmospheric inversion. Above the inversion level, the air is much cleaner.

found: Ozone in the troposphere reacts with automobile exhaust to produce smog, and therefore it is a pollutant. Ozone in the stratosphere is beneficial, and the destruction of the ozone layer creates serious problems.

smog Smoky fog; a word used loosely to define visible air pollution.

Ozone irritates the respiratory system, causing loss of lung function and aggravating asthma in susceptible individuals. Ozone also increases susceptibility to heart disease and is a suspected carcinogen. High ozone concentrations slow the growth of plants, which is a particularly serious problem in the rich, agricultural areas of California.

Nationwide, ozone levels have decreased by about 25 percent between 1980 and 2005. However, local problems remain. According to a recent EPA report, 49 counties, where 42 million people live, had poor ozone air quality in 2004, and scientists recorded unhealthy ozone levels in 7 counties, where 18 million people live, primarily in California and Texas.

In the 1950s, air pollution experts worked mostly in the industrialized cities of the East Coast and the Midwest. When they were called to diagnose the problem in southern California, they looked for the sources of air pollution they knew well, especially sulfur dioxide. But the smog was nothing like the pollution they were familiar with. These researchers eventually learned that incompletely burned gasoline in automobile exhaust reacts with nitrogen oxides and atmospheric oxygen in the presence of sunlight to form ozone (O_3). The ozone then reacts further with automobile exhaust to form smog (Figure 17.16).

As you learned in Section 17.5, ozone in the stratosphere absorbs ultraviolet radiation and protects life on Earth. Yet excessive ozone in the air can be harmful. Is ozone a pollutant to be eliminated, or a beneficial component of the atmosphere that we want to preserve? The answer is that it is both, depending on *where* it is

Toxic Volatiles

A volatile compound is one that evaporates readily and therefore easily escapes into the atmosphere. Whenever chemicals are manufactured or petroleum is refined, some volatile by-products escape into the atmosphere. When metals are extracted from ores, gases such as sulfur dioxide are released. When pesticides are sprayed onto fields and orchards, some of the spray is carried off by wind. When you paint your house, the volatile parts of the paint evaporate into the air. As a result of all these processes, tens of thousands of volatile compounds are present in polluted air: some are harmless, others are poisonous, and many have not been studied. Consider the case of dioxin.

Very little dioxin is intentionally manufactured. It is not an ingredient in any herbicide, pesticide, or other industrial formulation. You cannot buy dioxin at your local hardware store or pharmacy. Dioxin forms as an unwanted by-product in the production of certain chemicals and when specific chemicals are burned. For example, in the United States today, garbage incineration is the most common source of dioxin. When a compound containing chlorine, such as the plastic polyvinyl chloride (PVC), is burned, some of the chlorine reacts with organic compounds to form dioxin. The dioxin then goes up the smokestack of

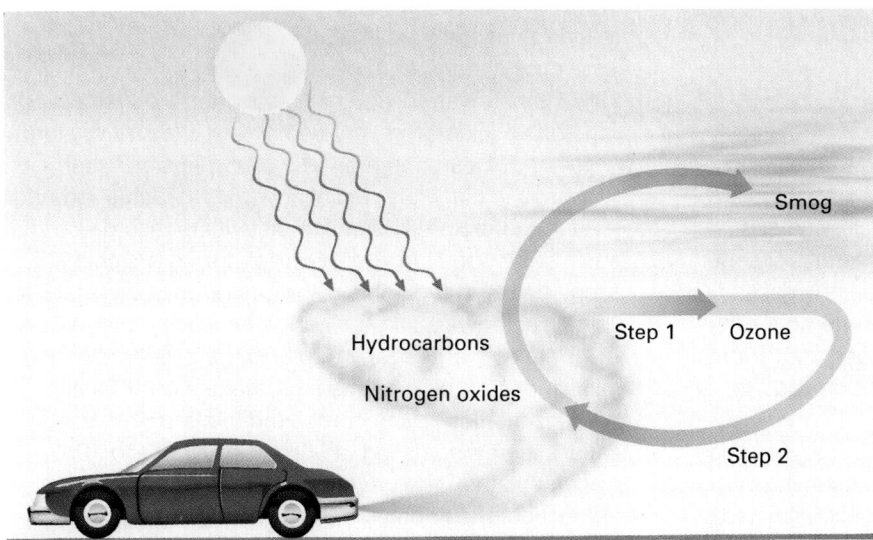

FIGURE 17.16 Smog forms in a sequential process. Step 1: Automobile exhaust reacts with air in the presence of sunlight to form ozone. Step 2: Ozone reacts with automobile exhaust to form smog.

particle *or* **particulate** In pollution terminology, any small piece of solid matter larger than a molecule, such as dust or soot.

aerosol In pollution terminology, a particle or particulate that is suspended in air.

fly ash Noncombustible minerals that escape into the atmosphere when coal burns, eventually settling as gritty dust.

chlorofluorocarbons (CFCs) Organic compounds containing chlorine and fluorine, which rise into the upper atmosphere to destroy the ozone layer.

halons Compounds containing bromine and chlorine, which rise into the upper atmosphere to destroy the ozone layer.

ozone hole An unusually low ozone concentration in the stratosphere.

the incinerator, diffuses into the air, and eventually falls to Earth. Cattle eat grass lightly dusted with dioxin and store the dioxin in their fat. Humans ingest the compound mostly in meat and dairy products. The EPA estimates that the average U.S. citizen ingests about 0.0000000001 gram (100 picograms) of dioxin in food every day. Although this is a minuscule amount, the EPA has argued that dioxin is the most toxic chemical known and that even these low background levels may cause adverse effects such as cancer, disruption of regulatory hormones, reproductive and immune system disorders, and birth defects. Others disagree. The Chemical Manufacturers Association has written: "There is no direct evidence to show that any of the effects of dioxins occur in humans in everyday levels."

No one knows whether very small doses of potent poisons are harmful. Environmentalists argue that it is "better to be safe than sorry," and that therefore we should reduce ambient concentrations of volatiles like dioxin. Others counter that because the harmful effects are unproven, we should not burden our economy with the costs of control.

Particulates and Aerosols

A **particle**, or **particulate**, is any small piece of solid matter, such as dust or soot. An **aerosol** is any small particle that is larger than a molecule and suspended in air. These three terms are used interchangeably to discuss air pollution. Many natural processes release aerosols. Windblown silt, pollen, volcanic ash, salt spray from the oceans, and smoke and soot from wildfires are all aerosols. Industrial emissions add to these natural sources.

Smoke and soot are carcinogenic aerosols formed whenever fuels are burned. Coal always contains clay and other noncombustible minerals that accumulated when the coal formed in the muddy bottoms of ancient swamps. When the coal burns, some of these minerals escape from the chimney as **fly ash**, which settles as gritty dust. When metals are mined, the drilling, blast-

ing, and digging raise dust, and this, too, adds to the total load of aerosols.

In 1988, EPA epidemiologists noted that whenever atmospheric aerosol levels rose above a critical level in Steubenville, Ohio, the number of fatalities from all causes—car accidents to heart attacks—rose. After several studies substantiated the Steubenville report, the EPA proposed additional reductions of the ambient aerosol levels in the United States. Opponents argued that it is unfair to target all aerosols because the term covers a wide range of substances from a benign grain of salt to a deadly mist of toxic volatiles.

17.7 Depletion of the Ozone Layer

As described in Section 17.5, solar energy breaks oxygen molecules (O_2) apart in the stratosphere, releasing free oxygen atoms (O). The free oxygen atoms combine with oxygen molecules to form ozone (O_3). Ozone absorbs high-energy ultraviolet light. This absorption protects life on Earth because ultraviolet light causes skin cancer, inhibits plant growth, and otherwise harms living tissue.

In the 1970s, scientists learned that organic compounds containing chlorine and fluorine, called **chlorofluorocarbons (CFCs)**, or compounds containing bromine and chlorine, called **halons**, rise into the upper atmosphere, react with, and destroy ozone (Figure 17.17). At that time, CFCs were used as cooling agents in almost all refrigerators and air conditioners, as propellants in some aerosol cans, as cleaning solvents, and in plastic foam coffee cups and some building insulation.

In 1985, scientists observed an unusually low ozone concentration in the stratosphere over Antarctica, called the **ozone hole**. The ozone concentration over Antarctica continued to decline between 1985 and 1993, until it was 65 percent below normal over 23 million square kilometers, an area almost the size of North America. Research groups also reported significant increases in ultraviolet radiation from the Sun at ground level in the region. In addition, scientists recorded ozone depletion in the Northern Hemisphere. In March 1995, ozone concentration above the United States was 15 to 20 percent lower than during March 1979.

Data on global ozone depletion persuaded the industrial nations of the world to limit the use of CFCs and other ozone-destroying compounds. In a series of international agreements signed between 1978 and 1992, many nations of the world agreed to reduce or curtail production of compounds that destroy atmospheric ozone. Most industrialized countries stopped production of CFCs on January 1, 1996.

The international bans have had positive results. The concentration of ozone-destroying chemicals

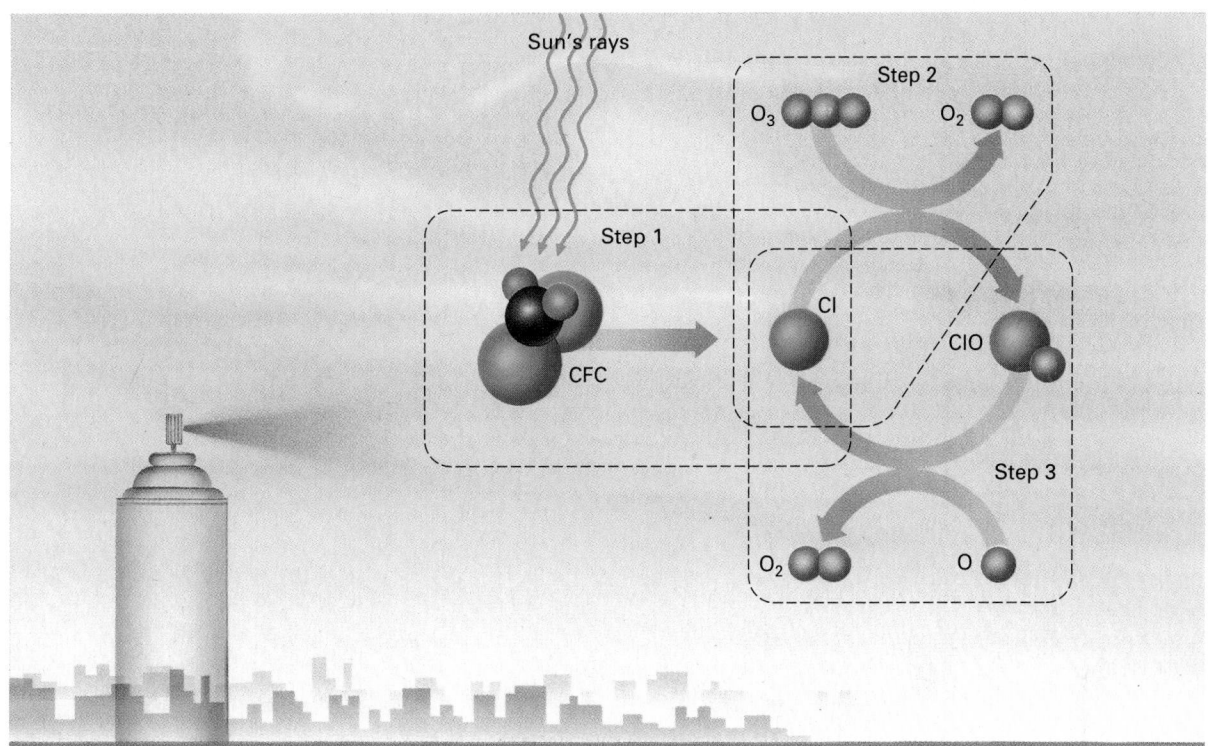

FIGURE 17.17 CFCs destroy the ozone layer in a three-step reaction. Step 1: CFCs rise into the stratosphere. Ultraviolet radiation breaks the CFC molecules apart, releasing chlorine atoms. Step 2: Chlorine atoms react with ozone, O3, to destroy the ozone molecule and release oxygen, O2. The extra oxygen atom combines with chlorine to produce ClO. Step 3: The ClO sheds its oxygen to produce another free chlorine atom. Thus, chlorine is not used up in the reaction, and one chlorine atom reacts over and over again to destroy many ozone molecules.

peaked in the troposphere (lower atmosphere) in 1994 and has been declining ever since. As a result, fewer CFCs and halons have been drifting into the stratosphere. The CFCs and halons that are already in the stratosphere break down slowly, but the concentration of ozone-destroying chemicals in the stratosphere has peaked and is beginning to decline. In a review article published in May 2006, the authors conclude that "ozone abundances have at least not decreased . . . for most of the world," although natural cycles make

it difficult to judge the relative effects of human and natural influences. In summary, they write, "it is therefore unlikely that ozone will stabilize at levels observed before 1980, when a decline in ozone concentrations was first observed"[3] (Figure 17.18).

---

3. Elizabeth C. Weatherhead and Signe Bech Andersen, "The Search for Signs of Recovery of the Ozone Layer," *Nature* 441 (May 4, 2006), 39; online at http://www.nature.com/nature/journal/v441/n7089/abs/nature04746.html.

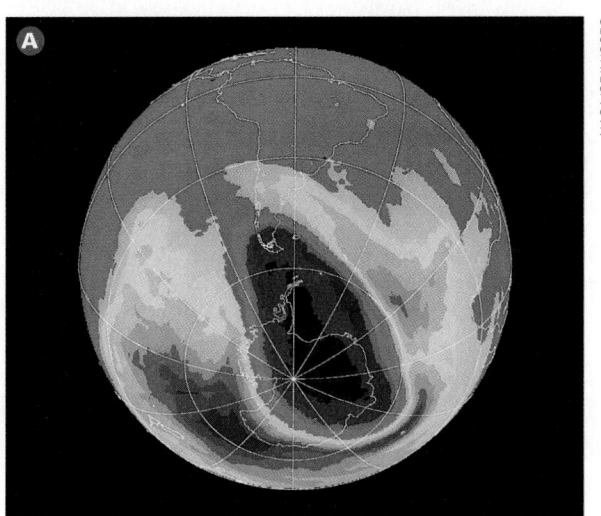

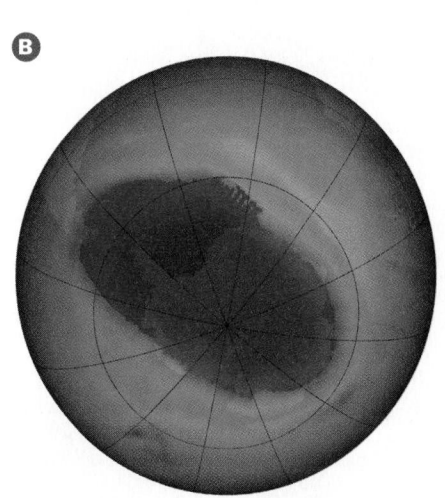

NASA/GRIN/GSFC

NASA/GODDARD SPACE FLIGHT CENTER SCIENTIFIC VISUALIZATION STUDIO

FIGURE 17.18
(A) A satellite image of stratospheric ozone over Antarctica in 1994. (B) Similar data for 2004. In both images, dark purple shows the lowest ozone concentration.

18

ENERGY BALANCE IN THE ATMOSPHERE

Glaciers, snowfields, and snow-covered mountains of the Japanese Alps reflect light and cool the Earth.

> # How glorious a greeting the sun gives the mountains!
> *John Muir*

Weather is the state of the atmosphere at a given place and time. Temperature, wind, cloudiness, humidity, and precipitation are all components of weather. The weather changes frequently, from day to day or even from hour to hour.

Climate is the characteristic weather of a region, particularly the temperature and precipitation, averaged over several decades. Miami and Los Angeles have warm climates; summers are hot and even the winters are warm. In contrast, New York and Chicago experience much greater temperature extremes. Even though winters are cool and often snowy, summers can be almost as hot as those in Miami. Seattle experiences moderate temperatures with foggy, cloudy winters. In this and the following two chapters, we discuss the atmosphere, weather, and climate. In Chapter 21, we will consider global climate change.

Many of the processes that drive energy balance in the atmosphere vary in gradual and predictable manners. But the weather and climate don't always vary gradually and predictably because they are affected by many complex, overlapping system interactions, and by threshold and feedback mechanisms. As a simple example: Incoming solar radiation is most intense at the equator and decreases predictably toward the poles. But some high-latitude locations are warmer than regions closer to the equator, and on any given day, it can be warmer in Montreal than in Houston. In this and subsequent chapters, we will explain the fundamental processes that drive weather and climate, and how they interact.

18.1 Incoming Solar Radiation

Solar energy streams from the Sun in all directions, and Earth receives only one two-billionth of the total solar output. However, even this tiny fraction warms Earth's surface and makes it habitable.

The space between Earth and the Sun is nearly empty. How does sunlight travel through a vacuum? In the late 1600s and early 1700s, light was poorly understood. Isaac Newton postulated that light consists of streams of particles that he called "packets" of light. Two other physicists, Robert Hooke and Christian Huygens, argued that light travels in waves. Today we know that Hooke and Huygens were correct—light behaves as a wave; but Newton was also right—light acts as if it is composed of particles. But how can light be both a wave and a particle at the same time? In a sense this is an unfair question because light is fundamentally different from familiar objects. Light is unique; it behaves as a wave *and* a particle simultaneously.

Particles of light are called **photons**. In a vacuum, photons travel only at one speed, the speed of light, never faster and never slower. The speed of light is 3×10^8 meters per second. At that rate a photon covers the 150 million kilometers between the Sun and Earth in about 8 minutes. Photons are unlike ordinary matter in that they appear when they are emitted and disappear when they are absorbed.

Light also behaves as an electrical and magnetic wave, called **electromagnetic radiation**. Terminology describing a light wave is the same as for sound waves or ocean waves (Figure 18.1). Its wavelength is the distance between successive wave crests. The **frequency** of a wave is the number of complete wave cycles, from crest to crest, that pass by any point in a second. (Think of how *frequently* the waves pass by.) Electromagnetic radiation occurs in a wide range of wavelengths and frequencies. The **electromagnetic spectrum** is the continuum of radiation of different wavelengths and frequencies

weather The state of the atmosphere at a given place and time as characterized by temperature, wind, cloudiness, humidity, and precipitation.

climate The characteristic weather of a region, averaged over several decades; refers to yearly cycles of temperature, wind, rainfall, etc., and not to daily variations.

photon A particle of light; the smallest particle or packet of electromagnetic energy.

electromagnetic radiation Radiation consisting of an oscillating electric and magnetic field, including radio waves, infrared, visible light, ultraviolet, x-rays, and gamma rays.

frequency The number of complete wave cycles, from crest to crest, that pass by any point in a second.

electromagnetic spectrum The entire range of electromagnetic radiation from very-long-wavelength (low frequency) radiation to very-short-wavelength (high frequency) radiation.

(Figure 18.2). At one end of the spectrum, radiation given off by ordinary household current has a long wavelength (5,000 kilometers) and low frequency (60 cycles per second). At the other end, cosmic rays from outer space have a short wavelength (about one-trillionth of a centimeter, or 10^{-14} meter) and very high frequency (10^{22} cycles per second). Visible light is only a tiny portion (about one-millionth of 1 percent) of the electromagnetic spectrum.

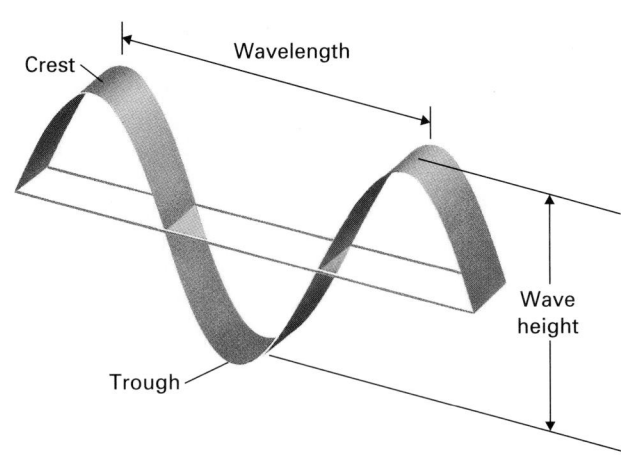

FIGURE 18.1 The terms used to describe a light wave are identical to those used for water, sound, and other types of waves.

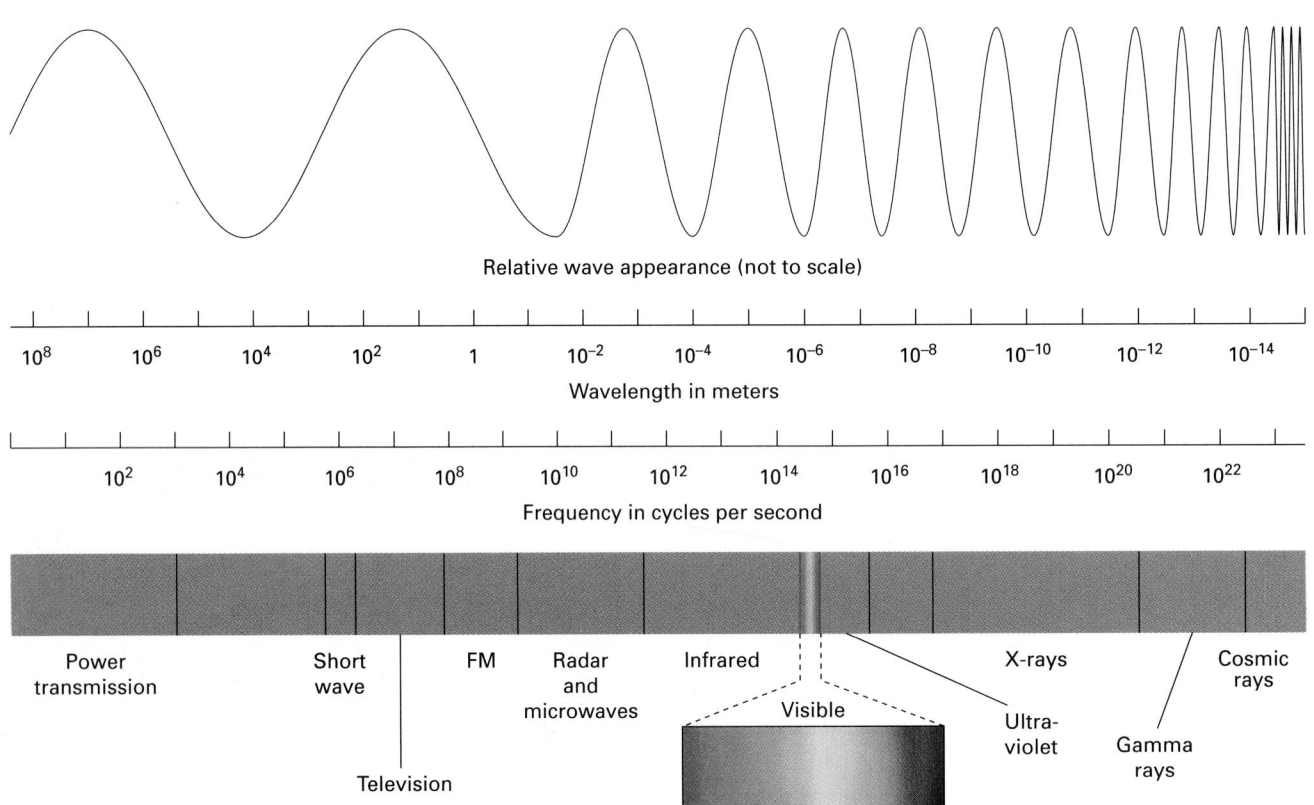

Relative wave appearance (not to scale)

Wavelength in meters

Frequency in cycles per second

Power transmission · Short wave · Television · FM · Radar and microwaves · Infrared · Visible · Ultra-violet · X-rays · Gamma rays · Cosmic rays

FIGURE 18.2 The electromagnetic spectrum. The wave shown is not to scale. In reality the wavelength varies by a factor of 10^{22}, and this huge difference cannot be shown.

Absorption and Emission

If you go outside on a cold, snowy, February day, you wear a heavy jacket, hat, and gloves. Yet the sunlight warms your face. If the sky is clear, even though the temperature may be well below freezing, you must wear sunglasses to prevent snow blindness, and if you have unprotected fair skin you will suffer sunburn.

Each photon is a tiny packet of concentrated energy. The energy of a single photon is related only to the frequency of the light, and *not* to the ambient temperature. When a photon strikes your face, it may be absorbed. During **absorption of radiation**, the energy of the photon may initiate chemical and physical reactions in the molecules of your skin. One possible reaction is roughly analogous to cooking, and this reaction causes sunburn or tan. A photon may also cook a molecule in the retina of your eye, and if this reaction occurs frequently enough, you will become snow-blind.

Some absorbed photons do not cause chemical reactions; instead they cause molecules to vibrate or rotate more rapidly. This rapid motion makes your skin feel warm.

Alternatively, a photon's energy may be transferred to an electron in a molecule. In this case, the electron jumps to what scientists call an **excited state**—it has a higher level of energy than its *ground state* (lowest level). But electrons do not remain in their excited states forever. Frequently they fall back to a lower energy state, initiating **emission of radiation** in the process. In other words, the photon's energy was transferred to the electron and eventually the electron releases that energy back in the form of a photon. All objects emit radiant energy at some wavelength (except at a temperature of absolute zero).

Let's shift our focus from your skin to an iron bar. An iron bar at room temperature emits *infrared* radiation. This radiation has low energy and long wavelengths. It is invisible because the energy is too low to activate the sensors in our eyes. Infrared radiation is sometimes called *radiant heat*, or *heat rays*. Thus even at room temperature an iron bar radiates heat. If you place the bar in a hot flame, it begins to glow with a dull, red color when it becomes hot enough. The heat has excited electrons in the iron bar, and the excited electrons then emit visible, red, electromagnetic radiation (Figure 18.3). If you heat the bar further, it gradually changes color until it becomes white. This demonstration shows another property of emitted electromagnetic radiation: the wavelength (color) of the radiation is determined by the temperature of the source. With increasing temperature, the energy level of the radiation increases, the wavelength decreases, and the color changes progressively.

The Sun's surface temperature is about 6,000°C. Because of its high temperature, the Sun emits relatively high-energy (short-wavelength) radiation, primarily in the ultraviolet and visible portions of the spectrum. When this radiation strikes Earth, it is absorbed by rock and soil. After the radiant energy is absorbed, the rock and soil reemit it. But Earth's surface is much cooler than the Sun's. Therefore Earth emits low-energy, infrared heat radiation, which has a relatively long wavelength and low frequency. Thus Earth absorbs high-energy, visible light and emits low-energy, invisible, infrared heat radiation.

excited state A state of physical energy higher than the lowest energy level (or *ground state*) of an electron in an atom or molecule.

emission of radiation The process that occurs when energy, in the form of a photon, is emitted as an electron falls out of an excited state, with the equivalent loss of energy from the emitting substance.

albedo The reflectivity of a surface; surfaces that reflect more light have a higher albedo.

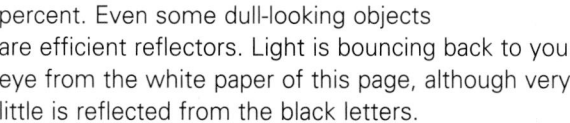

Reflection

Radiation reflects from many surfaces. We are familiar with the images reflected by a mirror or the surface of a still lake. Some surfaces are better reflectors than others. The reflectivity of a surface is referred to as its **albedo** (from the Latin for "whiteness") and is often expressed as a percentage. A mirror reflects nearly 100 percent of the light that strikes it and has an albedo close to 100 percent. Even some dull-looking objects are efficient reflectors. Light is bouncing back to your eye from the white paper of this page, although very little is reflected from the black letters.

Snowfields and glaciers have high albedos and reflect 80 to 90 percent of sunlight. Clouds have the second-highest albedo and reflect 50 to 55 percent of sunlight. On the other hand, city buildings and dark pavement have albedos of only 10 to 15 percent (Figure 18.4). Forests, with many independent surfaces of dark

FIGURE 18.3 Iron glows red, then orange, then heats to white as a welder works in his shop.

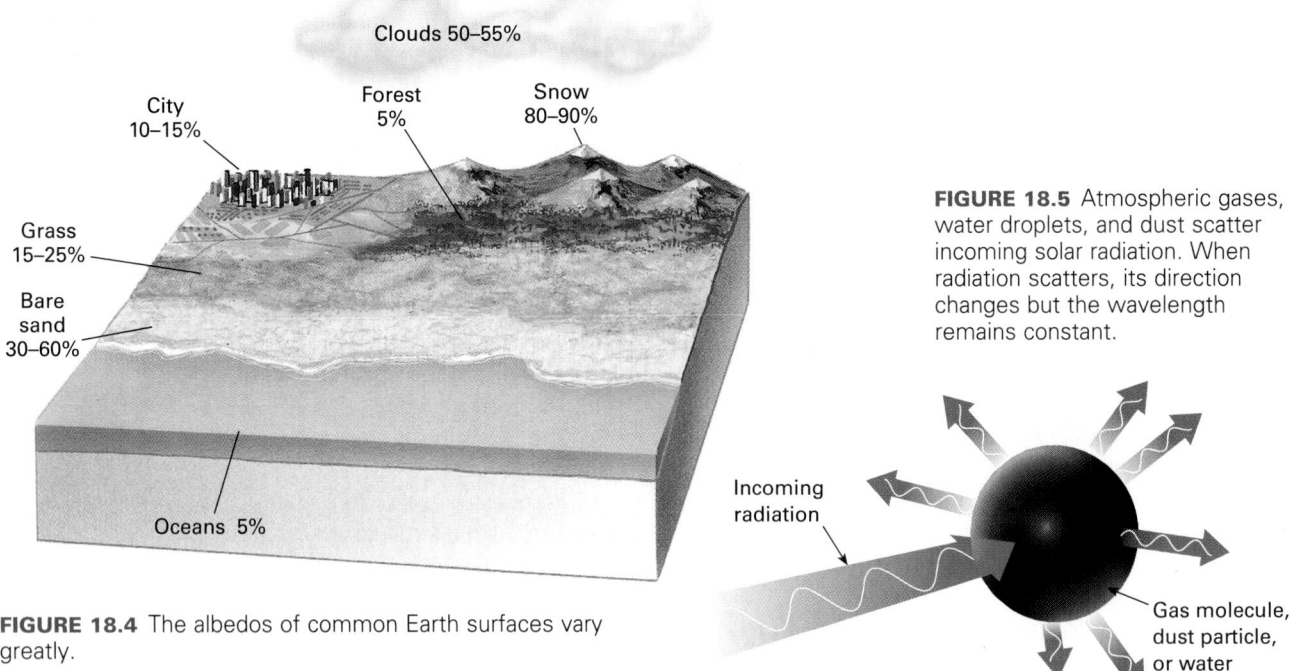

FIGURE 18.4 The albedos of common Earth surfaces vary greatly.

FIGURE 18.5 Atmospheric gases, water droplets, and dust scatter incoming solar radiation. When radiation scatters, its direction changes but the wavelength remains constant.

leaves, have an even lower albedo of about 5 percent. The oceans, which cover about two-thirds of Earth's surface, also have a low albedo. As a result, they absorb considerable solar energy and strongly affect Earth's radiation balance. Thus the temperature balance of the atmosphere is profoundly affected by the albedos of the hydrosphere, geosphere, and biosphere. If Earth's albedo were to rise by growth of glaciers or cloud cover, the surface of our planet would cool. Alternatively, a decrease in albedo (caused by the melting of glaciers) would cause warming. Thus the growth and shrinking of Earth's snow and ice is a classic example of a feedback mechanism. If Earth were to cool by a small amount, snowfields and glaciers would grow. But snow and ice reflect solar radiation back into space and lead to additional cooling—which causes further expansion of the snow and ice—and so on.

Scattering

On a clear day the Sun shines directly through windows on the south side of a building, but if you look through a north-facing window, you cannot see the Sun. Even so, light enters through the window, and the sky outside is blue. If sunlight were only transmitted directly, a room with north-facing windows would be dark and the sky outside the window would be black. Atmospheric gases, water droplets, and dust particles scatter sunlight in all directions, as shown in Figure 18.5. It is this scattered light that illuminates a room with north-facing windows and turns the sky blue.

The amount of scattering is inversely proportional to the wavelength of light. Short-wavelength blue light,

therefore, scatters more than longer-wavelength red light. The Sun emits light of all wavelengths, which combine to make up white light. Consequently, in space the Sun appears white. The sky appears blue from Earth's surface because the blue component of sunlight scatters more than other frequencies and colors the atmosphere. The Sun appears yellow from Earth because yellow is the color of white light with most of the blue light removed.

18.2 The Radiation Balance

With this background, let us examine the fate of sunlight as it reaches Earth (Figure 18.6). Of all the sunlight that reaches Earth, 50 percent is absorbed, scattered, or reflected by clouds and atmosphere, 3 percent is reflected by Earth's surface, and 47 percent is absorbed by Earth's surface. The absorbed radiation warms rocks, soil, and water.

If Earth absorbs radiant energy from the Sun, why doesn't Earth's surface get hotter and hotter until the oceans boil and the rocks melt? The answer is that rocks, soil, and water reemit virtually all the energy they absorb. As explained previously, most solar energy that reaches Earth is short-wavelength, visible and ultraviolet radiation. Earth's surface absorbs this radiation and then reemits the energy mostly as long-wavelength, invisible, infrared (heat) radiation. Some of this infrared heat escapes directly into space, but some is absorbed by the atmosphere. The atmosphere traps this heat radiating from Earth and acts as an insulating blanket.

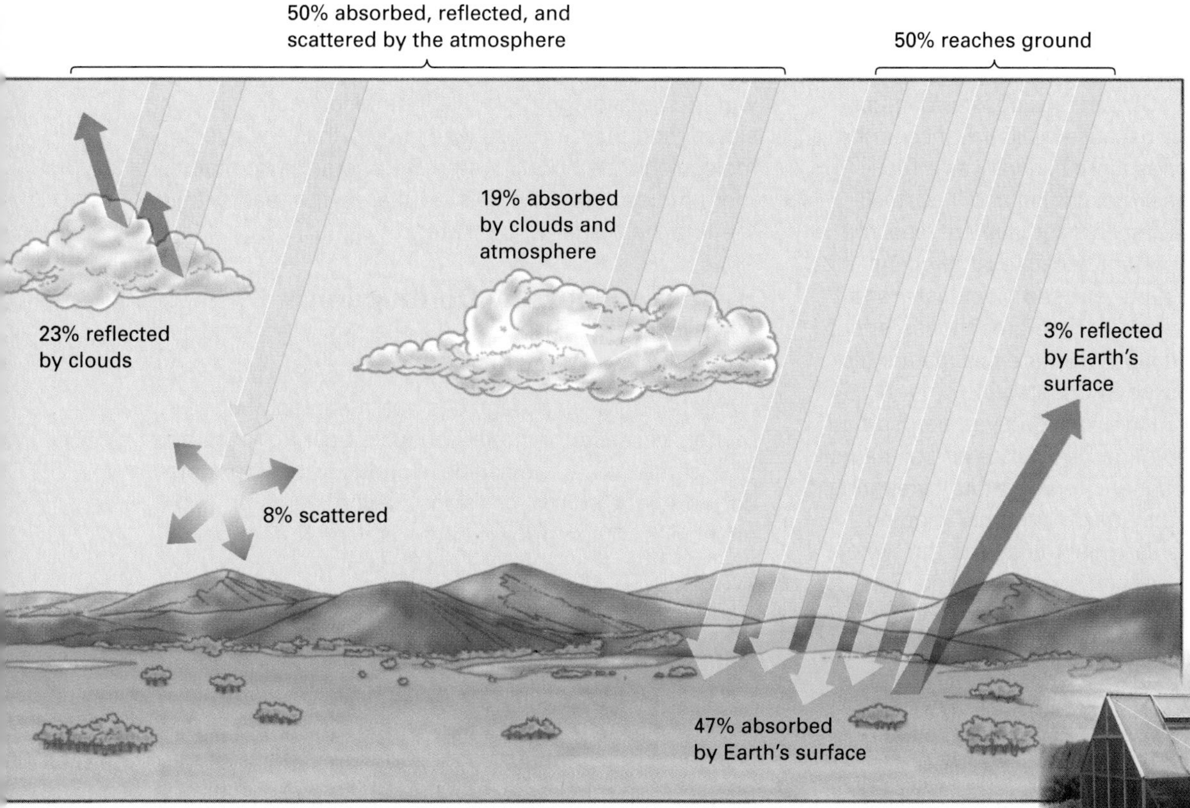

50% absorbed, reflected, and
scattered by the atmosphere

50% reaches ground

19% absorbed
by clouds and
atmosphere

23% reflected
by clouds

3% reflected
by Earth's
surface

8% scattered

47% absorbed
by Earth's surface

© NICOLE GORDINE/SHUTTERSTOCK

FIGURE 18.6 Half of the incoming solar radiation reaches Earth's surface. The atmosphere scatters, reflects, and absorbs the other half. All of the radiation absorbed by Earth's surface is reradiated as long-wavelength heat radiation.

If Earth had no atmosphere, radiant heat loss would be so rapid that Earth's surface would cool drastically at night. Earth remains warm at night because the atmosphere absorbs and retains much of the radiation emitted by the ground. If the atmosphere were to absorb even more of the long-wavelength radiant heat from Earth, the atmosphere and Earth's surface would become warmer. This warming process is called the **greenhouse effect** (Figure 18.7).[1]

Some gases in the atmosphere absorb infrared radiation and others do not. Oxygen and nitrogen, which together make up almost 99 percent of dry air at ground level, do not absorb infrared radiation; water, carbon dioxide, methane, and a few other gases do. Thus,

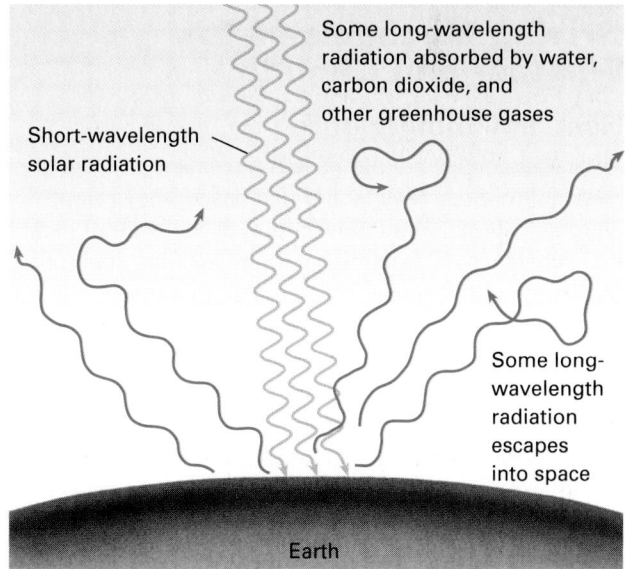

Some long-wavelength
radiation absorbed by water,
carbon dioxide, and
other greenhouse gases

Short-wavelength
solar radiation

Some long-
wavelength
radiation
escapes
into space

Earth

FIGURE 18.7 The greenhouse effect can be viewed as a three-step process. Step 1: Rocks, soil, and water absorb short-wavelength solar radiation, and become warmer (orange lines). Step 2: The Earth reradiates the energy as long-wavelength infrared heat rays (red lines). Step 3: Molecules in the atmosphere absorb some of the heat, and the atmosphere becomes warmer.

---

1. The comparison between the atmosphere and a greenhouse is only partially correct. Both the glass in a greenhouse and Earth's atmosphere are transparent to incoming, short-wavelength radiation and partially opaque to emitted, long-wavelength, heat radiation. However, the glass in a greenhouse is also a physical barrier that prevents heat loss through air movement, whereas the atmosphere is not. Yet in the Earth, the absorption of radiation is sufficient to warm the atmosphere and cause significant changes in Earth's climate. We use the term *greenhouse effect* because it has become common, both in atmospheric science and in everyday use.

temperature A measure of heat in a substance, proportional to the average speed of atoms and molecules in a sample.

heat A measure of the total energy in a sample.

conduction The transport of heat by direct collision among atoms or molecules.

they are called *greenhouse gases*. Water is the most abundant greenhouse gas in Earth's atmosphere.

Carbon dioxide is also important because its abundance in the atmosphere can vary as a result of several natural and industrial processes. Methane has become important because large quantities are released by industry and agriculture. The greenhouse effect and global climate change are discussed in Chapter 21.

Dust, cloud cover, aerosols, and other particulate air pollutants also affect Earth's atmospheric temperature by altering the amount of sunlight that is absorbed or reflected. Large volcanic eruptions can inject dust into the upper atmosphere and reflect enough sunlight to cool Earth. The effect of man-made pollution is harder to interpret. Dust and aerosols close to the surface of Earth reflect incoming solar radiation but also absorb infrared radiation emitted by the ground. The net result depends on a complex balance of factors, including particle size, particle composition, and natural cloudiness. Whatever the outcome, it is certain that whenever people change the composition of air, they run the risk of altering weather and climate.

18.3 Energy Storage and Transfer: The Driving Mechanisms for Weather and Climate

Heat and Temperature

All matter consists of atoms and molecules that are in constant motion. They fly through space, they rotate, and they vibrate. The **temperature**, or measure of heat in a substance, is proportional to the average speed of the atoms or molecules in a sample.[2] In a teacup full of boiling water, the water molecules are racing around rapidly, smashing into each other, spinning like so many boomerangs, and vibrating like spheres connected by pulsating springs. In a bathtub full of ice water, the water molecules are also moving, but much more slowly. Molecules in the hot water are moving faster than those in the cold water, so the hot water has a higher temperature.

In contrast, **heat** is a measure of the total energy in a sample. It is related to the average energy of every

molecule multiplied by the total number of molecules. It may seem counterintuitive, but there is more heat in a bathtub full of ice water than in a teacup full of boiling water. The average molecule in the ice water is moving slower and therefore has less energy than the average molecule in the boiling water. But there are so many more molecules in the bathtub than in the teacup that the total heat energy is greater.

Heat Transport by Conduction and Convection

If you place a metal frying pan on the stove, the handle gets hot, even though it is not in contact with the burner, because the metal conducts heat from the bottom of the pan to the handle. **Conduction** is the transport of heat by direct collisions among atoms or molecules. When the frying pan is heated from below, the metal atoms on the bottom of the pan move more rapidly. They then collide with their neighbors and transfer energy to them. Like a falling row of dominoes, energy is passed from one atom to another throughout the pan until the handle becomes hot.

Metals conduct heat rapidly and efficiently, but air is a poor conductor. To understand how air transports heat, imagine that a heater is placed in one corner of a cold room. The heated air in the corner expands, becoming less dense. This light, hot air rises to the ceiling. It flows along the ceiling, cools, falls, and returns to the stove, where it is reheated (Figure 18.8A). Recall from Chapter 6 that *convection* is the upward and downward flow of fluid material in response to heating and cooling. Convection thus involves the transport of heat by the movement of currents. Convection occurs

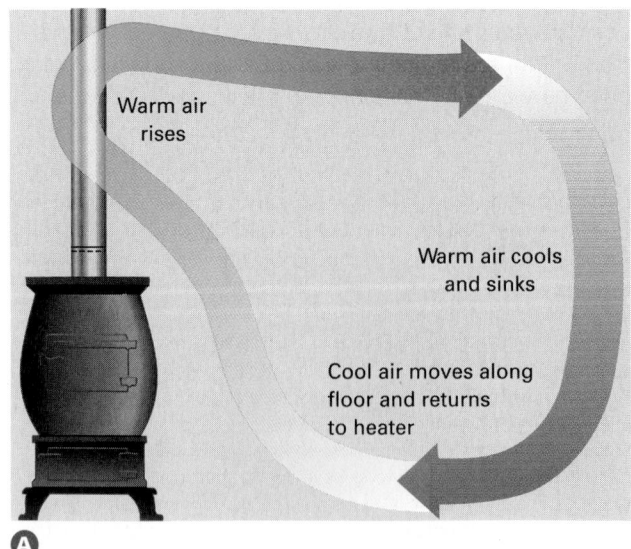

A

FIGURE 18.8 (A) Convection currents distribute heat throughout a room.

---

2. More precisely, temperature is proportional to the kinetic energy of the atoms and molecules. In turn, the kinetic energy equals (mass) × (velocity)[2] for a monatomic gas.

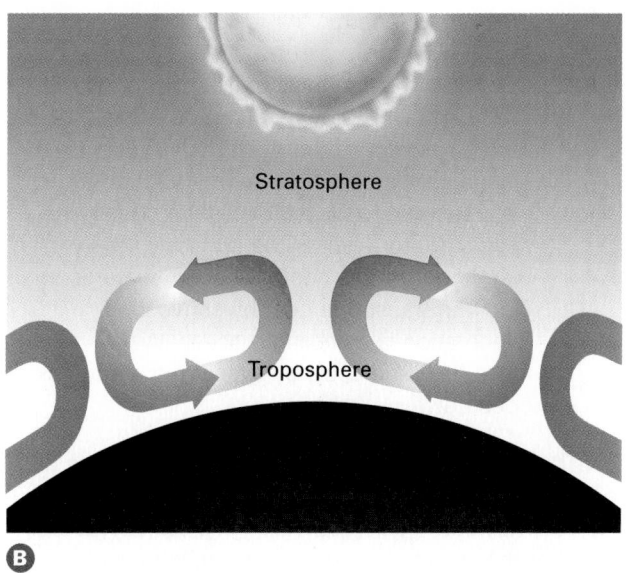

FIGURE 18.8 (B) Convection also distributes heat through the atmosphere when the Sun heats Earth's surface. In this case the ceiling is the boundary between the troposphere and the stratosphere.

readily in liquids and gases. Recall from Chapter 16 that ocean currents transport large quantities of heat northward or southward, thereby altering climate. For example, the Gulf Stream carries tropical water northward and warms the west coast of Europe.

Similar currents occur in the atmosphere (Figure 18.8B). If air in one region is heated above the temperature of surrounding air, this warm air becomes less dense and rises, just as a log floats to the surface of a pond. As the warm air rises, cooler, denser air in another portion of the atmosphere sinks. Air then flows along the surface to complete the cycle. In common speech, this horizontal airflow is called *wind*.

Changes of State

Given the proper temperature and pressure, most substances can exist in three states: solid, liquid, and gas. However, at Earth's surface, many substances commonly exist in only one state. In our experience, rock is almost always solid and molecular oxygen is almost always a gas. Water commonly exists in all three states—as solid ice, as liquid, and as gaseous water vapor.

Latent heat (stored heat) is the energy released or absorbed when a substance changes from one state to another. As shown in Figure 18.9, about 80 calories are required to melt a gram of ice at a constant temperature of 0°C. As a comparison, 100 calories are

needed to heat the same amount of water from freezing to boiling (from 0°C to 100°C). Another 540 to 600 calories are needed to evaporate a gram of water at constant temperature and pressure.[3] The energy transfers also work in reverse. When 1 gram of water vapor condenses to liquid, 540 to 600 calories are released. When 1 gram of water freezes, 80 calories are released. *Sublimation* is the transformation directly from solid ice to water vapor, without passing through an intermediate the liquid phase. For water, this process requires about 620 calories per gram.

If you walk onto the beach after swimming, you feel cool, even on a hot day. Your skin temperature drops because water on your body is evaporating, and evaporation absorbs heat, cooling your skin. Similarly, evaporation from any body of water cools the water and the air around it. Conversely, condensation releases heat. The energy released when water condenses into rain during a single hurricane can be as great as the energy released by several atomic bombs.

The energy absorbed and released during freezing, melting, evaporation, and condensation of water is important in the atmospheric energy balance. For example, in the northern latitudes March is usually colder than September, even though equal amounts

latent heat Stored heat; the energy released or absorbed when a substance changes from one state to another, by melting, freezing, vaporization, condensation, or sublimation.

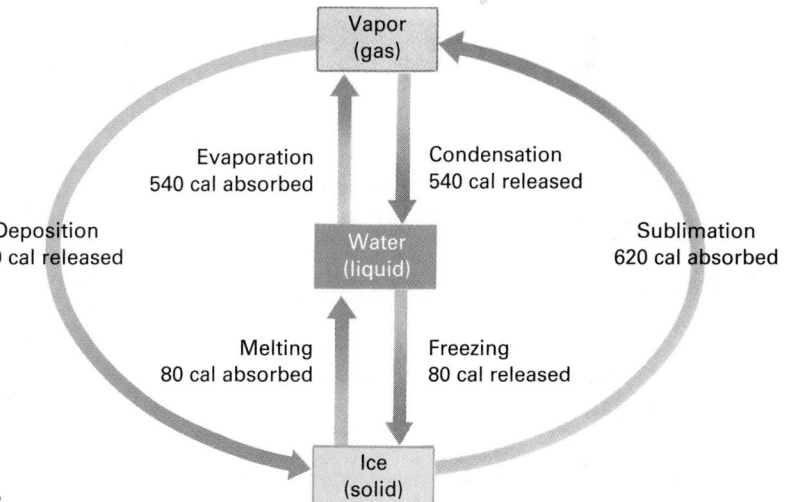

FIGURE 18.9 Water releases or absorbs latent heat as it changes among its liquid, solid, and vapor states. (Calories are given per gram at 0°C and 100°C. The values vary with temperature. Red arrows show processes that absorb heat; blue arrows show those that release heat.)

---

3. The heat of vaporization varies with temperature and pressure; at 100°C and 1 atmosphere pressure, the value is 539.5 calories per gram.

specific heat The amount of energy required to raise the temperature of 1 gram of a substance by 1°C.

of sunlight are received in both months. However, in March much of the solar energy is absorbed by melting snow. Snow also has a high albedo and reflects sunlight efficiently. Evaporation cools seacoasts, and the energy of a hurricane comes, in part, from the condensation of massive amounts of water vapor.

Heat Storage

If you place a pan of water and a rock outside on a hot summer day, the rock becomes hotter than the water. Both have received identical quantities of solar radiation.[4]

Why is the rock hotter?

1. **Specific heat** is the amount of energy needed to raise the temperature of 1 gram of material by 1°C. Specific heat is different for every substance, and water has an unusually high specific heat. Thus, if water and rock absorb equal

---

4. Water actually absorbs a bit more solar energy than rock because water has a lower albedo than rock, but this difference is overwhelmed by the other factors.

amounts of energy, the rock becomes hotter than the water.

2. Rock absorbs heat only at its surface, and the heat travels slowly through the rock. As a result, heat concentrates at the surface. Heat disperses more effectively through water for two reasons. First, solar radiation penetrates several meters below the surface of the water, warming it to this depth. Second, water is a fluid and transports heat by convection.

3. Evaporation is a cooling process. Water loses heat and cools by evaporation, but rock does not.

Think of the consequences of the temperature difference between rock and water. On a hot summer day you may burn your feet walking across dry sand or rock, but the surface of a lake or ocean is never burning hot. Suppose that both the ocean and the adjacent coastline are at the same temperature in spring. As summer approaches, both land and sea receive equal amounts of solar energy. But the land becomes hotter, just as rock becomes hotter than water. Along the seacoast the cool sea moderates the temperature of the land. The interior of a continent is not cooled in this manner and is generally hotter than the coast. In winter

Latitude and Longitude

If someone handed you a perfectly smooth ball with a dot on it and asked you to describe the location of the dot, you would be at a loss to do so because all positions on the surface of a sphere are equal. How, then, can locations on a spherical Earth be described? Even if we ignore irregularities, continents, and oceans, Earth has points of reference because it rotates on its axis and has a magnetic field that nearly (but not exactly) coincides with the axis of rotation.

The North Pole and South Pole lie on the rotational axis, and lines of latitude form imaginary horizontal rings around the axis. Mathematicians measure distance on a sphere in degrees. Using this system, the equator is defined as 0° latitude, the North Pole is 90° north latitude, and the South Pole is 90° south latitude (Figure A).

No natural east–west reference exists, so a line running through Greenwich, England, was arbitrarily chosen as the 0° line. The planet was then divided by lines of longitude, also measured in degrees. On a globe with the rotational axis vertically oriented, lines of longitude also run vertically. To a navigator they measure east–west angular distance from Greenwich, England.

The system is easy to use. Minneapolis–St. Paul lies at 45° north latitude and 93° west longitude. The latitude tells us that the city lies halfway between the equator (0°) and the North Pole (90°). Because a circle has 360°, the longitude tells us that Minneapolis–St. Paul is about one-quarter of the way around the world in a westward direction from Greenwich, England.

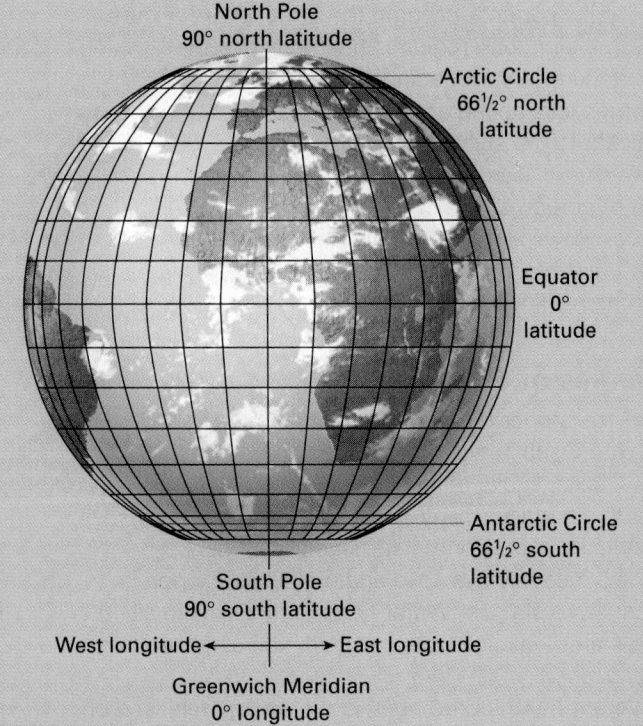

FIGURE A Latitude and longitude allow navigators to identify a location on a spherical Earth.

the opposite effect occurs, and inland areas are generally colder than the coastal regions. Thus, coastal areas are commonly cooler in summer and warmer in winter than continental interiors. The coldest temperatures recorded in the Northern Hemisphere occurred in central Siberia and not at the North Pole, because Siberia is landlocked, whereas the North Pole lies in the middle of the Arctic Ocean.[5] In summer, however, Siberia is considerably warmer than the North Pole. In fact, the average temperature in some places in Siberia ranges from −50°C in winter to +20°C in summer, the greatest range on Earth.

18.4 Temperature Changes with Latitude and Season

Temperature Changes with Latitude

The region near the equator is warm throughout the year, whereas polar regions are cold and ice-bound even in summer. To understand this temperature difference, consider first what happens if you hold a flashlight above a flat board. If the light is held directly overhead and the beam shines straight down, a small area is brightly lit. If the flashlight is held at an angle to the board, a larger area is illuminated. However, because the same amount of light is spread over a larger area, the intensity is reduced (Figure 18.10).

Now consider what happens when the Sun shines directly over the equator. The equator, analogous to

the flat board under a direct light, receives the most concentrated radiation. The Sun strikes the rest of the globe at an angle and thus radiation is less concentrated at higher latitudes (Figure 18.11). Because the equator receives the most concentrated solar energy, it is generally warm throughout the year. Average atmospheric temperature becomes progressively cooler poleward (north and south of the equator). But, as mentioned in the introduction to this chapter, the average temperature does not change steadily with latitude, because many other factors—such as winds, ocean currents, albedo, and proximity to the oceans—also affect atmospheric temperature in any given region.

The Seasons

Earth circles the Sun in a planar orbit, while simultaneously spinning on its axis. This axis is tilted at 23.5° from a line drawn perpendicular to the orbital plane.

Earth revolves around the Sun once a year. As shown in Figure 18.12, the North Pole tilts toward the Sun in summer and away from it in winter. June 21 is the summer **solstice** in the Northern Hemisphere because at this time, the North Pole leans the full 23.5°

solstice Either of two times per year when the Sun is furthest from the equator and shines directly overhead at either 23.5° north latitude (on or about June 21) or 23.5° south latitude (on or about December 22). The solstice on or about June 21 marks the longest day of the year in the Northern Hemisphere and the shortest day in the Southern Hemisphere. The solstice on or about December 22 marks the longest day in the Southern Hemisphere and the shortest day in the Northern Hemisphere.

---

5. Even though the Arctic Ocean is covered with ice, water lies only a few meters below the surface and it still influences climate.

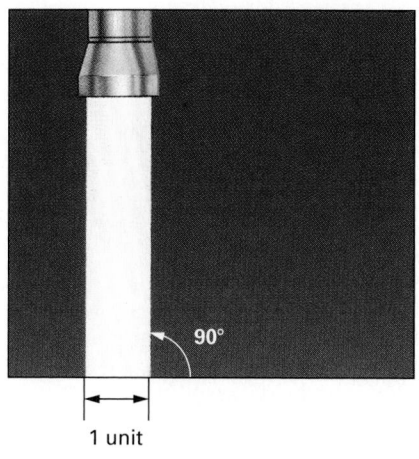

90°

1 unit

One unit of light is concentrated over 1 unit of surface

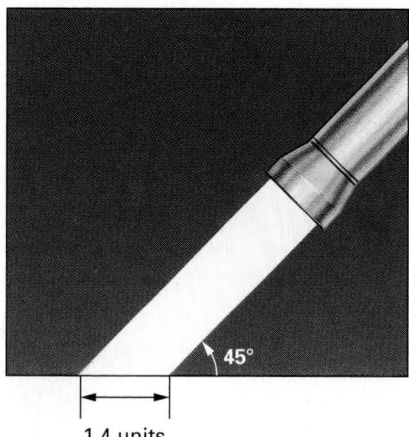

45°

1.4 units

One unit of light is dispersed over 1.4 units of surface

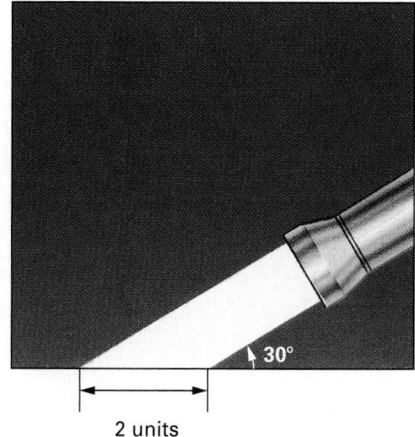

30°

2 units

One unit of light is dispersed over 2 units of surface

FIGURE 18.10 If a light shines from directly overhead, the radiation is concentrated on a small circular area. However, if the light shines at an angle, or if the surface is tilted, the radiant energy is dispersed over a larger, elliptical area.

FIGURE 18.11
When the Sun shines directly over the equator, the equator receives the most intense solar radiation, and the poles receive little.

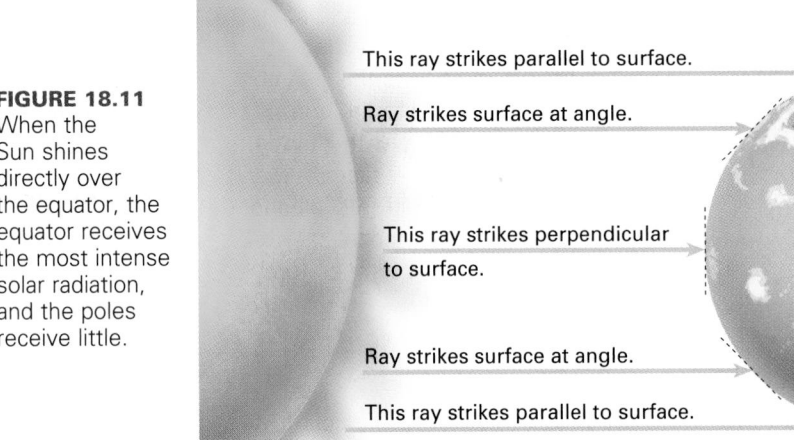

This ray strikes parallel to surface.

Ray strikes surface at angle.

This ray strikes perpendicular to surface.

Ray strikes surface at angle.

This ray strikes parallel to surface.

FIGURE 18.12
Weather changes with the seasons because Earth's axis is tilted relative to the plane of its orbit around the Sun. As a result, the Northern Hemisphere receives more direct sunlight during summer but less during winter.

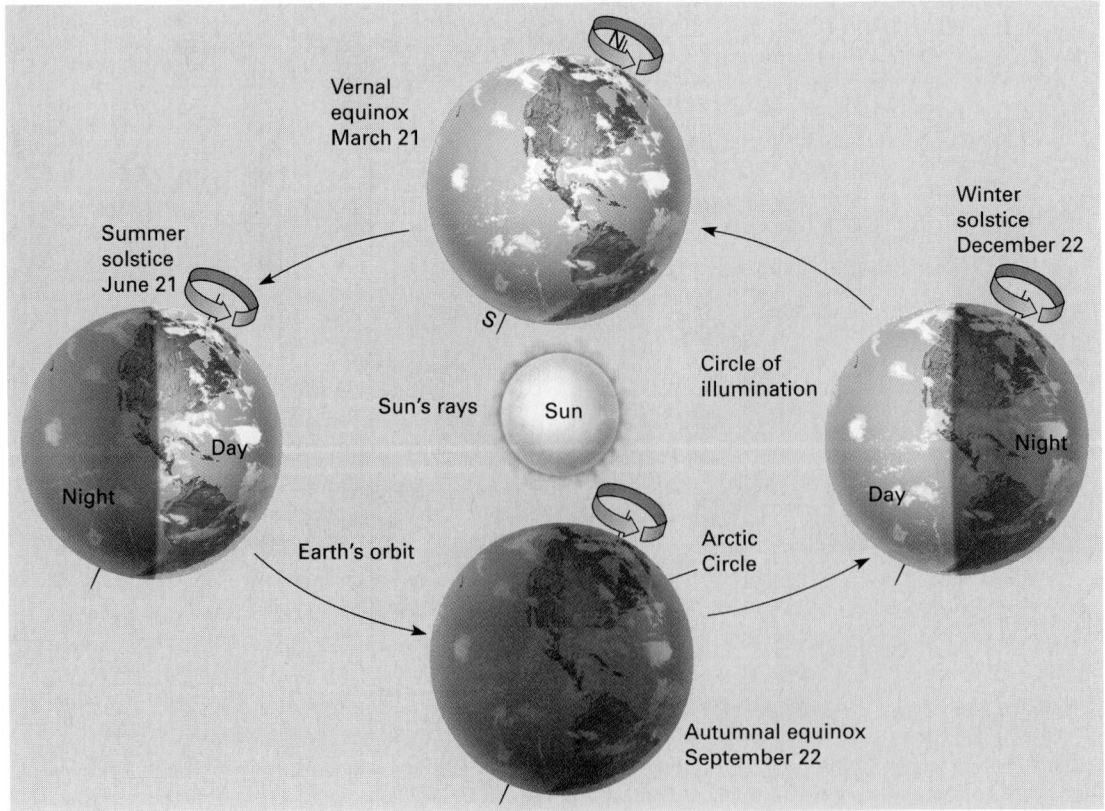

Vernal equinox March 21

Winter solstice December 22

Summer solstice June 21

Circle of illumination

Sun's rays

Sun

Day

Night

Night

Day

Earth's orbit

Arctic Circle

Autumnal equinox September 22

toward the Sun. As a result, sunlight strikes Earth from directly overhead at a latitude 23.5° north of the equator. This latitude is called the **tropic of Cancer**. If you stood on the tropic of Cancer at noon on June 21, you would cast no shadow. June is warm in the Northern Hemisphere for two reasons: (1) when the Sun is high in the sky, sunlight is more concentrated than it is in winter; (2) when the North Pole is tilted toward the Sun, it receives 24 hours of daylight. Polar regions are called "lands of the midnight Sun" because the Sun never sets in the summertime (Figure 18.13). Below the Arctic Circle the Sun sets in the summer, but the days are always longer than they are in winter (Table 18.1).

When it is summer in the Northern Hemisphere, the South Pole tilts away from the Sun and thus the Southern Hemisphere receives low-intensity sunlight and has short days. June 21 is the first day of winter in the Southern Hemisphere. Six months later, on December 21 or 22, the seasons are reversed. The North Pole tilts away from the Sun, giving rise to the winter solstice in the Northern Hemisphere, while it is summer in the Southern Hemisphere. On this day sunlight strikes Earth directly overhead at the **tropic of Capricorn**, latitude 23.5° south. At the North Pole, the Sun never rises and it is continuously dark, while the South Pole is bathed in continuous daylight.

On March 21 and September 22, Earth's axis lies at right angles to a line drawn between Earth and the Sun. As a result, the poles are not tilted toward or away from the

tropic of Cancer The latitude 23.5° north of the equator. On or about June 21, the summer solstice in the Northern Hemisphere, sunlight strikes Earth from directly overhead at noon at this latitude.

tropic of Capricorn The latitude 23.5° south of the equator. On or about December 22, the summer solstice in the Southern Hemisphere, sunlight strikes Earth from directly overhead at noon at this latitude.

COPYRIGHT AND PHOTOGRAPH BY DR. PARVINDER S. SETHI

FIGURE 18.13 The Sun shines down brightly on this Norwegian lake at 2 o'clock in the morning.

TABLE 18.1 Hours of Sunlight per Day

Latitude	Geographic Reference	Summer Solstice	Winter Solstice	Equinoxes
0° north	Equator	12 hr	12 hr	12 hr
30° north	New Orleans	13 hr 56 min	10 hr 04 min	12 hr
40° north	Denver	14 hr 52 min	9 hr 08 min	12 hr
50° north	Vancouver	16 hr 18 min	7 hr 42 min	12 hr
90° north	North Pole	24 hr	0 hr 00 min	12 hr

equinox Either of two times during the year—on or about March 21 and September 22—when the Sun shines directly overhead at the equator and every portion of Earth receives 12 hours of daylight and 12 hours of darkness.

isotherms Lines on a weather map connecting areas with the same average temperature.

Sun and the Sun shines directly overhead at the equator at noon. If you stood at the equator at noon on either of these two dates, you would cast no shadow. But north or south of the equator, a person casts a shadow even at noon. In the Northern Hemisphere, March 21 is the first day of spring and September 22 is the first day of autumn, whereas the seasons are reversed in the Southern Hemisphere. On the first days of spring and autumn, every portion of the globe receives 12 hours of direct sunlight and 12 hours of darkness. For this reason, March 21 and September 22 are called the **equinoxes**, meaning "equal nights."

All areas of the globe receive the same total number of hours of sunlight every year. The North Pole and South Pole receive direct sunlight in dramatic opposition, 6 months of continuous light and 6 months of continuous darkness, whereas at the equator, each day and night are close to 12 hours long throughout the year. Although the poles receive the same number of sunlight hours as do the equatorial regions, the sunlight reaches the poles at a much lower angle and therefore delivers much less total energy per unit of surface area.

18.5 Temperature Changes with Geography

Even though all locations at a given latitude receive equal amounts of solar radiation, some places have cooler climates than others at the same latitude. Figure 18.14 shows temperatures around Earth in January and in July. Lines called **isotherms** connect areas of the same average temperature. Note that the isotherms loop and dip across lines of latitude. For example, the January 0°C line runs through Seattle, Washington, dips southward across the center of the United States, and then swings northward to northern Norway. Such variations occur with latitude because winds and ocean currents transport heat from one region of Earth to another. Other factors, discussed in Section 18.3, include the different heat storage properties of oceans and continents, evaporative cooling of the oceans, and the latent heat of snow and ice.

Altitude

Recall from Chapter 17 that at higher elevations in the troposphere, the atmosphere is thinner and absorbs less energy. In addition, lower parts of the troposphere have absorbed much of the heat radiating from Earth's surface. Consequently, temperature decreases with elevation.

Mount Everest lies at 28° north, about the same latitude as Tampa, Florida. Yet, at 8,000 meters, climb-

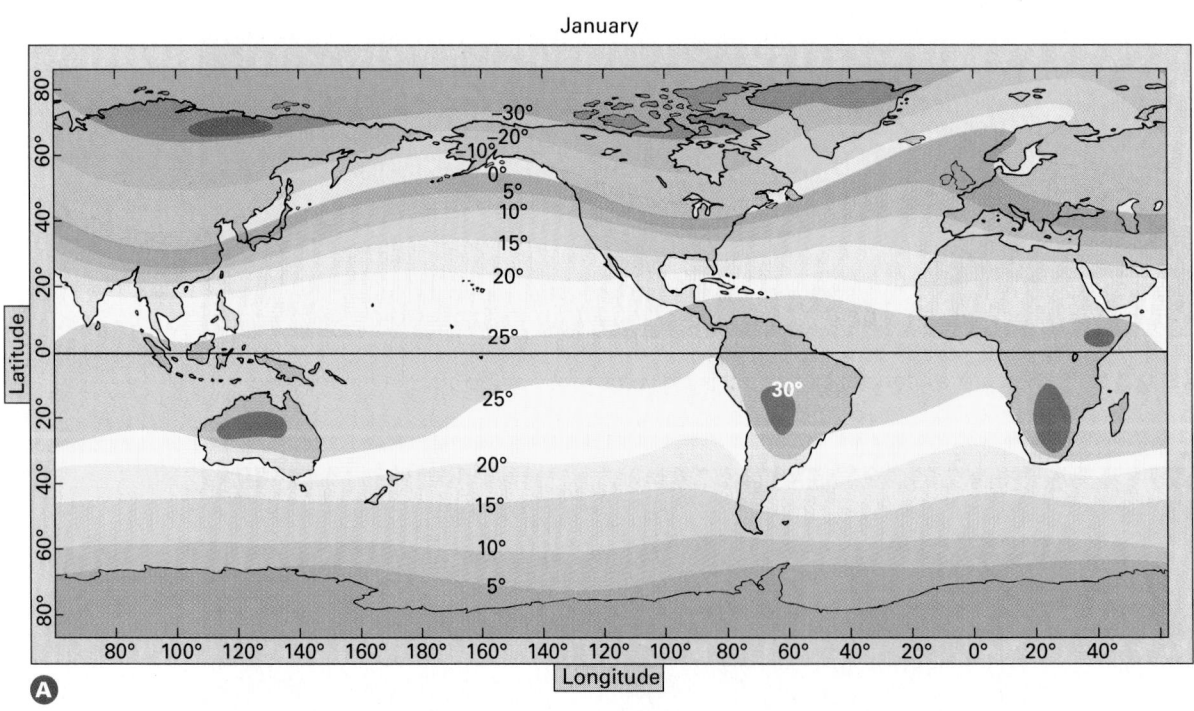

FIGURE 18.14 (A) Global temperature distributions in January. Isotherm lines connect places with the same average temperatures.

Unit 5: The Atmosphere

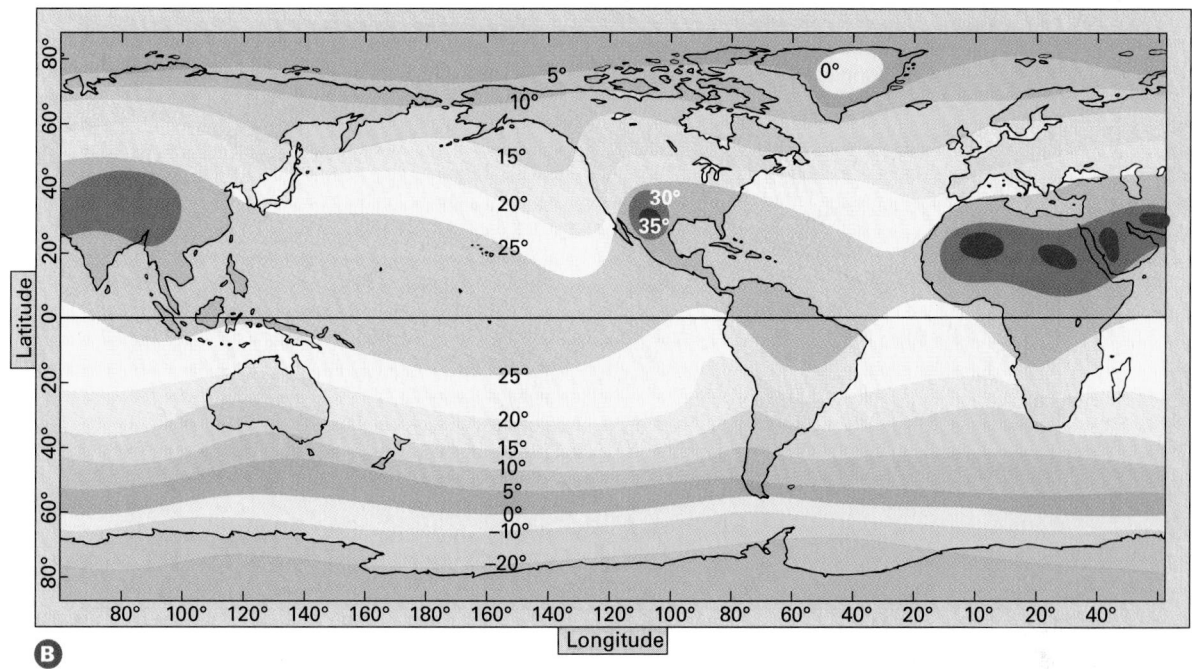

FIGURE 18.14 (B) Global temperature distributions in July. Isotherm lines connect places with the same average temperatures.

ers on Everest wear heavy down clothing to keep from freezing to death, while during the same season sunbathers in Tampa lounge on the beach in bikinis.

Ocean Effects

Recall from Section 18.3 that land heats more quickly in summer and cools more quickly in winter than ocean surfaces do. As a result, continental interiors show greater seasonal extremes of temperature than coastal regions. For example, San Francisco and St. Louis both lie at approximately 38° latitude, but St. Louis is in the middle of the continent and San Francisco lies on the Pacific coast. On average, St. Louis is 9°C cooler in winter and 9°C warmer in summer than San Francisco (Figure 18.15).

FIGURE 18.15 The average temperature of continental St. Louis (red line) is colder in winter and warmer in summer than that of coastal San Francisco (blue line).

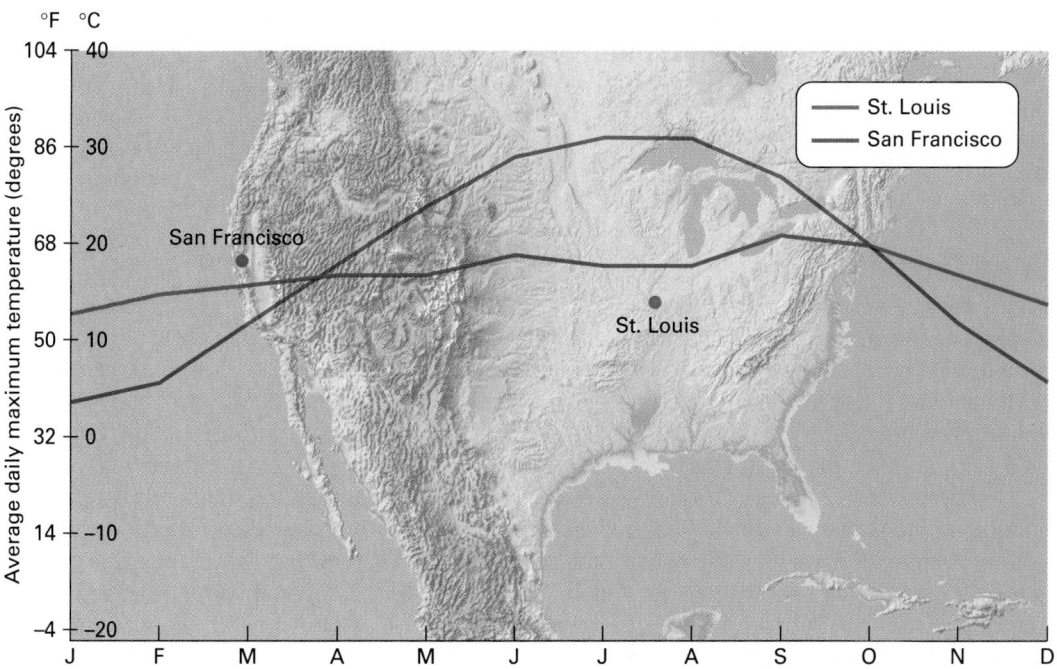

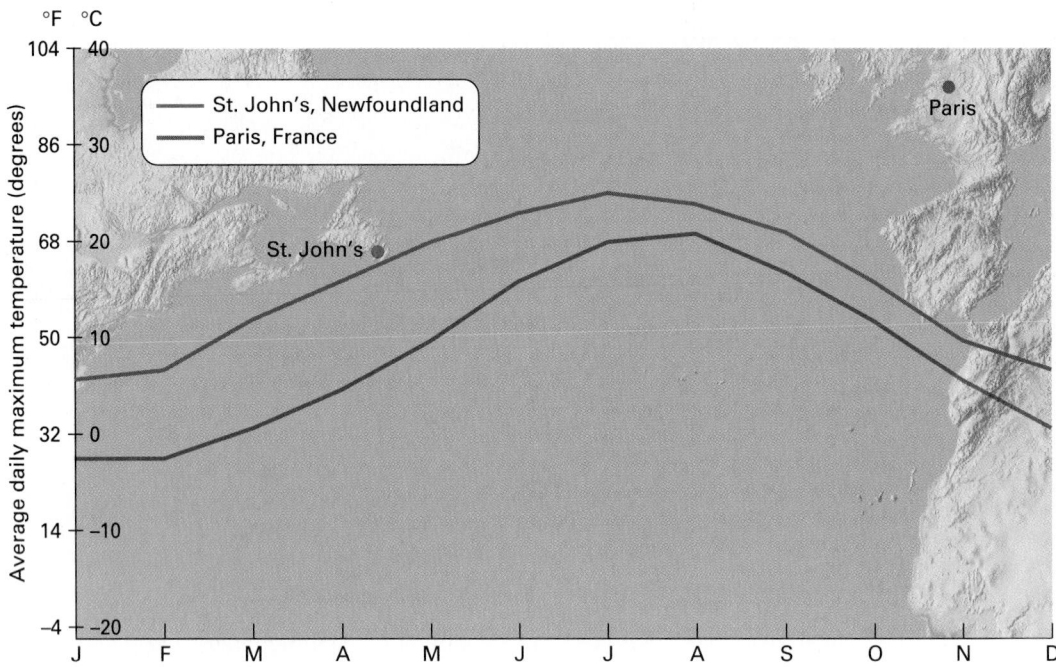

°F °C

104 ─ 40

86 ─ 30

68 ─ 20

50 ─ 10

32 ─ 0

14 ─ −10

−4 ─ −20

Average daily maximum temperature (degrees)

— St. John's, Newfoundland
— Paris, France

Paris

St. John's

J F M A M J J A S O N D

FIGURE 18.16 Paris is warmed by the Gulf Stream and the North Atlantic Drift. On the other hand, St. John's, Newfoundland, is alternately warmed by the Gulf Stream and cooled by the Labrador Current. The cooling effect of the Labrador Current depresses the temperature of St. John's year-round.

Ocean currents also play a major role in determining average temperature. Paris, France, lies at 48° north latitude, north of the U.S.–Canadian border on the other side of the Atlantic Ocean. Although winters are dark because the Sun is low in the sky, the warm Gulf Stream carries enough heat northward so the average minimum temperature in Paris in January is 21°C, just below freezing. In contrast, St. John's, Newfoundland, is at about the same latitude but under the influence of the Labrador Current that flows from the North Pole. As a result, the average minimum January temperature in St. John's is 28°C, significantly colder than it is in Paris (Figure 18.16).

Wind Direction

Winds carry heat from region to region just as ocean currents do. Vladivostok is a Russian city on the west coast of the Pacific Ocean at about 43° north latitude. It has a near-Arctic climate with frigid winters and heavy snow. Forests are stunted by the cold, and trees are generally small.

In comparison, Portland, Oregon, lies at 45° north latitude on the east coast of the Pacific Ocean. Its temperate climate supports majestic cedar forests, and rain is more common than snow in winter (Figure 18.17). The main difference in this case is that Vladivostok is cooled by frigid Arctic winds from Siberia.

Cloud Cover and Albedo

If you are sunbathing on the beach and a cloud passes in front of the Sun, you feel a sudden cooling. During the day, clouds reflect sunlight and cool the surface. In contrast, clouds have the opposite effect and warm the surface during the night. Recall that after the Sun sets, Earth cools because radiant heat from soil and rock escape into the air. Clouds act as an insulating blanket by absorbing outgoing radiation and reradiating some of it back downward (Figure 18.18). Thus cloudy nights are generally warmer than clear nights. Because clouds cool Earth during the day and warm it at night, cloudy regions generally have a narrower daily temperature range than regions with predominantly blue sky and starry nights.

Snow also plays a major role in affecting regional temperature. Imagine that it is raining and the temperature is 0.5°C, or just above freezing. After the rain stops, the Sun comes out. The bare ground absorbs solar radiation and the surface heats up. Now imagine that the temperature had dropped to −0.5°C during the storm. In this case, the precipitation would fall as snow. After the storm passed and the Sun came out, the surface would be covered with a sparkling white cover with 80 to 90 percent albedo. Most of the incoming solar radiation would reflect back into the atmosphere and the surface

© ADISA/SHUTTERSTOCK

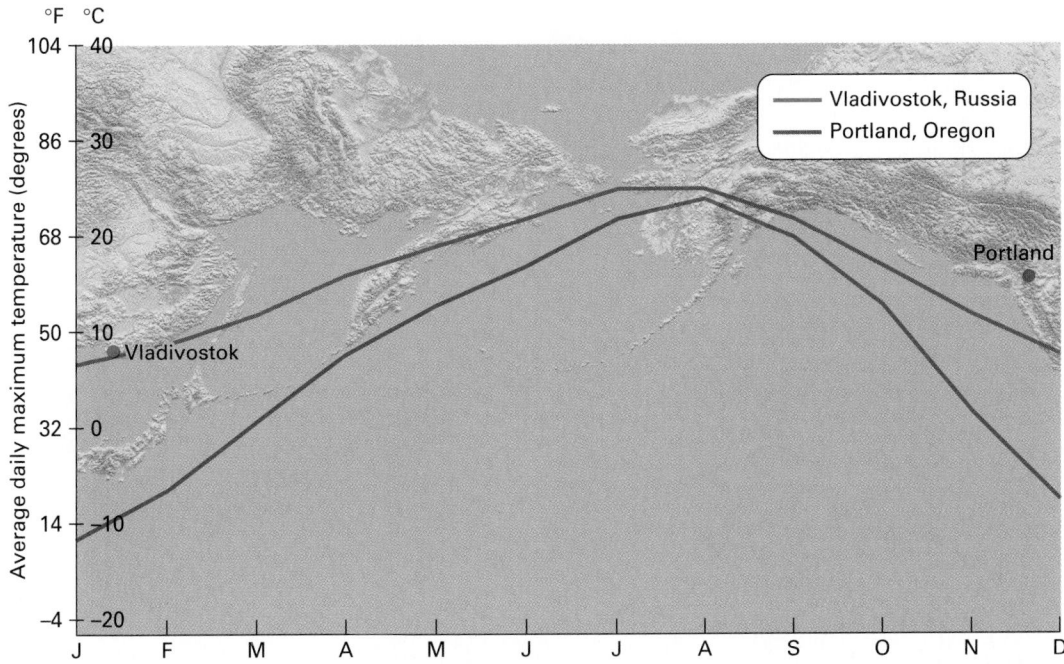

FIGURE 18.17 During summer, temperatures in Vladivostok, Russia, and Portland, Oregon, are nearly identical. However, frigid Arctic wind from Siberia cools Vladivostok during the winter (red line) so that the temperature is significantly colder than that of Portland (blue line).

would remain much cooler. If the surface temperature did rise above freezing, the snow would have to melt before the ground temperature would rise. But snow has a very high latent heat; it takes a lot of energy to melt the snow without raising the temperature at all. In this example, a seemingly insignificant 1° initial temperature difference during a storm would cause a much larger temperature difference when the Sun came out after the storm.

This type of threshold mechanism is very important in weather and climate and will be discussed in Chapters 19 and 20.

FIGURE 18.18 Clouds cool Earth's surface during the day but warm it during the night.

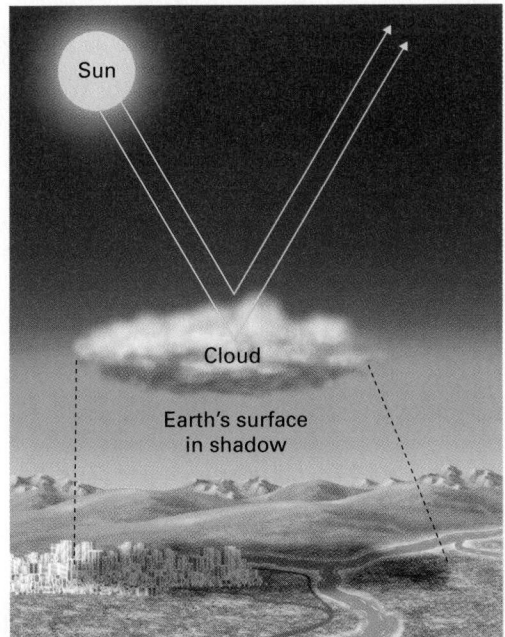

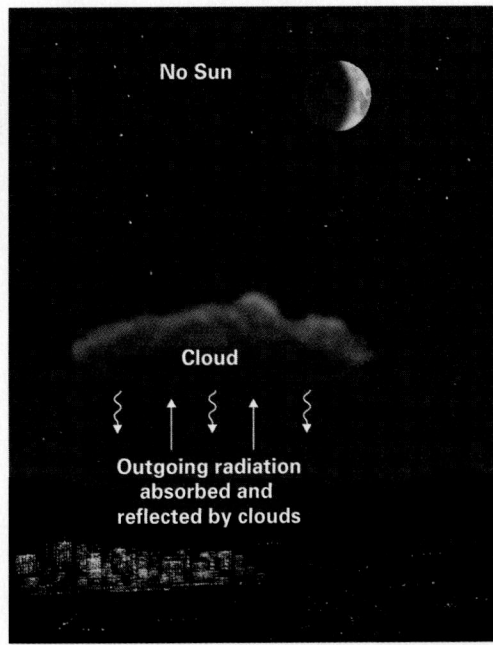

19

MOISTURE, CLOUDS, AND WEATHER

Heavy rain falls in the distance from a thick cumulonimbus cloud cover.

visit **4ltrpress.cengage.com**

19.1 Moisture in Air

Precipitation occurs only when there is moisture in the air. Therefore to understand precipitation, we must first understand how moisture collects in the atmosphere and how it behaves.

Humidity

When water boils on a stove, a steamy mist rises above the pan and then disappears into the air. The water molecules have not been lost; they have simply become invisible. In the pan, water is liquid, and in the mist above, the water exists as tiny droplets. These droplets then evaporate, and the invisible water vapor mixes with air. Water also evaporates into air from the seas, streams, lakes, and soil. Winds then distribute this moisture throughout the atmosphere. Thus, all air contains some water vapor—even over the driest deserts.

Humidity is the amount of water vapor in air. **Absolute humidity** is the mass of water vapor in a given volume of air, expressed in grams per cubic meter (g/m³).

Air can hold only a certain amount of water vapor, and warm air can hold more water vapor than cold air can. For example, air at 25°C can hold 23 g/m³ of water vapor, but at 12°C, it can hold only half that quantity, 11.5 g/m³ (Figure 19.1). **Relative humidity** is the amount of water vapor in air relative to the maximum it can hold at a given temperature. It is expressed as a percentage:

$$\text{Relative humidity (\%)} = \frac{\text{Actual quantity of water per unit of air}}{\text{Maxiumum quantity at the same temperature}} \times 100$$

If air contains half as much water vapor as it can hold, its relative humidity is 50 percent. Suppose that air at 25°C contains 11.5 g/m³ of water vapor. Since air at that temperature can hold 23 g/m³, it is carrying half of its maximum, and the relative humidity is 11.5 g/23 g × 100 = 50 percent.

Now let us take some of this air and cool it without adding or removing any water vapor. Because cold air holds less water vapor than warm air holds, the relative humidity increases even though the *amount* of water vapor remains constant. If the air cools to 12°C, and it still contains 11.5 g/m³, the relative humidity reaches 100 percent because air at that temperature can hold only 11.5 g/m³.

When relative humidity reaches 100 percent, the air is *saturated*. The temperature at which

humidity The amount of water vapor in the air.

absolute humidity The mass of water vapor in a given volume of air, expressed in grams per cubic meter (g/m³).

relative humidity The ratio of the amount of water vapor in a given volume of air divided by the maximum amount of water vapor that can be held by that air at a given temperature, expressed as a percentage.

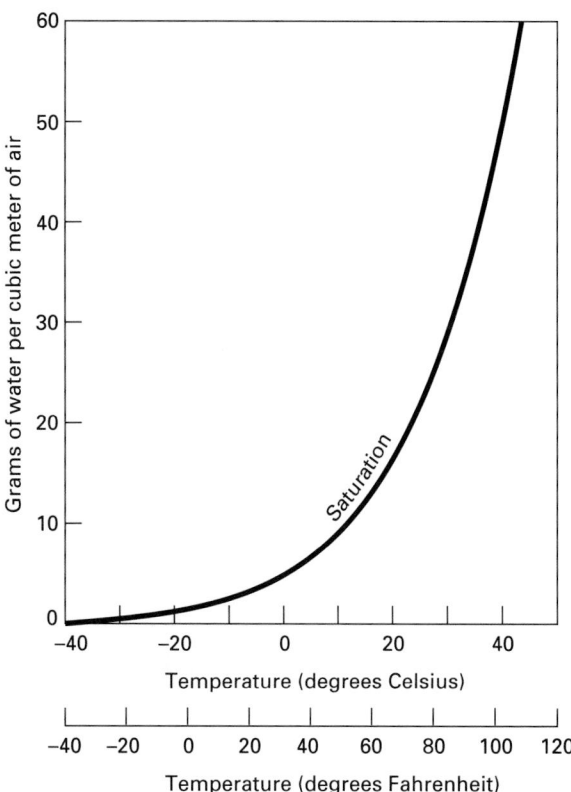

FIGURE 19.1 Warm air can hold more water vapor than cold air can.

saturation The
maximum amount
of water vapor
that air can hold.

saturation The maximum amount of water vapor that air can hold.

dew point The temperature at which the relative humidity of air reaches 100 percent and the air becomes saturated.

supersaturation A condition in which the relative humidity of the air exceeds 100 percent.

supercooling A condition in which water droplets in air do not freeze even when the air cools below the freezing point.

dew Moisture that is condensed onto objects from the atmosphere, usually during the night, when the ground and leaf surfaces become cooler than the surrounding air.

frost Ice crystals formed directly from vapor when the dew point is below freezing.

saturation occurs, 12°C in this example, is the **dew point**. If saturated air cools below the dew point, some of the water vapor may condense into liquid droplets (although, as discussed next, under special conditions in the atmosphere, the relative humidity can rise above 100 percent).

Supersaturation and Supercooling

When the relative humidity reaches 100 percent (at the dew point), water vapor condenses quickly onto solid surfaces such as rocks, soil, and airborne particles. Airborne particles such as dust, smoke, and pollen are abundant in the lower atmosphere. Consequently, water vapor may condense easily at the dew point in the lower atmosphere, and there the relative humidity rarely exceeds 100 percent. However, in the clear, particulate-free air high in the troposphere, condensation occurs so slowly that for all practical purposes it does not happen. As a result, the air commonly cools below its dew point but water remains as vapor. In that case, the relative humidity rises above 100 percent, and the air reaches a point of **supersaturation**.

Similarly, liquid water does not always freeze at its freezing point. Small droplets can remain liquid in a cloud even when the temperature is −40°C. Such water has undergone **supercooling**.

19.2 Cooling and Condensation

Moisture condenses to form water droplets or ice crystals when moist air cools below its dew point. Clouds and fog are visible concentrations of this airborne water and ice. Three atmospheric processes cool air to its dew point and cause condensation: (1) Air cools when it loses heat by radiation. (2) Air cools by contact with a cool surface such as water, ice, rock, soil, or vegetation. (3) Air cools when it rises.

Radiation Cooling

As described in Chapter 18, the atmosphere, rocks, soil, and water absorb the Sun's heat during the day and then radiate some of this heat back out toward space at night. As a result of heat lost by radiation, air, land, and water become cooler at night, and condensation may occur.

Contact Cooling: Dew and Frost

You can observe condensation on a cool surface with a simple demonstration. Heat water on a stove until it boils and then hold a cool drinking glass in the clear air just above the steam. Water droplets will condense on the surface of the glass because the glass cools the hot, moist air to its dew point.

The same effect occurs in a house on a cold day. Water droplets or ice crystals appear on windows as warm, moist, indoor air cools on the glass (Figure 19.2).

In some regions, the air on a typical summer evening is warm and humid. After the Sun sets, plants, houses, windows, and most other objects lose heat by radiation and therefore become cool. During the night, water vapor condenses on the cool objects. This condensation is called **dew**. If the dew point is below freezing, **frost** forms. Frost is not frozen dew, but ice crystals formed directly from vapor.

Cooling of Rising Air

Radiation and contact cooling close to Earth's surface form dew, frost, and some types of fog. However, clouds and precipitation normally form at higher elevations where the air is not cooled by direct contact with the ground. Almost all cloud formation and precipitation occur when air cools as it rises (Figure 19.3).

FIGURE 19.2 Ice crystals condense on a window on a frosty morning.

© PATRIZIA TILLY/SHUTTERSTOCK

© SHERRI R. CAMP/SHUTTERSTOCK

COURTESY OF GRAHAM R. THOMPSON/JONATHAN TURK

FIGURE 19.3 Most clouds form as rising air cools. The cooling causes invisible water vapor to condense into visible water droplets, or ice crystals, which we see as a cloud.

COURTESY OF GRAHAM R. THOMPSON/JONATHAN TURK

Work and heat are both forms of energy. Work can be converted to heat or heat can be converted to work, but energy is never lost. If you pump up a bicycle tire, you are performing work to compress the air. This energy is not lost; much of it converts to heat. Therefore, both the pump and the newly filled tire feel warm. Conversely, if you puncture a tire, the air rushes out. It must perform work to expand, so the air rushing from a punctured tire cools. Variations in temperature caused by compression or expansion of gas are called **adiabatic temperature changes**. *Adiabatic* means without gain or loss of heat. During adiabatic warming, air warms up because work is done on it, not because heat is added. During adiabatic cooling, air cools because it performs work, not because heat is removed.

Air pressure decreases with elevation. When dense surface air rises, it expands because the atmosphere around it is now of lower density, just as air expands when it rushes out of a punctured tire. Rising air performs work to expand, and therefore it cools adiabatically. Dry air cools by 10°C for every 1,000 meters it rises (5.5°F/1,000 ft). This cooling rate is called the **dry adiabatic lapse rate**. Thus, if dry air were to rise from sea level to 9,000 meters (about the height of Mount Everest), it would cool by 90°C (162°F).

Almost all air contains some water vapor. As moist air rises and cools adiabatically, its temperature may eventually decrease to the dew point. At the dew point, moisture may condense as droplets, and a cloud forms. But recall that condensing vapor releases latent heat. As the air rises through the cloud, its temperature now is affected by two opposing processes. It cools adiabatically, but at the same time it is heated by the latent heat released by condensation. However, the warming caused by latent heat is generally less than the amount of adiabatic cooling. The net result is that the rising air continues to cool, but more slowly than at the dry adiabatic lapse rate. The **wet adiabatic lapse rate** is the cooling rate after condensation has begun. It varies from 5°C for every 1,000 meters it rises (2.7°F/1,000 ft) for air with a high moisture content, to 9°C for every 1,000 meters it rises (5°F/1,000 ft) for relatively dry air (Figure 19.4). Thus, once clouds start to form, rising air no longer cools as rapidly as it did lower in the atmosphere. Rising air cools at the dry adiabatic lapse rate until it cools to its dew point

adiabatic temperature changes Temperature changes caused by compression or expansion of gas, that occur without gain or loss of heat.

dry adiabatic lapse rate The rate at which dry air cools adiabatically as it rises—10°C for every 1,000 meters above sea level. Rising air cools at the dry adiabatic lapse rate until it cools to its dew point and condensation begins.

wet adiabatic lapse rate The rate at which rising moist air cools adiabatically after it has reached its dew point and condensation has begun—varying depending on moisture content from 5°C to 9°C for every 1,000 meters it rises.

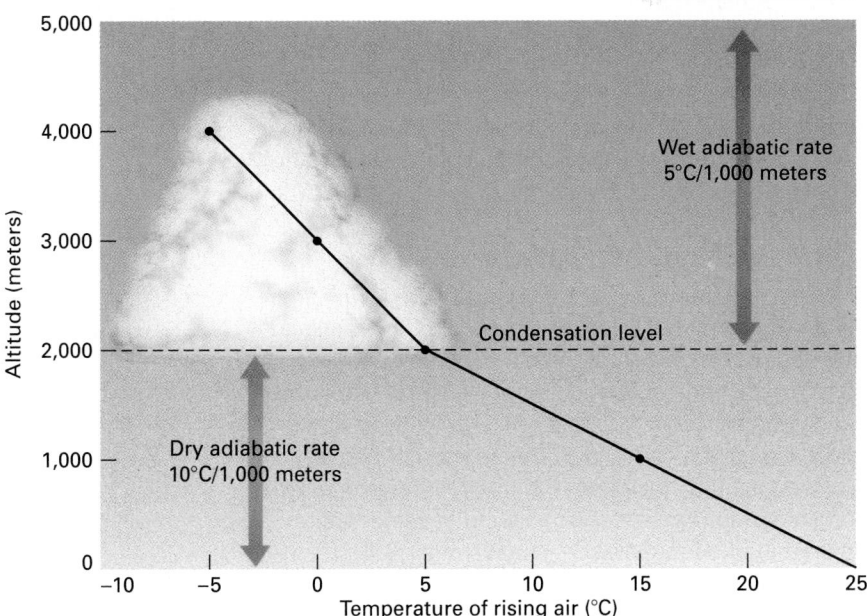

FIGURE 19.4 A rising air mass initially cools rapidly at the dry adiabatic lapse rate. Then, after condensation begins and clouds start to form, it cools more slowly at the wet adiabatic lapse rate.

and condensation begins. Then, as it continues to rise, it cools at the lesser, wet adiabatic lapse rate as condensation continues.

In contrast, sinking air becomes warmer because of adiabatic compression. Warm air can hold more water vapor than cool air can. Consequently, water does not condense from sinking, warming air, and the latent heat of condensation does not affect the rate of temperature rise. As a result, sinking air always becomes warmer at the dry adiabatic rate.

19.3 Rising Air and Precipitation

To summarize: When moist air rises, it cools and forms clouds. Three mechanisms cause air to rise (Figure 19.5): orographic lifting, frontal wedging, and convection–convergence.

Orographic Lifting

When air flows over mountains, it is forced to rise by a mechanism called **orographic lifting**. This rising air frequently causes rain or snow over the mountains, as will be explained in Section 19.8.

Frontal Wedging

A moving mass of cool, dense air may encounter a mass of warm, less-dense air. When this occurs, the cool, denser air slides under the warm air mass, forcing the warm air upward to create a weather front. This process is called **frontal wedging**. We will discuss weather fronts in more detail in Sections 19.7, 19.9, and 19.10.

Convection–Convergence

Recall that *convection* is the upward, and downward, and horizontal flow of fluids in response to heating and cooling. If one portion of the atmosphere becomes warmer than the surrounding air, the warm air expands, becomes less dense, and rises. Thus a hot-air balloon rises because it contains air that is warmer and less dense than the air around it. If the Sun heats one parcel of air near Earth's surface to a warmer temperature than that of surrounding air, the warm air will rise, just as the hot-air balloon rises.

Convective Processes and Clouds

On some days clouds hang low over the land and obscure nearby hills. At other times clouds float high in

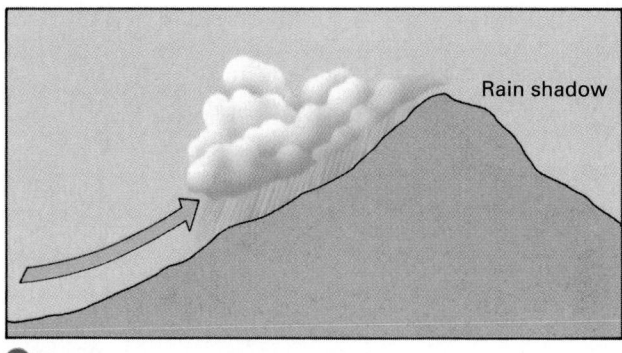

A Orographic lifting

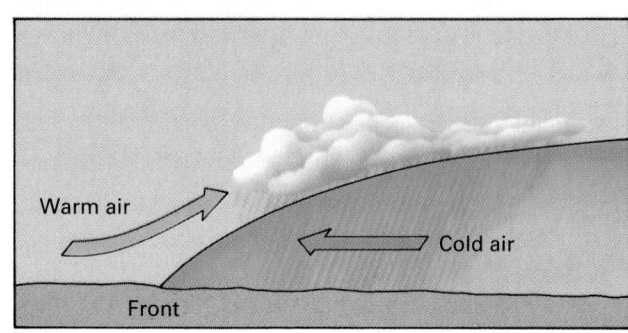

B Frontal wedging

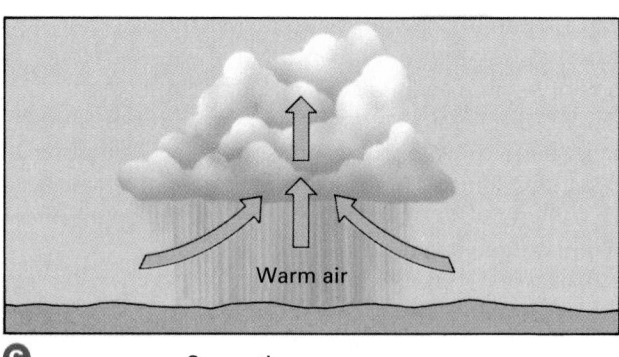

C Convection–convergence

FIGURE 19.5 Three mechanisms cause air to rise and cool: (A) orographic lifting, (B) frontal wedging, and (C) convection–convergence.

the sky, well above the mountain peaks. What factors determine the height and shape of a cloud?

Recall that air is generally warmest at Earth's surface and cools with elevation throughout the troposphere. The rate at which air that is neither rising nor falling cools with elevation is called the **normal lapse rate**. The average normal lapse rate is 6°C for every 1,000 meters of elevation (3.3°F/1,000 ft) and thus is less than the dry adiabatic lapse rate. However, the normal lapse rate is variable. Typically, it is greatest near Earth's surface and decreases with altitude. The normal lapse rate also varies with latitude, the time of day, and the seasons. It is important to note that the normal lapse rate is simply the vertical temperature structure of the atmosphere. In contrast, rising air cools because of adiabatic cooling.

Figure 19.6 shows two rising, warm-air masses, one consisting of dry air and the other of moist air. The central part of the figure shows that the normal lapse rate is the same for both air masses: the temperature of the atmosphere decreases rapidly in the first few thousand meters and then more slowly with increasing elevation. However, the two air masses behave differently because of their different moisture contents.

The dry air mass (part A of the figure) rises and cools at the dry adiabatic lapse rate of 10°C/1,000 m. As a result, in this example the mass's temperature and density become equal to that of surrounding air at an elevation of 3,000 meters. Because the density of the rising air is the same as that of the surrounding air, the rising air is no longer buoyant, and it stops rising. No clouds form because the air has not cooled to its dew point.

In part B of the figure, the rising moist air initially cools at the dry adiabatic rate of 10°C/1,000 m, but only until the air cools to its dew point, at an elevation of 1,000 meters. At that point, moisture begins to condense, and clouds form. But the condensing moisture releases latent heat of condensation. This additional heat causes the rising air to cool more slowly, at the

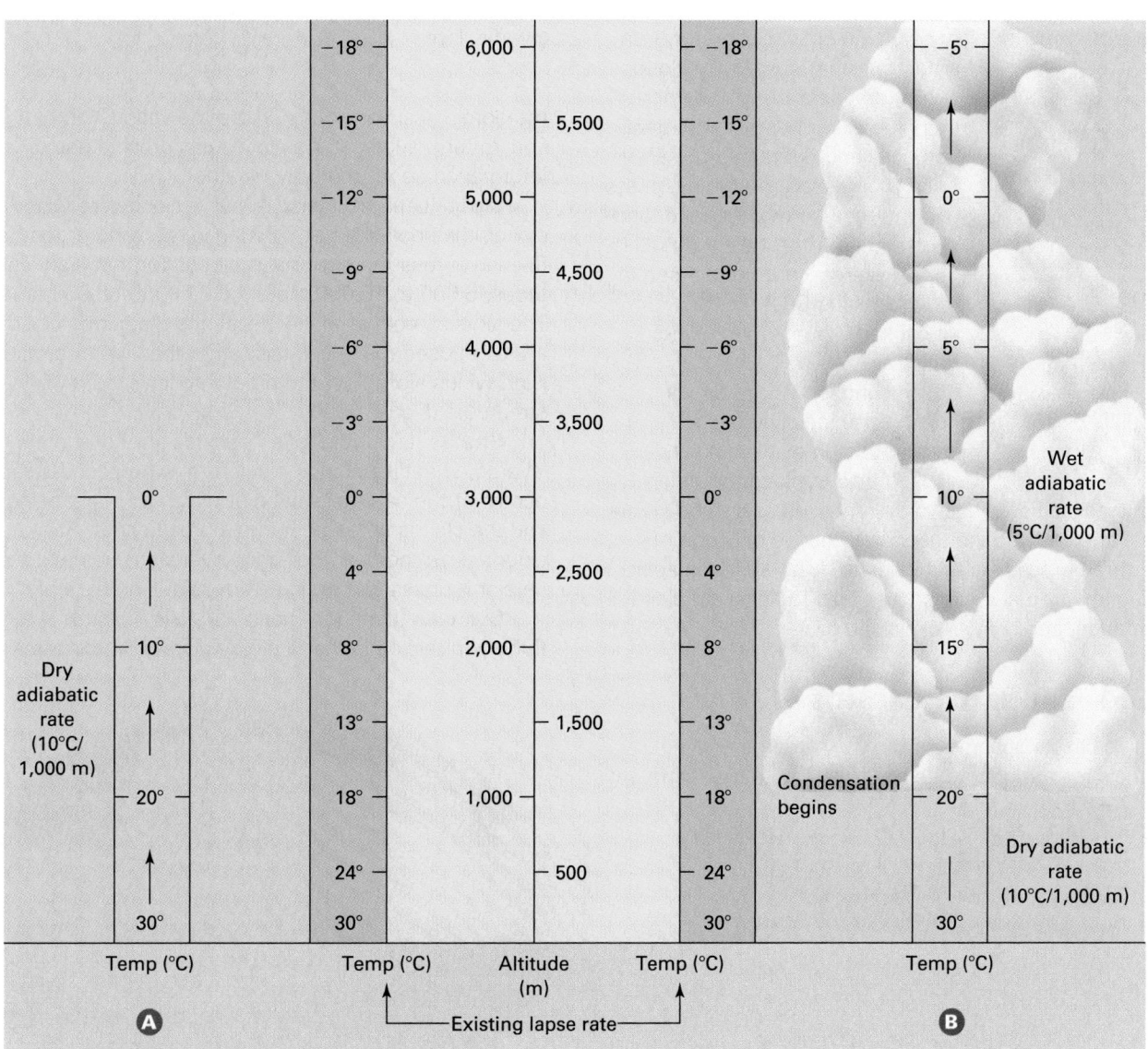

FIGURE 19.6 (A) As dry air rises, it expands and cools at the dry adiabatic lapse rate. Thus, it soon cools to the temperature of the surrounding air, and it stops rising. (B) As moist air rises, initially it cools at the dry adiabatic lapse rate. It soon cools to its dew point, and clouds form. Then, it cools more slowly at the wet adiabatic lapse rate. As a result, it remains warmer than surrounding air and continues to rise for thousands of meters. It stops rising when all moisture has condensed, and the air again cools at its dry adiabatic rate.

unstable air A parcel of warm, moist air that rises rapidly, ascends to high elevations, and leads to formation of towering clouds and heavy rainfall.

stable air A parcel of warm, dry air that does not rise rapidly, does not ascend to high elevations, and does not lead to cloud formation and precipitation.

cirrus Wispy, high-altitude clouds composed of ice crystals.

stratus Horizontally layered clouds that spread out into a broad sheet, usually creating dark, overcast skies.

cumulus Fluffy white clouds with flat bottoms and billowy tops.

wet adiabatic rate of 5°C/1,000 m. As a result, the rising air remains warmer and more buoyant than surrounding air, and it continues to rise for thousands of meters, creating a towering, billowing cloud with the potential for heavy precipitation.

In simple terms, warm, moist air is **unstable air** because it rises rapidly, forming towering clouds and heavy rainfall. Also, as shown in Figure 19.5C, air rushes along the ground to replace the rising air, thus generating surface winds. Most of us have experienced a violent thunderstorm on a hot, summer day. Puffy clouds seem to appear out of nowhere in a blue sky. These clouds grow vertically and darken as the afternoon progresses. Suddenly, gusts of wind race across the land, and shortly thereafter, heavy rain falls. These events, to be described in more detail in Section 19.7, are all caused by unstable, rising, moist air.

In contrast, warm, dry air doesn't rise rapidly, doesn't ascend to high elevations, and doesn't lead to cloud formation and precipitation. Thus warm, dry air is said to be **stable air**. Yet, keep in mind that convection is only one of the three processes that leads to rising air. Orographic lifting and frontal wedging also lead to rising air, cloud formation, and rain, as explained earlier.

19.4 Types of Clouds

Even a casual observer of the daily weather will notice that clouds are quite different from day to day. Different

FIGURE 19.7 Cirrus clouds are high, wispy clouds composed of ice crystals.

RALPH F. KRESGE/NOAA

FIGURE 19.8 Stratus clouds spread out across the sky in a low, flat layer.

meteorological conditions create the various cloud types and, in turn, a look at the clouds can provide useful information about the daily weather.

Cirrus (Latin for "wisp of hair") clouds are wispy clouds that look like hair blowing in the wind or feathers floating across the sky. Cirrus clouds form at high altitudes, 6,000 to 15,000 meters (20,000 to 50,000 feet). The air is so cold at these elevations that cirrus clouds are composed of ice crystals rather than water droplets. High winds aloft blow them out into long, gently curved streamers (Figure 19.7).

Stratus (Latin for "layer") clouds are horizontally layered, sheetlike clouds. They form when condensation occurs at the same elevation at which air stops rising and the clouds spread out into a broad sheet. Stratus clouds form the dark, dull-gray, overcast skies that may persist for days and bring steady rain (Figure 19.8).

Cumulus (Latin for "heap" or "pile") clouds are fluffy, white clouds that typically display flat bottoms

COPYRIGHT AND PHOTOGRAPH BY DR. PARVINDER S. SETHI

COURTESY OF GRAHAM R. THOMPSON/JONATHAN TURK

FIGURE 19.9 Cumulus clouds are fluffy white clouds with flat bottoms and billowy tops.

Unit 5: The Atmosphere

and billowy tops (Figure 19.9). On a hot, summer day the top of a cumulus cloud may rise 10 kilometers or more above its base in cauliflower-like masses. The base of the cloud forms at the altitude at which the rising air cools to its dew point and condensation starts. However, in this situation the rising air remains warmer than the surrounding air and therefore continues to rise. As it rises, more vapor condenses, forming the billowing columns.

Other types of clouds are named by combining these three basic terms (Figure 19.10). **Stratocumulus** clouds are low, sheet-like clouds with some vertical structure. The term *nimbo* refers to a cloud that precipitates; thus a **cumulonimbus** cloud is a towering rain cloud. If you see one, you should seek shelter, because cumulonimbus clouds commonly produce intense rain, thunder, lightning, and sometimes hail. A **nimbostratus** cloud is a stratus cloud from which rain or snow falls. Other prefixes are also added to cloud names. For example, *Alti* is derived from the Latin root *altus*, meaning "high." An **altostratus** cloud is simply a high stratus cloud.

Types of Precipitation
Rain

Why does rain fall from some clouds, whereas other clouds float across a blue sky on a sunny day and produce no rain? The droplets in a cloud are small, about 0.01 millimeter in diameter (about one-seventh the diameter of a human hair). In still air, such a droplet would require 48 hours to fall from a cloud 1,000 meters above Earth. But these tiny droplets never reach Earth because they evaporate faster than they fall.

If the air temperature in a cloud is above freezing, the tiny droplets may collide and coalesce. You can observe similar behavior in droplets sliding down

© SERGIEIEV/SHUTTERSTOCK

stratocumulus Low, sheetlike clouds with some vertical structure.

cumulonimbus Towering storm clouds that form in columns and produce intense rain, thunder, lightning, and sometimes hail.

nimbostratus Stratus clouds from which rain or snow falls.

altostratus High-altitude stratus clouds.

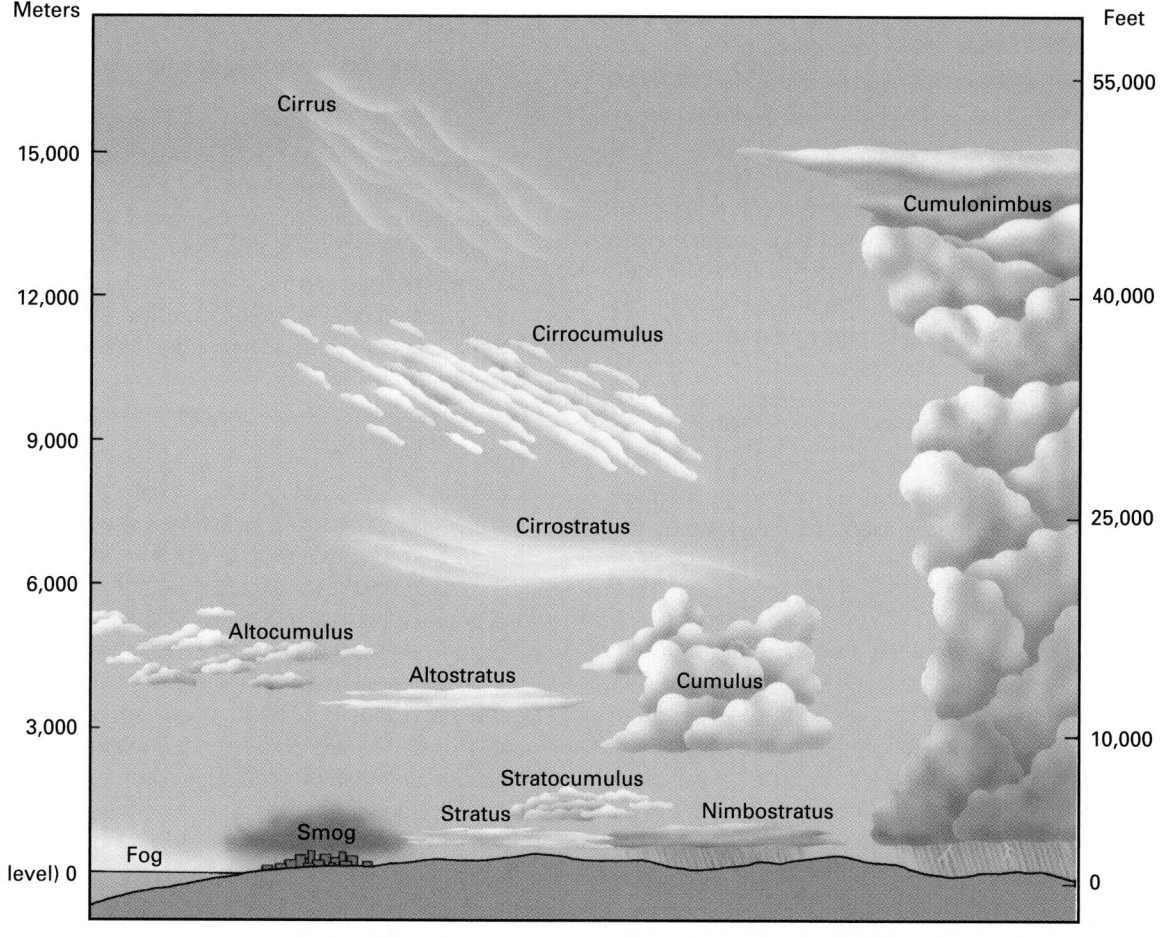

FIGURE 19.10 Cloud names are based on the shape and altitude of the clouds.

sleet Small spheres of ice that develop when raindrops form in a warm cloud and freeze as they fall through a layer of cold air at lower elevation.

glaze An ice coating that forms when rain falls on subfreezing surfaces.

hail Large ice globules varying from 5 millimeters to a record 14 centimeters in diameter that fall from cumulonimbus clouds.

a window pane on a rainy day. If two droplets collide, they merge to become one large drop. If the droplets in a cloud grow large enough, they fall as drizzle (0.1 to 0.5 millimeter in diameter) or light rain (0.5 to 2 millimeters in diameter). About 1 million cloud droplets must combine to form an average-size raindrop.

In many clouds, however, water vapor initially forms ice crystals rather than condensing as tiny droplets of supercooled water. Part of the reason for this is that the temperature in clouds is commonly below freezing, but another factor also favors ice formation. At temperatures near or below freezing, air that is slightly undersaturated with respect to water is slightly supersaturated with respect to ice. For example, if the relative humidity of air is 95 percent with respect to water, it is about 105 percent with respect to ice. Thus, as air cools toward its dew point, all the vapor forms ice crystals rather than supercooled water droplets. The tiny ice crystals then grow larger as more water vapor condenses on them, until they are large enough to fall. The ice then melts to form raindrops as it falls through warmer layers of air.

If you have ever been caught in a thunderstorm, you may remember raindrops large enough to be painful as they struck your face or hands. Recall that a cumulus cloud forms from rising air and that its top may be several kilometers above its base. The temperature in the upper part of the cloud is commonly below freezing. As a result, ice crystals form and begin to fall.

Condensation continues as the crystal falls through the towering cloud, and the crystal grows. If the lower atmosphere is warm enough, the ice melts before it reaches the surface. Raindrops formed in this manner may be 3 to 5 millimeters in diameter, large enough to hurt when they hit.

Snow, Sleet, and Glaze

As explained previously, when the temperature in a cloud is below freezing, the cloud is composed of ice crystals rather than water droplets. If the temperature near the ground is also below freezing, the crystals remain frozen and fall as snow. In contrast, if raindrops form in a warm cloud and fall through a layer of cold air at lower elevation, the drops freeze and fall as small spheres of ice called **sleet**. Sometimes the freezing zone near the ground is so thin that raindrops do not have time to freeze before they reach Earth. However, when they land on subfreezing surfaces, they form a

DAVE PUTNAM

FIGURE 19.11 Glaze forms when rain falls on a surface that is colder than the freezing temperature of water.

coating of ice called **glaze** (Figure 19.11). Glaze can be heavy enough to break tree limbs and electrical transmission lines. It also coats highways with a dangerous icy veneer. In the winter of 1997–1998, a sleet and glaze storm in eastern Canada and the northeastern United States caused billions of dollars in damage. The ice damaged so many electric lines and power poles that many people were without electricity for a few weeks.

Hail

Occasionally, precipitation takes the form of very large ice globules called **hail**. Hailstones vary from 5 millimeters in diameter to a record 14 centimeters in diameter; that record breaker weighed 765 grams (more than 1.5 pounds) and fell in Kansas. A 500-gram (1-pound) hailstone crashing to Earth at 160 kilometers (100 miles) per hour can shatter windows, dent car roofs, and kill people and livestock. Even small hailstones can damage crops. Hail falls only from cumulonimbus clouds. Because cumulonimbus clouds form in columns with distinct boundaries, hailstorms occur in local, well-defined areas. Thus, one farmer may lose an entire crop while a neighbor is unaffected.

A hailstone consists of ice in concentric shells, like the layers of an onion. Two mechanisms have been proposed for their formation. In one, turbulent winds blow

FIGURE 19.12 Radiation fog is seen as a morning mist in this field in Idaho.

advection fog Fog that forms when warm, moist air from the sea blows onto cooler land, where the air cools and water vapor condenses at ground level.

radiation fog Fog that occurs when Earth's surface and the air near the surface cool by radiation during the night, and water vapor in the air condenses because it cools below its dew point.

evaporation fog Fog that forms when air is cooled by evaporation from a body of water, commonly a lake or river and typically in late fall or early winter when the air is cool but the water is still warm. The water evaporates, but the vapor cools and condenses to fog.

upslope fog Fog that forms when air cools as it rises along a land surface.

falling ice crystals back upward in the cloud. New layers of ice accumulate as additional vapor condenses on the recirculating ice grain. An individual particle may rise and fall several times until it grows so large and heavy that it drops out of the cloud. In the second mechanism, hailstones form in a single pass through the cloud. During their descent, supercooled water freezes onto the ice crystals. The layering develops because different temperatures and amounts of supercooled water exist in different portions of the cloud, and each layer forms in a different part of the cloud.

19.5 Fog

Fog is a cloud that forms at or very close to ground level, although most fog forms by processes different from those that create higher-level clouds. **Advection fog** occurs when warm, moist air from the sea blows onto cooler land. The air cools to its dew point, and water vapor condenses at ground level. San Francisco and Seattle, as well as Vancouver, British Columbia, all experience foggy winters as warm, moist air from the Pacific Ocean is cooled first by the cold California current and then by land. The foggiest location in the United States is Cape Disappointment, Washington, on the Pacific Ocean at the mouth of the Columbia River, where visibility is obscured by fog 29 percent of the time.

Radiation fog occurs when Earth's surface and air near the surface cool by radiation during the night (Figure 19.12). Water vapor condenses as fog when the air cools below its dew point. Often the cool, dense, foggy air settles into valleys. If you are driving late at night in hilly terrain, beware, because a sudden dip in the roadway may lead you into a thick fog where visibility is low. A ground fog of this type typically "burns off" in the morning. The rising Sun warms the land or water surface which, in turn, warms the low-lying air. As the air becomes warmer, its capacity to hold water vapor

increases, and the fog droplets evaporate. Radiation fog is particularly common in areas where the air is polluted because water vapor condenses readily on the tiny particles suspended in the air.

Recall that vaporization of water absorbs heat and, therefore, cools both the surface and the surrounding air. In addition, vaporization adds moisture to the air. The cooling and the addition of moisture combine to form conditions conducive to fog. **Evaporation fog** occurs when air is cooled by evaporation from a body of water, commonly a lake or river. Evaporation fogs are common in late fall and early winter, when the air has become cool but the water is still warm. The water evaporates, but the vapor cools and condenses to fog almost immediately upon contact with the cold air.

Upslope fog occurs when air cools as it rises along a land surface. Upslope fogs occur both on gradually sloping plains and on steep mountains. For example, the Great Plains rise from sea level at the Mississippi Delta to 1,500 meters (5,000 feet) at the Rocky Mountain front. When humid air moves northwest from the Gulf of Mexico toward the Rockies, it rises and cools adiabatically to form upslope fog. The rapid rise at the mountain front also forms fog.

19.6 Pressure and Wind

Warm air is less dense than cold air. Thus warm air exerts a relatively low atmospheric pressure and cold air exerts a relatively high atmospheric pressure. Warm air rises because it is less dense than the surrounding cool air (Figure 19.13A). Air rises slowly above a typical low-pressure region, at a rate of about 1 kilometer per day. In contrast, if air in the upper atmosphere cools, it becomes denser than the air beneath it and sinks (Figure 19.13B).

Air must flow inward over Earth's surface toward a low-pressure region to replace a rising air mass. But a sinking air mass displaces surface air, pushing it outward from a high-pressure region. Thus vertical airflow in both high- and low-pressure regions is accompanied by horizontal airflow, called **wind**. Winds near Earth's surface always flow away from a region of high pressure and toward a low-pressure region. Ultimately, all wind is caused by the pressure differences resulting from unequal heating of Earth's atmosphere.

Pressure Gradient

Wind blows in response to differences in pressure. Imagine that you are sitting in a room and the air is still. Now you open a can of vacuum-packed coffee and hear the hissing as air rushes into the can. Because the pressure in the room is higher than that inside the coffee can, wind blows from the room into the can. But if you

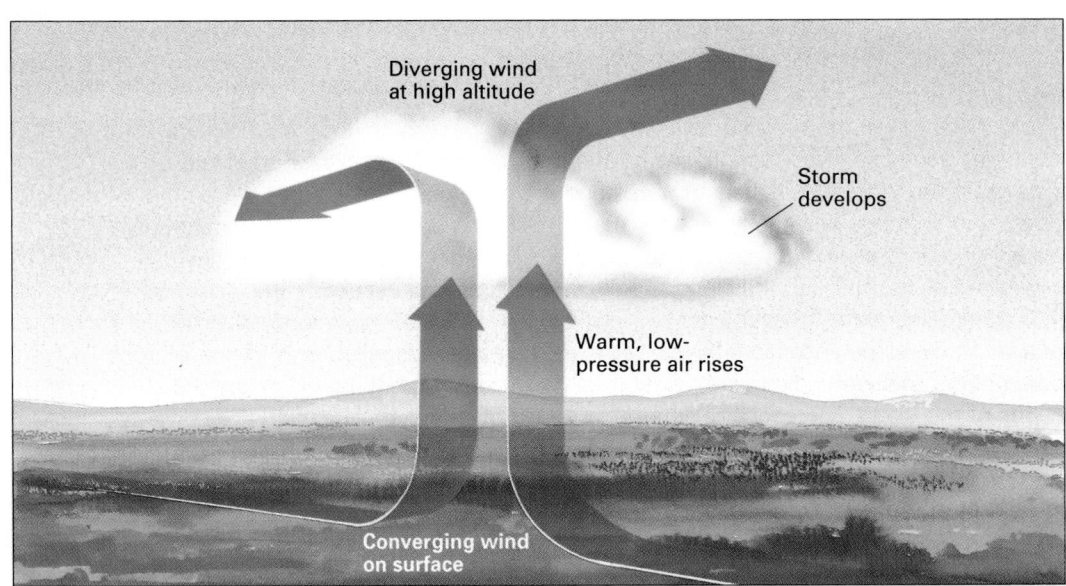

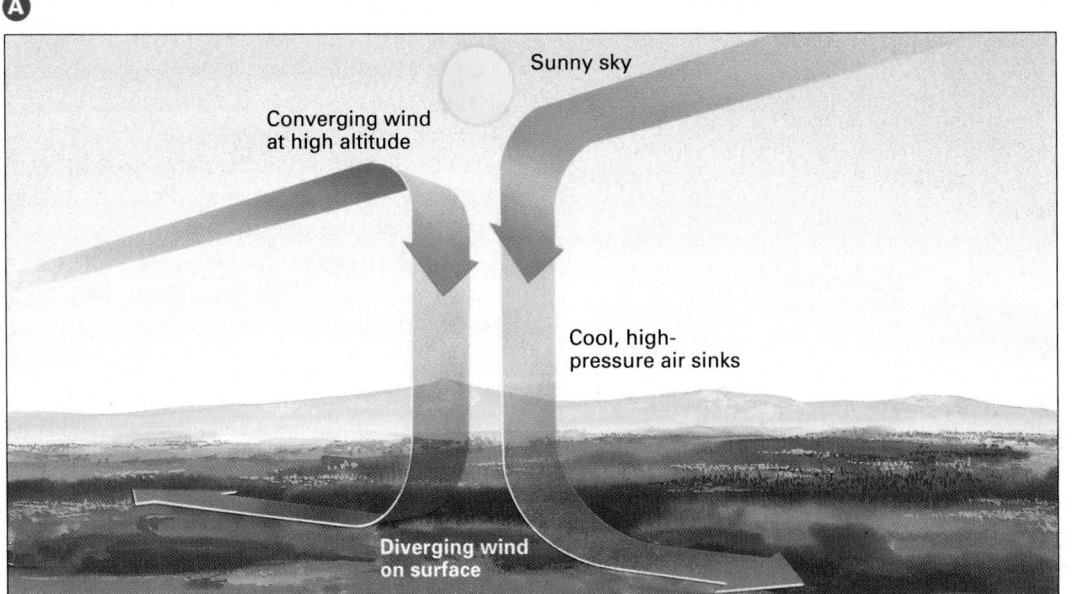

FIGURE 19.13 (A) Rising, low-pressure air creates clouds and precipitation. Air flows inward toward the low-pressure zone, creating surface winds. (B) Sinking, high-pressure air creates clear skies. Air flows outward from the high-pressure zone and also creates surface winds.

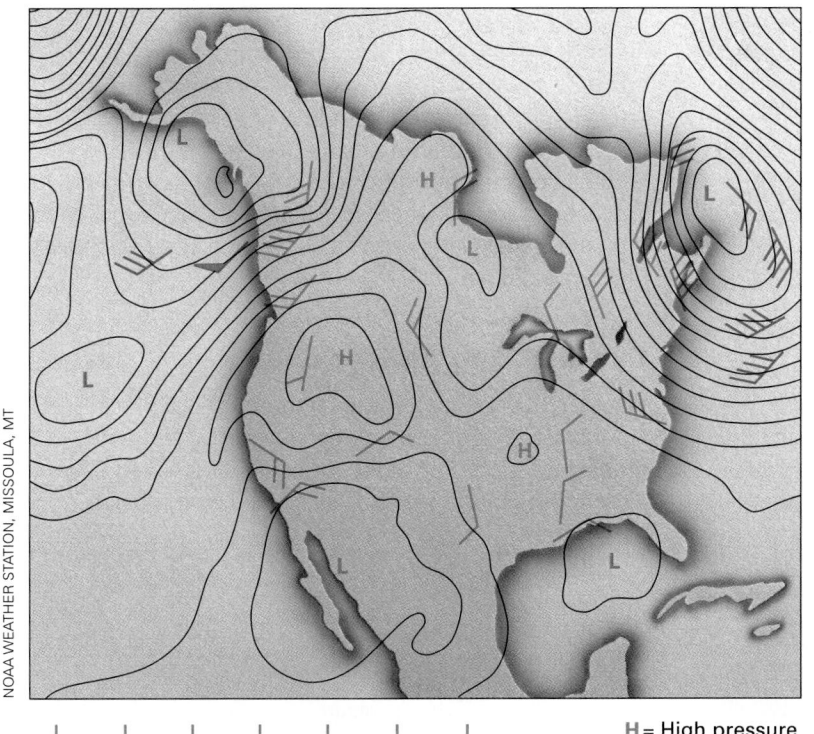

5	10	15	20	30	40	50

H = High pressure

L = Low pressure

Wind flags represent wind speed in knots. The ends of the flags point in the direction the wind is blowing.

FIGURE 19.14 Pressure map and winds at 5,000 feet in North America on February 3, 1992. High-altitude data are shown because the winds are not affected by surface topography and thus the effect of pressure gradient is well illustrated. Note that in the Northeast and Northwest, steep pressure gradients, shown by closely spaced isobars, cause high winds that spiral counterclockwise into the low-pressure zones. Widely spaced isobars around high-pressure zones in the central United States cause weaker winds.

blow up a balloon, the air inside the balloon is at higher pressure than the air in the room. When the balloon is punctured, wind blows from the high-pressure zone of the balloon into the lower-pressure zone of the room.

Wind speed is determined by the magnitude of the pressure difference over distance, called the **pressure gradient**. Thus wind blows rapidly if a large pressure difference exists over a short distance. A steep pressure gradient is analogous to a steep hill. Just as a ball rolls quickly down a steep hill, wind flows rapidly across a steep pressure gradient. To create a pressure-gradient map, air pressure is measured at hundreds of different weather stations. Points of equal pressure are connected by map lines called **isobars**. A steep pressure gradient is shown by closely spaced isobars, whereas a weak pressure gradient is indicated by widely spaced isobars (Figure 19.14). Pressure gradients change daily, or sometimes hourly, as high- and low-pressure zones

move. Therefore, maps are updated frequently.

Coriolis Effect

Recall from Chapter 16 that the Coriolis effect, caused by Earth's spin, deflects ocean currents. The Coriolis effect similarly deflects winds. In the Northern Hemisphere wind is deflected toward the right, and in the Southern Hemisphere, to the left (Figure 19.15). The Coriolis effect alters wind direction but not its speed.

Friction

Rising and falling air generates wind both along Earth's surface and at higher elevations. Surface winds are affected by friction

wind Horizontal airflow caused by pressure differences resulting from unequal heating of Earth's atmosphere. Winds near Earth's surface always flow from a region of high pressure toward a low-pressure region.

pressure gradient A measure of the change in air pressure over distance, used to determine wind speed.

isobars Lines on a weather map connecting points of equal air pressure.

Winds are deflected to the right in the Northern Hemisphere

Rotation of Earth

Winds are deflected to the left in the Southern Hemisphere

FIGURE 19.15 The Coriolis effect deflects winds to the right in the Northern Hemisphere and to the left in the Southern Hemisphere. Only winds blowing due east or west are unaffected.

with Earth's surface, whereas high-altitude winds are not. As a result, wind speed normally increases with elevation. This effect was first noted during World War II. On November 24, 1944, U.S. bombers were approaching Tokyo for the first mass bombing of the Japanese capital. Flying between 8,000 and 10,000 meters (27,000 to 33,000 feet), the pilots suddenly found themselves roaring past landmarks 140 kilometers (90 miles) per hour faster than the theoretical top speed of their airplanes! Amid the confusion, most of the bombs missed their targets, and the mission was a military failure. However, this experience introduced meteorologists to **jet streams**, narrow bands of high-altitude wind. The jet stream in the Northern Hemisphere flows from west

to east at speeds between 120 and 240 kilometers per hour (75 and 150 mph). As a comparison, surface winds attain such velocities only in hurricanes and tornadoes. Airplane pilots traveling from Los Angeles to New York fly with the jet stream to gain speed and save fuel, whereas pilots moving from east to west try to avoid it.

Jet-stream influence on weather and climate will be discussed again later in this chapter and in more detail in Chapter 20.

Cyclones and Anticyclones

Figure 19.16A shows the movement of air in the Northern Hemisphere as it converges toward a low-pressure area. If Earth did not spin, wind would flow directly across the isobars, as shown by the black

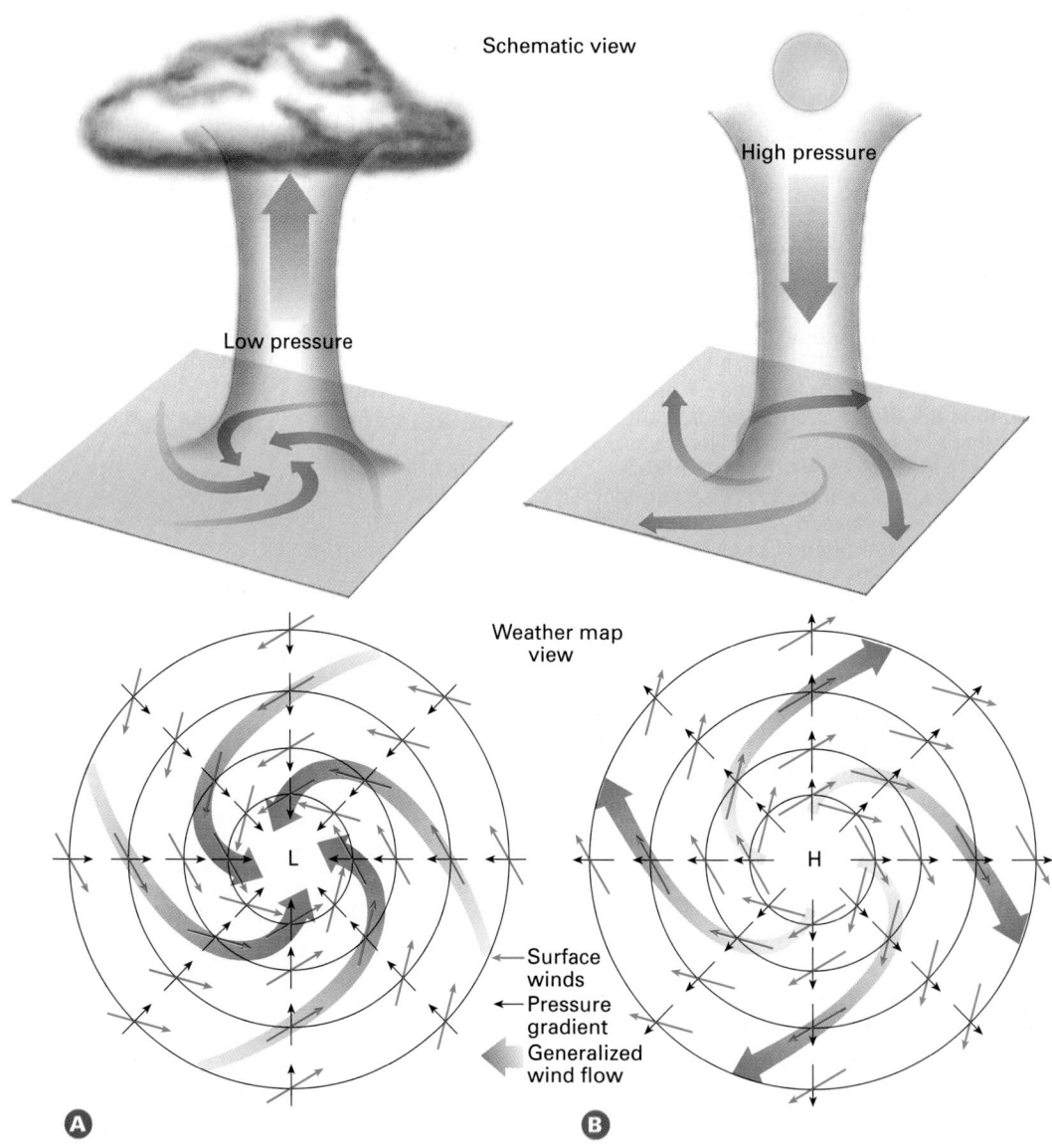

Schematic view

Low pressure

High pressure

Weather map view

L

H

Surface winds

Pressure gradient

Generalized wind flow

A   B

FIGURE 19.16 (A) In the Northern Hemisphere, a cyclone consists of winds spiraling counterclockwise into a low-pressure region. (B) An anticyclone in the Northern Hemisphere consists of winds spiraling clockwise out from a high-pressure zone.

arrows. However, Earth does spin, and the Coriolis effect deflects wind to the right, as shown by the small blue arrows. This rightward deflection creates a counterclockwise vortex near the center of the low-pressure region, as shown by the large, magenta arrows. In the Southern Hemisphere the direction is clockwise.

Such a low-pressure region with its accompanying surface wind is called a *cyclone*. In this usage, **cyclone** means a system of inwardly directed rotating winds, not the violent storms that are sometimes called cyclones, hurricanes, or typhoons. The opposite mechanism forms an **anticyclone** around a high-pressure region. When descending air reaches the surface, it spreads out in all directions. In the Northern Hemisphere, the Coriolis effect deflects the diverging winds of an anticyclone to the right, forming a pinwheel pattern, with the wind spiraling clockwise (Figure 19.16B). In the Southern Hemisphere, the Coriolis effect deflects winds leftward and creates a counterclockwise spiral.

Pressure Changes and Weather

As explained earlier, wind blows in response to any difference in pressure. However, low pressure generally brings clouds and precipitation with the wind, and sunny days predominate during high pressure. To understand this distinction, recall that warm air is less dense than cold air. If warm and cold air are in contact, the less dense, and therefore buoyant, warm air rises.

Rising air forms a region of low pressure. But rising air also cools adiabatically. If the cooling is sufficient, clouds form and rain or snow may fall. Thus, low barometric pressure is an indication of wet weather. Alternatively, when cool air sinks, it is compressed and the pressure rises. In addition, sinking air is heated adiabatically. Because warm air can hold more water

vapor than cold air can, the sinking air absorbs moisture, and thus clouds generally do not form over a high-pressure region. Thus, fair, dry weather generally accompanies high pressure.

19.7 Fronts and Frontal Weather

An **air mass** is a large body of air with approximately uniform temperature and humidity at any given altitude. Typically, an air mass is 1,500 kilometers or more across and several kilometers thick. Because air acquires both heat and moisture from Earth's surface, an air mass is classified by its place of origin. Temperature can be either polar (cold) or tropical (warm). Maritime air originates over water and has high moisture content, whereas continental air has low moisture content (Figure 19.17, Table 19.1).

Air masses move and collide. The boundary between a warmer air mass and a cooler one is a **front**. The term was first used during World War I because weather systems were considered analogous to armies that advance and clash along battle lines. When two air masses collide, each may retain its integrity for days before the two mix. During a collision, one of the air masses is forced to rise, which often results in cloudiness and precipitation. Frontal weather patterns are determined by the types of air masses that collide and their relative speeds and directions. The symbols commonly used on weather maps to describe fronts are shown in Figure 19.18.

Warm Fronts and Cold Fronts

Fronts are classified by whether a warm air mass moves toward a stationary (or more slowly moving) cold mass, or vice versa. A **warm front** forms when moving warm air collides with a stationary or

cyclone A low-pressure region with its accompanying system of inwardly directed rotating winds. In common, nonscientific usage the term often refers to a variety of different violent storms including hurricanes and tornados.

anticyclone A high-pressure region with its accompanying system of outwardly directed rotating winds that develop where descending air spreads over Earth's surface.

air mass A large body of air that has approximately the same temperature and humidity at any given altitude throughout.

front In meteorology, the boundary between a warmer air mass and a cooler one.

warm front A front that forms when moving warm air collides with a stationary or slower-moving cold air mass. The moving warm air rises over the denser cold air, cools adiabatically, and the cooling generates clouds and precipitation.

TABLE 19.1 Classification of Air Masses

Classification according to temperature (latitude):		
Polar (P) air masses originate in high latitudes and are cold.		
Tropical (T) air masses originate in low latitudes and are warm.		
Classification according to moisture content:		
Continental (c) air masses originate over land and are dry.		
Maritime (m) air masses originate over water and are moist.		
Symbol	Name	Characteristics
mP	Maritime polar	Moist and cold
cP	Continental polar	Dry and cold
mT	Maritime tropical	Moist and warm
cT	Continental tropical	Dry and warm

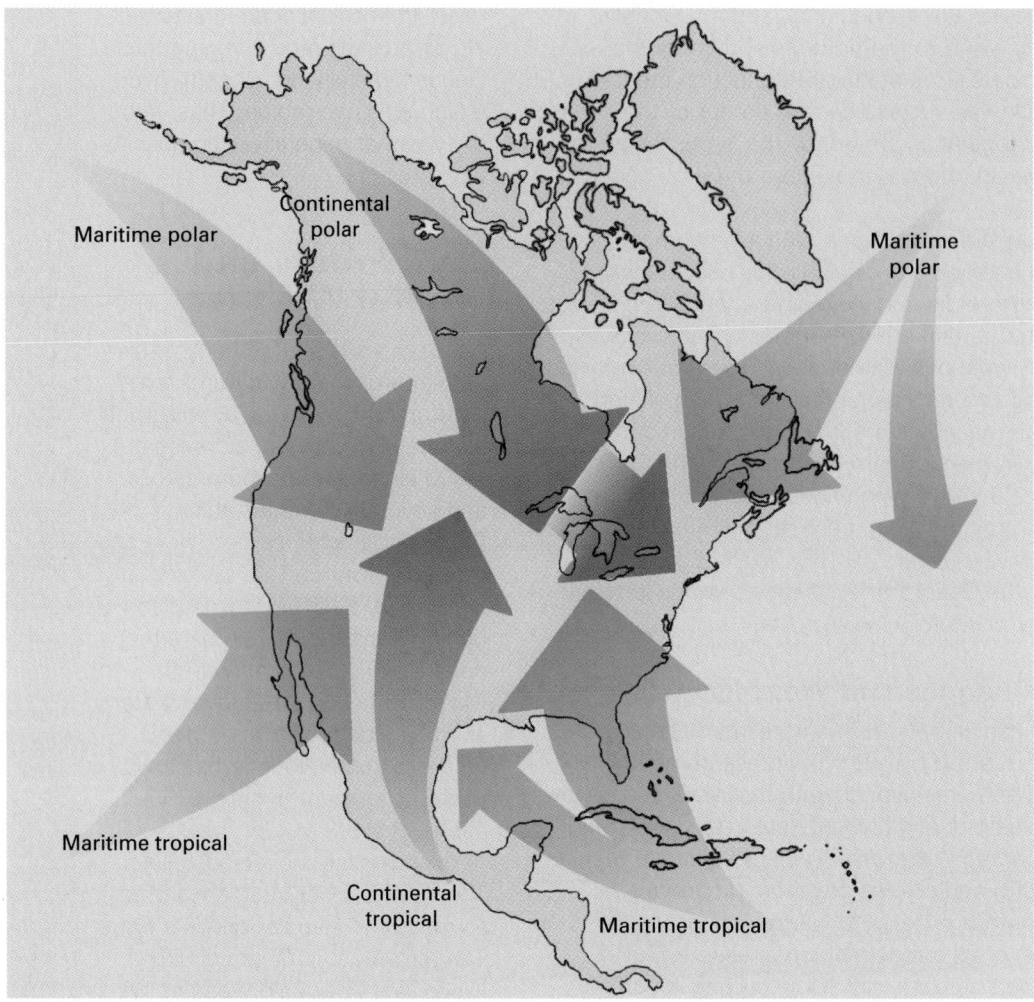

FIGURE 19.17 Air masses are classified by their source regions.

Showers   Rain   Snow

Fronts

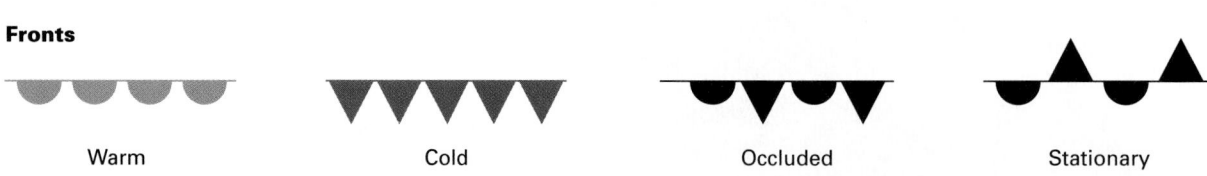

Warm   Cold   Occluded   Stationary

FIGURE 19.18 Symbols commonly used in weather maps. Warm and cold are relative terms. Air over the central plains of Montana at a temperature of 0°C may be warm relative to polar air above northern Canada but cold relative to a 20°C air mass over the southeastern United States.

slower-moving cold air mass. A **cold front** forms when moving cold air collides with stationary or slower-moving warm air.

In a warm front, the moving warm air rises over the denser cold air as the two masses collide (Figure 19.19). The rising warm air cools adiabatically and the cooling generates clouds and precipitation. Precipitation is generally light because the air rises slowly along the gently sloping frontal boundary. Figure 19.19 shows that a characteristic sequence of clouds accompanies a warm front. High, wispy cirrus and cirrostratus clouds develop near the leading edge of the rising warm air. These high clouds commonly precede a storm. They form as much as 1,000 kilometers ahead of an advancing band of precipitation that falls from thick, low-lying

nimbostratus and stratus clouds near the trailing edge of the front. The cloudy weather may last for several days because of the gentle slope and broad extent of the frontal boundary.

A cold front forms when faster-moving cold air overtakes and displaces warm air. The dense cold air forms a blunt wedge and pushes under the warmer air (Figure 19.20). Thus the leading edge of a cold front is much steeper than that of a warm front. The steep contact between the two air masses

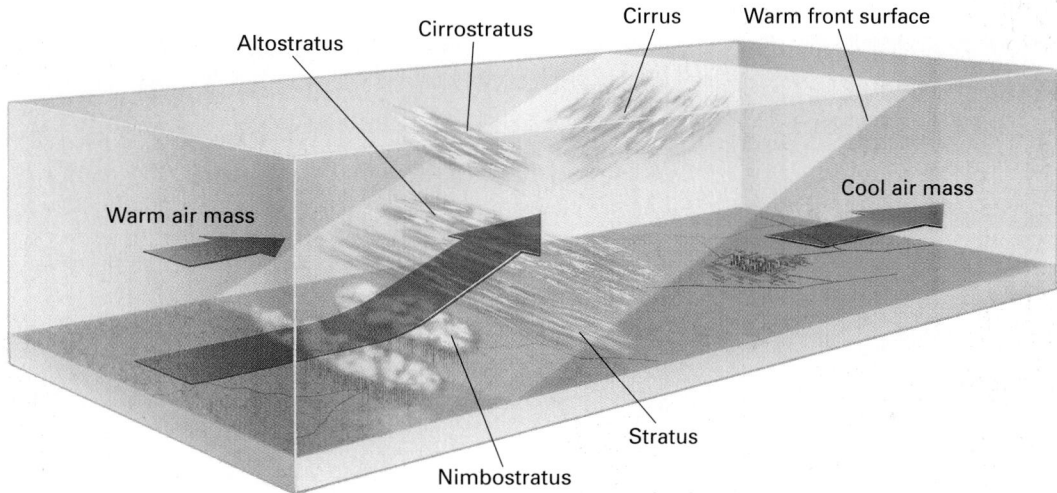

FIGURE 19.19 In a warm front, moving warm air rises gradually over cold air. High, wispy cirrus and cirrostratus clouds typically develop near the leading edge of the rising warm air, followed by descending altostratus, stratus, and nimbostratus clouds near the trailing edge of the front.

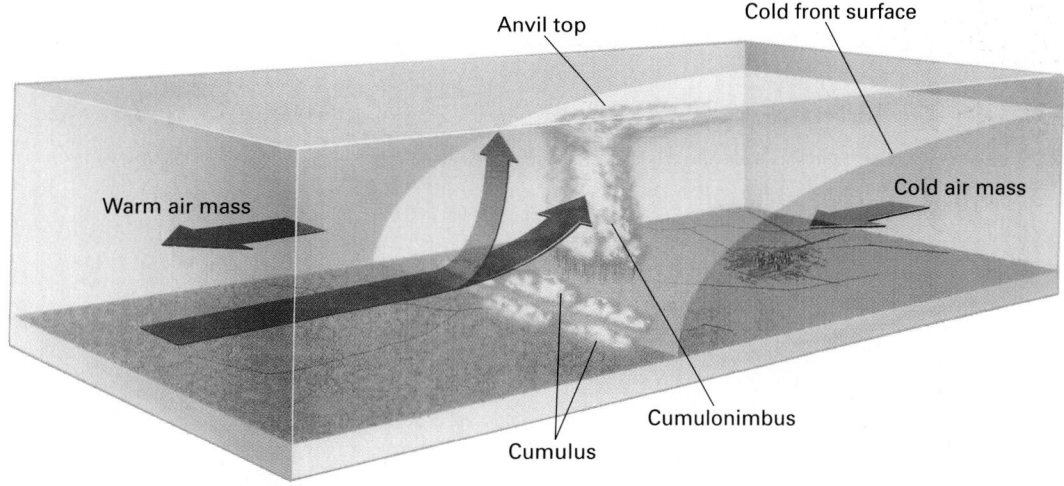

FIGURE 19.20 In a cold front, moving cold air slides abruptly beneath warm air, forcing it steeply upward and creating a narrow band of violent weather commonly accompanied by cumulus and cumulonimbus clouds.

causes the warm air to rise rapidly, creating a narrow band of violent weather commonly accompanied by cumulus and cumulonimbus clouds. The storm system may be only 25 to 100 kilometers wide, but within this zone downpours, thunderstorms, and violent winds are common.

Occluded Front

An **occluded front** forms when a faster-moving cold air mass traps a warm air mass against a second mass of cold air. Thus the warm air

mass becomes trapped between two colder air masses (Figure 19.21). The faster-moving cold air mass then slides beneath the warm air, lifting it completely off the ground. Precipitation occurs along both frontal boundaries, combining the narrow band of heavy precipitation of a cold front with the wider band of lighter precipitation of a warm front. The net result is a large zone of inclement weather. A storm of this type is commonly short-lived because the warm air mass is cut off from its supply of moisture evaporating from Earth's surface.

Stationary Front

A **stationary front** occurs along the boundary between two stationary air masses. Under these conditions, the front can remain over an area for several days. Warm

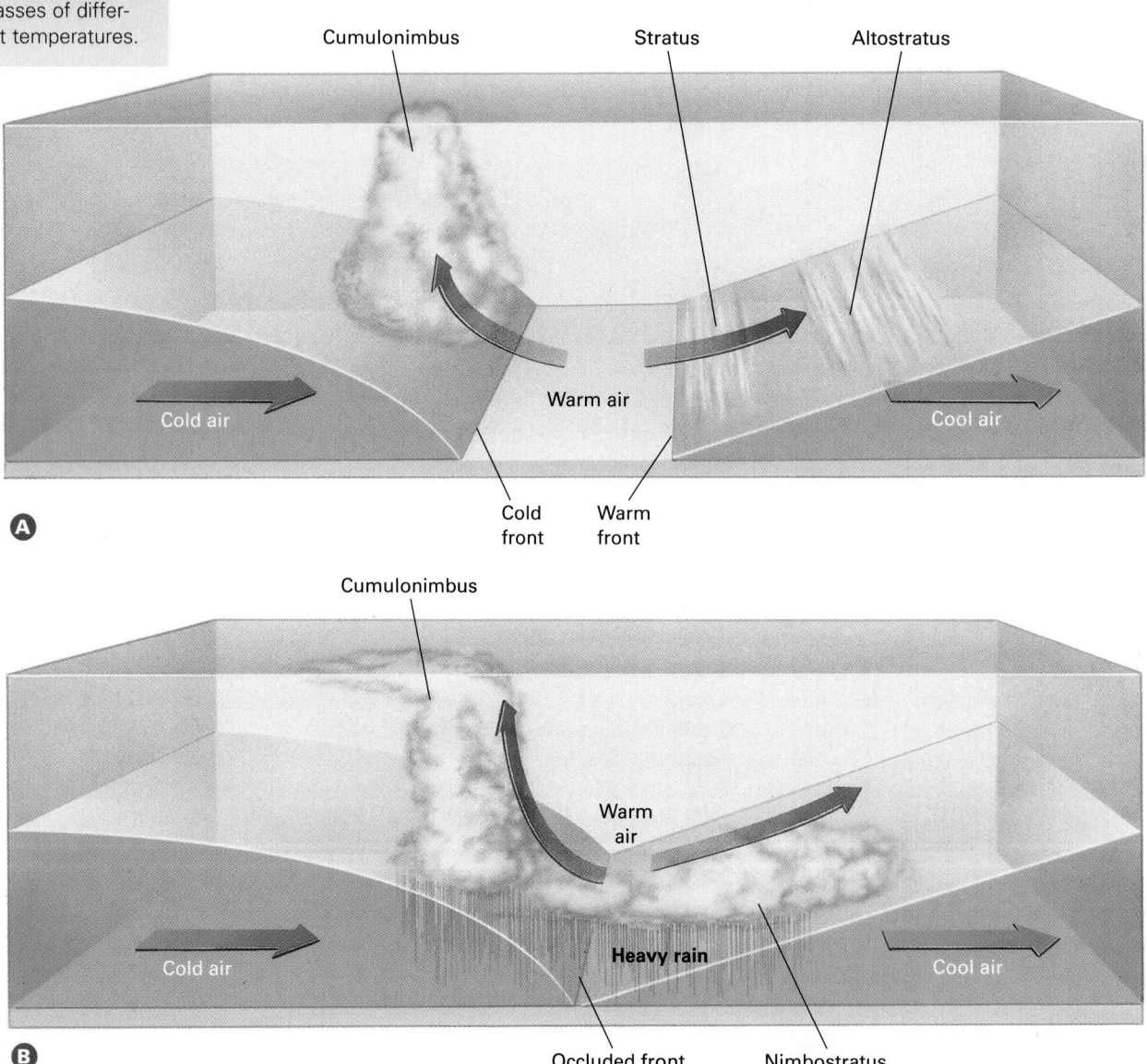

FIGURE 19.21 An occluded front forms where warm air is trapped and lifted between two cold air masses. Precipitation occurs along both frontal boundaries in a large zone of short-lived inclement weather.

air rises, forming conditions similar to those in a warm front. As a result, rain, drizzle, and fog may occur.

The Life Cycle of a Midlatitude Cyclone

As you learned in Section 19.6, a cyclone is a low-pressure system with rotating winds.[1] Most cyclones in the middle latitudes of the Northern Hemisphere develop along a front between polar and tropical air masses. The storm often starts with winds blowing in opposite directions along a stationary front between the two air masses (Figure 19.22A). In the figure, a warm air mass was moving northward and was deflected to the east by the Coriolis effect. At the same time, a cold air mass traveling southward was deflected to the west.

1. In common usage, the world cyclone is applied loosely to a variety of storm systems, and the usage varies from place to place. In this textbook, we apply the scientific usage.

In Figure 19.22B, the cold polar air continues to push southward, creating a cold front and lifting the warm air off the ground. Then, some small disturbance—a topographic feature such as a mountain range, airflow from a local storm, or perhaps a local temperature variation—deforms the straight frontal boundary, forming a wavelike kink in the front. Once the kink forms, the winds on both sides are deflected to strike the front at an angle. Thus, a warm front forms to the east and a cold front forms to the west.

Rising warm air then forms a low-pressure region near the kink (Figure 19.22C). In the Northern Hemisphere, the Coriolis effect causes the winds to circulate counterclockwise around the kink, as explained in Section 19.6. To the west, the cold front advances southward, and to the east, the warm front advances northward. At the same time, rain or snow falls from the rising warm air (Figure 19.22D). Over a period of one to three days, the air rushing into the low-pressure

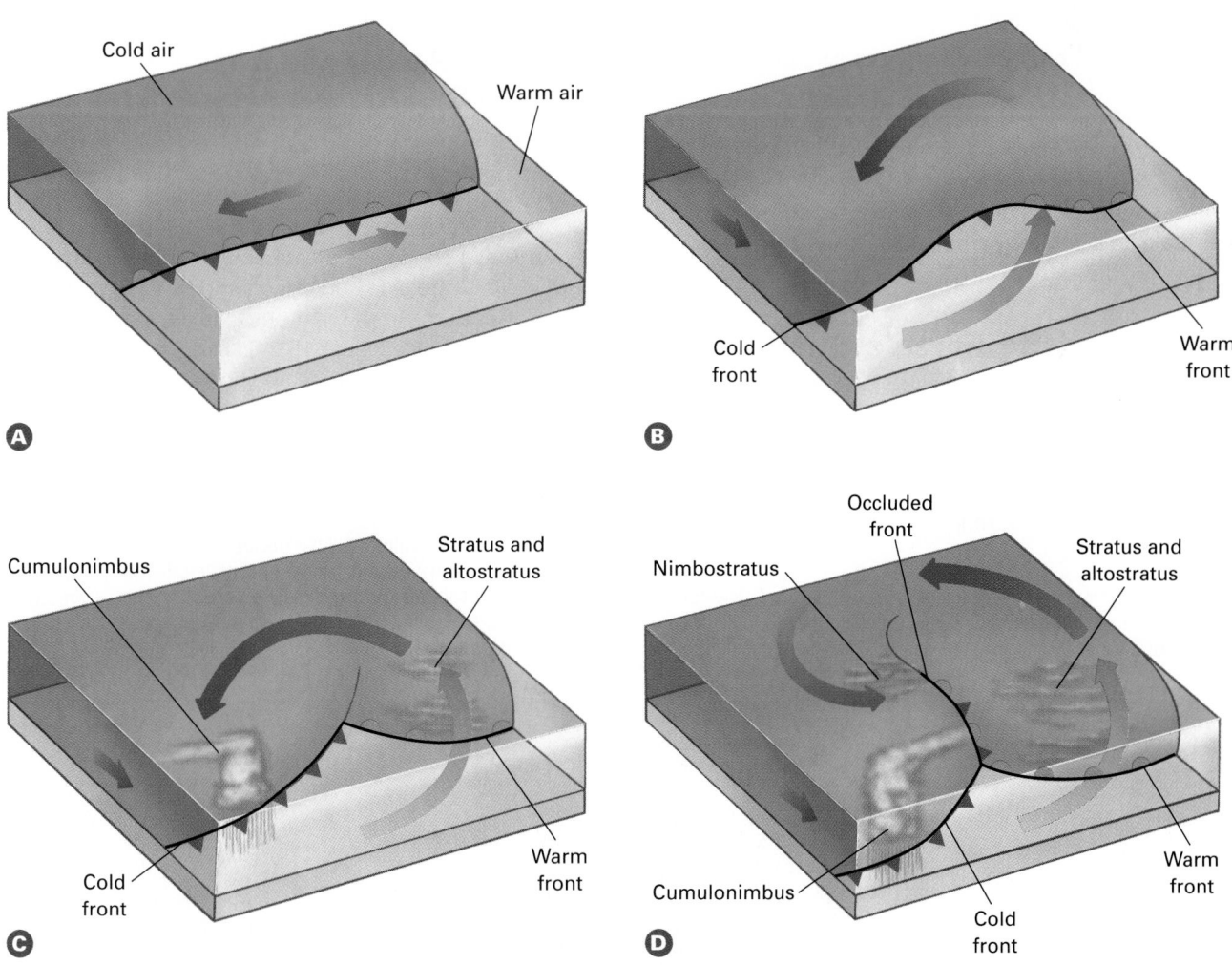

FIGURE 19.22 A midlatitude cyclone develops along a front between polar air and a tropical air mass. (A) The front develops. (B) Some small disturbance such as a topographical feature, nearby storm, or local temperature variation creates a kink in the front. (C) A low-pressure region and cyclonic circulation develop near the kink. (D) An occluded front forms.

Paths repeatedly followed by storms.

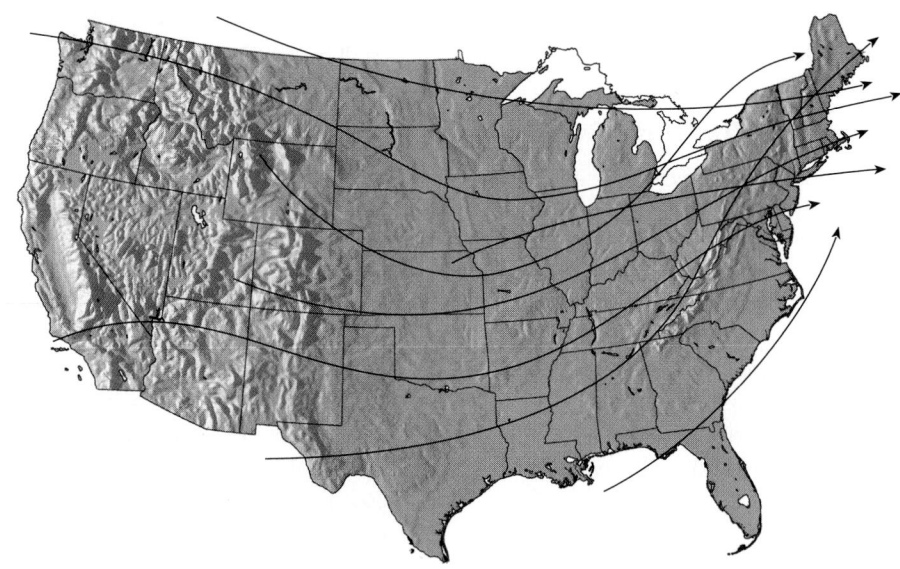

FIGURE 19.23 Most North American cyclones follow certain paths, called storm tracks, from west to east.

region equalizes pressure differences, and the storm dissipates.

Many of the pinwheel-shaped storms seen on weather maps are cyclones of this type. In North America, the jet stream and other prevailing, upper-level, westerly winds generally move cyclones from west to east along the same paths, called **storm tracks** (Figure 19.23).[2]

19.8 How the Earth's Surface Features Affect Weather

Earth's surface features—including mountain ranges, rainforests, proximity to the sea, and uneven heating and cooling of continents—can create conditions that affect the weather of a region.

_____
2. In common usage, the world cyclone is applied loosely to a wide variety of storm systems, and the usage varies from place to place. In this textbook, we apply the scientific usage.

Mountain Ranges and Rain-Shadow Deserts

As we described in Section 19.3, air rises in a process called *orographic lifting* when it flows over a mountain range. As the air rises, it cools adiabatically, and water vapor may condense into clouds that produce rain or snow. These conditions create abundant precipitation on the windward side and the crest of the range. When the air passes over the crest onto the leeward (downwind) side, it sinks (Figure 19.24). This air has already lost much of its moisture. In addition, it warms adiabatically as it falls, absorbing moisture and creating a rain-shadow desert on the leeward side of the range. For

FIGURE 19.24 A rain-shadow desert forms where moist air rises over a mountain range and precipitates most of its moisture on the windward side and crest of the range. The dry, descending air on the leeward side absorbs moisture, forming a desert.

Rising air generates low pressure, which leads to precipitation.

Warm, moist air rises.

Prevailing winds

Dry air descends, creating high-pressure zone.

Rain-shadow desert

example, Death Valley, California, is a rain-shadow desert and receives only 5 centimeters of rain a year, while the nearby west slope of the Sierra Nevada receives 178 centimeters of rain a year.

Forests and Weather

Recall that atmospheric moisture condenses when moist air cools below its dew point. Forests cool the air. Large quantities of water evaporate from leaf surfaces in the process called *transpiration* (Chapter 11), and evaporation cools the surrounding air. In addition, forests shade the soil from the hot sun, and tree roots and litter retain moisture. In an open clearing, rainwater evaporates quickly after a storm or runs off the surface. But forest soils remain moist long after the rain dissipates. Evaporation from soil litter combines with transpiration cooling from leaf surfaces to maintain relatively cool temperatures during times when there is no rain. In yet another feedback mechanism: Forests cool the air. Cool air promotes rainfall. Rainfall supports forests.

In today's tropical rainforests, local rainfall has decreased by as much as 50 percent when the forests were cut and replaced by farmland or pasture. When the rainfall decreases, wildfires ravage the boundary between the logged area and the remaining virgin forests. More forest is destroyed, establishing a negative feedback mechanism of increasing drought, fire, and forest loss.

Sea and Land Breezes

Anyone who has lived near an ocean or large lake has encountered winds blowing from water to land and from land to water. Sea and land breezes are caused by uneven heating and cooling of land and water. Recall that land surfaces heat up faster than adjacent bodies of water and cool more quickly. If land and sea are nearly the same temperature on a summer morning, during the day the land warms and heats the air above it. Hot air then rises over the land, producing a local low-pressure area. Cooler air from the sea flows inland to replace the rising air. Thus, on a hot sunny day, winds generally blow from the sea onto land. The rising air is good for flying kites or hang-gliding but often brings afternoon thunderstorms.

At night the reverse process occurs. The land cools faster than the sea, and descending air creates a local high-pressure area over the land. Then the winds reverse, and breezes blow from the shore out toward the sea.

Monsoons

A **monsoon** is a seasonal wind and weather system caused by uneven heating and cooling of continents and adjacent oceans. Just as sea and land breezes reverse direction with day and night, monsoons reverse direction with the seasons. In the summertime the continents become warmer than the sea. Warm air rises over land, creating a large low-pressure area and drawing moisture-laden maritime air inland. When the moist air rises as it flows over the land, clouds form and heavy monsoon rains fall. In winter the process is reversed. The land cools below the sea temperature, and as a result air descends over land, producing dry, continental, high pressure. At the same time, air rises over the ocean and the prevailing winds blow from land to sea. More than half of the inhabitants of Earth depend on monsoons because the predictable, heavy, summer rains bring water to the fields of Africa and Asia. If the monsoons fail to arrive, crops cannot grow and people starve.

monsoon A seasonal wind and weather system caused by uneven heating and cooling of land and adjacent sea, generally blowing from the sea to the land in the summer when the continents are warmer than the ocean, and from land to sea in winter when the ocean is warmer than the land.

19.9 Thunderstorms

An estimated 16 million thunderstorms occur every year, and at any given moment about 2,000 thunderstorms are in progress over different parts of Earth. A single bolt of lightning can involve several hundred million volts of energy and for a few seconds produces as much power as a nuclear power plant. It heats the surrounding air to 25,000°C or more, much hotter than the surface of the Sun. The heated air expands instantaneously to create a shock wave that we hear as thunder.

Despite their violence, thunderstorms are local systems, often too small to be included on national weather maps. A typical thunderstorm forms and then dissipates in a few hours and covers from about 10 to a few hundred square kilometers. It is not unusual to stand on a hilltop in the sunshine and watch rain squalls and lightning a few kilometers away. All thunderstorms develop when warm, moist air rises, forming cumulus clouds that develop into towering cumulonimbus clouds. Different conditions cause these local regions of rising air:

1. *Wind convergence.* Central Florida is the most active thunderstorm region in the United States. As the subtropical Sun heats the Florida peninsula, rising air draws moist air from both its east and west coasts. Where the two air masses converge, the moist air rises rapidly to create a thunderstorm. Thunderstorms also occur in other environments where moist air masses converge.

2. *Convection.* Thunderstorms also form in continental interiors during the spring or summer,

when afternoon sunshine heats the ground and generates cells of rising, moist air.

3. *Orographic lifting.* Moist air rises as it flows over hills and mountain ranges, commonly generating mountain thunderstorms.

4. *Frontal thunderstorms.* Thunderstorms commonly occur along frontal boundaries, particularly at cold fronts.

Lightning

Lightning is an intense discharge of electricity that occurs when the buildup of static electricity overwhelms the insulating properties of air (Figure 19.25). If you walk across a carpet on a dry day, the friction between your feet and the rug shears electrons off the atoms on the rug. The electrons migrate into your body and concentrate there. If you then touch a metal doorknob, a spark consisting of many electrons jumps from your finger to the metal knob.

In 1752 Benjamin Franklin showed that lightning is an electrical spark. He suggested that charges separate within cumulonimbus clouds and build until a bolt of lightning jumps from the cloud. In the more than 250 years since Franklin, atmospheric physicists have been unable to agree upon the exact mechanism of lightning. According to one hypothesis, friction between the intense winds and moving ice crystals in a cumulonimbus cloud generates both positive and negative electrical charges in the cloud, and the two types of charges become physically separated (Figure 19.26A). The positive charges tend to accumulate in the upper portion of the cloud, and the negative charges build up in the lower reaches of the cloud. When enough charge accumulates, the electrical potential exceeds the insulating properties of air, and a spark jumps from the cloud to

A

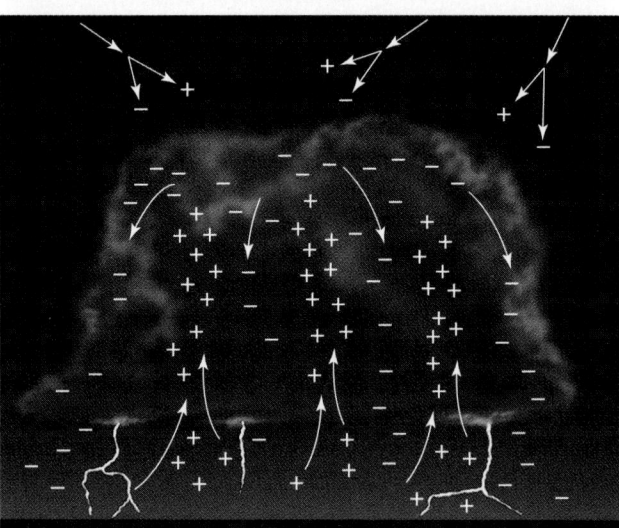

B

FIGURE 19.26 Two hypotheses for the origin of lightning. (A) Friction between intense winds and ice particles generates charge separation. (B) Charged particles are produced from above by cosmic rays and below by interactions with the ground. The particles are then distributed by convection currents.

NOAA PHOTO LIBRARY. OAR/ERL/NATIONAL SEVERE STORMS LABORATORY

FIGURE 19.25 This time-lapse photo captures multiple cloud-to-ground lightning strikes during a nighttime thunderstorm in Norman, Oklahoma, in March 1978.

the ground, from the ground to the cloud, or from one cloud to another.

Another hypothesis suggests that cosmic rays bombarding the cloud from outer space produce ions at the top of the cloud. Other ions form on the ground as winds blow over Earth's surface. The electrical discharge occurs when the potential difference between the two groups of electrical charges exceeds the insulating properties of air (Figure 19.26B).

Perhaps neither hypothesis is entirely correct and some combination of the mechanisms causes lightning.

19.10 Tornadoes and Tropical Cyclones

Tornadoes and tropical cyclones are both intense, low-pressure centers. Strong winds follow the steep pressure gradients and spiral inward toward a central column of rising air.

Tornadoes

A **tornado** is a small, short-lived, funnel-shaped storm that protrudes from the base of a cumulonimbus cloud (Figure 19.27A). The base of the funnel can be from 2 meters to 3 kilometers in diameter. Some tornadoes remain suspended in air while others touch the ground. After a tornado touches ground, it may travel for a few meters to a few hundred kilometers across the surface. The funnel travels at 40 to 65 kilometers per hour, and in some cases as much as 110 kilometers per hour, but the spiraling winds within the funnel are much faster. Few direct measurements have been made of pressure and wind speed inside a tornado. However, we know that a large pressure difference occurs over a very short distance. Meteorologists estimate that winds in tornadoes may reach 500 kilometers per hour or greater. These winds rush into the narrow, low-pressure zone and then spiral upward. After a few seconds to a few hours, the tornado lifts off the ground and dissipates.

Tornadoes are the most violent of all storms. One tornado in 1910 lifted a team of horses and then deposited it, unhurt, several hundred meters away. They were lucky. In the past, an average of 120 Americans were killed every year by these storms, and property damage costs millions of dollars. The death toll has decreased in recent years because effective warning systems allow people to seek shelter, but the property damage has continued to increase (Figure 19.27B). Tornado winds can lift the roof off a house and then flatten the walls. Flying debris kills people and livestock caught in the open. Even so, the total destruction from tornadoes is not as great as that from hurricanes because the path of a tornado is narrow and its duration short.

Although tornadoes can occur anywhere in the world, 75 percent of the world's twisters concentrate in the Great Plains, east of the Rocky Mountains. Approximately 700 to 1,000 tornadoes occur in the United States each year. They frequently form in the spring or early summer. At that time, continental polar (dry, cold) air from Canada collides with maritime tropical (warm, moist) air from the Gulf of Mexico. As explained previously, these conditions commonly create thunderstorms. Meteorologists cannot explain why most thunderstorms dissipate harmlessly but a few develop tornadoes. However, one fact is apparent: tornadoes are most likely to occur when large differences in temperature and moisture exist between the two air masses and the boundary between them is sharp.

Tropical Cyclones

A **tropical cyclone** or **tropical storm** is less intense than a tornado but much larger and longer-lived (Table 19.2). Tropical cyclones are circular disturbances that average 600 kilometers in diameter and persist for days or weeks. If the wind

> **tornado** A small, intense, short-lived, funnel-shaped storm that protrudes from the base of a cumulonimbus cloud.
>
> **tropical cyclone** *or* **tropical storm** A broad, circular storm with intense low pressure that forms over warm oceans.

FIGURE 19.27 (A) The dark funnel cloud of a tornado descends on Dimmitt, Texas, on June 2, 1995. (B) Tornadoes that accompanied Hurricane Andrew added to the terror and confusion at La Place, Louisiana, where two people died and many homes were destroyed.

hurricane A tropical storm occurring in North America or the Caribbean whose wind exceeds 120 kilometers per hour; called a *typhoon* in the western Pacific and a *cyclone* in the Indian Ocean.

typhoon A tropical storm occurring in the western Pacific Ocean whose wind exceeds 120 kilometers per hour; called a *hurricane* in North America and the Caribbean and a *cyclone* in the Indian Ocean.

storm surge Abnormally high coastal waters and flooding, created by a combination of strong onshore winds and the low atmospheric pressure of a storm that raises the sea surface by several meters.

TABLE 19.2 Comparison of Tornadoes and Tropical Cyclones

Feature	Range	
	Tornado	Tropical Cyclone
Diameter	2 to 3 km	400 to 800 km
Path length (distance traveled across terrain)	A few meters to hundreds of kilometers	A few hundred to a few thousand kilometers
Duration	A few seconds to a few hours	A few days to a week
Wind speed	300 to 800 km/hr	60 to 120 km/hr (tropical storm); 120 to 300 km/hr (hurricane)
Speed of motion	40 to 110 km/hr	20 to 30 km/hr
Pressure fall	20 to 200 millibars	20 to 60 millibars

NASA-GSC

FIGURE 19.28 This image shows Hurricane Katrina making landfall near New Orleans on August 28, 2005.

exceeds 120 kilometers per hour, a tropical cyclone is called a **hurricane** in North America and the Caribbean, a **typhoon** in the western Pacific, and a *cyclone* in the Indian Ocean (Figure 19.28). Intense low pressure in the center of a hurricane can generate wind as strong as 300 kilometers per hour.

The low atmospheric pressure created by a tropical cyclone can raise the sea surface by several meters. Often, as a tropical cyclone strikes shore, strong onshore winds combine with the abnormally high water level created by low pressure to create a **storm surge** that floods coastal areas. In 2005 during Hurricane Katrina, sea level rose about 8.5 meters above normal on the Gulf Coast as a result of a storm surge.

Tropical cyclones form only over warm oceans, never over cold oceans or land. Thus, moist, warm air is crucial to the development of this type of storm. A

midlatitude cyclone develops when a small disturbance produces a wavelike kink in a previously linear front. A similar mechanism initiates a tropical cyclone. In late summer, the Sun warms tropical air. The rising hot air creates a belt of low pressure that encircles the globe over the tropics. In addition, many local low-pressure disturbances move across the tropical oceans at this time of year. If a local disturbance intersects the global tropical low, it creates a bulge in the isobars. Winds are deflected by the bulge and, directed by the Coriolis effect, begin to spiral inward. Warm, moist air rises from the low. Water vapor condenses from the rising air, and the latent heat warms the air further, which causes even more air to rise. As the low pressure becomes more intense, strong surface winds blow inward to replace the rising air. This surface air also rises, and more condensation and precipitation occur.

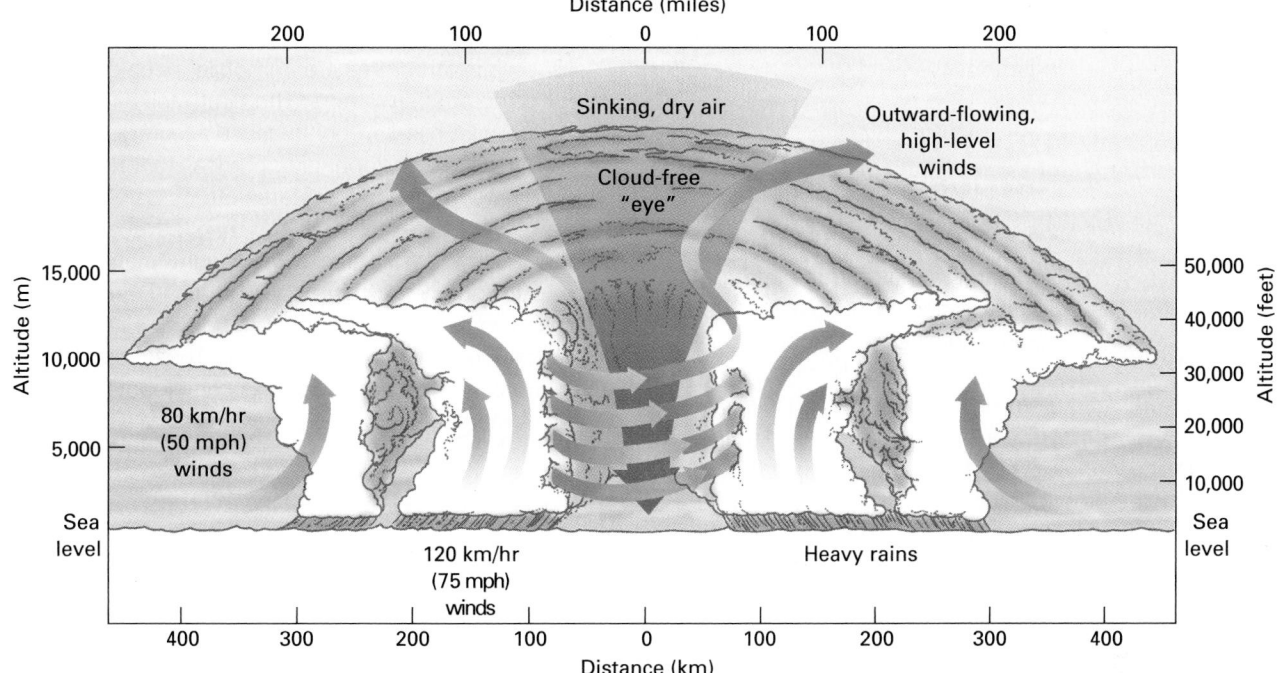

FIGURE 19.29 Surface air spirals inward toward a hurricane, rises through the towering wall of clouds, and then flows outward above the storm. Falling air near the storm's center creates the eerie calm in the eye of the hurricane.

But the additional condensation releases more heat, which continues to add energy to the storm.

The center of the storm is a region of vertical airflow, called the *eye*. In the outer, and larger, part of the eye, the air that has been rushing inward spirals upward. In the inner eye, air sinks. Thus, the horizontal wind speed in the eye is reduced to near zero (Figure 19.29). Survivors who have been in the eye of a hurricane report an eerie calm. Rain stops, and the Sun may even shine weakly through scattered clouds. But this is only a momentary reprieve. A typical eye is only 20 kilometers in diameter, and after it passes the hurricane rages again in full intensity.

Thus, a hurricane is powered by a classic feedback mechanism: the low-pressure storm causes conden-

sation; condensation releases heat; heat powers the continued low pressure. The entire storm is pushed by prevailing winds, and its path is deflected by the Coriolis effect. A hurricane usually dissipates only after it reaches land or passes over colder water because the supply of moist, warm air is cut off. Condensing water vapor in a single tropical cyclone releases as much latent heat energy as that produced by all the electric generators in the United States in a six-month period.

The numerical "category" of tropical cyclones is based on a rating scheme called the *Saffir–Simpson scale*, after its developers (Table 19.3). This scale, commonly mentioned in weather reports, rates the damage potential of a hurricane or other tropical storm, and typical values of atmospheric pressure, wind

TABLE 19.3 The Saffir–Simpson Hurricane Damage Potential Scale

Type	Category	Damage	Pressure (millibars)	Winds (km/h)	Storm surge (m)
Depression				> 56	
Tropical storm				63 to 117	
Hurricane	1	minimal	980	119 to 152	1.2 to 1.5
Hurricane	2	moderate	965 to 979	154 to 179	1.8 to 2.4
Hurricane	3	extensive	945 to 964	179 to 209	2.7 to 3.7
Hurricane	4	extreme	920 to 944	211 to 249	4.0 to 5.5
Hurricane	5	catastrophic	< 920	> 249	> 5.5

speed, and height of storm surge associated with storms of increasing intensity.

19.11 Hurricane Katrina

Scope of the Disaster

Hurricane Katrina formed in the South Atlantic on August 23, 2005, made landfall as a category 1 hurricane just north of Miami, then crossed the Florida peninsula and grew to a category 5 storm in the Gulf of Mexico with sustained winds of 280 kilometers per hour. By the time it made a second landfall on the Mississippi Delta, Hurricane Katrina had diminished to a category 3–4 storm, with winds hovering around 200 kilometers per hour. Yet despite the reduced winds, Katrina quickly became the most costly and one of the deadliest natural disasters in U.S. history. Parts of New Orleans were flooded to the rooftops of houses. In addition, a record 8.5-meter storm surge inundated the New Orleans and Mississippi coastlines, flattening and flooding homes over 233,000 square kilometers, an area almost as large as the United Kingdom. Approximately 1.3 million people were driven from their homes. The official death toll was 1,836 with more than 700 people still unaccounted for. The direct structural damage was estimated to cost the United States as much as $110 billion. Of course, indirect costs such as lost wages and disruption of human life are much higher still, running into the hundreds of billions of dollars.

Brief History of Gulf Hurricanes

Hurricanes have ravaged the southeast United States numerous times in the past and will certainly do so again. But while hurricanes are not preventable, hurricane deaths and costs result from a combination of factors, including the severity of the hurricane, building practices, and human responses. For example, from Table 19.4 we see that prior to Hurricane Katrina, the deadliest U.S. hurricanes all occurred before 1957. However, the costliest storms all occurred after 1955. With the exception of Katrina, none of the 10 deadliest storms are on the list of costliest storms, and vice versa. Why is there so little correlation between structural damage (cost) and death toll?

In the last half-century, structural damage has been high for three main reasons: (1) more Americans lived near the Atlantic and Gulf coasts in the late 1990s and early 2000s than they had previously; (2) homes and other structures have had an increased inflation-adjusted value; and (3) Americans now own more things than

ever before and, consequently, the dollar value of their possessions is at an all-time high.

But while structures are immobile, people can evacuate if given ample warning. The death toll has been low in the past half-century because accurate forecasting warns people of impending hurricanes. The deadliest hurricane in the United States struck Galveston, Texas, in September 1900. More than 8,000 people died because the population was caught unaware. A tropical cyclone has a sharp boundary, and even a few hundred kilometers outside that boundary, fluffy white clouds may be floating in a blue sky. Today, satellites track hurricanes and news reports give people ample time to evacuate.

19.12 El Niño

Hurricanes are common in the southeastern United States and on the Gulf Coast, but they rarely strike California. Consequently, Californians were taken by surprise in late September 1997 when Hurricane Nora ravaged Baja California and then, somewhat diminished by landfall, struck San Diego and Los Angeles. The storm brought the first rain to Los Angeles after a record 219 days of drought, then spread eastward to flood parts of Arizona, where it caused the evacuation of 1,000 people.

Other parts of the world also experienced unusual weather during the autumn of 1997. In Indonesia and Malaysia, fall monsoon rains normally douse fires intentionally set in late summer to clear the rainforest. The rains were delayed for two months in 1997, and as a result the fires raged out of control, filling cities with such dense smoke that visibility at times was no more than a few meters. Even an airliner crash was attributed to the smoke. Severe drought in nearby Australia caused ranchers to slaughter entire herds of cattle for lack of water and feed. At the same time, far fewer hurricanes than usual threatened Florida and the U.S. Gulf Coast. Floods soaked northern Chile's Atacama Desert, a region that commonly receives no rain at all for a decade at a time, while record snowfalls blanketed the Andes and heavy rains caused floods in Peru and Ecuador.

All of these weather anomalies have been attributed to El Niño, an ocean current that brings unusually warm water to the west coast of South America. But the current does not flow every year; instead, it occurs about every 3 to 7 years, and its effects last for about a year before conditions return to normal. Although meteorologists paid little attention to the phenomenon until the El Niño year of 1982–1983, many now think that El Niño affects weather patterns for nearly three-quarters of Earth.

TABLE 19.4 The Deadliest and Costliest Storms in U.S. History

Deadliest Storms				
Rank	Hurricane	Year	Category	Deaths*
1	Galveston, Texas	1900	4	8,000+
2	Florida	1928	4	2,500+
3	Katrina	2005	3–4	1,836+
4	Louisiana	1893	4	1,100+
5	South Carolina and Georgia	1893	3	1,000+
6	South Carolina and Georgia	1891	2	700
7	Florida Keys	1935	5	408
8	Louisiana	1856	4	400
9	Texas and Louisiana (Audrey)	1957	4	390
10	Florida, Alabama, and Mississippi	1926	4	372

Costliest Storms				
Rank	Hurricane	Year	Category	Damage (in billions)†
1	Katrina	2005	3–4	$75.0
2	Andrew (Florida, Louisiana)	1992	5	$43.7
3	Charley (Florida)	2004	4	$15.0
4	Ivan (Florida, Alabama)	2004	3	$14.2
5	Hugo (South Carolina)	1989	4	$12.3
6	Agnes (Florida, Northeast U.S.)	1972	1	$11.3
7	Betsy (Florida, Louisiana)	1965	3	$10.8
8	Frances (Florida)	2004	2	$8.8
9	Camille (Mississippi, Louisiana, Virginia)	1969	5	$8.9
10	Diane (Northeast U.S.)	1955	1	$7.0

*Does not include offshore deaths
†2004 dollars adjusted for inflation

Meteorologists first began recording El Niños in 1982–1983, although they were known to Peruvian fishermen much earlier because they warm coastal waters and diminish fish harvests. The fishermen called the warming events *El Niño* because they commonly occur around Christmas, the birthday of El Niño—the Christ Child.

To understand the El Niño effects, first consider interactions between southern Pacific sea currents and weather in a normal, non–El Niño year (Figure 19.30). Normally in fall and winter, strong trade winds blow westward from South America across the Pacific Ocean. The winds drag the warm, tropical surface water away from Peru and Chile and pile it up in the western Pacific near Indonesia and Australia. In the western Pacific, the warm water forms a low mound thousands of kilometers across. The water is up to 10°C warmer and as much as 60 centimeters higher than the surface of the ocean near Peru and Chile.

As the wind-driven surface water flows away from the South American coast, cold, nutrient-rich water rises from the depths to replace the surface water. The nutrients support a thriving fishing industry along the coasts of Peru and Chile.

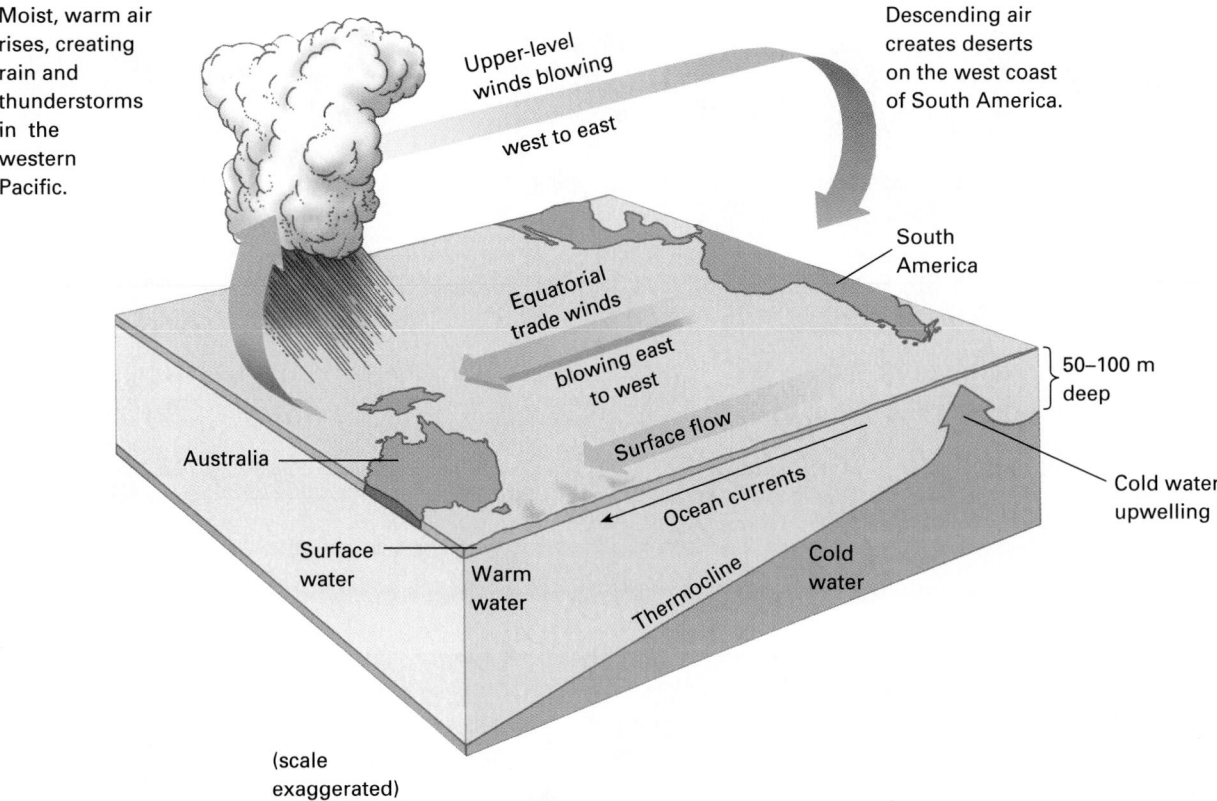

Moist, warm air rises, creating rain and thunderstorms in the western Pacific.

Upper-level winds blowing west to east

Descending air creates deserts on the west coast of South America.

South America

Equatorial trade winds blowing east to west

50–100 m deep

Cold water upwelling

Australia

Surface flow

Ocean currents

Surface water

Warm water

Thermocline

Cold water

(scale exaggerated)

FIGURE 19.30 In a normal year, trade winds drag warm surface water westward across the Pacific and pile it up in a low mound near Indonesia and Australia, where the warm water causes rain. The surface flow creates upwelling of cold, deep, nutrient-rich waters along the coast of South America.

Abundant moisture evaporates from the surface of a warm ocean. In a normal year, much of this water condenses to bring rain to Australia, Indonesia, and other lands in the southwestern Pacific, which are adjacent to the mound of warm water. On the eastern side of the Pacific, the cold, upwelling ocean currents cool the air above the coasts of Peru and northern Chile. This cool air becomes warmer as it flows over land. The warming lowers the relative humidity and creates the coastal Atacama Desert.

In an El Niño year, for reasons poorly understood by meteorologists, the trade winds slacken (Figure 19.31). The mound of warm water near Indonesia and Australia then flows downslope—eastward across the Pacific Ocean toward Peru and Chile. The anomalous accumulation of warm water off South America causes unusual rains in normally dry coastal regions, and heavy snowfall in the Andes. At the same time, the cooler water near Indonesia, Australia, and nearby regions causes drought. The mass of warm water that caused the 1997–1998 El Niño was about the size of the United States and created one of the strongest El Niño weather disturbances in history.

El Niño has global effects that go far beyond regional rainfall patterns. For example, El Niño deflects the jet stream from its normal path as it flows over North America, directing one branch northward over Canada and the other across southern California and Arizona. Consequently, those regions receive more winter precipitation and storms than usual, while fewer storms and warmer winter temperatures affect the Pacific Northwest, the northern plains, the Ohio River valley, the mid-Atlantic states, and New England. Southern Africa experiences drought, while Ecuador, Peru, Chile, southern Brazil, and Argentina receive more rain than usual.

Altered weather patterns created by El Niño wreak havoc. Globally, 2,000 deaths and more than $13 billion in damage are attributed to the 1982–1983 El Niño effects. In the United States alone, more than 160 deaths and $2 billion in damage occurred, mostly from storm and flood damage. In southern Africa, economic losses of $1 billion and uncounted deaths due to disease and starvation have been attributed to the 1982–1983 El Niño.

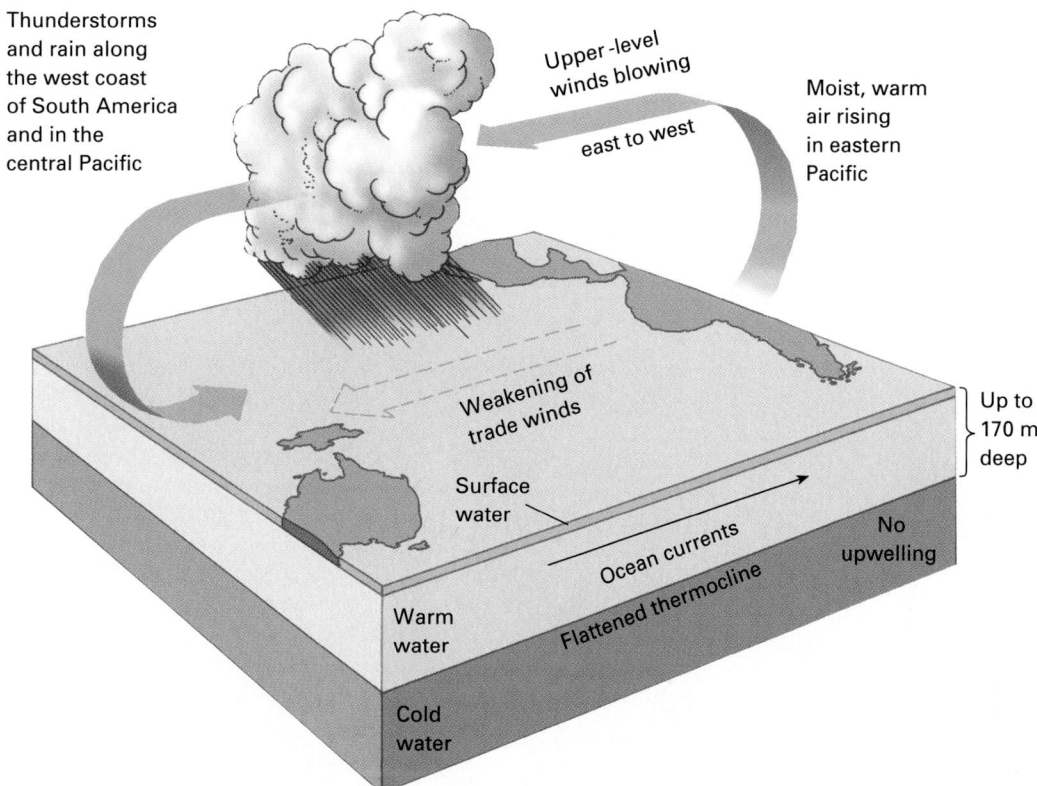

FIGURE 19.31 In an El Niño year, the trade winds slacken and the warm water flows eastward toward South America, causing the storms and rain to move over South America, and diminishing the upwelling currents.

20

CLIMATE

Summer in the Smoky Mountains, Tennessee.

> # Climate is what we expect, weather is what we get.
>
> *Mark Twain*

doldrums A vast low-pressure region of Earth near the equator with hot, humid air and where local squalls and rainstorms are common but steady winds are rare.

If you are planning a picnic for next Sunday, you hope for bright sunshine, but at the same time, you understand that your plans may be spoiled by rain. Next Sunday's temperature and precipitation are determined by daily fluctuations of weather. But weather varies only within a relatively narrow range. You are certain that it is not going to snow on the Fourth of July in Texas, and that if you live in Montana you are not going to bask outside in shorts and a T-shirt in January.

Over the time span of generations or centuries, climate is stable enough so that we plan our lives around it. Farmers in Kansas plant wheat and never attempt to raise bananas. In winter, hotel owners in Florida prepare for an influx of tourists escaping the northern winter. As a result, climate strongly influences many aspects of our lives: our outdoor recreation, the houses we live in, and the clothes we wear. Humans tend to migrate toward warm, sunny regions. In the United States, the populations of California and Texas have increased rapidly in the past generation, while North Dakota and Montana have seen little growth.

Yet such stability is only guaranteed over relatively short periods of geologic time. Dinosaur bones and coal deposits indicate that parts of Antarctica were once covered by warm, humid swamps. Desert dunes lie buried beneath the fertile wheat fields of Colorado. When geologists study the climate record in a specific region, they must separate the effects of tectonic movement from global climate change. For example, the coal deposits in Antarctica do not tell us that Earth was once warm enough for coal-producing swamps to grow near the South Pole. Instead, other evidence indicates that the continent of Antarctica was once close to the equator, as explained in our study of tectonics. Yet, even when the effects of tectonic plate movements are factored out, it is clear that global climate has changed dramatically and often abruptly over the long spans of Earth's history. In this chapter, we will examine today's global climate. Climate is regulated by many mechanisms that have already been discussed, including latitude (Chapter 18), wind (Chapters 18 and 19), oceans and ocean currents (Chapter 16 and 18), altitude (Chapter 17), and albedo (Chapter 18). We discuss global wind systems in more detail below.

20.1 Global Winds and Climate

The Sun is the ultimate energy source for winds and evaporation. The Sun shines most directly at or near the equator and warms the air near Earth's surface. The warm air gathers moisture from the equatorial oceans. The warm, moist, rising air forms a vast region of low pressure near the equator, with little horizontal airflow. As the rising air cools adiabatically, the water vapor condenses and falls as rain. Therefore, local squalls and thunderstorms are common, but steady winds are rare. This hot, still region was a serious barrier in the age of sailing ships. Mariners called the equatorial region the **doldrums** (from the Middle English for "dull"), and the old sailing literature is filled with stories recounting the despair and hardship of being becalmed on the vast, windless seas. On land, the frequent rains near the equatorial low-pressure zone nurture lush tropical rainforests.

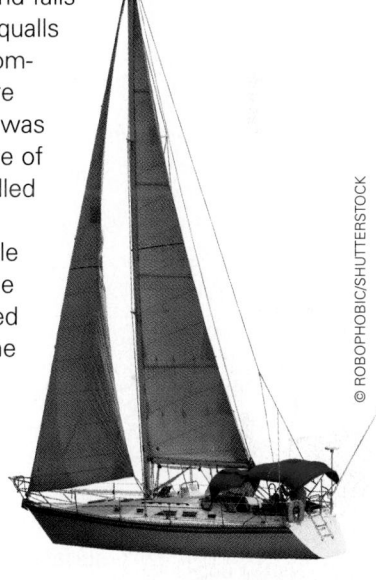

© ROBOPHOBIC/SHUTTERSTOCK

The air rising at the equator splits to flow north and south at high altitudes. However, these high-altitude winds do not continue to flow due north and south, but rather are deflected by the Coriolis effect. Thus their poleward movement is interrupted.

In both the Northern Hemisphere and Southern Hemisphere, this air veers until it flows due east at about 30° north and south latitudes (Figure 20.1). The air then cools enough to sink to the surface, creating subtropical high-pressure zones at 30° north and south latitudes. The sinking air warms adiabatically, absorbing water and forming clear, blue skies. At the center of the high-pressure area, the air moves vertically and not horizontally, and therefore few steady surface winds blow. This calm, high-pressure belt circling the

JACQUES DESCLOITRES, MODIS RAPID RESPONSE TEAM, NASA/GSFC

horse latitudes
A calm, high-pressure region of Earth lying at about 30° north and south latitudes, in which generally dry conditions prevail and steady winds are rare.

trade winds The winds that blow steadily toward the equator from the northeast in the Northern Hemisphere and from the southeast in the Southern Hemisphere, between 5° and 30° north and south latitudes.

prevailing westerlies The winds that blow steadily toward the poles from the southwest in the Northern Hemisphere and from the northwest in the Southern Hemisphere, between 30° and 60° north and south latitudes.

polar easterlies Persistent polar surface winds in the Northern Hemisphere that flow from east to west.

polar front The low-pressure boundary at about 60° latitude formed by warm air rising at the convergence of the polar easterlies and prevailing westerlies.

globe is called the **horse latitudes**. The region was so named because sailing ships were becalmed and horses transported as cargo often died of thirst and hunger. The warm, dry, descending air in this high-pressure zone forms many of the world's great deserts, including the Sahara in northern Africa, the Kalahari in southern Africa, and the Australian interior desert.

Descending air at the horse latitudes splits and flows over Earth's surface in two directions: toward the equator and toward the poles. The surface winds moving toward the equator are deflected by the Coriolis effect, so they blow from the northeast in the Northern Hemisphere and from the southeast in the Southern Hemisphere. Sailors depended on these reliable winds and called them the **trade winds**. The winds moving toward the poles are also deflected by the Coriolis effect, forming the **prevailing westerlies**. They flow from the southwest in the Northern Hemisphere and from the northwest in the Southern Hemisphere (Figure 20.1A).

The poles are cold year-round. The cold polar air sinks, creating yet another band of high pressure. The sinking air flows over the surface toward lower latitudes. In the Northern Hemisphere these surface winds are deflected by the Coriolis effect to form the **polar easterlies**. The polar easterlies and prevailing westerlies converge at about 60° latitude. Warm air rises at the convergence, forming a low-pressure boundary zone called the **polar front**.

These global wind patterns follow the **three-cell model**, which depicts three convection cells in each hemisphere. The cells are bordered by alternating bands of high and low pressure (Figure 20.1B). In the three-cell model, global winds are generated by heat-driven convection currents, and then their direction is altered by Earth's rotation (Coriolis effect).

Because the rising and falling air of global wind systems is dependent on temperature, these boundar-

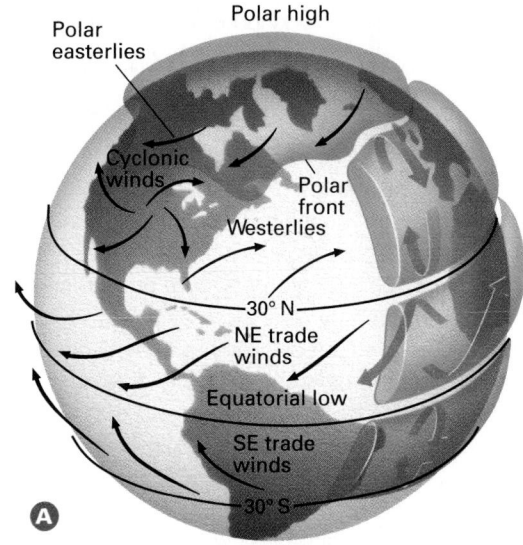

A

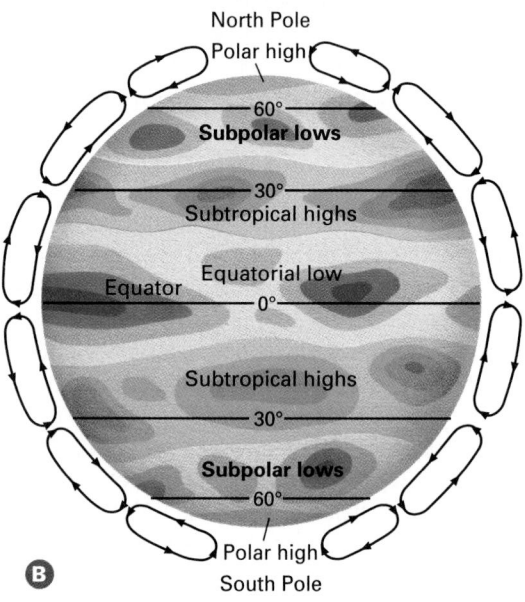

B

FIGURE 20.1 Global wind patterns predicted by the three-cell model. (A) Air rising at the equator moves poleward at high elevations, falls at about 30° north and south latitudes, and returns to the equator, forming the trade winds. The orange arrows show both upper-level and surface wind patterns. The black arrows show only surface winds. (B) High- and low-pressure belts are indicated on the sphere, with surface and upper-level wind patterns shown on the edges.

ies migrate north and south with the seasons. They are also distorted by surface topography and local air movement. For example, in the Northern Hemisphere, cyclones and anticyclones develop along the polar front as explained in Chapter 19. These storms bring alternating rain and sunshine, conditions that are favorable for agriculture. Thus the great wheat belts of the United States, Canada, and Russia all lie between 30° and 60° north latitude.

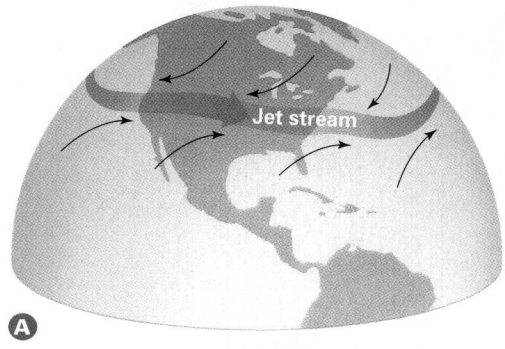

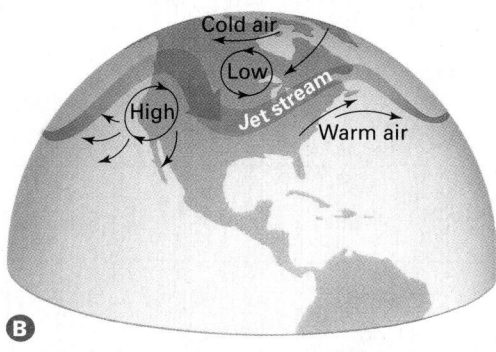

FIGURE 20.2 The polar front and the polar jet stream migrate with the seasons and with local conditions. Storms commonly occur along the jet stream.

Recall from Chapter 19 that a jet stream is a narrow band of fast-moving, high-altitude air. Jet streams form at boundaries between Earth's climate cells as high-altitude air is deflected by the Coriolis effect. The **subtropical jet stream** flows between the trade winds and the westerlies, and the **polar jet stream** forms along the polar front. When you watch a weather forecast on TV, the meteorologist may show the movement and direction of the polar jet stream as it snakes across North America. Storms commonly occur along this line because the jet stream marks the boundary between cold, polar air and the warm, moist, westerly flow that originates in the subtropics. The storms develop where the two contrasting air masses converge (Figure 20.2).

20.2 Climate Zones of Earth

Earth's major climate zones are classified primarily by temperature and precipitation. But an area with both wet and dry seasons has a different climate from one with moderate rainfall all year long, even though the two areas may have identical total annual precipitation. Therefore, climatic zones are also classified on the basis of seasonal variations in temperature and precipitation.

The **Koeppen Climate Classification**, used by climatologists throughout the world, is illustrated in Table 20.1. Subclassifications of these groups are shown in Table 20.2, and Earth's climate zones based on this classification system are shown on the map in Figure 20.3.

three-cell model A model of global wind patterns that depicts three convection cells in each hemisphere, bordered by alternating bands of high and low pressure.

subtropical jet stream A jet stream that flows between the trade winds and the westerlies.

polar jet stream A jet stream that flows along the polar front.

Koeppen Climate Classification A climate classification system describing Earth's principal climate zones, used by climatologists throughout the world.

TABLE 20.1 Koeppen Climate Classification

Climate Type	Name	Description
A	Humid tropical	In A climates, every month is warm with a mean temperature over 18°C (64°F). The temperature difference between day and night is greater than the difference between December and June averages. There is enough moisture to support abundant plant communities.
B	Arid	B climates have a chronic water deficiency; in most months evaporation exceeds precipitation. Temperatures vary according to latitude: some B climates are hot while others are frigid.
C	Humid mesothermal	C climates occur in midlatitudes, with distinct winter and summer seasons and enough moisture to support abundant plant communities. The winters are mild. Snow may fall but snow cover does not persist, with the average temperature in the coldest month above –3°C.
D	Humid microthermal	D climates are similar to C climates, with distinct summer and winter seasons, but D climates are colder. Winters are more severe, with persistent winter snow cover and an average temperature in the coldest month below –3°C.
E	Polar	In E climates, winters are extremely cold and even the summers are cool, with the average temperature in the warmest month below 10°C (50°F).
F	Highlands	F climates, found at latitudes worldwide, are determined by elevation above Earth's surface. Temperature changes with altitude, ranging from about –18°C to +10°C. Precipitation tends to decrease with altitude; windward sides of mountains usually receive more precipitation than leeward sides.

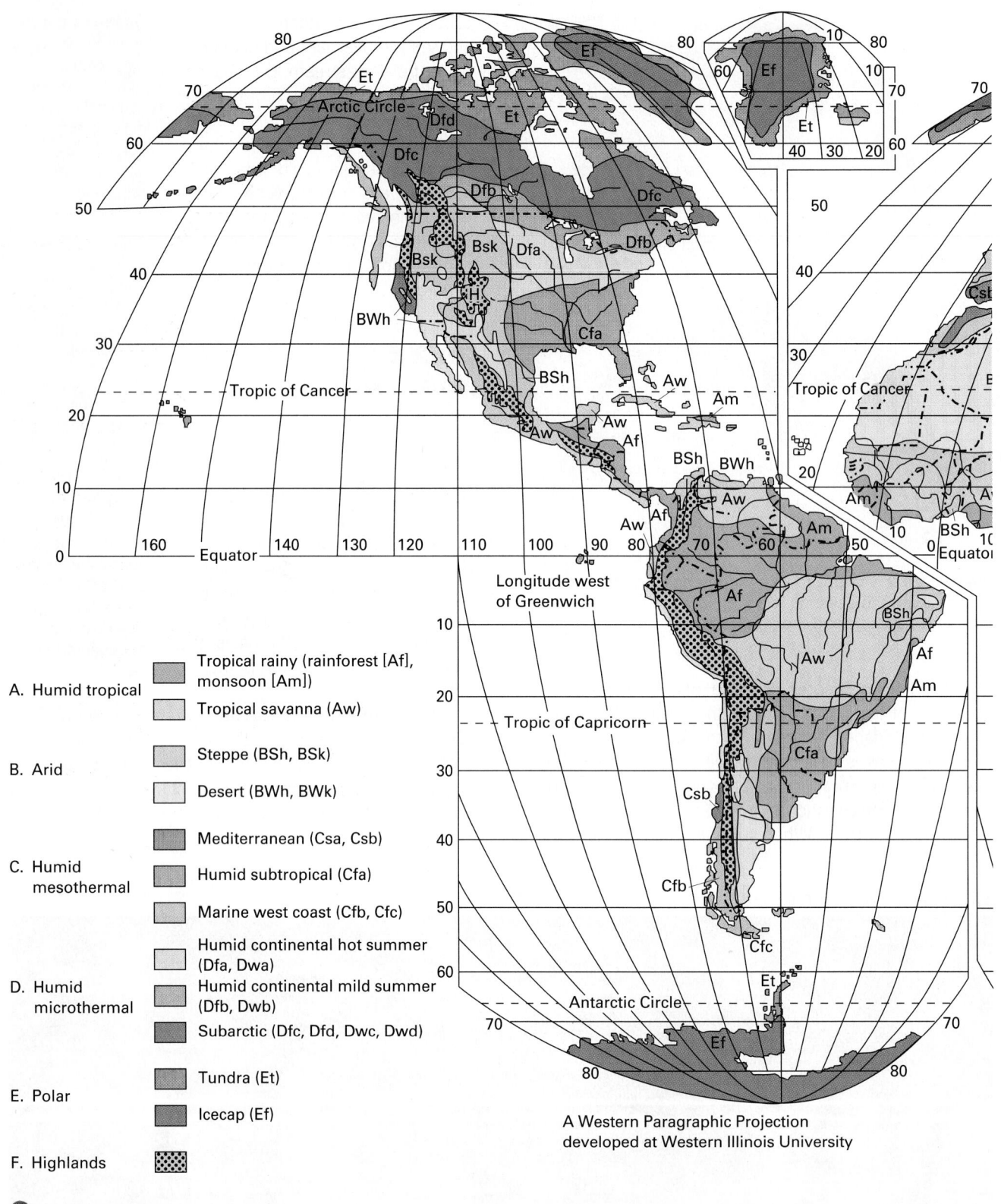

A. Humid tropical
- Tropical rainy (rainforest [Af], monsoon [Am])
- Tropical savanna (Aw)

B. Arid
- Steppe (BSh, BSk)
- Desert (BWh, BWk)

C. Humid mesothermal
- Mediterranean (Csa, Csb)
- Humid subtropical (Cfa)
- Marine west coast (Cfb, Cfc)

D. Humid microthermal
- Humid continental hot summer (Dfa, Dwa)
- Humid continental mild summer (Dfb, Dwb)
- Subarctic (Dfc, Dfd, Dwc, Dwd)

E. Polar
- Tundra (Et)
- Icecap (Ef)

F. Highlands

A Western Paragraphic Projection developed at Western Illinois University

A

FIGURE 20.3 (A) Global climate zones. Each climate zone supports a unique biome.

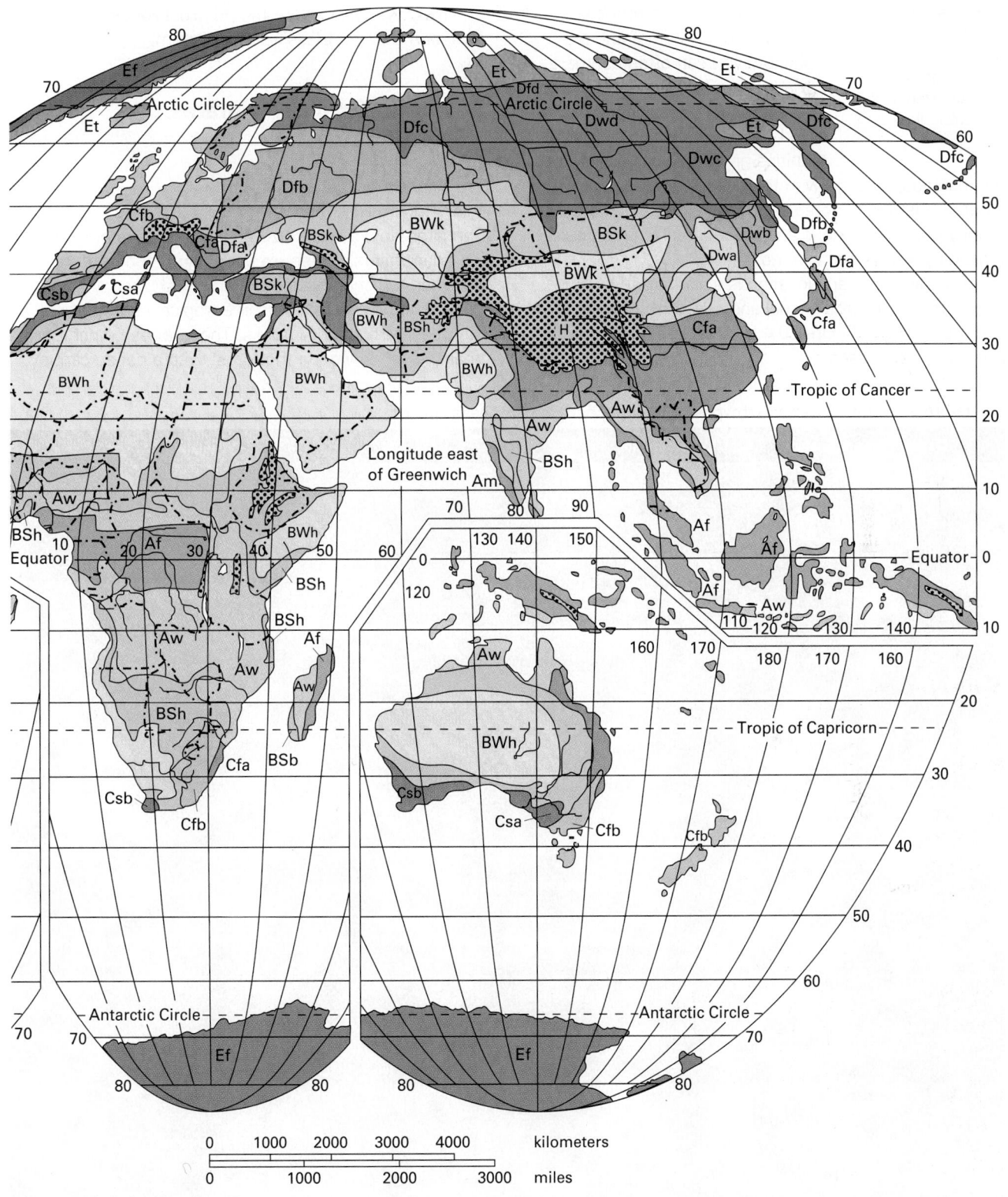

FIGURE 20.3 (B) Global climate zones. Each climate zone supports a unique biome.

biome A community of plants growing in a large geographic area characterized by a particular climate.

Although climate zones are defined by temperature and precipitation, you can often estimate climate types from a photograph of an area. Visual classification is possible because specific plant communities grow in specific climates. For instance, cacti grow in the desert, and trees grow where moisture is more abundant. A **biome** is a community of plants living in a large geographic area characterized by a particular climate.

The climate in any location is summarized by a *climograph* that records annual and seasonal temperature and precipitation. Figure 20.4 is a model climograph for Nashville, Tennessee.

Next we will explore the principal zones of the Koeppen Climate Classification system.

A. Humid Tropical Climates: No Winter

Tropical Climates with Abundant Year-Round Rainfall

The large, low-pressure zone near the equator causes abundant rainfall, often exceeding 400 centimeters per year, which supports **tropical rainforests** (Figure 20.5). The dominant plants in a tropical rainforest are tall trees with slender trunks. These trees branch only near the top, covering the forest with a dense canopy

TABLE 20.2 World Climates: Subtypes of the Koeppen Climate Classification

	Climate Type and Subtype	Description	Vegetation
A	Humid tropical	No winter	
	• Tropical rainforest	Abundant year-round rainfall	Rainforest
	• Tropical monsoon	Wet, with a short dry season	Rainforest
	• Tropical savanna	Summer rains, winter dry season	Grassland savanna
B	Arid	Dry; evaporation greater than precipitation	
	• Tropical steppe	Semiarid	Steppe grassland
	• Midlatitude steppe	Cool and dry; continental interior or rain-shadow location	Steppe grassland
	• Tropical desert	Very warm	Desert
	• Midlatitude desert	Cool and dry; continental interior or rain-shadow location	Desert
C	Humid mesothermal	Midlatitude, mild winter	
	• Humid subtropical	Warm, wet summer with winter cyclonic storms	Forest
	• Mediterranean	Dry summer and wet winter with cyclonic storms; subtropical	Shrubby plants and oak savanna
	• Marine west coast	Cool summer and mild winter; maritime, westerlies, and cyclonic storms	Forest to temperate rainforest
D	Humid microthermal	Midlatitude; severe winter	
	• Humid continental, hot summer	Midlatitude continental; warm summer, cold winter	Prairie and forest
	• Humid continental, mild summer	High-midlatitude continental; cool summer, cold winter	Prairie and forest
	• Subarctic	High-latitude continental	Forest taiga
E	Polar	Little warmth, even in summer	
	• Tundra	High latitude; short summer, harsh and long winter	Mosses, grasses, and flowers; some bushes and even a few small trees in protected areas on southern boundary of tundra
	• Ice cap	High latitude; short summer, harsh and long winter	No vegetation
F	Highlands	Cool to cold; found in mountain or high plateau areas; varies according to elevation	Changes as elevation increases

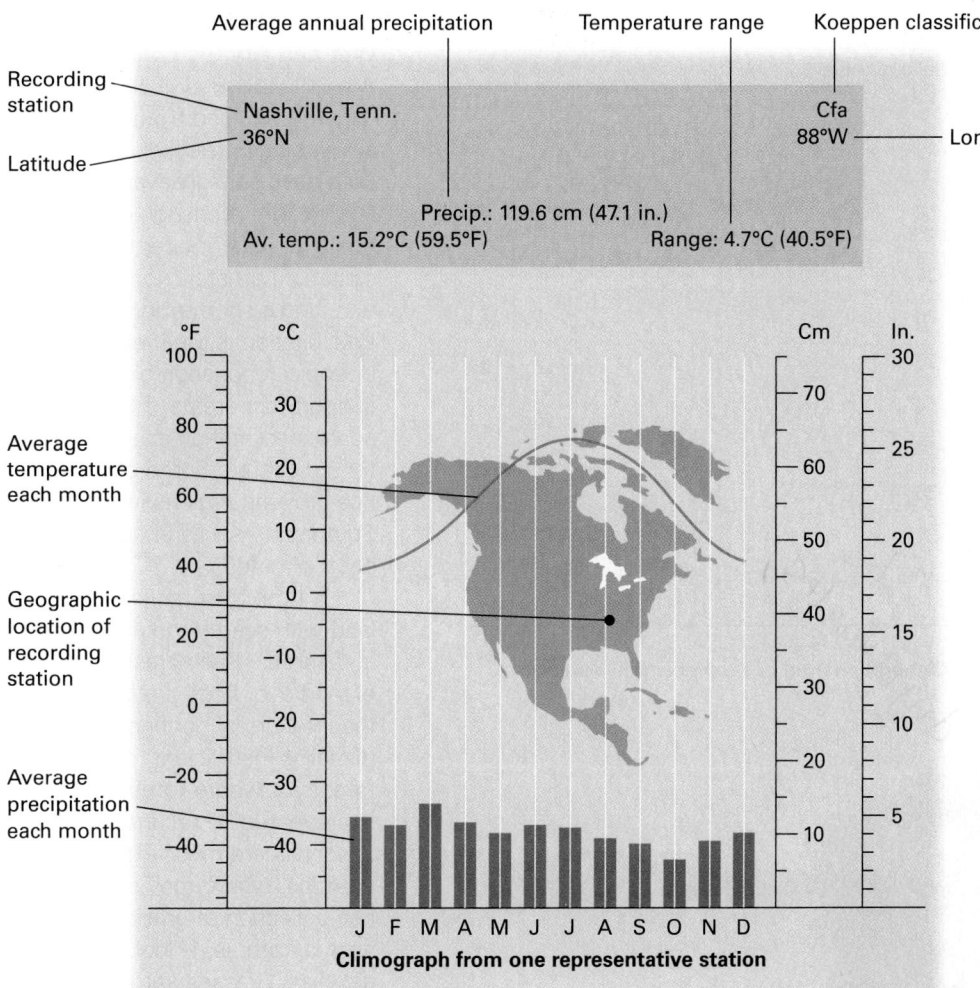

Recording station — Nashville, Tenn.
Latitude — 36°N

Average annual precipitation

Precip.: 119.6 cm (47.1 in.)
Av. temp.: 15.2°C (59.5°F)

Temperature range

Range: 4.7°C (40.5°F)

Koeppen classification — Cfa
88°W — Longitude

Average temperature each month

Geographic location of recording station

Average precipitation each month

°F °C
100   30
 80   20
 60   10
 40   0
 20 −10
  0 −20
−20 −30
−40 −40

Cm   In.
      30
 70
      25
 60
 50   20
 40   15
 30
 20   10
      5
 10

J F M A M J J A S O N D

Climograph from one representative station

tropical rain-forest A forest growing in a tropical climate with abundant, year-round rainfall; characterized by tall trees that branch only near the top, creating a dense canopy of leaves that block out most of the light, and with soggy ground and water dripping everywhere.

FIGURE 20.4 A model climograph. This graph shows the average monthly temperature (curved line) and average monthly precipitation (vertical bars) for Nashville, Tennessee. The labels highlight other features of the climograph.

FIGURE 20.5 Tropical rainforest.

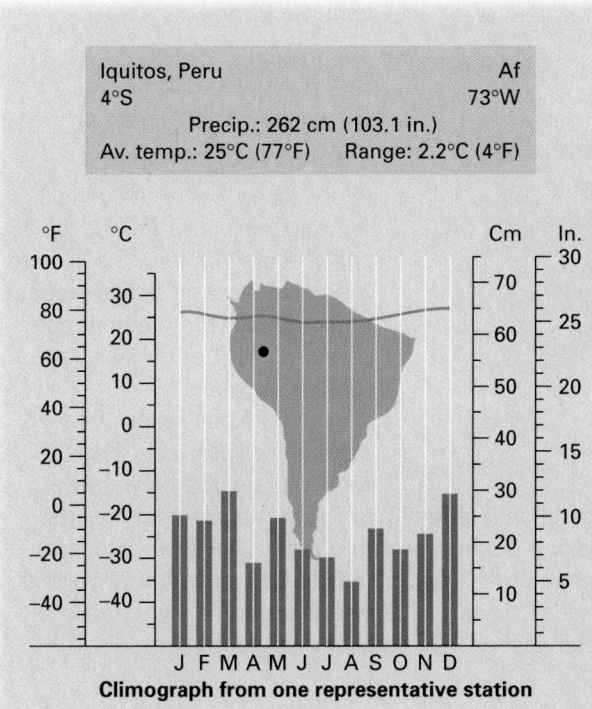

Iquitos, Peru   Af
4°S   73°W
Precip.: 262 cm (103.1 in.)
Av. temp.: 25°C (77°F)   Range: 2.2°C (4°F)

°F °C
100   30
 80   20
 60   10
 40   0
 20 −10
  0 −20
−20 −30
−40 −40

Cm   In.
      30
 70
      25
 60
 50   20
 40   15
 30
 20   10
      5
 10

J F M A M J J A S O N D

Climograph from one representative station

Tropical rainforest

Latitude: 0° to 15°

Av. temperature difference: 23°C to 28°C (winter to summer)

Av. annual precip.: 200 cm/yr. to 500 cm/yr.

General statistics for climate type

tropical monsoon climate A tropical climate with distinct wet and dry seasons, characterized by high rainfall during the wet months and a relatively short dry season.

tropical savanna climate A tropical climate with distinct wet and dry seasons but with relatively low annual rainfall, characterized by large areas of grassland supporting herds of grazing animals.

steppe A vast, semiarid, grass-covered plain found in dry climates such as those in southeast Europe, central Asia, and parts of central North America.

humid subtropical climate A midlatitude climate characterized by hot, humid summers but cooler winters and rainfall that falls throughout the year.

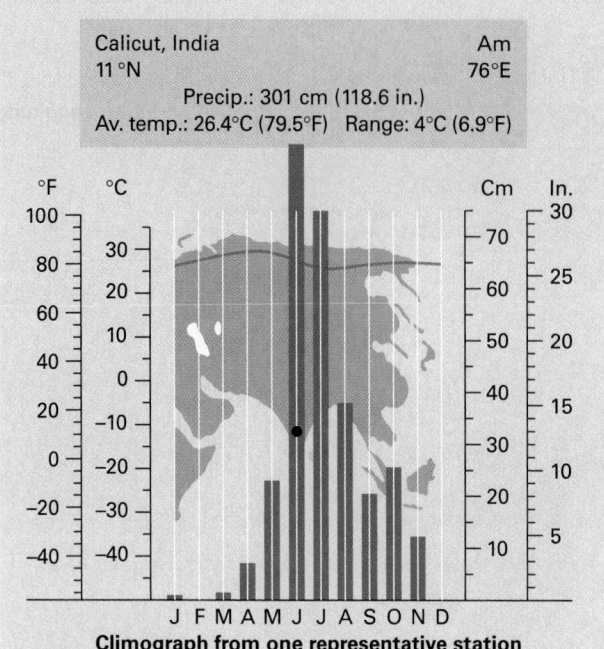

Calicut, India   Am
11 °N   76°E
Precip.: 301 cm (118.6 in.)
Av. temp.: 26.4°C (79.5°F)   Range: 4°C (6.9°F)

Climograph from one representative station

Tropical monsoon

Latitude: 5° to 30°

Av. temperature difference: 20°C to 30°C (winter to summer)

Av. annual precip.: 150 cm/yr. to 400 cm/yr.

General statistics for climate type

FIGURE 20.6 Tropical monsoon.

Tropical Climates with Distinct Wet and Dry Seasons

A **tropical monsoon climate** (Figure 20.6) and **tropical savanna climate** (Figure 20.7) both have seasonal variations in rainfall. The monsoon climate, however, has greater total precipitation, greater monthly variation, and a shorter dry season. Precipitation is great enough in tropical monsoon biomes to support rainforests. The seasonal precipitation is also ideal for agriculture. Some of the great rice-growing regions in India and Southeast Asia lie in tropical monsoon climates.

A tropical savanna is a grassland with scattered small trees and shrubs. Such grasslands extend over large areas, often in the interiors of continents, where rainfall is insufficient to support forests or where forest growth is prevented by recurrent fires. Savannas are most extensive in Africa, where they support a rich collection of grazing animals such as zebras, wildebeest, and gazelles. Grasses sprout and grow with the seasonal rains, and the great African herds migrate with the foliage.

B. Dry Climates: Evaporation Greater Than Precipitation

In dry zones where the annual precipitation varies from 25 to 35 centimeters per year, the climate is semiarid and grassland **steppes** predominate. The great steppe grasslands of Central Asia fall into these categories.

If the rainfall is less than 25 centimeters per year, deserts form and support only sparse vegetation (Figure 20.8). The world's largest deserts lie along the high-pressure zones of the 30° latitude, although rain-shadow and coastal deserts exist in other latitudes. Thus some deserts are torridly hot, while Arctic deserts are frigid for much of the year.

of leaves. The canopy blocks out most of the light, so as little as 0.1 percent of the sunlight reaches the forest floor, which consequently has relatively few plants. The ground in a tropical forest is soggy, the tree trunks are wet, and water drips everywhere.

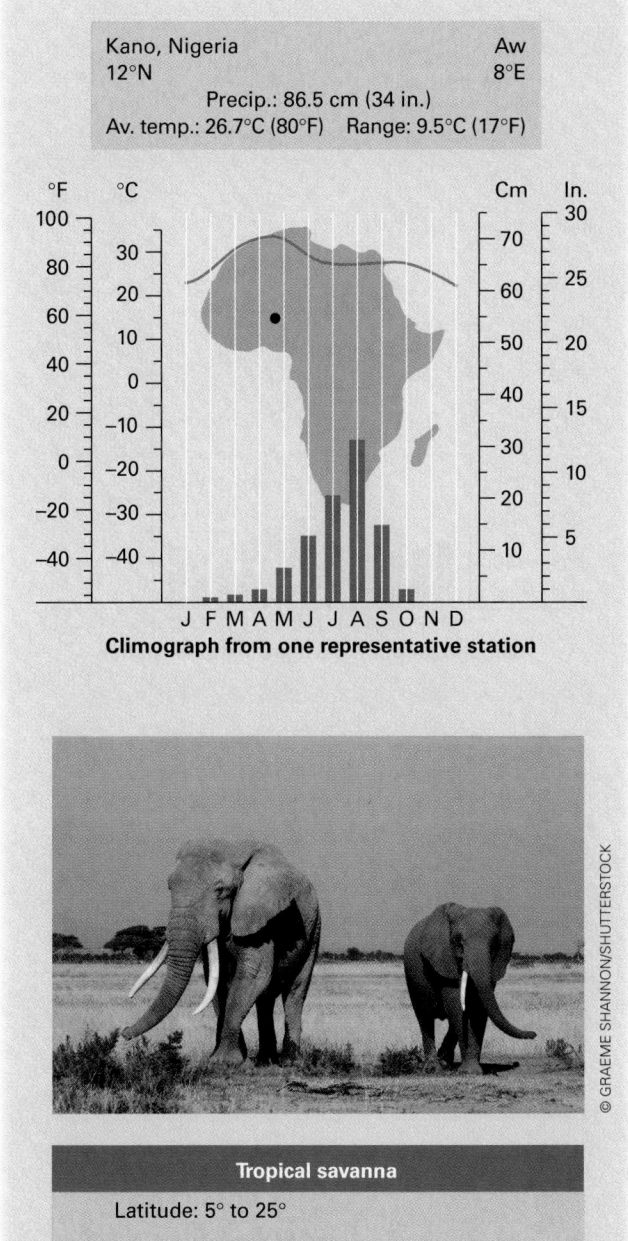

Kano, Nigeria   Aw
12°N   8°E
Precip.: 86.5 cm (34 in.)
Av. temp.: 26.7°C (80°F)   Range: 9.5°C (17°F)

Climograph from one representative station

Tropical savanna

Latitude: 5° to 25°

Av. temperature difference: 20°C to 30°C
(winter to summer)

Av. annual precip.: 100 cm/yr. to 180 cm/yr.

General statistics for climate type

© GRAEME SHANNON/SHUTTERSTOCK

FIGURE 20.7 Tropical savanna.

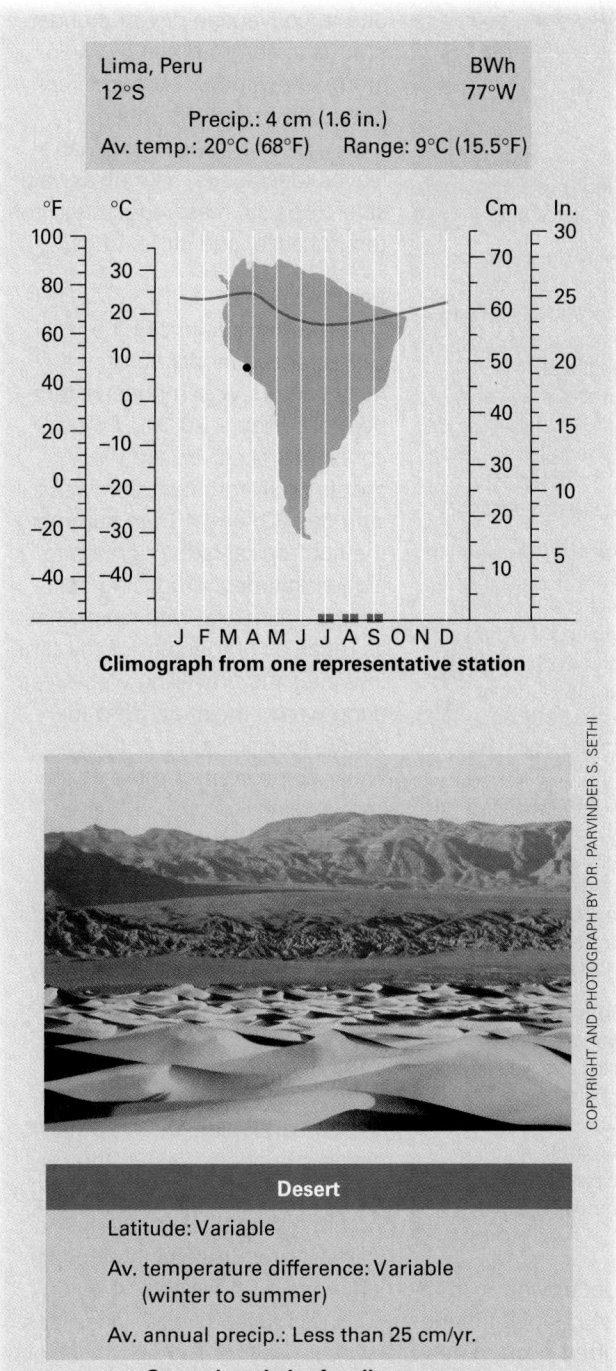

Lima, Peru   BWh
12°S   77°W
Precip.: 4 cm (1.6 in.)
Av. temp.: 20°C (68°F)   Range: 9°C (15.5°F)

Climograph from one representative station

Desert

Latitude: Variable

Av. temperature difference: Variable
(winter to summer)

Av. annual precip.: Less than 25 cm/yr.

General statistics for climate type

COPYRIGHT AND PHOTOGRAPH BY DR. PARVINDER S. SETHI

FIGURE 20.8 Desert.

C. Humid Midlatitude Climates with Mild Winters

Humid Subtropics
The southeastern United States has a **humid subtropical climate** (Figure 20.9). During the summer, conditions can be as hot and humid as in the tropics, and rain and thundershowers are common. However, during the winter, arctic air pushes southward, forming cyclonic storms. Although the average monthly temperature seldom falls below 7°C, cold fronts occasionally bring frost and snow. Precipitation is relatively constant year-round

Mediterranean climate A midlatitude climate characterized by dry summers, rainy winters, and moderate temperatures.

marine west coast climate A midlatitude climate characterized by relatively mild but wet winters and cool summers, with little temperature difference between seasons.

temperate rainforest A rainforest that grows in marine west coast climates where rainfall is more than 100 centimeters per year and is relatively constant throughout the year, particularly along the northwest coast of North America from Oregon to Alaska.

humid continental climate A midlatitude climate characterized by hot summers, cold winters, and precipitation throughout the year.

due to convection-driven thunderstorms in summer and cyclonic storms in winter. This zone supports both trees with needles (conifers) and trees with broad leaves (deciduous) as well as valuable crops such as vegetables, cotton, tobacco, and citrus fruits.

Mediterranean

The **Mediterranean climate** is characterized by dry summers, rainy winters, and moderate temperature (Figure 20.10). These conditions occur on the west coasts of all continents between latitudes 30° and 40°. In summer the subtropical high migrates to higher latitudes, producing near-desert conditions with clear skies as much as 90 percent of the time. In winter the prevailing westerlies bring warm, moist air from the ocean, leading to fog and rain. Thus, 75 percent or more of the annual rainfall occurs in winter. Although redwoods, the largest trees on Earth, grow in specific environments in central California, the summer heat and drought of Mediterranean climates generally retard the growth of large trees. Instead, shrubs and scattered trees dominate. Fires occur frequently during the dry summers and spread rapidly through the dense shrubbery. During the dry fall of 1991, a particularly devastating fire swept through the hills of Oakland and Berkeley, California, destroying 3,354 homes worth $1.5 billion and killing 24 people. If vegetation is destroyed by fire, landslides often occur when the winter rains return. Torrential winter rains frequently bring extensive flooding and landslides to southern California.

Marine West Coast

Marine west coast climate zones border the Mediterranean zones and extend poleward to 65° (Figure 20.11). They are severely influenced by ocean currents that moderate temperature and bring abundant precipitation. Thus summers are cool and winters warm, and the temperature difference between the seasons is small. For example, average monthly temperatures vary by only 15.5°C in Portland, Oregon. In contrast, Eau Claire, Wisconsin, which is a continental city at the

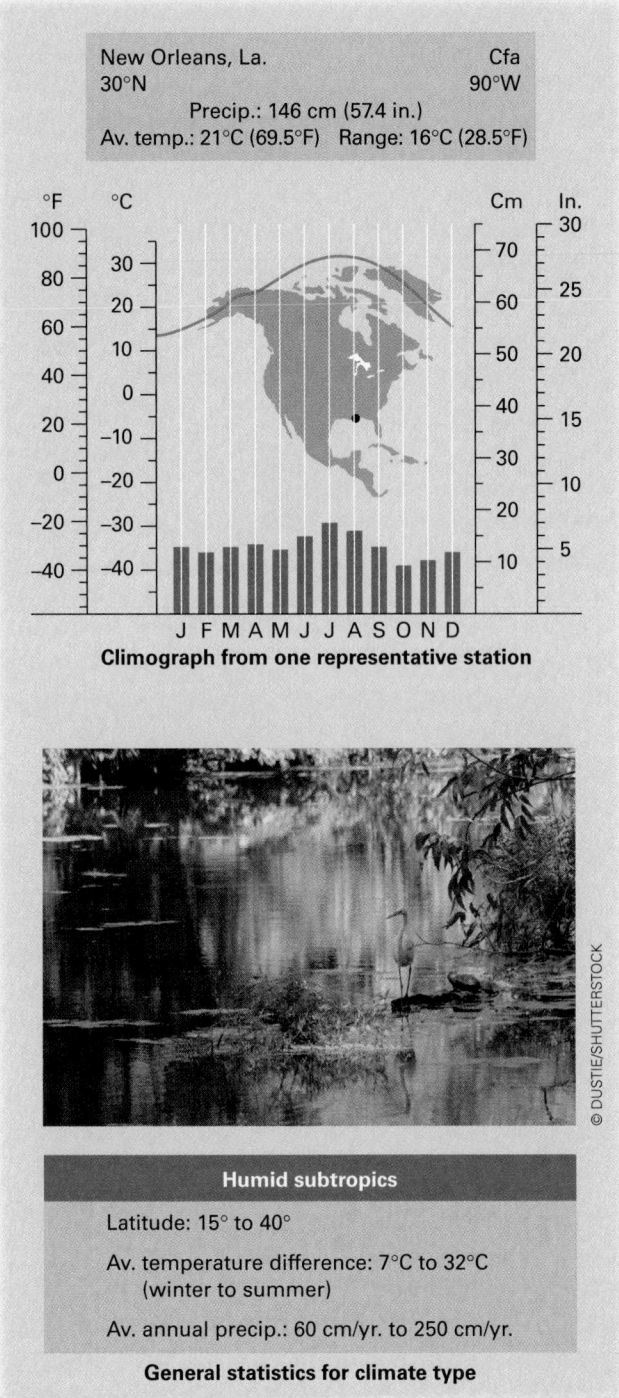

New Orleans, La.   Cfa
30°N   90°W
Precip.: 146 cm (57.4 in.)
Av. temp.: 21°C (69.5°F)   Range: 16°C (28.5°F)

J F M A M J J A S O N D
Climograph from one representative station

Humid subtropics

Latitude: 15° to 40°

Av. temperature difference: 7°C to 32°C (winter to summer)

Av. annual precip.: 60 cm/yr. to 250 cm/yr.

General statistics for climate type

FIGURE 20.9 Humid subtropics.

same latitude, experiences a 31.5°C annual temperature range. Seattle and other northwestern coastal cities experience rain and drizzle for days at a time, especially during the winter when the warm, moist, maritime air from the Pacific flows first over cool currents close to shore and then over cool land surfaces. The total rainfall varies from moderate, 50 centimeters per year (20

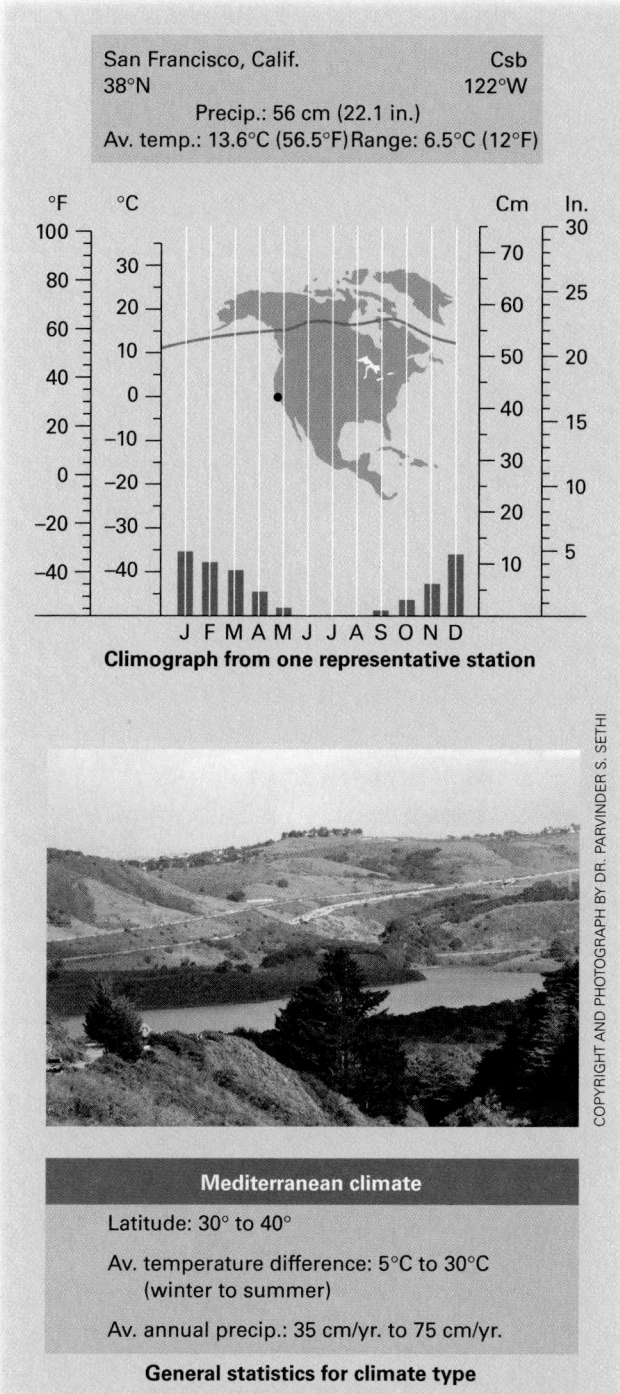

San Francisco, Calif.   Csb
38°N   122°W
Precip.: 56 cm (22.1 in.)
Av. temp.: 13.6°C (56.5°F) Range: 6.5°C (12°F)

Climograph from one representative station

COPYRIGHT AND PHOTOGRAPH BY DR. PARVINDER S. SETHI

Mediterranean climate

Latitude: 30° to 40°

Av. temperature difference: 5°C to 30°C
(winter to summer)

Av. annual precip.: 35 cm/yr. to 75 cm/yr.

General statistics for climate type

FIGURE 20.10 Mediterranean climate.

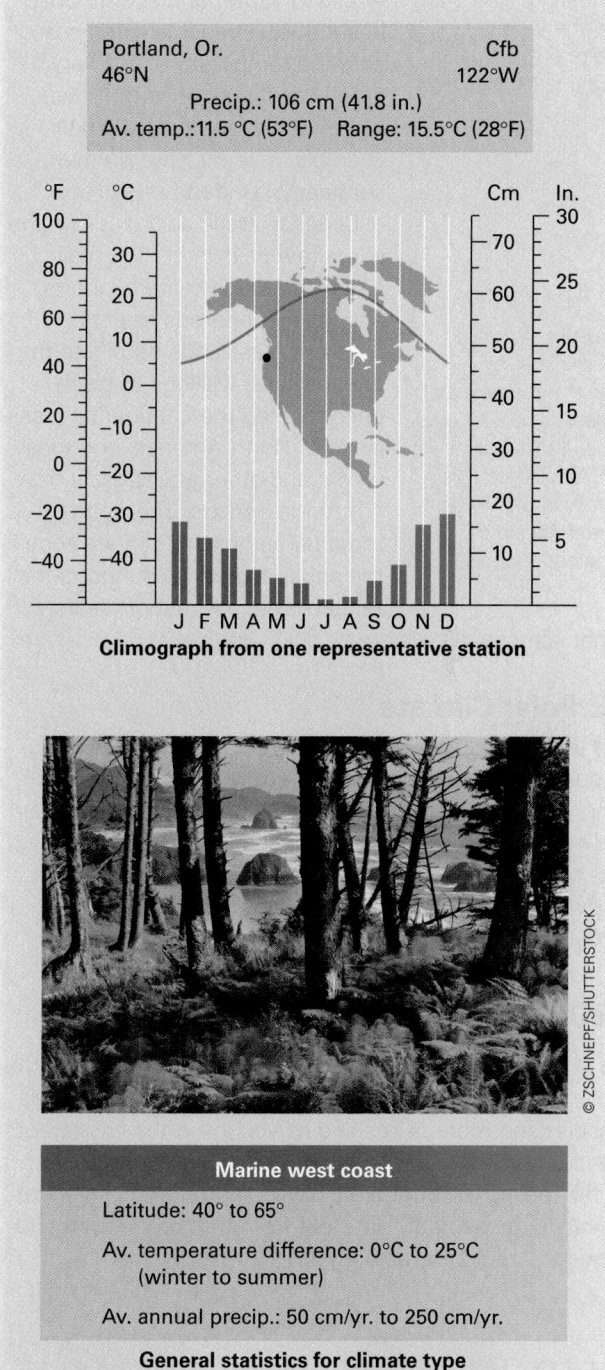

Portland, Or.   Cfb
46°N   122°W
Precip.: 106 cm (41.8 in.)
Av. temp.:11.5 °C (53°F)   Range: 15.5°C (28°F)

Climograph from one representative station

© ZSCHNEPF/SHUTTERSTOCK

Marine west coast

Latitude: 40° to 65°

Av. temperature difference: 0°C to 25°C
(winter to summer)

Av. annual precip.: 50 cm/yr. to 250 cm/yr.

General statistics for climate type

FIGURE 20.11 Marine west coast.

inches per year), to wet, 250 centimeters per year (100 inches per year). The wettest climates occur where mountains interrupt the maritime air. **Temperate rainforests** grow where rainfall is greater than 100 centimeters per year and is constant throughout the year. Temperate rainforests are common along the northwest coast of North America, from Oregon to Alaska.

D. Humid Midlatitude Climate with Severe Winters

Continental interiors in the midlatitudes are characterized by hot summers and cold winters, giving rise to **humid continental climates** (Figure 20.12). Thus in the northern Great Plains the temperature can drop to

subarctic The northernmost edge of the humid continental climate zone, lying just south of the arctic.

taiga A biome of conifers that can survive the extreme winters of the subarctic.

tundra A biome dominated by low-lying grasses, mosses, flowers, and a few small bushes that grow in the Arctic and high alpine regions.

−40°C in winter and soar to 38°C in summer. Even in a given season, the temperature may vary greatly as the polar front moves northward or southward. For example, in winter the northern continental United States may experience arctic cold one day and rain a few days later. If rainfall is sufficient, this climate supports abundant coniferous forests, whereas grasslands dominate the drier regions. Millions of bison once roamed the continental grasslands of North America, and today wheat and other grains grow from horizon to horizon. The northernmost portion of this climate zone is the **subarctic**, which supports the **taiga** biome, a forest of conifers that can survive extremely cold winters.

E. Polar Climate

In the Arctic and Antarctic, winters are harsh and long, and the temperature remains above freezing only during a short summer. The climate is classified as a *polar climate*. Trees cannot survive, and low-lying plants such as mosses, grasses, flowers, and a few small bushes cover the land. This biome is called **tundra** (Figure 20.13).

20.3 Urban Climates

If you ride a bicycle from the center of a city toward the countryside, you may notice that the air gradually feels cooler and more refreshing as you leave the city streets and enter the green fields or hills of the outlying area. This feeling is not entirely psychological; the climate of a city is measurably different from that of the surrounding rural regions (Table 20.3).

© DIGITALDEPTH/SHUTTERSTOCK

Duluth, Minn. Dfb
46°N 92°W
Precip.: 76 cm (29.9 in.)
Av. temp.: 4°C (39°F) Range: 31°C (56°F)

Climograph from one representative station

J F M A M J J A S O N D

COURTESY OF GRAHAM R. THOMPSON/JONATHAN TURK

Midlatitude with severe winters

Latitude: 40° to 65°

Av. temperature difference: −15°C to 25°C (winter to summer)

Av. annual precip.: 50 cm/yr. to 150 cm/yr.

General statistics for climate type

FIGURE 20.12 Midlatitude with severe winters.

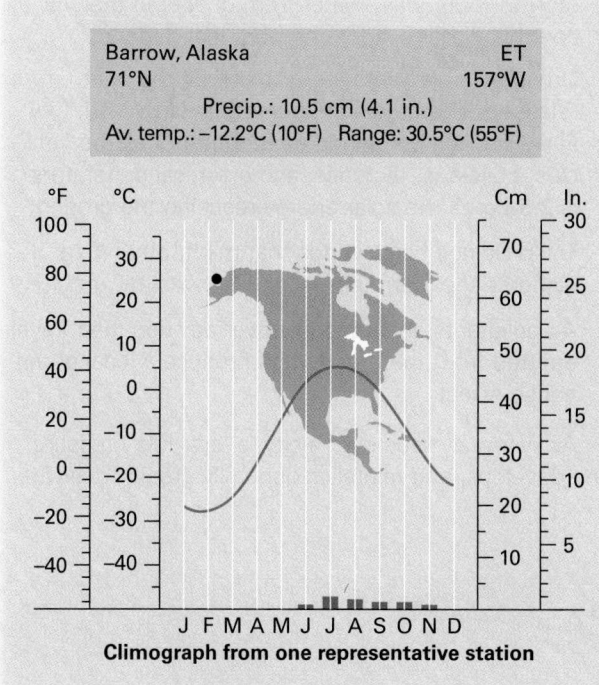

Barrow, Alaska
71°N
ET
157°W

Precip.: 10.5 cm (4.1 in.)
Av. temp.: –12.2°C (10°F) Range: 30.5°C (55°F)

Climograph from one representative station

© EKATERINA/SHUTTERSTOCK

Arctic tundra

Latitude: 65° to 80°

Av. temperature difference: –35°C to +10°C
(winter to summer)

Av. annual precip.: generally low, 10 cm/yr.
to 35 cm/yr.

General statistics for climate type

FIGURE 20.13 Arctic tundra.

TABLE 20.3 Average Changes in Climatic Elements Caused by Urbanization

Weather Element	Comparison with Rural Environment
Cloudiness:	
Cloud cover	5 to 10% more
Fog, winter	100% more
Fog, summer	30% more
Precipitation, total	5 to 10% more
Relative humidity:	
Winter	2% lower
Summer	8% lower
Radiation:	
Total	15 to 20% less
Direct sunshine	5 to 15% less
Temperature:	
Annual mean	0.5 to 1.0°C higher
Winter minimum (average)	1.0 to 3.0°C higher
Wind speed:	
Annual mean	20 to 30% lower
Extreme gusts	10 to 20% lower
Calms	5 to 20% higher

urban heat island effect Warmer temperatures in a city compared to the surrounding countryside, caused by factors in the urban environment itself.

As shown in Figure 20.14, the average winter minimum temperature in the center of Washington, D.C., is more than 3°C warmer than in outlying areas. This temperature difference, called the **urban heat island effect**, is caused by numerous factors:

• Stone and concrete buildings and asphalt roadways absorb solar radiation and reradiate it as infrared heat.

• Cities are warmer because little surface water exists and, as a result, little evaporative cooling occurs. In contrast, in the countryside, water collects in the soil and evaporates for days after a storm. Roots draw water from deeper in the soil, and this water evaporates from leaf surfaces.

• Urban environments are warmed by the heat released when fuels are burned. In New York City in winter, the combined heat output of all the vehicles, buildings, factories, and electrical generators is 2.5 times the solar energy reaching the ground.

• Tall buildings block winds that might otherwise disperse the warm air.

• Air pollutants absorb long-wave radiation (heat rays) emitted from the ground and produce a local greenhouse effect.

As warm air rises over a city, a local low-pressure zone develops, and rainfall is generally greater over the

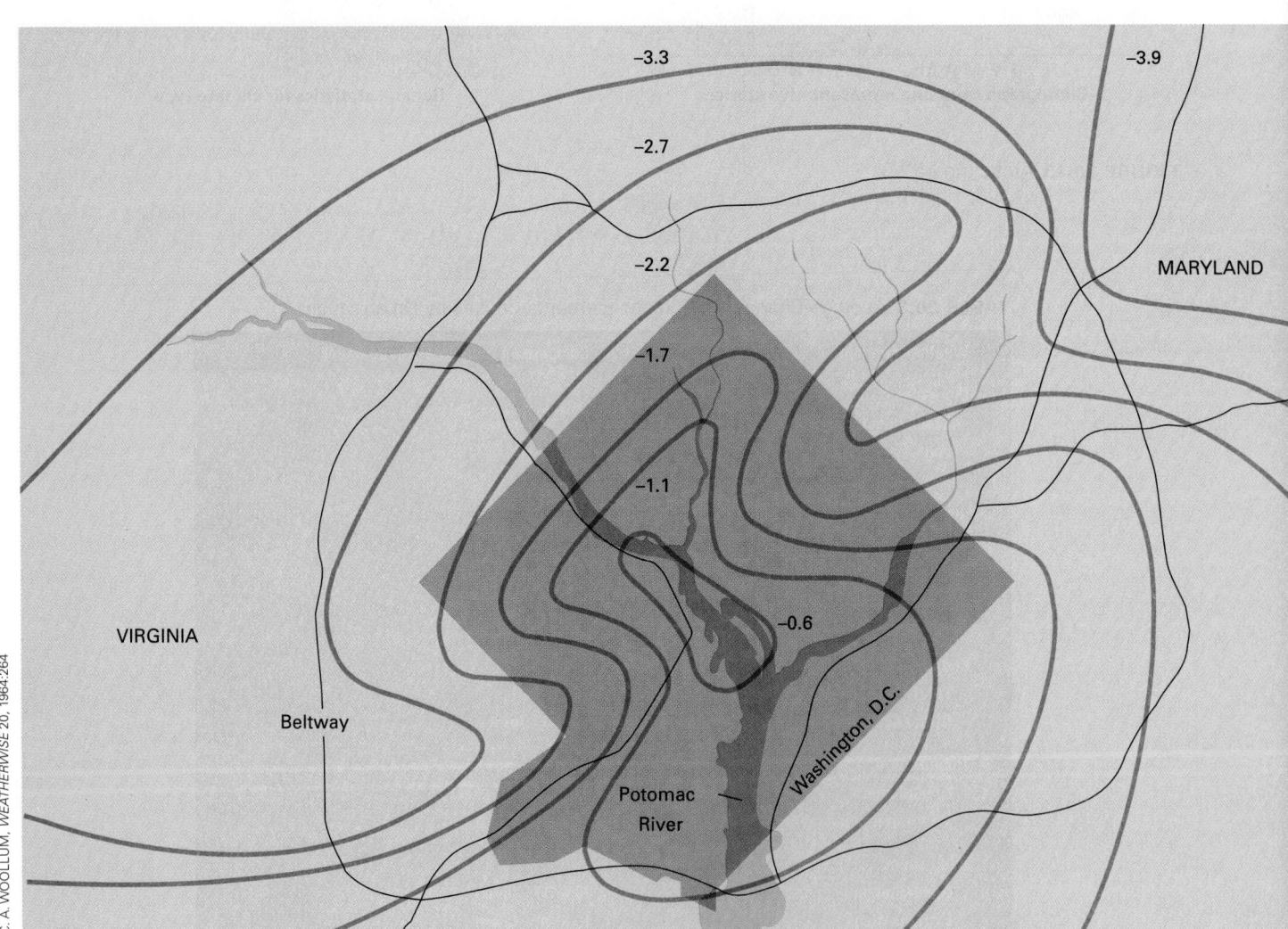

C. A. WOOLLUM, *WEATHERWISE* 20, 1964:264

FIGURE 20.14 The urban heat island effect. The average minimum temperature in and around Washington, D.C., during the winter; contours represent relative T minima in degrees Celsius.

city than in the surrounding areas (Figure 20.15). Water condenses on dust particles, which are abundant in polluted urban air. Weather systems collide with the city buildings and linger, much as they do on the windward side of mountains. Thus, a front that might pass quickly over rural farmland remains longer over a city and releases more precipitation.

In 1600, less than 1 percent of the global population lived in cities. By 1950, 30 percent of the world's population was urban, and by 2000 that ratio had grown to 55 percent. Therefore, although urban climate change may not affect global climate, it affects the lives of the many people living in cities.

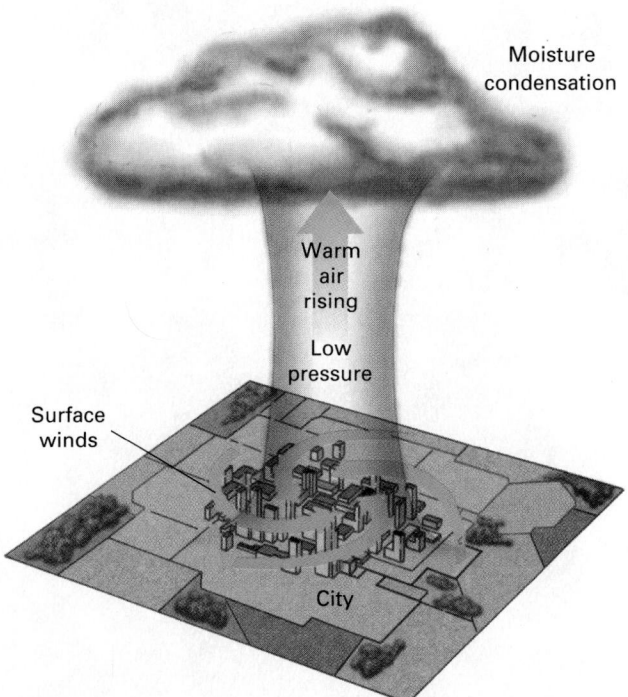

FIGURE 20.15 Warm air rising over a city creates a low-pressure zone. As a result, precipitation is greater in the city than over the surrounding countryside.

21

CLIMATE CHANGE

Larsen B Ice Shelf 1/31/2002

Larsen B Ice Shelf 3/5/2002

FIGURE 21.1 The Larsen B Ice Shelf rims a small portion of the eastern coastline of the Antarctic Peninsula. Between January 31 and March 5 of 2002, 3,250 square kilometers of ice, about the size of the state of Rhode Island, broke off the ice shelf and floated northward.

visit 4ltrpress.cengage.com

{ **The warnings about global warming have been extremely clear for a long time. We are facing a global climate crisis.** }

Al Gore

In March 2000, an 11,000-square-kilometer iceberg as large as the state of Connecticut broke free from the Ross Ice Shelf in Antarctica and drifted north, where it melted in warmer water. Two months later, a massive chunk of ice cracked off the nearby Ronne Ice Shelf. In September, a second Connecticut-sized chunk of the Ross Ice Shelf disintegrated. Then in January of 2002, a Rhode Island–sized section of the Larsen Ice Shelf splintered into millions of small fragments, which you can see in Figure 21.1. Since the height of the last ice age, 18,000 years ago, the Ross Ice Shelf has receded 700 kilometers, shedding 5.3 million cubic kilometers of ice. And, as evidenced by the recent disintegration, the process continues to this day.

In order to have a balanced perspective on what is happening today, we must first study historical climate change. The geological record in almost every locality on Earth provides evidence that past regional climates were different from modern climates. Geologists have discovered sand dunes beneath prairie grasslands near Denver, Colorado, indicating that this semiarid region was recently desert. Moraines on Long Island, New York, tell us that this temperate region was once glaciated. Fossil ferns in nearby Connecticut indicate that, before the glaciers, the northeastern United States was warm and wet.

Many of these regional climatic fluctuations resulted from global climate change. Thus, 18,000 years ago, Earth was cooler than it is today and glaciers descended to lower latitudes and altitudes. During Mississippian time, from 360 to 325 million years ago, Earth was warmer than it is today. Vegetation grew abundantly, and some of it collected in huge swamps to form coal.

During the past 10,000 years, global climate has been mild and stable as compared with the preceding 100,000 years. During this time, humans have developed from widely separated bands of hunter-gatherers, to agrarian farmers, and then moved into crowded communities in huge industrial megalopolises. Today, with a global population of more than 6 billion, people have stressed the food-producing capabilities of the planet. If temperature or rainfall patterns were to change, even slightly,

crop failures could lead to famines. In addition, cities and farmlands on low-lying coasts could be flooded under rising sea level. As a result, climate change may be one of the most important issues that we face in the 21st century.

21.1 Climate Change in Earth's History

Recall from Chapter 17 that Earth's primordial atmosphere contained high concentrations of carbon dioxide (CO_2) and water vapor (H_2O). Both greenhouse gases absorb infrared radiation in the atmosphere. Astronomers have calculated that the Sun was 20 to 30 percent fainter early in Earth's history than it is today, yet oceans did not freeze. The high concentrations of atmospheric carbon dioxide and water vapor retained enough of the Sun's diminished radiation to warm Earth's atmosphere and surface to temperatures that kept the oceans liquid. Luckily for us, the concentration of carbon dioxide and water in the atmosphere declined gradually as the Sun warmed.

Figure 21.2 on the next page is a graph of mean global temperature and precipitation throughout Earth's history. Note, for example, that the planet plunged into a deep freeze on at least five occasions. There is evidence that the cold periods 600 to 700 million years ago were so extreme that the oceans are thought to have frozen from pole to pole, accompanied by near-total land coverage of glaciers, leading to a "Snowball Earth," described in Chapter 13. In contrast, our planet was relatively warm for 248 million years, from the start of the Mesozoic Era almost to the present. The last 2 million years have witnessed the most recent ice age. According to many climatologists, the current period is an interglacial warming episode and the ice sheets are likely to return in the geologically near future.

Many additional climate changes occurred during Earth's history, but they were too short in duration to be apparent on the graph. For example, ice core records from Greenland, Antarctica, and alpine glaciers show that between 110,000 and 10,000 years ago the mean annual global

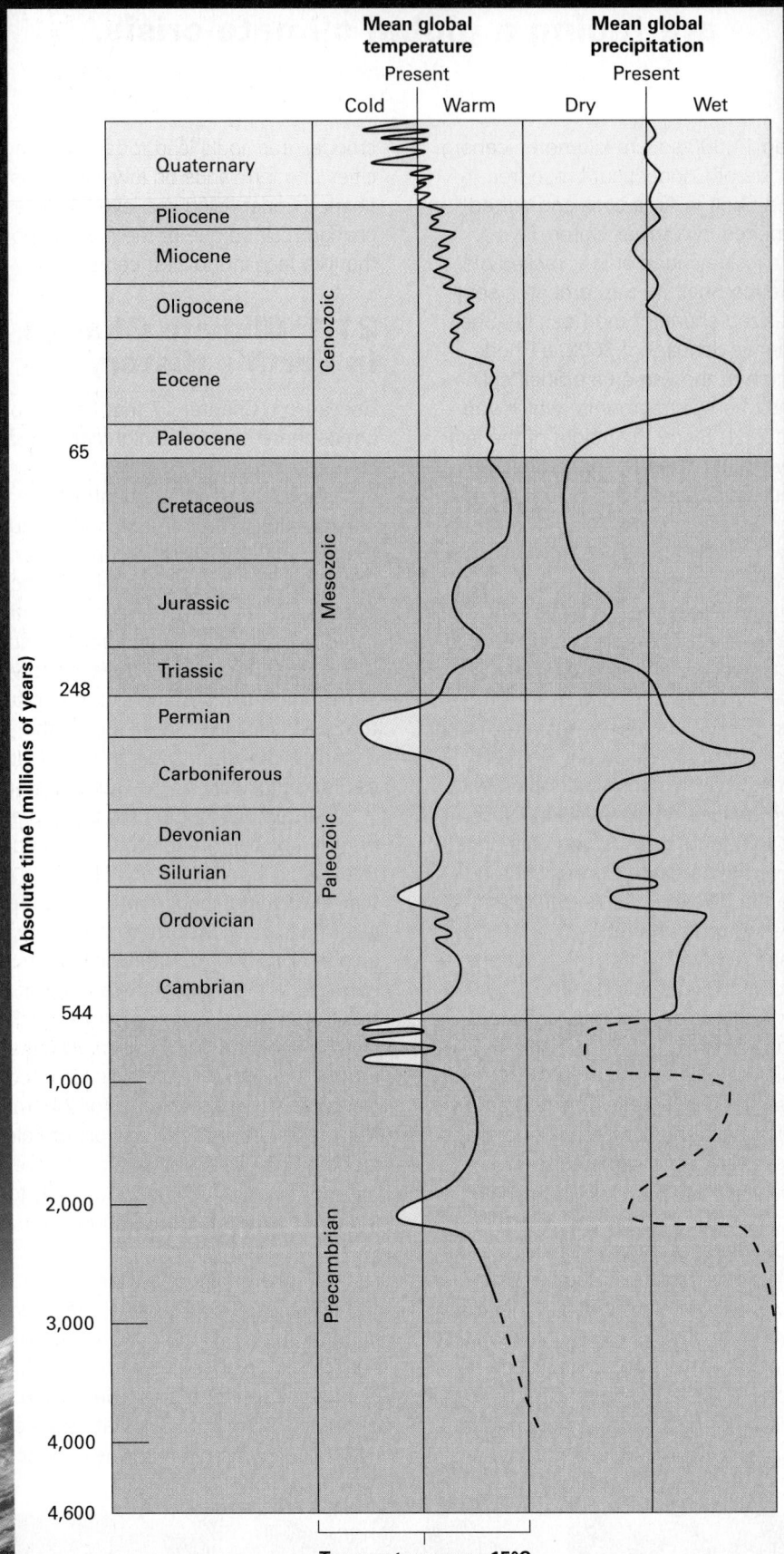

Source: Redrawn from L. A. Frakes, *Climates throughout Geologic Time* (New York: Elsevier Scientific Publishing, 1989).

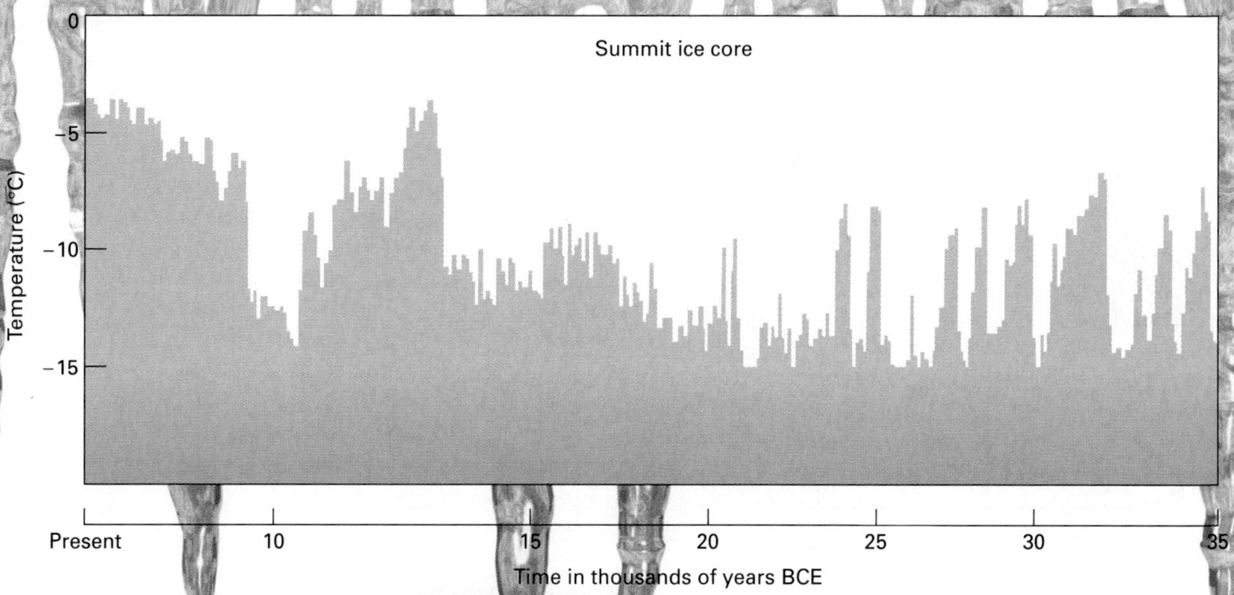

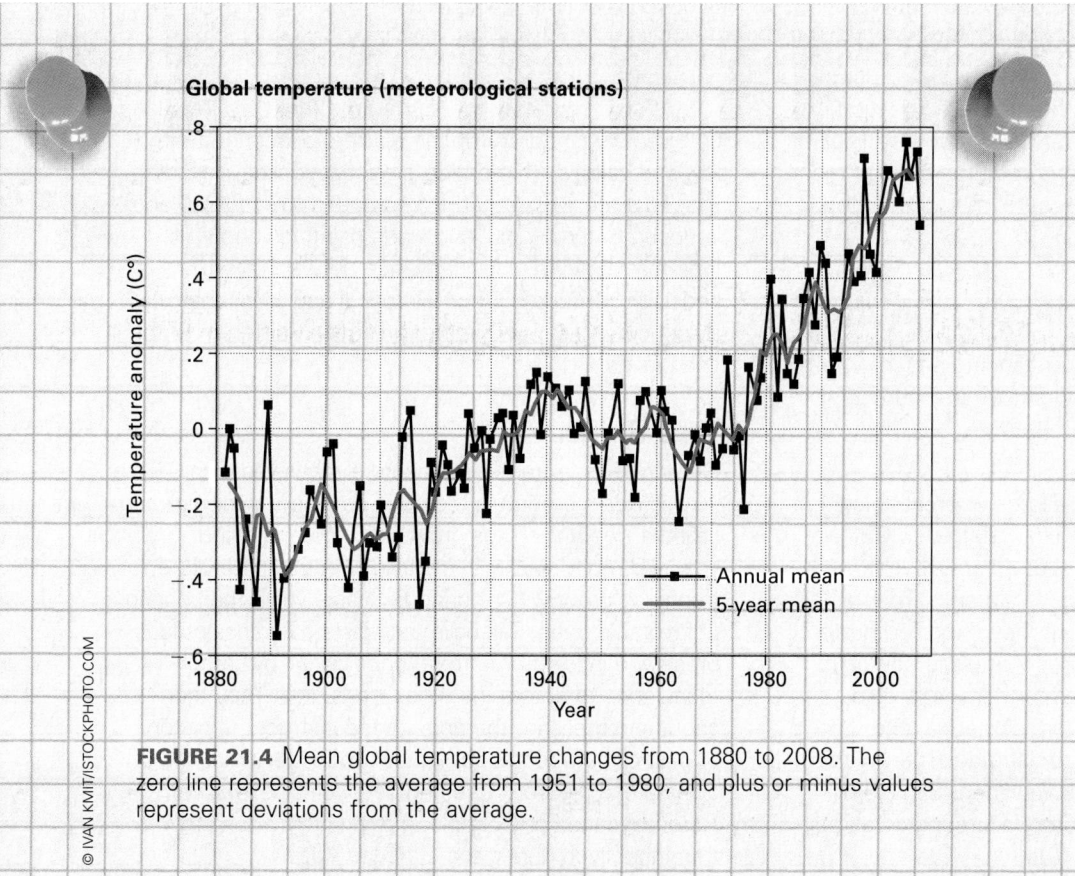

FIGURE 21.3 Atmospheric temperature fluctuations during the past 35,000 years in Greenland. The data were collected by the Greenland Ice Core Project.

temperature changed frequently and dramatically. Figure 21.3 gives a closer look at a portion of the temperature data. The graph shows that the atmospheric temperature in Greenland changed several times by 5°C to 10°C within 5 to 10 years. Many of the cold intervals persisted for 1,000 years or more. As mentioned earlier, the past 10,000 years, during which civilization developed, witnessed anomalously stable climate.

Figure 21.4 gives us an even more detailed look at a 128-year span from 1880 to 2008. Note that the temperature rose slowly from 1880 to 1970 and then increased dramatically. During the past century, people burned large quantities of fuel, thereby injecting carbon dioxide into the atmosphere. The correlation between carbon dioxide emission and global temperature implies—but does not prove—that human activities have caused the current global warming. The temperature rise has been a little less than 0.8°C; temperature changes much larger than this have occurred repeatedly throughout Earth's history. For the past 100 years, meteorologists have used

FIGURE 21.4 Mean global temperature changes from 1880 to 2008. The zero line represents the average from 1951 to 1980, and plus or minus values represent deviations from the average.

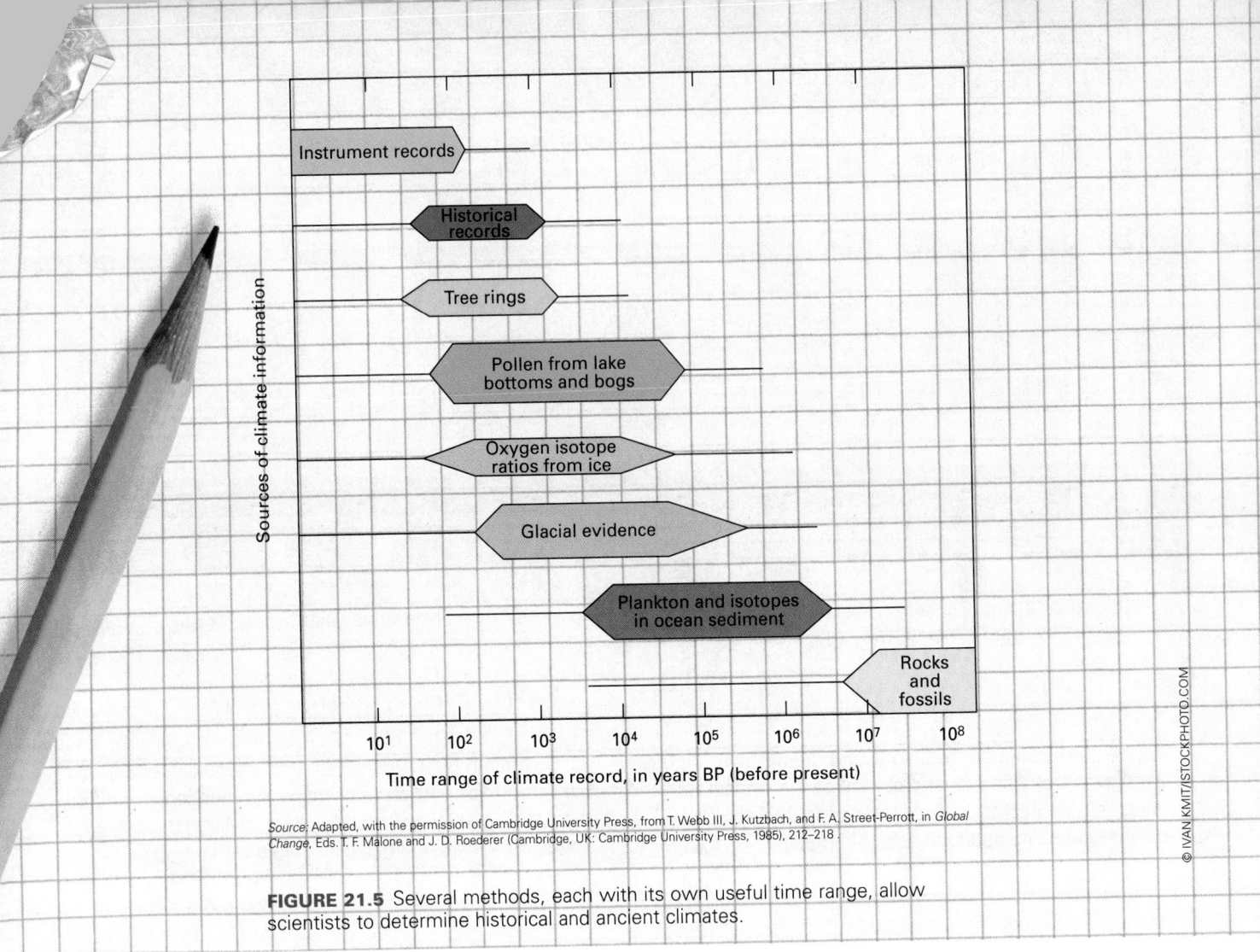

Source: Adapted, with the permission of Cambridge University Press, from T. Webb III, J. Kutzbach, and F. A. Street-Perrott, in *Global Change*, Eds. T. F. Malone and J. D. Roederer (Cambridge, UK: Cambridge University Press, 1985), 212–218 .

FIGURE 21.5 Several methods, each with its own useful time range, allow scientists to determine historical and ancient climates.

instruments to measure temperature, precipitation, wind speed, and humidity. But how do we interpret prehistoric climates? Figure 21.5 reviews several techniques for determining past climate, as represented in Figure 21.2.

21.2 Measuring Climate Change

This chapter is devoted to understanding and interpreting the 125-year warming trend shown in Figure 21.4.

Historical Records

Historians search for written records or archaeological data that chronicle climate change. In 985 CE, the Viking explorer Eric the Red sailed to southwest Greenland with a few hundred immigrants. They established two colonies, from which they exported butter and cheese to Iceland and Europe, and the population flourished. Some Vikings sailed farther west, colonized the Labrador Coast of North America, and visited Ellesmere Island, near 80° north latitude. Then within 300 to 400 years, the colonies vanished. Sagas tell of heavy sea ice in summer, crop failures, starvation, and death.

At approximately the same time that the Greenland colonies vanished, European glaciers descended into lowland valleys. This period of global cooling, called the *Little Ice Age*, lasted from about 1450 to 1850 and is documented by old landscape paintings and writings that depict a glacial advance between the 15th and 19th centuries. Other historical and archaeological evidence chronicles climate changes at different times and places.

Tree Rings

Growth rings in trees also record climatic variations. Each year, a tree's growth is recorded as a new layer of wood called a *tree ring*. Trees grow slowly during a cool, dry year and more quickly in a warm, wet year; therefore, tree rings grow wider during favorable years than during unfavorable ones. Paleoclimatologists date ancient logs preserved in ice, permafrost, or glacial till by carbon-14 techniques to determine when trees died. They then count and measure the rings to reconstruct the history of past climate recorded in the wood. Interpretations of climate change from tree-ring data coincide well with historical data. For example, growth rings are narrow in trees that lived during the Little Ice Age.

Plant Pollen

Plant pollen is widely distributed by wind and is coated with a hard, waxy cover that resists decomposition. As a result, pollen grains are abundant and well preserved in sediment in lake bottoms and bogs. For example, 11,000 years ago, spruce was the most abundant tree species in a Minnesota bog. In modern forests, spruce dominates in colder Canadian climates but is less abundant in Minnesota. Therefore, scientists deduce that the climate in Minnesota was colder 11,000 years ago than it is at present. Pollen in younger layers of sediment shows that about 10,500 years ago, pines displaced the spruce, indicating that the temperature became warmer.

Oxygen Isotope Ratios in Glacial Ice

Oxygen consists mainly of two isotopes, abundant ^{16}O and rare ^{18}O. Both isotopes are incorporated into water, H_2O. Water molecules containing ^{16}O are lighter and evaporate more easily than those containing ^{18}O. At high temperatures, however, evaporating water vapor contains a higher proportion of ^{18}O than it does at lower temperatures. Therefore, the ratio of $^{18}O/^{16}O$ in vapor from warm water is higher than that in vapor from cool water. Some of the water vapor condenses as snow, which accumulates in glaciers. Thus the $^{18}O/^{16}O$ ratios in glacial ice reflect water temperature at the time the water evaporated. Because most of the atmospheric water vapor that falls as snow originated from evaporation of ocean water, scientists then use the $^{18}O/^{16}O$ data from glacial ice to estimate mean ocean surface temperatures. Because the sea surface and the atmosphere are in close contact, mean ocean surface temperature reflects mean global atmospheric temperature.

Geologists have drilled deep into Greenland and Antarctic glaciers, where the ice is up to 110,000 years old, and have carefully removed ice cores. The age of the ice at any depth is determined by counting annual ice deposition layers or by carbon-14 dating of wind-blown pollen within the glacier. The oxygen isotope ratios in each layer reflect the air temperature at the time the snow fell. The temperature data in Figure 21.3 were obtained from Greenland ice cores.

Glacial Evidence

Erosional and depositional features created by glaciers, such as the tills, tillites, and glacial striations described in Chapter 13, are evidence of the growth and retreat of alpine glaciers and ice sheets, which in turn reflect climate. The timing of recent glacial advances can be determined by several methods. One effective technique is carbon-14 dating of logs preserved in glacial till. Unfortunately, carbon-14 dating can only be used to obtain ages spanning about 10 half-lives, from 50,000 years ago to the present.

Plankton and Isotopes in Ocean Sediment

In a technique that parallels pollen studies, scientists estimate climate by studying fossils in deep-sea sediment. The dominant life-forms in the ocean are microscopic plankton that float near the sea surface. Just as pollen ratios change with air temperature, plankton species ratios change with sea-surface temperature.

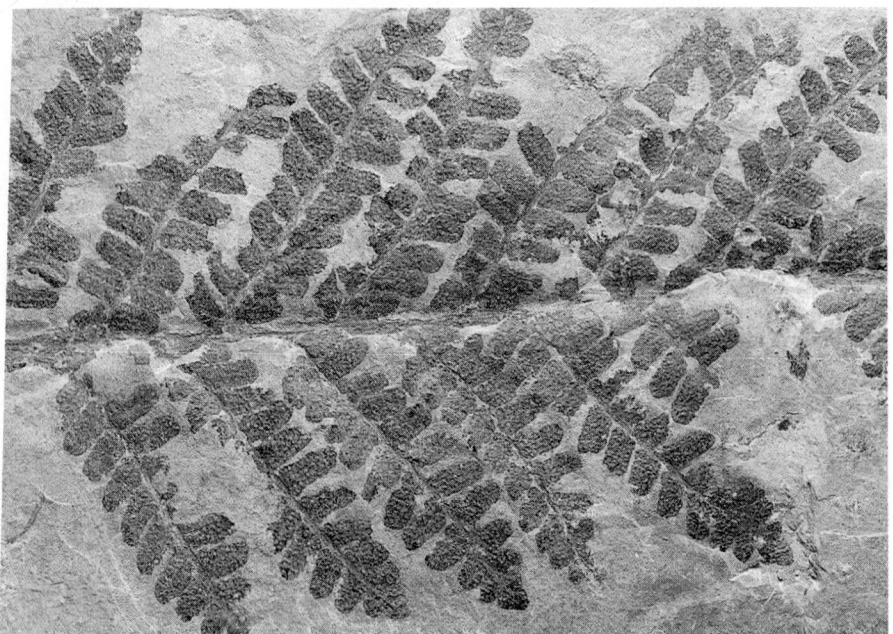

FIGURE 21.6 A fossil fern indicates that a region was wet and warm at the time the fern grew.

Thus, fossil plankton assemblages found in sediment cores reflect sea-surface temperature.

Most hard tissues formed by animals and plants, such as shells, exoskeletons, teeth, and bone, contain oxygen. Many organisms absorb a high ratio of $^{18}O/^{16}O$ at low temperatures, but the ratio decreases with increasing temperature. For example, consider foraminifera, tiny marine organisms. During a time of Pleistocene cooling and glacial growth, their shells contain an average of 2 percent more ^{18}O than similar shells formed during a warm interglacial interval. Thus, just as scientists estimate paleoclimate by measuring oxygen isotope ratios in glacial ice, they can estimate ancient climate by measuring oxygen isotope ratios in fossil corals, plankton, teeth, and the remains of other organisms. Oxygen is also incorporated into soil minerals, so isotope ratios in soil and sea-floor sediment also reflect paleoclimate.

The Rock and Fossil Record

Fossils are abundant in many sedimentary rocks of Cambrian age and younger.

Geologists can approximate climate in ancient ecosystems by comparing fossils with modern relatives of the ancient organisms (Figure 21.6). For example, modern coral reefs grow only in tropical water. Therefore, we infer that fossil reefs also formed in the tropics. Coal deposits and ferns, like the one in Figure 21.6, formed in moist tropical environments; cacti indicate that the region was once desert.

Looking backward even farther—into the Proterozoic Eon, before life became abundant—it is difficult to measure climate with fossils. Thus, geologists search for clues in rocks. Tillite is a sedimentary rock formed from glacial debris and thus indicates a cold climate. Lithified dunes formed in deserts or along coasts.

Sedimentary rocks form in water, so their existence tells us that the temperature was above freezing and below boiling. Carbonate rocks precipitate from carbon dioxide dissolved in seawater. Geochemists know the chemical conditions under which carbon dioxide dissolves and precipitates, so they can calculate a range of atmospheric and oceanic compositions and temperatures that would have produced limestone and other carbonate rocks. Most thick limestones formed in warm, shallow seas. Some ancient mineral deposits, such as the banded iron deposits described in Chapter 17, also reflect the chemistry of the ancient atmosphere.

21.3 Astronomical Causes of Climate Change

After scientists learned that climates have changed, they began to search further to understand how climates change. Despite its immense

When the Earth was formed, the sun only produced 70% of the energy that it produces today.

complexity, Earth's atmospheric temperature is determined by three basic processes:

- **The heat of the Sun.** Solar output fluctuates, so in the absence of mitigating factors, when the Sun becomes hotter, Earth warms. When the Sun cools, so does our planet.

- **Albedo.** If the incoming solar radiation is absorbed at Earth's surface, the planet becomes warmer; if heat is reflected back into space, the planet becomes cooler. Therefore, the reflectivity, or albedo, of Earth's surface is a critical factor in determining the planet's temperature.

- **Heat retention by the atmosphere.** Heat that is reflected back into space may be absorbed by gases in the atmosphere, in a process called the *greenhouse effect*. Since different gases retain heat differently, atmospheric composition is the third basic factor that regulates our planet's atmospheric temperature.

These three factors determine the amount of heat available to drive the planet's climate and weather engines. Multitudinous feedback loops and threshold effects—involving atmosphere, hydrosphere, geosphere, and biosphere—then perturb the basic system to produce the resultant climate.

Changes in Solar Radiation

Recall from Chapter 13 that variations in Earth's orbit and rotational pattern may have caused the climate fluctuations responsible for the glacial advances and retreats of the Pleistocene Ice Age. Other astronomical factors may also cause climate change.

A star the size of our Sun produces energy by hydrogen fusion for about 10 billion years. During this time, its energy output increases slowly. When the Earth first formed, the Sun produced only 70 percent of the energy that it produces today. If sunlight were the only factor that influenced temperature, our planet would have been much cooler and the earth would have been covered with ice. However, sedimentary rocks formed in this early earth, indicating that water was present. The Earth's primordial atmosphere contained abundant methane, which is a potent heat-trapping greenhouse gas that warmed the early Earth. As the sun heated up, the concentration of methane declined and the composition of the atmosphere changed. Within a relatively narrow range, processes leading to warming balanced those leading to cooling, and for the most part the Earth's temperature remained mostly in the range where water was abundant in its liquid form.

Within the past few hundred million years, solar output has changed by only one fifty-millionth of 1

COPYRIGHT AND PHOTOGRAPH BY DR. PARVINDER S. SETHI

Face of the Portage Glacier in June 2009. Each year, more of this particular rock surface is visible at this location, as the glacier continues its retreat upslope, changing the delicate balance created by the Earth's albedo.

percent per century, and therefore the variation had no measurable influence on climate change over thousands, to even millions, of years.

Bolide Impacts

Evidence strongly suggests that a bolide crashed to Earth about 65 million years ago. The impact blasted enough rock and dust into the sky to block out sunlight and cool the planet. According to one current hypothesis, this cooling led to the extinction of the dinosaurs. Other bolide impacts may have caused rapid and catastrophic climate changes throughout Earth's history.

21.4 Water and Climate

Water is abundant in all four of Earth's spheres. It occurs in rocks and soil of the geosphere, comprises over 90 percent of the living organisms of the biosphere, constitutes essentially the entire hydrosphere, and exists in the atmosphere as vapor, liquid droplets,

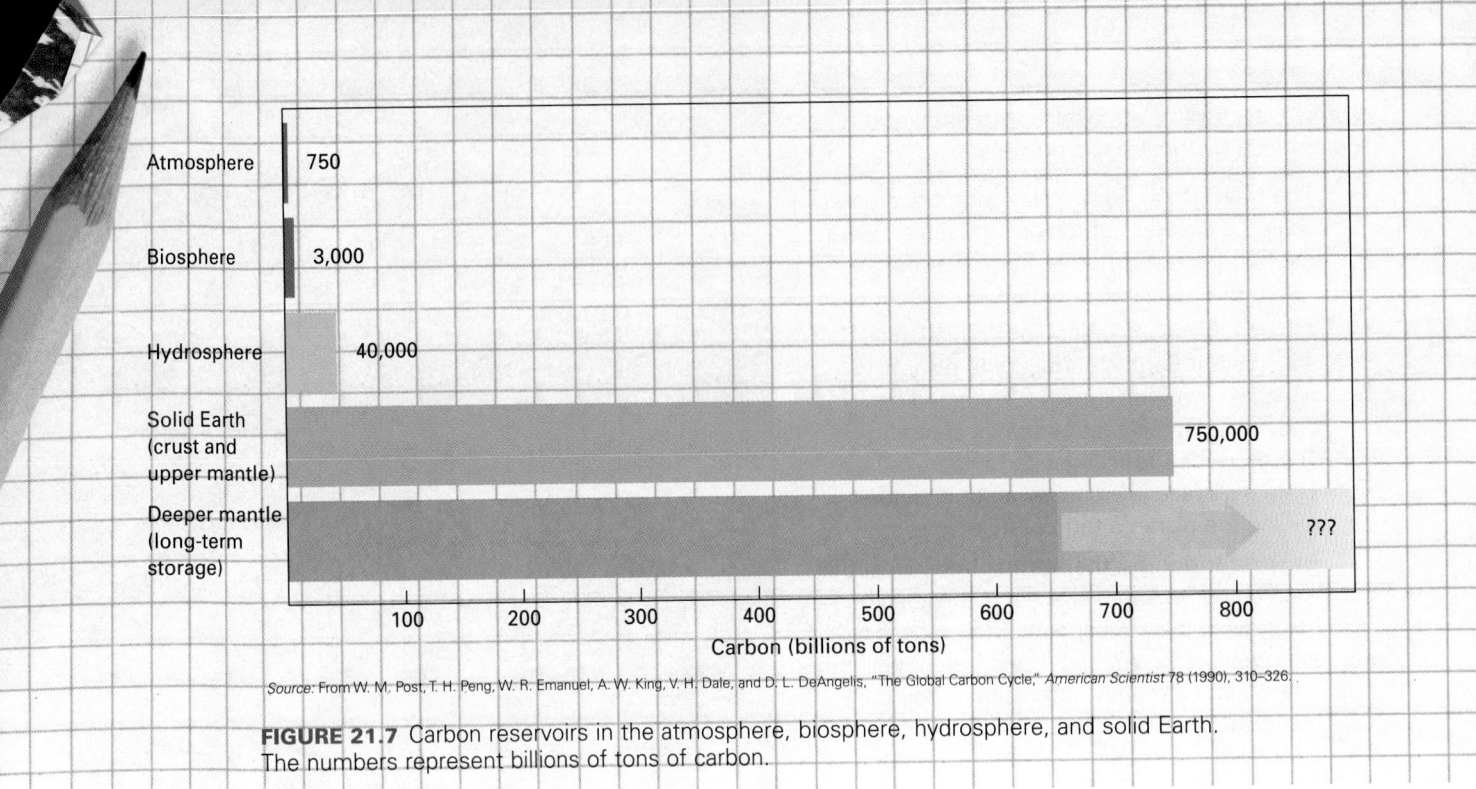

Atmosphere   750

Biosphere   3,000

Hydrosphere   40,000

Solid Earth (crust and upper mantle)   750,000

Deeper mantle (long-term storage)   ???

Carbon (billions of tons)

Source: From W. M. Post, T. H. Peng, W. R. Emanuel, A. W. King, V. H. Dale, and D. L. DeAngelis, "The Global Carbon Cycle," *American Scientist* 78 (1990), 310–326.

FIGURE 21.7 Carbon reservoirs in the atmosphere, biosphere, hydrosphere, and solid Earth. The numbers represent billions of tons of carbon.

and solid crystals of snow and ice. Moreover, water moves freely from one system to another. As a result, "water acts as the Venetian blind of our planet, as its central heating system, and as its refrigerator, all at the same time."[1] We have already discussed many of the components of the complex, and often conflicting, relationship between water and climate. As a brief review:

- Water vapor is the most abundant greenhouse gas. It warms the atmosphere and Earth's surface.

- Clouds reflect sunlight and therefore cool the atmosphere and Earth's surface. But clouds also absorb heat radiating from Earth's surface. Thus, water in the atmosphere causes both warming and cooling.

- Glaciers and snowfields have a high albedo (80 to 90 percent); they reflect sunlight and also cool Earth's climate. Conversely, surface water has a very low albedo, only about 5 percent. As a result, any change from surface water to glaciers, or vice versa, can have a dramatic effect on the planet's albedo and temperature.

- When water evaporates from the ocean surface, solar energy is stored as latent heat of the resultant vapor. The vapor moves vast distances, transporting this heat. The heat is then released when the water vapor condenses to form rain or snow.

- Similarly, heat is released when water freezes to form snow and ice. But once a glacier forms or a portion of Earth's surface becomes snow covered, a lot of heat is required to melt the ice.

- Flowing water weathers rocks and initiates chemical reactions that alter the carbon dioxide concentration in the atmosphere. Carbon dioxide is a greenhouse gas that warms Earth's surface.

- Ocean currents move heat to and from the polar regions. But because more currents flow from the equator toward the poles than the other way around, there is a net transport of heat toward the polar regions. This polar warming effect is counterbalanced by the fact that water evaporates from warm ocean currents. Some of this vapor condenses and falls as snow, thereby increasing the albedo and cooling the polar regions.

Climate models attempt to quantify all of these factors, but clearly the balances are challenging to unravel. For this reason, it is difficult to predict climate changes.

21.5 The Natural Carbon Cycle and Climate

Carbon circulates among the atmosphere, the hydrosphere, the biosphere, and the geosphere and is stored in each of these reservoirs, in proportions you can see in Figure 21.7.

Carbon in the Atmosphere

Oxygen and nitrogen, the most abundant gases in the atmosphere, are transparent to infrared radiation and are not greenhouse gases. Carbon exists in the atmosphere mostly as carbon dioxide (CO_2), and in smaller amounts as methane (CH_4).

1. Heike Langenberg, commentary introducing the special Insight section "Climate and Water," *Nature* 419 (September 12, 2002), 187.

Although only 0.1 percent of the total carbon near Earth's surface is in the atmosphere, this reservoir plays an important role in controlling atmospheric temperature because carbon dioxide and methane are greenhouse gases; they absorb infrared radiation and heat the lower atmosphere. If either of these compounds is removed from the atmosphere, the atmosphere cools; if they are released into the atmosphere, the air becomes warmer.

Carbon in the Biosphere

Carbon is the fundamental building block for all organic tissue. Plants extract carbon dioxide from the atmosphere and build their body parts predominantly of carbon and hydrogen. This process occurs both on land and in the sea. Most of the aquatic fixation of carbon is conducted by microscopic photoplankton. Therefore, healthy terrestrial and aquatic ecosystems play a vital role in removing carbon from the atmosphere.

Most of the carbon is released back into the atmosphere by natural processes, such as respiration, fire, or rotting, which are all part of the carbon cycle as demonstrated in Figure 21.8. However, at certain times and places, organic material does not decompose completely and is stored as fossil fuels—coal, oil, and gas. Thus, plants transfer carbon from the biosphere to rocks of the upper crust.

Carbon in the Hydrosphere

Carbon dioxide dissolves in seawater. Most of it then reacts to form bicarbonate, HCO_3^- (commonly found in your kitchen as baking soda or bicarbonate of soda, $NaHCO_3$) and carbonate $(CO_3)^{2-}$.

The amount of carbon dioxide dissolved in the oceans depends in part on the temperature of the atmosphere and of the oceans. When seawater warms, it releases dissolved carbon dioxide into the atmosphere, causing greenhouse warming. In turn, greenhouse warming further heats the oceans, causing more carbon dioxide to escape. Warmth evaporates seawater as well, and water vapor also absorbs infrared radiation. Clearly, such a feedback mechanism can escalate. A runaway greenhouse effect may be responsible for the high temperature on Venus.

Carbon in the Crust and Upper Mantle

As shown in Figure 21.7, the atmosphere contains about 750 billion tons of carbon. In contrast, the crust and upper mantle contain 1,000 times as much, or 750 trillion tons of carbon. The upper geosphere, combined with the hydrosphere and biosphere, contain almost 800 trillion tons of carbon. Thus, if only a minute fraction of the carbon in the geosphere, hydrosphere, and biosphere is released, the atmospheric concentration of

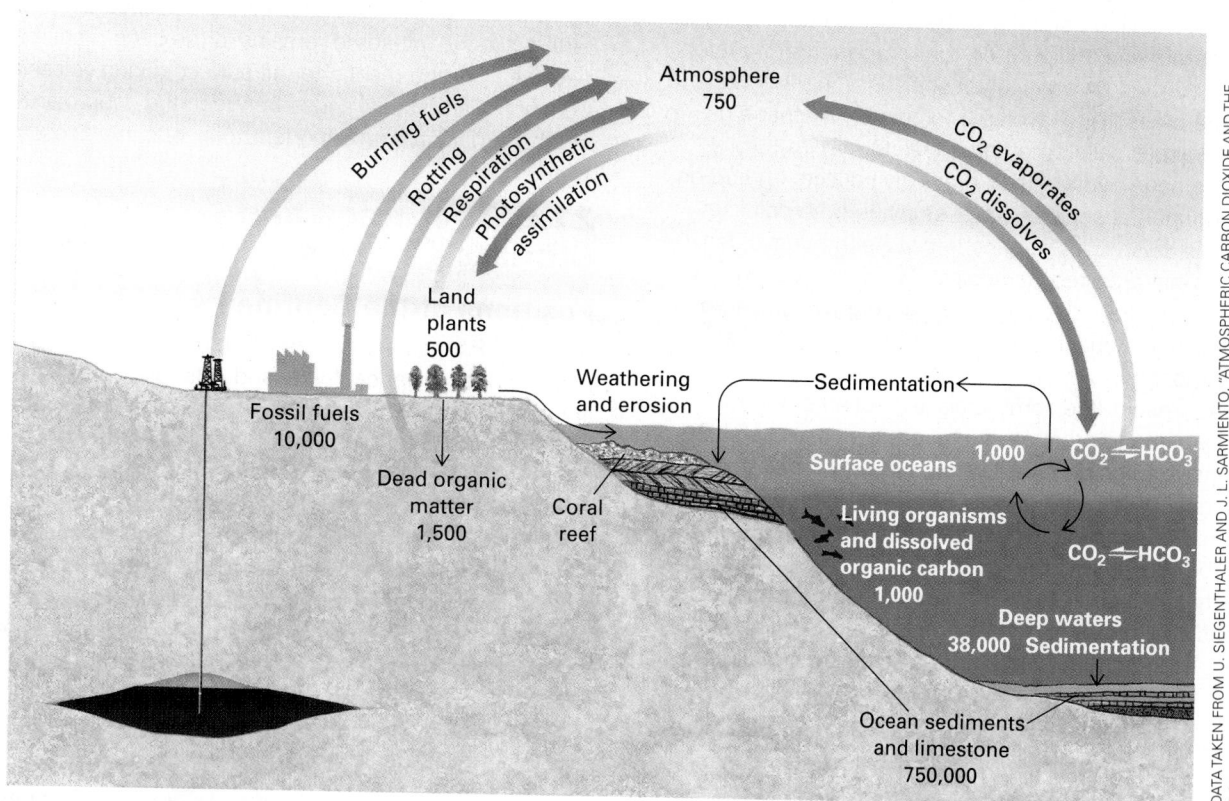

DATA TAKEN FROM U. SIEGENTHALER AND J. L. SARMIENTO, "ATMOSPHERIC CARBON DIOXIDE AND THE OCEAN," NATURE 365 (SEPTEMBER 9, 1993): 119.

FIGURE 21.8 The carbon cycle. The numbers show the size of the reservoirs in billions of tons of carbon.

carbon dioxide can change dramatically, with a draconian effect on climate.

Carbonate Rocks

Marine organisms absorb calcium and carbonate ions from seawater and convert them into calcium carbonate ($CaCO_3$) in shells and other hard parts. This process removes carbon from seawater and causes more atmospheric carbon dioxide to dissolve into the seawater. Thus, formation of shells removes carbon dioxide from the atmosphere. The shells and skeletons of these organisms gradually collect to form limestone.

When sea level falls or tectonic processes raise portions of the sea floor above sea level, limestone and silicate rocks weather by processes that extract additional carbon dioxide from the atmosphere.[2] The exception to that is with acid rain reactions like $CaCO_3 + H_2SO_4$ (aq) $\rightarrow$ Ca (aq) + (SO$_4$) + H_2O + CO_2 (gas), which return carbon dioxide to the air.

Carbon in Fossil Fuels

Carbon is stored in fossil fuels, and carbon dioxide is released when these fuels are burned. Recoverable fossil fuels contain about 4 trillion tons of carbon, 5 times the amount in the atmosphere today. For this reason, scientists are concerned that burning fossil fuels will raise atmospheric carbon dioxide levels.

Methane in Sea-Floor Sediment

When organic material falls to the sea floor and is buried with mud, bacteria decompose it, releasing methane, commonly called natural gas. Between a depth of about 500 meters and 1 kilometer, the temperature of water-saturated mud on continental shelves is low enough and the pressure is favorable to convert methane gas to a frozen solid called methane hydrate.

Methane hydrate then gradually collects in mud on the continental shelves. After studying both drill samples and seismic data, geochemist Keith Kvenvolden of the United States Geological Survey estimated that methane hydrate deposits hold twice as much carbon as all conventional fossil fuels—10 times more than is in the atmosphere.

At present, the commercial extraction of methane hydrates to produce natural gas is impractical. It is expensive to drill in deep water, and a thin layer spread throughout the continental shelves would be prohibitively expensive to exploit. However, scientists are studying links between methane hydrates and climate. Tectonic activity at subduction zones or landslides on continental slopes could release methane from hydrate deposits. Changes in bottom temperatures on continental shelves

resulting from warming of seawater could also release methane from the frozen hydrates. In turn, increased atmospheric methane could trigger global warming. Ice core studies show that global atmospheric methane concentration has changed rapidly in the past, perhaps by sudden releases of oceanic methane hydrates.

About 55 million years ago, near the end of the Paleocene Epoch, climate suddenly warmed and many aquatic and terrestrial species became extinct. According to one model, a change in sea-surface circulation caused equatorial waters to remain in low latitudes. High equatorial temperatures evaporated enough water to increase the salinity of the sea surface. When the salinity reached a threshold value where the surface water was denser than the cold deep water, the warm, salty water sank. The warm, sinking water melted the methane hydrates and released the methane. Aquatic species were poisoned by the methane in the water, and many terrestrial species succumbed to the rapid greenhouse warming.[3]

Carbon in the Deeper Mantle

During subduction, oceanic crust sinks into the mantle (Figure 21.9). The descending plate may carry carbonate rocks and sediment. As this material sinks to greater depths, the carbonate minerals become hot and release carbon dioxide, which is carried back to the surface by volcanic eruptions. Some carbonate rock may be carried into deeper regions of the mantle during subduction. Large quantities of carbon were trapped within Earth during its formation. Much of this carbon escaped early in Earth's history, but some remains in the deep mantle and may rise from the mantle to the surface during volcanic eruptions. Carbon exchanges between the deep mantle and the surface are an important topic of current research.

21.6 Tectonics and Climate Change

Positions of the Continents

A map of Pangea shows that 200 million years ago Africa, South America, India, and Australia were all clustered near the South Pole (Figure 21.10). Because climate is colder at high latitudes than near the equator, continental position alters continental climate.

In addition, continental interiors generally experience colder winters and hotter summers than coastal areas do. When all the continents were joined into a supercontinent, the continental interior was huge, and regional climates must have been different from the climates on many smaller continents with extensive coastlines.

---

2. The net reaction is:
   $CaCO_3 + CO_2 + H_2O = Ca(HCO_3)_2$
   Limestone plus carbon dioxide plus water react to form calcium bicarbonate (soluble).

3. Gerald R. Dickens, Maria M. Castillo, and James C. G. Walker, "A Blast of Gas in the Latest Paleocene: Simulating First-Order Effects of Massive Dissociation of Oceanic Methane Hydrate," *Geology* (March 1997), 259–262.

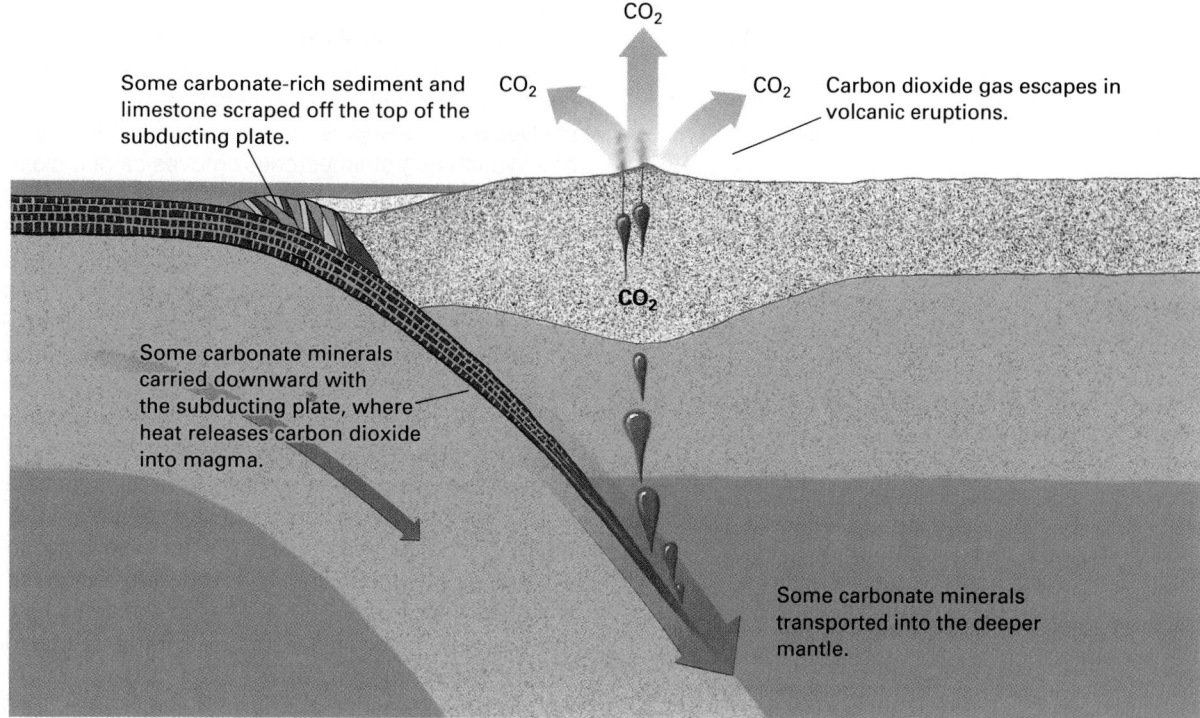

FIGURE 21.9 A subducting oceanic plate carries limestone and other carbonate-rich sediment into the mantle. Some of the carbonate minerals are heated to produce carbon dioxide, which escapes during volcanic eruptions. Some of the carbonate minerals may be stored in the mantle.

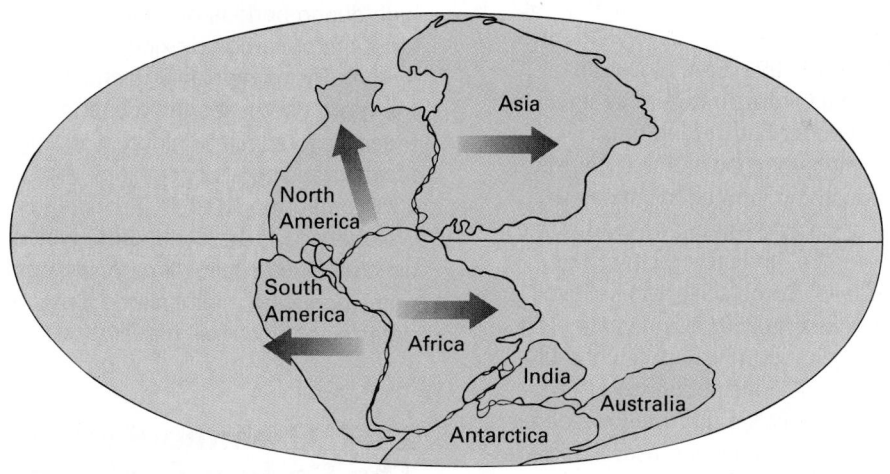

FIGURE 21.10 Two hundred million years ago, when the Pangea supercontinent was assembled, Africa, South America, India, and Australia were all positioned close to the South Pole.

The positions of the continents also influence wind and sea currents, which, in turn, affect climate. For example, today the Arctic Ocean is nearly landlocked, with three straits connecting it with the Atlantic and Pacific oceans. The Bering Strait between Alaska and Siberia is 80 kilometers across, Kennedy Channel between Ellesmere and Greenland is only 40 kilometers across, and a third, wider seaway runs along the east coast of Greenland. Presently, cold currents run southward through Kennedy Channel and the Bering Strait,

and the North Atlantic Drift carries warm water northward along the coast of Norway. If any of these straits were to widen or close, global heat transfer would be affected. Deep-sea currents also transport heat and are affected by continental positions.

Tectonic plates move from 1 to 16 centimeters per year. A plate that moves 5 centimeters per year travels 50 kilometers in a million years. Thus continental motion can change global climate within a geologically short period of time by opening or closing a crucial strait.

(However, even this geologically "short" period of time is extremely long when compared with the rise of human civilization.) Much longer times are required to modify climate by altering the proximity of a continent to the poles or by creating a supercontinent.

Mountains and Climate

Global cooling during the past 40 million years coincided with the formation of the Himalayas and the North American Cordillera,[4] the system of mountain ranges that form the backbone of the continent. Mountains interrupt airflow, altering regional winds. Air cools as it rises and passes over high, snow-covered peaks. However, it is unclear whether this regional cooling could account for the global cooling that accompanied this episode of mountain formation.

Large portions of the Himalayas and the North American Cordillera are composed of marine limestone. Recall from Section 21.5 that when marine limestone weathers, carbon dioxide is removed from the atmosphere. When sea-floor rocks are thrust upward to form mountains, they become exposed to the air. Rapid weathering then may remove enough atmospheric carbon dioxide to cause global cooling.

Volcanoes and Climate

Volcanoes emit ash and sulfur compounds that reflect sunlight and cool the atmosphere. For two years after Mount Pinatubo erupted in 1991, Earth cooled by a few tenths of a degree Celsius. Temperature rose again in 1994 after the ash and sulfur had settled out.

Volcanoes also emit carbon dioxide that warms the atmosphere by absorbing infrared radiation. The net result—warming or cooling—depends on the size of the eruption, its violence, and the proportion of solids and gases released. Some scientists believe that a great eruption in Siberia 250 million years ago cooled the atmosphere enough to cause or contribute to the Permian extinction. A huge sequence of eruptions 120 million years ago, called the *mid-Cretaceous superplume*, may have emitted enough carbon dioxide to warm the atmosphere by 7°C to 10°C. Dinosaurs flourished in huge swamps, and some of the abundant vegetation collected to form massive coal deposits.

How Tectonics, Sea Level, Volcanoes, and Weathering Interact to Regulate Climate

Tectonics, sea level, volcanoes, and weathering are all part of a tightly interconnected Earth system that affects both global and regional climates. When tectonic plates spread slowly, the Mid-Oceanic Ridge system is so narrow that it displaces relatively small amounts of seawater. As a result, sea level falls. When sea level falls, large marine limestone deposits on the continental shelves are exposed as dry land. The limestone weathers. Weathering of limestone removes carbon dioxide from the atmosphere, leading to global cooling. At the same time, when sea-floor spreading is slow, subduction is also slow. Volcanic activity at both the spreading centers and the subduction zones slows down, so relatively small amounts of carbon dioxide are emitted. With small additions of carbon dioxide from volcanic eruptions and removal of atmospheric carbon dioxide by weathering, the atmospheric carbon dioxide concentration decreases and the global temperature cools. In addition, dropping sea level decreases the surface area of the oceans and increases the surface area of the higher-albedo continents. This results in an increase of average global albedo and, consequently, reinforces the global cooling (Figure 21.11). These conditions may have caused the cooling at the end of the Carboniferous Period shown in Figure 21.2.

In contrast, during periods of rapid sea-floor spreading, a high-volume Mid-Oceanic Ridge system raises sea level. Marine limestone beds are submerged, weathering slows, and weathering removes less carbon dioxide from the atmosphere. Volcanic activity is high during periods of rapid plate movement, so large amounts of carbon dioxide are released into the atmosphere. Rising sea level decreases continental area and therefore decreases the average global albedo. All of these factors lead to global warming (Figure 21.12). But rapid spreading also coincides with rapid subduction and accelerated mountain-building, leading to accelerated weathering on the continents, which consumes carbon dioxide. Once again, climate systems are driven by so many opposing mechanisms that it is often difficult to determine which will prevail.

21.7 Greenhouse Effect: The Carbon Cycle and Global Warming

We have learned that the amount of carbon in the atmosphere is determined by many natural factors, including rates of plant growth, mixing of surface ocean water and deep ocean water, growth rates of marine organisms, weathering, the movement of tectonic plates, and volcanic activity. Within the past few hundred years, humans have become an important part of the carbon cycle. About 20 percent of this change occurs by cutting forests and other urbanization of land surfaces, thus reducing the carbon uptake from the atmosphere to plants. The remaining 80 percent occurs when gases are emitted by industry and agriculture.

4. William F. Ruddiman and John Kutzbach, "Plateau Uplift and Climatic Change," *Scientific American* (March 1991), 66ff.

Figure 21.11 Slow subduction = Narrow Mid-Oceanic Ridge = Cooler Earth

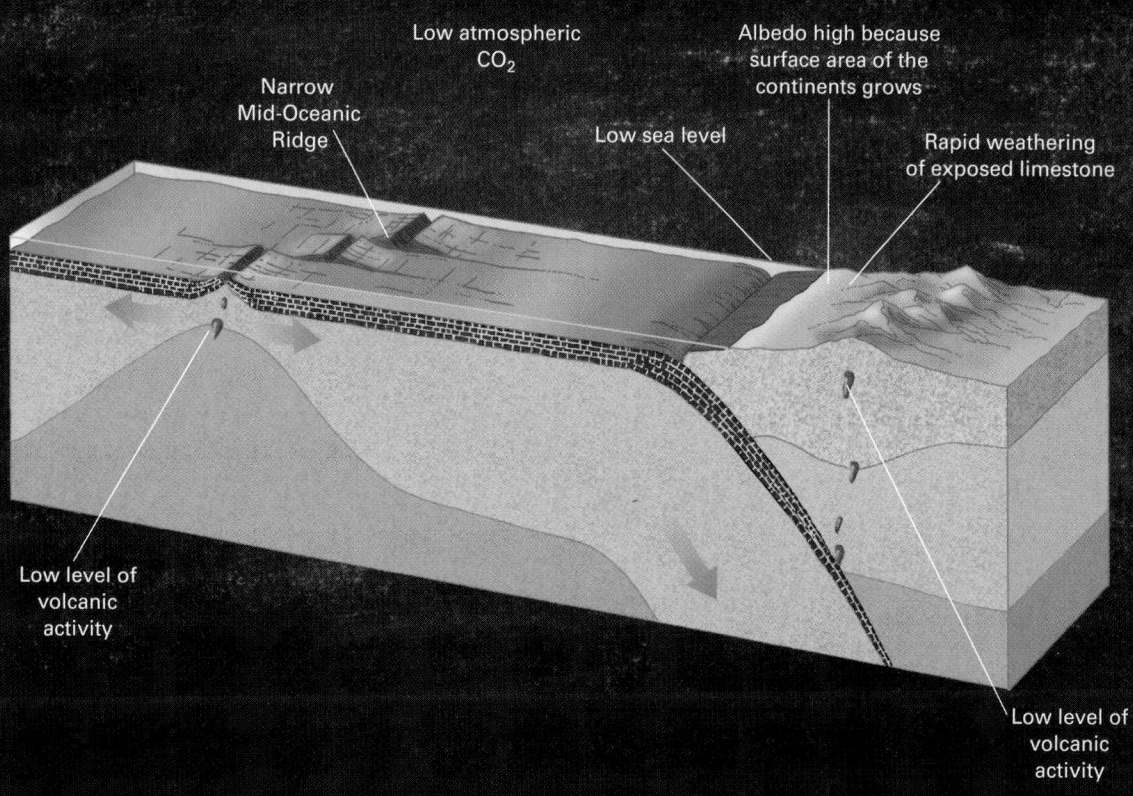

Narrow Mid-Oceanic Ridge

Low atmospheric CO$_2$

Low sea level

Albedo high because surface area of the continents grows

Rapid weathering of exposed limestone

Low level of volcanic activity

Low level of volcanic activity

Figure 21.12 Fast subduction = Wide Mid-Oceanic Ridge = Warmer Earth

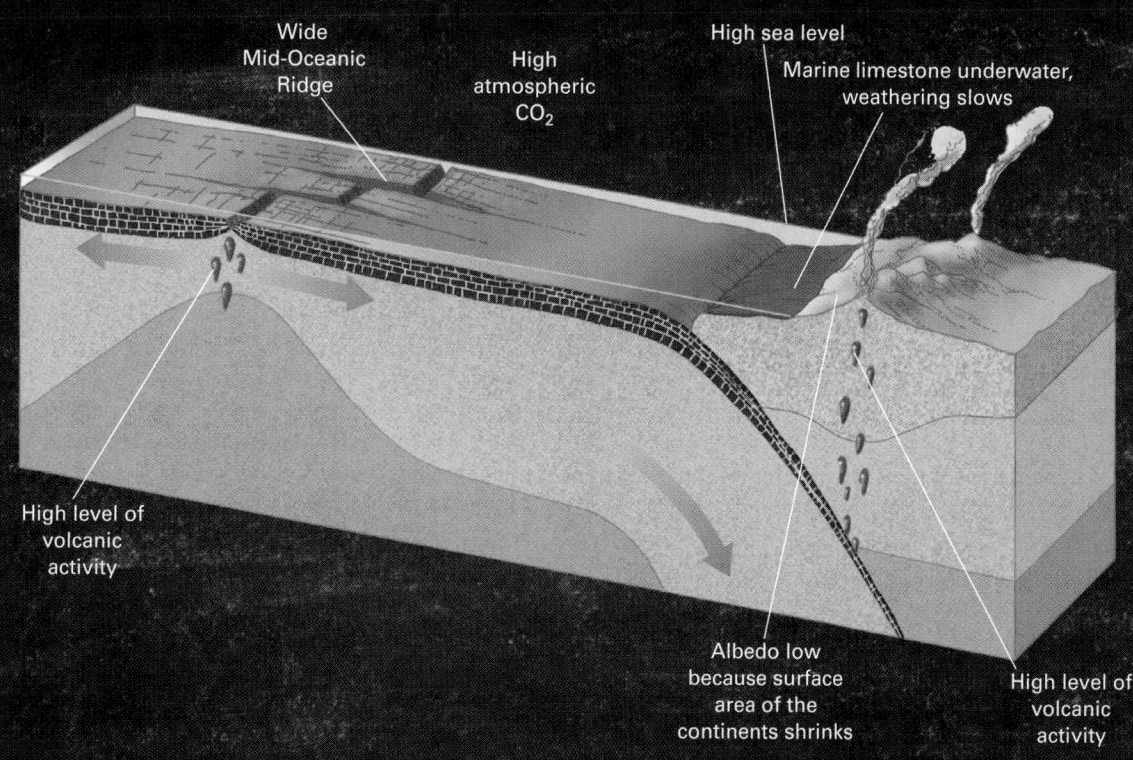

Wide Mid-Oceanic Ridge

High atmospheric CO$_2$

High sea level

Marine limestone underwater, weathering slows

High level of volcanic activity

Albedo low because surface area of the continents shrinks

High level of volcanic activity

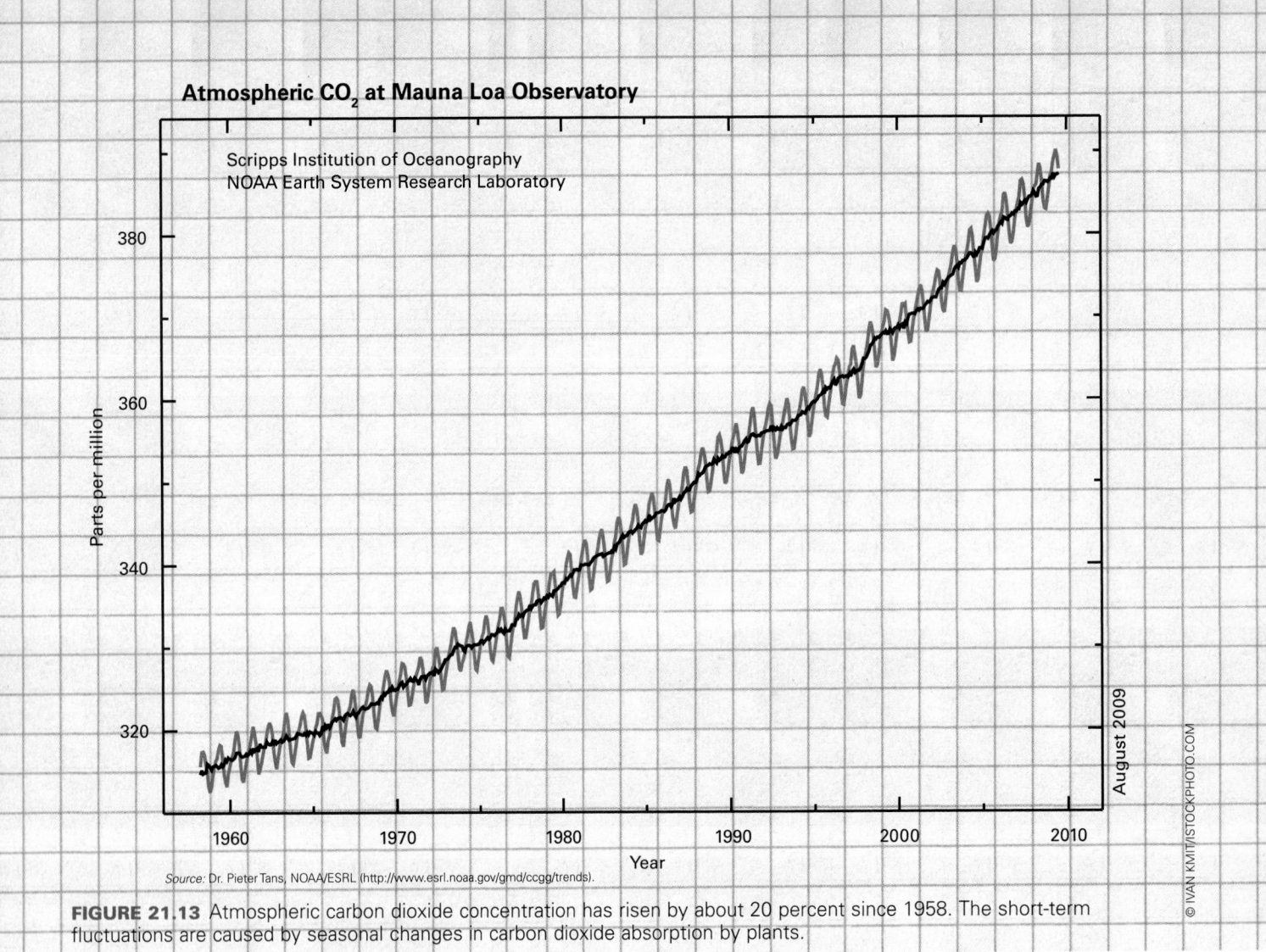

Atmospheric CO$_2$ at Mauna Loa Observatory

Scripps Institution of Oceanography
NOAA Earth System Research Laboratory

Source: Dr. Pieter Tans, NOAA/ESRL (http://www.esrl.noaa.gov/gmd/ccgg/trends).

FIGURE 21.13 Atmospheric carbon dioxide concentration has risen by about 20 percent since 1958. The short-term fluctuations are caused by seasonal changes in carbon dioxide absorption by plants.

Modern industry releases four greenhouse gases—carbon dioxide, methane, chlorofluorocarbons (CFCs), and nitrogen oxides.

People release carbon dioxide whenever they burn fossil fuels or biofuels. This release is inherent in the chemistry of combustion. Carbon in the fuel reacts with oxygen in the air to produce carbon dioxide. Furthermore, once carbon dioxide is released, it is, for all practical purposes, impossible to remove this gas from the atmosphere. If you drive your car to town today, the carbon dioxide released will remain in the atmosphere for centuries. Logging also frees carbon dioxide because stems and leaves are frequently burned and forest litter rots more quickly when it is disturbed by heavy machinery. The recent rise in the concentration of atmospheric carbon dioxide has attracted considerable attention because it is the most abundant industrial greenhouse gas (Figure 21.13).

In addition, several other greenhouse gases are released by modern agriculture and industry. Small amounts of methane are released during some industrial processes. Larger amounts are released from the guts of cows, other animals, and termites, and from rotting that occurs in rice paddies. Today, industry and agriculture combined add about 37×10^{12} grams of methane into the atmosphere every year. Chlorofluorocarbons were, until recently, used as refrigerants and as propellants in aerosol cans. This source is rapidly diminishing because international treaties have banned CFC production. N$_2$O, yet another greenhouse gas, is released from the manufacture and use of nitrogen fertilizer, some industrial chemical syntheses, and from the exhaust of high-flying jet aircraft.

Over the past few decades, a few scientists and some politicians and writers have argued that climate change is not really happening. However, today the data are overwhelming; the Intergovernmental Panel on Climate Change (IPCC) concluded that "warming in the climate system is unequivocal." Henry Pollack, a member of the IPCC, wrote that "use of the word 'unequivocal' leaves no wiggle room. 'Maybe, maybe not' is over. Significant climate change is happening."

The next question is: Are we sure that humans are causing this climate change?

JODHPUR INDIA

An increased demand for manufactured goods means more industry, which spews greater amounts of greenhouse emissions in our atmosphere.

CONSEQUENCES OF GLOBAL WARMING

Agriculture

- Shifts in food-growing areas
- Changes in crop yields
- Increased irrigation demands
- Increased pests, crop diseases, and weeds in warmer areas

Biodiversity

- Extinction of some plant and animal species
- Loss of habitats
- Disruption of aquatic life

Weather Extremes

- Prolonged heat waves and droughts
- Increased flooding
- More intense hurricanes, typhoons, tornadoes, and violent storms

Water Resources

- Changes in water supply
- Decreased water quality
- Increased drought
- Increased flooding

Melting of Arctic Sea Ice and Glaciers

- Changes in albedo
- Rise in sea level
- Changes in temperature and salinity of ocean
- Altered water balance
- Species extinction

Human Population

- Increased deaths
- More environmental refugees
- Increased migration

Forests

- Changes in forest composition and locations
- Disappearance of some forests
- Increased fires from drying
- Loss of wildlife habitats and species

Sea Level and Coastal Areas

- Rising sea levels
- Flooding of low-lying islands and coastal cities

- Flooding of coastal estuaries, wetlands, and coral reefs
- Beach erosion
- Disruption of coastal fisheries
- Contamination of coastal aquifers with saltwater

Human Health

- Increased deaths from heat and disease
- Disruption of food and water supplies
- Spread of tropical diseases to temperate areas
- Increased respiratory disease
- Increased water pollution from coastal flooding

KEY TERMS

atmosphere The gaseous layer above the Earth's surface, mostly nitrogen and oxygen, with smaller amounts of argon, carbon dioxide, and other gases. The atmosphere is held to Earth by gravity and thins rapidly with altitude.

biosphere The zone of Earth comprising all forms of life in the sea, on land, and in the air.

catastrophism The principle stating that infrequent catastrophic geologic events alter the course of Earth history and modify the path of slow geologic time.

core The dense, metallic, innermost region of Earth's geosphere, consisting mainly of iron and nickel. The outer core is molten, but the inner core is solid.

crust The outermost layer of Earth's geosphere, about 7 to 70 kilometers thick and composed of relative low-density silicate rocks.

ecosystem A complex community of individual organisms, interacting with each other and with their physical environment, functioning as an ecological unit in nature.

feedback mechanism A reaction whereby a small initial perturbation in one component affects a different component of Earth's systems, which amplifies the original effect, which perturbs the system even more, which leads to an even greater effect, and so on.

geosphere The solid Earth, consisting of the entire planet from the center of the core to the outer crust.

gradualism A principle stating that geological change occurs as a consequence of slow or gradual accumulation of small events, such as the deposition of sand in a river delta to eventually form a large land mass. More recently, scientists studying biological evolution use the term to describe a theory of evolution that proposes that species change gradually in small increments.

ground water Subsurface water contained in the soil and bedrock of the upper few kilometers of the geosphere, comprising about 0.63 percent of all water in the hydrosphere.

hydrosphere All of Earth's water, which circulates among oceans, continents, glaciers, and atmosphere.

1.1 The Earth's Four Spheres

Earth consists of four spheres: The geosphere is composed of a dense, hot, central core, surrounded by a large mantle that comprises most of Earth's volume, with a thin, rigid crust. The crust and mantle are composed of rock while the core is composed of iron and nickel. The hydrosphere is mostly ocean water. Most of Earth's freshwater is locked in glaciers. Most of the liquid freshwater lies in ground water reservoirs; streams, lakes, and rivers account for only 0.01 percent of the planet's water. The atmosphere is a mixture of gases, mostly nitrogen and oxygen. Earth's atmosphere supports life and regulates climate. The biosphere is the thin zone that life inhabits.

FIGURE 1.1 All of Earth's cycles and spheres are interconnected.

lithosphere The cool, rigid, outer part of Earth, about 100 kilometers thick, which includes the crust and the uppermost mantle.

mantle The rocky, mostly solid layer of Earth's geosphere lying beneath the crust and above the core. The mantle extends from the base of the crust to a depth of about 2,900 kilometers.

system Any combination of interacting components that form a complex whole.

tectonic plates The seven segments of Earth's outermost, cool, rigid shell, consisting of the lithosphere and overlying crust. Tectonic plates float on the weak, plastic rock beneath.

threshold effect A reaction whereby the environment initially changes slowly (or not at all) in response to a small perturbation, but after the threshold is crossed, an additional small perturbation causes rapid change.

uniformitarianism A principle stating that the geologic processes operating today also operated in the past.

1. Explain how water vapor and carbon dioxide influenced the evolution of the climate and atmosphere on Venus, Earth, and Mars.
2. List and briefly describe each of Earth's four spheres.
3. List the three major layers of Earth. Which is/are composed of rock, which is/are metallic? Which is the largest; which is the thinnest?
4. List six types of reservoirs that collectively contain most of Earth's water.
5. What is ground water? Where in the hydrosphere is it located?
6. What two gases compose most of Earth's atmosphere?
7. How thick is Earth's atmosphere?
8. Briefly discuss the size and extent of the biosphere.
9. Define a system and explain why a systems approach is useful in earth science.
10. Briefly explain the statement: "Matter can be recycled, but energy cannot."
11. How old is Earth? When did life first evolve? How long have humans and their direct ancestors been on this planet?
12. Compare and contrast gradualism and catastrophism. Give an example of each type of geologic change.
13. Briefly explain the threshold and feedback effects.
14. Briefly outline the magnitude of human impact on the planet.

1.2 Earth Systems

A system is composed of interrelated, interacting components that form a complex whole. Earth's four major systems—the four spheres—are subdivided into a great many interacting smaller ones. All the spheres continuously exchange matter and energy.

1.3 Time and Rates of Change in Earth Science

Earth is about 4.6 billion years old; life formed at least 3.8 billion years ago; abundant multicellular life evolved about 544 million years ago, and hominids have been on this planet for a mere 5 to 7 million years. The principle of gradualism states that geologic change occurs over a long period of time by a sequence of almost imperceptible events. In contrast, the principle of catastrophism postulates that geologic change occurs mainly during infrequent catastrophic events.

1.4 Threshold and Feedback Effects

A threshold effect occurs when the environment initially changes slowly (or not at all) in response to a small perturbation—but after the threshold is crossed, an additional small perturbation causes rapid and dramatic change. A feedback mechanism occurs when a small initial perturbation in one component affects a different component of Earth systems, amplifying the original effect, which perturbs the system even more and leads to an even greater effect.

1.5 Humans and Earth Systems

Due to our sheer numbers and technological prowess, humans have become a significant engine of change in Earth systems.

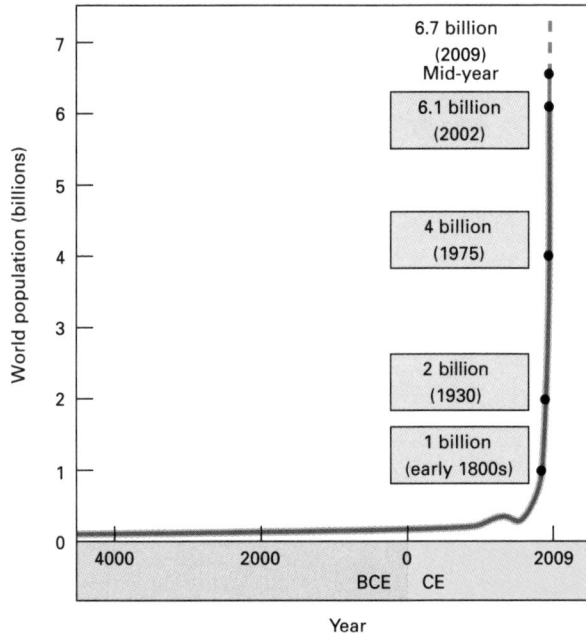

FIGURE 1.8 The human population has increased rapidly since the 1700s. For most of human history, there were fewer than a half-billion people on Earth. In mid-2009, 6.7 billion people inhabited our planet.

*remember

There are more review tools for this course online at 4ltrpress .cengage.com.

KEY TERMS

anion An ion that has a negative charge.

carbonates Minerals whose chemical elements include carbon and oxygen as a major part of their chemical composition; an example is calcite ($CaCO_3$), containing CO_3^{-2} (the "carbonate atom").

cation A positively charged ion.

cleavage The tendency of some minerals to break along flat surfaces, which are planes of weak bonds in the crystal. When a mineral has excellent cleavage, sheet after sheet can be peeled from the crystal, like peeling layers from an onion.

crystal A solid element or compound whose atoms are arranged in a regular, periodically repeated pattern.

crystal face The flat surface that develops if a crystal grows freely in an uncrowded environment. Under perfect conditions, the crystal that forms will be symmetrical.

crystal habit The characteristic shape of an individual crystal, and the manner in which aggregates of crystals grow.

crystalline structure The orderly, repetitive arrangement of atoms in a crystal.

element A substance that cannot be broken down into other substances by ordinary chemical means.

fracture The manner in which minerals break, other than along planes of cleavage.

gem A mineral that is prized for its rarity and beauty rather than for industrial use.

hardness The resistance of a mineral to scratching, controlled by the bond strength between its atoms.

industrial minerals Rocks or minerals that have economic value exclusive of metal ores, fuels, and gems.

ion An atom with an electrical charge, either positive or negative.

2.1 What Is a Mineral?

Minerals are the substances that make up rocks. A mineral is a naturally occurring, inorganic solid with a definite chemical composition and a crystalline structure.

2.2 The Chemical Composition of Minerals

Each mineral consists of chemical elements bonded together in definite proportions, so that its chemical composition can be given as a chemical formula.

TABLE 2.1 The Eight Most Abundant Chemical Elements in the Earth's Crust*

Element	Symbol	Weight Percent	Atom Percent	Volume Percent[†]
Oxygen	O	46.60	62.55	93.80
Silicon	Si	27.72	21.22	0.90
Aluminum	Al	8.13	6.47	0.50
Iron	Fe	5.00	1.92	0.40
Calcium	Ca	3.63	1.94	1.00
Sodium	Na	2.83	2.64	1.30
Potassium	K	2.59	1.42	1.80
Magnesium	Mg	2.09	1.84	0.30
Totals		98.59	100.00	100.00

*Abundances are given in percentages by weight, by number of atoms, and by volume.
[†]These numbers will vary somewhat as a function of the ionic radii chosen for the calculations.
Source: From *Principles of Geochemistry* by Brian Mason and Carleton B. Moore. © 1982 by John Wiley & Sons, Inc.

2.3 The Crystalline Nature of Minerals

Every mineral has a crystalline structure—an orderly, periodically repeated arrangement of its atoms—and therefore every mineral is a crystal. The shape of a crystal is determined by the shape and arrangement of its cells. Every mineral is distinguished from others by its chemical composition and crystal structure.

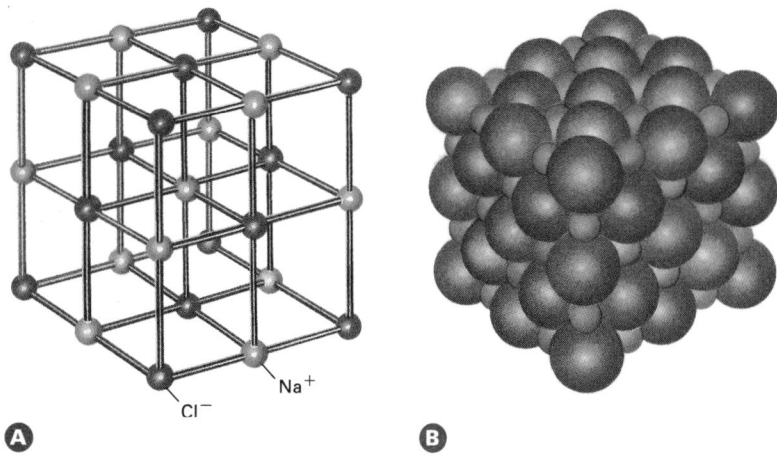

FIGURE 2.2 The orderly arrangement of sodium and chlorine ions in NaCl, or halite. The crystal model in (A) is exploded so that you can see into it; the ions are actually closely packed, as in (B).

luster The quality and intensity of light reflected from the surface of a mineral.

mineral A naturally occurring inorganic solid with a definite chemical composition and a crystalline structure.

Mohs hardness scale A scale, based on a series of 10 fairly common minerals and numbered 1 to 10 (from softest to hardest), used to measure and express the hardness of minerals.

native elements Minerals that consist of only one element and thus the element occurs in the native state (not chemically bonded to other elements). With the exception of atmospheric gases, only about 20 elements occur as native elements.

ore minerals Minerals from which metals or other elements can be profitably recovered.

rock-forming minerals The nine minerals or mineral groups that are most abundant in the Earth's crust and that combine to make most rocks. They are olivine, pyroxene, amphibole, mica, the clay minerals, quartz, feldspar, calcite, and dolomite.

silicate tetrahedron The fundamental building block of all silicate minerals, a pyramid-shaped structure of a silicon ion bonded to four oxygen ions, $(SiO_4)^{4-}$.

silicates Minerals whose chemical elements include silicon and oxygen and whose crystal structures contain silicate tetrahedra. All rocks are composed principally of silicate minerals.

specific gravity The weight of a substance relative to the weight of an equal volume of water.

streak The color of the fine powder of a mineral, usually obtained by rubbing the mineral on an unglazed, porcelain streak plate.

sulfides Minerals whose chemical elements include sulfur (S) bonded to a metal ion; an example is pyrite (FeS_2).

2.4 Physical Properties of Minerals

Most common minerals are easily recognized and identified visually. Identification is aided by observing a few physical properties, including crystal habit, cleavage, fracture, hardness, specific gravity, color, streak, and luster.

2.5 Mineral Classes and the Rock-Forming Minerals

Although more than 3,500 minerals are known in Earth's crust, only the nine rock-forming minerals make up most of Earth's crust. They are feldspar, quartz, pyroxene, amphibole, mica, clay, olivine, calcite, and dolomite. The first seven on this list are silicates; their structures and compositions are based on the silicate tetrahedron, in which a silicon atom is surrounded by four oxygen atoms to form a pyramid-shaped structure. Two carbonate minerals, calcite and dolomite, are also sufficiently abundant to be called rock-forming minerals.

2.6 Commercially Important Minerals

Ore minerals, industrial minerals, and gems are important for economic reasons.

2.7 Harmful and Dangerous Rocks and Minerals

Most minerals and rocks are environmentally safe in their natural states, but some can release environmentally hazardous materials when they are mined, milled, or smelted.

REVIEW QUESTIONS

1. What properties distinguish minerals from other substances?
2. Explain why oil and coal are not minerals.
3. What does the chemical formula for quartz, SiO_2, tell you about its chemical composition? What does $KAlSi_3O_8$ tell you about orthoclase feldspar?
4. What is an atom? An ion? A cation? An anion? What roles do they play in minerals?
5. Every mineral has a crystalline structure. What does this mean?
6. What are the factors that control the shape of a well-formed crystal?
7. What is a crystal face?
8. What conditions allow minerals to grow well-formed crystals? What conditions prevent their growth?
9. List and explain the physical properties of minerals that are most useful for identification.
10. Why do some minerals have cleavage and others do not? Why do some minerals have more than one set of cleavage planes?
11. Why is color often an unreliable property for mineral identification?
12. List the rock-forming minerals. Why are they called rock-forming? Which are silicates? Why are so many of them silicates?
13. Draw a three-dimensional view of a single silicate tetrahedron. Draw the five arrangements of tetrahedra found in the rock-forming silicate minerals. How many oxygen ions are shared between adjacent tetrahedra in each of the five configurations?
14. Make a table with two columns. List the basic silicate structures in the left column. In the right column, list one or more examples of rock-forming minerals for each structure.
15. Explain how mining can release harmful or poisonous materials from rocks that were benign in their natural environment.

KEY TERMS

basement rock The older igneous and metamorphic rock that lies beneath the thin layer of sedimentary rocks and soil covering much of Earth's surface, forming the "basement" of the crust.

bedding Layering that develops as sediments are deposited; also called *stratification*.

bedrock The solid rock that lies beneath soil or unconsolidated sediments; it can be igneous, metamorphic, or sedimentary.

carbonate rocks Bioclastic sedimentary rocks composed of the carbonate minerals (minerals whose chemical elements include carbon and oxygen).

compaction Tight packing of sedimentary grains, usually resulting from the weight of overlying sediment, causing weak lithification and a decrease in porosity.

conglomerate A clastic sedimentary rock that consists of lithified gravel.

country rock The older rock already in an area, cut into by a younger igneous intrusion or mineral deposit.

cross-bedding A sedimentary structure in which wind or water deposits small beds at an angle to the main sedimentary layering.

extrusive igneous rock Igneous rock formed from material that has erupted through the crust onto the surface of Earth; usually fine grained. Also called *volcanic rock*.

foliation The layering in micas and other minerals created by metamorphism.

fossil The imprint, remains, or any other trace of a plant or animal preserved in rock.

igneous rock Rock that forms when magma rises to Earth's surface, cools, and solidifies.

intrusive igneous rock A rock formed when magma solidifies within Earth's crust, without erupting to the surface; usually medium to coarse grained. Also called *plutonic rock*.

lava Fluid magma that flows onto Earth's surface from a volcano or fissure. Also, the rock formed by solidification of the same material.

lithification The process by which loose sediment is converted to solid rock.

magma Molten rock generated from below Earth's crust, from which igneous rock is formed.

3.1 Rocks and the Rock Cycle

Geologists divide rocks into three groups, depending upon how the rocks formed. Igneous rocks solidify from magma. Sedimentary rocks form from clay, sand, gravel, and other sediment that collects at Earth's surface. Metamorphic rocks form when any rock is altered by temperature, pressure, or an influx of hot water. The rock cycle summarizes processes by which rocks continuously recycle in the outer layers of Earth, forming new rocks from old ones. Rock cycle processes exchange energy and materials with the atmosphere, the hydrosphere, and the biosphere.

Solidification · Weathering · Lithification

Igneous rock · Sediment

Magma · Sedimentary rock

Melting · Metamorphism

Metamorphic rock

FIGURE 3.1 The rock cycle shows that rocks change over geologic time. The arrows show paths that rocks can follow as they change.

3.2 Igneous Rocks

Extrusive (or volcanic) igneous rocks are fine-grained rocks that solidify from magma that has erupted onto Earth's surface. Granite and basalt are the two most common igneous rocks. Intrusive (or plutonic) igneous rocks are medium- to coarse-grained rocks that solidify within Earth's crust.

metamorphic grade The intensity of metamorphism that formed a rock; the maximum temperature and pressure attained during metamorphism.

metamorphic rock A rock formed when igneous, sedimentary, or other metamorphic rocks recrystallize in response to elevated temperature, increased pressure, chemical change, and/or deformation.

metamorphism The process by which rocks and minerals change form in response to changes in temperature, pressure, chemical conditions, and/or deformation.

mud cracks Irregular polygonal fractures that develop when mud dries, forming patterns that may be preserved when the mud is lithified.

parent rock Any original rock before it is changed by weathering, metamorphism, or other geological processes.

peat A loose, unconsolidated, brownish mass of partially decayed plant matter; a precursor to coal.

pore space The empty space between particles of rock, sediment, or soil.

precipitation A chemical reaction that produces a solid salt, called a *precipitate*, from a liquid solution.

ripple marks Small, parallel ridges and toughs formed in sediment by wind, water currents, or waves, which are often preserved when the sediment is lithified.

rock cycle The sequence of events in which rocks are formed, destroyed, altered, and reformed by geological processes.

sediment Solid rock or mineral fragments that are transported and deposited by wind, water, gravity, or ice; that are weathered by natural forces, precipitated by chemical reactions, or secreted by organisms; and that accumulate in loose, unconsolidated layers.

sedimentary rock Rock formed when sediment becomes cemented or compacted through the process of lithification.

sedimentary structure Any feature of sedimentary rock formed during deposition or by later sedimentary processes—for example, layering, ripple marks, or fossils.

shale A clastic sedimentary rock that consists of lithified tiny clay minerals and smaller amounts of quartz and organic particles. The organic material in shale is the source of most oil and natural gas.

slaty cleavage A metamorphic foliation producing a parallel fracture pattern that cuts across original sedimentary bedding in micas or other metamorphic rocks.

texture The size, shape, and arrangement of mineral grains, or crystals, in a rock.

weathering The decomposition and disintegration of rocks and minerals at Earth's surface by chemical and physical processes.

3.3 Sedimentary Rocks

Sediment forms by weathering of rocks and minerals. It includes all solid particles such as rock and mineral fragments, organic remains, and precipitated minerals. It is transported by streams, glaciers, wind, and gravity; is deposited in layers; and eventually is lithified to form sedimentary rock. Shale, sandstone, and limestone are the most common kinds of sedimentary rock.

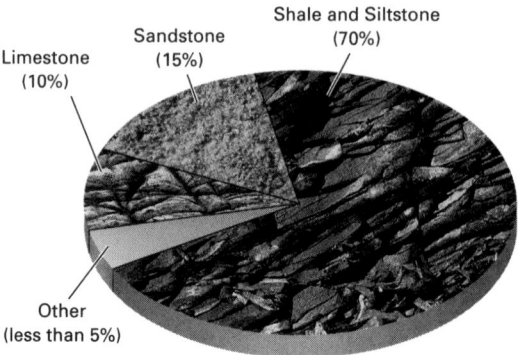

FIGURE 3.5 Earth's sedimentary rocks: Shale, siltstone, and sandstone are clastic rocks that make up more than 85 percent of all sedimentary rocks. Limestone and some "other" sedimentary rocks make up less than 15 percent.

3.4 Metamorphic Rocks

When a rock is heated, when pressure increases, or when hot water alters its chemistry, both its minerals and its textures change in a process called metamorphism. Contact metamorphism affects rocks heated by a nearby igneous intrusion. Burial metamorphism alters rocks as they are buried deeply within Earth's crust. In regions where tectonic plates converge, high temperature and deformation from rising magma and plate movement all combine to cause regional dynamothermal metamorphism. Hydrothermal metamorphism is caused by hot solutions soaking through rocks and is often associated with emplacement of ore deposits. Slate, schist, gneiss, and marble are common metamorphic rocks.

REVIEW QUESTIONS

1. Explain what the rock cycle tells us about Earth processes.
2. Describe ways in which rock cycle processes exchange energy and materials with the atmosphere, the hydrosphere, and the biosphere.
3. What are the three main kinds of rock in Earth's crust?
4. How do the three main types of rock differ from each other?
5. What is magma? Where does it originate?
6. Describe how igneous rocks are classified and named.
7. What are the two most common kinds of igneous rock?
8. Describe and explain the differences between plutonic and volcanic rock.
9. What is sediment? How does it form?
10. How do sedimentary grains become rounded?
11. Where in your own area would you look for rounded sediment?
12. Describe how sediment becomes lithified.
13. What are the differences among shale, sandstone, and limestone?
14. Explain why almost all sedimentary rocks are layered, or bedded.
15. How does cross-bedding form?
16. What is metamorphism? What factors cause metamorphism?
17. What kinds of changes occur in a rock as it is metamorphosed?
18. What is metamorphic foliation? How does it differ from sedimentary bedding?
19. How do contact metamorphism and regional metamorphism differ, and how are they similar?

KEY TERMS

absolute age Time measured in years.

angular unconformity An unconformity in which younger sediment or sedimentary rocks rest on the eroded surface of tilted or folded older rocks.

Archean Eon A division of geologic time 3.8 to 2.5 billion years ago. The oldest known rocks formed at the beginning of, or just prior to, the start of the Archean Eon.

Cenozoic Era The latest of the four eras into which geologic time is subdivided, 65 million years ago to the present.

conformable A term describing sedimentary layers that were deposited continuously without detectable interruption.

correlation The process of establishing the age relationship of rocks or geologic features from different locations on Earth; can be done by comparing characteristics of the layers or the types of fossils found in those layers. There are two types of correlation: *time correlation* (age equivalence) and *lithologic correlation* (continuity of the rock unit).

disconformity A type of unconformity in which the sedimentary layers above and below the unconformity are parallel.

eon The largest unit of geologic time. The most recent eon, the Phanerozoic Eon, is further subdivided into eras.

epoch The smallest unit of geologic time. Periods are divided into epochs.

era A geologic time unit. Eons are divided into eras and, in turn, eras are subdivided into periods.

evolution The biological theory that life-forms have changed in their physical and genetic characteristics over time.

geologic column A composite, columnar diagram that shows the sequence of rocks at a given place or region, arranged to show their position in the geologic time scale.

geologic time scale A chronological arrangement of geologic time subdivided into eons, eras, periods, and epochs.

4.1 Earth Rocks, Earth History, and Mass Extinctions

At least six catastrophic mass extinctions have occurred in geologic history. The greatest, 248 million years ago, annihilated 90 percent of all marine species, two-thirds of reptile and amphibian species, and 30 percent of insect species.

4.2 Geologic Time

Scientists measure geologic time in two ways. Relative age measurement refers only to the order in which events occurred. Absolute age is measured in years.

4.3 Relative Geologic Time

Determinations of relative time are based on geologic relationships among rocks and the evolution of life-forms through time. The criteria for relative dating are summarized in a few simple principles: the principle of original horizontality, the principle of superposition, the principle of crosscutting relationships, and the principle of faunal succession.

4.4 Unconformities and Correlation

Layers of sedimentary rock are conformable if they were deposited without major interruptions. An unconformity represents a major interruption of deposition and a significant time gap between formation of successive layers of rock. In a disconformity, layers of sedimentary rock immediately above and below the unconformity are parallel. An angular unconformity forms when lower layers of rock are tilted and partially eroded prior to deposition of the upper beds. In a nonconformity, sedimentary layers lie on top of an erosion surface developed on igneous or metamorphic rocks. Correlation is the process of establishing the age relationship of rocks from different locations on Earth by comparing characteristics of the layers or the fossils found in those layers. Index fossils and key beds are important tools in time correlation—the demonstration that sedimentary rocks found in different geographic localities formed at the same time.

Conglomerate

Sandstone

COURTESY OF GRAHAM R. THOMPSON/JONATHAN TURK

FIGURE 4.12
A disconformity separates parallel layers of sandstone and overlying conglomerate in Wyoming. Some sandstone layers were eroded away before the conglomerate was deposited.

Hadean Eon The earliest time in Earth's history, ranging from 4.6 billion years ago to 3.8 billion years ago.

half-life The time it takes for half of the atoms of a radioactive isotope in a sample to decompose.

index fossil A fossil that dates the layers where it is found because it came from an organism that is abundantly preserved in rocks, was widespread geographically, and existed as a species or genus for only a relatively short time.

isotopes Atoms of the same element that have the same number of protons but different numbers of neutrons.

key bed A thin, widespread, easily recognized sedimentary layer that can be used for correlation because it was deposited rapidly and simultaneously over a wide area.

mass extinction A sudden, catastrophic event during which a significant part of all life-forms on Earth become extinct.

Mesozoic Era The part of geologic time roughly from 248 to 65 million years ago. Dinosaurs rose to prominence and became extinct during this era.

nonconformity A type of unconformity in which layered sedimentary rocks lie on an erosion surface cut into igneous or metamorphic rocks.

Paleozoic Era The part of geologic time occurring during a period from 543 to 248 million years ago. During this era invertebrates, fishes, amphibians, reptiles, ferns, and cone-bearing trees were dominant.

period A geologic time unit longer than an epoch and shorter than an era.

Phanerozoic Eon The most recent 543 million years of geologic time, including the present, represented by rocks that contain evident and abundant fossils.

Precambrian A term referring to all of geologic time before the Paleozoic Era, encompassing approximately the first 4 billion years of Earth's history. Also refers to all rocks formed during that time.

principle of crosscutting relationships The obvious principle that a rock or feature must first exist before anything can happen to it; thus, if an intrusion of rock cuts across an existing rock, the dike (intrusion) is younger.

principle of faunal succession The principle that species succeeded one another through time in a definite order, so that sedimentary rocks of the same age contain identical fossils and rocks of different ages contain different fossils; therefore, the relative ages of rocks can be identified from their fossils.

principle of original horizontality The principle that most sediment is deposited as nearly

4.5 Absolute Geologic Time

Absolute time is measured by radiometric dating, which relies on the fact that radioactive parent isotopes decay to form daughter isotopes at a fixed known rate as expressed by the half-life of the isotope. The cumulative effects of the radioactive decay process can be determined because the daughter isotopes accumulate in rocks and minerals.

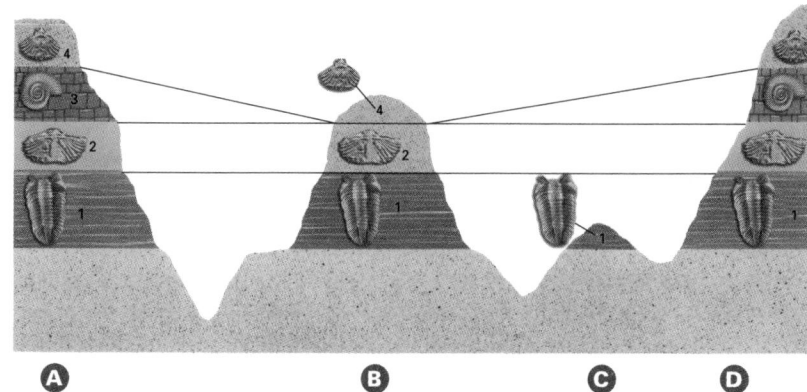

FIGURE 4.17 Index fossils demonstrate age equivalency of sedimentary rocks from widely separated localities. Sedimentary beds containing the same index fossils are interpreted to be of the same age. In this figure, the fossils show that at locality B, layer 3 is missing because layer 4 directly lies on top of layer 2. Either layer 3 was deposited and then eroded away before 4 was deposited at locality B, or layer 3 was never deposited here. At locality C, all layers above 1 are missing, either because of erosion or because they were never deposited.

4.6 The Geologic Column and Time Scale

Worldwide correlation of rocks of all ages has resulted in the geologic column, a composite record of rocks formed throughout the history of Earth. The major units of the geologic time scale are eons, eras, periods, and epochs. The Phanerozoic Eon, the most recent 543 million years of geologic time, is finely and accurately subdivided because sedimentary rocks deposited at this time are often well preserved and they contain abundant well-preserved fossils. In contrast, Precambrian rocks and time are only coarsely subdivided because fossils are scarce and poorly preserved and the rocks are often altered.

REVIEW QUESTIONS

1. Describe the two ways of measuring geologic time. How do they differ?
2. Give an example of how the principle of original horizontality might be used to determine the order of events affecting a sequence of folded sedimentary rocks.
3. Explain a conformable relationship in sedimentary rocks.

horizontal beds, and therefore most sedimentary rocks started out with nearly horizontal layering.

principle of superposition The principle that in any undisturbed layers of sediment or sedimentary rock, the age becomes progressively younger from bottom to top; younger layers always accumulate on top of older layers.

Proterozoic Eon The portion of geological time occurring during a period from 2.5 billion to 543 million years ago.

radiometric dating The process of measuring the absolute age of rocks, minerals, and fossils by measuring the concentrations of radioactive isotopes and their decay products.

relative age An approach to measuring geologic time based on the order

in which events occurred, but not measured in years.

unconformity An interruption in sediment deposits or a break between eroded igneous and overlying sedimentary layers, causing a gap in the geological record for that place. Types of unconformities include disconformity, angular unconformity, and nonconformity.

KEY TERMS

banded iron formation Iron-rich sedimentary rocks composed of alternating iron-rich and silica-rich layers; source of most of the world's supply of iron.

bauxite A gray, yellow, or reddish-brown rock, composed of a mixture of aluminum oxides and hydroxides, that formed as a residual deposit; the principle source of aluminum.

bitumen A thick, sticky, oil-like substance that permeates tar sands and can be converted to crude oil.

black smoker A jet of black water spouting from a fracture in the sea floor, commonly near the Mid-Oceanic Ridge. The black color is caused by precipitation of fine-grained metal sulfide minerals as the hydrothermal solutions cool on contact with seawater.

coal bed methane Methane that is chemically bonded to coal. The methane can be recovered by removing the ground water in a coal bed, which decreases the pressure and allows the methane to separate from the coal as a gas.

crystal settling A process in which the crystals that solidify first from a cooling magma settle to the bottom of the magma chamber because the minerals are more dense than magma; the ultimate result is a layered body of rock, each layer containing different minerals.

disseminated ore deposit A large, low-grade hydrothermal deposit in which generally metal-bearing minerals are widely scattered throughout a rock body; not as concentrated as a hydrothermal vein.

energy resources Geologic resources, including petroleum, coal, natural gas, and nuclear fuels, used for heat, light, work, and communication.

5.1 Mineral Resources

Useful rocks and minerals are called mineral resources; they include both nonmetallic mineral resources and metals. All mineral resources are nonrenewable. Ore is rock sufficiently enriched in one or more minerals to be mined profitably; geologists usually use the term to refer to metallic mineral deposits.

5.2 Ore and Ore Deposits

Four types of geologic processes concentrate elements to form ore: (1) Magmatic processes form ore as magma solidifies. (2) Hydrothermal processes transport and precipitate metals from hot water. (3) Sedimentary processes form placer deposits, evaporite deposits, and banded iron formations. (4) Weathering removes easily dissolved elements from rocks and minerals, leaving behind residual ore deposits such as bauxite.

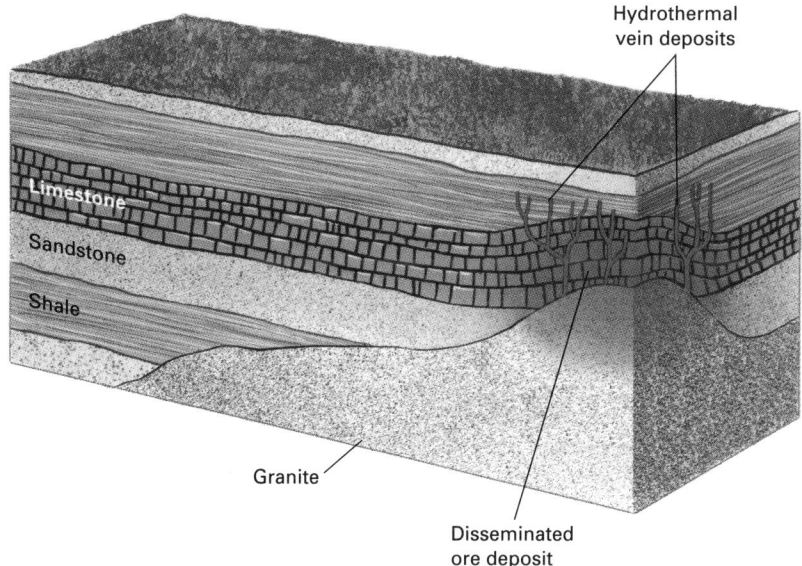

FIGURE 5.3 Hot water scavenges metals from country rock and deposits metallic minerals in ore-rich veins that fill fractures in bedrock. It also deposits low-grade disseminated metal ore in large volumes of rock surrounding the veins.

5.3 Mineral Reserves

Mineral reserves are the known amount of ore in the ground.

5.4 Mines and Mining

Metal ores and coal are extracted from underground mines and surface mines.

fossil fuels Energy resources including petroleum, coal, and natural gas, which formed from the partially decayed remains of plants and animals; they are nonrenewable and unrecyclable.

fuel cell An electrochemical energy-conversion device that produces electricity from an external supply of fuel, such as hydrogen.

hydrogen economy An energy economy in which electricity is used to dissociate water into hydrogen and oxygen, and the hydrogen is used as a fuel.

hydrothermal processes Geologic processes in which hot water or steam dissolves metals and minerals from rocks or magma; the solutions then seep through cracks before cooling, to create ore deposits.

hydrothermal vein deposit A rich, sheet-like mineral deposit that forms when dissolved minerals, precipitated from hot water solutions, fill a fault or other fracture.

kerogen The waxy, solid organic material in

oil shales that yields oil when the shales are heated and distilled; the precursor of liquid petroleum.

magmatic processes Geologic processes that form ore deposits as liquid magma solidifies into igneous rock.

manganese nodule A manganese-rich, potato-shaped rock found on the ocean floor.

mineral reserves A term to describe the known supply of ore in the ground; can be used on a local, national, or global scale.

mineral resources Geologic resources including both metal ore and nonmetallic minerals.

natural gas A mixture of naturally occurring light hydrocarbons composed mainly of methane, CH_4, that is used for home heating, cooking, and to fuel large electrical generating plants.

nonmetallic mineral resources Economically useful rocks or minerals that are not metals—such as salt, building stone, sand, and gravel.

nuclear fuels Radioactive isotopes, such as those of uranium, used to generate electricity in nuclear reactors.

oil shale A kerogen-bearing sedimentary rock that yields liquid or gaseous hydrocarbons when heated.

petroleum A complex liquid mixture of hydrocarbons, formed from decayed plant and animal matter, that can be extracted from sedimentary strata and refined to produce propane, gasoline, and other fuels. Also called *crude oil* or simply *oil*.

placer deposit A surface mineral deposit formed along streambeds, beneath waterfalls, or on beaches when water currents slow down and deposit high-density minerals.

reservoir Oil-saturated permeable rock in which liquid petroleum or gas accumulates.

residual ore deposit A mineral deposit formed from relatively insoluble ions left in the soil near Earth's surface after most of the soluble ions were dissolved and removed by abundant water.

secondary and tertiary recovery techniques Methods of extracting oil or natural gas by artificially augmenting the reservoir energy, as by injection of water, detergent, pressurized gas, or other fluid.

solar cell A device that produces electricity directly from sunlight; also sometimes called a *photovoltaic (PV) cell*.

source rock The shale or other sedimentary rock in which oil or natural gas originates.

submarine hydrothermal ore deposit An ore deposit that forms when hot seawater dissolves metals from sea-floor rocks and then, as it rises through the upper layers of oceanic crust, cools and precipitates the metals.

5.5 Energy Resources: Coal, Petroleum, and Natural Gas

One important energy resource is fossil fuels: coal, oil, and natural gas. Fossil fuels are nonrenewable and unrecyclable. Plant matter decays to form peat. Peat converts to coal when it is buried and subjected to elevated temperature and pressure. Petroleum forms from the remains of organisms that settle to the ocean floor or lake bed and are incorporated into source rock. The organic matter converts to liquid oil when it is buried and heated. The petroleum then migrates to a reservoir, where an oil trap retains it. Natural gas forms in source rock or an oil reservoir subjected to high temperature, and consequently many oil fields contain a mixture of oil with natural gas floating above the heavier liquid petroleum.

5.6 Energy Resources: Tar Sands and Oil Shale

Secondary and tertiary recovery can extract additional supplies of petroleum from old wells and from tar sands and oil shale.

5.7 Energy Resources: Renewable Energy

Solar, wind, geothermal, hydroelectric, and biomass fuels are renewable sources of energy.

5.8 Energy Resources: Nuclear Fuels and Reactors

Nuclear power is expensive, and questions about the safety and disposal of nuclear wastes have diminished its future in the United States. Nuclear fuels, like mineral resources, are nonrenewable, although uranium is abundant. Inexpensive uranium ore will be available for a century or more.

5.9 Conservation as an Alternative Energy Resource

The single quickest and most effective way to decrease energy consumption and to prolong the availability of fossil fuels is to conserve energy.

5.10 Energy for the 21st Century

Alternative energy resources currently supply a small fraction of our energy needs but have the potential to provide abundant renewable energy.

REVIEW QUESTIONS

1. Describe the two major categories of geologic resources.
2. Describe the differences between nonrenewable and renewable resources. List one example of each.
3. Discuss the formation of hydrothermal ore deposits.
4. Describe the unique advantages of hydrogen fuel.
5. What is ore? What are mineral reserves? Describe three factors that can change estimates of mineral reserves.

AGRICULTURAL STABILIZATION AND CONSERVATION SERVICE/USDA

FIGURE 5.8 The Bingham Canyon, Utah, open-pit copper mine is the largest human-created excavation on Earth. It is 4 kilometers in diameter and 0.8 kilometer deep.

surface mine A hole excavated into Earth's surface for the purpose of recovering mineral or fuel resources.

tar sands Sand deposits permeated with heavy oil and an oil-like substance called bitumen.

underground mine A mine consisting of subterranean passages that commonly follow ore veins or coal seams.

KEY TERMS

asthenosphere The portion of the upper mantle just beneath the lithosphere, extending from a depth of about 100 kilometers to about 350 kilometers below the surface of Earth and consisting of weak, plastic rock where magma may form.

continental drift The theory proposed by Alfred Wegener that Earth's continents were once joined together and later split and drifted apart. The continental drift theory has been replaced by the more complete plate tectonics theory.

continental rifting The process by which a continent is pulled apart at a divergent plate boundary.

convection The upward and downward flow of fluid material in response to heating and cooling. Convection occurs slowly in Earth's mantle and much more quickly in the oceans and the atmosphere.

convergent boundary A plate boundary where two tectonic plates move toward each other or collide head-on.

divergent boundary A plate boundary where tectonic plates move apart from each other and new lithosphere is continuously forming; also called a *spreading center* or a *rift zone*.

Gondwanaland The southern part of Pangea, consisting of what is now South America, Africa, Antarctica, India, and Australia.

hot spot A persistent volcanic center thought to be located directly above a rising plume of hot mantle rock.

isostasy The concept that the lithosphere floats on the asthenosphere as an iceberg floats on water.

Laurasia The northern part of Pangea, consisting of what is now North America and Eurasia.

6.1 Alfred Wegener and the Origin of an Idea: The Continental Drift Hypothesis

Alfred Wegener's hypothesis of continental drift foreshadowed the theory of plate tectonics, which provides a unifying framework for much of modern geology.

6.2 The Earth's Layers

Earth is a layered planet. The crust is its outermost layer and varies from 4 to 70 kilometers thick. The mantle extends from the base of the crust to a depth of 2,900 kilometers, where the core begins. The lithosphere is the cool, hard, strong outer 75 to 125 kilometers of Earth; it includes all of the crust and the uppermost mantle. The hot, plastic asthenosphere extends to 350 kilometers in depth. The core is mostly iron and nickel and consists of a liquid outer layer and a solid inner sphere.

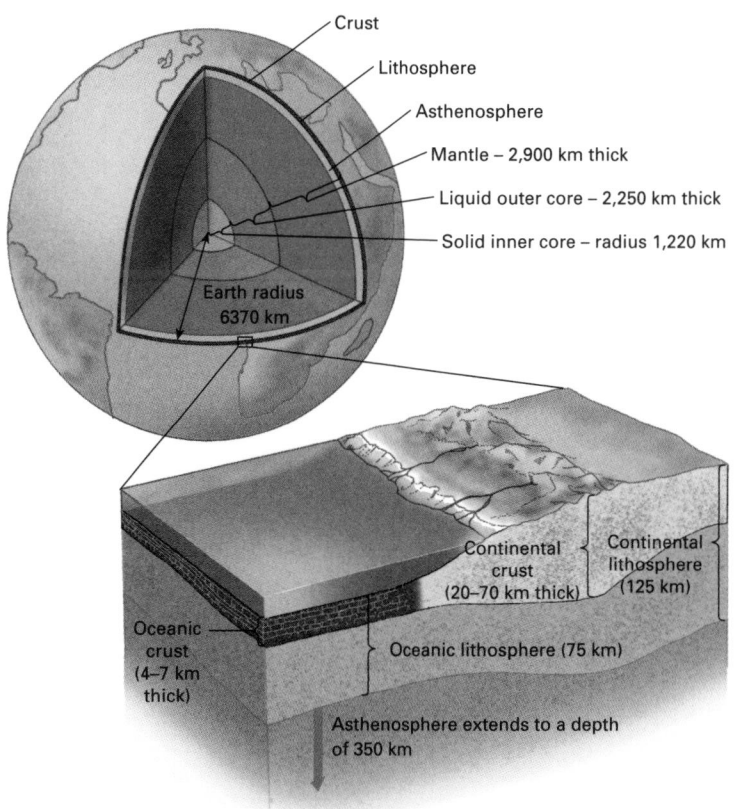

FIGURE 6.4 Earth is a layered planet. The insert is drawn on an expanded scale to show near-surface layering. Note that the average thickness of the lithosphere varies from about 75 kilometers beneath the oceans to about 125 kilometers beneath continents.

6.3 The Sea-Floor Spreading Hypothesis

The hypothesis of sea-floor spreading was proposed as a general model for the origin of all oceanic crust.

magnetic reversal A change in Earth's magnetic field in which the north magnetic pole becomes the south magnetic pole and vice versa; has occurred on average every 500,000 years over the past 65 million years.

mantle plume A relatively small rising column of mantle rock that is hotter than surrounding rock. As pressure decreases in a rising plume, magma forms at a hot spot.

Mid-Atlantic Ridge The portion of the Mid-Oceanic Ridge system that lies in the middle of the Atlantic Ocean, halfway between North America and South America to the west, and Europe and Africa to the east.

Mid-Oceanic Ridge system The undersea mountain chain that forms at the boundary between divergent tectonic plates within oceanic crust. It circles the planet like the seam on a baseball, forming Earth's longest mountain chain.

normal magnetic polarity A magnetic orientation the same as that of Earth's current magnetic field.

Pangea The supercontinent that existed when all Earth's continents were joined together, about 300 million to 200 million years ago, first identified and named by Alfred Wegener.

plate boundary A fracture or boundary that separates two tectonic plates.

plate tectonics A theory of global tectonics stating that the lithosphere is segmented into several plates that move about relative to one another by floating on and gliding over the plastic asthenosphere. Seismic and tectonic activity occur mainly at the plate boundaries.

reversed magnetic polarity Magnetic orientations in rock that are opposite to the current orientation of Earth's magnetic field.

sea-floor spreading The hypothesis that segments of oceanic crust are separating at the Mid-Oceanic Ridge.

subduction The process in which two lithospheric plates of different densities converge and the denser one sinks into the mantle beneath the other.

subduction zone A long narrow region at a convergent boundary where a lithospheric plate is sinking into the mantle during subduction; also referred to as *subduction boundary*.

supercontinent A continent, such as Alfred Wegener's Pangea, consisting of all or most of Earth's continental crust joined together to form a single, large landmass. At least three supercontinents are thought to have existed during the past 2 billion years, and each broke apart after a few hundred million years.

transform boundary A plate boundary where two tectonic plates slide horizontally past one another.

6.4 The Theory of Plate Tectonics

Plate tectonics theory is the concept that the lithosphere floats on the asthenosphere and is segmented into seven major tectonic plates, which move relative to one another by gliding over the asthenosphere. Most of Earth's major geological activity occurs at plate boundaries. Three types of plate boundaries exist: (1) new lithosphere forms and spreads outward at a divergent boundary, or spreading center; (2) two lithospheric plates move toward each other at a convergent boundary; and (3) two plates slide horizontally past each other at a transform boundary.

6.5 The Anatomy of a Tectonic Plate

Volcanoes, earthquakes, and mountain building occur near plate boundaries. Interior parts of lithospheric plates are tectonically stable. Tectonic plates move horizontally at rates that vary from 1 to 16 centimeters per year. Plate movements carry continents across the globe, cause ocean basins to open and close, and affect climate and the distribution of plants and animals.

6.6 Why Plates Move: The Earth as a Heat Engine

Mantle convection and movement of lithospheric plates can occur because the mantle is hot, plastic, and capable of flowing. The entire mantle, from the top of the core to the crust, convects in huge cells. Horizontally moving tectonic plates are the uppermost portions of convection cells. Convection occurs because (1) the mantle is hottest near its base, (2) new lithosphere glides downslope away from a spreading center, and (3) the cold leading edge of a plate sinks into the mantle and drags the rest of the plate along.

6.7 Supercontinents

Supercontinents may assemble, split apart, and reassemble every few hundred million years.

6.8 Isostasy: Vertical Movement of the Lithosphere

The concept that the lithosphere floats on the asthenosphere is called isostasy. When weight, such as a glacier, is added to or removed from Earth's surface, the lithosphere sinks or rises. This vertical movement in response to changing burdens is called isostatic adjustment.

6.9 How Plate Movements Affect Earth Systems

The movements of tectonic plates help shape Earth's surface by generating volcanic eruptions and earthquakes, building mountain ranges, and changing the global distributions of continents and oceans. Tectonic activities also impact climate.

REVIEW QUESTIONS

1. Briefly describe Alfred Wegener's theory of continental drift. What evidence supported his ideas? How does Wegener's theory differ from the modern theory of plate tectonics?
2. Briefly describe the sea-floor spreading hypothesis and the evidence used to develop the hypothesis. How does this idea differ from Wegener's theory and the theory of plate tectonics?
3. Draw a cross-sectional view of Earth. List all the major layers and the thickness of each.
4. Describe the physical properties of each of Earth's layers.
5. Describe and explain the important differences between the lithosphere and the asthenosphere.
6. What properties of the asthenosphere allow the lithospheric plates to glide over it?
7. Describe some important differences between the crust and the mantle.
8. Describe some important differences between oceanic crust and continental crust.

KEY TERMS

660-kilometer discontinuity A boundary in the mantle, at a depth of about 660 kilometers, where seismic wave velocities increase because pressure is great enough that the minerals in the mantle recrystallize to form denser minerals.

Benioff zone A zone of earthquake activity along the upper portion of a sinking plate, as it scrapes past the opposing plate, in a subduction zone.

body waves Seismic waves that travel through the interior of Earth, carrying energy from the earthquake's focus to the surface.

earthquake A sudden motion or trembling of Earth caused by the abrupt release of slowly accumulated elastic energy in rocks.

epicenter The point on Earth's surface directly above the initial rupture point (focus) of an earthquake.

fault creep A continuous, slow movement of solid rock along a fault, resulting from a constant stress acting over a long time. Creeping faults do not usually have large earthquakes.

focus The initial rupture point of an earthquake, typically lying below Earth's surface.

foreshocks Small earthquakes that precede a large quake by a few seconds to a few weeks.

liquefaction A geological process in which a soil loses its shear strength during an earthquake and becomes a fluid.

Mercalli scale A scale of earthquake intensity that expresses the strength of an earthquake based on its destructive power and its effects on buildings and people, without accurately measuring the energy released by the quake.

Mohorovičić discontinuity The boundary between the crust and the mantle, identified by a change in the velocity of seismic waves; also called the *Moho*.

moment magnitude An earthquake scale that measures the amount of movement and the surface area of a fault during a quake, closely reflecting the total amount of energy released.

7.1 Anatomy of an Earthquake

An earthquake is a sudden motion or trembling of Earth caused by the abrupt release of slowly accumulated energy in rocks. Most earthquakes occur along tectonic plate boundaries. Earthquakes occur either when the elastic energy accumulated in rock exceeds the friction that holds rock along a fault, or when the elastic energy exceeds the strength of the rock and the rock breaks.

7.2 Earthquake Waves

Seismology is the study of earthquakes and the nature of Earth's interior based on evidence from seismic waves. An earthquake starts at the initial point of rupture, called the focus, typically lying below Earth's surface. The location on Earth's surface directly above the focus is the epicenter. Seismic waves include body waves, which travel through the interior of Earth, and surface waves, which travel on the surface. P waves are compressional body waves that cause alternate compression and expansion of the rock. They are the first or "primary" waves to reach an observer. S waves travel slower than P waves and are the "secondary" body waves to reach an observer. They consist of a shearing motion and travel through solids but not liquids. Surface waves travel more slowly than either type of body wave. Seismic waves are recorded on a seismograph; a record of earth vibration is called a seismogram. Early in the 20th century, geologists used the Mercalli scale to record the extent of earthquake damage. The Richter scale expresses earthquake magnitudes. Modern geologists use the moment magnitude scale to record the energy released during an earthquake. The distance from a seismic station to an earthquake is calculated by constructing a time-travel curve, which calibrates the difference in arrival times between S and P waves. The epicenter can be located by measuring the distance from three or more seismic stations.

7.3 Earthquakes and Tectonic Plate Boundaries

Earthquakes are common at all three types of plate boundaries. The San Andreas Fault zone is an example of a strike-slip fault along a transform plate boundary. Along some portions of the fault zone, rocks slip past one another at a continuous, snail-like pace called fault creep. In other regions, friction prevents slippage until elastic deformation builds and is eventually released in a large earthquake. Subduction zone earthquakes occur along the Benioff zone when the subducting plate slips suddenly. Earthquakes occur at divergent plate boundaries as blocks

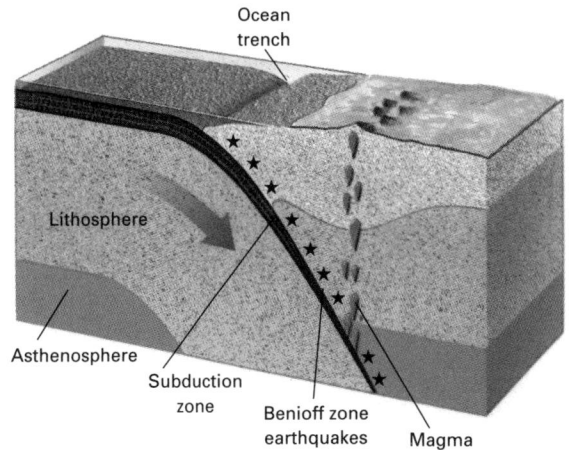

FIGURE 7.17 A descending lithospheric plate generates magma and earthquakes in a subduction zone. Earthquake foci, shown by stars, concentrate along the upper portion of the subducting plate, called the Benioff zone.

P waves Body waves that travel faster than other seismic waves and are the first or "primary" waves to reach an observer; formed by alternate compression and expansion of rock, shaking the ground with a back-and-forth movement parallel to the direction of the wave.

plastic deformation Deformation that occurs without fracture after a rock's elastic limit is reached, when it continues to deform, like putty, while still solid. The rock keeps its new shape and does not store the energy used to deform it; thus, earthquakes do not occur when rocks deform plastically.

Richter scale A scale of earthquake magnitude that expresses the amount of energy released; calculated from the amplitude of the largest body wave on a standardized seismograph, although not a precise measure of earthquake energy.

S waves Seismic waves that travel slower than P waves and are the "secondary" waves to reach an observer; sometimes called *shear waves* because of the shearing motion in which they shake the ground, perpendicular to the direction of wave travel.

San Andreas Fault zone A zone of strike-slip faults, extending from San Francisco to San Diego, which forms the transform boundary between the Pacific plate and the North American plate. Frequent earthquakes occur along this fault zone.

seismic wave An elastic wave that travels through rock, produced by an earthquake or explosion.

seismogram The record of Earth vibration made by a seismograph.

seismograph An instrument that records seismic waves.

seismology The study of earthquakes and the nature of Earth's interior based on evidence from seismic waves.

strike-slip fault A fault whose surface is vertical but on which the rocks on opposite sides move horizontally.

surface waves Seismic waves that radiate from the earthquake's epicenter and travel along the surface of Earth or along a boundary between layers within Earth.

time-travel curve A graph that records arrival times of P and S earthquake waves, used to measure the distance from a recording station to an earthquake epicenter.

tsunami A large destructive sea wave, produced by an undersea earthquake or volcano; sometimes called a *tidal wave*, although it has nothing to do with tides.

of lithosphere along the fault drop downward. Earthquakes occur in plate interiors along old faults.

7.4 Earthquake Prediction

Long-term earthquake prediction is based on the observation that most earthquakes occur on preexisting faults at tectonic plate boundaries. Short-term prediction is based on occurrences of foreshocks, release of radon gas, and changes in the land surface, the water table, and electrical conductivity.

7.5 Earthquake Damage and Hazard Mitigation

Earthquake damage is influenced by rock and soil type, construction design, and the likelihood of fires, landslides, and tsunamis.

7.6 Studying the Earth's Interior

Earth's internal structure and properties are known by studies of earthquake wave velocities and refraction and reflection of seismic waves as they pass through Earth. The boundary between the crust and the mantle, called the Mohorovičić discontinuity (or the Moho), lies at a depth ranging from 4 to 70 kilometers. Oceanic crust is thinner than continental crust, and continental crust is thicker under mountain ranges than it is under plains. The mantle is almost 2,900 kilometers thick and composes about 80 percent of Earth's volume. The upper portion of the mantle is part of the hard and rigid lithosphere. Beneath the lithosphere, the asthenosphere is plastic and partially melted. Beneath the lithosphere, the mantle is stronger and less plastic; at the 660-kilometer discontinuity, the mineral composition of the mantle changes. Wave and density studies show that the outer core is liquid iron and nickel and the inner core is solid iron and nickel.

7.7 Earth's Magnetism

Flowing metal in the outer core generates Earth's magnetic field.

REVIEW QUESTIONS

1. Explain how energy is stored prior to, and then released during, an earthquake.
2. Give two mechanisms that can release accumulated elastic energy in rocks.
3. Why do most earthquakes occur at the boundaries between tectonic plates? Are there any exceptions?
4. Define *focus* and *epicenter*.
5. Discuss the similarities and differences among P waves, S waves, and surface waves.
6. Explain how a seismograph works. Sketch what an imaginary seismogram would look like before and during an earthquake.
7. Describe the similarities and differences between the Richter and moment magnitude scales. What is actually measured and what information is obtained?
8. Describe how the epicenter of an earthquake is located.
9. Discuss earthquake mechanisms at the three types of tectonic plate boundaries.
10. What is the Benioff zone? At what type of tectonic boundary does it occur?
11. Why do only shallow earthquakes occur along the Mid-Oceanic Ridge?
12. Discuss earthquake mechanisms at plate interiors.

KEY TERMS

aa Lava that has a jagged, rubbly, broken surface.

ash flow A mixture of volcanic ash, larger pyroclastic particles, and gas that flows rapidly along Earth's surface as a result of an explosive volcanic eruption.

ash-flow tuff A pyroclastic rock formed when an ash flow solidifies.

batholith A large pluton, exposed over more than 100 square kilometers of Earth's surface.

caldera A large circular depression created by the collapse of the magma chamber after an explosive volcanic eruption.

cinder cone A small volcano, typically less than 300 meters high, made up of loose, pyroclastic fragments blasted out of a central vent; usually active for only a short time.

cinders Glassy, pyroclastic volcanic fragments 4 to 32 millimeters in size.

columnar joints Regularly spaced cracks that commonly develop in lava flows, growing downward, forming five- or six-sided columns.

composite cone *or* **stratovolcano** A steep-sided volcano formed by an alternating series of lava flows and pyroclastic eruptions and marked by repeated eruption.

crater A bowl-like depression at the summit of a volcano, created by volcanic activity.

dike A sheetlike igneous rock, cutting through layers of country rock, that forms when magma oozes into a fracture; as the country rock erodes, the dike is left standing on the surface.

fissures Breaks, cracks, or fractures in rocks.

flood basalt Basaltic lava that erupts gently in great volume from cracks at Earth's surface to cover large areas of land and form lava plateaus.

8.1 Magma

Rocks of the asthenosphere partially melt to produce basaltic magma as a result of three processes: rising temperature, pressure-release melting, and addition of water. These processes occur beneath spreading centers, in mantle plumes, and in subduction zones to form both volcanoes and plutons.

8.2 Basalt and Granite

Basalt makes up most of the oceanic crust, and granite is the most abundant rock in continents. Basaltic magma forms by partial melting of mantle peridotite. Granitic magma forms when basaltic magma rises into and melts granitic rocks of the lower continental crust.

8.3 Partial Melting and the Origin of Continents

Earth's earliest continents were probably formed by partial melting of the original peridotite crust to form basalt, and then by further partial melting of the basalt to form andesite and then granite.

8.4 Magma Behavior

Basaltic magma usually erupts in a relatively gentle manner onto Earth's surface from a volcano. In contrast, granitic magma typically solidifies within Earth's crust. When granitic magma does erupt onto the surface, it often does so violently. These contrasts in behavior of the two types of magma are caused by differences in silica and water content.

8.5 Plutons

A pluton is any intrusive mass of igneous rock. A batholith is a pluton with more than 100 square kilometers of exposure at Earth's surface. A dike and a sill are both sheetlike plutons. Dikes cut across layering in country rock, and sills run parallel to layering.

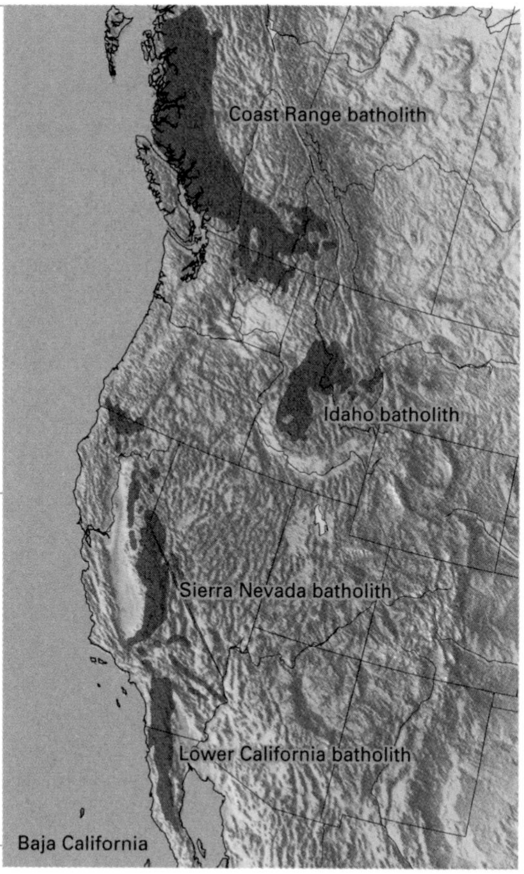

8.6 Volcanoes

Magma may flow onto Earth's surface as lava or may erupt explosively as pyroclastic material. Fluid lava forms lava plateaus and shield volcanoes. A pyroclastic eruption may form a cinder cone. Alternating eruptions of fluid lava and pyroclastic material from the same vent create a composite cone.

FIGURE 8.9 The large batholiths in western North America, shown here in dark gray, form high mountain ranges.

TABLE 8.1 Characteristics of Different Types of Volcanic Features

Type of Volcanic Feature	Physical Form	Size	Type of Magma	Style of Activity	Examples
Basalt plateau	Flat to gentle slope	100,000 to 1,000,000 km2 in area; 1 to 3 km thick	Basalt	Formed by gentle fissure eruptions	Columbia River plateau
Shield volcano	Slightly sloped, 6° to 12°	Up to 9,000 m high	Basalt	Gentle; some fire fountains	Hawaii
Cinder cone	Moderate slope	100 to 400 m high	Basalt or andesite	Ejections of pyroclastic material	Parícutin (Mexico)
Composite volcano	Alternate layers of flows and pyroclastics	100 to 3,500 m high	Variety of types of magmas and ash	Often violent	Vesuvius (Italy); Mount St. Helens; Aconcagua (Argentina)
Caldera	Circular depression, sometimes with steep walls	Less than 40 km in diameter	Granite	Formed by a violent cataclysmic explosion; potential for violent eruption remains	Yellowstone; San Juan Mountains

lava plateau *or* **basalt plateau** A broad plateau covering thousands of square kilometers, formed by lava deposited during a rapid sequence of fissure eruptions.

pahoehoe Lava with a smooth, billowy, or ropy surface.

partial melting The process in which a silicate rock only partly melts as it is heated, forming magma that is more silica-rich than the original rock.

pluton A body of intrusive igneous rock.

pressure-release melting Melting of asthenosphere rock that occurs when pressure in the asthenosphere decreases and the rock expands and melts.

pyroclastic rock Rock made up of liquid magma and solid rock fragments that were ejected explosively from a volcanic vent.

shield volcano A large, gently sloping volcanic mountain formed by successive flows of basaltic magma.

sill A sheetlike igneous rock, parallel to the grain or layering of country rock, that forms when magma oozes between layers.

stock A pluton exposed over less than 100 square kilometers of Earth's surface; similar to a batholith, but smaller.

vent An opening in a volcano, typically in the crater, through which lava and rock fragments erupt.

vesicles Holes in lava rock that formed when the lava solidified before bubbles of gas or water could escape.

volcanic ash The smallest pyroclastic particles, less than 2 millimeters in diameter.

volcano A hill or mountain formed from lava and rock fragments ejected through a volcanic vent.

8.7 Volcanic Explosions: Ash-Flow Tuffs and Calderas

When granitic magma rises to Earth's surface, it may erupt explosively, forming ash-flow tuffs and calderas.

8.8 Risk Assessment: Predicting Volcanic Eruptions

Volcanic eruptions are common near a subduction zone, near a spreading center, and at a hot spot over a mantle plume, but are rare in other environments. Eruptions on a continent are often violent, whereas those in oceanic crust are gentle. Such observations form the basis of regional predictions of volcanic hazards. Short-term predictions are made on the basis of earthquakes caused by magma movements, swelling of a volcano, increased emissions of gas and ash from a vent, and other signs that magma is approaching the surface.

8.9 Volcanic Eruptions and Global Climate

Large volcanic episodes affect the atmosphere, climate, and living organisms.

REVIEW QUESTIONS

1. Describe several ways in which volcanoes and volcanic eruptions can threaten human life and destroy property.
2. Describe three processes that generate magma in the asthenosphere.
3. Describe magma formation in a spreading center, a hot spot, and a subduction zone.
4. Describe the origin of granitic magma.
5. How much silica does average granitic magma contain? How much does basaltic magma contain?
6. How much water does average granitic magma contain? How much does basaltic magma contain?
7. Why does magma rise soon after it forms?
8. Explain why basaltic magma and granitic magma behave differently as they rise toward Earth's surface.
9. Many rocks and even entire mountain ranges at Earth's surface are composed of granite. Does this observation imply that granite forms at the surface?
10. Explain the difference between a dike and a sill.
11. How do a shield volcano, a cinder cone, and a composite cone differ from one another? How are they similar?
12. How does a composite cone form?
13. How does a caldera form?
14. Explain why additional eruptions in Yellowstone Park seem likely. Describe what such an eruption might be like.

KEY TERMS

Andean margin A continental margin characterized by subduction of an oceanic lithospheric plate beneath a continental plate; also called an *active continental margin*.

anticline A fold in rock that arches upward; the oldest rocks are in the middle.

basin A bowl-shaped syncline structure, commonly filled with sediment.

dome A circular or elliptical anticline structure, resembling an inverted cereal bowl.

fault A fracture in rock along which one side has moved relative to the other side. Compare with *joint*.

fault zone An area of numerous, closely spaced faults.

fold A bend in rock.

footwall A term to describe the lower side of an inclined fault or vein (i.e., the rock one would walk on).

forearc basin A sedimentary basin between the oceanic trench and the magmatic arc, either in an island arc or at an Andean margin.

geologic structure Any feature produced by rock deformation, such as a fold or a fault; also refers to the combination of all such features of an area or region.

graben A wedge-shaped block of rock that has dropped downward between two normal faults, forming a valley.

hanging wall A term to describe the top side of an inclined fault or vein (i.e., the rock hanging above one's head).

horst The block of rock between two grabens, which has moved relatively upward along normal faults as the grabens have settled downward.

island arc A gently curving chain of volcanic islands in the ocean formed by convergence of two plates, each bearing ocean crust, and the resulting subduction of one plate beneath the other.

9.1 Folds and Faults: Geologic Structures

When stress is applied to rocks, the rocks can deform in an elastic or a plastic manner, or they may rupture by brittle fracture. The nature of the material, temperature, pressure, and rate at which stress is applied all affect rock behavior under stress. A geologic structure is any feature produced by deformation of rocks. Folds usually form when rocks are compressed.

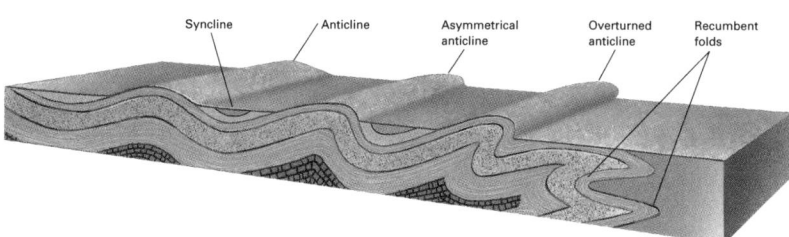

FIGURE 9.5 Five common types of folds. Folds can be symmetrical, as shown on the left, or asymmetrical, as shown in the center. If a fold has tilted beyond the perpendicular, it is overturned, as shown on the right.

A fault is a fracture along which rock on one side has moved relative to rock on the other side. Normal faults are usually caused when rocks are pulled apart; reverse and thrust faults are caused by compression; and strike-slip faults form where blocks of crust slip horizontally past each other along vertical fractures. A joint is a fracture where the rock on either side has not moved.

9.2 Mountains and Mountain Ranges

Mountains form when the crust thickens and rises isostatically. Their peak heights decrease when crustal rocks flow outward or are worn away by erosion.

9.3 Island Arcs: Subduction Where Two Oceanic Plates Converge

If two converging plates carry oceanic crust, a volcanic island arc forms.

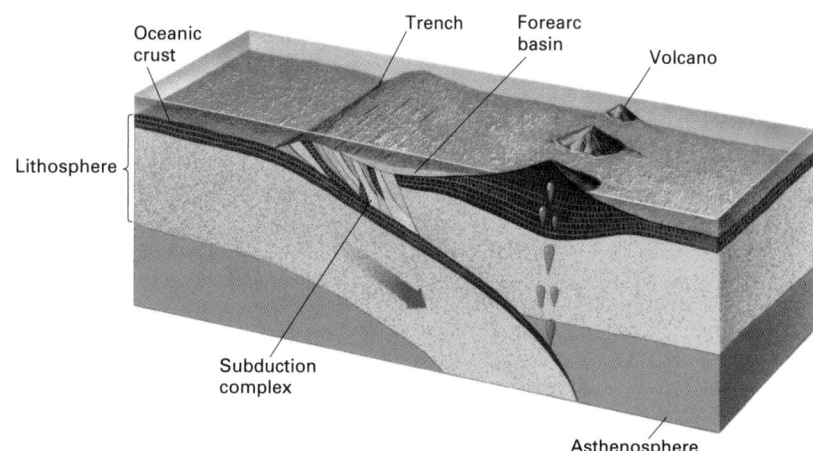

FIGURE 9.17 Formation of an island arc. A subduction complex contains slices of oceanic crust and upper mantle scraped from the top of a subducting plate. Magma forms in the subduction zone and rises to build submarine volcanoes, which eventually grow above sea level to form the volcanic mountain chain.

joint A fracture along which the rock on either side of the break does not move. Compare with *fault*.

limbs The sides of a fold in rock.

normal fault A fault in which the hanging wall has moved downward relative to the footwall.

orogeny The process of mountain building; all tectonic processes associated with mountain building.

reverse fault A fault in which the hanging wall has moved up relative to the footwall.

slip The distance that rocks on opposite sides of a fault have moved.

stress The force exerted against an object, usually measured as force per unit area or pressure.

subduction complex The accumulation of sea-floor sediment, basaltic oceanic crust, and slices of the upper mantle that are scraped from the upper layers of the subducting slab in a subduction zone and added to the opposite plate.

syncline A fold in rock that arches downward, and whose center contains the youngest rocks.

thrust fault A type of reverse fault that is nearly horizontal, with a dip of 45° or less over most of its extent.

underthrusting The process by which one continent moves beneath the other during a continent–continent collision.

9.4 The Andes: Subduction at a Continental Margin

The Andes are a mountain chain consisting predominantly of igneous rocks formed at a type of continental margin called an Andean margin. An Andean margin is characterized by subduction of an oceanic lithospheric plate beneath a continental plate. Andean margins are dominated by granitic plutons and andesitic volcanoes. They also contain rocks of a subduction complex and sedimentary rocks deposited in a forearc basin.

9.5 The Himalayas: A Collision between Continents

The Himalayas began with formation of an Andean-type continental margin along the edge of Tibet. Later, a continental plate carrying India collided with Tibet, producing compressional forces that created the mountain range. The geology of mountain ranges formed by continent–continent collisions, such as the Himalayas, is dominated by vast regions of folded and thrust-faulted sedimentary and metamorphic rocks and by earlier-formed plutonic and volcanic rocks.

9.6 Mountains and Earth Systems

Mountains exert profound effects on climate, weather, and plant and animal habitats.

REVIEW QUESTIONS

1. What is tectonic stress? Explain the main types of stress.
2. Explain the different ways in which rocks can respond to tectonic stress. What factors control the response of rocks to stress?
3. What is a geologic structure? What are the three main types of structures? What type(s) of rock behavior does each type of structure reflect?
4. At what type of tectonic plate boundary would you expect to find normal faults?
5. Explain why folds accommodate crustal shortening.
6. Draw a cross-sectional sketch of an anticline–syncline pair and label the limbs.
7. Draw a cross-sectional sketch of a normal fault. Label the hanging wall and the footwall. Use your sketch to explain how a normal fault accommodates crustal extension. Sketch a reverse fault and show how it accommodates crustal shortening.
8. Explain the similarities and differences between a fault and a joint.
9. In what tectonic environment would you expect to find a strike-slip fault, a normal fault, and a thrust fault?
10. What mountain chain has formed at a divergent plate boundary? What are the main differences between this chain and those developed at convergent boundaries? Explain the differences.
11. Explain why erosion initially causes a mountain range to rise and then eventually causes the peak heights to decrease.
12. Describe the similarities and differences between an island arc and the Andes. Why do the differences exist?
13. Describe the similarities and differences between the Andes and the Himalayan chain. Why do the differences exist?
14. Draw a cross-sectional sketch of an Andean-type plate margin to a depth of several hundred kilometers.
15. Draw a sequence of cross-sectional sketches showing the evolution of a Himalayan-type plate margin. Why does this type of boundary start out as an Andean-type margin?

KEY TERMS

abrasion A mechanical weathering process that consists of the grinding and rounding of rock surfaces by friction and impact.

A horizon The layer of soil below the O horizon, composed of a mixture of humus, sand, silt, and clay; combines with the O horizon as to form topsoil.

angle of repose The maximum slope or steepness at which loose material remains stable. If the slope becomes steeper than the angle of repose, the material slides.

aspect The orientation of a slope with respect to the Sun; the direction toward which it faces.

B horizon The soil layer just below the A horizon, containing less organic matter, and where ions leached from the A horizon accumulate; also called *subsoil*.

caliche A hard crust on the soil of arid and semiarid regions, formed when calcium carbonate precipitates and cements the soil particles together.

capillary action The process by which water is pulled upward through the soil due to the natural attraction of water molecules to soil particles.

chemical weathering The decomposition of rock when it chemically reacts with air, water, or other agents in the environment, altering its chemical composition and mineral content.

C horizon The lowest soil layer.

creep A subcategory of flow mass wasting; the very slow movement of loose material downslope, usually at a rate of only about 1 centimeter per year and usually on land with vegetation. Trees on a creeping block tilt downhill.

dissolution A chemical weathering process in which mineral or rock dissolves, forming a solution.

erosion The removal of weathered rocks that occurs when water, wind, ice, or gravity transports the material to a new location.

10.1 Weathering and Erosion

Weathering is the decomposition and disintegration of rocks and minerals at Earth's surface. Erosion is the removal of weathered rock or soil by flowing water, wind, glaciers, or gravity. After being eroded from the immediate environment, rock or soil may be transported large distances and eventually deposited.

10.2 Mechanical Weathering

Mechanical (or physical) weathering can occur by pressure-release fracturing, frost wedging, abrasion, organic activity, and thermal expansion and contraction.

10.3 Chemical Weathering

The most important processes of chemical weathering are dissolution, hydrolysis, and oxidation. A few minerals dissolve readily in water. Acids and bases enhance the solubility of minerals. Rainwater is slightly acidic due to reactions between water and atmospheric carbon dioxide. During hydrolysis, water reacts with one mineral to form a new mineral. Oxidation is weathering in which a mineral decomposes when it reacts with oxygen. Chemical weathering and mechanical weathering often operate together. For example, saltwater seeps through cracks in rock; the dissolved salts crystallize when the water evaporates, and the growing crystals break the rock apart. Hydrolysis may also fracture granite to form exfoliation slabs.

10.4 Soil

Soil is the layer of weathered material overlying bedrock. Sand, silt, clay, and humus are commonly found in soil. Water leaches soluble ions downward through the soil. Clays are also transported downward by water. The uppermost layer of soil, called the O horizon, consists mainly of litter and humus. The amount of organic matter decreases downward. Leaching removes dissolved ions and clay from the A horizon and deposits them in the B horizon. Six major factors control soil characteristics: parent rock, climate, rates of plant growth and decay, slope aspect and steepness, time, and transport of soil materials. Dry climates have salt-encrusted soils called pedocals, which often form a hard cement crust called caliche. In moist climates, pedalfer soils develop. In these regions, soluble ions are removed from the soil, leaving high concentrations of lesssoluble aluminum and iron. Laterite soils form in very warm, moist climates, where all of the more soluble ions are removed.

10.5 Erosion

Interactions with flowing water, wind, and glaciers erode soil as it forms.

10.6 Landslides

Mass wasting is the downhill movement of rock and soil under the influence of gravity. Landslide is a general term for mass wasting and the landforms it creates. The stability of a slope and the severity of a landslide depend on (1) steepness of the slope, (2) type of rock and orientation of rock layers, (3) nature of unconsolidated materials, (4) water and vegetation, and (5) earthquakes or volcanic eruptions.

10.7 Types of Landslides

Landslides fall into three categories: flow, slide, and fall. During flow, a mixture of rock, soil, and water moves as a viscous fluid. Creep is a slow type of flow that occurs at a

exfoliation A weathering process resulting in fracture when concentric plates or shells split away from a main rock mass like the layers of an onion; frequently explained as a form of pressure-release fracturing, but many geologists believe it is caused by hydrolysis-expansion.

fall Mass wasting in which unconsolidated material falls freely or bounces down steep slopes or cliffs.

flow Mass wasting (landslide) in which loose soil or sediment moves downslope as a fluid, not as a consolidated mass; may occur slowly (less than 1 centimeter per year) or rapidly (like water over a waterfall).

frost wedging A mechanical weathering process in which water freezes in a crack in rock, and then the expansion wedges the rock apart.

humus The dark, organic component of soil consisting of litter that has decomposed enough so that the origin of the individual pieces cannot be determined.

hydrolysis A chemical weathering process in which a mineral reacts with water to form a new mineral that has water as part of its crystal structure.

landslide A general term for mass wasting (the downslope movement of rock and regolith under the influence of gravity) and the landforms it creates.

laterite A highly weathered soil rich in oxides of iron and aluminum that usually develops in warm, moist, tropical regions.

leaching The downward movement of water and dissolved ions from the O and A soil horizons into the B horizon, where they accumulate.

litter Leaves, twigs, and other plant or animal material that have fallen to the surface of the soil but have not decomposed.

loam The most fertile soil, a mixture especially rich in sand and silt with generous amounts of organic matter.

mass wasting The downslope movement of earth material, primarily caused by gravity. *See also* landslide.

mechanical weathering *or* **physical weathering** The disintegration of rock into smaller pieces by physical processes without altering the chemical composition of the rock.

mudflow A subcategory of flow mass wasting that involves the downslope movement, usually on unvegetated land, of fine-grained soil particles mixed with a large amount of water; can be slow-moving, as slow as 1 meter per year, or as fast as a speeding car.

O horizon The uppermost layer of soil, named for its organic component; the combined O and A horizons are called *topsoil.*

organic activity A mechanical weathering process in which a crack in a rock is expanded by tree or plant roots growing there.

oxidation A chemical weathering process in which a mineral decomposes when it reacts with oxygen.

pedalfer A common soil type that forms in moist environments, characterized by abundant iron and aluminum oxides and a concentration of clay in the B horizon.

rate of about 1 centimeter per year. A mudflow is a fluid mass of sediment and water. Slide is the movement of a coherent block of material along a fracture. Slump is a type of slide in which the moving mass slips downhill over a gently curved fracture. In a rockslide, a newly detached segment of bedrock slides along a tilted bedding plane or fracture. Fall occurs when particles fall or tumble down a steep cliff.

10.8 Predicting and Avoiding Landslides

Earthquakes and volcanic eruptions trigger devastating mass wasting. Proper planning and engineering can avert much damage to human habitation.

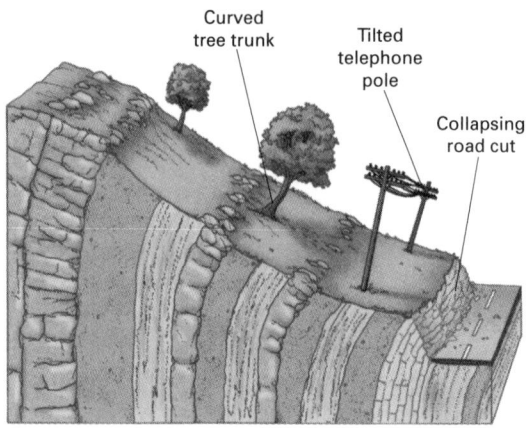

FIGURE 10.26 During creep, the land surface moves more rapidly than deeper layers, so objects embedded in rock or soil tilt downhill.

REVIEW QUESTIONS

1. Explain the differences between mechanical weathering and chemical weathering.
2. What is a talus slope? What conditions favor the formation of talus slopes?
3. Characterize the four major horizons of a mature soil.
4. List and describe each of the factors that control slope stability.

pedocal A salt-encrusted soil in arid and semiarid climates characterized by an accumulation of calcium carbonate.

pressure-release fracturing A mechanical weathering process in which tectonic forces lift deeply buried rocks close to the surface and then erosion removes overlying rock and sediment—the net result of which is to remove the pressure from overlying material, causing the rock to expand and fracture.

regolith The thin layer of loose, unconsolidated, weathered material that overlies bedrock. Some earth scientists and engineers use the terms *regolith* and *soil* interchangeably; soil scientists include soil as only the upper layers of regolith.

rockslide A subcategory of slide mass wasting in which a segment of

bedrock slides downslope along a fracture and the rock breaks into fragments and tumbles down the hillside; also called a *rock avalanche.*

salt cracking A chemical weathering process in which salts that are dissolved in water in the pores of rock crystallize, widening cracks and pushing the mineral grains apart.

slide Mass wasting in which the rock or soil initially moves as a consolidated unit along a fracture surface.

slump A subcategory of slide mass wasting in which blocks of material slide downslope as a consolidated unit over a fracture in rock or regolith; trees on the slumping blocks tilt uphill.

soil The upper layers of regolith that support plant growth. Some earth sci-

entists and engineers use the terms *soil* and *regolith* interchangeably.

soil horizon A layer of soil that is distinguishable from other layers because of differences in appearance and in physical and chemical properties.

talus An accumulation of loose, angular rocks at the base of a cliff, created when the rocks broke off as a result of frost wedging.

thermal expansion and contraction A mechanical weathering process that fractures rock when temperature changes rapidly, causing the surface of the rock to heat or cool faster than its interior and to expand or contract faster than the interior.

topsoil The fertile, dark-colored surface soil; the combined O and A soil horizons.

KEY TERMS

alluvial fan A fan-shaped accumulation of sediment created where a steep mountain stream rapidly slows down as it reaches a relatively flat plain.

aquifer A body of rock that can yield economically significant quantities of ground water; should be both porous and permeable.

artesian aquifer An inclined aquifer sandwiched between layers of impermeable rock, and where the water in the lower part is under pressure from the weight of water above.

artesian well A well drilled into an artesian aquifer, in which the water rises without pumping and in some cases spurts to the surface.

artificial levee A wall built along the banks of a stream to prevent rising floodwater from spilling out of the channel onto the flood plain.

banks The rising slopes bordering the sides of a stream channel.

base level The deepest level to which a stream can erode its bed. The ultimate base level is usually sea level, but this is seldom attained.

bed The floor of a stream channel.

bed load The portion of a stream's load—boulders, cobbles, and sand—that is transported along the bottom, immediately above the streambed.

braided stream A stream that flows in many shallow, interconnecting channels; formed because more sediment was supplied to the stream than it could carry, accumulating in the channel and forcing the stream to overflow its banks and erode new channels.

capacity The maximum quantity of sediment that a stream can carry past a given point in a given amount of time.

cavern An underground cavity or series of chambers created when groundwater dissolves large amounts of rock, usually limestone; also called a *cave*.

channel characteristics Features describing the shape and roughness of a stream channel.

11.1 The Water Cycle

Only about 0.64 percent of Earth's water is fresh. The rest is salty seawater and glacial ice. The constant circulation of water among the sea, land, the biosphere, and the atmosphere is called the hydrologic cycle, or the water cycle. Most of the water that evaporates from the seas and from land returns to the surface as rain or snow. The precipitation that falls on the continents returns to the sea via runoff and moving ground water, or it returns to the atmosphere via evaporation and transpiration.

11.2 Streams

A stream is any body of water flowing in a channel. The velocity of a stream is determined by its gradient (steepness), discharge (amount of water), and channel characteristics (shape and roughness of the stream channel). Streams shape Earth's surface by eroding soil and bedrock. Streams transport sediment as dissolved load, suspended load, and bed load. Downcutting, lateral erosion, and mass wasting combine to form a stream valley. Base level is the lowest elevation to which a stream can erode its bed; it is usually sea level.

11.3 Stream Deposition

When streams slow down they deposit their bed load first. If water slows sufficiently, the suspended load may also fall to the bottom, but the dissolved load remains in the water until the chemical environment changes. A stream feeding a delta or fan splits into many channels called distributaries.

11.4 Floods

A flood occurs when a stream overtops its banks and flows over its flood plain. Artificial levees and channels can contain small floods but they may exacerbate large ones.

column A cave deposit formed when a stalactite and a stalagmite meet and fuse together.

competence A measure of the largest particles that a stream can transport.

delta A nearly flat, fan-shaped accumulation of sediment, forming a tract of land where a stream enters a lake or ocean.

discharge The volume of water flowing downstream, usually measured in units of cubic meters per second (m³/sec).

dissolved load The portion of a stream's sediment load that consists of ions dissolved in water.

distributaries Channels that split from the main stream feeding a delta or alluvial fan, spreading out to build the area covered by the delta or fan.

downcutting Downward erosion by a stream into its bed, usually cutting a V-shaped valley along a relatively straight path.

drainage basin The region that is drained by a single river.

eutrophic lake A relatively shallow lake characterized by abundant plant nutrients, thus sustaining multiple living organisms.

flood A relatively high stream flow that overtops the stream banks, covering land that is not usually under water.

flood plain That portion of a river valley adjacent to the channel; it is built by sediment deposited during floods and is covered by water during a flood.

geothermal energy Energy extracted from Earth's heat.

geyser A type of hot spring that intermittently erupts with violent jets of hot water and steam when ground water comes in contact with hot rock.

graded stream A stream with a smooth, concave profile, in equilibrium with its sediment supply; it transports all the sediment supplied to it with neither erosion nor deposition in the streambed.

gradient The steepness or vertical drop of a stream over a specific distance.

hot spring A spring formed where hot ground water flows to the surface.

hydrologic cycle The continuous circulation of water among the hydrosphere, the atmosphere, the biosphere, and the geosphere; also called the *water cycle*.

karst topography A type of irregular landscape that forms over limestone or other soluble rock and is characterized by caverns, sinkholes, and underground streams.

kettle lake A lake that forms in a depression created by a receding glacier, filled with the water from the melting glacier.

lake A large, inland body of standing water that occupies a depression in the land surface.

lateral erosion The action of a low-gradient stream as it cuts into and erodes its banks, swinging from one side of its channel to the other, forming a wide flat valley.

meanders A series of twisting curves or loops in the course of a stream.

oligotrophic lake A deep lake characterized by nearly pure water but with low concentrations of plant nutrients, thus sustaining relatively few living organisms.

oxbow lake A crescent-shaped lake created where a meander loop is cut off from a stream when the ends of the meander became plugged with sediment.

perched water table A localized water table above the main water table, formed where a layer of impermeable rock or clay lies above the main water table, creating a locally saturated zone.

permeability The ability of a material to transmit water, as measured by the speed at which fluid can travel through the material.

point bar A deposit of sediment in the slower water on the inside of a meander.

porosity The proportional volume of a material that consists of pores or open spaces, indicating the amount of water it can hold.

recharge To replenish an aquifer by the addition of water.

runoff Surface water that flows to the oceans in streams and rivers.

sinkhole A circular depression on Earth's surface caused by the collapse of a cavern roof or by the dissolution of surface rocks.

spring A place where the water table intersects the land surface and ground water flows or seeps onto the surface.

stalactite An icicle-like dripstone of dissolved calcite precipitated from drops of water, which hangs from the ceiling of a cavern.

stalagmite A cone-shaped deposit of dissolved calcite precipitated from drops of water that have fallen to the floor of a cavern.

stream A moving body of water, confined in a channel and flowing downslope; a *river* is a large stream fed by smaller ones.

11.5 Lakes

Lakes are short-lived landforms because streams fill them with sediment. Recent glaciers created many modern lakes; as a result, we live in an unusual time of abundant lakes. The life history of a lake commonly involves a progression from an oligotrophic lake to a eutrophic lake to a swamp and, finally, to a flat meadow or forest.

11.6 Ground Water

Much of the rain that falls on land seeps into soil and bedrock to become ground water. Ground water saturates the upper few kilometers of soil and bedrock to a level called the water table, which separates the zone of saturation from the zone of aeration. Most ground water moves slowly, about 4 centimeters per day. Springs occur where the water table intersects the land surface and water flows or seeps onto the surface. An inclined layer of permeable rock sandwiched between layers of impermeable rock can produce an artesian aquifer. Caverns form where ground water dissolves limestone. A sinkhole forms when the roof of a limestone cavern collapses. Karst topography, with numerous caves, sinkholes, and subterranean streams, is characteristic of limestone regions.

11.7 Hot Springs, Geysers, and Geothermal Energy

Hot springs and geysers develop when hot ground water rises to the surface. Hot ground water has been tapped to produce geothermal energy.

11.8 Wetlands

Wetlands are among the most biologically productive environments on Earth. In addition, they are natural water purification systems, and they mitigate flood effects by absorbing floodwaters. Despite these facts, government and private efforts have eliminated more than half of the wetlands that existed in the lower 48 states when European settlers first arrived.

REVIEW QUESTIONS

1. Describe the movement of water through the hydrologic cycle.
2. Describe the factors that determine the velocity of stream flow.
3. What is karst topography? How can it be recognized? How does it form?

submarine delta The portion of a river delta that lies underwater.

suspended load The portion of a stream's sediment load that is carried for a considerable time in suspension, free from contact with the streambed.

thermocline The boundary between the upper warm layers and deeper cool layers of water in a lake.

transpiration Direct evaporation of water into the atmosphere from the leaf surfaces of plants.

tributary A stream that feeds water into another stream or river.

turnover A process, occurring in fall and spring in temperate climates, in which a lake's surface water changes temperature in response to seasonal weather changes and convection mixes the water to equalize temperature throughout the lake.

water table The top level of subsurface ground water, at the top of the zone of saturation and below the zone of aeration.

well A hole dug or drilled into Earth, generally for the production of water, petroleum, natural gas, brine, sulfur, or for exploration.

wetlands Regions that are water soaked or flooded for all or part of the year; also known as *swamps, bogs, marshes, sloughs, mud flats,* and *flood plains*.

zone of aeration A subsurface zone above the water table where the rock or soil may be moist but not saturated, with air occupying some or all of the pore space; also called the *unsaturated zone*.

zone of saturation A subsurface zone below the water table in which the soil and bedrock are completely saturated with water.

KEY TERMS

anaerobic Without oxygen; anaerobic bacteria are bacteria that live without oxygen.

biodegradable pollutants Pollutants that decay naturally in a reasonable amount of time, being consumed or destroyed by organisms that live naturally in soil or water.

bioremediation The use of microorganisms to decompose an environmental contaminant.

chemical remediation Treatment of a contaminated area by injecting it with a chemical compound that reacts with the pollutant to produce harmless products.

Clean Water Act A federal law passed in 1972, mandating the cleaning of the nation's rivers, lakes, and wetlands and forbidding the discharge of pollutants into waterways.

cone of depression A cone-shaped depression in the water table, created when water is pumped out of a well more rapidly than it can flow through the aquifer. If water continues to be pumped from the well at this rate, the water table will drop.

consumption Any process that uses water and then returns it to Earth far from its source.

diversion system A pipe, canal, or other infrastructure that transports water from its natural place and path in the hydrologic cycle to a new place and path to serve human needs.

externalities Additional or unintended consequences of an activity or condition, such as the indirect costs of environmental degradation, including reduction in tourism and lowered land values.

nonbiodegradable pollutants Pollutants that do not decay naturally in a reasonable amount of time, including some industrial compounds,

12.1 Water Supply and Demand

Growing populations have created demands for water that stretch, and in some cases exceed, the amount of water that is available both globally and in the United States. Most of the water used by homes and industry is withdrawn and then returned to streams or ground water reservoirs near the site of withdrawal. But most of the water used by agriculture is consumed because it evaporates. Water use falls into three categories: (1) domestic use accounts for 10 percent of U.S. water consumption; (2) industrial use accounts for 49 percent; and (3) agricultural use accounts for 41 percent.

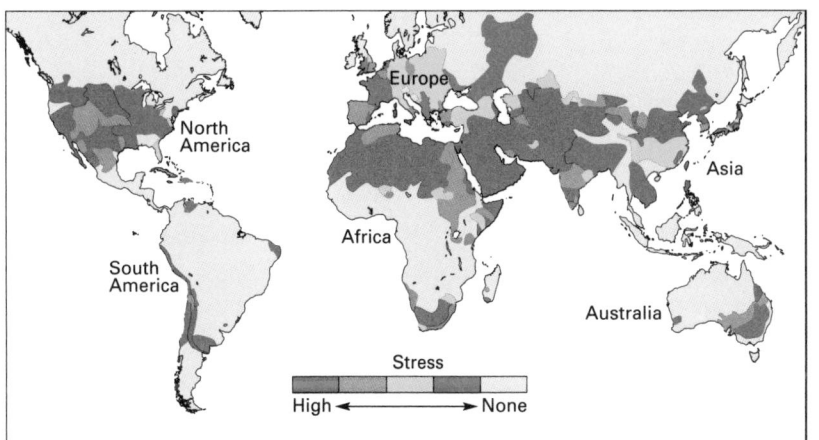

DATA FROM WORLD COMMISSION ON WATER USE IN THE 21ST CENTURY.

FIGURE 12.8 As humans use water for irrigation and other purposes, surface water resources become stressed. (Stress evaluation is based on the amount of water available compared with the amount used by people.)

12.2 Dams and Diversion

The United States receives about 3 times more water from precipitation than it uses, but some of the driest regions use the greatest amounts of water. Water diversion projects collect and transport surface water and ground water from places where water is available to places where it is needed. Dams are especially useful in regions that receive seasonal rainfall, and some dams also generate hydroelectric energy. But they can create undesirable effects, including water loss, salinization, silting, erosion, disaster when a dam fails, recreational and aesthetic losses, and ecological disruptions. Ground water projects pump water to Earth's surface for human use, but because ground water flows so slowly, the extraction often creates a cone of depression that may lead to a drop in the water table. The chapter's discussion of the Ogallala aquifer illustrates ground water depletion. Ground water withdrawal may also cause subsidence and saltwater intrusion.

12.3 The Great American Desert

The Great American Desert is a mostly arid region of the western United States. Americans have built great cities and extensive farms and ranches in the region, all supplied by irrigation systems. Diminishing water reserves and increasing costs of water diversion projects suggest that their future is uncertain. The chapter outlined the Owens River and Colorado River diversion projects.

12.4 Water and International Politics

Water conflict leads to serious international conflict between hostile nations.

toxic inorganic compounds, and nontoxic sediment that muddies streams and habitats.

nonpoint source pollution Pollution that is generated over a broad area, such as fertilizers and pesticides spread over agricultural fields.

persistent bioaccumulative toxic chemicals (PBTs) Nonbiodegradable toxins that are released into and accumulate in the environment.

plume of contamination The slow-growing, three-dimensional zone within an aquifer or of surface water affected by a dispersing pollutant.

point source pollution Pollution that arises from a specific site such as a septic tank or a factory.

pollution The reduction of the quality of a resource by the introduction of impurities.

remediation The treatment of a contaminated area, such as an aquifer, to remove or decompose a pollutant.

salinization A process whereby salts accumulate in soil that is irrigated heavily, lowering soil fertility.

saltwater intrusion A condition in coastal regions in which excessive pumping of fresh ground water causes salty ground water to invade an aquifer.

subsidence The irreversible sinking or settling of Earth's surface.

withdrawal Any process that uses water and then returns it to Earth locally.

REVIEW QUESTIONS

1. The United States receives 3 times more water in the form of precipitation than it uses. Why do water shortages exist in many parts of the country?
2. Describe the three main categories of water use. What proportion of total U.S. water use falls into each category?
3. Explain the differences among water use, water withdrawal, and water consumption.
4. Why is agriculture responsible for the greatest proportion of water consumption?
5. What are the two main sources of water exploited by water diversion projects?
6. Describe the beneficial effects of dams and their associated water delivery systems.
7. Describe the negative effects of dams and their associated water delivery systems.
8. Why does salinization commonly result from desert irrigation?
9. Describe the factors that make ground water a valuable resource.
10. Describe problems caused by excessive pumping of ground water.
11. Why does ground water depletion cause longer-term problems than might result from the draining of a surface reservoir?

12.5 Water Pollution

Water pollution is the reduction in the quality of water by the introduction of impurities. In the United States, the Clean Water Act was passed in 1972 in response to serious water pollution disasters such as the Cuyahoga River fire and Love Canal disaster. Water pollutants include biodegradable materials such as sewage, disease organisms, and fertilizers; nonbiodegradable materials (also called persistent bioaccumulative toxic chemicals, or PBTs) such as industrial organic compounds, toxic inorganic compounds, and sediment; radioactive materials; and heat. Pollution can originate from both point sources and nonpoint sources.

12.6 How Sewage, Detergents, and Fertilizers Pollute Waterways

Biodegradable pollutants nourish an ecosystem and lead to oxygen depletion and growth of anaerobic bacteria, with resultant degradation of the ecosystem.

12.7 Toxic Pollutants, Risk Assessment, and Cost–Benefit Analysis

It is difficult to determine the health effects of low doses of water pollutants. Cost–benefit analysis compares the cost of pollution control with the cost of externalities.

12.8 Ground Water Pollution

Ground water pollutants normally spread slowly into an aquifer as a plume of contamination. Because contaminants permeate all the tiny pores' spaces in an aquifer, cleanup, or remediation, is expensive and difficult. Techniques include elimination of the source, monitoring, modeling, bioremediation, chemical remediation, and removal of contaminated rock and soil.

12.9 Nuclear Waste Disposal

Radioactive wastes must be isolated from water resources because they are impossible to destroy and persist for long times. In the United States, Yucca Mountain is the designated site for spent fuels from nuclear power plants.

12.10 The Clean Water Act: A Modern Perspective

Since the Clean Water Act was passed in 1972, waterways in the United States have been cleaned considerably—but many of our streams still remain polluted.

12. If land subsides when an underlying aquifer is depleted, will it rise to its original level when pumping stops and the aquifer is recharged? Explain your answer.
13. Why does saltwater intrusion affect only coastal areas?
14. Describe the geographic region that John Wesley Powell called the Great American Desert.
15. List four categories of water pollutants. What is/are the source(s) of each, and what are the harmful effects?
16. How does a nontoxic substance, such as cannery waste, become a water pollutant?
17. Discuss the difficulties in assessing the health risks of low doses of toxic water pollutants.
18. What is cost–benefit analysis and how does it work?
19. Explain why pollutants persist in ground water longer than they do in surface water.
20. Outline possible procedures for cleaning polluted ground water.
21. Discuss the controversy over the Yucca Mountain nuclear waste repository.

KEY TERMS

alpine glacier A glacier that forms in mountainous terrain.

arête A sharp, narrow rib of rock forming a border between adjacent valleys or between two cirques, created when two alpine glaciers moved along opposite sides of the mountain ridge and eroded both sides.

basal slip Movement of a glacier in which the entire mass slides over bedrock.

cirque A steep-walled spoon-shaped depression eroded into a mountain peak by a glacier.

crevasse A fracture or crack in the brittle upper 40 meters of a glacier, formed when the glacier flows over uneven bedrock.

drift Any rock or sediment transported and deposited by a glacier or by glacial meltwater.

drumlins Elongate hills, usually occurring in clusters, formed when a glacier flows over and reshapes a mound of till or stratified drift.

eccentricity A term referring to the elliptical shape of Earth's orbit around the Sun; the more elliptical the orbit, the more eccentric it is said to be.

end moraine A ridge of till that forms at the end, or terminus, of a glacier that is neither advancing nor retreating and whose terminus has remained in the same place for years.

erratics Boulders, usually different from bedrock in the immediate vicinity, that were transported to their present location by a glacier.

esker A long, snake-like ridge formed as the channel deposit of a stream that flowed within or beneath a melting glacier.

fjord A deep, narrow, glacially carved valley on a high-latitude seacoast that was later flooded by encroaching seas as the glaciers melted.

glacial striations Parallel grooves and scratches in bedrock that form as rocks are dragged along at the base of a glacier.

glacier A massive, long-lasting accumulation of compacted snow and ice that forms on land and moves downslope or spreads outward under its own weight.

13.1 Formation of Glaciers

If snow survives through one summer, it becomes a relatively hard, dense material called firn. A glacier is a massive, long-lasting accumulation of compacted snow and ice that forms on land and creeps downslope or outward under the influence of its own weight. Alpine glaciers form in mountainous regions; continental glaciers cover vast regions.

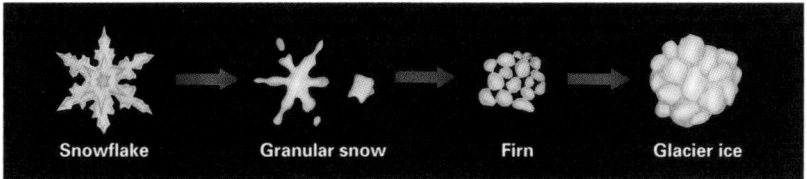

FIGURE 13.1 Newly fallen snow changes through several stages to form glacier ice.

13.2 Glacial Movement

Glaciers move by two mechanisms: basal slip and plastic flow. The upper 40 meters of a glacier is too brittle to flow, and large cracks called crevasses develop in this layer. In the zone of accumulation, the annual rate of snow accumulation is greater than the rate of melting, whereas in the zone of ablation, melting exceeds accumulation. The snow line is the boundary between permanent snow and seasonal snow. The end of a glacier is called its terminus.

13.3 Glacial Erosion

Glaciers erode bedrock, forming U-shaped valleys, cirques, and other landforms.

13.4 Glacial Deposits

Drift is rock or sediment transported and deposited by a glacier. The unsorted drift deposited directly by a glacier is called till. Most glacial terrain is characterized by large mounds of till known as moraines. Terminal moraines, ground moraines, recessional moraines, lateral moraines, medial moraines, and drumlins are all depositional features formed by glaciers. Stratified drift consists of sediment first carried by a glacier and then transported, sorted, and deposited by streams. Valley trains, outwash plains, kames, and eskers are composed of stratified drift. A kettle is a depression created when a large block of ice left behind by a retreating glacier melts.

13.5 The Pleistocene Ice Age

During the past 1 billion years, at least six major ice ages have occurred. The most recent is the Pleistocene Ice Age. One hypothesis contends that Pleistocene advances and retreats were caused by climate change induced by variations in Earth's orbit and the orientation of its rotational axis. Sea level falls when continental ice sheets form and rises again when the ice melts.

ground moraine The moraine formed when a glacier recedes steadily and deposits till in a relatively thin layer over a broad area.

hanging valley A small glacial valley lying high above the floor of the main valley.

horn A sharp, pyramid-shaped rock summit where three or more cirques intersect near the summit.

ice age A time of extensive glacial activity, when alpine glaciers descended into lowland valleys and continental glaciers spread over the higher latitudes.

ice fall A section of a glacier consisting of numerous crevasses and towering ice pinnacles.

ice sheet or **continental glacier** A glacier that covers an area of 50,000 square kilometers or more and spreads outward in all directions under its own weight.

iceberg A large chunk of ice that breaks from a glacier into a body of water.

kame A small mound or ridge of stratified drift deposited by a stream that flows on top of, within, or beneath a glacier.

lateral moraine A ridge-like moraine that forms from sediment on or adjacent to the sides of a mountain glacier.

medial moraine A moraine formed in or on the middle of a glacier by the merging of lateral moraines as two glaciers flow together.

moraine A mound or ridge of till deposited directly by glacial ice.

outwash Sediment deposited by streams flowing from the terminus of a melting glacier.

outwash plain A broad, level surface formed when outwash spreads onto a wide valley or plain beyond a glacier.

paternoster lakes A series of lakes in a glacial valley, strung out like beads and connected by short streams and waterfalls.

plastic flow Movement of a glacier in which the ice flows as a viscous fluid.

Pleistocene Ice Age The most recent ice age, which began roughly 2 or 3 million years ago, characterized by several advances and retreats of glaciers. Most climate models indicate that Earth is still in the Pleistocene Ice Age.

precession The circling or wobbling of Earth's axis as the planet travels in its orbit, like that of a wobbling top.

recessional moraine A moraine that forms at the new terminus of a glacier as the glacier stabilizes temporarily during retreat.

snow line The boundary between permanent snow and seasonal snow.

stratified drift Glacial drift that was first carried by a glacier and then transported and deposited in layers by a stream.

tarn A small lake at the base of a cirque.

terminal moraine An end moraine that forms when a glacier is at its greatest advance before beginning to retreat.

terminus The end, or foot, of a glacier.

till Glacial drift that was deposited directly by glacial ice.

tillite Till that was deposited by glaciers so long ago that it became lithified into solid rock.

tilt The angle of Earth's axis with respect to a line perpendicular to the plane of its orbit around the Sun. Earth's axis is tilted by about 23.5°.

U-shaped valley A glacially eroded valley with a broad, characteristic U-shaped cross section.

valley train Outwash deposited in a narrow mountain valley by the streams flowing from an alpine glacier.

zone of ablation The lower-altitude part of an alpine glacier, where more snow melts in summer than accumulates in winter, and where the melting snow leaves behind a surface of old, hard glacial ice.

zone of accumulation The higher-elevation upper end of an alpine glacier, where more snow falls in winter than melts in summer and snow accumulates from year to year, and where the glacier's surface is covered by snow year-round.

13.6 Snowball Earth: The Greatest Ice Age in Earth's History

The greatest ice age in Earth's history occurred during late-Precambrian time. Glaciers covered all continents and the seas froze over completely, entombing the globe in a shell of ice.

13.7 The Earth's Disappearing Glaciers

Glaciers and sea ice are shrinking in more places and at more rapid rates than at any other time since scientists began keeping records. The melting of the Arctic and Antarctic ice has profound effects for the planet.

REVIEW QUESTIONS

1. List the major steps in the metamorphism of newly fallen snow to glacial ice.
2. Differentiate between alpine glaciers and continental glaciers. Where are alpine glaciers found today? Where are continental glaciers found today?
3. Distinguish between basal slip and plastic flow.
4. Why are crevasses only about 40 meters deep, even though many glaciers are much thicker?
5. Describe the surface of a glacier in the summer and in the winter in the zone of accumulation and in the zone of ablation.
6. Describe how glacial erosion can create a cirque, a paternoster lake, and striated bedrock.
7. Describe the formation of arêtes, horns, and hanging valleys.
8. Distinguish among ground, recessional, terminal, lateral, and medial moraines.
9. Why are kames and eskers features of receding glaciers? How do they form?
10. What topographic features were left behind by the continental ice sheets? Where can they be found in North America today?
11. Discuss the evidence for the ice age known as Snowball Earth.

*remember

There are more review tools for this course online at 4ltrpress.cengage.com.

KEY TERMS

bajada A broad, gently sloping depositional surface formed by the merging of alluvial fans from closely spaced canyons, extending outward into a desert valley.

barchan dune A crescent-shaped dune, highest in the center, with the tips facing downwind; generally forms in rocky deserts where there is little sand.

butte A flat-topped mountain, smaller and more tower-like than a mesa, characterized by steep cliff faces.

deflation Erosion by wind.

desert Any region that receives less than 25 centimeters (10 inches) of rain per year and consequently supports little or no vegetation.

desertification A process by which semiarid land is converted to desert, by human mismanagement or by climate change.

desert pavement A continuous cover of closely packed stones left behind when wind erodes smaller particles such as silt and sand.

dune A mound or ridge of wind-deposited sand.

flash flood A rapid, intense, local flood of short duration, usually following a rainstorm.

loess A homogenous, porous deposit of wind-blown silt, typically unlayered, that forms vertical bluffs and cliffs.

longitudinal dune A long, symmetrical dune that is parallel with the direction of the prevailing wind; forms where sand is limited and wind direction is erratic but from the same general direction.

mesa A flat-topped mountain, shaped like a table, that is smaller than a plateau and larger than a butte.

parabolic dune A crescent-shaped dune with tips pointing into the wind; forms in moist semidesert regions and along seacoasts where sparse vegetation is present to anchor the tips of the dune.

pediment A broad, gently sloping erosional surface that forms along the front of desert mountains uphill from a bajada, usually covered by a patchy veneer of gravel only a few meters thick.

14.1 Why Do Deserts Exist?

Deserts have an annual precipitation of less than 25 centimeters. The world's largest deserts occur near 30° north and south latitudes, where warm, dry, descending air absorbs moisture from the land. Deserts also occur on the leeward side (in rain shadows) of mountains, in continental interiors, and in coastal regions adjacent to cold ocean currents.

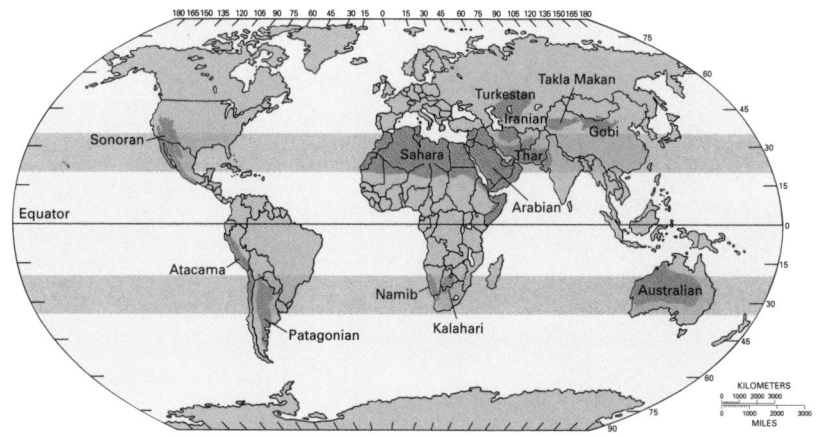

FIGURE 14.2 The major deserts of the world are concentrated at approximately 30° north and south latitudes.

14.2 Water and Deserts

Desert streams are often dry for much of the year, but flash floods may occur during a rainstorm. Playa lakes are desert lakes that dry up periodically, leaving abandoned lake beds called playas. Alluvial fans in desert environments may be several kilometers across and rise a few hundred meters above the surrounding valley floor. A bajada is a broad depositional surface formed by merging alluvial fans. A pediment is a broad, gently sloping erosional surface that forms along the front of desert mountains and that merges imperceptibly with a bajada.

14.3 Two American Deserts

The Colorado River drains the Colorado Plateau desert. Thus, streams carry sediment away from the region, forming canyons and eroding the plateaus to form mesas and buttes. Death Valley and the Great Basin have no external drainage and, as a result, the valleys are filling with sediment eroded from the surrounding mountains.

FIGURE 14.10
The Colorado Plateau and the Great Basin together make up a large desert and semiarid region of the western United States. The Colorado River flows through the Colorado Plateau to the Gulf of California, but no streams flow out of the Great Basin.

plateau A large, elevated area of relatively flat land.

playa The dry desert lake bed of a playa lake.

playa lake A temporary desert lake that dries up during the dry season.

rain-shadow desert A desert formed on the downwind side of a mountain range.

saltation The bouncing, hopping movement of sand grains as the wind blows them along the ground or as they are carried in suspension by water or some other fluid.

slip face The steep leeward side of a dune, typically at the angle of repose for loose sand, so that the sand slides or slips down the face, where it is deposited.

transverse dune A relatively long, straight dune with a gently sloping windward side and a steep lee face that is perpendicular to the prevailing wind; forms where sand is plentiful and evenly dispersed.

wash A streambed that is dry for most of the year.

14.4 Wind

Deflation is erosion by wind. Silt and sand are removed selectively, leaving larger stones on the surface and creating desert pavement. Sand grains are relatively large and heavy and are carried only short distances by saltation, seldom rising more than a meter above the ground. Silt can be transported great distances at higher elevations. Wind erosion forms blowouts. Windblown particles are abrasive, but because the heaviest grains travel close to the surface, abrasion occurs mainly near ground level. A mound or ridge of wind-deposited sand is called a dune. Most dunes are asymmetrical, with gently sloping, windward sides and steeper slip faces on the lee sides. Dunes migrate. The various types of dunes include barchan dunes, transverse dunes, longitudinal dunes, and parabolic dunes. Wind-deposited silt is called loess, which typically forms vertical cliffs and bluffs.

14.5 Desertification

The Sahara is the largest desert on the planet. South of the Sahara lies the semiarid Sahel. Low rainfall has been responsible for the expansion of the Sahara into the Sahel. Still, while a desert ecosystem is created by low rainfall, human management can significantly alter the productivity of the land.

REVIEW QUESTIONS

1. Why are many deserts concentrated along zones at 30° latitude in both the Northern Hemisphere and the Southern Hemisphere?
2. List three factors that control global distribution of deserts.
3. Why do flash floods and mudflows occur in deserts?
4. Why are alluvial fans more conspicuous in deserts than in humid environments?
5. Compare and contrast floods in deserts with floods in more humid environments.
6. Compare and contrast pediments and bajadas.
7. Why is wind erosion more prominent in desert environments than it is in humid regions?
8. Describe the formation of desert pavement.
9. Describe the evolution and shape of a dune.
10. Describe the differences among barchan dunes, transverse dunes, parabolic dunes, and longitudinal dunes. Under what conditions does each type of dune form?
11. Compare and contrast desert plateaus, mesas, and buttes. Describe the formation of each.
12. Compare the effects of stream erosion and deposition in the Colorado Plateau and Death Valley.

KEY TERMS

abyssal fan A large, fan-shaped accumulation of sediment deposited at the base of many submarine canyons adjacent to the deep sea floor; also called a *submarine fan*.

abyssal plains The flat, level, largely feature-less parts of the ocean floor between the Mid-Oceanic Ridge and the continental rise.

accreted terrane A landmass that originated as an island arc or a microcontinent and was later added onto a continent.

active continental margin A continental margin that occurs at a convergent plate boundary, characterized by subduction of an oceanic litho-spheric plate beneath a continental plate; also called an *Andean margin*.

atoll A circular coral reef that forms a ring of islands around a central lagoon and is bounded on the outside by deep water of the open sea.

bolide A large piece of space debris, such as an asteroid, that crashes into a planet.

carbonate platforms Extensive accumulations of limestone, such as the Florida Keys and the Bahamas, formed on a continental shelf in warm regions where sediment does not muddy the water and reef-building organisms thrive.

chemosynthesis A process in which bacteria produce energy from hydrogen sulfide and thus are not dependent on photosynthesis.

continental rise An apron of sediment at the foot of the continental slope where it merges with the deep sea floor.

continental shelf A shallow, nearly level area of continental crust covered by sediment and sedimentary rocks, submerged below sea level at the edge of a continent between the shore-line and the continental slope.

continental slope The relatively steep (averaging 3° but varying between 1° and 10°) under-water slope between the continental shelf and the continental rise.

echo sounder An instrument that emits sound waves and then records them as they reflect off the sea floor; the data are then used to record the topography of the sea floor.

guyot A flat-topped seamount, formed when the top of a sinking island is eroded by sea waves.

15.1 The Origin of Oceans

Ocean water, atmospheric gases, and the molecular building blocks of life were carried to Earth by bolides early in Earth's history.

15.2 The Earth's Oceans

Oceans cover about 71 percent of Earth's surface. Continental lithosphere is thicker and less dense than oceanic lithosphere. Consequently, continents float isostatically to high elevations, whereas oceanic lithosphere sinks to low elevations. The ocean basins form topographic depressions on Earth's surface, which fill with water to form oceans.

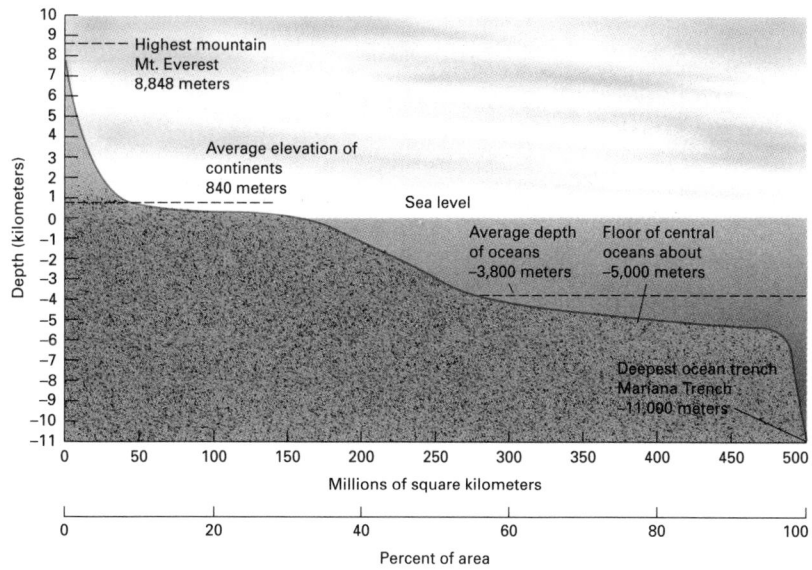

FIGURE 15.3 A schematic cross section of the continents and ocean basins. The vertical axis shows elevations relative to sea level. The horizontal axis shows the relative areas of the types of topography. Thus, 30 percent, or about 150 million square kilometers, of Earth's surface lies above sea level.

15.3 Studying the Sea Floor

Because of the great depth and remoteness of the ocean floor and oceanic crust, knowledge of them comes mainly from sampling and remote sensing.

15.4 Features of the Sea Floor

The Mid-Oceanic Ridge system is a submarine mountain chain that extends through all of Earth's major ocean basins. A rift valley runs down the center of many parts of the ridge, and the ridge and rift valley are both offset by numerous transform faults. The Mid-Oceanic Ridge forms at a spreading center where new oceanic crust is added to the sea floor. The Mid-Oceanic Ridge system supports thriving ecosystems based on chemosynthesis. Oceanic trenches are the deepest parts of ocean basins. Island arcs (chains of volcanoes formed at subduction zones where two oceanic plates collide) are common features of some ocean basins, particularly the southwestern Pacific. Seamounts and oceanic islands form in oceanic crust as a result of volcanic activity over mantle plumes. An atoll is a circular coral reef growing on a sinking volcanic island.

layer 1 The uppermost layer of oceanic crust, composed of terrigenous and pelagic sediment.

layer 2 The second layer of oceanic crust, consisting of pillow basalt.

layer 3 The deepest and thickest layer of oceanic

crust, directly overlying the mantle, consisting of basalt dikes and gabbro.

magnetometer An instrument that measures a magnetic field.

microwave radar A satellite-based instrument that measures the echo of microwave pulses to detect subtle swells and depressions on the sea surface, which reflect sea-floor topography and can be used to map the sea floor.

oceanic island A submarine mountain (seamount) that rises above sea level.

oceanic trench A long, narrow, steep-sided depression of the sea floor formed where a subducting oceanic plate sinks into the mantle, dragging the sea floor downward.

passive continental margin A margin that occurs where continental and oceanic crust are firmly joined together and where little tectonic activity occurs.

pelagic sediment Muddy ocean sediment that consists of a mixture of clay carried from continents and the skeletons of tiny marine organisms.

pillow basalt Molten lava that solidified under water, forming spheroidal lumps that resemble a stack of pillows.

rift valley An elongate depression that develops at a divergent plate boundary. Examples include continental rift valleys and the rift valley along the center of the Mid-Oceanic Ridge system.

rock dredge An open-mouthed steel net dragged along the sea floor behind a research ship for the purpose of sampling rocks from submarine outcrops.

sea-floor drilling A process in which drill rigs mounted on offshore platforms or on research vessels cut cylindrical cores from both sediment and rock of the sea floor, which are then brought to the surface for study.

seamount A submarine mountain, usually of volcanic origin, that rises 1 kilometer or more above the surrounding sea floor.

seismic profiler A device that emits a high-energy signal that penetrates and reflects from layers in sediment and rock beneath the sea floor; the data are used to construct a topographic profile of the sea floor.

submarine canyon A deep, V-shaped, steep-walled trough eroded into a continental shelf and slope.

terrigenous sediment Sea-floor sediment composed of sand, silt, and clay eroded from the continents.

transform fault A strike-slip fault between two offset segments of a mid-ocean ridge.

turbidity current A rapidly flowing submarine current laden with suspended sediment, which develops when loose, wet sediment tumbles down the continental slope in a submarine landslide.

15.5 Sediment and Rocks of the Sea Floor

Oceanic crust varies from about 4 to 7 kilometers thick and consists of three layers. The top layer (layer 1) is sediment, which varies from zero to 3 or more kilometers thick. Beneath this, in layer 2, lies about 1 to 2 kilometers of pillow basalt. Layer 3, the deepest layer of oceanic crust, is from 3 to 5 kilometers thick and consists of basalt dikes on top of gabbro. The base of this layer is the boundary between oceanic crust and mantle. The age of sea-floor rocks increases regularly away from the Mid-Oceanic Ridge. No oceanic crust is older than about 200 million years because it recycles into the mantle at subduction zones.

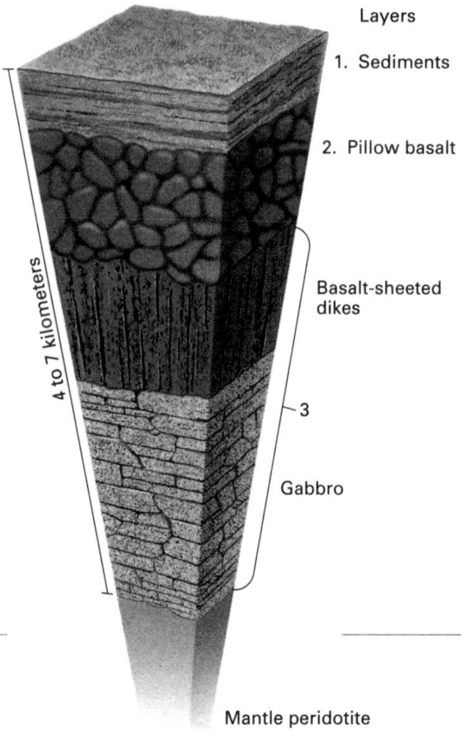

15.6 Continental Margins

A passive continental margin includes a continental shelf, a continental slope, and a continental rise formed by accumulation of terrigenous sediment. Submarine canyons eroded by turbidity currents notch continental margins and commonly lead into abyssal fans, where the turbidity currents deposit sediments on the continental rise. An active continental margin, where oceanic crust sinks in a subduction zone beneath the margin of a continent, usually includes a narrow continental shelf and a continental slope that steepens abruptly into an oceanic trench.

FIGURE 15.21 The three layers of oceanic crust. Layer 1 consists of sediment. Layer 2 is pillow basalt. Layer 3 consists of vertical dikes overlying gabbro. Below layer 3 is the upper mantle.

REVIEW QUESTIONS

1. Describe the main differences between oceans and continents.
2. Sketch a cross section of the central rift valley in the Mid-Oceanic Ridge.
3. Describe the dimensions of the Mid-Oceanic Ridge system.
4. Explain why the Mid-Oceanic Ridge is topographically elevated above the surrounding ocean floor. Why does its elevation gradually decrease away from the ridge axis?
5. Explain the origin of the rift valley in the center of the Mid-Oceanic Ridge.
6. Why are the abyssal plains characterized by such low relief?
7. Sketch a cross section of oceanic crust from a deep sea basin. Label, describe, and indicate the approximate thickness of each layer.
8. Describe the two main types of sea-floor sediment. What is the origin of each type?
9. Compare the ages of oceanic crust with the ages of continental rocks. Why are they so different?
10. Sketch a cross section of both an active continental margin and a passive continental margin. Label the features of each. Give approximate depths below sea level of each of the features.
11. Explain how a continental shelf–slope–rise complex forms on a continental margin.
12. Why does an active continental margin typically have a steeper continental slope than a passive margin?
13. Why do oceanic islands sink after they form?

KEY TERMS

backshore The upper zone of a beach that is usually dry but is washed by waves during storms.

barrier island A long, narrow, low-lying island that extends parallel to the shoreline.

baymouth bar A spit that extends partially or completely across the entrance to a bay.

beach Any strip of shoreline washed by waves and tides.

beach drift The gradual movement of sediment (usually sand) along a beach, parallel to the shoreline, when waves strike the beach obliquely.

break To collapse or crash, as when a wave approaches a beach and the front of the wave rises over its base, growing steeper until it collapses forward.

Coriolis effect A deflection of air or water currents, caused by the rotation of Earth.

crest The highest part of a wave.

current A continuous flow of water in a particular direction.

deep-sea current Vertical and horizontal flow of water below a depth of 400 meters in the oceans, caused mainly by gravity.

Ekman transport The natural process by which wind causes movement of water near the ocean surface. Each layer of water in the ocean drags with it the layer beneath; thus the movement of each layer of water is affected by the movement of the layer above

emergent coastline A coastline that was previously under water but has become exposed, because either the land has risen or sea level has fallen.

equatorial upwelling Oceanic upwelling in which surface currents flowing westward on both sides of the equator are deflected poleward and are replaced by upward flow of deep, nutrient-rich waters.

16.1 Geography of the Oceans

All of Earth's oceans are connected, so Earth has a single global ocean with several distinct ocean basins existing within it.

16.2 Seawater

Seawater contains about 3.5 percent dissolved salts. The upper layer of the ocean, about 450 meters thick, is relatively warm. In the thermocline, below the warm surface layer, temperature drops rapidly with depth. Deep ocean water is consistently around 1°C to 2.5°C.

16.3 Tides

Tides are caused by gravitational pull of the Moon and Sun. Two high tides and two low tides occur approximately every day.

16.4 Sea Waves

Wave size depends on (1) wind speed, (2) the length of time the wind has blown, and (3) the distance that the wind has traveled. The highest part of a wave is the crest; the lowest is the trough. The distance between successive crests is called the wavelength. Wave height is the vertical distance from the crest to the trough. The water in a wave moves in a circular path.

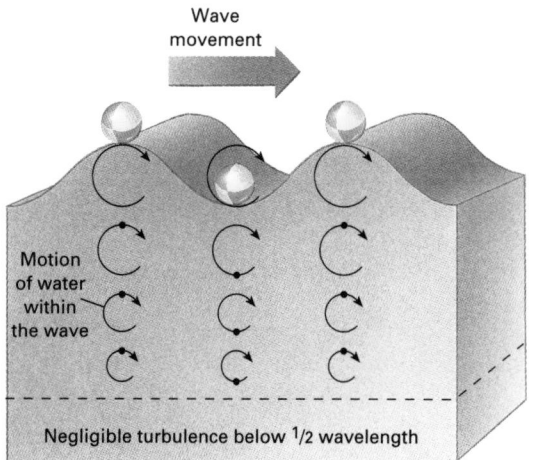

FIGURE 16.8 While a wave moves horizontally across the sea surface, the water itself moves only in small circles.

16.5 Ocean Currents

A current is a continuous flow of water in a particular direction. Surface currents are driven by wind and deflected by the Coriolis effect, Ekman transport, and sea-surface topography. Deep-sea currents are driven by differences in seawater density. Cold, salty water is dense and therefore sinks and flows along the sea floor. When ocean water sinks in some places, it must rise in others to maintain mass balance. This upward flow of water is called upwelling.

eutastic sea-level change A global sea-level change, caused by three different processes: the growth or melting of glaciers, changes in water temperature, and changes in the volume of the Mid-Oceanic Ridge.

foreshore *or* **intertidal zone** The part of a beach that lies between the high-tide and low-tide lines, and is exposed to the air at low tide but is covered by water at high tide.

groin A narrow barrier or wall built on a beach, perpendicular to the shoreline, to trap sand transported by currents and waves.

Gulf Stream One of many oceanic surface currents. It begins near the Gulf of Mexico and travels northward up the Atlantic coast, growing wider and slower as it moves to the northeast.

gyre A circular or elliptical current in either water or air.

lagoon A sheltered body of water separated from the sea by a reef or barrier island.

longshore current A current flowing parallel and close to the coast and transporting sand for great distances along the coastline, generated when waves strike a shore at an angle.

neap tides The relatively small tides that occur when the Moon is 90° out of alignment with the Sun and Earth.

phytoplankton Plankton that conduct photosynthesis like land-based plants and that are the base of the food chain for aquatic animals.

plankton Small marine organisms that live mostly within a few meters of the sea surface, where sunlight is available, and that conduct most of the photosynthesis and nutrient consumption in the ocean and form the base of the marine food web.

reef A wave-resistant ridge or mound built by corals or other marine organisms.

refraction The bending of a wave that occurs when it approaches the shore at an angle; the end of the wave in shallow water slows down while the end in deeper water continues at a faster speed.

salinity The total quantity of dissolved salts in seawater, expressed as a percentage.

sea arch An arch created when a seashore cave is eroded all the way through a narrow headland, leaving behind only the cave's entranceway.

sea stack A pillar of rock left when a sea arch collapses or when the inshore portion of a headland erodes faster than the tip.

spit A small ridge of sand or gravel extending from a beach into a body of water.

spring tides The very high and very low tides that occur when the Sun, Moon, and Earth are aligned and their gravitational fields combine to create a strong tidal bulge.

submergent coastline A coastline that was previously above sea level but has been drowned, because either the land has sunk or sea level has risen.

surf The chaotic, turbulent waves breaking along the shore.

surface current Horizontal flow of water in the upper 400 meters of the oceans, caused by wind blowing over the sea surface.

thermohaline circulation The force behind deep-sea currents, created by differences in water temperature and salinity (and therefore differences in water density).

tidal current A current, channeled by a bay with a narrow entrance or by closely spaced islands, caused by the rise and fall of the tides.

tides The cyclic rise and fall of ocean water caused by the gravitational force of the Moon and, to a lesser extent, of the Sun.

trough The lowest part of a wave.

upwelling A rising ocean current that transports cold water (and nutrients) from the depths to the surface.

16.6 The Seacoast

When a wave nears the shore, the bottom of the wave slows and the wave breaks, creating surf. Ocean waves weather and erode coastlines by hydraulic action, abrasion, solution, and salt cracking. The bending of a wave as it strikes shore is called refraction. Refracted waves often form longshore currents that transport sediment along a shore. Tidal currents also transport sediment in some areas. Irregular coastlines are straightened by erosion and deposition.

16.7 Emergent and Submergent Coastlines

If land rises or sea level falls, the coastline migrates seaward and old beaches are abandoned above sea level, forming an emergent coastline. In contrast, a submergent coastline forms when land sinks or sea level rises.

16.8 Beaches

A beach is a strip of shoreline washed by waves and tides. Most coastal sediment is transported to the sea by rivers. Glacial drift, reefs, and local erosion also add sand in certain areas. Coastal emergence may expose large amounts of sand. Spits, baymouth bars, and barrier islands are common on sandy coastlines. Human intervention such as the building of groins may upset the natural movement of coastal sediment and alter patterns of erosion and deposition on beaches. A rocky coast is dominated by wave-cut cliffs, wave-cut platforms, sea arches, and sea stacks. A fjord is a steep-sided, narrow, glacially carved, submerged valley on a high-latitude seacoast.

16.9 Life in the Sea

Most of the photosynthesis and nutrient consumption in the ocean is carried out by plankton, tiny organisms that live mostly within a few meters of the sea surface where light is available. Phytoplankton conduct photosynthesis, just as land-based plants do, and are the base of the food chain for aquatic animals. Zooplankton are tiny animals that feed on the phytoplankton. The shallow water of a continental shelf supports large populations of marine organisms. Commercial overfishing and habitat destruction threaten world fish harvests. A reef is a wave-resistant ridge or mound built by corals, oysters, algae, or other organisms.

16.10 Global Warming and Sea-Level Rise

Ocean water expands when it warms or when glaciers melt and freshwater flows into the sea—both mechanisms causing sea level to rise. Global sea level has risen over the past century and may continue into the next. Many poor countries cannot afford coastal protection for even a small sea-level rise; a dramatic sea-level rise could inundate coastal cities.

REVIEW QUESTIONS

1. Name the major ocean basins and give their locations.
2. Describe the temperature profile of the open oceans.
3. Explain why two high tides occur every day, even though the Moon lies directly above any portion of Earth only once a day.
4. List the three factors that determine the size of a wave.
5. Explain the Coriolis effect. What is a gyre?
6. What is refraction? How does it affect coastal erosion?
7. Explain how coastal processes straighten an irregular coastline.

wave height The vertical distance from the crest to the trough of a wave.

wave-cut cliff A cliff created when waves erode the headland of a rocky coastline into a steep profile.

wave-cut platform The flat or gently sloping platform left when a wave-cut cliff erodes.

wavelength The distance between successive wave crests (or troughs).

zooplankton Tiny marine animals that live mostly within a few meters of the sea surface and feed on phytoplankton.

KEY TERMS

acid precipitation *or* **acid rain** Rain, snow, fog, or mist that has become acidic after reacting with air pollutants.

aerosol In pollution terminology, a particle or particulate that is suspended in air.

atmospheric pressure *or* **barometric pressure** The pressure of the atmosphere at any given location and time.

bar Unit of measurement for atmospheric pressure. One bar is equal to sea-level atmospheric pressure.

barometer A device used to measure barometric pressure.

chlorofluorocarbons (CFCs) Organic compounds containing chlorine and fluorine, which rise into the upper atmosphere to destroy the ozone layer.

cyanobacteria Blue-green bacteria that were among the earliest photosynthetic life-forms on Earth.

First Great Oxidation Event The process, occurring approximately 2.4 billion years ago, when the oxygen concentration in Earth's atmosphere increased suddenly from trace amounts to appreciable quantities, probably because of a combination of biological and geochemical processes.

fly ash Noncombustible minerals that escape into the atmosphere when coal burns, eventually settling as gritty dust.

Gaia The term (Greek for "Earth") used by James Lovelock to refer to our planet, which he likened to a living creature due to the interconnectivity of all of Earth's systems.

halons Compounds containing bromine and chlorine, which rise into the upper atmosphere to destroy the ozone layer.

mesopause The ceiling of the mesosphere; the boundary between the mesosphere and the thermosphere.

mesosphere The layer of air that lies above the stratopause, extending upward from about

17.1 Earth's Early Atmospheres

Earth's primordial atmosphere was composed almost entirely of hydrogen and helium. After these light elements boiled off into space, bolides carried gases to create a secondary atmosphere, which was composed predominantly of carbon dioxide, with lesser amounts of water and nitrogen, and trace gases. By 4 billion years ago, outgassing, combined with reactions of gases with rocks of the geosphere, created an atmosphere rich in hydrogen and carbon dioxide.

17.2 Life, Iron, and the Evolution of the Modern Atmosphere

Photosynthesis by primitive cyanobacteria began producing oxygen about 2.7 billion years ago. However, for 300 million years, the oxygen concentration did not increase dramatically in the atmosphere due to two processes: (1) as the oxygen concentration increased, it initially reacted with dissolved iron in seawater to form banded iron formations; oxygen concentration rose to its present level only after nearly all the dissolved iron had been removed by precipitation; and (2) when life first evolved, the hydrogen concentration of the atmosphere remained high and oxygen reacted with the hydrogen to produce water; bacteria converted the hydrogen to methane, and the oxygen concentration in the atmosphere rose only after the hydrogen concentration declined. The First Great Oxidation Event occurred about 2.4 billion years ago. After that time, further production of oxygen was roughly balanced by respiration. About 600 million years ago, numerous processes reduced decay and respiration, leading to the Second Great Oxidation Event.

17.3 The Modern Atmosphere

Today, dry air is roughly 78 percent nitrogen (N_2), 21 percent oxygen (O_2), and 1 percent other gases. Air also contains water vapor, dust, liquid droplets, and pollutants.

17.4 Atmospheric Pressure

Atmospheric pressure (or barometric pressure) is the weight of the atmosphere per unit area. Pressure varies with weather and decreases with altitude.

17.5 Atmospheric Temperature

Atmospheric temperature decreases with altitude in the troposphere. Then the temperature rises in the stratosphere because ozone absorbs solar radiation; at an altitude of about 50 kilometers, the stratosphere is as warm as at Earth's surface. The temperature decreases again in the mesosphere where it is extremely cold, and then in the uppermost layer, the thermosphere, temperature increases to just below freezing as high-energy radiation is absorbed.

17.6 Air Pollution

The increasing ill effects of air pollution prior to and just after Word War II convinced lawmakers to pass legislation such as the 1963 Clean Air Act. Incomplete combustion of coal and petroleum produces carcinogenic hydrocarbons such as benzene as well as carbon monoxide and methane, which is a greenhouse gas. Impurities in these fuels burn to produce oxides of nitrogen and sulfur. Nitrogen and sulfur oxides react in the atmosphere to produce acid rain, which damages health, weathers materials, and reduces growth of crops and forests.

Incompletely burned fuels in automobile exhaust react with nitrogen oxides in the presence of sunlight and atmospheric oxygen to form ozone; ozone then further reacts with automobile exhaust to form smog. Dioxin is an example of a compound that is produced inadvertently during chemical manufacture and when certain materials are burned. Some scientists argue that even tiny amounts of dioxin and other toxic volatiles may be harmful to human health, but others disagree. Scientific studies show that industrial aerosols are harmful to health, but aerosols are so varied that it is difficult to know which ones are most harmful.

55 kilometers to about 80 kilometers above Earth's surface.

outgassing The release of volatiles from Earth's mantle to the surface in volcanic eruptions.

ozone hole An unusually low ozone concentration in the stratosphere.

particle *or* **particulate** In pollution terminology, any small piece of solid matter larger than a molecule, such as dust or soot.

pH scale A logarithmic scale that measures the acidity of a solution. A pH of 7 is neutral; numbers lower than 7 represent acidic solutions, and numbers higher than 7 represent basic ones.

photosynthesis The process by which chlorophyll-bearing plant cells convert carbon dioxide and water to organic sugars, using sunlight as an energy source; oxygen is released in the process.

Second Great Oxidation Event The process, occurring about 600 million years ago, when the oxygen concentration in Earth's atmosphere increased abruptly a second time in response to interactions among chemical, physical, and biological processes.

smog Smoky fog; a word used loosely to define visible air pollution.

stratopause The ceiling of the stratosphere; the boundary between the stratosphere and the mesosphere.

stratosphere The layer of air above the tropopause, extending upward to about 55 kilometers.

thermosphere An extremely high and diffuse region of the atmosphere lying above the mesosphere, from about 80 kilometers upward.

tropopause The top of the troposphere; the boundary between the troposphere and the stratosphere.

troposphere The layer of air that lies closest to Earth's surface and extends upward to about 17 kilometers.

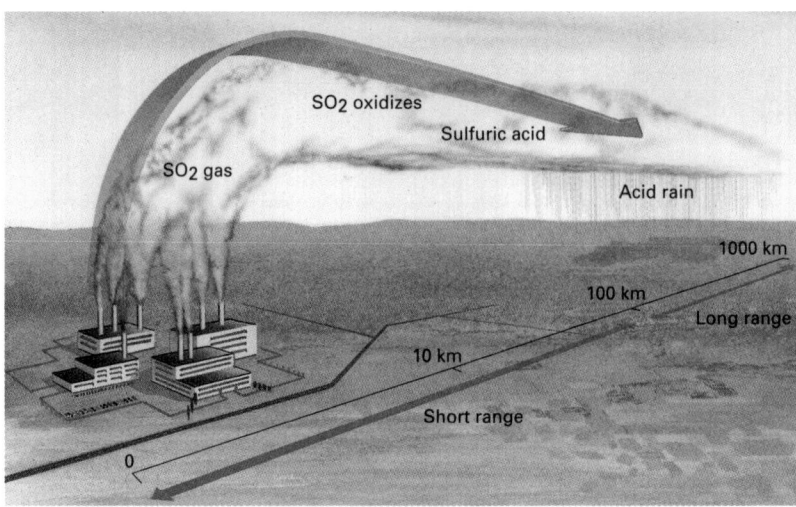

FIGURE 17.13 Acid rain develops from the addition of sulfur compounds to the atmosphere by industrial smokestacks.

17.7 Depletion of the Ozone Layer

Chlorofluorocarbons and halons, compounds containing chlorine and bromine, rise into the stratosphere and deplete the ozone that filters out harmful UV radiation and protects Earth. Ozone-destroying chemicals have been regulated by international treaty, and their concentration in the troposphere is slowly diminishing. Overall, Earth's stratospheric ozone level has been increasing, but serious ozone holes remain at the poles.

REVIEW QUESTIONS

1. Discuss the formation and composition of Earth's earliest atmosphere.
2. How and when did Earth's secondary atmosphere form? Compare the composition of this atmosphere with that of the modern one.
3. How did outgassing affect the composition of the Hadean atmosphere?
4. What process first began producing oxygen about 2.7 billion years ago? Why did the oxygen concentration remain low for 300 million years?
5. Briefly describe the First Great Oxidation Event.
6. How did banded iron formations form? How did their formations affect the atmosphere?
7. Briefly describe the Second Great Oxidation Event.
8. List the two most abundant gases in the atmosphere. List three other, less abundant, gases. List three nongaseous components of natural air.
9. Draw a graph of the change in pressure with altitude. Explain why the pressure changes as you have shown.
10. What is a barometer and how does it work?
11. Draw a figure showing temperature changes with altitude. Label all the significant layers in Earth's atmosphere.
12. Discuss the air pollution disaster in Donora. What pollutants were involved? Where did they come from? How did they become concentrated?
13. Briefly list the major sources of air pollution.
14. What air pollutants are generated when coal and gasoline burn?
15. What is acid rain, how does it form, and how does it affect people, crops, and materials?
16. What is smog, how does it form, and how does it differ from automobile exhaust?
17. What is a volatile? Why are volatiles hard to control?
18. Discuss the production, dispersal, and health effects of dioxin.
19. What is an aerosol? Briefly discuss the health effects of aerosols.
20. Explain how CFCs deplete the ozone layer.
21. Why is ozone in the troposphere harmful to humans, while ozone in the stratosphere is beneficial?
22. Discuss the effects of the international ban on ozone-destroying chemicals.

KEY TERMS

absorption of radiation The process that occurs when energy is absorbed: the energy of a photon is converted to electrical, chemical, vibrational, or heat energy, and the photon disappears.

albedo The reflectivity of a surface; surfaces that reflect more light have a higher albedo.

climate The characteristic weather of a region, averaged over several decades; refers to yearly cycles of temperature, wind, rainfall, etc., and not to daily variations.

conduction The transport of heat by direct collision among atoms or molecules.

electromagnetic radiation Radiation consisting of an oscillating electric and magnetic field, including radio waves, infrared, visible light, ultraviolet, x-rays, and gamma rays.

electromagnetic spectrum The entire range of electromagnetic radiation from very-long-wavelength (low frequency) radiation to very-short-wavelength (high frequency) radiation.

emission of radiation The process that occurs when energy, in the form of a photon, is emitted as an electron falls out of an excited state, with the equivalent loss of energy from the emitting substance.

equinox Either of two times during the year—on or about March 21 and September 22—when the Sun shines directly overhead at the equator and every portion of Earth receives 12 hours of daylight and 12 hours of darkness.

excited state A state of physical energy higher than the lowest energy level (or *ground state*) of an electron in an atom or molecule.

frequency The number of complete wave cycles, from crest to crest, that pass by any point in a second.

greenhouse effect An increase in the temperature of the planet's surface caused when infrared-absorbing gases in the atmosphere trap energy from the Sun.

heat A measure of the total energy in a sample.

isotherms Lines on a weather map connecting areas with the same average temperature.

18.1 Incoming Solar Radiation

Light is a form of electromagnetic radiation and exhibits properties of both waves and particles. Wavelength is the distance between wave crests, and frequency is the number of cycles that pass in a second. The electromagnetic spectrum is the continuum of radiation of different wavelengths and frequencies.

When radiation is absorbed by matter, the electrons may be promoted into an excited state in which they have an increased level of energy. When they fall back to their lower energy state, they reemit the energy, but often at a dif-

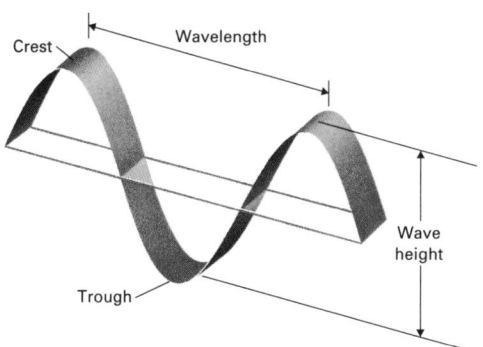

FIGURE 18.1 The terms used to describe a light wave are identical to those used for water, sound, and other types of waves.

ferent frequency from the incoming photons. The Sun's high-energy (short-wavelength) radiation is absorbed by rock and soil, which reemit low-energy (long-wavelength) radiation. Radiation reflects from many surfaces; the reflectivity of a surface is referred to as its albedo. Scattering of sunlight makes the sky appear blue.

18.2 The Radiation Balance

About 8 percent of solar radiation is scattered back into space; 19 percent is absorbed in the atmosphere; 26 percent is reflected; and 47 percent is absorbed by soil, rocks, and water. However, the absorbed radiation is eventually reemitted into space. The greenhouse effect is the warming of the atmosphere due to the absorption of this long-wavelength radiation by water vapor, carbon dioxide, methane, and other greenhouse gases.

18.3 Energy Storage and Transfer: The Driving Mechanisms for Weather and Climate

Temperature is a measure of heat proportional to the average speed of atoms or molecules in a sample, while heat measures the total thermal energy in a sample. Conduction is the transport of heat by direct collisions among atoms or molecules. Winds and ocean currents transfer heat from one region of the globe to another by convection and advection. Latent heat (stored heat) is the energy released or absorbed when a substance changes from one state to another. Large quantities of latent heat are absorbed or emitted when water freezes, melts, vaporizes, or condenses.

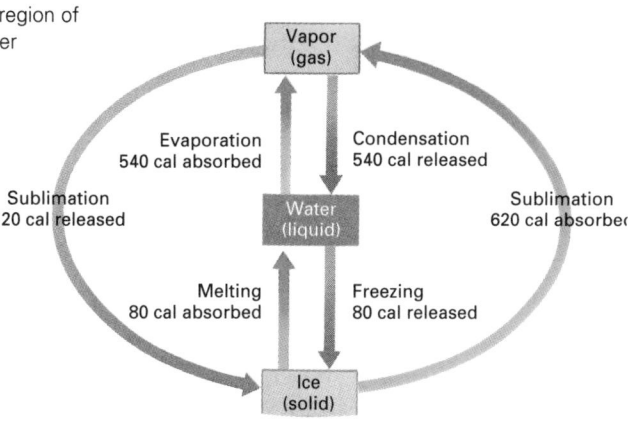

FIGURE 18.9 Water releases or absorbs latent heat as it changes among its liquid, solid, and vapor states. (Calories are given per gram at 0°C and 100°C. The values vary with temperature. Red arrows show processes that absorb heat; blue arrows show those that release heat.)

latent heat Stored heat; the energy released or absorbed when a substance changes from one state to another, by melting, freezing, vaporization, condensation, or sublimation.

photon A particle of light; the smallest particle or packet of electromagnetic energy.

solstice Either of two times per year when the Sun is furthest from the equator and shines directly overhead at either 23.5° north latitude (on or about June 21) or 23.5° south latitude (on or about December 22). The solstice on or about June 21 marks the longest day of the year in the Northern Hemisphere and the shortest day in the Southern Hemisphere. The solstice on or about December 22 marks the longest day in the Southern Hemisphere and the shortest day in the Northern Hemisphere.

specific heat The amount of energy required to raise the temperature of 1 gram of a substance by 1°C.

temperature A measure of heat in a substance, proportional to the average speed of atoms and molecules in a sample.

tropic of Cancer The latitude 23.5° north of the equator. On or about June 21, the summer solstice in the Northern Hemisphere, sunlight strikes Earth from directly overhead at noon at this latitude.

tropic of Capricorn The latitude 23.5° south of the equator. On or about December 22, the summer solstice in the Southern Hemisphere, sunlight strikes Earth from directly overhead at noon at this latitude.

weather The state of the atmosphere at a given place and time as characterized by temperature, wind, cloudiness, humidity, and precipitation.

Oceans affect weather and climate because water has low albedo and a high specific heat. In addition, currents transport heat both vertically and horizontally. Finally, evaporation cools the sea surface. As a result of all of these factors, coastal areas are generally cooler in summer and warmer in winter than continental interiors at the same latitude.

18.4 Temperature Changes with Latitude and Season

The general decrease in temperature from the equator to the poles results from the decreasing intensity of solar radiation from the equator to the poles. Changes of seasons are caused by the tilt of Earth's axis relative to the Earth–Sun plane.

18.5 Temperature Changes with Geography

Temperature changes with geography as a result of altitude, differences in temperature between ocean and land, ocean currents, wind direction, cloud cover, and albedo.

REVIEW QUESTIONS

1. Briefly outline the nature of light. What is a photon? What is the electromagnetic spectrum?
2. What happens to light when it is absorbed? When it is emitted? Give an example of each of these phenomena.
3. What is albedo? How does the albedo of a surface change when snow melts? When forests are converted to agricultural fields? When fields are paved into parking lots?
4. What is the fate of the solar radiation that reaches the Earth?
5. Explain the greenhouse effect.
6. Explain the difference between heat and temperature.
7. Explain the difference between conduction and convection.
8. What is latent heat? How does it affect Earth's surface temperature?
9. What is specific heat? How does it affect Earth's surface temperature?
10. Explain how the tilt of Earth's axis affects climate in the temperate and polar regions.
11. Discuss temperature and lengths of days at the poles, the midlatitudes, and the equator at the following times of year: June 21, December 21, and the equinoxes.
12. Explain how altitude, wind, the ocean, cloud cover, and albedo affect regional temperature.

*remember

There are more review tools for this course online at 4ltrpress .cengage.com.

KEY TERMS

absolute humidity The mass of water vapor in a given volume of air, expressed in grams per cubic meter (g/m³).

adiabatic temperature changes Temperature changes caused by compression or expansion of gas, that occur without gain or loss of heat.

advection fog Fog that forms when warm, moist air from the sea blows onto cooler land, where the air cools and water vapor condenses at ground level.

air mass A large body of air that has approximately the same temperature and humidity at any given altitude throughout.

altostratus High-altitude stratus clouds.

anticyclone A high-pressure region with its accompanying system of outwardly directed rotating winds that develop where descending air spreads over Earth's surface.

cirrus Wispy, high-altitude clouds composed of ice crystals.

cold front A front that forms when moving cold air collides with stationary or slower-moving warm air. The dense cold air distorts into a blunt wedge and pushes under the warmer air, creating a narrow band of violent weather.

cumulonimbus Towering storm clouds that form in columns and produce intense rain, thunder, lightning, and sometimes hail.

cumulus Fluffy white clouds with flat bottoms and billowy tops.

cyclone A low-pressure region with its accompanying system of inwardly directed rotating winds. In common, nonscientific usage the term often refers to a variety of different violent storms including hurricanes and tornados.

dew Moisture that is condensed onto objects from the atmosphere, usually during the night, when the ground and leaf surfaces become cooler than the surrounding air.

dew point The temperature at which the relative humidity of air reaches 100 percent and the air becomes saturated.

dry adiabatic lapse rate The rate at which dry air cools adiabatically as it rises—10°C for every 1,000 meters above sea level. Rising air cools at the dry adiabatic lapse rate until it cools to its dew point and condensation begins.

El Niño An episodic weather pattern occurring every 3 to 7 years in which the trade winds slacken in the Pacific Ocean and warm water accumulates off the coast of South America and causes unusual rains and heavy snowfall in the Andes.

evaporation fog Fog that forms when air is cooled by evaporation from a body of water.

front In meteorology, the boundary between a warmer air mass and a cooler one.

frontal wedging A process by which a moving mass of cool, dense air encounters a mass of warm, less-dense air; the cool, denser air slides under the warm air mass, forcing the warm air upward to create a weather front.

frost Ice crystals formed directly from vapor when the dew point is below freezing.

glaze An ice coating that forms when rain falls on subfreezing surfaces.

hail Large ice globules varying from 5 millimeters to a record 14 centimeters in diameter that fall from cumulonimbus clouds.

humidity The amount of water vapor in the air.

hurricane A tropical storm occurring in North America or the Caribbean whose wind exceeds 120 kilometers per hour; called a *typhoon* in the western Pacific and a *cyclone* in the Indian Ocean.

isobars Lines on a weather map connecting points of equal air pressure.

19.1 Moisture in Air

Relative humidity is the amount of water vapor in air compared to the amount the air could hold at that temperature. When relative humidity reaches 100 percent, the air is saturated. Condensation occurs when saturated air cools below its dew point.

19.2 Cooling and Condensation

Three atmospheric processes cool air to its dew point and cause condensation: (1) radiation, (2) contact with a cool surface, and (3) adiabatic cooling of rising air. Radiation and contact cooling cause the formation of dew, frost, and some types of fog. However, clouds and precipitation normally form as a result of the cooling that occurs when air rises.

19.3 Rising Air and Precipitation

Three mechanisms cause air to rise: orographic lifting, frontal wedging, and convection –convergence. Warm, moist air is said to be unstable because it rises rapidly, forming towering clouds and heavy rainfall. Warm, dry air is said to be stable because it doesn't rise rapidly, doesn't ascend to high elevations, and doesn't lead to cloud formation and precipitation.

19.4 Types of Clouds

A cloud is a concentration of water droplets or ice crystals in air. The three fundamental types of clouds are cirrus, stratus, and cumulus. Precipitation occurs when the water droplets or ice crystals in a cloud coalesce until they become large enough to fall.

19.5 Fog

Fog is a cloud that forms at or very close to ground level, although most fog forms by processes different from those that create higher-level clouds. Types of fog include advection fog, radiation fog, evaporation fog, and upslope fog.

jet streams Narrow bands of high-altitude, fast-moving wind.

monsoon A seasonal wind and weather system caused by uneven heating and cooling of land and adjacent sea.

nimbostratus Stratus clouds from which rain or snow falls.

normal lapse rate The vertical temperature structure of the atmosphere; in other words, the rate at which air that is neither rising nor falling cools with elevation.

occluded front A front that forms when a faster-moving cold air mass traps a warm air mass against a second mass of cold air.

orographic lifting Lifting of air that occurs when air flows over a mountain.

pressure gradient A measure of the change in air pressure over distance, used to determine wind speed.

radiation fog Fog that occurs when Earth's surface and the air near the surface cool by radiation during the night, and water vapor in the air condenses because it cools below its dew point.

relative humidity The ratio of the amount of water vapor in a given volume of air divided by the maximum amount of water vapor that can be held by that air at a given temperature, expressed as a percentage.

saturation The maximum amount of water vapor that air can hold.

sleet Small spheres of ice that develop when raindrops form in a warm cloud and freeze as they fall through a layer of cold air at lower elevation.

stable air A parcel of warm, dry air that does not rise rapidly, does not ascend to high elevations, and does not lead to cloud formation and precipitation.

stationary front A front at the boundary between two stationary air masses of different temperatures.

storm surge Abnormally high coastal waters and flooding, created by a combination of strong onshore winds and the low atmospheric pressure of a storm that raises the sea surface by several meters.

storm tracks Paths repeatedly followed by storms.

stratocumulus Low, sheetlike clouds with some vertical structure.

stratus Horizontally layered clouds that spread out into a broad sheet, usually creating dark, overcast skies.

supercooling A condition in which water droplets in air do not freeze even when the air cools below the freezing point.

supersaturation A condition in which the relative humidity of the air exceeds 100 percent.

tornado A small, intense, short-lived, funnel-shaped storm that protrudes from

19.6 Pressure and Wind

When air is heated, it expands and rises, creating low pressure. Cool air sinks, exerting a downward force and forming high pressure. Uneven heating of Earth's surface causes pressure differences, which, in turn, cause wind. Wind speed is determined by the pressure gradient.

19.7 Fronts and Frontal Weather

A front is the boundary between a warmer air mass and a cooler one. When two air masses collide, the warmer air rises along the front, forming clouds and often precipitation.

19.8 How Earth's Surface Features Affect Weather

Air cools adiabatically when it rises over a mountain range, often causing precipitation. Sea breezes and monsoons arise because ocean temperature changes slowly in response to daily and seasonal changes in solar radiation, whereas land temperature changes quickly. In the case of forests, a feedback mechanism operates: forests cool the air, cool air promotes rainfall, and rainfall supports forests.

19.9 Thunderstorms

A thunderstorm is a small, short-lived storm from a cumulonimbus cloud. Lightning occurs when charged particles separate within the cloud.

19.10 Tornadoes and Tropical Cyclones

A tornado is a small, short-lived, funnel-shaped storm that protrudes from the bottom of a cumulonimbus cloud. A tropical cyclone (or tropical storm) is a larger, longer-lived storm that forms over warm oceans and is powered by the energy released when water vapor condenses to form clouds and rain.

19.11 Hurricane Katrina

Death and property loss from Hurricane Katrina was caused by the intensity of the storm and by a variety of human factors, including human-caused erosion of the delta and levee failure.

19.12 El Niño

El Niño is a weather pattern caused by shifting motion of ocean currents every 3 to 7 years.

the base of a cumulonimbus cloud.

tropical cyclone *or* **tropical storm** A broad, circular storm with intense low pressure that forms over warm oceans.

typhoon A tropical storm occurring in the western Pacific Ocean whose wind exceeds 120 kilometers per hour; called a *hurricane* in North America and the Caribbean and a *cyclone* in the Indian Ocean.

unstable air A parcel of warm, moist air that rises rapidly, ascends to high elevations, and leads to formation of towering clouds and heavy rainfall.

upslope fog Fog that forms when air cools as it rises along a land surface.

warm front A front that forms when moving warm air collides with a stationary or slower-moving cold air mass. The moving warm air rises over the denser cold air, cools adiabatically, and the cooling generates clouds and precipitation.

wet adiabatic lapse rate The rate at which rising moist air cools adiabatically after it has reached its dew point and condensation has begun—varying depending on moisture content from 5°C to 9°C for every 1,000 meters it rises.

wind Horizontal airflow caused by pressure differences resulting from unequal heating of Earth's atmosphere. Winds near Earth's surface always flow from a region of high pressure toward a low-pressure region.

KEY TERMS

biome A community of plants growing in a large geographic area characterized by a particular climate.

doldrums A vast low-pressure region of Earth near the equator with hot, humid air and where local squalls and rainstorms are common but steady winds are rare.

horse latitudes A calm, high-pressure region of Earth lying at about 30° north and south latitudes, in which generally dry conditions prevail and steady winds are rare.

humid continental climate A midlatitude climate characterized by hot summers, cold winters, and precipitation throughout the year.

humid subtropical climate A midlatitude climate characterized by hot, humid summers but cooler winters and rainfall that falls throughout the year.

Koeppen Climate Classification A climate classification system describing Earth's principal climate zones, used by climatologists throughout the world.

marine west coast climate A midlatitude climate characterized by relatively mild but wet winters and cool summers, with little temperature difference between seasons.

Mediterranean climate A midlatitude climate characterized by dry summers, rainy winters, and moderate temperatures.

polar easterlies Persistent polar surface winds in the Northern Hemisphere that flow from east to west.

polar front The low-pressure boundary at about 60° latitude formed by warm air rising at the convergence of the polar easterlies and prevailing westerlies.

polar jet stream A jet stream that flows along the polar front.

prevailing westerlies The winds that blow steadily toward the poles from the southwest in the Northern Hemisphere and from the northwest in the Southern Hemisphere, between 5° and 30° north and south latitudes.

steppe A vast, semiarid, grass-covered plain found in dry climates such as those in southeast Europe, central Asia, and parts of central North America.

subarctic The northernmost edge of the humid continental climate zone, lying just south of the arctic.

20.1 Global Winds and Climate

The three-cell model of global wind circulation shows three cells of global airflow bordered by alternating bands of high and low pressure. Warm air rises near the equator, forming a low-pressure region called the doldrums. This air flows north and south at high altitude until it cools and falls at 30° latitude and forms a high-pressure region called the horse latitudes, where most of the world's largest deserts prevail. The falling air splits. A portion flows back toward the equator along the surface, forming the trade winds and completing the tropical cell.

The remaining air that falls at 30° north and south latitudes flows poleward along the surface. In the Northern Hemisphere, this air is deflected to form the prevailing westerlies. Air rises at a low-pressure region near 60° latitude and returns at high altitude to complete the cell. A second region of high pressure exists over the poles, where the descending air spreads outward to form the polar easterlies. The polar front forms where the polar easterlies and the prevailing westerlies converge. The jet stream blows at high altitude along the polar front and along the boundaries between cells at the horse latitudes.

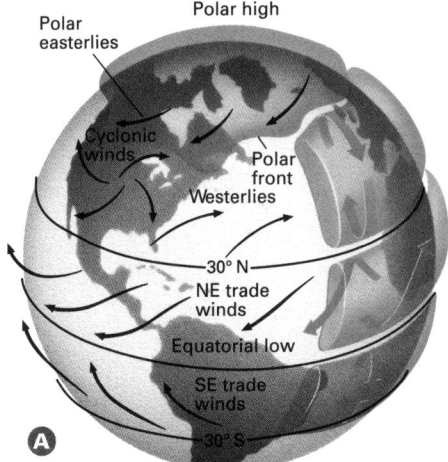

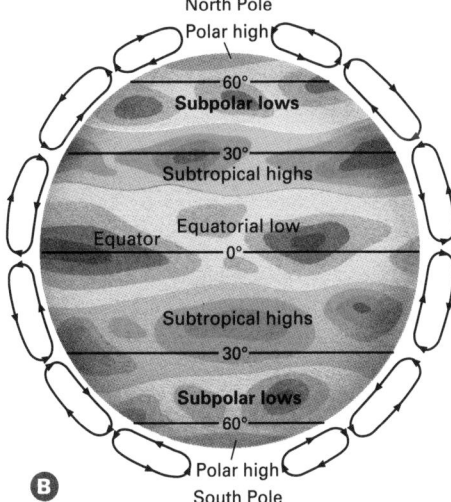

FIGURE 20.1 Global wind patterns predicted by the three-cell model. (A) Air rising at the equator moves poleward at high elevations, falls at about 30° north and south latitudes, and returns to the equator, forming the trade winds. The orange arrows show both upper-level and surface wind patterns. The black arrows show only surface winds. (B) High- and low-pressure belts are indicated on the sphere, with surface and upper-level wind patterns shown on the edges.

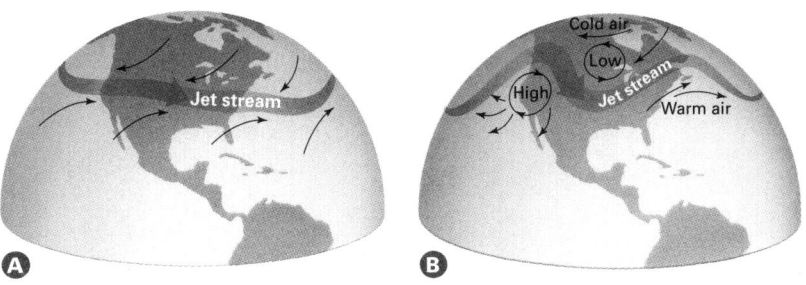

FIGURE 20.2 The polar front and the polar jet stream migrate with the seasons and with local conditions. Storms commonly occur along the jet stream.

subtropical jet stream A jet stream that flows between the trade winds and the westerlies.

taiga A biome of conifers that can survive the extreme winters of the subarctic.

temperate rainforest A rainforest that grows in marine west coast climates where rainfall is more than 100 centimeters per year and is relatively constant throughout the year, particularly along the northwest coast of North America from Oregon to Alaska.

three-cell model A model of global wind patterns that depicts three convection cells in each hemisphere, bordered by alternating bands of high and low pressure.

trade winds The winds that blow steadily toward the equator from the northeast in the Northern Hemisphere and from the southeast in the Southern Hemisphere, between 5° and 30° north and south latitudes.

tropical monsoon climate A tropical climate with distinct wet and dry seasons, characterized by high rainfall during the wet months and a relatively short dry season.

tropical rainforest A forest growing in a tropical climate with abundant, year-round rainfall; characterized by tall trees that branch only near the top, creating a dense canopy of leaves that block out most of the light, and with soggy ground and water dripping everywhere.

tropical savanna climate A tropical climate with distinct wet and dry seasons but with relatively low annual rainfall, characterized by large areas of grassland supporting herds of grazing animals.

tundra A biome dominated by low-lying grasses, mosses, flowers, and a few small bushes that grow in the Arctic and high alpine regions.

urban heat island effect Warmer temperatures in a city compared to the surrounding countryside, caused by factors in the urban environment itself.

20.2 Climate Zones of Earth

Climate zones are classified according to annual temperature, annual precipitation, and variability in either of these factors from month to month or season to season. World climate types include humid tropical, arid, humid mesothermal, humid microthermal, polar, and highland climates.

20.3 Urban Climates

Urban areas are generally warmer and wetter than surrounding countryside.

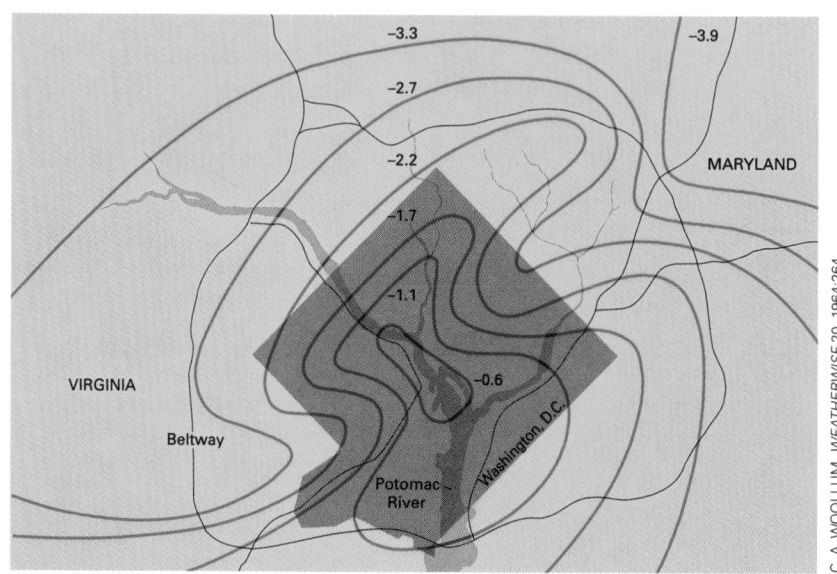

C. A. WOOLLUM, *WEATHERWISE* 20, 1964:264

FIGURE 20.14 The urban heat island effect. The average minimum temperature in and around Washington, D.C., during the winter; contours represent relative T minima in degrees Celsius.

REVIEW QUESTIONS

1. Discuss the factors that control the average temperature in any region.
2. Discuss how oceans affect coastal climate.
3. Explain the three-cell model for global wind circulation.
4. Describe the trade winds. Why are they so predictable?
5. Why is the doldrums region relatively calm and rainy? Why are the horse latitudes calm and dry?
6. Describe the polar front and the jet stream. How do they affect weather?
7. Explain how horizontal and vertical ocean currents affect temperature.
8. Discuss and describe the major climate zones of the Koeppen Climate Classification and the plant communities associated with each.
9. Explain how urban climate differs from that of the surrounding countryside.

REVIEW QUESTIONS

1. Explain why early Earth was warm, even though the Sun emitted about 30 percent less energy at that time.
2. Briefly outline climate changes both from 100,000 to 10,000 years ago and from 10,000 years ago to the present. List some of the methods used to gather data to develop this time line.
3. Explain how volcanic eruptions can cause either a cooling or a warming of the atmosphere.
4. List and explain four feedback loops that affect climate. For each one, outline the contributions of geosphere, atmosphere, biosphere, and hydrosphere.
5. Discuss the consequences of a warmer Earth. How would temperature changes affect precipitation? How would changes in temperature and precipitation affect agriculture, ecosystems, and human society?

21.1 Climate Change in Earth's History

Global climate has changed throughout Earth's history, from extreme cold with extensive glaciation to a 248-million-year warm period from the start of the Mesozoic Era almost to the present. During the last 100,000 years, the mean annual global temperature has changed frequently and dramatically, although the past 10,000 years, during which civilization developed, witnessed anomalously stable climate.

21.2 Measuring Climate Change

Past climates can be studied through historical records, tree rings, pollen assemblages in sediment, oxygen isotope ratios in glacial ice, erosional and depositional features created by glaciers, plankton assemblages and isotope studies in ocean sediment, and fossils in sedimentary rocks. The data can be compared with current data to measure climate change.

21.3 Astronomical Causes of Climate Change

Despite its immense complexity, Earth's atmospheric temperature is determined by the heat of the Sun, albedo, and heat retention by the atmosphere, which had great impact on early Earth in particular. When the Earth first formed, the Sun produced about 70 percent of the energy it produces today. Other components of the primordial atmosphere warmed early Earth; as the composition of the atmosphere changed, so did climate. Bolide impacts may also have great impact and caused rapid, catastrophic changes in Earth's climate.

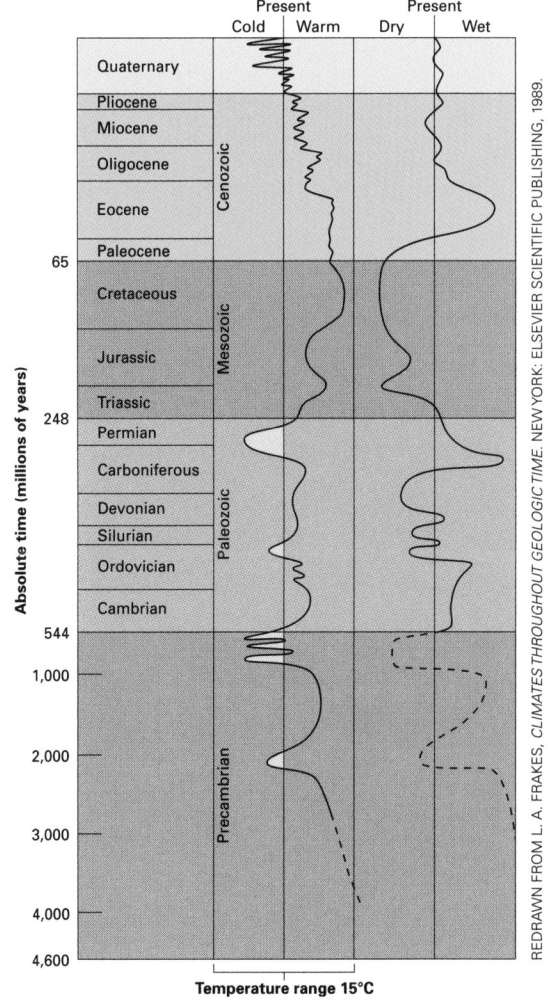

FIGURE 21.2 Mean global temperature and precipitation have both fluctuated throughout Earth's history.

REDRAWN FROM L. A. FRAKES, *CLIMATES THROUGHOUT GEOLOGIC TIME*. NEW YORK: ELSEVIER SCIENTIFIC PUBLISHING, 1989.

21.4 Water and Climate

Water in its various forms can cause warming or cooling. Water vapor warms the atmosphere, but clouds can cause either warming or cooling. Glaciers and snowfields reflect sunlight and cause cooling. Evaporation is a cooling process, while condensation releases heat. The evaporation–condensation cycle transports heat from the equatorial regions to the poles. Weathering alters the carbon dioxide concentration in the atmosphere. Ocean currents transport heat and moisture.

21.5 The Natural Carbon Cycle and Climate

Carbon circulates among all four of Earth's spheres. Carbon dioxide is a greenhouse gas that warms the atmosphere. Carbon exists in the biosphere as the fundamental building block for organic tissue. Carbon dioxide dissolves in seawater to form bicarbonate and carbonate ions. Carbon exists in the crust and mantle in several forms: (1) Marine organisms absorb calcium and carbonate ions and convert them to solid calcium carbonate (shells and skeletons), and as a result large quantities of carbon exist in limestone and marine sediment. (2) Fossil fuels are largely composed of carbon. (3) Large quantities of carbon exist as methane hydrates on the sea floor. (4) Carbon also exists in the deeper mantle; some of this is primordial carbon trapped during Earth's formation and some is carried into the mantle on subducting plates.

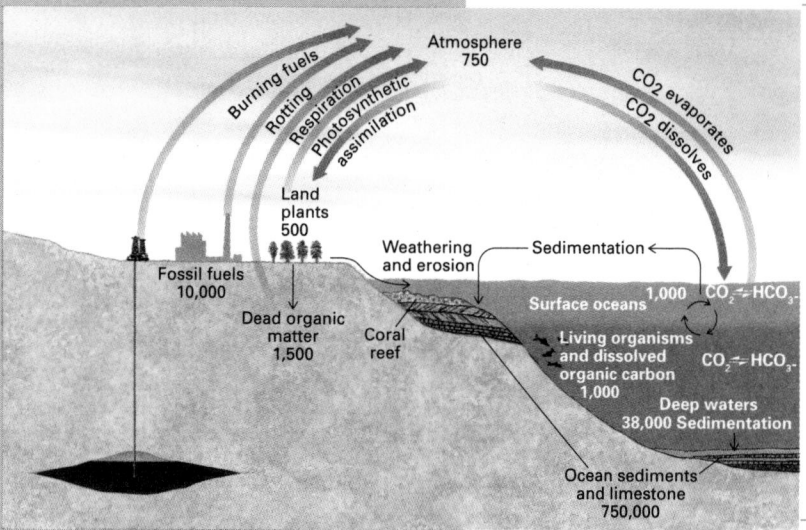

Source: Data taken from U. Siegenthaler and J. L. Sarmiento, "Atmospheric Carbon Dioxide and the Ocean," *Nature* 365 (September 9, 1993), 119.

FIGURE 21.8 The carbon cycle. The numbers show the size of the reservoirs in billions of tons of carbon.

21.6 Tectonics and Climate Change

Natural climate change also results from changes in the positions of continents, the growth of mountains, and volcanic eruptions. The positions of the continents influence wind and sea currents, which affect climate. Mountains interrupt airflow, changing regional winds. The limestone in several large mountain chains removes carbon dioxide from the air in the weathering process, possibly causing global cooling. Volcanic eruptions emit ash and sulfur compounds that reflect sunlight and cool the atmosphere, but also emit carbon dioxide and ash into the atmosphere, which warms the planet by absorbing infrared radiation. All of these factors interact to maintain a specific global temperature.

21.7 Greenhouse Effect: The Carbon Cycle and Global Warming

People release carbon dioxide into the atmosphere when they burn fuel. Logging, some industrial processes, and some aspects of agriculture contribute greenhouse gases to the atmosphere. Most scientists agree that human introduction of greenhouse gases has altered climate in the past century. Global warming could lead to a longer growing season in the high latitudes, but warmth affects plants in ways to decrease crop yields. One study has shown that rice yields decline in a warmer world, and the ratio of respiration to photosynthesis increases, leading to more carbon dioxide in the atmosphere. Global warming will affect precipitation and snowmelt, cause extreme weather, alter biodiversity, melt arctic sea ice and glaciers, lead to acidification of the oceans, and cause sea level to rise.

21.8 Feedback and Threshold Mechanisms in Climate Change

Climate feedback loops include snowmelt and albedo and changes in rates of plant respiration and photosynthesis. The Younger Dryas Event was a sudden cool climate period that may have been caused by changes in the ocean currents due to freshwater influx. The collapse of the West Antarctic Ice Sheet could lead to rising sea level and further global warming. Melting permafrost could release trapped methane gas, which is 20 times as effective at trapping heat as carbon dioxide. Such release could create catastrophic feedback warming.

KEY TERMS

absorption spectrum A spectrum of radiation created when light passes through a substance, such as a star, that absorbs some of the wavelengths; visible as dark lines crossing the bands of colors in a spectrum.

blue shift A shift toward shorter (blue) wavelengths, observed in the spectrum of a distant galaxy or other object that is moving toward Earth; caused by the Doppler effect.

celestial spheres The hypothetical series of concentric transparent spheres surrounding Earth in Aristotle's model of the Universe. Aristotle postulated that the Sun, Moon, planets, and stars are imbedded in the spheres.

constellation A group of stars that seem to form a pattern when viewed from Earth, to which astronomers have given a name for convenience in referring to the positions of objects in the night sky.

corona The outer atmosphere of the Sun, normally invisible but appearing as a halo around a black Moon during a solar eclipse.

crescent moon A lunar phase in which the Moon appears as a thin crescent, when either waxing or waning.

Doppler effect The observed change in frequency of light or sound that occurs when the source of the wave is moving either toward or away from the observer.

emission spectrum A spectrum created when absorbed radiation is reemitted from its source.

eyepiece The lens, closest to the eye in a refracting telescope, that magnifies the image.

full moon The lunar phase when the Moon appears round and fully illuminated, because it is on the opposite side of Earth from the Sun and its entire sunlit area is visible.

geocentric A model that places Earth at the center of the Universe.

gibbous moon A bright moon, either waxing or waning, with only a sliver of dark visible.

heliocentric A model that places the Sun at the center of the Solar System.

inertia The tendency of an object to resist a change in motion.

22.1 The Motions of the Heavenly Bodies

Constellations are groups of stars that seem to form a pattern when viewed from Earth and serve as points of reference for the positions of objects in the night sky. For most of the year, planets appear to drift eastward with respect to the stars, but sometimes they seem to reverse direction and drift westward. This apparent reverse movement is called retrograde motion.

22.2 Aristotle and the Earth-Centered Universe

Aristotle proposed a geocentric, or Earth-centered, Universe in which a stationary, central Earth is surrounded by celestial spheres that contain the Sun, the Moon, the planets, and the stars. Ptolemy modified the geocentric model to explain the retrograde motion of the planets. In Ptolemy's model, each planet moves in small circles within its larger orbit.

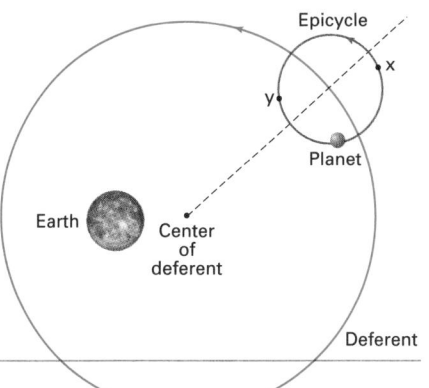

22.3 The Renaissance and the Heliocentric Solar System

Copernicus believed that the Universe should operate in the simplest manner possible and showed that a heliocentric Solar System best explains the movement of planets including retrograde motion. Kepler calculated the elliptical orbits of the planets. Galileo used observation and experimentation to discredit the geo-

FIGURE 22.5 Ptolemy's explanation of retrograde motion. Each planet revolves in a small orbit (the *epicycle*) around the larger orbit (the *deferent*). When the planet is in position x, it appears to be moving eastward. When it is in position y, it appears to reverse direction and to be moving westward. Ptolemy did not realize that the planets moved in elliptical orbits. To compensate for this error, he placed Earth away from the center of the deferent.

centric model and show that the planets revolve around the Sun. Newton proved that gravity holds the planets in elliptical orbits rather than flying off into space.

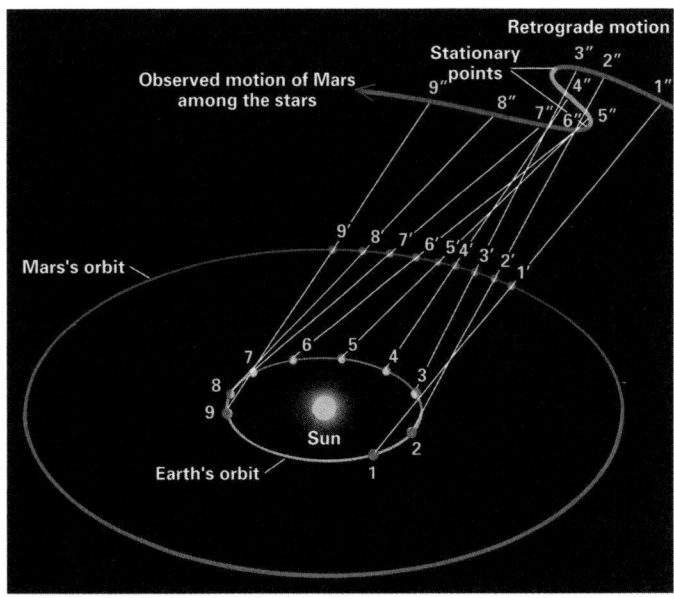

FIGURE 22.6 Retrograde motion is explained within the heliocentric model by changes in the relative positions of Earth and the planet we are observing.

light pollution The nighttime glow of city lights that competes with the light of distant stars and reduces the vision of a telescope.

lunar eclipse A phenomenon that occurs when Earth lies directly between the Sun and the Moon, causing Earth's shadow to fall on the Moon and darken it.

new moon The lunar phase during which the Moon is dark when viewed from Earth, because it has moved to a position between Earth and the Sun and its sunlit side faces away from us.

objective lens The lens, farthest from the eyepiece in a refracting telescope, that collects light from a distant object.

parallax The apparent change in position of an object due to the change in position of the observer.

penumbra A wide band outside of the umbra, where only a portion of the Sun is hidden from view during a solar eclipse.

planet A celestial body that revolves in a fixed orbit around a star.

red shift A shift toward longer (red) wavelengths, observed in the spectrum of a distant galaxy or other object that is moving away from Earth; caused by the Doppler effect.

reflecting telescope A type of telescope that uses a mirror or mirrors to collect and focus an image, which is then reflected to the eyepiece by another mirror.

refracting telescope A type of telescope that uses two lenses: one to collect light from a distant object and another to magnify the image.

resolution The capability of distinguishing individual parts, such as the ability of an optical telescope to distinguish details.

retrograde motion The apparent motion of planets when viewed from Earth, in which they temporarily appear to move backwards (westward) with respect to the stars, before resuming their original eastward motion.

revolution Orbiting around a central point. A satellite revolves around Earth, and Earth revolves around the Sun.

rotation Turning or spinning on an axis. Tops and planets rotate on their axes.

solar eclipse A phenomenon that occurs when the Moon passes directly between Earth and the Sun; the Moon casts its shadow on Earth, thus blocking the Sun's light.

spectrum An ordered array of colors; a pattern of wavelengths into which a beam of light or other electromagnetic radiation is separated.

umbra The narrow band of the Moon's shadow where the Sun is completely blocked out during a solar eclipse.

waning Becoming smaller; the 14 to 15 days after a full moon and before a new moon

when the visible portion of the Moon decreases every day.

waxing Becoming full; the 14 to 15 days after a new moon and before a

full moon when the visible portion of the Moon increases every day.

22.4 The Motions of the Earth and the Moon

The revolution of the Moon around Earth causes the phases of the Moon. A lunar eclipse occurs when Earth lies directly between the Sun and the Moon. A solar eclipse occurs when the Moon lies directly between the Sun and Earth.

22.5 Modern Astronomy

Objects in space are studied with both optical telescopes and telescopes sensitive to invisible wavelengths. Emission and absorption spectra provide information about the chemical compositions and temperatures of stars and other objects. The Doppler effect causes light from an object moving away from Earth to reach us at a lower frequency than it had when it was emitted. Light from an object moving toward Earth has a higher frequency than it had when it was emitted. Instruments carried aloft by spacecraft eliminate interference by Earth's atmosphere and allow close inspection of objects in the Solar System.

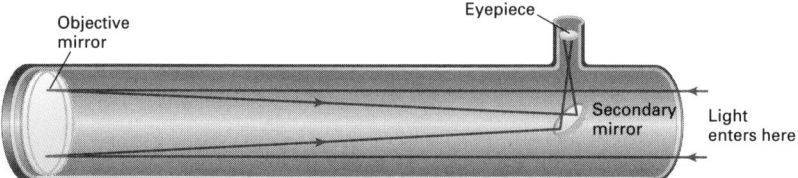

FIGURE 22.17 A Newtonian reflecting telescope. Incoming light (right) is collected and focused by a curved objective mirror. The light is then reflected back to the eyepiece by a secondary mirror.

REVIEW QUESTIONS

1. List and explain Aristotle's observations and the reasoning he used to support his geocentric model of the Solar System.
2. Why did Aristotle and other ancient astronomers fail to notice a parallax shift when they observed the stars?
3. What is retrograde motion? How is it explained in Ptolemy's and Copernicus's models?
4. Explain how Galileo's studies of physics contributed to his rejection of Aristotle's geocentric model.
5. Explain how Galileo's observations of the Milky Way, the Sun, and the Moon led him to question Aristotle's geocentric model.
6. What evidence convinced Galileo that Earth revolves around the Sun?
7. What was the astronomical significance of Newton's studies of gravity?
8. How is the Moon positioned with respect to Earth and the Sun when it is full, new, gibbous, and crescent?
9. Explain the difference between rotation and revolution as they relate to planetary motion.
10. Draw a picture of the Sun, Earth, and Moon as they will be during an eclipse of the Sun. Draw a picture of the Sun, Earth, and Moon as they are aligned during an eclipse of the Moon. Explain how these positions produce eclipses.
11. Explain how optical telescopes work. How is a refracting telescope different from a reflecting telescope?
12. Discuss the relative advantages and disadvantages of space telescopes versus ground-based observatories.
13. Describe the differences between absorption spectra and emission spectra.
14. Explain the Doppler effect.

KEY TERMS

asteroids Small celestial bodies in orbit around the Sun, primarily in the region between Mars and Jupiter.

blob tectonics Tectonic activity dominated by rising and sinking of the mantle and crust, believed to predominate on Venus, as opposed to the horizontal movement of plates associated with tectonic activity on Earth.

Callisto Jupiter's outermost Galilean moon; marked by a heavily cratered surface.

chondrule A small grain about 1 millimeter in diameter embedded in a meteorite, often containing amino acids or other organic molecules.

coma The bright outer sheath of a comet, surrounding the nucleus.

comet An interplanetary orbiting body composed of loosely bound rock and ice, which forms a bright head and extended fuzzy tail when it approaches the Sun. It appears to be fiery hot, but it is actually a cold object and its "flame" is reflected light.

dwarf planet A body (for example, Pluto or Eris) that orbits the Sun, is not a satellite of a planet, and is massive enough to pull itself into a spherical shape, but is not massive enough to clear out other bodies in and near its orbit.

escape velocity The speed that an object must attain to escape the gravitational field of a planet or other object in space.

Europa The second-closest of Jupiter's moons; similar to Earth in that much of its interior is composed of rock and much of its surface is covered with water, although the water is frozen into a vast planetary ice crust.

Ganymede A moon of Jupiter marked by a convecting, metallic core and a brittle water/ice crust that behaves much like rock.

Great Red Spot A giant hurricane-like storm on the surface of Jupiter that has existed for centuries.

Io The innermost moon of Jupiter and the most active volcanic body in the Solar System.

23.1 The Solar System: A Brief Overview

Table 23.1 provides an overview of the eight major planets. Due to the differences in composition, the terrestrial planets are much denser than the Jovian planets. However, the Jovian planets are much larger and more massive than the Earth and its neighbors.

TABLE 23.1 Comparison of the Eight Major Planets

Planet	Distance from Sun (millions of kilometers)	Radius (compared to radius of Earth = 1)	Mass (compared to mass of Earth = 1)	Density (compared to density of water = 1)	Composition of Planet	Density of Atmosphere (compared to Earth's atmosphere = 1)	Number of Moons
Terrestrial Planets							
Mercury	58	0.38	0.06	5.4		One-billionth	0
Venus	108	0.95	0.82	5.2	Rocky with metallic core	90	0
Earth	150	1	1	5.5		1	1
Mars	229	0.53	0.11	3.9		0.01	2
Jovian Planets							
Jupiter	778	11.2	318	1.3	Liquid hydrogen surface with liquid metallic mantle and solid core	Dense and turbulent	62
Saturn	1,420	9.4	94	0.7			62
Uranus	2,860	4.0	15	1.3	Hydrogen and helium outer layers with solid core	Similar to Jupiter except that some compounds that are gases on Jupiter are frozen on the outer planets	27
Neptune	4,490	3.9	17	1.7			13

23.2 The Terrestrial Planets

Mercury is the closest planet to the Sun. It has virtually no atmosphere, and it rotates slowly on its axis; therefore, it experiences extremes of temperature. Mercury's surface is heavily cratered from meteorite bombardment that occurred early in the history of the Solar System. Venus has a hot, dense atmosphere as a result of a runaway greenhouse effect. Its surface shows signs of recent tectonic activity, probably resulting from vertical upwelling, called blob tectonics. Mars is a dry, cold planet with a thin atmosphere, but its surface bears signs of tectonic activity and ancient water erosion. The mobile robots Spirit and Opportunity detected clear signs that water once existed on Mars.

23.3 The Moon: Our Nearest Neighbor

Earth's Moon probably formed from the debris of a collision between a Mars-sized body and Earth. The Moon was heated by the energy released during condensation of the debris, by radioactive decay, and by meteorite bombardment. Evidence of ancient volcanism exists, but the Moon is cold and inactive today. In 2009 NASA instruments confirmed the presence of water molecules in the Moon's polar regions, although in relatively small amounts.

Jovian planets The outer planets—Jupiter, Saturn, Uranus, and Neptune—all massive, with relatively small rocky or metal cores surrounded by swirling liquid and gaseous atmospheres.

Jupiter The largest planet in the Solar System and fifth from the Sun.

Kuiper Belt objects Tiny ice dwarfs, similar to Pluto, orbiting in a disk-shaped region at the outer reaches of the Solar System.

liquid metallic hydrogen A form of hydrogen under extreme temperature and pressure, which forces the atoms together

so tightly that the electrons move freely throughout the packed nuclei, and as a result the hydrogen conducts electricity.

maria Dry, barren, flat expanses of volcanic rock on the Moon, first thought to be seas.

Mars The fourth planet from the Sun.

Mercury The closest planet to the Sun.

meteor A falling meteoroid that enters Earth's atmosphere and glows as it vaporizes; colloquially called a *shooting star*.

meteorite A meteor that does not completely vaporize and that strikes Earth's surface.

meteoroid A small interplanetary body, most often an asteroid or comet fragment, traveling in an irregular orbit through the inner Solar System.

Neptune The eighth planet from the Sun; similar to Uranus in size, composition, and atsmosphere.

nucleus The dense, solid core of a comet.

Opportunity One of two mobile robots that landed on Mars in 2004.

Pluto Once considered to be the ninth planet from the Sun in our Solar System, reclassified in 2006 as a dwarf planet.

Saturn The second-largest planet and sixth from the Sun; marked by its distinctive rings.

solar wind A stream of electrons and positive ions radiating outward from the Sun at high speed.

Spirit One of two mobile robots that landed on Mars in 2004.

stony meteorite A meteorite with a mass ratio of rock to metal similar to the mass ratio of Earth's mantle to Earth's core, thus reflecting the primordial composition of the Solar System and representing a window into its past. Most meteorites are stony meteorites.

tail The long trailing portion of a comet, always pointing away from the Sun, formed when solar winds blow away lighter particles from the comet's head. What appears to be a fiery arrow is actually reflected light from the Sun.

terrestrial planets The four Earth-like planets closest to the Sun—Mercury, Venus, Earth, and Mars—which are composed primarily of non-volatile metals and silicate rocks.

Titan Saturn's largest moon; the only moon in the Solar System with an appreciable atmosphere.

Triton The largest of Neptune's moons; marked by craters filled with ice or frozen methane.

Uranus The seventh planet from the Sun; similar to Neptune in size, composition, and atmosphere.

Venus The second planet from the Sun; resembles Earth in size and density.

23.4 The Jovian Planets: Size, Compositions, and Atmospheres

Jupiter, Saturn, Uranus, and Neptune are all large planets with low densities. Jupiter and Saturn have dense atmospheres, surfaces of liquid hydrogen, inner zones of liquid metallic hydrogen, and cores of rock and metal. Uranus and Neptune have higher proportions of rock and ice than Jupiter and Saturn do, and their magnetic fields are not in line with their axes of rotation—a phenomenon for which no satisfactory hypothesis has yet been developed.

23.5 Moons of the Jovian Planets

The largest moons of Jupiter are Io, which is heated by gravitational forces and exhibits extensive volcanic activity; Europa, which is ice covered; and Ganymede and Callisto, which are large spheres of rock and ice. Titan, the largest moon of Saturn, has an atmosphere rich in nitrogen and methane. Photos of its surface show evidence that landscape was sculpted by wind, flowing liquids, and tectonic activity. Triton, Neptune's largest moon, is about 75 percent rock and 25 percent ice.

23.6 Planetary Rings

All the Jovian planets have one or more rings, but the most spectacular are those of Saturn, which are visible from Earth using just a small telescope. Saturn's rings are made up of many small particles of dust, rock, and ice. They might have formed from a moon that was fragmented or from rock and ice that never coalesced to form a moon.

23.7 Pluto and Other Dwarf Planets

Pluto has a low density and is a small planet composed of ice and rock. It may be related to numerous similar icy bodies known as Kuiper Belt objects, residual planetesimals left over from the formation of the Solar System.

23.8 Asteroids, Comets, and Meteoroids

Asteroids are small, planet-like bodies in solar orbits between the orbits of Mars and Jupiter. A comet is an object that orbits the Sun in an elongated, elliptical orbit. Comets are composed mainly of water/ice and other frozen volatiles, mixed with smaller concentrations of dust, silicate rock, and metals. When a comet is in the inner part of the Solar System, it consists of a small, dense nucleus composed of ice and rock; an outer sheath or coma composed of gases, water vapor, and dust; and a long tail made up of particles blown outward by the solar wind. A meteorite is a meteoroid (a piece of matter orbiting through the inner Solar System) that falls to Earth. Most meteorites are stony, and some contain organic molecules. About 10 percent of all meteorites are metallic.

REVIEW QUESTIONS

1. List the eight planets in order of distance from the Sun, and distinguish between the terrestrial and the Jovian planets.
2. Give a brief description of Mercury. Include its atmosphere, surface temperature, surface features, and speed of rotation about its axis.
3. Compare and contrast the atmospheres of Mercury, Venus, the Moon, Earth, and Mars.
4. Venus boiled, life evolved on Earth under moderate temperatures, and Mars froze. Discuss the evolution of the atmospheres and climates on these three planets. (Refer back to Chapter 1 for additional information to answer this question.)
5. Discuss how the information from Spirit and Opportunity creates a picture of the climate history of Mars.
6. Discuss the evidence that the Martian atmosphere was once considerably different from what it is today.
7. Why are there fewer meteorite craters visible on Venus than on Mercury?
8. Explain how Venusian tectonics differs from tectonics on Earth.

KEY TERMS

absolute brightness or luminosity The brightness of a star as it would appear if it were a fixed distance from Earth.

absorption nebula A cold, dark nebula that absorbs light.

accelerating Universe A model in which an unidentified force is overpowering gravity and causing the expansion of the Universe to accelerate.

apparent brightness or luminosity The luminosity of a star as seen from Earth.

barred spiral galaxy A spiral galaxy with a straight bar of stars, gas, and dust extending across the nucleus.

big bang An event 10 to 20 billion years ago, thought to mark the beginning of the Universe when all matter exploded from a single infinitely dense point.

black hole An infinitesimally small region of space, created after the supernova explosion of a huge star, that contains matter packed so densely that light cannot escape from its intense gravitational field.

chromosphere A turbulent, diffuse, gaseous layer of the Sun that lies above the photosphere.

closed Universe A model in which the gravitational force of the Universe is so great that all the galaxies will eventually slow down, reverse direction, and fall back to the center, forming another point of infinite density.

convective zone The subsurface zone in a star where energy is transmitted primarily by convection.

core The center of the Sun, where hydrogen fusion takes place.

24.1 In the Beginning: The Big Bang

The big bang theory states that the Universe began with a cataclysmic explosion that instantly created space and time.

24.2 The Nonhomogeneous Universe

The nonhomogeneity of the Universe was generated by quantum mechanical interactions very early in its infancy.

24.3 The Birth of a Star

Stars form from condensing nebulae, clouds of gas and dust in interstellar space. When the condensing gases become hot enough, fusion begins.

24.4 The Sun

The central core of the Sun has a temperature of about 15 million K and a density 150 times that of water. Hydrogen fusion occurs in this region. The visible surface of the Sun, called the photosphere, is only 5,700 K and has a pressure of 1/100 of the Earth's atmosphere at sea level.

24.5 Stars: The Main Sequence

A star can appear luminous either because it really is bright or because it is close. The absolute brightness or luminosity of a star is how bright it would appear if it were a fixed distance of 32.6 light-years away. A Hertzsprung–Russell (H–R) diagram is a graph that plots absolute stellar luminosity against temperature. Most stars fall within a band called the main sequence.

24.6 The Life and Death of a Star

When the hydrogen in the core of a mature star is exhausted and hydrogen fusion ends, gravity compresses the core and the temperature rises. Fusion then begins in the outer shell, and the star expands to become a red giant. After helium fusion in an average-mass star ends, the star releases a planetary nebula and then shrinks to become a white dwarf. In a more massive star, enough heat is produced to fuse heavier elements; the iron-rich core then explodes to become a supernova.

cosmic background radiation Low-energy, microwave radiation that began traveling through space when the Universe was only 300,000 years old and now pervades all space in the Universe.

dark energy One component of dark matter, comprising the poorly understood repulsive force that causes the modern Universe to accelerate.

dark mass One component of dark matter, comprising a collection of as-yet-undetected particles, large enough to make up 23 percent of the Universe, that exert conventional gravitational attraction.

dark matter Mysterious invisible matter that may comprise up to 96 percent of the mass of the Universe.

elliptical galaxy A galaxy with an oval shape and no spiraling arms; can be either giant or dwarf.

emission nebula A hot nebula that glows because it is emitting light energy.

first generation energy Cosmic background radiation that emanated from the primordial sea of atoms, particles, and energy of the big bang.

galactic halo A spherical cloud of dust and gas that surrounds the galactic disk of a spiral galaxy such as the Milky Way.

galaxy A large volume of space containing many billions of stars, held together by mutual gravitational attraction.

gamma ray burst A burst of high-energy radiation emanating from space. Longer bursts come from the death throes of massive stars, while shorter bursts form when a neutron star is sucked into a black hole or when two neutron stars collide to form a black hole.

globular cluster A concentration of older stars that shared a common origin and are held together by gravity, lying within the galactic halo of a spiral galaxy such as the Milky Way.

Hertzsprung–Russell (H–R) diagram A graph that plots absolute stellar luminosity against temperature.

Hubble's law A law that states that the velocity of a galaxy is proportional to its distance from Earth. Thus the most distant galaxies are traveling at the highest velocities.

light-year The distance traveled by light in one year, approximately 9.5×10^{12} kilometers.

main sequence A band running across a Hertzsprung–Russell diagram that contains most of the stars, which are fueled by hydrogen fusion.

Milky Way The barred spiral galaxy in which our Sun and Solar System are located.

nebula A cloud of interstellar gas and dust; plural *nebulae*.

neutron star A small, extremely dense star, created from remnants of the supernova explosion of a large star and composed almost entirely of compressed neutrons.

open Universe A model in which the gravitational force of the Universe is not sufficient to stop the current expansion and the galaxies will continue to fly apart forever.

parsec A distance used in astronomy, equal to about 3.26 light-years.

photosphere The surface of the Sun visible from Earth; also called the solar atmosphere.

planetary nebula A nebula created when a star about the size of our Sun explodes and blows a shell of gas out into space.

population I star A relatively young star formed from material ejected by an older, dying star; composed mainly of primordial hydrogen and helium, with a small percentage of heavier elements. Our Sun is a population I star.

population II star An old star, with lower concentration of heavy elements than a population I star.

primordial nucleosynthesis The formation of helium nuclei that occurred by nuclear fusion during the first 8.5 minutes in the life of the Universe.

prominence A red, flame-like jet of gas rising from the Sun's corona.

pulsar A neutron star that emits a pulsating radio signal.

quasar An object, less than 1 light-year in diameter and very distant from Earth, that emits an extremely large amount of energy as radio waves.

radiative zone An inner zone of a star surrounding the core where radiation energy is

24.7 Neutron Stars, Pulsars, and Black Holes

The remnant of a supernova can contract to become a neutron star. If a neutron star emits pulses of radio waves, it is called a pulsar. If the mass of a dying star is great enough, it collapses into a black hole.

24.8 Galaxies

Galaxies and clusters of galaxies form the basic structure of the Universe. Commonly, galaxies are elliptical, spiral, or barred spiral, although other shapes also exist. Hubble's law states that the most distant galaxies are moving away from Earth at the greatest speeds, while closer ones are receding more slowly.

24.9 The Milky Way

Our Sun lies in the Milky Way, a barred spiral galaxy. In addition to stars in the main disk, the galaxy contains captured dwarf galaxies, diffuse clouds of gas and dust between the stars, and a galactic halo and globular clusters of stars surrounding the disk.

24.10 Quasars

Quasars are very far away, emit as much as 100 times the energy of an entire galaxy, and are small compared with a galaxy. A black hole may lie in the center of a quasar.

24.11 Dark Matter

Up to 96 percent of the Universe is invisible and is called dark matter.

24.12 The End of the Universe

In trying to determine whether the Universe is open or closed, astronomers learned that the expansion of the Universe is accelerating.

24.13 Why Are We So Lucky?

Life could have formed under only a very narrow range of fundamental physical constants.

REVIEW QUESTIONS

1. Outline the three main forms of evidence to support the big bang theory.
2. Describe the Universe when it was a trillionth of a trillionth of a billionth of a second old.

transmitted by absorption and emission.

red giant A stage in the life of a star when its core is composed of helium that is not undergoing fusion, but a hot shell of hydrogen around the core is fusing at a rapid rate, producing enough energy to cause the star to expand. Its core is hotter, but its expanded surface is cooler and emits red light.

second generation energy Radiation emitted by stars, galaxies, quasars, and other forms of concentrated matter.

solar mass A standard unit for expressing the mass of a star relative to the mass of our Sun; 1 solar mass equals the mass of the Sun.

spicule A jet of gas at the edge of the Sun, shooting upward from the chromosphere.

spiral galaxy A galaxy characterized by arms that radiate out from the center like a pinwheel.

sunspot A comparatively cool, dark region on the Sun's surface caused by a magnetic disturbance.

supernova An exploding star that is releasing massive amounts of energy.

white dwarf A stage in the life of a star when fusion has halted and the star glows solely from the residual heat produced during past eras. White dwarfs are very small stars.

Let us first summarize the data:

- Human activities release greenhouse gases.

- The concentration of these gases in the atmosphere has risen since the beginning of the Industrial Revolution.

- Greenhouse gases absorb infrared radiation and trap heat.

- The atmosphere has warmed by almost 0.8°C during the last 50 years.

Of the 20 hottest years on record, 19 have occurred after 1980, and 2005 was the hottest year in more than a century.

The obvious connection is that rising atmospheric concentrations of industrial greenhouse gases have caused the recent global temperature rise. Again, some government and industry sources have disagreed, stating that the current warming trend may be an unrelated natural event that just happened to coincide with the atmospheric increase in carbon dioxide. Although this debate has raged for a few decades, today scientists are essentially unanimous in concluding that industrial emissions are causing global warming. Of 928 peer-reviewed papers appearing in scientific journals between 1993 and 2003, every author and every study concluded that anthropogenic greenhouse gas emissions are causing the observed warmer temperatures.

Consequences of Greenhouse Warming

Many people ask, "What's the big deal? What difference will it make if the planet is a few degrees warmer than it is today?"

Temperature Effects on Agriculture

A warmer global climate would mean a longer frost-free period in the high latitudes, which would benefit agriculture. In parts of North America, the growing season is now a week longer than it was a few decades ago. On the negative side, warmth affects plants in many ways that decrease crop yields. In one study, researchers found that rice yields decrease by 10 percent for every 1°C increase in nighttime temperatures. The scientists concluded that warmer nighttime temperatures increased plant respiration and therefore decreased the energy available for storage in the rice kernels. In another study, scientists estimated that heat stress, increased virulence of parasites and disease organisms, and loss of soil moisture led to a 30 percent reduction in forest and grassland growth during the 2003 European drought. The effect on crops was variable. Winter wheat yields were barely affected, because the maximum growth period occurred before the heat wave. On the other end of the spectrum, the Italian corn yields decreased by 36 percent, despite extensive irrigation.

Precipitation and Soil-Moisture Effects on Agriculture

In a warmer world, both precipitation and evaporation would increase. Computer models show that the effects would differ from region to region. Between 1900 and 2000, rainfall increased in Pakistan and in the great grain-growing regions in North America and Russia. However, parts of North Africa, India, and Southeast Asia received less rainfall over the same time. The resulting drought has led to famine during recent decades.

In a hotter world, more soil moisture will evaporate. In the Northern Hemisphere, scientists predict that global warming will lead to increased rainfall, adding moisture to the soil. At the same time, warming will increase evaporation that removes soil moisture. Most computer models forecast a net loss of soil moisture and depletion of ground water, despite the prediction that rainfall will increase. Drought and soil moisture depletion would increase demands for irrigation. But, irrigation systems are already stressing global water resources, causing both shortages and political instability in many parts of the world. As a result of all these factors, most scientists predict that higher mean global temperature would decrease global food production, perhaps dramatically.

In many regions of the world, winter precipitation accumulates as snow in the high mountains. Snowmelt feeds rivers during the summer and this river water is used to irrigate crops. In a warmer world, more precipitation falls as rain, and the snow that does fall is more likely to melt in early spring rather than in late summer. Also, glaciers recede, thus diminishing the amount of water in long-term storage. As a result of all these factors, river levels during mid- to late summer are lower in a warmer world than in a cooler world. Thus, unless people build expensive dams, there is less water available for irrigation.

Extreme Weather Events

Computer models predict that weather extremes such as intense rainstorms, flooding, heat waves, prolonged droughts, and violent storms—hurricanes, typhoons,

ANNA LISA YODER/ISTOCKPHOTO.COM

and tornadoes—will become more common in a warmer, wetter world. In the United States, the number of heavy downpours (defined as 5 centimeters, or 2 inches, of rain in a single day) increased by 25 percent from 1900 to 2000. Worldwide, there were 10 times as many catastrophic floods in the decade from 1990 to 2000 as there were in an average decade between 1950 and 1985. Hurricanes are driven by warm sea surface temperatures, and many climate models predict that extreme tropical storms and hurricanes, like Katrina, will become more frequent in a warmer world. Thus, we face the paradox of both more floods and more droughts. Hurricane intensity is likely to increase in a warmer world.

Changes in Biodiversity

According to a study published by the World Wildlife Fund in 2000, global warming could alter one-third of the world's wildlife habitats by 2100. In some northern-latitude regions, 70 percent of habitats will be significantly affected. As soil moisture decreases, trees will die and wildfires will become more common, changing forests into savannas. Plants and animals that thrive in cold temperatures will die off. Several recent studies have shown that many plant and animal pathogens thrive better in warmer temperatures than in a colder world. Thus, the pine beetle in northern North America has flourished in recent years, decimating huge swaths of forests in the Rocky Mountains of western United States and Canada. Populations of frogs have declined dramatically as they have succumbed to virulent disease organisms. In extreme cases, species that are unable to adapt to the warmer temperatures might become extinct. For example, as Arctic sea ice diminishes, polar bears are losing their hunting grounds and migration pathways. As a result, these majestic creatures are threatened. Undoubtedly, other species will flourish. Ecological systems will change, but it is important to remember that terrestrial ecosystems have adjusted to far greater perturbations than we are experiencing today.

Melting of Arctic Sea Ice and Glaciers

About 3.5 million years ago, in the mid-Pliocene, the average global temperature was about 2°C to 3.3°C warmer than it is today. During this time, the Northern Hemisphere was ice free and there was significantly less ice in Antarctica than exists there today. Recall that in the past 50 years, Earth's temperature has increased by 0.8°C, and now is only 1.5°C to 2.5°C cooler than that prehistorical warm period. In recent decades, as the Earth has warmed, glacial and sea ice has melted. Between 1979 and 2009, the Arctic Ocean lost approximately 7 percent of its spring ice cover. But more alarming, the ice that remains is significantly thinner than it was 30 years ago, causing concern that the rate of loss could quickly accelerate with a small additional temperature increase. Ice loss in Antarctica increased by 75 percent in the last 10 years due to a speed-up in the flow of its glaciers. Greenland Ice Sheet's annual loss rose from 90 cubic kilometers in 1996 to 150 cubic kilometers in 2007. Stephen Schneider, a climatologist at Stanford, says that there is a "few percent chance" that meltwater that has already percolated into the ice cap will "obliterate the Greenland ice cover irrevocably."[5] He goes on to add that if the Northern Hemisphere warms by 3°C, there is a 90 percent chance that the Greenland ice cap will disappear. Note that this estimate corresponds with what actually happened 3.5 million years ago.

When land-based glaciers melt, sea level rises, affecting civilizations throughout the world. (Melting of sea ice has no effect on sea level.) Changes of sea level and loss of ice cover also affect the Earth's climate balance in many ways, as will be discussed in Section 21.8.

Acidification of the Oceans

About one-third of all the carbon dioxide that enters the atmosphere dissolves in the ocean. On one hand, this buffer mechanism reduces the concentration of carbon dioxide in the atmosphere, thus maintaining a cooler planet. On the other hand, the dissolved carbon dioxide alters the chemistry of the oceans. When carbon dioxide dissolves in seawater, it reacts to form carbonic acid, H_2CO_3. Since preindustrial times, the oceans have become more acidic, showing a decline of about 0.1 pH units. When the oceans become more acidic, marine organisms have reduced ability to build hard shells or exoskeletons out of calcium carbonate. For example, along the Great Barrier Reef in Australia, the calcification rate decreased by 14 percent between 1990 and 2008. Thus, a small change in pH leads directly to a large biological disruption. The resultant changes affect species distributions that could adversely affect ocean food webs.

Sea-Level Change

When water is warmed, it expands slightly. From 1900 to 2000, mean global sea level rose 10 to 25 centimeters, or roughly the thickness of a dime, every year. Oceanographers suggest that even this modest rise may be affecting coastal estuaries, wetlands, and coral

---

5. Stephen H. Schneider, "The Worst-Case Scenario," *Nature* 458 (April 30, 2009), 1104–1105.

Increasingly destructive storms threaten homes along the California coastline. Expensive reinforcement measures offer only short-term protection from the waves' erosion.

reefs. However, sea-level rise is accelerating dramatically because the polar ice sheets of Antarctica and Greenland are melting. In 2005, 150 cubic kilometers of Antarctic ice melted and the water flowed into the oceans. To exacerbate the problem, the flow of ice from Greenland has more than doubled over the past decade (Figure 21.14).

Effects on People

Humans evolved during a period of rapid climate change in the savannas and forests of Africa. In fact, many anthropologists argue that we flourished because we adapted to climate change faster and more efficiently than other species in the African ecosystem. A few million years later, we survived the Pleistocene Ice Age. We are a clever and resourceful species. Yet the consequences of global warming could be severe. In a warmer world, tropical diseases such as malaria have been spreading to higher latitudes. But the biggest problem arises because expanding human population is already stressing global systems. Because of high population density, land ownership, and political boundaries, people cannot migrate as freely as they once could. As a result, small shifts in food production or water availability could lead to mass starvation, water scarcity, and political instability. In addition, rising sea level could flood coastal cities and farmlands, causing trillions of dollars worth of damage throughout the world. There is already great stress on our water supply, as you read about in Chapter 12.

21.8 Feedback and Threshold Mechanisms in Climate Change

If the concentration of greenhouse gases and the mean global temperature were related linearly, a small change in one would produce a corresponding small change in the other. However, climate change does not follow linear relationships.

For example, the melting of ice is a threshold phenomenon. If the air above a glacier warms from −1.5°C to −0.5°C, the ice does not melt and the warming may cause only minimal environmental change. However, if the air warms another degree to +0.5°C, the ice warms beyond the threshold defined by its melting point. Melting ice can cause sea-level rise and a host of cascading effects.

Recall that a feedback mechanism occurs when a small initial perturbation affects another component of Earth's systems, which amplifies the original effect, which perturbs the system even more, which leads to a greater effect, and so on. Four important climate feedback mechanisms are discussed next.

Albedo Effects

Sparkling snowfields and icy glaciers have high albedos and cool Earth by reflecting sunlight. When Earth warms a little, snow and ice covers decrease, and the

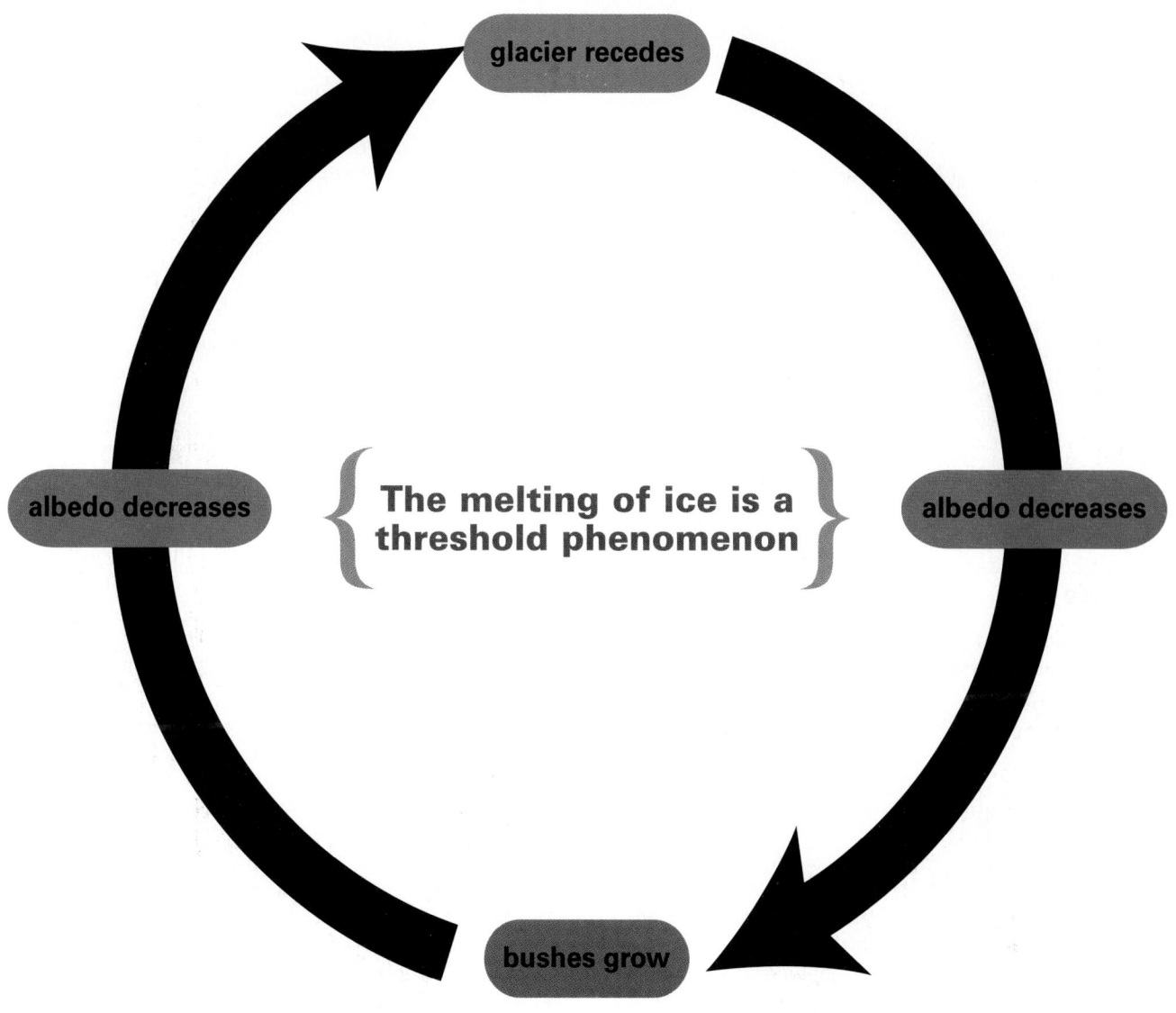

glacier recedes

{ **The melting of ice is a threshold phenomenon** }

albedo decreases

albedo decreases

bushes grow

albedo decreases. But when the albedo decreases, the atmosphere gets warmer, which melts more snow, which warms the atmosphere even more. A recent study has shown that a second, interlocking feedback mechanism exacerbates the first. In the Arctic, warmer temperatures have promoted the growth of bushes that often protrude above the spring snow and trap solar heat—thereby depressing the albedo. Thus snow melts faster, providing favorable conditions for bushes to grow, which decreases the albedo even further, which warms the atmosphere, as you can see in the figure above.

The accelerating nature of this double feedback loop has contributed to the fact that between 1960 and 1990, Arctic landscapes warmed by roughly 0.15°C per decade. However, between 1990 and 2004, this warming more than doubled to 0.3°C to 0.4°C per decade.

Imbalances in Rates of Plant Respiration and Photosynthesis

Any process that increases the total photosynthesis in an ecosystem will remove carbon dioxide from the air and sequester the carbon in plant tissue. In contrast, any process that increases respiration over photosynthesis will introduce more carbon dioxide into the air and increase global warming. Numerous studies have shown that decay respiration in forest and grassland soils accelerates in a warmer world, thus warming the world more by introducing carbon dioxide into the atmosphere.

In another study, scientists showed that both respiration and photosynthesis decreased during the 2003 European drought, but photosynthesis declined more than respiration, leading to a net increase in carbon emissions into the atmosphere.

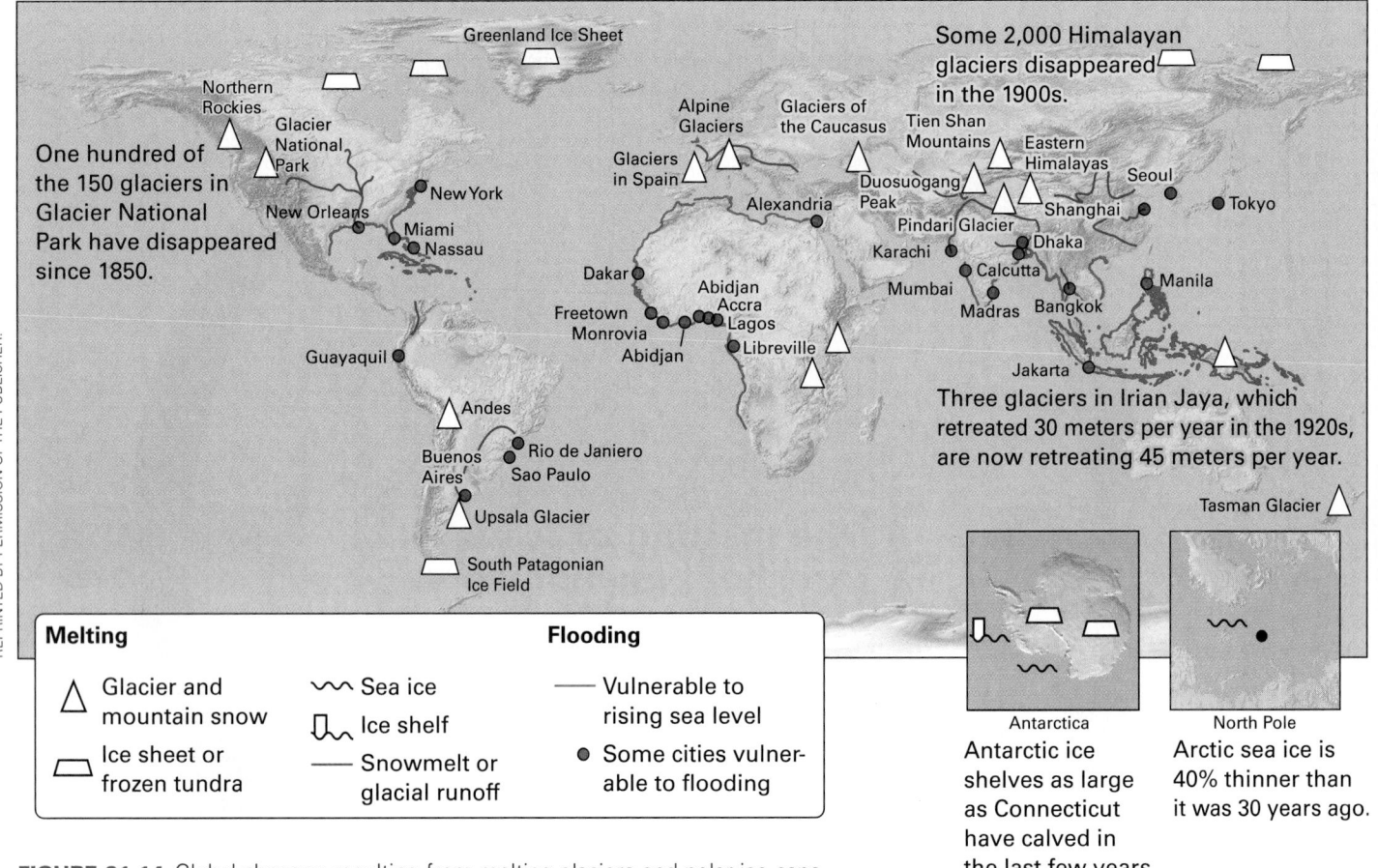

FROM "THE MERCURY'S RISING," FROM NEWSWEEK, 12/4/2000, P.52. REPRINTED BY PERMISSION OF THE PUBLISHER.

One hundred of the 150 glaciers in Glacier National Park have disappeared since 1850.

Some 2,000 Himalayan glaciers disappeared in the 1900s.

Three glaciers in Irian Jaya, which retreated 30 meters per year in the 1920s, are now retreating 45 meters per year.

Melting

△ Glacier and mountain snow

▱ Ice sheet or frozen tundra

〰 Sea ice

⬓〰 Ice shelf

— Snowmelt or glacial runoff

Flooding

— Vulnerable to rising sea level

● Some cities vulnerable to flooding

Antarctica
Antarctic ice shelves as large as Connecticut have calved in the last few years.

North Pole
Arctic sea ice is 40% thinner than it was 30 years ago.

FIGURE 21.14 Global changes resulting from melting glaciers and polar ice caps.

Changes in Ocean Currents

Approximately 12,800 years ago, air temperature over the North Atlantic dropped suddenly by more than 10°C, a climate change called the Younger Dryas Event, which you can see on the graph in Figure 21.15.

The speed of this change implies that some threshold or feedback mechanisms occurred, most probably relating to changes in ocean currents. Recall that in the Atlantic Ocean, warm-water currents carry heat northward, thus heating southern Greenland and northern Europe. One argument proposes that after the ice age maximum, as the planet warmed, massive amounts of ice melted, creating a huge lake, called Lake Agassiz in North America, about where the Great Lakes are now. Initially, Lake Agassiz flowed southward into the Gulf of Mexico. But when the North American ice sheet retreated sufficiently, a huge volume of water suddenly flowed through what is now the St. Lawrence River valley into the North Atlantic. By one estimate, 9,500 cubic kilometers of water cascaded into the ocean. Many scientists argue that the huge influx of freshwater caused the Younger Dryas Event. However, there is significant debate over how this freshwater influx caused the recorded temperature change. According to one argument, the freshwater lowered the surface density of the seawater. Due to this lowered density,

seawater in the northeast Atlantic failed to sink, as it had previously. This disruption of vertical movement disrupted the horizontal currents. Others argue that ocean currents are driven primarily by wind systems, and the change in water flow must have altered wind patterns, which in turn altered the currents in the Atlantic. A third argument proposes that the cold freshwater was simply a physical barrier, just as a tributary will alter the flow of the main current in a river system. On the other hand, maybe the emptying of Lake Agassiz was not the cause of the temperature change at all. This argument is supported by the observation that a second meltwater pulse from central North America occurred at the end of the Younger Dryas, when the planet was warming and this influx of meltwater did not cause a cooling trend.

While the mechanism remains uncertain, ice core data tell us that the Younger Dryas cooling, while the most dramatic, was not the only rapid cooling in the Pleistocene Epoch. In fact it was only the most recent of a series of large and abrupt climate swings that occurred repeatedly during the last ice age, and were recorded in many places around the globe. Thus, it is clear that sequences of threshold and feedback mechanisms have changed climate rapidly and dramatically in the recent geologic past.

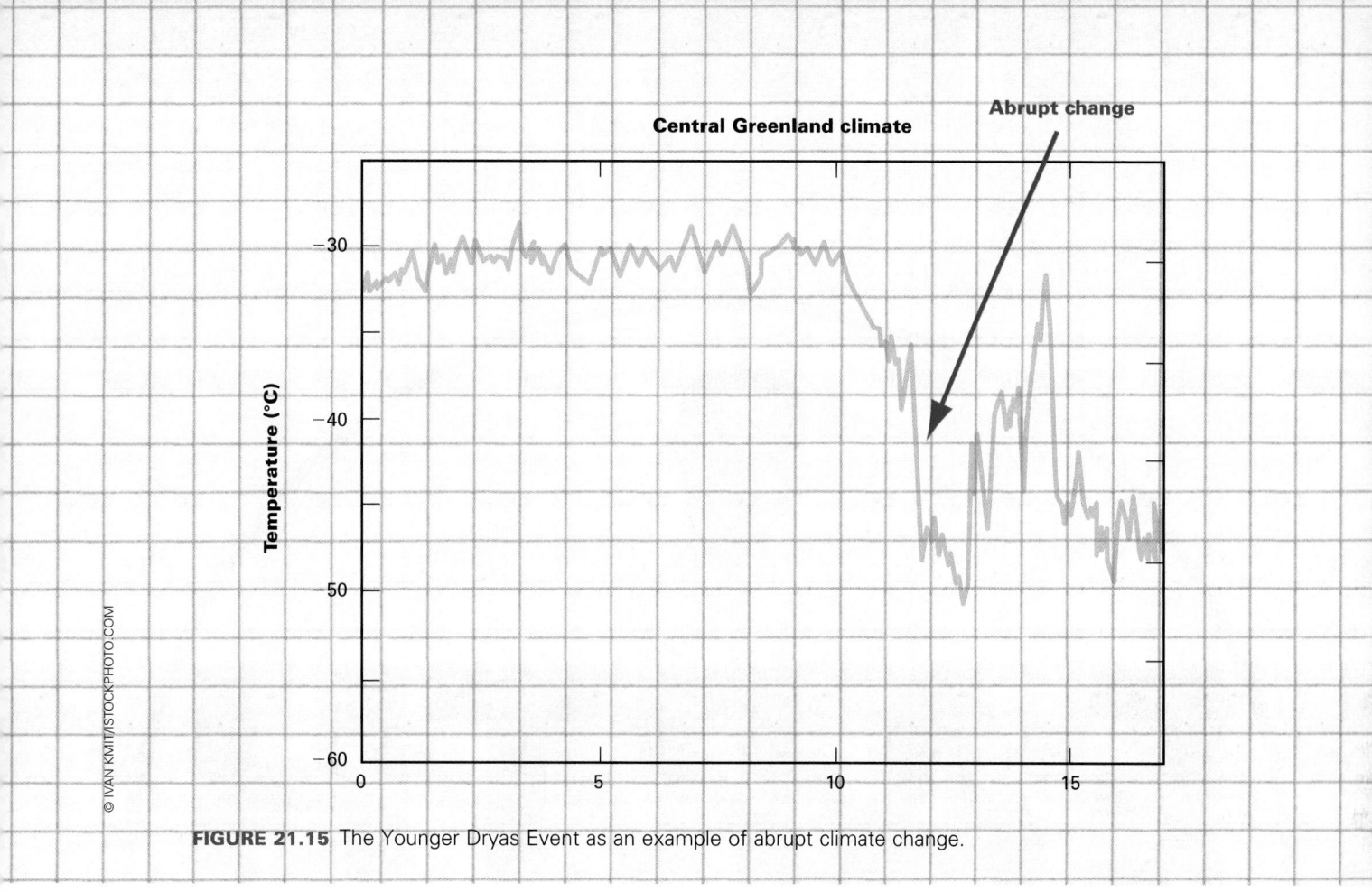

FIGURE 21.15 The Younger Dryas Event as an example of abrupt climate change.

Permafrost and Deep-Sea Methane Deposits

Large deposits of methane gas are trapped both in permafrost and in sediments beneath oceans. Methane is 20 times as effective in trapping heat in the atmosphere as carbon dioxide. Warming temperatures can release methane from both sources, thus causing a rapid and possibly catastrophic feedback warming. Vast areas of permafrost contain stores of carbon, which accumulated during multiple glacial advances and retreats over millions of years. Some of this carbon already exists as methane, produced by microbial decay in the past. Additional amounts can be released quickly when the ice melts and microbes have lush, carbon-rich material to feast on. In a similar manner, methane is also trapped in sea-floor sediments as methane hydrates, in which methane is trapped in a molecular cage of ice. If ocean water warms sufficiently to melt the ice, then the methane bubbles to the surface. In both of these cases, initial warming from an unrelated cause, such as human release of carbon dioxide, can release the methane, which leads to additional and more rapid warming. Many scientists hypothesize that about 55 million years ago, a 100,000-year hot spell called the Paleocene-Eocene Thermal Maximum was caused by such a release of methane-bearing sea-floor deposits. This was the last time that the Earth was entirely without ice.

No one knows what the future will bring, but we do know that climates have changed radically in the past, from the frigid cold of Snowball Earth to the warmth of the Cretaceous. Today, by altering the atmospheric composition, we are undergoing a great and unintentional experiment with the climate system that sustains us. By making artificial changes to the natural order, such as the flow of water, we are altering not only the Earth's topography but also its natural cycles in ways that we may not be prepared to handle. Furthermore, global climate change has already affected the Earth, its ecosystems, and the humans who call this planet home. Are we prepared to account for the consequences linked to the way we use our planet? By understanding how human activity can affect the Earth's natural cycles, perhaps we can work with our environment and not against it to maintain a habitable planet.